THE PHYSIOLOGY OF TROPICAL FISHES

This is Volume 21 in the

FISH PHYSIOLOGY series

Edited by William S. Hoar, David J. Randall, and Anthony P. Farrell

A complete list of books in this series appears at the end of the volume

THE PHYSIOLOGY OF TROPICAL FISHES

Edited by

ADALBERTO L. VAL
Laboratory of Ecophysiology and Molecular Evolution
Department of Ecology,
National Institute for Research in the Amazon-INPA

VERA MARIA F. DE ALMEIDA-VAL
Laboratory of Ecophysiology and Molecular Evolution
Department of Ecology,
National Institute for Research in the Amazon-INPA

DAVID J. RANDALL
Department of Biology and Chemistry
City University of Hong Kong, SAR China

Amsterdam • Boston • Heidelberg • London • New York • Oxford
Paris • San Diego • San Francisco • Singapore • Sydney • Tokyo
Academic Press is an imprint of Elsevier

ELSEVIER

ACADEMIC
PRESS

Academic Press is an imprint of Elsevier
84 Theobald's Road, London WC1X 8RR, UK
30 Corporate Drive, Suite 400, Burlington, MA 01803, USA
525 B Street, Suite 1900, San Diego, California 92101-4495, USA

This book is printed on acid-free paper ∞

Copyright © 2006, Elsevier (USA).

Library of Congress Catalog Number: 2005923101

British Library Cataloguing in Publication Data
A catalogue record for this book is available from the British Library

ISBN–13: 978-0-12-350445-6
ISBN–10: 0-12-350445-7

For information on all Academic Press publications
visit our web site at http://books.elsevier.com

Printed and bound by CPI Group (UK) Ltd, Croydon, CR0 4YY

Transferred to Digital Print 2011

Working together to grow
libraries in developing countries

www.elsevier.com | www.bookaid.org | www.sabre.org

ELSEVIER   BOOK AID
International   Sabre Foundation

CONTENTS

4. Biological Rhythms
Gilson Luiz Volpato and Eleonora Trajano

5. Feeding Plasticity and Nutritional Physiology in Tropical Fishes
Konrad Dabrowski and Maria Celia Portella

6. The Cardiorespiratory System in Tropical Fishes: Structure, Function, and Control
Stephen G. Reid, Lena Sundin, and William K. Milsom

7. Oxygen Transfer
    Colin J. Brauner and Adalberto L. Val

8. Nitrogen Excretion and Defense Against Ammonia Toxicity
    Shit F. Chew, Jonathan M. Wilson, Yuen K. Ip,
    and David J. Randall

9. Ionoregulation in Tropical Fishes from Ion-Poor,
    Acidic Blackwaters
    Richard J. Gonzalez, Rod W. Wilson, and Christopher M. Wood

10. Metabolic and Physiological Adjustments to Low
     Oxygen and High Temperature in Fishes of the Amazon
     *Vera Maria F. de Almeida-Val, Adriana Regina Chippari Gomes,
     and Nívia Pires Lopes*

11. Physiological Adaptations of Fishes to Tropical
     Intertidal Environments
     *Katherine Lam, Tommy Tsui, Kazumi Nakano,
     and David J. Randall*

12. Hypoxia Tolerance in Coral Reef Fishes
     Göran E. Nilsson and Sara Östlund-Nilsson

CONTRIBUTORS

The numbers in parentheses indicate the pages on which the authors' contributions begin.

VERA MARIA F. DE ALMEIDA-VAL *(1, 443), National Institute for Research in the Amazon, Laboratory of Ecophysiology and Molecular Evolution, Ave André Araújo 2936, CEP 69083–000, Manaus, AM, Brazil*

COLIN J. BRAUNER *(277), Department of Zoology, University of British Columbia, 6270 University Blvd, Vancouver, BC, Canada V6T 124*

SHIT F. CHEW *(307), Natural Sciences, National Institute of Education, Nayang Technological University, 1 Nanyang Walk, Singapore 637616, Republic of Singapore*

ADRIANA REGINA CHIPPARI GOMES *(443), National Institute for Research in the Amazon, Laboratory of Ecophysiology and Molecular Evolution, Ave André Araújo 2936, CEP 69083–000, Manaus, AM, Brazil*

KONRAD DABROWSKI *(155), School of Natural Resources, Ohio State University, Columbus, Ohio*

RICHARD J. GONZALEZ *(397), Department of Biology, University of San Diego, 5998 Alcala Park, San Diego, CA 92110*

PETER A. HENDERSON *(85), Pisces Conservation Ltd, The Square, Pennington, Lymington, Hampshire, UK*

YUEN K. IP *(307), Department of Biological Sciences, National University of Singapore, 10 Kent Ridge Road, Singapore 117543, Republic of Singapore*

KATHERINE LAM *(501), Marine Laboratory, Department of Biology and Chemistry, City University of Hong Kong, Hoi Ha Wan, Hong Kong*

ix

NÍVIA PIRES LOPES *(443)*, National Institute for Research in the Amazon, Laboratory of Ecophysiology and Molecular Evolution, Ave André Araújo 2936, CEP 69083–000, Manaus, AM, Brazil

WILLIAM K. MILSOM *(225)*, Department of Zoology, University of British Columbia, Vancouver, BC, Canada, V6T1Z4

KAZUMI NAKANO *(501)*, Food Processing and Preservation Division, National Research Institute of Fisheries Science, Yokohama, Japan

GÖRAN E. NILSSON *(583)*, Physiology Programme, Department of Molecular Biosciences, University of Oslo, P.O. Box 1041, N-0316 Oslo, Norway

SARA ÖSTLUND-NILSSON *(583)*, Department of Biology, University of Oslo, P.O. Box 1066, N-0316 Oslo, Norway

MÁRIO C. C. DE PINNA *(47)*, Department of Vertebrates, Museu de Zoologia, Av. Nazaré 481, Caixa Postal 42594, São Paulo-SP 04299–970, Brazil

MARIA CELIA PORTELLA *(155)*, Fisheries Institute, Regional Pole for Technological Agribusiness Development, Ribeirao Preto, SP, and University of Sao Paulo State, Jaboticabal, Brazil

DAVID J. RANDALL *(1, 307, 501)*, Department of Biology and Chemistry, City University of Hong Kong, Tat Chee Avenue, Hong Kong, China

STEPHEN G. REID *(225)*, The Centre for the Neurobiology of Stress, Department of Life Sciences, University of Toronto, Toronto, Ontario, Canada

LENA SUNDIN *(225)*, Department of Zoology, Göteborg University, Box 463, SE 405 30, Göteborg, Sweden

ELEONORA TRAJANO *(101)*, Department of Zoology, Institute of Biosciences, University of São Paulo, C.P. 11461, 05422–970 São Paulo, Brazil

TOMMY TSUI *(501)*, Department of Biology and Chemistry, City University of Hong Kong, Kowloon Tong, Hong Kong

ADALBERTO L. VAL *(1, 277)*, National Institute for Research in the Amazon, Laboratory of Ecophysiology and Molecular Evolution, Ave André Araújo 2936, CEP 69083–000, Manaus, AM, Brazil

GILSON LUIZ VOLPATO *(101)*, Research Center on Animal Welfare – RECAW, Laboratory of Animal Physiology and Behavior, Department of Physiology, IB, UNESP, CP. 510, CEP 18618–000, Botucatu, SP, Brazil

JONATHAN M. WILSON *(307)*, *Centro Interdisciplinar de Investigacão Marinhae Ambiental-CIIMAR, Rua do Campo Alegre 823, 4150–180 Porto, Portugal*

ROD W. WILSON *(397)*, *Department of Biological Sciences, Hatherly Laboratories, University of Exeter, Prince of Wales Road, Exeter, EX4 4PS, UK*

CHRISTOPHER M. WOOD *(397)*, *Department of Biology, McMaster University, 1280 Main St West, Hamilton, Ontario L8S 4K1, Canada*

PREFACE

The present volume was conceived to cover the most recent advances in the physiology of tropical fishes. The readers can find information about the physiology of tropical fishes in many of the first 20 volumes of the Fish Physiology series. However, *The Physiology of Tropical Fishes* is the first volume to specifically gather information about the large and important group of fishes that live in the tropics. Tropical environments are as diverse as are the groups of fishes living there. Rather than trying to cover all areas of the physiology of tropical fishes, this book brings together the subjects related to their physiological adaptations to tropical environments, which they have shaped during their evolutionary history, and what make tropical fishes an amazing group to study. *The Physiology of Tropical Fishes* hopes to broaden our understanding of what is so special about freshwater and marine tropical habitats that makes tropical fishes one of the most diverse groups of vertebrates in the world. Indeed, subjects such as Growth, Biological Rhythms, Feeding Plasticity and Nutrition, Cardiorespiration, Oxygen Transfer, Nitrogen Excretion, Ionoregulation, Biochemical and Physiological adaptations are all presented and discussed in the light of their specific fitness to tropical environments such as intertidal pools, coral reefs, and the Amazon's different types of waters, all of them typically hypoxic and warm water bodies. These subjects have been developed by top scientists studying specific characteristics of tropical species and also their many interactions with their ever-changing environments. The voyage through this volume brings us the conviction that tropical fishes are barely studied and much more needs to be done before we have a clear picture of the adaptive characteristics that allow them to survive extreme tropical environmental and biological conditions. We are very grateful to all colleagues who contributed to this volume, for their enthusiasm and their dedication to this project. Also, we are grateful to the many reviewers for their constructive comments. We thank Claire Hutchins for her support and the staff of Elsevier for providing the proofreading formats and helping with the final editing of the volume. At last, but not least, we thank the Editors of the Series Fish Physiology, Bill Hoar, David Randall, and Tony Farrell for their invitation and for keeping such an important subject updated for the many generations to come.

Adalberto Luis Val
Vera Maria Fonseca de Almeida-Val
David John Randall

1

TROPICAL ENVIRONMENT

ADALBERTO L. VAL
VERA MARIA F. DE ALMEIDA-VAL
DAVID J. RANDALL

I. INTRODUCTION

The tropical climate zone occupies *ca.* 40% of the surface of the earth and is located between the Tropics of Cancer (latitude 23.5 °N) and Capricorn (latitude 23.5 °S). The main ecological driving forces within this zone are relatively stable high temperatures and air humidity. Although there are variations in climate within the tropics, 90% of tropical ecosystems are hot and humid, whether permanent or seasonal, and the remaining 10% are hot and dry and include mainly desert-like ecosystems. These variations are determined by altitude, topography, wind patterns, ocean currents, the proportion of land to water masses, geomorphology, vegetation patterns, and more recently by large-scale/man-made environmental changes.

1

Various attempts have been made to classify the climates of the earth into climatic regions. Köppen Climate Classification System, proposed by the Köppen in 1936 (Köppen, 1936), is based on five inputs: (i) average temperature of the warmest month; (ii) average temperature of the coldest month; (iii) average thermal amplitude between the coldest and warmest months; (iv) number of months with temperature exceeding $10\,^{\circ}C$; and (v) winter summer rains. Two other classifications followed this classification: the classification of Holdridge, which takes into account temperature, evapotranspiration, and annual rainfall, being also known as life zone classification (Holdridge, 1947), and the classification of Thornthwaite, which takes into account moisture and temperature indexes (Thornthwaite, 1948). The "empirical, and somewhat obsolete, albcit fairly efficient" characteristics of the Köppen classification have driven some rejection to this and some other similar classifications (Le Houérou *et al.*, 1993). However, the Köppen classification is widely used with a variety of amendments, and based on this system five major climatic groups are recognized in the world, plus a sixth for highland climates. They are: Tropical humid (A); Dry (B); Mild mid-latitude (C); Severe mid-latitude (D); Polar (E); and Highlands (F). These main types are further classified into various subtypes, as reviewed by McKnight (1992). For the purpose of this book and based on Köppen's classification, three broad categories of tropical climates are distinguished: *Af*, tropical humid climate with relatively abundant rainfall every month of the year; *Am*, tropical humid climate with a short dry season; and *Aw*, tropical climate characterized by a longer dry season and prominent, but not extraordinary, wet season (Figure 1.1).

An intricate relationship does exist between soils and water bodies. Under many circumstances water composition and its major characteristics are determined by the surrounding soil that in tropics is highly diverse, and so are the water bodies, as we shall see later on this chapter. In addition, as the anthropogenic pressure increases dramatically on land, more and more aquatic environments are experiencing significant challenges. These water bodies, however, often have an amazingly high ability to neutralize the large quantity of diversified chemical products reaching them (Val *et al.*, 2004, in press; van der Oost *et al.*, 2003). This chapter aims to depict the major aquatic habitats of tropical fishes, with emphasis on their physical, chemical, and biological characteristics and effects of man-made and global changes on these environments.

Diversity is the keyword defining tropical aquatic ecosystems as they include hundreds of different types of water bodies, with different water composition, and different biological and physical characteristics. In addition, a single water body undergoes significant changes throughout the year, even disappearing, in some cases, during the dry season. Disappearance of water

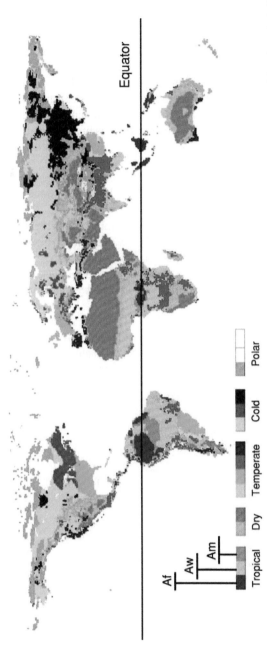

Fig. 1.1 Main types of tropical climates (Köppen's climate classification), including all climates that are controlled by equatorial and tropical air masses; *Af* – tropical moist climates, or **rainforest**, characterized by relatively abundant rainfall every month of the year; *Am* – tropical humid climate, characterized by a short dry season; and *Aw* – wet–dry tropical climate, or **savanna**, which is characterized by a longer dry season and prominent, but not extraordinary, wet season. This last type gets a little cooler during the dry season but will become very hot just before the wet season. Modified from Strahler, A. N., Strahler, A. H., *Elements of Physical Geography*. John Wiley & Sons, 1984.

bodies during the dry season is best exemplified by many shallow lakes in the Amazon (Junk *et al.*, 1989; Sioli, 1984; Val and Almeida-Val, 1995), and by the ephemeral lakes in Niger (Verdin, 1996). In many cases, water bodies of the same climatic region or even located close together may have different chemical composition and behave differently. In other words, a water body is a unique ecosystem, without parallel in the world. Each water body can be visualized as a "living tissue" that responds accordingly to each environmental factor. Thus, the biological, chemical, and physical characteristics of the different water systems presented in the following sections are roughly generalizations.

II. TROPICAL MARINE ENVIRONMENTS

The marine environment contains approximately 98% of the water of the planet, with the atmosphere being the smallest water compartment with only 0.001% of the total existing water (Table 1.1). The marine environment is not quiet, stable and uniform as it seems to a casual observer; in fact, it is a moving and changing environment with a large variety of biotopes, inhabited by

Table 1.1
Stocks of Water in the Different Compartments of the Earth

	Volume ($1000\,km^3$)	% of total water	% of total freshwater
Salt water			
Oceans	1 338 000	96.54	
Saline/brackish groundwater	12 780	0.93	
Salt water lakes	85	0.006	
Inland waters			
Glaciers, permanent snow cover	24 064	1.74	68.70
Fresh groundwater	10 530	0.76	30.06
Ground ice, permafrost	300	0.022	0.86
Freshwater lakes	91	0.007	0.26
Soil moisture	16.5	0.001	0.05
Atmospheric water vapor	12.9	0.001	0.04
Marshes, wetlands	11.5	0.001	0.03
Rivers	2.12	0.0002	0.006
Incorporated in biota	1.12	0.0001	0.003
Total water	**1 386 000**	**100**	
Total freshwater	**35 029**		**100**

Source: Shiklomanov (1993).

almost all animal Phyla on the planet (Angel, 1997). In a small area around India, for example, 167 biotopes have been mapped and identified for conservation and sustainable use (Singh, 2003). A marine biotope can be envisioned as an area in which habitat conditions and organic diversity are quite similar. Marine biotopes differ from place to place according to local geology, currents, temperature, depth, light, dissolved gases, transparency, and levels of ions and nutrients, among other parameters.

Basically, these marine biotopes are either pelagic or benthic, in general the environmental quality of conditions for life decreases with closeness to land and the surface of the water, so biotopes are most numerous in inshore waters. The pelagic environment is further divided into (1) neritic zone, a designation for waters over the continental shelf, i.e., from low tide mark up to 100 fathoms (about 200 meters) offshore, and (2) oceanic zone, a designation for all waters beyond the edge of the continental shelf. The area between the lowest and highest tide mark is known as the littoral or intertidal zone and is highly influenced by the supralittoral region (see next section in this chapter for further details). The oceanic zone is further divided into epipelagic, a designation for surface waters away from continental shelf up to about 200 meters in depth; mesopelagic, for waters between 200 and 1000 meters; bathypelagic, for water between 1000 and 4000 meters; and abyssopelagic, for waters roughly below 4000 meters. While the neritic zone biotopes experience seasonal variations in chemical, physical, and biological parameters, the oceanic biotopes are relatively less productive but are much more stable environments with a wide range of living conditions (Lagler et al., 1977). These differences are the major determinants of the fish fauna inhabiting each of these biotopes (see Chapter 11, this volume).

Temperature is a major driving force controlling the distribution of marine fish fauna. It can be as high as 55 °C in small intertidal pools during the summer but is normally between 26 and 32 °C in the superficial water of tropical marine environments (see Levinton, 1982, for relationship between temperature of ocean water surface and latitude). Indeed, temperature differences between water surface and deeper water layers are not uniform among the different climatic zones (Figure 1.2). In the tropics, a stable thermocline develops between 100 and 300 meters depth that restricts plankton biomass to the upper warm layers of water and consequently reduces the amount of food for fish living below the photic zone.

Temperature and salinity are independent variables. Between the Tropics of Cancer and Capricorn, salinity decreases towards the Equator while temperature increases, reaching the highest values at the Equator (see Thurman, 1996). This variation is dependent on the balance of evaporation and

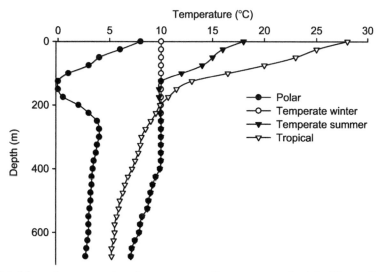

Fig. 1.2 Schematic representation of temperature profiles *versus* ocean depth in different climatic zones of the world. Notice that the deep and bottom waters of all latitudes are uniformly cold. Well-developed thermoclines are at low and mid-summer latitudes. Data compiled from several sources. See text for details.

precipitation that, in some instances, are dependent on atmospheric circulation. In general, surface water salinity is higher than deep water salinity due to evaporation. Below the surface, a halocline is formed where rapid changes in salinity are related to water depth. Near the Equator the halocline extends down to 1000 meters depth. Salinity is, on average, 35.5 ppm in the Pacific Ocean, 35.5 ppm in the Atlantic Ocean and near 40 ppm in the Red Sea. However, salinity can vary and be as high as 90 ppm in the Araruama Lagoon, at Rio de Janeiro, or even as high as 155 ppm in small tide pools subjected to intense evaporation. On the other hand, it can be as low as 8 ppm in the Baltic sea due to precipitation and river discharges (Soares-Gomes and Figueiredo, 2002).

Light is also a key factor that shapes marine biotopes as photosynthesis is entirely dependent on this physical parameter. Generally, three light zones can be distinguished: (1) photic zone, from water surface down to about 100 meters, where light is enough for photosynthesis, i.e., more than 5% of sunlight is available; (2) dysphotic zone, between *ca.* 100 and 200 meters, where light is weak for photosynthesis, less than 5% of the sunlight; and (3) aphotic zone, where no light is available at all. These major marine environmental characteristics determine the nature of the organic interactions in these environments.

A. Neritic Zone

1. SUPRALITTORAL

Supralittoral zone is the spray zone, extremely variable and very difficult to inhabit, requiring considerable specialized adaptations of animals. The few fish species inhabiting this habitat include, among others, mainly gobies, eels and clingfishes (Bone *et al.*, 1995).

2. INTERTIDAL

The intertidal environment, also known as littoral zone, is characterized by extreme conditions occurring during short periods of time, aggravated by intermittent drying periods that require from the inhabiting fishes extreme ability to overcome temperature, ionic and respiratory disturbances (see Chapter 11). The intertidal zone is, in fact, a demanding environment where the animals are knocked by waves and isolated in pools and mudflats. The most commonly known intertidal fishes are the mudskippers and the blennies that are truly amphibious because they emerge from water to graze on mud or rock in or above the splash zone (Bone *et al.*, 1995). Many intertidal fishes move in from and out to the sublittoral zone, e.g., species of stingrays (Dasyatidae), flounders (Bothidae and Pleuronectidae), soles (Soleidae), bonefish (*Albula*), eels (*Anguilla*), morays (Muraenidae), clingfish (Gobiesocidae), sculpins (Cottidae), searobins (Triglidae), snailfish and lumpfish (Cyclopteridae), midshipmen (*Porichthys*), blennies (Blenniidae), gobies (Gobiidae), pipefish and seashore (Syngnathidae), and cusk-eels (Ophidiidae) (see Bone *et al.*, 1995). There have been a number of reviews on the biology of intertidal fishes (Bone *et al.*, 1995; Graham, 1970; Horn *et al.*, 1998; Horn and Gibson, 1988) and their ecophysiology (Berschick *et al.*, 1987; Bridges, 1993).

3. SUBLITTORAL

The conditions for fish life are still good in the inner littoral zone, where seasonal variations are near maximum and light conditions support high productivity, in turn supporting a highly diverse group of fishes. In addition to the groups found in the littoral zone, the sublittoral zone also includes species of surfperch (Embiotocidae), skates (Rajidae), sharks (Squalidae), bonefish (Albulidae), croackers, kingfish and drums (Sciaenidae), hakes and pollocks (Gadiidae), rockfish (Scorpaenidae), wrasses (Labridae), butterflyfish and angelfish (Chaetodontidae), parrotfish (Scaridae), filefish and triggerfish (Balistidae), trunkfish (Ostraciidae), puffers (Tetraodontidae), porcupinefish (Diodontidae), and kelpfish (*Gibbonsia*) that migrate back and forth to outer littoral zone and to Coral Reefs, to which fish diversity from this sublittoral zone is somehow related.

The coral reefs constitute a distinct formation occurring in warm tropical seas that link sublittoral and littoral zones. By far, the major fish diversity living in shallow seas is found associated with coral reefs and atolls, which is the paragon of a rich marine community (Cornell and Karlson, 2000), including many small fishes. Coral reefs are mainly found in the Indian and Western Pacific oceans, in the Caribbean and around the West Indies. Coral reefs provide a wide diversity of habitats due to their physical structure and spatial coral arrangements. Despite that, there is a striking difference in the number of coral reef fish species in different regions from the richest central Indo-West Pacific reefs of the Philippines with more than 2000 species to the less rich reefs around Florida, which house between 500 and 700 fish species (Figure 1.3) (Sale, 1993). Recently, an elegant analysis of speciation of reef fishes shows how dispersal from a major center of origin can simultaneously account for both large-scale gradients in species richness and structure of local communities (Mora *et al.*, 2003). In addition, these authors succeeded in showing that the Indo-Pacific Region stands out as the major center of endemism in the Indian and Pacific Ocean and that the number of fish species decreases from the center (lower latitude) to the borders (high latitude, 30 °N, 30 °S), something that has been already demonstrated for other biological groups.

An analysis of coral reef assemblages is likely to be influenced by both coral diversity and substratum complexity; many studies have provided evidences that fish abundance and species richness are correlated with coral cover, availability of shelter, structural complexity, and biological characteristics, such as territoriality (Caley and John, 1996; Letourneur, 2000; McCormick, 1994; Munday, 2000; Nanami and Nishihira, 2003; Steele, 1999). Corals show nocturnal hypoxia and the effects of hypoxia on one of its inhabitants, *Gobiodon histrio*, has been described (Nilsson *et al.*, 2004; see Chapter 12). As coral reefs may be continuously distributed over a large area, widely spaced, and patchily distributed, there is increasing attention on the relationship of connectivity and species diversity and richness (Mora and Sale, 2002; Nanami and Nishihira, 2003), as this information is relevant for environmental management and conservation.

The outer sublittoral zone is comparatively less productive and, therefore, conditions for fish life vary seasonally. Light, ranging from blue to violet, reaches the bottom of this zone, further limiting its productivity. Fish community is poor and includes species of haddock, cod, hake, halibut, chimaera, hagfish and eel. Beyond this point is the abyssal zone, an essentially stable, dark and cold zone even within the tropical oceans (Lagler *et al.*, 1977; Lowe McConnell, 1987), and includes an almost unknown fish community (see Fish Physiology, Volume 16).

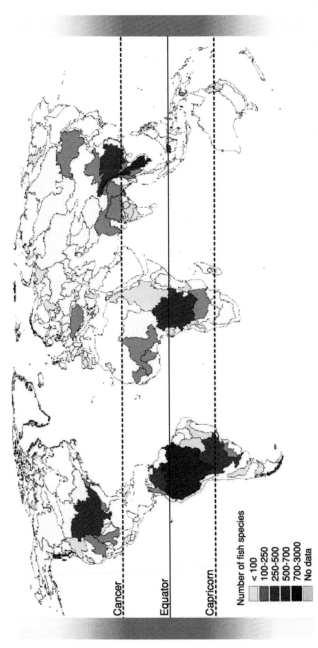

Fig. 1.3 Freshwater fish species diversity over the world showing the hotspots where fish species occur at high densities: hotspots are closer to the Equator, and the most-diversity site is located in the Amazon region (see text for further details). Modified from the map compiled by the World Conservation Monitoring United Nations Environmental Programme (UNEP-WCMC), (www.iucn.org).

Number of fish species
< 100
100-250
250-500
500-700
700-3000
No data

Cancer

Equator

Capricorn

B. Oceanic Zone

The open ocean covers nearly two-thirds of the world's surface and is the habitat of some 2500 fish species; half of them are pelagic. The Oceanic zone is relatively uniform; seasonal fluctuations affect only some areas, although conditions change with depth (Lowe McConnell, 1987). It is much less productive than the Neritic zone and Estuaries. As above mentioned, this zone is divided into (1) epipelagic, (2) mesopelagic and (3) bathypelagic zones. They refer, in fact, to the different stratum layers of the oceans that differ in depth and light availability, and so in biomass productivity.

1. EPIPELAGIC ZONE

This zone is an euphotic zone, where photosynthesis takes place. Despite the warm water and higher solar irradiance, the primary productivity of tropical oceans ranges from 18 to 50 $g.cm^{-2}.year^{-1}$ while its temperate counterparts ranges from 70 to 120 $g.cm^{-2}.year^{-1}$ (reviewed by Lourenço and Marquez Junior, 2002), a significantly higher primary productivity. In fact, primary productivity tends to increase from the latitude 0, the Equator, towards higher latitudes, with values more or less homogeneous up to the Tropics of Cancer and Capricorn and then peaking between this point and 60 °N and 60 °S (Field *et al.*, 1998). Inspecting the maps generated by the global CO_2 survey, JGOFS (Joint Global Ocean Flux Study) it becomes evident that the warm equatorial Pacific Ocean is the largest continuous and natural source of CO_2 to atmosphere while, in contrast, the cold North Atlantic, North Pacific and the Southern Ocean are important CO_2 sinks, i.e., the ocean regions where large amounts of CO_2 are physically absorbed and biologically assimilated (Takahashi *et al.*, 1999). Fish life depends ultimately on primary production by algae at the base of the food web, a condition met in tropical oceans, although at a relatively low level, throughout the year as a consequence of the constant temperature and solar incidence. The epipelagic fish fauna is richer in warm compared to cold regions (Bone *et al.*, 1995) and includes many species that feed on the neritic zone. The epipelagic fauna include mackerels, tuna (migrate to cold water during reproduction), sharks, marlin and others.

2. MESOPELAGIC ZONE

Inhabitants of the mesopelagic zone, also known as the twilight zone, depend on the plankton and other corpses dropping from the epipelagic zone. Many fish species living in this zone migrate upwards at night to feed in the upper zone, sinking again before dawn. They are adapted to dark, to save energy as food is scarce, and to pressure, since pressure increases by one atmosphere with every 10 meters of depth. In general, as depth increases, size,

Table 1.2
Diversity of Fish of the Brazilian Coast

Order	Families	Species	POZ*	Occurrence
Anguilliformes	4	5	M (60%)	E – M – M/B
Clupeiformes	2	2	M (100%)	M
Osmeriformes	4	6	B (57%)	M – B – M/B
Stomiiformes	8	19	M (76%)	M – B – M/B
Aulopiformes	5	11	M (54%)	E/M – M – M/B – E/M/B
Myctophiformes	2	38	M (95%)	M – B
Lampridiformes	3	4	E (67%)	E – E/M
Polymixiiformes	1	1	M (100%)	M
Gadiformes	6	10	M (60%)	E/M – M – M/B
Batrachoidiformes	1	1	E/M (100%)	E/M
Lophiiformes	3	3	M (60%)	E/M – M – M/B
Beloniformes	2	2	E (100%)	E
Beryciformes	3	3	M (50%)	E – M – M/B
Zeiformes	3	3	M (60%)	E – E/M – M
Gasterosteiformes	2	2	E (67%)	E – E/M
Scorpaeniformes	3	7	M (57%)	E/M – M
Perciformes	27	59	E/M (59%)	E – E/M – M
Pleuronectiformes	1	1	E/M (100%)	E/M
Tetraodontiformes	4	6	E (71%)	E – E/M
Total diversity observed and total percentage of each occupied zone				
TOTAL	84	183	-	E (16%)
				E/M (30%)
				M (40%)
				M/B (9%)
				B (5%)

*Abbreviations: POZ, Predominantly occupied zone (E = Epipelagic; M = Mesopelagic; B = Bathypelagic; E/M and M/B = transition zones).
Source: Compiled from Figueiredo *et al.* (2002).

abundance and fish diversity decreases. Fish diversity of the mesopelagic zone of tropical oceans is poorly known. Studying the fishes of the southern Atlantic, between Cabo de São Tomé (22 °S) and Arroio do Chuí (34 °S), within 200 nautical miles of the Brazilian coast, and using pelagic trawling, Figueiredo *et al.* (2002) collected a total of 28 357 specimens belonging to 185 species, 84 Families and 19 Orders (Table 1.2). Despite the gear type used, 86% of the sampled families had representatives inhabiting the epi and/or mesopelagic zone and only 14% inhabiting the meso and/or bathypelagic zones.

3. BATHYPELAGIC ZONE

Inhabitants of this zone depend entirely on the food gravitating from the zones above. Animals are adapted to high pressure, to darkness, to food limitations, and to energy economy. Most animals are bioluminescent. Fishes are greatly reduced in number and diversity. Interestingly, all the five strictly bathypelagic species collected in the Brazilian coast (*Bathylagus bericoides*, *Dolichopteryx anascopa*, *D. binocularis*, *Chauliodus sloani*, *C. atlanticus* and *Ceratoscopelus warmingii*) have large geographic distribution. All collected specimens were small in size and have bioluminescent organs (Figueiredo *et al.*, 2002).

C. Estuaries

An estuary is an area of interaction between oceanic salt water and freshwater from a stream. The estuary definition proposed by Cameron and Pritchard (1963) stating that an estuary is a semi-enclosed coastal body of water which has a free connection with the open sea and within which seawater is measurably diluted with fresh water derived from land drainage, is the most common and widely used. However, by restricting estuaries to semi-enclosed water bodies, the authors do not recognize the salinity gradient caused by the interaction of both types of waters that extends away from the land masses. In other words, Cameron and Pritchard's definition fails to consider the drainage of the Amazon and the Mississippi rivers as estuaries. Functionally, an estuary can be envisaged as an ecotone (Lagler *et al.*, 1977) and so it includes the boundaries of the salinity gradient in both the upstream and open ocean. The world's great estuaries are situated in the tropics: the Amazon, Orinoco, Congo, Zambezi, Niger, Ganges and Mekong; all very large rivers draining enormous geographical regions. The Amazon River discharges 20% of all freshwater entering the oceans of the world (see Sioli, 1984), with a flow of 0.2 Sv ($1 \text{ Sv} = 10^6 \text{ m}^3/\text{s}$) that accounts in large part for the sea surface salinity in the west tropical Atlantic Ocean (Masson and Delecluse, 2001). Tropical estuarine environments include, indeed, seasonally flowing streams and lacustrine water bodies intermittently connected to sea. Chemical, physical and biological conditions of these tropical estuaries are far from uniform, and greatly influence the estuarine life, including fish fauna. In general, estuaries are characterized by extreme changes in salinity, tidal and stream-current turbulence, turbidity and siltation. No other water systems undergo extreme seasonal fluctuations as observed in tropical estuaries. In addition, estuaries bordered by cities (13 out of 16 largest cities in the world are on the coast) and industries may also experience extremes of pollution. Roughly, tropical estuaries can be divided in four categories: (a) open estuaries; (b) estuarine coastal waters; (c) coastal lakes; and (d) blind estuaries.

1. OPEN ESTUARIES

All medium and large tropical rivers draining into the oceans form open estuaries and among them are all well-known tropical estuaries referred to above. They are never isolated from the sea and experience all the major environmental estuarine oscillations. These river-mouth estuaries exhibit layering, with freshwater overlaying the salt water beneath, that can extend for long distances, as observed for the Amazon River (the "Pororoca"). The water layers may have distinct ichthyofauna and may serve as a route for diadromous fishes or even as a route for casual freshwater invasion, as occur with elasmobranches that can be found in the middle Amazon River, near Manaus (Santos and Val, 1998; Thorson, 1974).

2. ESTUARINE COASTAL WATERS

The effects of Amazon drainage into the Atlantic are felt up to 400 km from the mouth, a distance that depends on several variables, including tidal cycles and seasonal changes in river water level. Similar situations occur with many other large and small rivers, such as the Orinoco (Venezuela), Ganges (India), Paraná (Brazil), and in general it is difficult to establish the boundaries of such environments. The shallow nature of these tropical waters and their lowered salinities, in conjunction with high turbidities, make them only partly estuarine from a fish fauna perspective (Baran, 2000; Blaber, 2002; Blaber *et al.*, 1990; Pauly, 1985).

3. COASTAL LAKES

Coastal lakes, also known as coastal lagoons, are lacustrine bodies behind tropical shorelines. They are relatively large water bodies, which is what makes them unique. They undergo high seasonal fluctuation that in the end determines the form and regularity of the lake–sea connection. Four main subtypes have been recognized: isolated lakes; percolation lakes, silled lakes and lagoonal inlets. Their fish fauna is mixed, marine and estuarine, depending on the salinity.

4. BLIND ESTUARIES

Blind estuaries are small water bodies, both in length and catchment, regularly formed by a sandbank across the sea mouth. When it is closed, freshwater enters from the river and fills the system. The salinity is dependent on tidal regimes, freshwater inflow, sandbank draining rates and wind. In general, blind estuaries are exploited for local subsistence.

Undoubtedly tropical estuaries are highly complex and variable aquatic ecosystems. They are among the most productive ecosystems, contributing effectively to maintain marine life. In addition, estuaries are the nursery grounds for many important fisheries, which often depend on one of its components that are the mangroves. Mangroves are composed of salt-tolerant trees and shrub that grow in shallow warm water; their muddy waters are rich in nutrients and serve as shelter for many types of marine organisms. A recent study has shown that mangroves in the Caribbean are unexpectedly important for neighboring coral reefs as well (Mumby *et al.*, 2004). Mangroves dominate the border of almost all known tropical estuaries, covering a quarter of the world's tropical coastline (Blaber, 2002; Wolanki, 1992).

III. TROPICAL FRESHWATER ENVIRONMENTS

Freshwater environments are many times smaller than the oceans both in area and in volume of water even though they are equivalent in terms of habitat diversity. They represent only 0.8% of total habitats in the Earth while marine environments represent 70.8% (Table 1.1). High and constant temperatures through the year and an almost absent seasonal variation in day length contrast tropical freshwater environments with their temperate counterparts. Water and land are distributed unevenly over the globe and so are freshwater bodies over the land and this is unrelated to population spread or economic development. In contrast to marine environments, freshwater bodies are many and vary in size, shape, depth and location. Thus, extensive water–terrestrial transition zones, so-called ecotones, are formed and play a central role in freshwater life. Ecotones, in fact, constitute habitats for some fish species, at least during part of their life (Agostinho and Zalewski, 1995). Indeed, external and internal processes influence the energy flow between the two interacting systems, the function of the ecotone, and, therefore, life and landscape interaction (Bugenyi, 2001; Johnson *et al.*, 2001). Freshwater bodies are broadly divided in two groups of environments: (a) standing water and (b) flowing water environments – lentic and lotic environments, respectively. Basically, lakes, reservoirs and wetlands are lentic, and rivers and streams are lotic environments. Two other types of freshwater bodies deserve some attention as habitats for tropical fishes: springs and caves.

A. Lakes and Ponds

Tropical lakes are far less numerous than temperate lakes because glacial lakes are rare in the tropics. Tropical lakes vary in size, from minute ponds to lakes with gigantic proportions, such as Lake Victoria, with a surface area of 68 635 km². However, the great majority of them are relatively small water bodies – only 88 lakes in the world have a surface area larger than 1000 km² and just 19 are larger than 10 000 km², six of them are located in the tropics: four in Africa (Victoria, Nyasa, Chad and Turkana), one in South America (Maracaibo) and one in Australia (Lake Eyre). Some tropical lakes are located in high mountains and most have tectonic origin, for example Lake Titicaca, which is 3812 m above sea level in South America, or Lake Victoria, located at

1136 m and Lake Tanganyika, at 773 m above sea level in Africa (Babkin, 2003). Table 1.3 shows the morphological characteristics of the major tropical lakes.

Roughly, tropical and temperate lakes are not different in total annual solar irradiance but they do differ in minimum annual irradiance (Lewis Jr, 1996). Changes in solar irradiance induce gradients in water temperature that further lead to water column mixing. Light controls photosynthesis that is further moderated by temperature and nutrient supply and these are different between both temperate and tropical lakes and among Tropical lakes. Clearly, mean temperature decreases from the Equator up to the Tropics of Cancer and Capricorn though there is no difference in the annual maximum temperature and, thus, temperate and tropical lakes are differentiated mainly by minimum rather than by maximum temperatures. Seasonal changes in temperature are associated with water mixing and stratification that are clearly present in temperate lakes but by no means absent in tropical lakes. Deep tropical lakes stratify and tend to mix predictably at a particular time of the year (Lewis, 1987). In contrast, floodplain lakes are destabilized annually by hydraulic forces as observed for some lakes of the Amazon. In this case, floodplain lakes are annually inundated by lateral overflows of rivers. Wind may also affect stratification and water mixing in tropical lakes more readily than in their temperate counterparts; this more dynamic process has been related to the efficiency of recycling nutrients and the productivity of tropical lakes compared with temperate lakes (Lewis, 1987).

Continuously high temperature throughout the water column and continuously high solar irradiance make the basic conditions for a high rate of annual photosynthesis in tropical lakes. This often results in hyperoxia during the day and hypoxia during the night. The chemical and biological demand for oxygen is high in tropical lakes. Together these conditions result in a hypoxic or even anoxic hypolimnion that has consequences for oxygen concentrations all through the water column, and dramatically affects biogeochemical cycles of carbon, nitrogen and phosphorus. Because chemical weathering of phosphorus is more efficient at higher temperatures and because denitrification is higher in tropical waters, tropical lakes experience low nitrogen:phosphorus ratios in the hypolimnion, and not rarely, a nitrogen deficit takes place throughout the water column when deep waters mix with surface waters. Thus, the deficit of nitrogen is more critical for tropical lakes than the amount of phosphorus (Lewis Jr, 2000).

The hypolimnion of tropical lakes is more prone to anoxia and chemical stratification than temperate lakes, due to the reduced oxygen solubility at high temperatures and to the increased oxygen consumption by a variety of biological and chemical processes. Oxygen scarcity is widespread in tropical freshwater, particularly in floodplain lakes, inundated forests and permanent

Table 1.3
Morphological Characteristics of Major Tropical Lakes

Lake	Country	Major characteristics	Depth maximum (m)	Volume (km^3)	Surface area (km^2)	Watershed area (km^2)	Residence time (years)	Annual fish catch (ton/year)
Chad	Chad, Cameroon, Niger, Nigeria	Known for its yield of natural soda, an activity that contributes to keeping the lake water fresh	10.5	72	1540	24 264	NA	135 500
Eyre	Australia	A great salt lake of tectonic origin. The vast catchment area is only marginally desert and as such is very responsive to even slight variations of rainfall	5.7	30.1	9690	1 140 000	NA	NA
Maracaibo	Venezuela	Largest lake of South America, semi-arid in the north and has an average rainfall of 127 cm in the south	60	280	13 010	NA	NA	NA
Nyasa	Mozambique, Malawi and Tanzania	Most southerly of the great African Rift Valley lakes, consisting of a single basin	706	8400	6400	6593	NA	21 000

Tanganyika	Tanzania, Zaire, Zambia and Burundi	Second largest of the African lakes; second deepest (next to L. Baikal) and the longest lake of the world. Its very ancient origin is only rivalled by such old lakes as Baikal	1430	17 800	32 890	263 000	NA	518 400
Titicaca	Bolivia and Peru	Largest lake in South America, highest elevation large lake in the world, one of the oldest lakes in the world	281	893	8372	58 000	1343	6327
Turkana	Ethiopia and Kenya	Tertiary volcanic rocks are found in the south and along most of the western side of the lake, while a later lava flow (Pleistocene) forms a barrier in the southern end of the lake	109	203.6	6750	130 860	12.5	15 000
Victoria	Kenya, Tanzania and Uganda	Second largest freshwater lake by surface area; one of the oldest lakes in the world	84	2750	68 800	184 000	23	120 000

Abbreviations: MY, million years; NA, data not available.
Source: Borre *et al.* (2001); ILEC (2004).

swamps (Carter and Beadle, 1930; Chapman *et al.*, 1999; Junk, 1996; Kramer *et al.*, 1978; Townsend, 1996; Val and Almeida-Val, 1995). In many of these habitats, dissolved oxygen exists only in the first few millimeters of the top of water column with levels close to zero or even zero below this water layer (Figure 1.4). In floodplain lakes of the Amazon, this oxygen is the sole source for many fish species that have evolved an extraordinary set of adaptations to explore this zone of the water column (Junk *et al.*, 1983; Val, 1995; see Chapters 6 and 7). Habitat diversity, structure and function of river floodplains of the Amazon have been reviewed elsewhere (Junk, 1997).

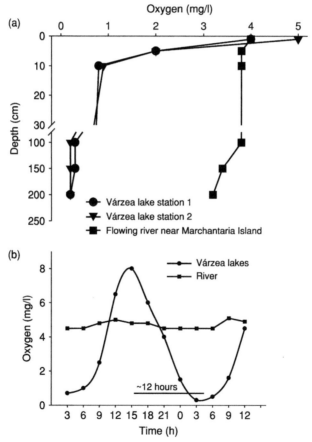

Fig. 1.4 Changes in dissolved oxygen in water bodies of the Amazon: (a) diurnal changes in oxygen levels according to water depth in a *várzea* lake; (b) comparison of dissolved oxygen during 24 hours in a *várzea* lake and in the river.

Seasonal variations in dissolved oxygen are observed for tropical shallow lakes but they are not as extreme as those described for temperate lakes. Extreme variations in dissolved oxygen do occur in tropical lakes but they tend to occur in much shorter periods of time, e.g., 24 hours. At floodplain lakes, dissolved oxygen can drop from oversaturated levels at noon to values close to zero at night (Junk et al., 1983; Val, 1996). After periods of nutrient unloading, the situation may be worsened, as an extensive cover of aquatic plants, most of them macrophytes, is developed limiting irradiance of the water column and further limiting the already weak photosynthesis capacity. Subsequently, such aquatic plant biomass increases oxygen demand due to organic decomposition. Under these situations, stratification develops resulting in hypoxic or even anoxic conditions in the hypolimnion. When the thermocline is disrupted by the end of the day or by winds, water layers mix and hypoxic conditions occur even in the water surface layer. Hydrogen sulfide is displaced throughout the water column when bottom water mixes into the water column above; this poses an extra challenge to fish (Affonso et al., 2002; Brauner et al., 1995). Adaptations of tropical fishes to hypoxia occur at all levels of their biological organization and will be discussed elsewhere in this volume (Chapters 6, 7 and 10).

Shallow lakes have small stock of water per unit of area but not necessarily a corresponding reduction of water fluxes and therefore are sensitive to water surface processes (Talling, 2001). This contrasts to deep tropical lakes. Lake Tanganyika is the second largest tropical lake and houses a great species richness primarily accounted for by endemic fishes. Its high productivity, interestingly, comes mainly from off-shore, open-water food web, which is biologically poor. The reduced temperature gradient between water surface and water bottom (1470 m) assure wind-driven mixing in bringing nutrients for primary production from the deep water layers. However, it seems that increased air and surface-water temperatures enhance the water density differences, reducing the effectiveness of water mixing with already mapped clear effects on primary production and fish yields.

Many tropical shallow lakes are located in arid and semi-arid zones over dryland. They are large and fill erratically and then recede and dry until the next major inflow, so that water levels may fluctuate widely often in accord with fluctuations in salinity. With increasing aridity, dryland lakes experience increased spatio-temporal variability of rainfall, i.e., in semi-arid and sub-humid regions rain falls on a seasonal basis while within arid zones rain falls unpredictably and episodically. As a large amount of water evaporates from these types of water bodies, leaving behind salts carried in, many of them experience increases in salinity above the limit, now widely accepted as 3 g/l, to be considered as freshwater lakes. Salinities of these so-called salt lakes may vary from 3 up to 300 g/l on a seasonal basis, depending on inflows and rainfall. However, a number of factors give rise to salinization and are related to anthropogenic effects such as excessive clearance of natural vegetation, overuse of water for irrigation, and changes in the nature of groundwater/ surface water interaction. Despite the causes of increases and changes in salinity, it demands significant physiological adjustments of the biological communities inhabiting these environments to maintain ionic homeostasis (see Chapter 9, and Timms, 2001; Williams, 2000; Williams *et al.*, 1998).

Salt lakes may be highly alkaline; for example, the soda lakes of East Africa. The water of Lake Magadi in the Kenyan rift valley is highly alkaline (pH10) and highly buffered ($CO_2 = 180\,mmol^{-1}$). Water with these character- istics would rapidly kill most teleost fishes as they are unable to excrete ammonia under these conditions, with the exception of the Lake Magadi tilapia, *Alcolapia grahami* (Randall *et al.*, 1989; Wood *et al.*, 1989). Recently, a review of cichlids inhabiting Lake Magadi and Lake Natron, another rep- resentative example of a rift valley soda lake in East Africa, indicates the presence of four species of tilapiines cichlids (Turner *et al.*, 2001), possibly all having the same ability to excrete nitrogen as urea at high rates (Narahara *et al.*, 1996) or an unknown system to avoid neurotoxicity caused by increased body levels of ammonia.

In an opposite situation, though not less challenging, are the acidic ion poor lakes and *igapós* of the Amazon. These water bodies are rich in dissolved organic carbon (DOC), very low in ions (resembling distilled water), acidic (pH 3–3.5) and often hypoxic (Furch and Junk, 1997; Matsuo and Val, 2003). However, fish fauna inhabiting these waters are relatively rich. Rio Negro harbors more than 1000 fish species (Ragazzo, 2002; Val and Almeida-Val, 1995) that are able to maintain ion homeostasis under the dominant environ- mental conditions (see Chapter 9, this volume and Gonzalez *et al.*, 1998; Gonzalez *et al.*, 2002; Matsuo and Val, 2002; Wilson *et al.*, 1999; Wood *et al.*, 1998). This fish diversity contrasts with that found in alkaline salt lakes that harbor a reduced number of fish species, which initially suggests that acidic conditions present fewer challenges than alkaline conditions, although this has yet to be proven. The presence of specific compounds in the blackwaters of Rio Negro, such as humic and fulvic acids, may provide additional protection against the dominant ion-poor acidic condition of these environments. Many intermediary water conditions and characteristics appear in areas where different types of primary water mix, for example in open lakes formed along the confluence of black and white waters of the Amazon.

Volcanic lakes contrast with these water bodies. Two types of volcanic lakes can be distinguished: (a) volcanic crater lakes with steep-sided walls that are, in general, deep lakes; and (b) volcanic barrier lakes, formed by the blockage of steep-sided river valleys by volcanic lava flows that provide wind-sheltered conditions conducive to long-term stratification (Beadle, 1966; Beadle, 1981). Volcanic lakes make up to 10% of all natural lakes in the tropics, as estimated by Lewis Jr (1996). Lakes of volcanic origin are abundant in parts of Africa and Central America, the Costa Rican and Nicaraguan volcanic lakes being the most studied (Chapman *et al.*, 1998; Umana *et al.*, 1999). Based on salinity, volcanic lakes are separated into saline and dilute groups that are further divided into several sub-groups (Pasternack and Varekamp, 1997). The environmental conditions of crater lakes are in general more challenging than barrier lakes and this is reflected in fish diversity. Cichlids and cyprinids are among the groups living in volcanic lakes (Bedarf *et al.*, 2001; Danley and Kocher, 2001).

Cave lakes are as challenging an environment as volcanic lakes for fish. Caves are formed by dissolution of large areas of underground limestone that, in general, interconnect with several chambers. These caves are usually open to the atmosphere and water percolating through supplies the material to form stalagtites, stalagmites, dripstones and flowstones. Cave habitats are intrinsically fragile and are among the most unknown environments (Culver, 1982). Fishes inhabiting caves depend on sinking materials as dark conditions prevent primary production. In general, cave fishes are blind, their populations are small and present a precocial life style, leading to slow population turnover, as already described for cave fishes (see Chapter 4; Romero, 2001; Trajano, 1997).

Equally challenging are tropical high altitude lakes that, in general, present thermal characteristics intermediate between temperate and tropical lakes (Chacon-Torres and Rosas-Monge, 1998). Lake Titicaca is the largest freshwater lake in South America and the highest of the world's large lakes, sitting 3810 m above sea level. Covering 8400 km^2 and having a volume of 932 km^3, Lake Titicaca consists of three basins: *Lago Grande, Bahia del Puno* and *Lago Pequeno*, which are all part of a large endorheic basin that drains into the Amazon basin. Despite draining into the hottest fish diversity spot, Lake Titicaca is itself relatively poor in fish diversity. In general, all high altitude lakes demand special management rules to preserve their biological characteristics (Borre *et al.*, 2001).

B. Reservoirs

In contrast to lakes, reservoirs are bodies of fresh water artificially created by humans by the establishment of dams or excavations across rivers, streams or run-off channels. In many rivers, damming is constructed in cascade, consisting of a series of sequential dams. The number of dams and reservoirs has increased from 5000 in 1950s to 40 000 in 1980s. Today, the total number of dams and reservoirs, including small ones, is estimated to be 800 000,

Table 1.4
Comparative Characteristics of Lakes and Reservoirs

Lakes	Reservoirs		
Main characteristics	Main characteristics	Positive benefits	Negative effects
Especially abundant in glaciated areas; orogenic areas are characterized by deep, ancient lakes; riverine and coastal plains are characterized by shallow lakes and lagoons	Located worldwide in most landscapes, including tropical forests, tundra and arid plains; often abundant in areas with a scarcity of shallow lakes and lagoons	Production of energy (hydropower)	Displacement of local populations following inundation of reservoir water basin and excessive human immigration into reservoir region, with associated social, economic and health problems
Generally circular water basin	Elongated and dendritic water basin	Increased low-energy water quality improvement	Deterioration of conditions for original population and increased health problems from increasing spread of waterborne disease and vectors
Drainage: surface area ratio usually <10:1	Drainage: surface area ratio usually >10:1	Retention of water resources in the drainage basin	Loss of edible native river fish species and loss of agricultural and timber lands
Stable shoreline (except for shallow lakes in semi-arid zones)	Shoreline can change because of ability to artificially regulate water level	Creation of drinking water and water supply resources	Loss of wetlands and land/water ecotones and loss of natural floodplains and wildlife habitats
Water level fluctuation generally small (except for shallow lakes in semi-arid zones)	Water level fluctuation can be great	Creation of representative biological diversity reserves	Loss of biodiversity, and displaced wildlife populations
Long water flushing time in deeper lakes	Water flushing time often short for depth	Increased welfare for local population	Need for compensation for loss of agricultural lands, fishery grounds and housing

Rate of sediment deposition in water basin is usually slow under natural conditions	Rate of sediment deposition often rapid	Enhanced recreational possibilities	Degradation of local water quality
Variable nutrient loading	Usually large nutrient loading	Increased protection of downstream river from flooding events	Decreased river flow rates below reservoir; increased flow variability; and decreased downstream temperatures, transport of silt and nutrients
Slow ecosystem succession Stable flora and fauna (often includes endemic species under undisturbed conditions)	Ecosystem succession often rapid Variable flora and fauna	Increased fishery possibilities Storage of water for use during low-flow periods	Barrier to upstream fish migration Decreased concentrations of dissolved oxygen and increased concentrations of hydrogen sulfide and carbon dioxide in reservoir bottom water layer and dam discharges
Water outlet is at surface	Water outlet is variable, but often at some depth in water column	Enhancement of navigation possibilities	Loss of valuable historic or cultural resources (e.g., burial grounds, relic sites, temples)
Water inflow typically from multiple, small tributaries	Water inflow typically from one or more large rivers	Increased potential for sustained agricultural irrigation	Decreased aesthetic values, and increased seismic activity

occupying an area equivalent to 400 000 km^2. The number of river dams of more than 15 m high is estimated to be 45 000, distributed unevenly worldwide. Approximately 1700 larger dams are currently under construction. Worldwide, dams and reservoirs are used for irrigation (48%), hydropower generation (20%) and for flood control (32%). The small reservoirs are more productive than large reservoirs as result of the greater area:volume ratio. Also, small reservoirs are characterized by thermal instability, with rapid exchange of nutrients within the water column and water–sediment interface (Mwaura *et al.*, 2002). Lakes and reservoirs differ in respect to many ecological parameters (Table 1.4) but nothing is more destructive to riverine and riparian fish species than dams that alter the conditions to which the local ecosystem has adapted and, therefore, almost certainly reduce species diversity.

Most dams and reservoirs are concentrated in the temperate and subtropical zones in developed countries. In South America, however, there is a high concentration of major dams and reservoirs constructed in cascade in many of the tributaries of Paraná and in the Paraná River itself. This basin has more than 50 dams and reservoirs, 14 of them are large (more than 15 m high). This concentration is paralleled only by that on the US Pacific coast. No other place within the tropics has a significant concentration of major dams and reservoirs than in tropical Africa. In particular, on the Zambezi River basin six major dams have been built, and all other African river basins have one or two major dams. Currently, Asian river basins experience a similar situation: six major dams will barricade the River Ganges and another 11 will be constructed on the Yangtze in the near future. The largest yet is the Yangtze Three Gorges dam, which will plug one of the largest rivers of the world (Chen, 2002).

In contrast to the Paraná basin, which will be the subject of a further analysis, the Amazon flows unplugged over its 6000 km. There are only five dams located on secondary tributaries in the Amazon (Table 1.5). These dams flood large areas, with the total flooded area by Tucuruí dam being by far the

Table 1.5
Dams of the Amazon

Dam	River basin	First operational date	Capacity (MW)	Reservoir area (km^2)
Tucuruí	Tocantins	1984	4000	2400
Balbina	Uatumã	1989	250	2300
Samuel	Jamari	1989	216	560
Curuá-Una	Curuá-Una	1977	30	78
Paredão	Araguari	1975	40	23
Total			**2536**	**5361**

largest at 2400 km^2, followed by the Balbina dam that flooded an area equivalent to 2300 km^2. In the Amazon region, river damming assumes special importance as it profoundly affects or even eliminates the natural flood pulses that are, as mentioned above, the major environmental driving force in this region, shaping all relationships of living beings with their environment, including all fish species (Gunkel *et al.*, 2003; Junk *et al.*, 1989; Middleton, 2002).

Most rivers without dams provide free corridors for aquatic fauna, including fish, and this contrasts to regulated rivers, as dams disrupt fish migrations. As large as the diversity of fish is the diversity of movements and behavior of tropical fishes and, comparatively, the biological consequences posed by dams in the tropics are likely to differ from the effects of dams on fishes in temperate regions. Thus, to better design the structures to reduce the disruption of fish movements in the tropics caused by river damming, information about such diversity is needed (Holmquist *et al.*, 1998). These structures should restore both upstream and downstream connectivity. Upstream connectivity is much more common and includes a number of structures to facilitate fish passage, contrasting with downstream connectivity and this has only just begun to be addressed (Larinier, 2000). An extensive analysis of the impact of large dams on freshwater fishes revealed that in 27% of the cases the impacts were positive against 73% that were negative, of which, 53% were downstream of the dam. Most of the negative impacts were related to obstruction of upwards migration and to changes in connections to the floodplain, though these kind of impacts were more frequent in temperate (56%) than in tropical (27%) zones (Craig, 2000).

C. Major River Basins

Rivers and streams are bodies of flowing water moving in one specific direction. Because of this characteristic they are classified as lotic environments. In general, their headwaters are located at springs, snowmelt places or even lakes from where they travel all the way down to their mouth, usually another water channel, a lake, or the ocean. Biological, chemical and physical characteristics change during this journey and are major determinants of fish fauna of these water bodies. Indeed, these characteristics are basin-specific keeping a strict relationship with regional weather conditions, and rivers are almost a "moving living tissue" that responds to local conditions. Taken together, all rivers of the world carry only 0.006% of total freshwater, as much as 68% of all freshwater is locked in glaciers and permanent snow cover and another 30% in groundwater. Though representing a small portion of all freshwater, rivers have been heavily threatened worldwide, mainly by changes in their hydrology due to constructions of dams and reservoirs

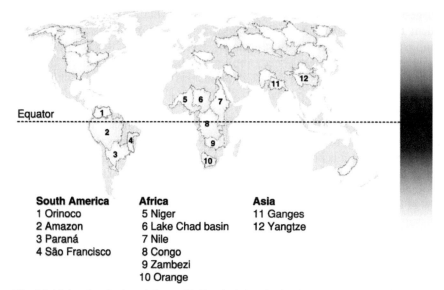

South America	Africa	Asia
1 Orinoco	5 Niger	11 Ganges
2 Amazon	6 Lake Chad basin	12 Yangtze
3 Paraná	7 Nile	
4 São Francisco	8 Congo	
	9 Zambezi	
	10 Orange	

Fig. 1.5 Major river basins of the world. Tropical river basins (1–12) are unevenly distributed along the tropics. Fish diversity increases towards the Equator (right bar). Modified from the map organized by United Nations Environmental Programme (UNEP); World Conservation Monitoring Centre (WCMC), World Research Institute (WRI), American Association for the Advancement of Science (AAAS), Atlas of Population and Environment 2001 (www.unep.org/vitalwater).

causing irreversible changes in many of the ecosystems closely associated with them.

The major tropical river basins are distributed unevenly between the Tropics of Cancer and Capricorn (Figure 1.5).

1. SOUTH AMERICA

In South America, four major river basins are recognized: Amazon, Orinoco, Paraná and São Francisco. The Amazon basin is by far the largest basin in the world, draining an area equivalent to the continental United States of America. The Amazon River itself discharges 175 thousand cubic meters of water into the Atlantic Ocean every second, representing 20% of the total freshwater entering all oceans, creating a phenomenon known locally as the "Pororoca" (from Tupi language, "great roar"). This discharge is five times that of the Congo River in Africa and 12 times that of the Mississipi River (Amarasekera *et al.*, 1997; Oltman, 1967; Sioli, 1984). A number of anastomosing water bodies create a very low topographic relief and an extensive aquatic "landscape" (Marlier, 1967; Sioli, 1991; Val and Almeida-Val, 1995) that undergoes a predictable annual flood cycle, the major ecological driving force in the Amazon. The difference between highest and lowest water level ranges from 4 meters near the mouth to 17 meters at the upper Japurá River; at the Port of Manaus, a crest of 10 meters, occasionally more, occurs in the main channel. Nearly all organic-water habitat interactions in the Amazon are fashioned by this flood pulse, in particular fish biology.

Environmental diversity in the Amazon is further magnified by the presence of different types of waters that are documented by the names of the rivers themselves, e.g., Rio Negro (black water), Rio Branco (white water), Rio Claro (clear water), among others. The colors of the waters have been related to specific conditions of the catchment areas. Sioli (1950) was the first to describe these three main types based, not only on color, but also on the amount and type of suspended solids, pH and dissolved minerals derived from specific conductance measurements. Sioli recognized three basic types of river water: (a) white water rivers, as the Amazon River, which are rich in suspended silt and dissolved minerals, carried from the Andes and Andean foothills, having a near-neutral pH; (b) black water rivers, which are black *in situ*, present high levels of humic and fulvic acids, leached from podzolic soils, have acidic pH, are poor in ions and present high transparency compared to the white water rivers; and (c) clear water rivers, drained from highly weathered tertiary sediments, that have pH varying from acidic to neutral, are low in dissolved minerals and present high transparency but may be greenish (Furch and Junk, 1997; Sioli, 1950). The Amazon basin harbors 3000 fish species (see Chapter 2), out of which 1000 or more thrive in Rio Negro waters, despite its low levels of minerals and acidic conditions (Table 1.6) that require special adjustments of ion regulation (see Chapter 9).

The Cassiquiare canal connects the upper part of the Rio Negro with the Orinoco basin, making possible fish exchange between these two basins. The Orinoco River is 2150 km long and is the third largest river in the world in discharge, after the Amazon and Congo rivers, with a mean discharge of 36 000 m^3/s (DePetris and Paolini, 1991). Again, most of the large tributaries of the Orinoco, such as the Apure, Meta and Guaviare rivers, have their origin in the Andes and join the Orinoco from its western side, carrying large amounts of Andean sediments. For fish, this river basin has represented the possibility of colonization of new environments, both the Amazon and the southern South American basins.

The Paraná river basin is the second largest basin in South America. The Paraná River rises in the south-east central highlands of Brazil and flows generally southward. The Paraná's major tributary is the Paraguay River, which rises in the Mato Grosso state of Brazil and flows into the Paraná state near the northern border of Argentina. The Paraná runs 4695 km (nearly 3000 miles), discharging 15000 m^3/s of water into the La Plata estuary near Buenos Aires and Montevideo. Suspended sediments in the Paraná

Table 1.6

Main Characteristics of Water of the Rio Negro in Comparison with
Solimões/Amazonas River, and Forest Streams

	Negro river	Solimões Amazon river	Forest streams
Na (mg/l)	0.380 ± 0.124	2.3 ± 0.8	0.216 ± 0.058
K (mg/l)	0.327 ± 0.107	0.9 ± 0.2	0.150 ± 0.108
Mg (mg/l)	0.114 ± 0.035	1.1 ± 0.2	0.037 ± 0.015
Ca (mg/l)	0.212 ± 0.066	7.2 ± 1.6	0.038 ± 0.034
Cl (mg/l)	1.7 ± 0.7	3.1 ± 2.1	2.2 ± 0.4
Si (mg/l)	2.0 ± 0.5	4.0 ± 0.9	2.1 ± 0.5
Sr (μg/l)	3.6 ± 1.0	37.8 ± 8.8	1.4 ± 0.6
Ba (μg/l)	8.1 ± 2.7	22.7 ± 5.9	6.9 ± 2.9
Al (μg/l)	112 ± 29	44 ± 37	90 ± 36
Fe (μg/l)	178 ± 58	109 ± 76	98 ± 47
Mn (μg/l)	9.0 ± 2.4	5.9 ± 5.1	3.2 ± 1.2
Cu (μg/l)	1.8 ± 0.5	2.4 ± 0.6	1.5 ± 0.8
Zn (μg/l)	4.1 ± 1.8	3.2 ± 1.5	4.0 ± 3.3
Total P (μg/l)	25 ± 17	105 ± 58	10 ± 7
Total C (mg/l)	10.5 ± 1.3	13.5 ± 1.3	8.7 ± 3.8
HCO_3-C (mg/l)	1.7 ± 0.5	6.7 ± 0.8	1.1 ± 0.4
pH	5.1 ± 0.6	6.9 ± 0.4	4.5 ± 0.2
Conductance (μS)	9 ± 2	57 ± 8	10 ± 3

Source: Furch (1984); Furch and Junk (1997).

River are carried at a mean of *ca.* 80 mega tons per year, mainly supplied by the
Bermejo River, the main Andean tributary of the Paraguay River that is itself
loaded with 4500 mg/l of sediment. Thus, in addition to the sediments carried
to the Atlantic in the North by the Amazon River, the Andes also contributes
with a significant amount of sediment carried to the Atlantic in the South. A
peculiar morphological characteristic of this river is the ample inner delta,
some 320 km long and over 60 km wide, which is made of many different
anastomosing water bodies, including numerous channels and streams,
ponds, ox-bows, lakes, some containing blackwater, separated by levees
and bars of variable extensions and height. Most of the sedimentological,
morphological and hydrological (see also Table 1.7) characteristics
of the Paraná River basin have been reviewed elsewhere (DePetris
and Paolini, 1991). Because this river basin drains a highly populated area
that creates an enormous demand for energy, the Paraná River is the highest
fragmented river of the world with 14 major dams (major dam = height greater
than 150 meters, volume greater than 15 mega cubic meters, reservoir
storage capacity with at least 25 cubic kilometers or generating capacity
greater than 1000 megawatts). Only the Colorado and the Columbia rivers,

Table 1.7
Sedimentological, Morphological and
Hydrological Characteristics of the Paraná River Basin

Parameter	Min.–Max.	Discharge weighted mean	Mass transport rate ($\times 10^6$ t/year)
Water gauge (m)	1.78–6.69	-	-
Secchi disk (m)	0.09–0.39	-	-
TSS (mg/l)	49–302	101	-
pH	6.26–7.92	7.19	-
Eh (mV)	342–502	398	-
Conductivity (μS/cm)	32–115	57.6	-
Alkalinity (meq/l)	0.21–1.5	0.69	21.5 (assuming all $CaCO_3$)
Chloride (mg/l)	4.2–12.5	6.47	4.2
Hardness (meq/l)	0.24–0.72	0.45	14.0 (assuming all $CaCO_3$)
Calcium (mg/l)	2.18–11.7	6.92	4.2
Magnesium (mg/l)	1.13–2.7	2.09	1.2
Sodium (mg/l)	1.27–10.1	5.32	3.3
Potassium (mg/l)	1.64–6.3	3.65	2.2
Total phosphate (mg/l)	0.06–2.5	1.1	0.7
Dissolved silica (mg/l)	16.1–19.7	17.1	11.0
Oxygen (mg/l)	4.43–10.8	8.14	4.9
O_2 saturation (%)	46.6–115	90.3	-

Source: DePetris and Paolini (1991).

both in United States, have more than 10 dams along their river courses. River damming generates an enormous impact on fish ecology as we shall see later. Despite this, more than 500 fish species still inhabit the Paraná River basin (Brasil, 1998).

The São Francisco River basin is located in the northeast of Brazil. The river flows 1609 km from the south–central region, traversing diverse climatic zones before discharging into the Atlantic Ocean. The São Francisco basin is as large as that of the Danube and Colorado and faces similar water problems as a result of haphazard development projects such as mining, irrigation, hydropower and damming. As many of these projects did not take environmental considerations into account fish diversity has been heavily threatened. This river basin harbors 150 freshwater fish species, many of them endemic to this watershed (Britski *et al.*, 1988; Menezes, 1996).

2. AFRICA

The major river basins of Africa contrast with those of South America in regard to three main issues: (a) the major rivers are largely international, requiring an international effort for effective management; (b) the spatial distribution of surface water is uneven, resulting in many areas being water stressed or dependent on other external sources – at least 14 countries suffer water stress or scarcity; and (c) reduced water quality and sanitation (UNEP, 2002). River fragmentation is another important issue in Africa – the region has more than 1200 dams and reservoirs, 60% of which are located in South Africa (539) and Zimbabwe (213), most of them constructed to facilitate irrigation. Among them are four out of the five biggest dams in the world (see below). River basins of Africa are divided into inland, which refers basically to the Lake Chad basin, and peripheral systems. The major peripheral rivers drain into the Atlantic Ocean (Congo, Niger and Orange), into the Indian Ocean (Zambezi) and into the Mediterranean Sea (Nile).

There are two major rivers in the Lake Chad basin, the Rivers Chari and Lagone which, together, provide near 95% of the water flowing into Lake Chad. These two rivers have their origins in the Cameroon Adamawa Highlands. This river basin has experienced a massive reduction in the past decades as a consequence of poor land use in the catchment, deforestation and natural drought (Birkett, 2000). The Chari and its extensive floodplains support a rich terrestrial fauna with many endemic species. Over 100 species of fish have been described and are under severe environmental pressure. Many of these fish species are found only in this basin.

The Congo river basin contains the most diverse and distinctive group of animals adapted to a large-river environment in tropical Africa – an exceptional group of endemic species adapted to large rapids in the lower Congo. In the Pliocene age, about 12 million years ago, this basin was a large lake that had no outlet to the ocean. Over time, Congo river water, which was a coastal river at that time, broke through the margins of the huge lake and flowed down a series of rocky large rapids to enter the Atlantic Ocean (Beadle, 1981). Even during the dry periods of the Pleistocene, the Congo basin remained relatively stable, compared with the surrounding terrestrial environments. This stability of the aquatic environment and the isolation from other eco-regions are thought to be the grounds for the appearance and evolution of a rich biota, in particular the rich endemic freshwater fauna within the Congo River basin (Beadle, 1981). Fortunately, nowadays, the Congo River basin is almost an undisturbed environment with reduced pollution, low deforestation rates and reduced number of dams. With regard to flood pulses, rainfall and animal migrations, the Congo River basin operates within most of their original conditions. In terms of fish diversity, the Congo River basin is the second richest environment, harboring nearly 700 species of fish, many of them endemic to this environment (see Chapter 2) and highly adapted to local conditions, for example the fishes and spiny-eels that live in the lower Congo River rapids.

The Niger river basin covers nine countries in West Africa; the Niger River itself is 4100 km in length, the third largest river in Africa, and traverses four countries. The singularities of this river were well expressed in the original name of the river, which was "Egerou n-egerou," which means "river of rivers." The headwaters of the Niger are located at Fouta Djalon and Mount Nimba in Guinea, ending in the Atlantic Ocean with an inter-annual discharge of 4800 m^3/s. Because most of the catchment area of this river basin is located within arid zones, with low rainfall (except for headwaters in Guinea and the low Niger) and because of dams constructed on the Nile and Benue rivers, the region has been seriously impacted by drought during the past 40 years. Drought is particularly alarming because of the associated changes in salinity in the Niger delta that affects the biota of the coastal environment. Sedimentation, which has increased significantly within the area, also represents a new environmental stress for this river basin. Two major areas of this basin should be pin-pointed: the inland delta in Mali and the Niger Delta in Nigeria. These two areas present high biodiversity, comparable only with that found in the Amazon basin. The floodplains of the Niger River and delta wetlands harbor a specialized flora that is adapted to extreme fluctuations in water levels. The inner delta is a wetland of major international importance, covering an area of *ca.* 3.2 million ha, producing almost 100 000 tones of fish annually. The Niger delta is equally environmentally complex. The distinctive ecological attributes of this delta include sandy ridge barriers, brackish and saline mangroves, seasonal swamps that harbor a variety of aquatic animals adapted to the main ecological forces of these ecosystems – the flood pulses and the tidal inversion of the Atlantic. The flood pulse also has a major effect on dissolved solutes. The Niger contains 250 fish species, 20 of them endemic to this basin (Abe *et al.*, 2003; Martins and Probst, 1991).

The Orange River basin, another peripheral basin draining into the Atlantic Ocean, is the most important one for South Africa, stretching over an area of more than one million square kilometers and including four countries: Botswana, Lesotho, Namibia and South Africa. The Orange River originates in the Lesotho highlands, flows 2300 km, forming the border between South Africa and Namibia, before entering the Atlantic Ocean. On its way from the headwaters to the ocean, the Orange River has a rainfall decreasing from 1800 mm at its source to a mere 25 mm westward. This river is seen as the most "developed" river in Africa having in total 29 dams, 22 of them in South Africa, and the most stressed water system in the world as water abstraction in the Orange River basin in southern Africa is close to the maximum amount available – a water stress indicator of 0.8 to 0.9 (water stress indicator is the ratio of water abstracted from a basin to total water available). Currently, the mean average discharge of the Orange River is 360 m^3/s, which represents about 20% of the mean annual runoff of South Africa. The second main river composing this basin is the Vaal, which occupies the old Witwatersrand basin, formed nearly 4 billion years ago (the Earth is about 4.5 billion years old).

Thus, this river is possibly the oldest river on Earth. Despite the major anthropogenic pressures, this river basin harbors 27 fish species, seven of them endemic to this river basin (Martins and Probst, 1991).

The Zambezi River basin is shared by Angola, Botswana, Malawi, Mozambique, Namibia, Tanzania, Zambia and Zimbabwe. The Zambezi is the fourth in size of the rivers of Africa and the largest flowing eastward, with a total length of 2650 km from its source in north-western Zambia to its delta into the Indian Ocean in central Mozambique. The basin, including the associated wetlands, is the largest intact block of wildlife habitat in Africa even though, in many parts, the river is highly fragmented. The average flow of this river as it enters the sea is $3600 \, m^3/s$. A peculiar formation along the Zambezi River is the Victoria Falls, which mark the end of what is known as the upper Zambezi. The middle and lower Zambezi are very similar as regard fish diversity, both differing from the upper part. A total of 122 fish species have been described for this river basin, 15 of them endemic to this environment (Lamore and Nilsson, 2000; Tumbare, 1999; WCD, 2000; Winemiller and Kelso-Winemiller, 2003).

The Nile River basin is unique, covering about 3 million square kilometers. Ten countries share this unique river basin – Burundi, Democratic Republic of Congo, Egypt, Eritrea, Ethiopia, Kenya, Rwanda, Sudan, Tanzania and Uganda, where 160 million people live within the boundaries of the basin. The Nile is the world's longest river; it traverses 6671 km from its farthest sources at the headwaters of the Kagera River to its delta in Egypt on the Mediterranean Sea, where it discharges an inter-annual mean of $3100 \, m^3/sec$ of water. The Ruvyironza River of Burundi, one of the upper branches of the Kagera, is regarded as the ultimate source of the Nile. During this journey a sequence of unique ecological systems is displayed, hosting a number of varied landscapes, with high mountains, tropical forests, woodlands, lakes, savannas, wetlands, arid lands and deserts, varying from rain forests and mountains in the south to savannas and swamps in southern Sudan to barren deserts in the north. In total 129 fish species have been described for this water body, 26 of them endemic to the Nile River (Sutcliffe and Parks, 1999).

3. ASIA

The Yangtze and the Ganges River basins are the two major tropical river basins that deserve attention in Asia. The Yangtze basin has a seasonal semitropical climate, lying between 25 °N and 35 °N, covering an area of $1.81 \times 10^{-6} \, km^2$. Thus, the south part of this basin is highly influenced by tropical climate; in fact, some tributaries of the south meander along the Tropic of Cancer. The Yangtze River, also known as the Changjiang, is often called as the "equator" of China, dividing the country into two parts: the humid south and the dry north. In the dry north, in the Qinghai Tibet Plateau, are the headwaters of the Yangtze River, from where it flows for 6300 km until emptying into the East China Sea at Shanghai. This river course is divided in three sections: (a) the upper Yangtze, referring to the mountainous section from the headwaters to the city of Yichang; (b) the middle Yangtze, that flows on a flat plain down to the city of Huckou; and (c) the lower Yangtze, that stretches from Huckou to the East China Sea. More than 3000 tributaries and 4000 lakes form a complex riverine–lacustrine network that collect most of the Yangtze water from rain, despite the important contribution of the melting water coming from the headwater glacier. Severe environmental problems including erosion, pollution, water abstraction, downstream sedimentation and river fragmentation are among the concerns related to this basin. The Three Gorges Dam, being constructed in the Upper Yangtze and scheduled to be completed by 2009, is causing a major concern with regard to fish diversity and conservation. Despite these environmental pressures, fish species richness is high in the Yangtze River basin which harbors 361 fish species, belonging to Cypriniformes (273 species, 75.6% of the total), to Siluriformes (43 species, 11.9%), to Perciformes (23 species, 6.4%) and to 9 other Orders (22 species, 6.1%). Out of the 361 recognized species, 177 species or subspecies are endemic to the Yangtze basin (Chen et al., 2001; Cheng, 2003; Fu et al., 2003; Zhang et al., 2003).

The other Asian hydrographic basin includes the Ganges basin, which, together with the Ganges–Brahmapur delta and the Bengal fan, composes what is locally known as the Ganges River system. The Ganges River basin encompasses three distinct sections: the Himalayan belt with its deep valleys and glaciers in the north; the Ganges alluvial region in the middle, representing about 56% of the entire basin; and the plateau and hills in the south. Part of the basin has a semi-arid climate and the river-flow depends largely on highly erratic monsoonal rains. The Ganges and its tributaries flow through three countries: Nepal, India and Bangladesh, remaining as the main source of freshwater for half of the population of India and Bangladesh, and for nearly the entire population of Nepal. The Ganges River rises in Uttar Kashi District, in the Himalayan Mountains, at an elevation of 4100 m above sea level, draining an area of $1.1 \times 10^6 \, km^2$. From the Uttar Kashi District, the Ganges River meanders for 2525 km, flowing through three states, Uttar Pradesh, Bihar and West Bengal, before emptying into the Bay of Bengal. During its long course, many small and large tributaries, side lakes and floodplain areas make a number of complex ecosystems that have been severely threatened. The Yamuna River, also originating in the Himalayas, is the largest tributary of the Ganges, flowing into it at Allahabad. The Yamuna river sub-basin represents 42% of the Ganges River basin. Ecological studies of this main tributary have revealed a rapid deterioration of water quality, loss of fisheries, increased pollution and significant changes in the biotic communities. A special feature of the Ganges basin is the Ganges–Brahmaputra delta that is a vast swamp forest, called Sunderbans, where the sediments collected from Himalaya and from the respective draining basins of both Ganges and Brahmaputra rivers are transported, being deposited in the delta itself and in the Bengal fan, Bengal Bay. The Ganges River systems contribute as much as 20% of the global sediment input into the world's oceans, which places the Ganges River as the third largest sediment-transporting river in the world. Its annual mean water discharge into the ocean is $15\,000\,\mathrm{m}^3/\mathrm{s}$, i.e., a mere 2% of the global water flow into the oceans. The upper Ganges River harbors 83 fish species belonging to 20 Families, Cyprinidae being the most specious one, with 32 species. The whole basin is home to 141 freshwater fish species, with many endangered populations (Gopal and Sah, 1993; Rao, 2001; Singh *et al.*, 2003).

Several secondary river basins are not included in this review. They are, indeed, important from the ecological point of view, and should be considered as global scale increases from major river basins to regional river basins. These secondary river basins harbor a number of fish species which are, in many cases, endemic to these secondary basins, what gives them a special environmental status.

IV. WORLD FISH DISTRIBUTION

There is reasonable evidence that regional processes influence both regional and local diversity. However, geological influences are important in defining the initial evolutionary history of water bodies; for example, fish species inhabiting lakes depend on the conditions leading to the appearance of such water bodies and their initial colonization processes. Indeed, age of the system and its size play a major role – systems that are older and larger tend to have more species than young and small water systems. Nearly 4500 known species of fish inhabit freshwater habitats. This number could rise by an order of magnitude with the inclusion of new species found in newly sampled areas. Freshwater fishes comprise 40% of all fishes; for example an estimated 11 000 fish species, that together with 6000 species of mollusks make up the major freshwater animal groups. Undoubtedly, there is no single explanation for species richness in the different parts of the world. Ecology, evolution, biogeography, systematic, and paleontology are all required to understand global patterns of fish diversity (Rickelefs and Schluter, 1993).

Species richness increases strongly towards the Equator, i.e., moving from high to low latitudes, species richness within a sampling area of similar size increases, as has been documented for a wide variety of taxonomic groups, including freshwater and marine fishes. Indeed, these latitudinal gradients in species richness should be taken with some precaution as they are highly influenced by other positional and environmental variables, as longitude, elevation, depth, topography, aridity, types of soil, speciation and extinction rates, and immigration and emigration of species. Mechanisms as historical perturbation, environmental stability and heterogeneity, productivity and organic-environment interactions have been also listed among those generating systematic latitudinal variation in these processes and have been evoked to explain higher species richness within the tropics (Gaston, 2000). Because these processes and mechanisms interact in complex ways, it is not surprising that simple correlations with a specific abiotic factor are not always observed. The picture constitutes a Gordian knot for ecologists and biogeographers and the arguments are becoming more and more complex. With climate change, for example, non-native species may cross frontiers and become new elements of the biota (Walther *et al.*, 2002). In many cases, incursions of organisms, particularly fish, both from freshwater into marine habitats and vice-versa, occur and affect local species richness. This is the case, for example, for several marine-derived groups inhabiting the Amazon, such as stingrays, flatfish, pufferfish and anchovies (Lovejoy *et al.*, 1998). These incursions tend to expand as a consequence of global and local environmental changes caused by man. Therefore, there is no single or simple explanation for geographic variation in species diversity.

About 40% of the known fish species are freshwater forms. Given the distribution of water on Earth (see major river basins and lakes section), this is equivalent to one fish species per 15 km^3 of fresh water, compared with one species per 100 000 km^3 of seawater. Near isolated freshwater systems tend to provide the conditions for the appearance of new species rendering high diversity for many lineages of fish and invertebrates, quite different from the marine environments. In many cases, species richness and endemism are positively correlated (Watters, 1992). Even so, several hypotheses have been evoked to explain spatial variability in species richness: species-area hypothesis (Preston, 1962); species-energy hypothesis (Wright, 1983) and historical hypotheses, such as the refuge theory (Haffer, 1969; Weitzman and Weitzman, 1987). These hypotheses are under intense debate (Oberdorff *et al.*, 1997; Salo, 1987) and, again, it seems that there is no single or simple explanation for spatial gradients in species richness and many more places are to be sampled and mapped before we can have a clear picture.

An extensive analysis of fish species richness revealed that out of 108 watersheds 27 have particularly high fish species diversity, 56% of them located within the tropics, mostly in Central Africa, Southeast Asia and South America (see Figure 1.3). The analyzed tropical watersheds represent only about one-third of all tropical watersheds because data is lacking for many of these water bodies, including even medium and large ones. Therefore, fish diversity in tropical inland waters is almost certainly higher than what is known. In the north of South America, the Orinoco and Amazon River basins are high in number of fish species and high in number of endemic fish species. They are paralleled by the African watersheds, headed by the Congo River basin (Congo, Nile, Zambezi, Niger and Lake Chad, see above), with 700 fish species, 500 endemic to this basin, *ca.* one-quarter of the species richness and one-third of the fish endemism found in the Amazon (Nelson, 1994; Revenga *et al.*, 2000; Val and Almeida-Val, 1995).

V. GLOBAL CLIMATE CHANGES

Three main global environmental changes have direct effects on tropical water systems and drastically affect fish diversity and fish yields: global warming, hydrological changes and eutrophication.

A. Global Warming

Earth's climate has warmed by approximately $0.6\,°C$ over the past 100 years (Walther *et al.*, 2002) and is predicted to warm between 1.4 and $5.8\,°C$ over the next 100 years (IPCC, 2001). Global warming will, undoubtedly, bring many modifications to tropical water bodies in general. Lake warming already has had profound effects on fish in several large tropical lakes. The observed nutrient input from the anoxic zone of deep lakes will increase even further with global warming, strengthening the already critical seasonal stratification observed, for example, for lakes Tanganyika, Malawi, and Victoria. In addition to eutrophication, salinitization of shallow lakes, particularly those located in arid zones of the planet, is another consequence of global warming. Sea level rises as a result of both thermal expansion and partial melting of mountain and polar glaciers. Antarctic and Greenland ice caps will definitively affect the coastal ecosystems and estuaries, changing their limnological characteristics thus causing loss of species diversity. In summary, global warming will act as an ecosystem disruptor and will lead to changes in the major ecological driving forces, profoundly affecting the biological diversity and possibly reducing fish distribution within the tropics (IPCC, 2001; Verburg *et al.*, 2003; Walther *et al.*, 2002; Williams *et al.*, 2003).

B. Hydrological Changes

Water abstraction by upstream dams, reservoirs and irrigation can trigger dramatic hydrological changes worldwide but more remarkably within arid zones and among tropical water bodies. Dams and reservoirs provide unquestionable benefits but they do disrupt hydrological cycles as no other global environmental pressure; these effects include suppression of natural flood cycles, disconnection of river and marginal lakes from their wetlands and floodplains, changes in the extension of ecotones, changes in deposition of sediments downstream, as well as the disappearance of habitats such as water-falls, rapids, riparian vegetation and floodplains. In other words, water abstraction affects the ecological functions of the aquatic ecosystems, reducing the ability of these ecosystems to buffer the anthropogenic pressures on it. In addition, as many of these areas are used as nursery and feeding places by many fish species, reduction of fish yields follow these derived environmental changes. Tropical reservoirs are also prone to colonization by floating plants (Gunkel *et al.*, 2003; Ramírez and Bicudo, 2002; Tundisi, 1981). Among the above-mentioned impacts, damming a river will cause biological fragmentation, a great concern for conservation. Population fragments may or may not be able to replace their original genetic variability and, therefore, recover their original size. This is a great challenge for river fish assemblages and has just started to be addressed. Major dams may not be a problem for resident fishes; while smaller dams may cause huge disturbances for migrating species, unless provided with the so-called fish stairs or ladders, which allow them to complete their reproductive cycle.

C. Eutrophication

Production of nitrogen as fertilizer has increased dramatically during the past decades as a consequence of pressure to maximize food production, particularly in tropical countries. In general, tropical soils receive less phosphorus than required for sustainable agricultural use. However, part of both elements reaches tropical water bodies contributing significantly to their eutrophication. Eutrophication produces an excess of phytoplankton, algae and rooted aquatic plants (macrophytes) resulting in decreased oxygen content, accumulation of ammonia in the water column, as well as in a re-suspension of several chemical compounds from the sediments under the increased extent of anaerobic conditions occurring in the hypolimnion. Hyper-eutrophication ($>100\,\mu g/L$ PO_4 and $>25\,\mu g/L$ chlorophyll A) causes concerns in many water bodies across the world, in particular within the tropics where this phenomenon is causing the deterioration of extensive water areas and massive fish kills, often as a consequence of red tides. Biodiversity of most eutrophic and hyper-trophic water habitats decreases and these trophic stages are expanding, reaching coastal marine environments. All these habitats are in need of rigid control of phosphorus reaching freshwater bodies and nitrogen reaching marine environments (Wu, 2002). A recent review on the eutrophication of water bodies showed that this phenomenon clearly causes concerns, and some control policies are beginning to appear as a main issue for law makers, regarding environmental protection measures (Prepas and Charette, 2003).

VI. THE FUTURE

The tropics contain most of the world's fish species and are the source of much of the variation that has spread to other regions. By comparison the reduced fauna of temperate regions is much more studied than that in the tropics and, as a result, we know much more about temperate than tropical animals, including fish. For example, the effect of seasonal temperature changes, which dominates temperate biology, is well known but we know little of the impact of seasonal pulses of water on biota in tropical regions. In addition, much of the human population is located in the tropics, concentrated close to water, either by rivers or the coast. This large and increasing human population has, and is having, an enormous impact on the surrounding aquatic ecosystems, especially through eutrophication and expanded hypoxic conditions in both rivers and coastal areas. When this is coupled with the impacts of global warming and overfishing, due to the fact that the world's fishing capacity is far larger than the available fish stocks, it is clear that tropical aquatic ecosystems are under a clear and imminent threat. Unfortunately we may not even know what we are losing! On a more positive note, it is also clear that ecosystem recovery, although poorly described, is much more rapid in warm tropical environments than in temperate regions such that, as we begin to understand these systems, rapid recovery may be possible. In fact, because of the probable contracted time frame, the tropics may be the region to study large-scale recovery of ecosystems in detail.

ACKNOWLEDGMENTS

Part of the work reviewed here was supported by CNPq, INPA and FAPEAM. A. L. Val and V. M. F. Almeida-Val are recipients of research fellowships from CNPq.

REFERENCES

Abe, J., Wellens-Mensah, J., Diallo, O. S., and Mbuyil Wa Mpoyi, C. (2003). "Global International Waters Assessment. Guinea Current. GIWA Regional assessment 42." GIWA, Kalmar.
Affonso, E. G., Polez, V. L., Correa, C. F., Mazon, A. F., Araujo, M. R., Moraes, G., and Rantin, F. T. (2002). Blood parameters and metabolites in the teleost fish *Colossoma macropomum* exposed to sulfide or hypoxia. *Comp. Biochem. Physiol.* **133C**, 375–382.
Agostinho, A. A., and Zalewski, M. (1995). The dependence of fish community structure and dinamics on floodplain and riparian ecotone zone in Paraná River, Brazil. *In* "The

Importance of Aquatic-Terrestrial Ecotones for Freshwater Fish" (Schiemer, F., and Zalewski, M., Eds.), pp. 141–148. Kluwer Academic, Dordrecht.

Amarasekera, K. N., Lee, R. F., Williams, E. R., and Eltahir, E. A. B. (1997). ENSO and natural variability in the flow of tropical rivers. *J. Hydrol.* **200**, 24–39.

Angel, M. V. (1997). Pelagic biodiversity. *In* "Marine Biodiversity. Patterns and Processes" (Ormond, R. F. G., Gage, J. D. F., and Angel, M. V., Eds.), p. 449. Cambridge University Press, Cambridge.

Babkin, V. I. (2003). The Earth and its physical feature. *In* "World Water Resources at the Beginning of the Twenty-first Century" (Shiklomanov, I. A., and Rodda, J. C., Eds.), pp. 1–18. Cambridge University Press, Cambridge.

Baran, E. (2000). Biodiversity of estuarine fish faunas in west Africa. *Naga* **23**, 4–9.

Beadle, L. C. (1966). Prolonged stratification and deoxygenation in tropical lakes. I. Crater lake Nkugute, Uganda, compared with Lakes Bunyoni and Edward. *Limnol. Oceanogr.* **11**, 152–163.

Beadle, L. C. (1981). "The Inland Waters of Tropical Africa: An Introduction to Tropical Limnology." Longman, New York.

Bedarf, A. T., McKaye, K. R., Van Den Berghe, E. P., Perez, L. J. L., and Secor, D. H. (2001). Initial six-year expansion of an introduced piscivorous fish in a Tropical Central American lake. *Biol. Invas.* **3**, 391–404.

Berschick, P., Bridges, C. R., and Grieshaber, M. K. (1987). The influence of hyperoxia, hypoxia and temperature on the respiratory physiology of the intertidal rockpool fish *Gobius cobitis pallas. J. Environ. Biol.* **130**, 369–387.

Birkett, C. M. (2000). Synergistic remote sensing of Lake Chad: Variability of basin inundation. *Remote Sensing Environ.* **72**, 218–236.

Blaber, S. J. M. (2002). Fish in hot water: The challenges facing fish and fisheries research in tropical estuaries. *J. Fish Biol.* **61**, 1–20.

Blaber, S. J. M., Brewer, D. T., Salini, J. P., and Kerr, J. (1990). Biomass, catch rates and patterns of abundance of demersal fishes, with particular reference to penaeid prawn predators, in a tropical bay in the Gulf of Carpentaria, Australia. *Marine Biol.* **107**, 397–408.

Bone, Q., Marshall, N. B., and Blaxter, J. H. S. (1995). "Biology of Fishes." Chapman & Hall, New York.

Borre, L., Barker, D. R., and Duker, L. E. (2001). Institutional arrangements for managing the great lakes of the world: Results of a workshop on implementing the watershed approach. *Lakes Reservoirs Res. Manag.* **6**, 199–209.

Brasil (1998). Primeiro relatório nacional para a Convenção sobre Diversidade Biológica. Ministério do Meio Ambiente, dos Recursos hídricos e da Amazônia Legal, pp. 283. Brasília.

Brauner, C. J., Ballantyne, C. L., Randall, D. J., and Val, A. L. (1995). Air breathing in the armoured catfish (*Hoplosternum littorale*) as an adaptation to hypoxic, acid and hydrogen sulphide rich waters. *Canad. J. Zool.* **73**, 739–744.

Bridges, C. R. (1993). Ecophysiology of intertidal fish. *In* "Fish Ecophysiology" (Rankin, J. C., and Jensen, F. B., Eds.), pp. 375–400. Chapman & Hall, London.

Britski, H. A., Sato, Y., and Rosa, A. B. S. (1988). "Manual de identificação de peixes da região de Três Marias (com chaves de identificação para os peixes da bacia do São Francisco)." Câmara dos Deputados/CODEVASF, Brasília.

Bugenyi, F. W. B. (2001). Tropical freshwater ecotones: Their formation, functions and use. *Hydrobiologia* **458**, 33–43.

Caley, M. J., and John, J. S. (1996). Refuge availability structures assemblages of tropical reef fishes. *J. Animal Ecol.* **65**, 414–428.

Cameron, W. M., and Pritchard, D. W. (1963). Estuaries. *In* "The Sea" (Hill, M. N., Ed.), Vol. 2, pp. 306–324. Wiley & Sons, New York.

Carter, G. S., and Beadle, L. C. (1930). The fauna of the swampews of the Oaraguayan Chaco in relation to its environment. I. Physico-chemical nature of the environment. *J. Linnean Soc. Zool.* **37,** 205–258.

Chacon-Torres, A., and Rosas-Monge, C. (1998). Water quality characteristics of a high altitude oligotrophic Mexican lake. *Aquatic Ecosyst. Health Manag.* **1,** 237–243.

Chapman, L. J., Chapman, C. A., Brazeau, D. A., McLaughlin, B., and Jordan, M. (1999). Papyrus swamps, hypoxia and faunal diversification: Variation among populations of *Barbus meumayeri. J. Fish Biol.* **54,** 310–327.

Chapman, L. J., Chapman, C. A., Crisman, T. L., and Nordlie, F. G. (1998). Dissolved oxygen and thermal regimes of a Ugandan crater lake. *Hydrobiologia* **385,** 201–211.

Chen, C. T. A. (2002). The impact of dams on fisheries: Case of the Three Gorges Dam. *In* "Challenges of a Changing Earth" (Steffen, W., Jager, J., Carson, D. J., and Bradshaw, C., Eds.), pp. 97–99. Springer, Berlin.

Chen, X., Zong, Y., Zhang, E., Xu, J., and Li, S. (2001). Human impacts on the Changjiang (Yangtze) River basin, China, with special reference to the impacts on the dry season water discharges into the sea. *Geomorphology* **41,** 111–123.

Cheng, S. P. (2003). Heavy metal pollution in China: Origin, pattern and control. *Environ. Sci. Pollution Technol.* **10,** 192–198.

Cornell, H. V., and Karlson, R. H. (2000). Coral species richness: Ecological versus biogeographical influences. *Coral Reefs* **19,** 37–49.

Craig, J. F. (2000). Large dams and freshwater fish biodiversity. *In* "Dams, Ecosystem Functions and Environmental Restoration. Thematic Review II.1 prepared as an input to the World Commission on Dams" (Berkamp, G., McCartney, M., Dugan, P., McNeely, J., and Acreman, M., Eds.), pp. 1–58. WCD, Cape Town.

Culver, D. C. (1982). "Cave life: Evolution and Ecology." Harvard University Press, Cambridge.

Danley, P. D., and Kocher, T. D. (2001). Speciation in rapidly diverging systems: Lessons from Lake Malawi. *Mol Ecol.* **10,** 1075–1086.

DePetris, P. J., and Paolini, J. E. (1991). Biogeochemical aspects of South American Rivers: The Paraná and the Orinoco. *In* "Biogeochemistry of Major World Rivers. Scope 42" (Degens, E. T., Kempe, S., and Richey, J. E., Eds.), pp. 105–125. John Wiley & Sons, Chichester.

Field, C. B., Behrenfeld, M. J., Randerson, J. T., and Falkowski, P. (1998). Primary production of the biosphere: Integrating terrestrial and oceanic components. *Science* **281,** 237–240.

Figueiredo, J. L., Santos, A. P., Yamaguti, N., Bernardes, R. A., and Rossi-Wongtschowski, C. L. D. B. (2002). "Peixes da Zona econômica exclusiva da região sudeste-sul do Brasil. Levantamento com rede de meia água." EDUSP, São Paulo.

Fu, C., Wu, J., Chen, J. C., Wu, Q., and Lei, G. (2003). Freshwater fish biodiversity in the Yangtze River basin of China: patterns, threats and conservation. *Biodiversity Conserv.* **12.**

Furch, K. (1984). Water chemistry of the Amazon basin: the distribution of chemical elements among fresh waters. *In* "The Amazon. Limnology and Landscape Ecology of a Mighty Tropical river and Its Basin" (Sioli, H., Ed.), pp. 167–200. W. Junk, Dordrecht.

Furch, K., and Junk, W. J. (1997). Physicochemical conditions in the floodplains. *In* "The Central Amazon floodplain. Ecology of a Pulsing System" (Junk, W. J., Ed.), Vol. 126, pp. 69–108. Springer Verlag, Heidelberg.

Gaston, K. J. (2000). Global patterns in biodiversity. *Nature* **405,** 220–227.

Gibson, R. N. (1982). Recent studies on the biology of intertidal fishes. *Oceanogr. Marine Biol. Ann. Rev.* **20,** 363–414.

Gonzalez, R., Wood, C., Wilson, R., Patrick, M., Bergman, H., Narahara, A., and Val, A. (1998). Effects of water pH and calcium concentration on ion balance in fish of the rio Negro, Amazon. *Physiol. Zool.* **71,** 15–22.

Gonzalez, R. J., Wilson, R. W., Wood, C. M., Patrick, M. L., and Val, A. L. (2002). Diverse strategies for ion regulation in fish collected from the ion-poor, acidic Rio Negro. *Physiol. Biochem. Zool.* **75**, 37–47.

Gopal, S., and Sah, M. (1993). Conservation and management of river in India. Case-study of the River Yamuna. *Environ. Conserv.* **20**, 243–254.

Graham, J. B. (1970). Temperature sensitivity of two species of intertidal fishes. *Copeia* **1970**, 49–56.

Gunkel, G., Lange, U., Walde, D., and Rosa, J. W. (2003). The environmental and operational impacts of Curuá-Una, a reservoir in the Amazon region of Pará, Brasil. *Lakes Reservoirs Res. Manag.* **8**, 201–216.

Haffer, J. (1969). Speciation in Amazonian forest birds. *Science* **165**, 131–137.

Holdridge, L. R. (1947). Determination of world formations from simple climatic data. *Science* **105**, 367.

Holmquist, J. G., Schmidt-Gengenbach, J. M., and Yoshioka, B. B. (1998). High dams and marine-freshwater linkages: Effects on native and introduced fauna in the Caribbean. *Conserv. Biol.* **12**, 621–630.

Horn, M., Martin, K., and Chotkowski, M. (1998). "Intertidal Fishes: Life in Two Worlds." Academic Press, San Diego.

Horn, M. H., and Gibson, R. N. (1988). Intertidal fish. *Sci. Am.* **258**, 54–60.

ILEC (2004). World Lake Database. Survey of the state of the world lakes Vol. 2004. ILEC.

IPCC (2001). "Climate Change 2001: The Scientific Basis." Contribution of Working Group I to the Third Assessment Report of the Intergovernmental Panel on Climate Change (Houghton, J. T., Ding, Y., Griggs, D.J., Noguer, M., van der Linden, P.J., Dai, X., Maskell, K., and Johnson, C.A., Eds.). Cambridge University Press, Cambridge.

Johnson, N., Revenga, C., and Echeverria, J. (2001). Managing water for people and nature. *Science* **292**, 1071–1072.

Junk, W. J. (1997). Structure and function of the large central Amazonian river floodplains: Synthesis and discussion. *In* "The Central Amazon Floodplain" (Junk, W. J., Ed.), Vol. Ecological Studies 126, pp. 455–520. Springer Verlag, Heidelberg.

Junk, W. J., Bayley, P. B., and Sparks, R. E. (1989). The flood pulse concept in river–floodplain systems. *In* "Proceedings of the International Large River Symposium" (Dodge, D. P., Ed.), Vol. 106, pp. 110–127. Can. Spec. Publ. Fish. Aquat. Sci., Canada.

Junk, W. J., Soares, M. G., and Carvalho, F. M. (1983). Distribution of fish species in a lake of the Amazon river floodplain near Manaus (lago Camaleao), with special reference to extreme oxygen conditions. *Amazoniana* **7**, 397–431.

Junk, W. L. (1996). Ecology of floodplains - a challenge for tropical limnology. *In* "Perspectives in Tropical Limnology" (Schiemer, F., and Boland, E. J., Eds.), pp. 255–265. SBP Academic Publishing bv, Amsterdam.

Köppen, W. (1936). Das geographische system der klimate. *In* "Handbuch der Klimatologie" (Köppen, W., and Geiger, R., Eds.), Vol. Bd. 1, Teil C, pp. 1–44. Gerbrüder Borntraeger, Berlin.

Kramer, D. L., Lindsey, C. C., Moodie, G. E. E., and Stevens, E. D. (1978). The fishes and the aquatic environment of the central Amazon basin, with particular reference to respiratory patterns. *Canad. J. Zool.* **56**, 717–729.

Lagler, K. F., Bardach, J. E., Miller, R. R., and May Passino, R. R. (1977). "Icthyology." John Wiley & Sons, New York.

Lamore, G., and Nilsson, A. (2000). A process approach to the establishment of international river basin management in Southern Africa. *Phys. Chem. Earth (B)* **25**, 315–323.

Larinier, M. (2000). Dams and fish migration. *In* "Dams, Ecosystem Functions and Environmental Restoration. Thematic Review II.1 prepared as an input to the World Commission on

Dams" (Berkamp, G., McCartney, M., Dugan, P., McNeely, J., and Acreman, M., Eds.), pp. 1–30. WCD, Cape Town.

Le Houérou, H. N., Popov, G. F., and See, L. (1993). "Agro-bioclimatic Classification of Africa." FAO, Rome.

Letourneur, Y. (2000). Spatial and temporal variability in territoriality of a tropical benthic damselfish on a coral reef (Réunion Island). *Environ. Biol. Fishes* **57,** 377–391.

Levinton, J. S. (1982). "Marine Ecology." Prentice-Hall, Englewood Cliffs, NJ.

Lewis, W. M., Jr. (1996). Tropical lakes: How latitude makes a difference. *In* "Perspectives in Tropical Limnology" (Schiemer, F., and Boland, K. T., Eds.). SPB Academic Publishing bv, Amsterdam.

Lewis, W. M., Jr. (2000). Basis for protection and management of tropical lakes. *Lakes Reservoirs Res. Manag.* **5,** 35–48.

Lewis, W. M. (1987). Tropical limnology. *Ann. Rev. Ecol. Systemat.* **18,** 159–184.

Lourenço, S. O., and Marquez Junior, A. G. (2002). Produção primária marinha. *In* "Biologia Marinha" (Pereira, R. C., and Soares-Gomes, A., Eds.), pp. 195–227. Editôra Interciência, Rio de Janeiro.

Lovejoy, N. R., Bermingham, E., and Martin, A. P. (1998). Marine incursion into South America. *Nature* **396,** 421–422.

Lowe McConnell, R. H. (1987). "Ecological Studies in Tropical Fish Communities." Cambridge University Press, Cambridge.

Marlier, G. (1967). Ecological studies on some lakes of the Amazon valley. *Amazoniana* **1,** 91–115.

Martins, O., and Probst, J. L. (1991). Biogeochemistry of major African Rivers: Carbon and mineral transport. *In* "Biogeochemistry of Major World Rivers. Scope 42" (Degens, E. T., Kempe, S., and Richey, J. E., Eds.), pp. 127–156. John Wiley & Sons, Chichester.

Masson, S., and Delecluse, P. (2001). Influence of the Amazon River runoff on the tropical Atlantic. *Phys. Chem. Earth (B)* **26,** 136–142.

Matsuo, A. Y. O., and Val, A. L. (2002). Low pH and calcium effects on net Na^+ and K^+ fluxes in two catfishes species from the Amazon River (*Corydoras*: Callichthyidae). *Braz. J. Med. Biol. Res.* **35,** 361–367.

Matsuo, A. Y. O., and Val, A. L. (2003). Fish adaptations to Amazonian blackwaters. *In* "Fish Adaptations" (Val, A. L., and Kapoor, B. G., Eds.), pp. 1–36. Science Publishers, Inc., Enfield (NH), USA.

McCormick, M. I. (1994). Comparison of field methods for measuring surface topography and their associations with a tropical reef fish assemblage. *Marine Ecol. Progr. Ser.* **112,** 87–96.

McKnight, T. L. (1992). "Physical Geography, a Landscape Appreciation." Prentice Hall, Englewood Cliffs, NJ.

Menezes, N. A. (1996). Methods for assessing freshwater fish diversity. *In* "Biodiversity in Brazil: A First Approach" (Bicudo, C. E. M., and Menezes, N. A., Eds.), p. 326. CNPq, São Paulo.

Middleton, B. A. (2002). The flood pulse concept in wetland restoration. *In* "Flood Pulsing in Wetlands: Restoring the Natural Hydrological Balance" (Middleton, B. A., Ed.), pp. 1–10. John Wiley & Sons, New York.

Mora, C., Chittaro, P. M., Sale, P. F., Kritzer, J. P., and Ludsin, S. A. (2003). Patterns and processes in reef fish diversity. *Nature* **421,** 933–936.

Mora, C., and Sale, P. F. (2002). Are populations of coral reef fish open or closed? *Trends Ecol. Evol.* **117,** 422–428.

Mumby, P. J., Edwards, A. J., Arias-González, J. E., Lindeman, K. C., Blackwell, P. G., Gall, A., Gorczynska, M. I., Harbone, A. R., Pescod, C. L., Renken, H., Wabnitz, C. C. C., and Llewellyn, G. (2004). Mangroves enhance the biomass of coral reef fish communities in the Caribbean. *Nature* **427,** 533–536.

Munday, P. L. (2000). Interactions between habitat use and patterns of abundance in coral-dwelling fishes. *Environ. Biol. Fishes* **58**, 355–369.

Mwaura, F., Mavuti, K. M., and Wamicha, W. N. (2002). Biodiversity characteristics of small high-altitude tropical man-made reservoirs in the Eastern Rift Valley, Kenya. *Lakes Reservoirs Res. Manag.* **7**, 1–12.

Nanami, A., and Nishihira, M. (2003). Effects of habitat connectivity on the abundance and species richness of coral reef fishes: Comparison of an experimental habitat established at a rocky reef flat and at a sandy sea bottom. *Environ. Biol. Fishes* **68**, 186–193.

Narahara, A. B., Bergman, H. L., Laurent, P., Maina, J. N., Walsh, P. J., and Wood, C. M. (1996). Respiratory physiology of the Lake Magadi Tilapia (*Oreochromis alcalicus grahami*), a fish adapted to hot, alkaline and frequently hypoxic environment. *Physiol. Biochem. Zool.* **69**.

Nelson, J. S. (1994). "Fishes of the World." John Wiley & Sons, New York.

Nilsson, G. E., Hobbs, J.-P., Munday, P. L., and Östlund-Nilsson, S. (2004). Coward or braveheart: Extreme habitat fidelity through hypoxia tolerance in a coral-dwelling goby. *J. Exp. Biol.* **207**, 33–39.

Oberdorff, T., Hugueny, B., and Guégan, J. (1997). Is there an influence of historical events on contemporay fish species richness in rivers? Comparisons between Western Europe and North America. *J. Biogeogr.* **24**, 461–467.

Oltman, R. E. (1967). Reconnaissance investigations of the discharge water quality of the Amazon. *Atas Simposio sobre Biota Amazonica* **3**, 163–185.

Pasternack, G. B., and Varekamp, J. C. (1997). Volcanic lake systematics. I. Physical constraints. *Bull. Vulcanol.* **58**, 528–538.

Pauly, D. (1985). Ecology of coastal and estuarine fishes in Southeast Asia: A Phillipine case study. *In* "Fish Community, Ecology and Coastal Lagoons: Towards an Ecosystem Integration" (Yáñez-Arancibia, A., Ed.), pp. 499–514. UNAM Press, Mexico City.

Prepas, E. E., and Charette, T. J. V. (2003). Worldwide eutrophication of water bodies: causes, concerns, controls. *In* "Treatise on Geochemistry" (Lollar, B. S., Ed.), Vol. 9, pp. 311–331. Elsevier, Amsterdam.

Preston, F. W. (1962). The canonical distribution of commonness and rarity: I. *Ecology* **43**, 185–215.

Ragazzo, M. T. P. (2002). "Fishes of the Rio Negro. Alfred Russel Wallace." EDUSP. Imprensa Oficial do Estado., São Paulo.

Ramírez, J. J., and Bicudo, C. E. M. (2002). Variation of climatic and physical co-determinants of phytoplankton community in four nictemeral sampling days in a shallow tropical reservoir, southeastern Brazil. *Braz. J. Biol.* **62**, 1–14.

Randall, D. J., Wood, C. M., Perry, S. F., Bergman, H., Maloiy, G. M. O., Mommsen, T. P., and Wright, P. A. (1989). Urea excretion as a strategy for survival in a fish living in a very alkaline environment. *Nature* **337**, 165–166.

Rao, R. J. (2001). Biological conservation of the Ganga River, India. *Hydrobiol.* **458**, 159–168.

Revenga, C., Brunner, J., Henninger, N., Kassem, K., and Payne, R. (2000). "Pilot Analysis of Global Ecosystems: Freshwater Systems." WRI Publications, Washington, DC.

Rickelefs, R. E., and Schluter, D. (1993). Species diversity: Regional and historical influences. *In* "Species Diversity in Ecological Communities" (Ricklefs, R. E., and Schluter, D., Eds.), pp. 350–363. University of Chicago Press, London.

Romero, A. E. (2001). "The Biology of Hypogean Fishes." Kluwer Academic, Dordrecht.

Sale, P. (1993). "The Ecology of Fishes on Coral Reefs." Academic Press, New York.

Salo, J. (1987). Pleistocene forest refuges in the Amazon: evaluation of the biostratigraphical, lithostratigraphical and geomorphological data. *Ann. Zool. Fennici.* **24**, 203–211.

Santos, G. M., and Val, A. L. (1998). Ocorrência do peixe-serra (*Pristis perotteti*) no rio Amazonas e comentários sobre sua história natural. *Ciência Hoje* **23**, 66–67.

Shiklomanov, I. A. (1993). World fresh water resources. *In* "Water in Crisis: A Guide to the World's Fresh Water Resources" (Gleick, P. H., Ed.), pp. 13–24. Oxford University Press, New York.

Singh, H. S. (2003). Marine protected areas in India. *Ind. J. Marine Sci.* **32**, 226–233.

Singh, M., Müller, G., and Singh, I. B. (2003). Geogenic distribution and baseline concentration of heavy metals in sediments of the Ganges River, India. *J. Geochem. Explor.* **80**, 1–17.

Sioli, H. (1950). Das wasser im Amazonasgebiet. *Forschung Fortschritt.* **26**, 274–280.

Sioli, H. (1984). The Amazon and its main affluents: Hydrogeography, morphology of the river courses and river types. *In* "The Amazon. Limnology and Landscape Ecology of a Mighty Tropical River and Its Basin" (Sioli, H., Ed.), pp. 127–165. W. Junk, Dordrecht.

Sioli, H. (1991). "Amazônia. Fundamentos da ecologia da maior região de florestas tropicais." Vozes, Petrópolis.

Soares-Gomes, A., and Figueiredo, A. G. (2002). O ambiente marinho. *In* "Biologia Marinha" (Pereira, R. C., and Soares-Gomes, A., Eds.), pp. 1–33. Editôra Interciência, Rio de Janeiro.

Steele, M. A. (1999). Effects of shelter and predator on reef fishes. *J. Exp. Marine Biol. Ecol.* **233**, 65–79.

Sutcliffe, J. V., and Parks, Y. P. (1999). "The Hydrology of Nile." IAHS Special Publication No. 5, Wallingford.

Takahashi, T., Wanninkhof, R. H., Feely, R. A., Weiss, R. F., Chipmann, D. W., Bates, N., Olafsson, J. C. S., and Sutherland, S. C. (1999). Net sea-air CO_2 flux over the global oceans: An improved estimate based on the sea-air pCO_2 difference. *In* "Proceedings of the 2nd CO_2 in Oceans Symposium" (Nojiri, Y., Ed.), pp. 9–14. CGER/NIES, Tsukuba.

Talling, J. F. (2001). Environmnetal controls on the functioning of shallow tropical lakes. *Hydrobiol.* **458**, 1–8.

Thornthwaite, C. W. (1948). An approach towards a rational classification of climate. *Geogr. Rev.* **38**, 55–94.

Thorson, T. B. (1974). Occurrence of the sawfish, *Pristis perotteti*, in the Amazon river, with notes on *P. pectinatus. Copeia* **1974**, 560–564.

Thurman, H. V. (1996). "Introductory Oceanography." Prentice Hall, Englewood Cliffs, NJ.

Timms, B. V. (2001). Large freshwater lakes in arid Australia: A review of their limnology and threats to their future. *Lakes Reservoirs Res. Manag.* **6**, 183–196.

Townsend, A. S. (1996). Metalimnetic and hypolimnetic deoxygeneation in an Australian tropical reservoir of low trophic status. *In* "Perspectives in Tropical Limnology" (Schiemer, F., and Boland, E. J., Eds.), pp. 151–160. SPB Academic Publishing bv, Amsterdam.

Trajano, E. (1997). Population ecology of Trichomycterus itacarambiensis, a cave catfish from eastern Brazil (Siluriformes, Trichomycteridae). *Environ. Biol. Fishes* **50**, 357–369.

Tumbare, M. J. (1999). Equitable sharing of the water resources of the Zambezi River Basin. *Phys. Chem. Earth (B)* **24**, 571–578.

Tundisi, J. G. (1981). Typology of reservoirs in southern Brazil. *Verhandl. Int. Vereinigung Theoret. Angewandte Limnol.* **21**, 1031–1039.

Turner, G. F., Seehausen, O., Knight, M. E., Allender, C. J., and Robinson, R. L. (2001). How many species of cichlid fishes are there in African lakes? *Mol. Ecol.* **30**, 793–806.

Umana, V. G., Haberyan, K. A., and Horn, S. P. (1999). Limnology in Costa Rica. *In* "Limnology in Developing Countries" (Wetzel, R. G., and Gopal, B., Eds.), Vol. 2, pp. 33–62. International Scientific Publications, New Dehli.

UNEP (2002). "Global Environment Outlook 3 (GEO-3)." UNEP, Nairobi.

Val, A. L. (1995). Oxygen transfer in fish: Morphological and molecular adjustments. *Braz. J. Med. Biol. Res.* **28**, 1119–1127.

Val, A. L. (1996). Surviving low oxygen levels: Lessons from fishes of the Amazon. *In* "Physiology and Biochemistry of the Fishes of the Amazon" (Val, A. L., Almeida-Val, V. M. F., and Randall, D. J., Eds.), pp. 59–73. INPA, Manaus.

Val, A. L., and Almeida-Val, V. M. F. (1995). "Fishes of the Amazon and Their Environments. Physiological and Biochemical Features." Springer Verlag, Heidelberg.

Val, A. L., Chippari-Gomes, A. R., and Almeida-Val, V. M. F. (2004, in press). Hypoxia and petroleum: Extreme challenges for fish of the Amazon. *In* "Fish Physiology, Toxicology and Water Quality" (Rupp, G., Ed.). Environmental Protection Agency, USA, Montana.

van der Oost, R., Beyer, J., and Vermeulen, N. P. E. (2003). Fish bioaccumulation and biomarkers in environment risk assessment: A review. *Environ. Toxicol. Pharmacol.* **13**, 57–149.

Verburg, P., Hecky, R. E., and Kling, H. (2003). Ecological consequences of a century of warming in Lake Tanganyika. *Science* **301**, 505–507.

Verdin, J. P. (1996). Remote sensing of ephemeral water bodies in western Niger. *Int. J. Remote Sensing* **17**, 733–748.

Walther, G. R., Post, E., Convey, P., Menzel, A., Parmesan, C., Beebee, T. J. C., Fromentin, J. M., Hoegh-Guldberg, O., and Bairlein, F. (2002). Ecological responses to recent climate change. *Nature* **416**, 389–395.

Watters, G. T. (1992). Unionids, fishes and the species area curve. *J. Biogeogr.* **19**, 481–490.

WCD (2000). Kariba Dam Case Study. Prepared by Soils Incorporated (Pty) Ltd and Chalo Environmental and Sustainable Development Consultants, Cape Town.

Weitzman, S. H., and Weitzman, M. J. (1987). Biogeography and evolutionary diversification in neotropical freshwater fishes, with comments on the refuge theory. *In* "Biological Diversification in the Tropics" (Prance, G. T., Ed.), pp. 403–422. Columbia University Press, New York.

Williams, S. E., Bolitho, E. E., and Foc, S. (2003). Climate change in Australia tropical rainforest: An impending environmental catastrophe. *Proc. R. Soc. Lond. Ser. B (Biol. Sci.)* **270**, 1887–1892.

Williams, W. D. (2000). Dryland lakes. *Lakes Reservoirs Res. Manag.* **5**, 207–221.

Williams, W. D., De Deckker, P., and Shiel, R. J. (1998). The limnology of Lake Torrens, an episodic salt lake of central Australia, with particular reference to unique events in 1989. *Hydrobiologia* **384**, 101–110.

Wilson, R. W., Wood, C. M., Gonzalez, R. J., Patrick, M. L., Bergman, H. L., Narahara, A., and Val, A. L. (1999). Ion and acid-base balance in three species of Amazonian fish during gradual acidification of extremely soft water. *Physiol. Biochem. Zool.* **72**, 277–285.

Winemiller, K. O., and Kelso-Winemiller, L. C. (2003). Food habits of tilapiine cichlids of the Upper Zambezi River and floodplain during the descending phase of the hydrologic cycle. *J. Fish Biol.* **63**, 120–128.

Wolanki, E. (1992). Hydrodynamics of mangroves swamps and their coastal waters. *Hydrobiologia* **247**, 141–161.

Wood, C. M., Perry, S. F., Wright, P. A., Bergman, H. L., and Randall, D. J. (1989). Ammonia and urea dynamics in the lake Magadi tilapia, a ureotelic teleost fish adapted to an extremely alkaline environment. *Respir. Physiol.* **77**, 1–20.

Wood, C. M., Wilson, R. W., Gonzalez, R. J., Patrick, M. L., Bergman, H. L., Narahara, A., and Val, A. L. (1998). Responses of an Amazonian teleost, the tambaqui (*Colossoma macropomum*) to low pH in extremely soft water. *Physiol. Zool.* **71**, 658–670.

Wright, D. (1983). Species energy theory: an extension of species area theory. *Oikos* **41**, 495–506.

Wu, R. (2002). Hypoxia: From molecular responses to ecosystem responses. *Marine Pollution Bull.* **45**, 35–45.

Zhang, X., Zhang, Y., Wen, A., and Feng, M. (2003). Assessment of soil losses on cultivated land by using the ^{137}Cs technique in the Upper Yangtze River basin in China. *Soil Tillage Res.* **69**, 99–106.

2

DIVERSITY OF TROPICAL FISHES

MÁRIO C. C. DE PINNA

I. INTRODUCTION

Fishes of the tropical region constitute one of the most fascinating subjects of study in comparative biology. The myriad radiations and adaptive modifications displayed by different groups of tropical fishes cover practically all important phenomena in evolutionary biology. In many instances, tropical fishes provide unique case studies for specific processes.

Tropical fishes occupy practically all aquatic environments, either permanent or temporary, and it is nearly impossible to encounter a body of water in the tropics, however small or uninviting, where there are no fish living. They occur from high-mountain freezing streams to hot stagnant anoxic pools to subterranean waters. Across their wide range of habitats, tropical fishes display elaborate adaptations that adjust their mode of living to widely different and often extreme conditions. Such adaptations often

47

DOI: 10.1016/S1546-5098(05)21002-6

result in narrowly specialized species and groups of species which are found in very specific habitats and nowhere else. This is particularly evident in freshwater environments, where the level of endemism is particularly high.

As overwhelming as tropical fish diversity is known to be, its actual magnitude is still unknown, and certainly much larger than presently documented. New species are constantly being described, and current sampling coverage in most tropical areas is still limited. Collecting expeditions to regions already sampled normally yield a certain number of new species. Field efforts in poorly known or previously unsampled areas are certain to find numerous new taxa. It is reasonable to expect that important fish discoveries will be made in practically any previously unexplored tropical environment, especially those that are difficult to access by collectors. Tropical fishes excel at entering small marginal microhabitats which in other regions would be devoid of specialized fish life. The bottom of large rivers, interstitial leaf-litter water and temporary pools are just three examples where recent explorations have revealed a previously unsuspected ichthyological diversity (see Chapter 1).

In most groups of living organisms, species richness reaches its maximum in the tropical region, a pattern that has been known to biologists for centuries (at least since Humboldt and Bonpland, 1807). Fish are no exception (see Chapter 1). The reasons for that are multifaceted, and have been the focus of much discussion. Whatever the relative importance of each factor, it is generally recognized that biological diversity is a result of the interaction of two main forces: history and ecology. The biogeographical history of different biota determines the taxonomic composition of present-day areas, while ecology determines physical conditions for survival and other local biotic and abiotic parameters for the increase in diversification and its accumulation in time, the result of which we call biodiversity.

In this chapter I provide a general overview of the diversity of fish in the tropical region. The aim is to summarize current knowledge about the number and taxonomic composition of fish in freshwater and marine environments comprised in the areas of the world considered as tropical (see Chapter 1). The delimitation of the "tropical" region herein is broad, and follows geographical rather than ecological boundaries. The focus is on aquatic environments located within the circum-global tropical belt, regardless of mean annual temperature. Therefore, high-altitude environments with semi-tropical, temperate or even cold climates are included, as long as they are located within the tropical region. Likewise, the chapter often extends into the fish fauna of southern latitudes well beyond tropical borders, such as Austral South America and Southern South Africa. Those areas, although not strictly "tropical" under any meaning of the term, provide key information in understanding the diversity, history, and composition of fishes in the tropical region.

A discussion of hypotheses about geological evolution is included for certain areas, as are adaptations of fish life that are particularly interesting. Evolutionary and phylogenetic information is given priority in the presentation of information, as this is considered the only way to understand diversity in its intrinsically multidimensional nature. The amount of information and discussion provided in specific cases varies widely, and is not necessarily reflective of their intrinsic relevance. Rather, it is a result of the different state of current knowledge, which is highly uneven across different geographical areas and taxonomic boundaries. The underlying aim throughout this survey is to underscore the importance of phylogenetic patterns in understanding fish adaptations.

II. WHAT IS DIVERSITY?

The simplest measure of biological diversity, and one which first comes to mind for non-specialists, is number of species. One biota, region or area that harbors 100 species is considered to be more diverse than an equivalent unit which comprises only 50 species. The number of species can be expressed as a ratio, so that it expresses an average density of species per square kilometer, or some equivalent measure.

Measures of diversity based only on number of species, however, are incomplete. In fact, they measure only species richness, not diversity. Their non-dimensionality is oblivious to the fact that biological diversity is hierarchically organized. Diversity is an epiphenomenon of evolutionary history. Organisms are diverse because they have become diverse with time through a succession of lineage branching. This network of branching is an essential part of the structure of diversity. Biological diversification forms a hierarchical pattern that can be retrieved by phylogenetic analysis.

So, contrary to common notions, it is not species that are the fundamental units to understanding and measuring diversity. Taxa in general, rather, are the currency of diversity. Species are one kind of taxa, and one that many biologists consider as the most important of all because of its apparent taxonomic irreducibility and direct participation in biological processes. Elsewhere (Nelson, 1989; de Pinna, 1999), it has been argued that all properties that allegedly distinguish the species from other kinds of taxa are questionable. Although the species level is one that has special ecological and practical appeal, its near ubiquitousness in biodiversity assessments has been detrimental to a clear understanding of the subject.

The relevance of taxa, rather than simply species, in diversity assessments, is reflected in the different degrees of impact that different taxonomic ranks have when they represent new findings. The discovery of the coelacanth *Latimeria chalumnae* in 1938 attracted and still attracts much attention. That was not simply because it was a new species of fish. Tens of new species of fish are discovered annually. The discovery of the coelacanth became a legend in both scientific and non-scientific circles because it represented the discovery of a whole lineage long thought to be extinct. *Latimeria chalumnae*, although a single species, represented a whole basal clade of sarcopterygians, previously known only by fossils. The data obtained from that single species provided a comparative framework equivalent to that formed by all other sarcopterygians (a group that includes tetrapods) and is therefore of key importance in understanding the evolution of nearly half of all vertebrates. Clearly, the discovery of the coelacanth is a more important scientific breakthrough than the discovery of one more species of cichlid from Lake Victoria, where hundreds of similar species exist (or existed before their recent extinction).

The phylogenetic dimension of diversity explains why some particular areas are considered to contain such an important portion of biodiversity, despite a relatively small number of species. Regions that comprise a seemingly depauperate fish fauna are sometimes also the ones where "relicts" have survived. Such "relicts" are one or a few species which represent a hierarchically large clade. This is a result of highly asymmetrical phylogenies, where one or a few species constitute the sister group to hundreds or thousands of others (Stiassny and de Pinna, 1994). The reason for the asymmetry can be either different rate of differentiation or of extinction. Whatever the causing factor, sister groups have comparatively equivalent importance in understanding the structure of diversity, regardless of their relative number of constituent species. Therefore, clade diversity is as important as species diversity.

Examples of species-poor but clade-rich areas are many. For example, the gap of Dahomey (mostly in current Benin) comprises comparatively fewer species of freshwater fish than areas east and west of it. However, the gap of Dahomey is practically the only spot where *Denticeps clupeoides* is known to exist. That species is the single recent representative of the suborder Denticipitoidei, sister group to all other recent Clupeomorphs, a group that includes over 300 species (Grande, 1985). The austral trans-Andean region of South America, corresponding to the area of Chile, is also remarkable. It is the only region where various important clades of fish taxa survive today. It is the only place where one can find *Nematogenys inermis*, sole representative of the family Nematogenyidae and sister group to the Trichomycteridae. The latter is a large and diversified clade that includes some 200 species distributed throughout South America (including Chile) and part of Central America (de Pinna, 1998). The sister group to all other siluriforms, widely recognized as including the most "primitive" catfishes, is the family Diplomystidae. Species of that family occur only in central Chile and Argentina (Arratia, 1987; Azpelicueta, 1994; de Pinna, 1998). The same region is also home to various other fish groups that occur nowhere else in South America, such as percichthyids, galaxiids and mordaciids. Obviously, simply the number of fish species in austral South America is not reflective of its relevance in global fish diversity. One has to look for the clades they represent in order to have an idea of the significance of that region. Extinction of those taxa, although negligible in terms of species numbers, would be a tremendous loss for ichthyology and the understanding of fish biodiversity and evolution.

Another major example of species-poor but clade-rich area is Madagascar. That island is the only place where the catfish family Anchariidae (probably sister group to the Ariidae) exists. It is also home to several clades of basal cichlids, such as ptychochromines and *Paretroplus* (Stiassny, 1991; Stiassny and de Pinna, 1994; Stiassny and Raminosoa, 1994). Southern South Africa, again, has a remarkably depauperate fish fauna that comprises several unique clades, such as the catfish family Austroglanididae (with three species) and the only African galaxiid species (Skelton *et al.*, 1995).

III. THE EVOLUTIONARY IMPLICATIONS OF SALT TOLERANCE IN FRESHWATER FISHES

Freshwater fishes have been traditionally split into three divisions that are supposed to reflect their tolerance to saltwater: Primary, Secondary and Peripheral. The two former categories were originally proposed by Myers (1938) and have had enormous influence upon ichthyology, especially on the field of fish distribution and biogeography. The relevance of Myers' divisions has decreased in the vicariant biogeographical paradigm (cf. Rosen 1976), but they are still widely adopted in ichthyology. Primary division freshwater fishes are those that are strictly restricted to freshwater (with total dissolved salts less than 0.5 g/l) and which cannot survive saltwater for any significant length of time. Secondary freshwater fishes are those whose populations are normally restricted to freshwater, but which can tolerate and survive for some time in marine or brackish water. They can enter marine water voluntarily for short periods of time and may disperse across stretches of marine water which would constitute a barrier to Primary division freshwater fishes. Peripheral freshwater fishes (a term coined by Nichols, 1928) are those that live primarily in marine environments but which can enter and survive freshwater for a long time. The Peripheral division also usually includes diadromous species, which spend part of their lives in freshwater and part in saltwater. Although the definitions of Primary, Secondary and Peripheral hinge on a physiological characteristic, resistance to dissolved salt, actual application of the definitions are rarely based on actual experimentation of salt tolerance. Rather, salt tolerance is normally inferred on the basis of the environment where the fish is found in natural conditions. Information on the natural history of a species is therefore tantamount in determining whether it is a Primary, Secondary or Peripheral freshwater fish.

Primary freshwater fishes are considered to be of special interest in continental biogeography. Because of their inability to disperse through non-freshwater environments, their diversity and evolution are thought to be tightly connected to the history of drainages. Because of their limited vagility outside of aquatic environments, they are often more informative about the history of land masses than are most terrestrial organisms.

According to Myers' original essay, the difference between Primary, Secondary and Peripheral divisions are not simply physiological and ecological, but also embody an element of phylogenetic history. Primary freshwater fish groups are believed to have "carried down their physiological inability to survive in the sea, as family characters, from early times and probably since the origin of the groups concerned" (Myers, 1938). In current terminology, Myers' reasoning implies that Primary freshwater fish groups are those that share a homologous intolerance to saltwater, i.e., a primitive inability to survive marine or estuarine environments.

Proper categorization of the divisions of freshwater fishes is highly subjective, because it is not possible to apply them without resort to a taxonomic chart of reference. Also, there are many exceptions that do follow a clear pattern and the three categories freely cross taxonomic boundaries. It is often unclear in which category the exceptions fit. For example, otophysans are considered as a Primary freshwater fish group. Indeed, most species in that superorder are strictly restricted to freshwater. However, there are conspicuous exceptions. Among siluriforms, there are two families which include several marine species, Ariidae and Plotosidae (both families also include some entirely freshwater species as well). There are other siluriform families, such as Aspredinidae and Auchenipteridae, with some species that can certainly survive and reproduce in seawater, although they are not found far from estuaries. Within Cypriniformes, species of the East Asian cyprinid *Tribolodon* also live in saltwater. Therefore, the fact that otophysans, as a taxon, are a Primary freshwater fish group does not imply absolute salt tolerance of its component species.

Obviously, the divisions of freshwater fishes rely on a phylogenetic component that has so far been utilized in a rather subjective fashion. To further complicate the issue, several authors have used the term "Secondary" as a phylogenetic term, rather than in Myers' sense. In phylogenetic terminology, "Secondary" is sometimes used to refer to a condition that has evolved to resemble a phylogenetically precedent state. One can say that cetaceans are secondarily aquatic vertebrates, because their aquatic condition evolved from a terrestrial state, and is not primitively aquatic (as in primitive sarcopterygians and other "lower" vertebrates). The term "Secondary," as applied to a division of freshwater fishes, has not been necessarily used in a phylogenetic sense. Rather, it was used to describe a particular state of an ecological/physiological characteristic. It is true that this state has normally been applied to taxa which are indeed Secondary (in a phylogenetic sense) invaders derived from marine ancestor. Regardless, the terminological confusion is underlain by a conceptual one. It is possible to reform the definition of those categories so as to reflect more objective and explicitly phylogenetic criteria. Salt tolerance can be treated as a three-state character, ordered as 1 (Primary), 2 (Secondary) and P (Peripheral). The ordering reflects the assumption that a species does not directly jump from Primary freshwater to Peripheral and vice versa. The Secondary stage is probably a necessary intermediate stage in the evolution of habitat shift, and can be assumed to have existed even when the actual Secondary taxa are not observed. This is a working hypothesis which can be tested when numerous phylogenies and associated habitats have been examined. The character states can be mapped onto a known phylogeny, according to standard optimization procedures (Swofford and Maddison, 1987), and the optimization will provide a sequence of evolutionary habitat transitions which can be used as a basis for defining the divisions. Primary freshwater fishes can be considered as those which share a strict restriction to freshwater as a synapomorphy at some level. That means that their intolerance to saltwater is a homologous physiological feature. The same applies to the Secondary and Peripheral divisions. Fish in the Secondary division, being an intermediate character state, may have reached their condition from two directions. They may have derived either from a Primary or from a Peripheral (or marine) condition. In the first case (here called "Secondary division-1"), they represent a freshwater group that acquired some resistance to brackish water. In the second case ("Secondary division-2") they are part of a marine or Peripheral group that became adapted to freshwater environments.

The application of the proposed phylogenetic definitions is exemplified in Figure 2.1. The cladogram shows 16 taxa, A–P, each with their respective category of resistance to saltwater (1, Primary; 2, Secondary; P, Peripheral). The character is parsimoniously optimized as shown by the coded states for each node. Transitions between states are shown by a black rectangle. Monophyletic groups, each of which is represented by a node, can be categorized as

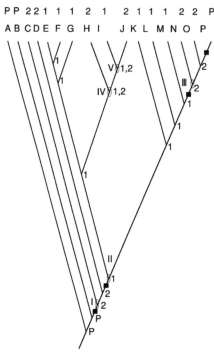

Fig. 2.1 Cladogram explaining phylogenetically based notions of primary (1), secondary (2) and peripheral (P) divisions of freshwater fish. The three conditions are treated as a multistate character ordered as 1-2-P, parsimoniously optimized on the tree. Node-optimized character-states are shown on the right of the respective node. Nodes numbered with Roman numerals and marked with an open circle are those discussed in text. Black rectangles represent character-state transitions. Terminal taxa are labelled A–P, with their respective character-states above them.

division-1, division-2 or Peripheral according to the optimization at that particular node. So, group "I" is Secondary division-2; group "II" is Primary division; group "III" is Secondary division-1; terminal "P" is a Peripheral taxon (be it a species or a monophyletic group). The root region of the tree indicates that the plesiomorphic division for the whole group is Peripheral. All other monophyletic groups in this example are Primary freshwater clades. Nodes corresponding to groups "IV" and "V" have ambiguous optimizations, which mean that there is more than one maximally-parsimonious sequence of state transitions. For example, node "IV" could be assigned state 1, inherited from the previous node, with transitions to state 2 in each of terminals "H" and "J." That would require a total of two steps on that portion of the tree. Alternatively, node "IV" could have state 2, which requires a transition at its base, plus a transition to state 1 in taxon "I." That alternative would amount to two steps also. Therefore, character-state optimization for node "IV" is uncertain and the characterization of that group as Primary or

Secondary is not determined. It cannot, however, be a Peripheral group, since that is not one of the alternatives among possible optimizations.

The phylogenetic definitions of freshwater divisions will not always agree with previous concepts. For example, the Cichlidae has long been considered as a Secondary freshwater fish group. The vast majority of cichlid species are never exposed to marine or brackish water in their natural environments and their tolerance to salt has never been tested. Their membership on the Secondary division has so far been based on the existence of some euryhaline cichlid species that live in estuaries plus the fact that they are labroids, a group otherwise composed entirely of marine families (Labridae, Pomacentridae and Embiotocidae; Stiassny and Jensen, 1987). Percoids considered being close to labroids are also typically marine fishes (Stiassny and Jensen, 1987). Although a detailed phylogeny of the entire Cichlidae is still not available, the most primitive cichlids seem to include the Asian genus *Etroplus* and the Malagasy *Paretroplus* and *Paratilapia* (Stiassny, 1991), all comprising species tolerant of saline water (Reithal and Stiassny, 1991). This fact optimizes the family as a Secondary division-2 freshwater group, because the marine environment of labroids and related percoids indicates that cichlids derive from saltwater ancestors. Of course, smaller subgroups of cichlids may be considered as other categories, depending on their tolerance to dissolved salts and their phylogenetic relationships.

The family Ariidae is another example of a marine group which is part of larger freshwater clades (Otophysi and Siluriformes). A recent phylogenetic study, the first to include the whole family in a global analysis, has hypothesized the genera *Galeichthys* and *Bagre* as two successive sister groups to all other ariids (Marceniuk, 2003). The two genera include marine and estuarine species only and this indicate that this is the primitive condition for the family. Therefore, Ariidae, as a group, is a Peripheral freshwater family (it cannot be considered as fully marine because at least part of their life cycle is always associated with estuaries). There are many freshwater ariid species. Those are either Secondary division-2 or Primary freshwater species (or groups of species), depending on their specific biology.

An example of the key role of phylogenetic structure in categorizing the division of freshwater fishes is the Autralasian catfish family Plotosidae. This family is usually considered a complicated case, because Siluriformes and Otophysans are widely recognized as comprising Primary division fishes. Species of Plotosidae are usually not included among Primary division fish fauna, because many species of that family are marine. On the other hand, there are also numerous genera and species which are exclusively freshwater in Australia and New Guinea. It is generally assumed that such freshwater species are derived from marine invaders, but in fact there is little concrete evidence for that. Phylogenetic relationships among Plotosidae are still unknown. In case the freshwater taxa turn out at basal positions in the cladogram of plotosids, e.g., forming a series of sister groups to the rest of the family, then the Plotosidae are a primarily freshwater group. Such hypothesis would imply that the family is primitively freshwater, and that the marine taxa are the result of subsequent adaptation(s) to saltwater. On the other hand, if plotosid phylogeny places marine taxa as basal, then it is the freshwater taxa which are Secondary invaders of continental waters. Until a hypothesis on the phylogeny of the Plotosidae is available, it is impossible to decide on which division of freshwater fish they should be included.

IV. THE IMPORTANCE OF PHYLOGENETIC INFORMATION FOR COMPARATIVE STUDIES

The past two decades have seen a surge of interest on phylogenetic patterns as a source of background information for evolutionary inferences. The hierarchical structure of branching diagrams (cladograms) provides a basis for understanding the temporal sequence of character states, their homology and level of generality. Phylogenetic hypotheses and their associated character set are powerful tools for understanding evolution. All biological attributes are, directly or indirectly, a result of phylogenetic history.

There are two reasons why two or more organisms have a trait in common. First, they may be similar because they are related and inherited the trait from a common ancestor which possessed the same condition. Second, they may have acquired the trait independently. Phylogenies allow biologists to discriminate between the two sources of similarity. The implications of that separation are fundamental for building evolutionary explanations for biological phenomena. When a trait is shared by organisms as a result of common ancestry, it is vain to search for causal explanations for the evolution of the trait among single species. The evolution of the trait, in that case, was a result of processes active on the ancestor of the group, not necessarily on its descendant species. Only an analysis of the group as a whole can shed light on common factors which may be significant for the understanding of the evolution of the trait. A different situation occurs when similar traits are shared by unrelated organisms. In that case, one may reasonably expect that there are common factors which are associated with the development of that condition. As cases of convergence multiply, it becomes increasingly less likely that the repetition of the pattern is a result of coincidence. The reduction of eyes and skin pigmentation seen in cave fishes, which occur in many unrelated groups, is obviously associated with the conditions under which such animals live. The fact that many different, unrelated, fish groups in many different caves shows similar reductions calls for a generalization. Conditions in the cave environment must be causally associated with the loss of eyes and pigment. If all cave fishes belonged to a single monophyletic group, the causal association (environment/trait) would be less compelling, because it might just be that they inherited their eyelessness and lack of dark pigment from a common ancestor which developed those traits due to factors unrelated to the cave environment (see Chapter 4). It is the repeated evolution of the trait in similar environments which indicates that a causal association may exist. The association may be refined on the basis of a more fine-grained comparative analysis. For example, similar reductions of eyes and pigment occur in fossorial psammophilic species (such as some glanapterygine trichomycterids and the freshwater ophichthyid eel *Stictorhinus potamius*). Comparisons of conditions in caves and psammic environments may pin down causal factors at a more specific (and informative) level.

V. FRESHWATER TROPICAL FISHES

Freshwater fishes have long been considered as a group of organisms potentially informative about continental relationships. This was particularly true for those considered to belong to primarily freshwater groups, physiologically incapable of crossing stretches of seawater (see above). Phylogenetic relationships among freshwater fishes have been intensively studied in recent years, a trend particularly pronounced in neotropical ostariophysans. Today, biogeographical inference for freshwater regions is far more richly provided with phylogenetic hypothesis than for marine areas. In part, this is a consequence of the more constrained geographical distribution of monophyletic groups in continental waters, which makes the task of gathering comparative material for phylogenetic studies less difficult. The elucidation of species-level phylogenies for marine groups usually requires examination of material from vast areas, often extending over different continents. This practical impediment apparently has been detrimental to the advance of phylogenetic understanding in marine fish taxa, which nowadays progresses at comparatively slower rates when compared to freshwater ones.

In the sections below, freshwater fishes are discussed according to the major land mass where they occur. In contrast to marine regions, tropical freshwater regions are highly segregated in terms of their fish-species composition. Few or no species are shared between major land masses. Therefore, inferences about transcontinental relationships based on freshwater fishes necessarily rely on hypotheses of phylogenetic kinship, rather than on pure faunal similarities. This factor has led to a deeper understanding of vicariant patterns in freshwater areas, when compared to marine regions.

A. The Neotropics

The neotropical region is the richest region in terms of number of freshwater fish species, with 4475 valid described species and at least another 1550 yet to be described (Reis *et al.*, 2003; see Chapter 1). Most of that diversity is represented by three orders of ostariophysan fish, Characifomes, Siluriformes and Gymnotiformes. The latter is exclusive to the neotropics. The Cypriniformes, massively represented in all other continents with the exception of Australia and Antarctica, is absent in South America. That fact is puzzling and represents one of the most intriguing and long-lasting controversies in continental biogeography. In addition to ostariophysan orders, South America has few other freshwater groups traditionally considered as Primary division: one genus of Lepidosirenidae, two genera of Osteoglossiformes and two genera of Nandidae.

A number of main areas of endemism have been recognized for freshwater fishes in South America, a pattern which resulted from over a century of work (summarized in Vari, 1988). Those, for the most part, follow the major river drainages in the continent: Paraguay, Upper Paraná, Coastal, São Francisco, Northeast, Amazon, Guianas, Orinoco, Western, Southwestern, Chilean and Patagonian (Weitzman and Weitzman, 1982; Vari, 1988; Menezes, 1996; Rosa *et al.*, 2003). The Amazon basin is by far the most speciose in the region (see also Chapter 1), which in the past has led to the assumption that it served as a center of dispersal for all the rest of South America (Darlington, 1957). This view is no longer accepted, and today the Amazon is seen as just the most diverse drainage in a region with a complex history of river basins dynamics, resulting in major vicariant patterns reflected in present-day distributions of monophyletic groups (Weitzman and Weitzman, 1982; Vari, 1988; Lundberg *et al.*, 1998).

Although much local diversification has occurred in South America, there are many known patterns of transcontinental relationships, indicating that diversity of freshwater fishes was substantial before the isolation of the continent. Relationships among major groups of Characiformes are still far from properly resolved. Nevertheless, there are various consistent hypotheses of relationships among characiform fishes which shed light on the biotic history of the continent. The African family Hepsetidae is known to be related to the neotropical families Erythrinidae, Lebiasinidae and Ctenoluciidae (Vari, 1995; Oyakawa, 1998). The Characidae, the most complex assemblage of characiforms, was formerly considered to include both neotropical and African forms. Recently, it was found that the neotropical characids are phylogenetically complex. One of its genera (*Chalceus*) is more closely related to African characids than to any neotropical taxa (Zanata and Vari, in press). The phenetic similarity between species of *Chalceus* and some alestids, such as those of *Arnoldichthys*, is striking. The elucidation of those relationships has led to the recognition of a separate family for former African characids, the Alestidae, and the transfer of *Chalceus* from the Characidae into the Alestidae (Zanata and Vari, in press). Two African families, Citharinidae and Distichodontidae, are considered as the sister group to all characiforms, both African and South American. All those complex patterns show that major portion of the diversification of characiforms predates the separation of Gondwana. At the same time, the group seems to be essentially Gondwanan, since no characiforms, recent or fossil, have been encountered out of Gondwanan land masses. While the complex intercontinental nature of characiform relationships indicates an ancient history of diversification, the presence of large monophyletic groups in both Africa and South America also demonstrate high levels of local diversification as well.

The number of characiform species differs markedly in Africa and South America. While there are over 1000 neotropical characiforms, the group is represented by less than 300 in Africa. This difference is made even more striking because the most primitive characiform clades are in Africa, indicating that the group has been in that area for a longer time. There are various speculations to explain this discrepancy. One of them is that cypriniforms in Africa occupy ecological niches similar to those of characiforms, especially in the small body-size ranges which normally concentrates much species diversity. Competition with cypriniforms would have restricted characiform diversity in Africa. Unquestionably, characiform geographical distribution in the Old World is smaller today than it was in the past, because there are characiform fossils in Southern Europe and in the Middle East, areas where there are no recent representatives of the group.

Neotropical fishes are represented by a rich assemblage of 12 endemic families, plus one present in North America (Ictaluridae) and the circumtropical and mostly marine Ariidae. Phylogenetically, the neotropics are the richest continent in catfish diversity. South America hosts the most "primitive" family of fishes, Diplomystidae, considered as the sister group to all other Siluriformes (Lundberg, 1970; Arratia, 1987; Mo, 1991; de Pinna, 1993, 1998). The Cetopsidae (including the former Helogenidae; de Pinna and Vari, 1995) seems to represent the sister group to all other recent nondiplomystid fishes (de Pinna *et al.*, submitted). The largest monophyletic group of catfishes that occurs on a single land mass is also South American, the superfamily Loricarioidea. That clade includes the families Nematogenyidae, Trichomycteridae, Callichthyidae, Scoloplacidae, Astroblepidae and Loricariidae (Baskin, 1973; de Pinna, 1998). The Loricarioidea comprises over 1000 species. A few neotropical families have been demonstrated to have closest relatives in other continents. The superfamily Doradoidea (including the families Doradidae and Auchenipteridae) is the sister group to the African Mochokidae. The Aspredinidae are part of an otherwise exclusively South and Southeast Asian group called superfamily Sisoroidea (de Pinna, 1996). Other neotropical families have not had their intraordinal affinities elucidated yet.

Perhaps the most striking specialization in neotropical catfishes is hematophagy, seen in some members of the family Trichomycteridae. All of the *ca.* 20 species (plus many not described) of the Trichomycterid subfamily Vandelliinae are exclusively blood-feeding as adults. This is a trophic specialization which is almost unique among gnathostomes, and occurs elsewhere only among vampire bats (coincidentally, also neotropical). Since the Vandelliinae are a monophyletic group, hematophagy is hypothesized to have evolved only once. Surprisingly little is known about vandelliine biology. It was only recently that juveniles of the group have been first discovered and found to be predators of small aquatic invertebrates (personal observation), a trophic condition shared with most other trichomycterids. Juvenile specimens also have a normal mouth structure, similar to that of nonparasitic members of the family. At some point, they undergo a metamorphosis during which their entire mouth structure changes profoundly, to reach the adult blood-feeding morphology (personal observation). The physiological specializations for hematophagy are entirely unknown. They may produce some type of anticoagulant, since the relatively enormous amount of blood per meal (approximately twice their body volume) remains liquid during digestion. It is also possible that some vandelliine species are carried by large migratory fishes to upriver spawning grounds. If that is the case, the large amount of blood ingested may even induce some hormonal fluctuations on the candiru, so that their reproductive periods may be synchronized with that of larger species. These possibilities are simply speculations at this point, but they indicate fascinating lines for future research.

Of special interest to fish physiologists is the order Gymnotiformes, the South American electric eels or American knifefishes. That is the only ostariophysan order restricted to a single land mass. The group includes five families and more than 100 species, distributed from the Río de La Plata in Argentina to Southwestern Chiapas in Mexico (Campos-da-Paz and Alberts, 1998; Alberts and Campos-da-Paz, 1998; Alberts, 2001). Their most remarkable characteristic is the ability to produce and detect electric fields, a phenomenon first reported by Lissmann (1958). Electric fields are generated by the emission of weak electric discharges (called electric organ discharges, or EOD in the literature), continuously emitted by electrogenic organs formed by modified muscle or nervous tissue. The body surface of gymnotiforms is covered with myriad electroreceptors, which are able to detect minor changes of shape in their self-generated field, as well as fields and discharges from other sources. Gymnotiforms use their electric field to orient them-selves in the environment (Heiligenberg, 1973) and to interact with other electrogenic/electro-sensitive fishes (Hopkins and Heiligenberg, 1978). Such abilities are associated with a host of fascinating adaptations which are collectively referred to as electro-communication. Details of wave form and frequency spectrum are diagnostic for different species (Hopkins, 1974), so that individuals can recognize themselves as con-specifics or not. Also, different electric signals are used in social communication, as signs of sexual interaction, dominance hierarchy, trophic and territorial behavior (Hage-dorn, 1986). Despite the diversity of gymnotiforms, there is a single instance of the evolution of strong electric discharges in the group, in the family Electrophoridae. The single species in that family, *Electrophorus electricus*, is able to emit discharges of up to 650 volts, used in predation and defense. The species, which is the largest known gymnotiform (the largest specimens reach over 2 m in length), also has the regular weak discharges typical of other knifefishes. The Gymnotiformes are also well-known for their extraor-dinary ability to regenerate most or all of the post-coelomic part of their bodies, which in their case means most of the body (Ellis, 1913; Anderson, 1987). Apparently, their posterior region is often severed by predation in natural conditions. The interest in gymnotiform structure, behavior and physiology is intense, and special volumes have been dedicated to the subject (Bullock and Heiligenberg, 1986; Heiligenberg, 1991).

The South American Cyprinodontiformes include much of the range of biologically interesting facts to be found in neotropical freshwater fishes. Transcontinental relationships include the diminutive Amazonian poeciliid *Fluviphylax*. According to Costa (1996), the four species of *Fluviphylax* comprise the tribe Fluviphylacini, itself sister group to the African tribe Aplocheilichthyini and included in the poeciliid subfamily Aplocheilichtyi-nae. The genus *Orestias*, including approximately 40 species endemic to high-altitude rivers and lakes along the Andean range in central Peru and northern Chile, is the only South American cyprinodontiform whose rela-tionships lie outside of the tropical region. It is agreed that *Orestias* is related to northern-hemisphere taxa (Eigenmann, 1920; Parenti, 1981; Costa, 1997). There is still disagreement, however, on which is exactly the sister group of the genus. Parenti (1981, 1984) and Parker and Kornfield (1995) placed *Orestias* as sister group to species of the genus *Lebias* (formerly known as *Aphanius*), endemic to salt, brackish and freshwater environments around the Mediterranean, Black, Red and Arabian seas. Costa (1997), on the other hand, hypothesized *Orestias* as sister group to the tribe Cyprinodontini, which includes *Lebias* plus various other genera occurring on the northern-hemisphere part of the Americas. The magnitude of the species richness of the neotropical cyprinodontiform family Rivulidae was only recently realized. Fourteen of the 27 currently recognized genera and 78 of the 235 species were described by a single author (W. J. E. M. Costa) during the past 16 years (cf. Costa, 2003). Several rivulid species are annual, living in temporary pools which may dry out completely in the dry season. Although most or all of the adults die on those occasions, their eggs survive in the dried substrate into the next rainy season, when they give rise to a new generation. Finally, two species of *Rivulus* represent the only cases of self-fertilizing hermaphroditism known in vertebrates (Harrington, 1961).

South America includes freshwater representatives of fish groups not normally occurring outside of the marine environment. The only recent radiation of chondrichthyan fishes in freshwater is the stingray family Potamotrygonidae. The 18 species currently recognized in the family (Carvalho *et al.*, 2003), all Cis-Andean, are arranged in three genera, *Paratrygon* (monotypic), *Plesiotrygon* (monotypic) and *Potamotrygon* (16 species). *Taeniura* and *Himantura*, marine successive sister taxa to Potamotrygonidae, are sometimes included in the family too (Lovejoy, 1996). Potamotrygonids are widely distributed in South America, but absent in the São Francisco, upper Paraná and eastern Atlantic drainages.

Various other primarily marine groups have freshwater invaders in the neotropics. Ophichthyid eels, otherwise known only from tropical marine environments, are represented in South America by one endemic freshwater genus and species, *Stictorhinus potamius*, from the Tocantins and Orinoco basins (also reported from the State of Bahia, in northeastern Brazil). Little is known about that species, which has only rarely been collected. *Stictorhinus* seems to have fossorial habits on the bottom of rivers (Böhlke and McCosker, 1975), congregating in substrate that accumulates amongst boulders (G. M. Santos, personal communication). *Brotulas* of the family Bythidae, otherwise exclusive marine in reefs and deepwater, have six species in freshwater or slightly brackish water in caves in Cuba, Yucatán and the Galapagos Islands (Nielsen, 2003). Likewise, the mostly marine Batrachoididae include five freshwater species in South and Central America (Collette, 2003).

South America is the stage of some of the most extraordinary ichthyological findings in recent times. The past 30 years have seen the discovery of an entire new family of neotropical catfishes, Scoloplacidae (Bailey and Baskin, 1976; Schaefer *et al.*, 1989), and various new subfamilies (e.g., de Pinna, 1992). This trend continues into the present. A remarkable recent case which has received considerable publicity is the discovery of a species of leaf-litter adapted fish, possibly related to Characiformes, but whose relationships are still obscure. That fish has a body shape entirely unique for neotropical fishes and a puzzling combination of internal and external anatomical characteristics, which indicate that it may represent a new family of ostariophysans.

B. Africa

Most of the African continent might be considered as tropical, as far as freshwater fishes are concerned. The number of African freshwater fish species is approximately 2900. That figure, however, is deceiving about the actual density of species per area. Much of the African continent is extremely dry, lacking permanent water bodies. The large northern half is dominated by the Sahara desert in which the fish fauna is practically non-existent, except for a few opportunistic species in oasis. The freshwater fish fauna north of the Sahara, in the northwestern area called the Maghreb along the south coast of the Mediterranean, is depauperate. Also, the affinities of its species are closer to Europe than with the rest of Africa (Roberts, 1975; Greenwood, 1983). This is particularly evident in cypriniforms of the genera *Barbus*, *Pseudophoxinus* and *Cobitis* as well as in the cyprinidontiform *Aphanius*. Non-Primary freshwater fishes, such as the European eel (*Anguilla anguilla*), trout (*Salmo trutta*) and stickleback (*Gasterosteus aculeatus*), which have populations in the Maghreb, also indicate non-African affinities. Apparently, the Atlas Mountains have played a major role in biogeographical isolation of the Maghreb fish fauna from the rest of Africa.

Tropical African freshwater fish diversity is therefore restricted geographically to the region called sub-Saharan Africa. Even so circumscribed, there are large dry stretches of desert and savannah that lack permanent water necessary to sustain significant fish communities. These factors restrict the majority of African fish biodiversity to the moist equatorial region, mostly associated with rainforests. The richest region is drained for the most part by four large basins: Congo (Zaire), Niger, Nile and Zambezi, which are home to most of the freshwater fish species of Africa (see Chapter 1). Smaller drainages contribute to a lesser extent to the African fish fauna, but are of great importance in understanding the biogeography of the continent and in harboring endemics not found elsewhere.

Africa has been divided into a number of ichthyofaunal provinces since Boulenger (1905). Roberts (1975) recognizes ten such provinces, which are reflective of the main patterns of freshwater fish distribution in the continent: Maghreb, Abyssinian highlands, Nilo-Sudan (including Lakes Albert, Edward, George and Rudolf), Upper Guinea, Lower Guinea, Zaire (including Lakes Kivu and Tanganyika), East Coast (including Lakes Kioga, Victoria and those of the Eastern Rift Valley except Malawi and Rudolf), Zambesi (including Lake Malawi), Quanza, and Cape.

Despite the relatively compressed geographical boundaries, the African fish fauna is extraordinarily diverse in clades and ecological specializations. Africa comprises the largest number of ostariophysan orders of any continent, with a diverse assortment of Gonorynchiformes, Cypriniformes,

Characiformes and Siluriformes. Only Gymnotiformes, endemic to South America, are absent. Also, Africa is the only continent where living representatives of Polypteriformes are found today. The Polypteriformes includes primitive fishes which are the sister group to all other members of the Actinopterygii. Interestingly, the African polypteriforms, although considered "living fossils" in systematic ichthyology, are far from being ecological relicts. There are two valid genera (*Polypterus* and *Erpetoichthys*) with at least 11 species, several of which are locally abundant in equatorial Africa. The African polypterids are an extraordinary source of comparative information for understanding the evolution of bony fishes. In the field of vertebrate evolution, polypterids stand for bony fishes as the coelacanth stands for tetrapods.

Another remarkable component of the African fish fauna is the African lungfish, *Protopterus*. It forms the sister group to the South American lungfish, *Lepidosiren*, a relationship expressed in the inclusion of the two genera in the order Lepidosireniformes. In contrast to its monotypic South American sister group, however, *Protopterus* includes four well-differentiated species.

The Osteoglossomorpha are more diverse in Africa than in any other continents, both in number of species and of clades. The African bonytongue *Heterotis niloticus* is the sister group to the South American Pirarucu *Arapaima gigas*, both being included in the family Arapaimidae. The Pantodontidae, with its single species *Pantodon bucholzi*, is exclusive to Africa. The mostly Asian Notopteridae is represented by two species in Africa. A remarkable radiation, unparalleled in any other osteoglossomorphs, is represented by the family Mormyridae, endemic to Africa. The family, with over 200 species, is widespread in the continent, and absent only in the Maghreb and Cape regions. Mormyrids are particularly noteworthy because of their electro-sensory and electrogenic adaptations, which parallel those of South American Gymnotiforms. It is possible that such physiological particularities are associated, as a key innovation, with the great diversification of mormyrids. The Gymnarchidae is a monotypic family related to the Mormyridae and which shares with it the same electro-location/communication abilities.

African Characiformes, with *ca.* 210 species, are considerably less diverse than their neotropical counterparts. There are only four families in the continent: Distichodontidae, Citharinidae, Alestidae and Hepsetidae. The phylogenetic relationships of African characiforms, however, are complex and they do not form a monophyletic group. The exclusively African Distichodontidae and Citharinidae form a monophyletic group considered as the sister group to all other characiforms, both African and South American

(Vari, 1979). The Hepsetidae, with a single species also restricted to Africa, is included in an otherwise exclusively South American clade including the families Erythrinidae, Ctenoluciidae and Lebiasinidae (Vari, 1995; Oyakawa, 1998). Finally, the Alestidae is currently defined to include the South American genus *Chalceus* plus a variety of African genera (Zanata and Vari, in press). The group formed by Alestidae exclusive of *Chalceus* is the largest exclusively African characiform radiation. Despite their comparatively small number of species, African characiforms include the largest body size in the order, 1.3 m attained by *Hydrocynus goliath* (Weitzman and Vari, 1988).

Siluriformes display much diversification in African freshwaters, although in number of endemic species and higher clades it is less diverse than South America. The Mochokidae, Amphiliidae, Claroteidae, Malapteruridae and Austroglanididae are the only endemic African families. The families Schilbidae, Clariidae and Bagridae are shared with Asia, while the estuarine Ariidae are pantropical. Although there are no freshwater catfish families shared between Africa and South America, the Mochokidae is solidly positioned as sister group to neotropical doradoids (Lundberg, 1993; de Pinna, 1998). There is also evidence placing the Amphiliidae as sister group to neotropical Loricarioidea (de Pinna, 1993, 1998; Britto, 2003). The impoverished fish fauna of Somalia includes a remarkable subterranean catfish, *Uegitglanis zammaronoi*, which seems to be the sister group to all clariids (of both Africa and Asia) except for the Indian *Horaglanis krishnai* (de Pinna, 1993). The bizarre electric catfish family Malapteruridae is widespread in most West African drainages and the Nile. The family was previously considered to comprise a single genus, *Malapterurus*, and two or three species only, one of which was widespread throughout most of Africa (*M. electricus*). More detailed work (Norris, 2002) has unveiled a previously unsuspected diversity, including dwarfed species which were previously considered to be juveniles of other species. Norris (2002) recognized two genera (*Malapterurus* and *Paradoxoglanis*) and a total of 19 species, whose patterns of endemism match those recognized for other African freshwater fishes. According to the results of that work, no malapterurid species shows a widespread Pan-African distribution. The failure to recognize the species diversity within the family was in part a result of the wide morphological gap between malapterurids and other siluriforms, which has drawn attention away from the comparatively minor within-family variation (Norris, 2002). Phylogenetic relationships of malapterurids among siluriforms have been a long-standing controversy among ichthyologists. The profound autapomorphic specializations typical of the species of Malapteruridae have made morphological comparisons with other catfishes difficult or inconclusive. One proposal that has been a result of an extensive phylogenetic analysis of the order

Siluriformes places *Malapterurus* as the sister group of the bagrid subfamily Auchenoglanidinae (de Pinna, 1993). In addition to the characters proposed by de Pinna (1993), a few of the characters offered by Mo (1991) as diagnostic for the Auchenoglanidinae (a subfamily of Bagridae) are also seen in malapterurids. Those include a round caudal fin and the mottled integumentary coloration pattern. Although both characters are highly homoplastic within siluriforms as a whole, they are unusual among African catfishes and may, after more detailed analysis, provide further evidence that malapterurids and auchenoglanidines are related.

The Cypriniformes form one of the main components of the African freshwater fish fauna, with 475 valid species described so far. The vast majority of tropical African cypriniforms belong to the family Cyprinidae, although there are two genera of Cobitidae in a small area of the eastern reaches of the continent, close to the Red Sea straight. A peculiar subterranean cyprinid, *Phreatichthys*, occurs only in limestone underground waterways in the dry Somalian region. This fish is not pigmented, is eyeless and scaleless.

The most extraordinary case of explosive radiation in African freshwaters is represented by the Cichlidae, a family shared with South America. Interestingly, cichlid evolution and diversification in Africa seems to be closely associated with lacustrine environments, especially with the large lakes of the Great Rift Valley. The African lake cichlids constitute one of the most extraordinary cases of adaptive radiation in vertebrates, and have become the focus of study of an entire field in ichthyology.

The temperate extreme southern portion of Africa is faunistically quite different from the rest of the continent, forming a dramatically distinct zoogeographical region. An excellent review of the distribution patterns and biogeography of the fishes of that region and of the Southern part of Africa in general is provided by Skelton (1994). Temperate South Africa is noteworthy not only for its endemics, but also for the absence of major groups typical of the rest of Africa. The northern part of Southern Africa has a rich tropical fish community, with over 200 Primary and Secondary freshwater species representing 21 families. This diversity declines sharply southwards at the Natal region on the east coast and south of the Cunene River on the West Coast, so that only 38 species in five families are found there (Skelton, 1986). The catfish family Austroglanididae is endemic to that region. It comprises only three species restricted to the Orange and Olifants basins, which open into the Eastern Atlantic. The Austroglanididae, a family established by Mo (1991) may be related to non-African taxa, although more research is needed on the subject. The only galaxiid species in Africa, *Galaxias zebratus*, is restricted to the extreme south of the continent and it is a remarkable testimony to the austral affinities of Southern Africa.

C. Asia

Tropical Asia is the only region where recent freshwater fish discoveries rival in kind and degree those made in South America. As in the neotropics, this situation is a result of the combination of rich endemic diversity with the existence of unexplored areas. Tropical Asia comprises the region including the political boundaries of India, Sri Lanka, Bangladesh, Myanmar, Thailand, Laos, Cambodia, Vietnam, Malaysia, Singapore and Indonesia (which includes Sundaland, formed by the Islands of Java, Sumatra and Borneo). The latter eight countries are within the area collectively referred to as Indochina, or South and Southeast Asia (Kottelat, 1989). This zoogeographical region is limited on the east by Wallace's line (as modified by Huxley). The land masses located east of it belong to the Australian region. Although tropical South and Southeast Asia belong to the Asian continent, their historical connections are likely with Gondwana, rather than Laurasia (Audley-Charles, 1983, 1987). Geologically, it appears that most, if not all, of the South Asian border is formed by Gondwanan fragments which moved north and collided with the south Asian border. This implies that the phylogenetic relationships of freshwater fishes in South and Southeast Asia are to be sought among African, Australian and South American taxa, rather European, Central Asian and North American ones.

The fish fauna of tropical Asia is exceedingly rich in both clades and species, with approximately 3000 species and 121 families. Of cypriniforms alone, there are over 1600 species (of which over 1000 are in a single family, Cyprinidae). Interestingly, tropical Asia has more peripheral families than any other continent, with 87 families having freshwater invaders. This is probably a reflection of the extraordinary diversity of marine fishes in the region, the richest in the world (see below). As in other tropical regions, the diversity of freshwater fishes in South and Southeast Asia is still incompletely known. In the southern Indian State of Kerala, it has been estimated that up to 20% of the fish fauna is not described (Pethiyagoda and Kottelat, 1994). Concentrated fieldwork often results in spectacular increases in known species richness. For example, 11 weeks of collecting in Laos yielded an 80% increase in the number of freshwater fish species known for the country (Kottelat, 1998). As in other tropical areas, a major proportion of newly discovered taxa in Southeast Asia is represented by small species occurring in specialized habitats. Six of the nine species of Chauduriidae were described since 1991, largely as a result of more careful exploration of interstitial habitats and peat swamp forests. The discovery of more abundant material has allowed more detailed investigations into their peculiar anatomy and paedomorphic conditions (Britz and Kottelat, 2003). The family Sundasalangidae, previously considered as related to salangids but now recognized as highly paedomorphic freshwater clupeiforms, has had their species number tripled in the past few years (Siebert, 1997).

Many tropical Asian fish families are shared exclusively with Africa: Notopteridae, Bagridae, Clariidae, Schilbidae, Anabantidae, Chanidae and Mastacembelidae. The repeated pattern of shared exclusive taxa, all of which demonstrably monophyletic, is indicative that these families were already differentiated by the time India and parts of Southeast Asia separated from Africa. There is a single case of relationship between a Southeast Asian and South American taxa, involving the neotropical catfish family Aspredinidae and the otherwise exclusively Asian superfamily Sisoroidea (de Pinna, 1996; Diogo et al., 2003). This is a puzzling relationship, and one which requires rather unorthodox lines of biogeographical explanation. Sisoroids, as a trans-pacific group, are part of the same pattern encountered in several invertebrate taxa with similar distributions. It is possible that the whole trans-pacific biota was part of the hypothetical continent Pacifica, which fragmented and collided with different parts of the Pacific rim. Alternatively, sisoroids may have been present in Africa, but went subsequently extinct in that continent. If the latter alternative is correct, there should be fossil sisoroids in Africa, not yet encountered.

D. The Australian Region

The tropical Australian region includes Australia and New Guinea. The two land masses are closely related geologically, and have been conjoined for much of the history of the region. During the last glaciation, sea level was low enough for a mostly continuous land connection between Australia and New Guinea. The southern river basins in the latter joined with drainages in the northern vertent of the former as recently as 6000 MYA (Lundberg et al., 2000). This close relationship is reflected in shared species diversity. Approximately 50 freshwater fish species are shared between Southern New Guinea and Northern Australia. The taxonomic composition is also very similar in the two areas.

As with much of its vertebrate fauna, the freshwater fishes of Australia are quite unique. There are approximately 200 species, most belonging to gobioid families Gobiidae and Eleotrididae (with ca. 50 species), southern smelts of the super-family Galaxioidea (26 species), perch-like Teraponidae and Percichthyidae (43 species), atherinomorphs Atherinidae and Melano-teniidae (ca. 30 species) and the catfish family Plotosidae (ca. 15 species). Australia is the only continent where ostariophysans are not the dominant group of fishes in freshwater. Only the catfish families Plotosidae and Ariidae occur there, and both families also contain numerous marine species, in addition to the freshwater ones. There are few other Primary freshwater fishes, but they include a number of very important relicts. The first of these is the Australian lungfish, *Neoceratodus*, which is restricted to parts of two small river systems in Southeastern Queensland. *Neoceratodus* is the most primitive recent lungfish. It constitutes the sister group to the African *Protopterus* plus the South American *Lepidosiren*. Another remarkable Primary freshwater fish is the osteoglossomorph *Scleropages*, with two Australian species. Representatives of that genus, which is the closest relative to the South American *Osteoglossum*, occur also in New Guinea and Southeast Asia. The relatively depauperate freshwater fish fauna of Australia apparently results from a combination of the present-day dry climate over most of the country, combined with a long history of isolation from other freshwater-rich land masses. The historical component is certainly paramount, because the biogeographically close New Guinean region is humid and more species-rich, but otherwise has a taxonomic composition very close to that of Australia. Of course, part of the "impoverished" condition of the fish fauna may be artifactual, a result of incomplete taxonomic work at species level in the region (Lundberg *et al.*, 2000). While that is most likely true, there is little question that there is indeed some pauperization in the Australian and New Guinean freshwater fishes. The absence or near absence of such elements as ostariophysans and cichlids, alone, are major gaps in comparison with most other major land masses.

The salamanderfish, *Lepidogalaxias salamandroides*, is one of the most notable fish icons of Australia. That small species, described as late as 1961 (Mees, 1961), is often considered as the sole member of the family Lepidogalaxiidae, and has been a resilient puzzle in the elucidation of phylogenetic relationships among lower teleosts. A detailed study on the comparative anatomy of *Lepidogalaxias* (Rosen, 1974) has resulted in a hypothesis that the fish is related to northern hemisphere esocoids. Other hypotheses place *Lepidogalaxias* at an unresolved position between Salmonidae and Neoteleosts (Fink, 1984). More recent and taxonomically more encompassing work (Johnson and Patterson, 1996) has resulted in a very specific hypothesis that *Lepidogalaxias* is the sister group to *Lovettia*, a galaxioid genus endemic to Tasmania. The pronounced differences in body shape and general aspect between the two forms makes that relationship surprising, but the character support for the hypothesis is robust. An alternative hypothesis for the position of *Lepidogalaxias* within galaxioids also exist (Williams, 1997). In any event, *Lepidogalaxias* seems to have diverged quite dramatically from all of its close relatives, resulting in an aberrant morphology that makes elucidative comparisons difficult. The movable anterior part of the vertebral column (which results in a movable "neck"), the lack of extrinsic eye muscles, the aestivating habits in temporary pools and accessory aerial respiration are all features that set *L. salamandroides* apart from all other galaxioid fishes. It would be interesting to investigate whether the remarkable physiological specializations of *Lepidogalaxias* (Berra and Pusey, 1997) are present, in some version, also in *Lovettia* or other galaxioids.

The approximately 350 species of New Guinean freshwater fishes (Allen, 1991; Lundberg *et al.*, 2000) fit mostly into the dominant groups occurring in Australia. Australian groups absent in New Guinea include Dipnoi, galaxioids and Percichthyidae. New Guinea, on the other hand, includes a few groups not present in Australian freshwaters, such as the perch-like Chandidae. The fish fauna of New Guinea seems to be less thoroughly known than that of Australia, but a number of spectacular discoveries are sill regularly made in both areas (cf. Lundberg *et al.*, 2000).

The biogeographical relationships of Australia and New Guinea, as a whole, lie strongly in South America, especially its southern portion. Groups such as Percichthyids, galaxioids and Osteoglossidae are shared nearly exclusively between the two areas and are clear indicators of Austral former Gondwanan connections.

VI. MARINE TROPICAL FISHES

As with freshwater fishes, the tropical region also comprises an extraordinary diversity of marine fishes. There is a remarkable difference of species density in marine and freshwater environments (see above) per unit of water volume. A major difference in the patterns of diversity between marine and freshwater environments is reflected at the family-level representation. Every major land mass has various endemic families of freshwater fishes. Contrastingly, there are few endemic families in tropical marine regions, reflecting an evident homogeneity in familial composition of tropical shore fish faunas (Bellwood, 1998). The same phenomenon holds for higher taxonomic categories, with most of the reef-fish species belonging to the Perciformes (Robertson, 1998). In freshwater, most species belong to the Ostariophysi, which in turn have a very limited representation in marine environments.

Contingent historical factors in marine environments are no less important than those in freshwater in determining different levels of fish diversity. The dynamics and mechanisms of isolation of populations are different in detail in marine and freshwater environments, but their relative importance is similar. Some recent cases of faunal changes in historical times demonstrate that the geological history of a marine area is as important as local ecological factors in determining its fish faunal composition. The eastern portion of the Mediterranean is a warm-water region which nevertheless shares the same depauperate fish fauna as the rest of that Sea (fewer than 550 species). The western portion of the Mediterranean and its opening to the

Atlantic is located in a cold-water sector of the Atlantic, which provides a barrier to tropical species which might otherwise colonize the eastern Mediterranean from tropical areas. Proof that historical, rather than ecological, factors determined the scarce Mediterranean fish fauna came with the opening of the Suez Canal. Since 1869, human-made contact between the richer region of the Red Sea and the warm yet species-poor Eastern Mediterranean has resulted in over 50 Indo-West Pacific species entering and establishing populations in the latter (Golani, 1993). Clearly, as in all biotas, the composition and abundance of a fish fauna is a result of local ecological conditions and a long sequence of historical and geological events.

Diversity of fishes varies markedly among different tropical oceanic regions and will be treated separately below. Phylogenetic information for most tropical marine groups, as they are relevant for biogeographical inference, is far scantier than for freshwater taxa. For that reason, most of the hypotheses available as yet rely heavily on comparative information which is not strictly phylogenetic. Instead, similarities in faunal composition are the main basis for most discussions on marine fish biogeography. While such comparisons are often not fully conclusive, they offer a preliminary picture of the major patterns of similarity, and possibly historical relationship, among the main tropical marine regions.

A. Indo-West Pacific

Unquestionably the most diverse of all marine regions is the Indo-West Pacific, covering an area that extends from Southeast Africa across the Red Sea, Arabian Peninsula, South and Southeast Asia, New Guinea, Australia, Hawaii and the South Pacific islands. It is estimated that approximately 4000 species of fish occur in the Indo-West Pacific (Springer, 1982; Myers, 1989), a number which exceeds by far that of any other marine region. There are certainly ecological and physical factors which are associated with species richness of the Indo-West Pacific fishes. The region is known to harbor the most extensive and diverse communities of reef-building corals (Rosen, 1988). Tropical fish diversity is strongly associated with corals, as is, in fact, diversity of the whole marine animal biota (Briggs, 1974). Obviously, the question arises as to what is the cause of higher coral diversity in the region. One possibility is the extremely complex geographical structure of the region, which is split into tens of thousands of islands of various sizes. The subdivided shorelines not only represent a magnification of suitable shore areas for coral growth, but also increase landscape complexity and therefore opportunities for vicariance, speciation and specialization. Pleistocene sea level fluctuations may also represent a relevant factor in determining fish diversity in the Indo-Pacific. Sea-level changes may constitute dynamic barriers by producing isolation of marine areas by land barriers during low sea-level periods, with subsequent inundation of those barriers by rising sea levels (Randall, 1998). Naturally, the more complex a shoreline, the more likely such fluctuations are to form isolated pockets of shore fishes.

The center of species diversity in the Indo-Pacific is the area of the Philippines and Indonesia, with approximately 2500 species. Fish species diversity decreases sharply eastwards. New Guinea has 2000 species, Australia 1300, New Caledonia 1000, Samoa 915, Society 633, Hawaii 557, Pitcairn 250 and Easter 125 (Planes, 2002; Randall, 1995). The decrease in number of species clearly follows the diminishing land masses and associated shore areas. There are various theories that attempt to explain this gradient, and none of them has gained general acceptance. Several models evoke inferences about speciation rates and routes of dispersal, but analyses of genetic data for different animal groups have been quite contradictory in their derivative biogeographical implications (Planes, 2002). Detailed phylogenetic hypotheses, not yet available in the needed scale for most groups, will provide an important element in resolving those issues.

Clearly, colonization is an important factor in the markedly divergent species richness in Indo-Pacific localities. A correlation has been demonstrated between the biology of certain fish groups and their representation in insular localities. For example, Randall (1995) noticed that the familial representation in Hawaiian Islands does not exactly match the proportion expected for the Indo-Pacific region. The Gobiidae, usually the most speciose family in Indo-Pacific localities, is represented by only 27 species in Hawaii. Similarly limited representation happens also with the families Pomacentridae, Blenniidae and Apogonidae. Randall (1995) noticed that species of those families are usually either mouth brooders or lay demersal eggs, factors which are expected to be associated with a relatively short planktonic larval life. The vast distances that separate Hawaii from other coral reef areas would be a barrier for species without a long planktonic phase. In agreement with that, families such as Muraenidae and Acanthuridae, which have a long larval period and long sojourn in open waters, have a representation in Hawaiian waters larger than would be expected for Indo-Pacific localities (Randall, 1995).

An alternative interpretation of faunal gradients in the Indo-Pacific is offered by Springer (1982). That author exhaustively surveyed distributions for different taxa (mainly fish, but other groups as well) and concluded that the Pacific Plate constitutes a well-defined biogeographical region which cannot be understood as simply an impoverished offshoot of a larger Indo-Pacific region. The Pacific lithospheric plate underlies most of the Pacific basin, except for its Western portions (corresponding instead to a complex crossroads of the Philippine, Eurasian and Indian-Australian Plates).

According to Springer, the faunal composition of the region corresponding to the Pacific Plate may be explained by the history of the Pacific Plate itself, and not by dispersal from the more species-rich Indo-West Pacific.

The Red Sea belongs to the Indo-West Pacific biogeographic realm, but is usually considered as an appendix to it (Briggs, 1974). It contains a relatively depauperate fish fauna when compared to the Indian Ocean, with approximately 800 species. Diversity also decreases gradually to the north, a pattern which is not restricted to fishes (Kimor, 1973). The fact is curious, since the free flow of water through the Gulf of Aden would expectedly permit massive recruitment of pelagic organisms and larvae. Theories to explain the relative depauperate fauna of the Red Sea are various. One physical characteristic that must be considered is the bottleneck represented by the strait of Bab-el Mandeb, which is only 20 km broad and at one point only 100 m deep. That probably explains the markedly poor mesopelagic fish fauna in the Red Sea, which comprises eight species, less than 3% of the 300 species known from the Indian Ocean (Johnson and Feltes, 1984). The Red Sea today is considerably more saline than the Indian Ocean, because of the high evaporation rates coupled with little freshwater input. It has been suggested that conditions in the past, which include complete isolation of the Red Sea and lowering of its level of 90–200 m during the last glacial period (Sewell, 1948), may have resulted in hypersaline conditions unsuitable for life of most organisms. This resulted in the annihilation of any remaining elements of the Tethys Sea fauna, and would explain the prevalence of Indian Ocean elements in the Red Sea today, after reestablishment of contact. Several authors disagree with that scenario, arguing that there is no fossil evidence of abiotic periods in the Red Sea (Por, 1972; Klausewitz, 1980).

B. Eastern Pacific

The Eastern Pacific Region, also known as the Panamanian Region, is far less rich in fish species than is the Indo-West Pacific, with fewer than 900 fish species. In fact, the relationships of its fish fauna seem to lie more closely with the Western Atlantic (see below) than with the remainder of the Pacific Ocean. There are several sister species shared between the Western Atlantic and the Eastern Pacific, which indicates a previously continuous biota which was separated by the Isthmus of Panama between 3.5 and 3.1 MYA (Coates and Obando, 1996). Some species that occur on both sides of the isthmus have not undergone noticeable differentiation yet. The cold austral tip of South America represents an older barrier to tropical fish dispersal via a southern route.

The relative paucity of Indo-West Pacific species (or sister species) in the Eastern Pacific is a puzzling distributional fact, in view of the lack of apparent physical barriers. In fact, 86% of the Indo-West Pacific species do not reach the west coast of South America (Briggs, 1974). There is, however, an enormous stretch of open water between the Central and South Pacific islands and the Western South American shore. Briggs (1974) has proposed that such vast expanse without shores or coral reefs has acted as a barrier that prevents colonization of the Eastern Pacific region by species from the Indo-West Pacific. This is called the Eastern Pacific Barrier. The effectiveness of that barrier is of course not complete, and there are islands off the coasts of Mexico and Costa Rica where a number of West-Pacific species are found (Helfman *et al.*, 1997). One must also remember that the number of Indo-West Pacific species decreases sharply eastwards from the Indonesian-Philippine area, perhaps a result of the Pacific Plate having a history distinct from that of the Indo-West Pacific Region (see above). If that is the case, it is also possible that the Eastern Pacific fish fauna is somehow associated with the history of the Nazca plate, the lithospheric plate underlying the region.

C. Western Atlantic

The tropical portion of the Western Atlantic is the area that extends from the southern shores of North America, ranging south through the Gulf of Mexico, the Caribbean and the tropical shores of South America. This region is the second richest in fish diversity, with approximately 1200 species. The center of species richness in the tropical Western Atlantic, and in fact in the whole Atlantic, is the Caribbean. There are approximately 700 species of Caribbean fish, and that region has been seen as a center of speciation and of accumulation of fish species (Rocha, 2003). Other regions in the tropical Western Atlantic are normally seen, to varying degrees, as offshoots of the Caribbean center.

The mouth of the Amazon, associated with its huge outflow of freshwater, has long been considered as a major barrier for the inshore marine fauna in the coastal Western Atlantic. There is a large gap of coral reefs along the region around the mouth of the Amazon and for a long stretch of shore northwards until the Orinoco delta. The gap, extending for *ca.* 2500 km, is located to the north of the mouth of the Amazon (see Chapter 1), because the Southern Equatorial Current moves the discharge in a north-northwest direction. Lowered coastal salinities are felt as far north as Guyana (Eisma and Marel, 1971). Water- and bottom-type changes induced by the influx of the Amazon forms a gap in coral reef distribution, a gap reflected in a major portion of the shore marine biota. Because of that, the tropical West Atlantic has traditionally been divided into a northern and a southern half, supposedly reflective of isolated shore marine fish populations (Briggs, 1974). The significance of the mouth of the Amazon as a barrier to marine shore fishes is not absolute, however. The offshore reaches of Amazonian freshwater seem to be mostly restricted to surface layers. Below that, denser saltwater predominates. It has been shown (Collette and Rützler, 1977) that elements of typical coral reef benthic fish fauna exist right at the mouth of the Amazon, below the superficial freshwater layer. Of course, fishes in that region are not associated with corals, because the water is too dark and turbid to support coral growth. Fishes are instead associated with sponges, which provide a solid substrate with some structural similarity to a coral reef. Such consolidated substrate, however, seems to occur as small isolated patches (Moura, 2003), and there can be little doubt that most of the bottom is soft mud. Relatively continuous consolidated substrate does not exist in the region south of Trinidad to the Brazilian State of Maranhão (Moura et al., 1999). The Amazonian marine barrier seems indeed to be effective for several taxa, and specific differentiation exists in some fish groups. Some South Western Atlantic population has been demonstrated to be distinct species after detailed taxonomic analysis (e.g., Moura et al., 2001). It is becoming increasingly clear that the shore fish fauna of the South Western Atlantic cannot be considered simply as a depauperate offshoot of the Caribbean fish community (Moura and Sazima, 2003). In fact, it has recently been demonstrated even that some species from the South Western Atlantic have sister species on the Eastern Atlantic, rather than in the Caribbean (Heiser et al., 2000; Muss et al., 2001; Rocha et al., 2002). The less than complete barrier represented by the mouth of the Amazon is reflected in the relative similarity of species pairs. The Amazon barrier is dated to approximately 10 MYA, but species pairs separated by it look more similar than those separated by the Isthmus of Panama, an absolute barrier dated ca. 3 MYA (Rocha, 2003).

The coral reefs around the islands in the Great Barrier Reef in Australia harbor approximately twice the number of fish species as similar islands in the Caribbean. This large difference has been accounted for by differences in taxonomic composition between each region, rather than different rates of diversification (Westoby, 1985). That implies that historical constraints, rather than ecology, are the determinant factors in the difference in coral-reef species diversity between the Caribbean and the Great Barrier. It has been proposed that Pleistocene events of reef fish extinctions in the Caribbean have been more frequent and extensive than those in the Indo-Pacific (Ormond and Roberts, 1997; Bellwood, 1997). The same factor has been proposed to explain the fewer species in the South Western Atlantic when compared to the Caribbean (Moura, 2003).

Despite the long-standing paradigm of the barrier represented by the mouth of the Amazon, species from the South Western Atlantic have usually been considered as conspecific with their closest relatives in the Caribbean and Gulf of Mexico. More recent and detailed taxonomic work, however, is showing that there is specific differentiation in various cases (Moura, 2003).

D. Eastern Atlantic

The least species-rich of the marine tropical shore fish regions is the Eastern Atlantic, with only about 500 recorded species, less than half the number in the Western Atlantic. Part of that may be a result of incomplete sampling, since the region is also the least studied in terms of faunal surveys. The Eastern Atlantic extends from Senegal in the north to Angola in the South, and its most prominent physical feature is the Gulf of Guinea. It also includes a few major islands, such as Cape Verde, St Helena and Ascension. The relative paucity of fish species seems to be associated with a general scarcity of coral environments (Rosen, 1988). The latitudes which would be the most favorable for coral growth are affected by the outflows of several large rivers, such as the Congo, Volta and Niger. Such freshwater inflow results in elevated turbidity and sediment deposition which are unsuitable for reef-forming coral species. Nearly half of Eastern Atlantic species are endemic (Briggs, 1974), indicating that the region is effectively isolated from other marine areas. On the other hand, over 100 species are shared with the Western Atlantic, more than with any other marine region. This similarity may be a result of the vicariance of a common marine biota during the early stages of separation of Gondwana, when the margins of the Proto-Atlantic were still close. This hypothesis, though likely, still needs careful testing on the basis of phylogenetic hypotheses for different taxa on both sides of the Atlantic. There are also a number of Eastern Atlantic fish species shared with the Indo-Pacific. However, the majority of those are either pantropical or found also in the Western Atlantic and their value as evidence of relationships is questionable. It seems that the Indo-Pacific species found in the Eastern Atlantic are simply those which manage to cross the Cape of Good Hope and do not result from a common ancestral biota.

VII. CONCLUSIONS

The extraordinary diversity of tropical fishes is among the richest fields for studies on evolutionary biology. The complexity of that fauna, however, requires a multifaceted approach in its description and comprehension. Proper understanding of the diversity of tropical fishes requires first and utmost a temporal context that is provided only by hypotheses on phylogenetic relationships. Biological diversity is hierarchically organized, a reflection of the diverging structure of the evolution of species and other taxa.

The branching pattern of evolution is reflected in schemes of phylogenetic relationships (cladograms) and their derivative phylogenetic classifications. Historical factors, with their contingencies, are a primary factor determining net diversity and taxonomic composition of fish faunas. Ecological factors are explanatory only within the constraints posed by historical background, expressed as phylogenetic patterns coupled with geological information. A review of the fish faunas in different regions shows that the evolution of continents, river drainages and ocean basins are tantamount in canalizing fish diversity.

Assessment of diversity requires far more than simply counting the number of species per unit area. Taxonomic representation, or phyletic diversity, is a measure far more significant than number of species in assessing the diversity of a fauna. Many areas are species-poor but have extraordinarily relevant fish faunas. Such areas, as Austral South America and the Cape region in South Africa, contain unique representation of lineages that form the sister group to large clades, or reveal especially informative transcontinental relationships.

The physiological adaptations of tropical fishes are a biological attribute as dependent on evolutionary history as any other. Therefore, research on comparative physiology relies heavily on the results of phylogenetic investigations. Results of comparative physiology, in turn, also may represent important sources of information in elucidating phylogenetic relationships. The repeated occurrence of certain adaptations, such as accessory aerial respiration, is a response to similar demands in various different lineages (Graham, 1997; see Chapter 10). In some cases, the adaptations are homologous, and have been the result of a single event inherited by descendant taxa. In some other cases they are not, and therefore reflect cases of convergence that may be particularly elucidative about the triggering factors for the evolution of particular traits. The evaluation of the homology or non-homology of adaptations requires detailed knowledge about the evolution of the biological entities being compared. Knowledge of that sort is available by means of phylogenetic hypotheses. By plotting physiological attributes on a cladogram and optimizing its various states, it is possible to estimate the number of events which have resulted in each condition observed. That information allows discriminating which traits are the result of inheritance and which are the result of evolutionary change (Harvey and Pagel, 1991; Harvey et al., 1996). The separation between proximal and historical factors is fundamental to the comprehension of the evolution of physiological and other kinds of adaptations.

It is becoming increasingly clear that fish biodiversity in the tropics may be grossly underestimated. This situation stems from two sources. The first and more obvious one is that there are many uncharted areas and habitats in tropical regions, especially in freshwater habitats. In view of the level of endemism displayed by many fish taxa, especially Primary freshwater ones, previously unsampled areas have a high probability of yielding previously undocumented forms. The second and less evident reason is a general unsatisfactory situation of basic taxonomy, a result of the current scarcity of active professionals. It is alarming that basic revision works, when properly done, often reveal numbers of species vastly larger than previously suspected. In some cases, the increase in number is manifold or even one order of magnitude higher than the expected ones (cf. Vari and Harold, 2001; Norris, 2002).

ACKNOWLEDGMENTS

The author thanks Flávio T. de Lima, Naércio Menezes and Mario de Vivo for reading the manuscript and offering valuable suggestions. The chapter also benefited from conversations on fish diversity and distribution with Heraldo Britski, José Lima Figueiredo and Rodrigo Leão de Moura. Research funding is provided by CNPq (305713/2003-5) and FAPESP (1999/09781-6).

REFERENCES

Alberts, J. S. (2001). Species diversity and phylogenetic systematics of American knifefishes (Gymnotiformes, Teleostei). *Misc. Publ., Mus. Zool., Univ. Michigan* **190**, 1–27.
Alberts, J. S., and Campos-da-Paz, R. (1998). Phylogenetic systematics of gymnotiformes with diagnoses of 58 clades: A review of available data. *In* "Phylogeny and Classification of Neotropical Fishes" (Malabarba, L., Reis, R. E., Vari, R. P., Lucena, Z. M., and Lucena, C. A. S., Eds.), pp. 419–446. Edipucrs, Porto Alegre.
Allen, G. R. (1991). Field guide to the freshwater fishes of New Guinea. Christensen Research Institute (Madang, Papua New Guinea) Publ. **9**, 1–268.
Anderson, M. J. (1987). Molecular differentiation of neurons from ependyma-derived cells in tissue cultures of regenerating teleost spinal chord. *Brain Res.* **388**, 131–136.
Arratia, G. (1987). Description of the primitive family Diplomystidae (Siluriformes, Teleostei, Pisces): Morphology, taxonomy and phylogenetic implications. *Bonner Zool. Monogr.* **24**, 1–123.
Audley-Charles, M. G. (1983). Reconstruction of eastern Gondwanaland. *Nature* **306**, 48–50.
Audley-Charles, M. G. (1987). Dispersal of Gondwanaland: Relevance to evolution of the angiosperms. *In* "Biogeographical Evolution of the Malay Archipelago" (Whitmore, T. C., Ed.), pp. 5–25. Oxford Monographs in Biogeography, Clarendon Press, Oxford.
Azpelicueta, M. M. (1994). Three east-Andean species of *Diplomystes* (Siluriformes: Diplomystidae). *Ichthyol. Expl. Freshwaters* **5**, 223–240.
Bailey, R. M., and Baskin, J. N. (1976). *Scoloplax dicra*, a new armored catfish from the Bolivian Amazon. *Occ. Pap. Mus. Zool., Univ. Mich.* **674**, 1–14.
Baskin, J. N. (1973). Structure and relationships of the Trichomycteridae. Unpublished PhD Dissertation, City University of New York, New York.
Bellwood, D. R. (1997). Reef fish biogeography: Habitat association, fossils and phylogenies. Proceedings of the 8th International Coral Reef Symposium **1**, pp. 379–384.

Bellwood, D. R. (1998). What are reef fishes? Comment on the report by D. R. Robertson: Do coral-reef fish faunas have a distinctive taxonomic structure? *Coral Reefs* **17**, 187–189.

Berra, T. M., and Pusey, B. J. (1997). Threatened fishes of the world: *Lepidogalaxias salamandroides* Mees, 1961 (Lepidogalaxiidae). *Environm. Biol. Fishes* **50**, 201–202.

Böhlke, J. E., and McCosker, J. E. (1975). The status of the ophichthyid eel genera *Caecula* Vahl and *Sphagebranchus* Bloch, and the description of a new genus and species from freshwaters in Brazil. *Proc. Acad. Natl Sci. Philadelphia* **127**, 1–11.

Boulenger, G. A. (1905). The distribution of African fresh water fishes. *Rep. Meet. Br. Assoc. Adv. Sci (S. Afr.)* **75**, 412–432.

Briggs, J. C. (1974). "Marine Zoogeography." McGraw–Hill, New York.

Britto, M. R. (2003). Análise filogenética da ordem Siluriformes com ênfase nas relações da superfamília Loricarioidea (Teleostei: Ostariophysi). PhD Dissertation, Instituto de Biociências, Universidade de São Paulo.

Britz, R., and Kottelat, M. (2003). Descriptive osteology of the family Chaudhuriidae (Teleostei, Synbranchiformes, Mastacembeloidei), with a discussion of its relationships. *Am. Mus. Novitates* **3418**, 1–62.

Bullock, T. H., and Heiligenberg, W. (1986). "Electroreception." Wiley-Interscience, New York.

Campos-da-Paz, R., and Alberts, J. S. (1998). The gymnotiform "eels" of tropical America: A history of classification and phylogeny of the South American electric knifefishes (Teleostei: Ostariophysi, Siluriphysi). *In* "Phylogeny and Classification of Neotropical Fishes" (Malabarba, L., Reis, R. E., Vari, R. P., Lucena, Z. M., and Lucena, C. A. S., Eds.), pp. 401–417. Edipucrs, Porto Alegre.

Carvalho, M. R., Lovejoy, N. R., and Rosa, R. S. (2003). Family Potamotrygonidae (river stingrays). *In* "Checklist of the freshwater fishes of South and Central America" (Reis, R. E., Kullander, S. O., and Ferraris, C. J., Jr., Eds.), pp. 22–28. Edipucrs, Porto Alegre.

Coates, A. G., and Obando, J. A. (1996). The geologic evolution of the Central American isthmus. *In* "Evolution and Environment in Tropical America" (Jackson, J. B. C., Budd, A. F., and Coates, A. G., Eds.), pp. 21–56. University Chicago Press, Chicago.

Collette, B. B. (2003). Lampridae (p. 952), Batrachoididae (pp. 1026–1042), Belonidae (pp. 1104–1113), Scomberesocidae (pp. 1114–1115), Hemiramphidae (pp. 1135–1144). In: Carpenter 2003 [ref. 27006] Western Central Atlantic. CAS Ref No.: 26981.

Collette, B., and Rutzler, K. (1977). Reef fishes over sponge bottoms off the mouth of the Amazon River. *Proc. 3rd. Int. Coral Reef Symp.* 305–310.

Costa, W. J. E. M. (1996). Relationships, monophyly and three new species of the neotropical miniature poeciliid genus *Fluviphylax* (Cyprinodontiformes: Cyprinodontoidei). *Ichthyol. Explor. Freshwaters* **7**, 111–130.

Costa, W. J. E. M. (1997). Phylogeny and classification of the Cyprinodontidae revisited (Teleostei: Cyprinodontiformes): Are Andean and Anatolian killifishes sister taxa? *J. Comp. Biol.* **2**, 1–17.

Costa, W. J. E. M. (2003). Family Rivulidae (South American annual fishes). *In* "Checklist of the freshwater fishes of South and Central America" (Reis, R. E., Kullander, S. O., and Ferraris, C. J., Eds.), pp. 526–548. Edipucrs, Porto Alegre.

Darlington, P. J. (1957). "Zoogeography: the Geographical Distribution of Animals." John Wiley & Sons, New York.

Diogo, R., Chardon, M., and Vandewalle, P. (2003). Osteology and myology of the cephalic region and pectoral girdle of *Erethistes pusillus*, comparison with other erethistids, and comments on the synapomorphies and phylogenetic relationships of the Erethistidae (Teleostei: Siluriformes). *J. Fish Biol.* **63**, 1160–1175.

Eigenmann, C. H. (1920). On the genera *Orestias* and *Empetrichthys*. *Copeia* **89**, 103–106.

Eisma, D., and Marel, H. W. (1971). Marine muds along the Guyana coast and their origin from the Amazon basin. *Contr. Mineral. Petrol* **31**, 321–334.

Ellis, M. M. (1913). The gymnotoid eels of tropical America. *Mem. Carneg. Mus* **6**, 109–195.

Fink, W. L. (1984). Basal euteleosts: relationships. *In* "Ontogeny and Systematics of Fishes" (Moser, H. G., *et al.*, Eds.), pp. 202–206. *Am. Soc. Ichthyol. Herpetol,* Spec. Publ. no. 1.

Golani, D. (1993). The biology of the Red Sea migrant *Saurida undosquamis* in the Mediterranean and comparison with the indigenous confamilial *Synodus saurus* (Teleostei: Synodontidae). *Hydrobiologica* **27**, 109–117.

Graham, J. B. (1997). "Air-breathing Fishes – Evolution, Diversity and Adaptation." Academic Press, San Diego.

Grande, L. (1985). Recent and fossil clupeomorph fishes with materials for revision of the subgroups of clupeoids. *Bull. Am. Mus. Natl Hist.* **181**, 231–372.

Greenwood, H. P. (1983). The zoogeography of African freshwater fishes: Bioaccountancy or biogeography? *In* "Evolution, Time and Space: The Emergence of the Biosphere" (Sims, R. W., Price, J. H., and Whalley, P. E. S., Eds.), pp. 179–199. Systematics Association Special Vol. 23. Academic Press, London.

Hagedorn, M. (1986). The ecology, courtship and mating of gymnotiform electric fish. *In* "Electroreception" (Bullock, T. H., and Heiligenberg, W., Eds.), pp. 497–525. Wiley, New York.

Harrington, R. W. (1961). Oviparous hermaphroditic fish with internal self fertilization. *Science* **134**, 1749–1750.

Harvey, P. H., and Pagel, M. D. (1991). "The Comparative Method in Evolutionary Biology". Oxford Series in Ecology and Evolution. Oxford University Press, Oxford.

Harvey, P. H., Leigh-Brown, A. J., Maynard-Smith, J., and Nee, S. (1996). "New Uses for New Phylogenies." Oxford University Press, Oxford.

Heiligenberg, W. F. (1973). Electrolocation of objects in the electric fish *Eigenmannia* (Rhamphichthyidae, Gymnotoidei). *J. Comp. Physiol.* **91**, 223–240.

Heiligenberg, W. F. (1991). "Neural nets in electric fish." MIT Press, Cambridge.

Heiser, J. B., Moura, R. L., and Robertson, D. R. (2000). Two new species of creole wrasse (Labridae: *Clepticus*) from opposite sides of the Atlantic. *Aqua-J. Ichthyol. Aquatic Biol.* **4**, 67–76.

Helfman, G. S., Collette, B. B., and Facey, D. E. (1997). "The Diversity of Fishes." Blackwell Science, Inc., Malden, MA.

Hopkins, C. D. (1974). Electric communication in fish. *Am. Sci.* **62**, 426–437.

Hopkins, C. D., and Heiligenberg, W. (1978). Evolutionary design for electric signals and electoreceptors in gymnotoid fishes of Surinam. *Behav. Ecol. Sociobiol.* **3**, 113–134.

Humboldt, A., and Bonpland, A. (1807). "Essai sur la géographie des plantes accompagné d'un tableau physique des régions équinoxiales." Schoell, Paris; reprint: Arno Press, New York, 1977.

Johnson, G. D., and Patterson, C. (1996). Relationships of lower euteleostean fishes. *In* "Interrelationships of Fishes" (Stiassny, M. L. J., Parenti, L. R., and Johnson, G. D., Eds.), pp. 251–332. Academic Press, San Diego.

Johnson, R. K., and Feltes, R. M. (1984). A new species of *Vinciguerria* (Salmoniformes: Photichthyidae) from the Red Sea and Gulf of Aqaba, with comments on the Red Sea mesopelagic fish fauna. *Fieldiana, Zoology* **22**(new series), 1–35.

Kimor, B. (1973). Plankton relations of the Red Sea, Persian Gulf and Arabian Sea. *In* "The Biology of the Indian Ocean" (Zeitschel, B., and Gerlach, S. A., Eds.), Ecological Studies. Analysis and Synthesis, Vol. 3, pp. 221–255. Springer Verlag, New York.

Klausewitz, W. (1980). Tiefenwasser- und tiefsee fische aus dem Roten Meer. I. Einleitung und neunachweis für *Bembrops adenensis* Norman, 1939 und *Histiopterus spinifer* Gilchrist, 1904. *Senckenb. Biol.* **61**(1/2), 11–24.

Kottelat, M. (1989). Zoogeography of the fishes from Indochinese inland waters with an annotated check-list. *Bull. Zoölogisch Mus., Univ. Amsterdam* **12**, 1–54.

Kottelat, M. (1998). Fishes of the Nam Theum and the Xe Bangfai basins, Laos, with diagnoses of twenty-two new species (Teleostei: Cyprinidae, Balitoridae, Cobitidae, Coiidae and Odontobutidae. *Ichthyol. Expl. Freshwaters* **9**, 1–128.

Lissmann, H. W. (1958). On the function and evolution of electric organs in fish. *J. Exp. Biol.* **35**, 156–191.

Lovejoy, N. R. (1996). Systematics of myliobatoid elasmobranchs, with emphasis on the phylogeny and historical biogeography of neotropical freshwater stingrays (Potamotrygonidae, Rajiformes). *Zool. J. Linn. Soc.* **117**, 207–257.

Lundberg, J. G. (1970). The evolutionary history of North American catfishes, Family Ictaluridae. Unpublished PhD Dissertation (Zoology), University of Michigan, Ann Arbor.

Lundberg, J. G. (1993). African-South American freshwater fish clades and continental drift: Problems with a paradigm. *In* "Biological Relationships between Africa and South America" (Goldblatt, R., Ed.), pp. 156–199. Yale University Press, New Haven, CT.

Lundberg, J. G., Marshall, L. G., Guerrero, J., *et al.* (1998). The stage for Neotropical fish diversification: A history of tropical South American rivers. *In* "Phylogeny and Classification of Neotropical Fishes" (Malabara, L. R., *et al.*, Eds.), pp. 13–48. Editora da Pontificia Universidade Católica do Rio Grande do Sul, Porto Alegre.

Lundberg, J. G., Kottelat, M., Smith, G. R., Stiassny, M. L. J., and Gill, A. C. (2000). So many fishes, so little time: An overview of recent ichthyological discovery in continental waters. *Ann. Missouri Bot. Gard.* **87**, 26–62.

Marceniuk, A. P. (2003). Relações filogenéticas e revisão dos gêneros da família Ariidae (Ostariophysi, Siluriformes). Unpublished PhD Dissertation, Instituto de Biociências. Universidade de São Paulo, São Paulo.

Mees, G. F. (1961). Description of a new fish of the family Galaxiidae from Western Australia. *J. R. Soc. W. Australia* **44**, 33–38.

Menezes, N. A. (1996). Methods for assessing freshwater fish diversity. *In* "Biodiversity in Brazil: A First Approach" (Bicudo, C. E. M., and Menezes, N. A., Eds.), pp. 289–295. CNPq, São Paulo.

Mo, T. (1991). "Anatomy and Systematics of Bagridae (Teleostei), and Siluroid Phylogeny." Koeltz Scientific Books, Koenigstein.

Moura, R. L. (2003). Riqueza de espécies, diversidade e organização de assembléias de peixes em ambientes recifais: um estudo ao longo do gradiente latitudinal da costa brasileira. PhD Dissertation, Insituto de Biociências, Universidade de São Paulo.

Moura, R. L., and Sazima, I. (2003). Species richness and endemism levels of the Brazilian reef fish fauna. *Proc. 9th Int. Coral Reef Symp.* **9**, 956–959.

Moura, R. L., Figueiredo, J. L., and Sazima, I. (2001). A new parrotfish (Scaridae) from Brazil, and revalidation of *Sparisoma amplum* (Ranzani, 1842), *Sparisoma frondosum* (Agassiz, 1831), *Sparisoma axillare* (Steindachner, 1878) and *Scarus trispinosus* Valenciennes, 1840. *Bull. Marine Sci.* **68**, 505–524.

Moura, R. L., Gasparini, J. L., and Sazima, I. (1999). New records and range extensions of reef fishes in the Western South Atlantic, with comments on reef fish distribution along the Brazilian coast. *Rev. Bras. Zool.* **16**, 513–530.

Muss, A., Robertson, D. R., Stepien, C. A., Wirtz, P., and Bowen, B. W. (2001). Phylogeography of *Ophioblennius*: The role of ocean currents and geography in reef fish evolution. *Evolution* **55**, 561–572.

Myers, G. S. (1938). Fresh-water fishes and West Indian zoogeography. *Smith. Rep.* **3465**, 339–364.

Myers, R. S. (1999). "Micronesian Reef Fishes." Coral Graphics, Guam.

Nelson, G. J. (1989). Species and taxa: systematics and evolution. *In* "Speciation and Its Consequences" (Otte, D., and Endler, J., Eds.), pp. 60–81. Sinauer Associates, Sunderland.

Nichols, J. T. (1928). Fishes from the White Nile. *Am. Mus. Novitates* **319**.

Nielsen, J. G. (2003). Family Bythidae (viviparous brotulas). *In* "Checklist of the Freshwater Fishes of South and Central America" (Reis, R. E., Kullander, S. O., and Ferraris, C. J., Eds.), pp. 507–508. Edipucrs, Porto Alegre.

Norris, S. M. (2002). A revision of the African electric catfishes, family Malapteruridae (Teleostei, Siluriformes), with erection of a new genus and descriptions of fourteen new species, and an annotated bibliography. *Ann. Mus. r. Afr. Centr., Zool.* **289**, 1–155.

Ormond, R. F. G., and Roberts, C. M. (1997). The biodiversity of coral reef fishes. *In* "Marine Biodiversity: Patterns and Processes" (Ormond, R. F. G., Gage, J. D., and Angel, M. V., Eds.), pp. 216–257. Cambridge University Press, Cambridge.

Oyakawa, O. T. (1998). Relações filogenéticas das famílias Pyrrhulinidae, Lebiasinidae e Erythrinidae (Osteichthyes: Characiformes). PhD Dissertation, Instituto de Biociências Universidade de São Paulo, São Paulo.

Parenti, L. R. (1981). A phylogenetic and biogeographic analysis of cyprinodontiform fishes (Teleostei, Atherinomorpha). *Bull. Am. Mus. Natl Hist.* **168**, 335–557.

Parenti, L. R. (1998). A taxonomic revision of the Andean killifish genus *Orestias* (Cyprinodontiformes, Cyprinodontidae). *Bull. Am. Mus. Natl. Hist.* **178**, 107–214.

Parker, A., and Kornfield, I. (1995). Molecular perspective on the evolution and zoogeography of cyprinodontid killifishes (Teleostei, Atherinomorpha). *Copeia* **1995**, 8–21.

Pethiyagoda, R., and Kottelat, M. (1994). Three new species of fishes of the genera *Osteochilichthys* (Cyprinidae), *Travancoria* (Balitoridae) and *Horabagrus* (Bagridae) from the Chalakudy River, Kerala, India. *J. South Asia Natl Hist.* **1**, 97–116.

de Pinna, M. C. C. (1992). A new subfamily of Trichomycteridae (Teleostei, Siluriformes), lower loricarioid relationships and a discussion on the impact of additional taxa for phylogenetic analysis. *Zool. J. Linn. Soc.* **106**, 175–229.

de Pinna, M. C. C. (1993). Higher-level phylogeny of Siluriformes (Teleostei, Ostariophysi), with a new classification of the order. Unpublished PhD Dissertation, City University of New York, New York.

de Pinna, M. C. C. (1996). A phylogenetic analysis of the Asian catfish families Sisoridae, Akysidae, and Amblycipitidae, with a hypothesis on the relationships of the neotropical Aspredinidae. *Fieldiana, Zool.* **84**, 1–83.

de Pinna, M. C. C. (1998). Phylogenetic relationships of neotropical Siluriformes (Teleostei: Ostariophysi): Historical overview and synthesis of hypotheses. *In* "Phylogeny and Classification of Neotropical Fishes" (Malabarba, L., Reis, R. E., Vari, R. P., Lucena, Z. M., and Lucena, C. A. S., Eds.), pp. 279–330. Edipucrs, Porto Alegre.

de Pinna, M. C. C. (1999). Species concepts and phylogenetics. *Rev. Fish Biol. Fisheries* **9**, 353–373.

de Pinna, M. C. C., and Vari, R. P. (1995). Monophyly and phylogenetic diagnosis of the family Cetopsidae, with synonymization of the Helogenidae (Teleostei: Siluriformes). *Smiths. Contr. Zool.* **571**, 1–26.

de Pinna, M. C. C., Ferraris, C. J., and Vari, R. P. (submitted) A phylogenetic study of the neotropical catfish family Cetopsidae, with a new classification of the subfamily Cetopsinae (Ostariophysi, Siluriformes).

Planes, S. (2002). Biogeography and larval dispersal inferred from population genetic analysis. *In* "Coral Reef Fishes. Dynamics and Diversity on a Complex Ecosystem" (Seale, P. F., Ed.), pp. 201–220. Academic Press, New York.

Por, F. D. (1972). Hydrobiological notes on the high-salinity waters off the Sinai Peninsula. *Mar. Biol.* **14**(2), 111–119.

Randall, J. E. (1993). Zoogeographic analysis of the inshore Hawaiian fish fauna. *In* "Marine and Costal Biodiversity in the Tropical Island Pacific Region: Vol. 1, Species Systematics and Information Management Priorities" (Maragos, J. E., *et al.*, Eds.), pp. 193–203. Pacific Science Association, Bishop Museum, Honolulu.

Randall, J. E. (1998). Zoogeography of shore fishes of the Indo-Pacific region. *Zool. Studies* **37**, 227–268.

Reis, R. E., Kullander, S. O., and Ferraris, C. J. (2003). Checklist of the freshwater fishes of South and Central America EDIPUCRS, Porto Alegre.

Reithal, P. N., and Stiassny, M. J. L. (1991). The freshwater fishes of Madagascar: A study of an endangered fauna with recommendations for a conservation strategy. *Conserv. Biol.* **5**, 231–243.

Roberts, T. R. (1975). Geographical distribution of African freshwater fishes. *Zool. J. Linn. Soc.* **57**, 249–319.

Robertson, D. R. (1998). Do coral-reef fish faunas have a distinctive taxonomic structure? *Coral Reefs* **17**, 179–186.

Rocha, L. A. (2003). Patterns of distribution and processes of speciation in Brazilian reef fishes. *J. Biogeogr.* **30**, 1161–1171.

Rocha, L. A., Bass, A. L., Robertson, D. R., and Bowen, B. W. (2002). Adult habitat preferences, larval dispersal, and the comparative phylogeography of three Atlantic surgeonfishes (Teleostei: Acanthuridae). *Mol. Ecol.* **11**, 243–252.

Rosa, R. S., Menezes, N. A., Britski, H. A., Costa, W. J. E. M., and Groth, F. (2003). Diversidade, padrões de distribuição e conservação dos peixes da Caatinga. *In* "Ecologia e Conservação da Caatinga" (Leal, I. R., Tavarelli, M., and Cardoso daSilva, J. M., Eds.), pp. 135–180. Editora Universitária da UFPE, Recife.

Rosen, B. R. (1988). Process, problems and patterns in the biogeography of reef corals and other tropical marine organisms. *Helgolander wiss meeresunters* **42**, 269–301.

Rosen, D. E. (1974). Phylogeny and biogeography of salmoniform fishes and relationships of *Lepidogalaxias salamandroides. Bull. Am. Mus. Natl Hist.* **153**, 265–326.

Rosen, D. E. (1976). A vicariance model of caribbean biogeography. *Syst. Zool.* **24**, 431–464.

Schaefer, S. A., Weitzman, S. H., and Britski, H. A. (1989). Review of the neotropical catfish genus *Scoloplax* (Pisces: Loricarioidea: Scoloplacidae) with comments on reductive characters in phylogenetic analysis. *Proc. Acad. Natl Sci. Philadelphia* **141**, 181–211.

Sewell, R. B. S. (1948). The free-swimming planktonic Copepoda. Geographical distribution. *Sci. Rep. John Murray Exped. 1933–34* **8**(3), 317–592.

Siebert, D. (1997). Notes on the anatomy and relationships of *Sundasalanx* Roberts (Teleostei, Clupeidae), with descriptions of four new species from Borneo. *Bull. Natl Hist. Mus. Lond. (Zool.)* **63**, 13–26.

Skelton, P. H. (1986). Distribution patterns and biogeography of non-tropical southern African freshwater fishes. *In* "Palaeoecology of Africa and the Surrounding Islands" (Van Zinderen Bakker, E. M., Coetzee, J. A., and Scott, L., Eds.), pp. 211–230. A. A. Balkema, Rotterdam.

Skelton, P. H. (1994). Diversity and distribution of freshwater fishes in East and Southern Africa. *Ann. Mus. r. Afr. Centr., Zool.* **275**, 95–131.

Skelton, P. H., Cambray, J. A., Lombard, A, and Benn, G. A. (1995). Patterns of distribution and conservation status of freshwater fishes in South Africa. *S. Afr. J. Zool.* **30**, 71–81.

Springer, V. G. (1982). Pacific plate biogeography, with special reference to shore fishes. Smithson. *Contr. Zool.* **465**, 1–182.

Stiassny, M. L. J. (1991). Phylogenetic intrarelationships of the family cichlidae: an overview. *In* "Cichlid Fishes: Behaviour, Ecology and Evolution" (Keenleyside, M. H. A., Ed.), pp. 1–35. Chapman & Hall, London.

Stiassny, M. J. L., and Jensen, J. S. (1987). Labroid intrarelationships revisited: Morphological complexity, key innovations, and the study of comparative diversity. *Bull. Mus. Comp. Zool.* **151**, 269–319.

Stiassny, M. L. J., and de Pinna, M. C. C. (1994). Basal taxa and the role of cladistic patterns in the evaluation of conservation priorities: A view from freshwater. *In* "Systematics and Conservation Evaluation" (Forey, P. L., *et al.*, Eds.), pp. 235–249. The Systematics Association Special Vol. 50. Clarendon Press, Oxford.

Stiassny, M. J. L., and Raminosoa, N. (1994). The fishes of the inland waters of Madagascar. *Ann. Mus. r. Afr. Centr., Zool.* **275**, 133–149.

Swofford, D. L., and Maddison, W. P. (1987). Reconstructing ancestral character states under Wagner parsimony. *Math. Biosci.* **87**, 199–229.

Vari, R. P. (1979). Anatomy, relationships and classification of the families Citharinidae and Distichodontidae. *Bull. Br. Mus. Natl Hist. (Zool.)* **36**, 261–344.

Vari, R. P. (1995). The neotropical fish family Ctenoluciidae (Teleostei: Ostariophysi: Characiformes): Supra and intrafamilial phylogenetic relationships, with a revisionary study. *Smith. Contr. Zool.* **564**, 1–97.

Vari, R. P. (1988). The Curimatidae, a lowland Neotropical fish family (Pisces: Characiformes); distribution, endemism, and phylogenetic biogeography. *In* "Proceeding of a Workshop on Neotropical Distribution Patterns" (Vanzolini, P., and Heyer, W. R., Eds.), pp. 343–377. Academia Brasileira de Ciências, Rio de Janeiro.

Vari, R. P., and Harold, A. S. (2001). Phylogenetic study of the neotropical fish genera *Creagrutus* Günther and *Piabina* Reinhardt (Teleostei: Ostariophysi: Characiformes), with a revision of the Cis-Andean species. *Smith. Contr. Zool.* **613**, 1–239.

Weitzman, S. H., and Vari, R. P. (1988). Miniaturization in South American freshwater fishes; an overview and discussion. *Proc. Biol. Soc. Wash.* **101**, 444–465.

Weitzman, S. A., and Weitzman, M. (1982). Biogeography and evolutionary diversification in Neotropical freshwater fishes, with comments on the refuge theory. *In* "Biological Diversification in the Tropics" (Prance, G., Ed.), pp. 403–422. Columbia University Press, New York.

Westoby, M. (1985). Two main relationships among the components of species richness. *Proc. Ecol. Soc. Aust.* **14**, 103–107.

Williams, R. R. G. (1997). Bones and muscles of the suspensorium in the galaxioid *Lepidogalaxias salamandroides* (Teleostei: Osmeriformes) and their phylogenetic significance. *Rec. Aust. Mus.* **49**, 139–166.

Zanata A. M. and Vari, R. P. (in press) The family Alestidae (Ostariophys.i, Characiformes): A phylogenetic analysis of a trans-Atlantic clade. *Smith. Contr. Zool.*

3

THE GROWTH OF TROPICAL FISHES

PETER A. HENDERSON

I. INTRODUCTION

This chapter describes the equations that are commonly used to explain the growth of tropical fishes and presents the range of growth rates and maximum sizes that are achieved. By focusing on tropical fishes, we raise the question as to whether the growth of tropical fishes displays any particular features not shown by fishes from other regions. It might be assumed that they can grow faster and without seasonal variation in growth rate because of the lack of large annual variations in temperature. However, many other factors can also limit growth, including local productivity and oxygen availability, so it is far from clear if tropical fishes will actually grow faster or more consistently than colder water species.

To determine the rate of growth it is essential to know the age of the fish. In temperate waters hard structures such as scales, otoliths, spines, and bones often put down a clear annual winter growth check so that it is possible to determine the number of winters the fish has experienced. Even when this is difficult, many temperate fishes have a short spawning season so that length frequency graphs can be used to follow the growth of a cohort or year class through time. In the tropics a clear seasonal growth check may not occur. It is frequently found that fishes have experienced checks that can be related to changes in water conditions and spawning. However, the number of checks produced per year may vary, making it impossible to use growth checks to

The Physiology of Tropical Fishes: Volume 21
FISH PHYSIOLOGY

infer age. Further, the spawning season may be much more extended than in temperate waters so that length frequency graphs may show a number of modes within the same year and making it impossible to distinguish between year classes. Given these difficulties and the size of the tropical fish fauna it is unsurprising that we presently know far more about the growth of temperate than tropical fishes. In particular, we know little about the growth of large numbers of non-commercially exploited small marine and freshwater tropical species. The incompleteness and bias within the available growth data should be remembered when considering the relevance of the conclusions presented below.

Tropical fishes have been observed to produce a single annual growth check. For example, Fabre and Saint-Paul (1998), in a study of the abundant Amazonian anostomid *Schizodon fasciatus*, found that annulus formation on the scales occurred during January and February when the gonads were maturing. In this habitat the marked annual variation in water depth and food availability created maximum growth during the season when water levels were falling (July to November) and a marked seasonality in reproduction. The pronounced seasonal variation in water depth in the Amazon basin rivers probably produces seasonal patterns of growth and reproduction in many fishes, but if the constraint is relaxed clear annual checks may no longer occur. Jepsen *et al.* (1999) noted that *Cichla* species in the river system probably only spawned once per year but populations in reservoirs showing little seasonality in physico-chemical characteristics may spawn several times per year producing growth checks that were no longer interpretable.

In recent years, otolith microstructure has been used to age short-lived tropical fishes. For example, Kimura (1995) used daily growth increments to study the growth of the clupeids *Stolothrissa tanganicae* and *Limnothrissa miodon* in Lake Tanganyika. It was notable that growth was found to be far more rapid than earlier analyses based on length-frequency analysis. While daily growth increments are a powerful technique for ageing larval and small fishes it is often difficult to prove that all the growth checks counted are daily increments.

As fish have no, or only a limited, ability to control their body temperature, the growth of fish in temperate, Antarctic and Arctic waters is often temperature-limited. It is therefore unsurprising that in species that can survive in both warm temperate and tropical waters the highest growth rates are found in warmer tropical waters. For example, Lowe-McConnell (1987) notes that Chinese grass carp, *Ctenopharyngodon idella*, grew up to 10 g per day in ponds in Malacca, Malaysia and matured at only 1 year old. In comparison, in south China they grew at 3.5 g per day and matured at 3–5 years old. It is also known that tropical species that spend the beginning of their lives in cooler waters grow more slowly. However, such observations are only made for commercial species and it is unclear if the potential growth rates of warm tropical waters are frequently realized.

There are clear indications that the growth rates achieved by fish vary with the density of their populations, which may vary naturally, or by human exploitation. Unexploited or little-fished populations can be anticipated to have growth rates much below that observed in the same species in heavily exploited populations. By far the most important factor determining growth rates is likely to be the amount of available food. In tropical waters it is often striking that small fishes will be often far more willing to approach any disturbance or even an offered hand than their temperate counterparts. This is presumably because the need for food is more pressing and greater risks must be taken. Edwards (1985) suggested that the fact that the growth rates of three tropical Australian snapper species was lower than North Sea cod might be linked to their low level of exploitation. When comparing tropical and temperate growth rates it is impossible to allow for population size. However, it can be anticipated that food availability will tend to equalize the growth rates of temperate and tropical species.

II. DESCRIBING GROWTH

To compare the pattern of growth of different fish populations, it is necessary to describe the pattern of growth mathematically. The basic requirement is for a simple and generally applicable equation that gives the size in terms of average length or weight at any given age. When the average length of a fish is plotted against age in years the result is usually a curve for which the rate of increase in size continuously decreases with age (Figure 3.1) so that length approaches a maximum asymptotic length termed L_∞. Weight also increases asymptotically with age, but the shape of the curve is sigmoid (Figure 3.2) with a point of inflection often at about one-third of the asymptotic weight, W_∞.

There is a large literature on growth equations and there are many plausible equations that can describe the general features of fish growth, none of which is entirely satisfactory. Our requirement is for an equation that describes the basic features of the growth shown by fish as simply and with as few parameters as possible. Commonly applied forms include von Bertalanffy, logistic, and the more general Richardson model. The most popular function to describe fish growth is the von Bertalanffy growth equation (Bertalanffy and Müller, 1943):

$$L(t) = L_\infty(1 - e^{-Kt})$$

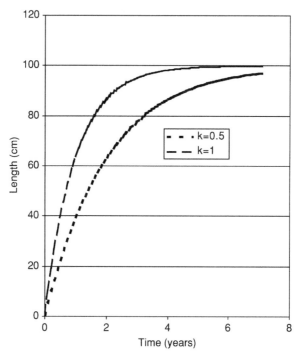

Fig. 3.1 The von Bertalanffy growth curve. Curves are shown for two hypothetical fish both of which have a maximum (asymptotic) length of 100 cm. They only differ in the value of the growth parameter, K. Note that the fish with the high K value initially grows faster and gets close to the asymptotic length quicker.

where $L(t)$ is the length at time or age t, L_∞ is called an asymptotic length, K is a growth constant and t time from birth. The greater the growth constant K, the faster the fish grows (Figure 3.1).

This equation has been extensively applied to fish populations and is the standard equation used. A list of references for the use of this equation can be found at http://homepage.mac.com/mollet/VBGF/VBGF_Ref.html. When fitted to field data a third term, t_0 is included in the von Bertalanffy equation. This is an adjustment parameter to allow for the size of the larvae at the time of hatching. For typical fish, this equation gives a good fit to the observed length at age and an adequate description of the decelerating rate of growth with size.

The growth of both tropical and colder water fishes has been found to give similarly adequate fits to the von Bertalanffy equation, indicating that there is no fundamental difference in the pattern of growth with size or age

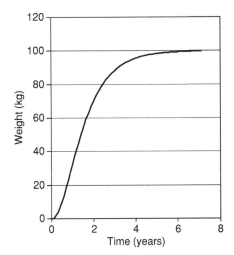

Fig. 3.2 The von Bertalanffy growth curve when weight is plotted rather than length.

between fish living in warm or cold waters. However, as will be discussed below, the equation takes no account of seasonality and therefore does not give a correct description of the temporal pattern of growth in highly seasonal environments.

It is important to remember that when comparing populations or species using a growth equation we are expressing some sort of generalization of the size at age. The actual pattern of growth shown by individual fish within the population may be very different. It can, for example, be altered by a whole range of events experienced by the individual, including lack of food, oxygen concentration, temperature, infections, parasites, and fin predators, plus individual characteristics determined by genetics.

A. The von Bertalanffy Parameters of Tropical Populations

The FishBase database was consulted (http://www.fishbase.org/search. cfm) to obtain the K and L_∞ values for individual fish populations. When multiple sets of parameters for a single species were available, these were all included in the analysis. This was appropriate because many species can show a wide range of variation in growth rate and maximum size over their geographical range and even between years.

Figure 3.3 shows the distribution of the von Bertalanffy growth constant K for tropical fishes. The distribution is highly skewed to the left with an

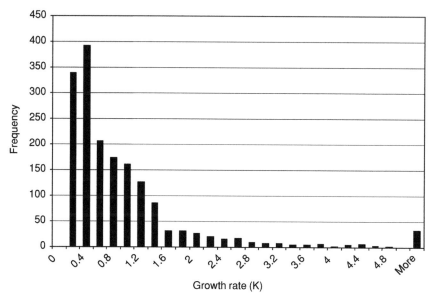

Fig. 3.3 The frequency distribution of the von Bertalanffy growth constant, K, for tropical fish species included in the FishBase database.

extremely long tail. About 90% of all populations studied had a *K* value of less than 1.6. An interesting feature is the lack of very low *K* values below 0.25. This would suggest that very slow average growth rates do not occur in tropical life history strategies.

Comparison of the distribution of growth rate, *K*, for tropical and colder water fishes is interesting. None tropical fishes show a left skew but there is a striking reduction of *K* values between about 0.8 and 1.6 (Figure 3.4) when compared with the distribution of tropical fish (Figure 3.3). It seems clear that a higher proportion of tropical than colder water fish have growth rates above 0.8.

The asymptotic length shows a similar left hand skew, reflecting the fact that the majority of tropical fishes are small (Figure 3.5). There is a general tendency for those fishes that only attain a small size to grow more rapidly. This is clearly seen in a scatter plot of the log L_∞ and log *K* (Figure 3.6). However, there is a considerable scatter in the data and there are many examples of rapidly growing tropical fishes that may achieve lengths of more than 1 m. The osteoglossid *Arapaima gigas,* which is the largest scaled fish in the Amazon basin, is a good example (Queiroz, 2000). Some species of tuna can also grow very rapidly and become sexually mature at only 3 or 4 years of age by which time they have reached a body length of more than 1 m.

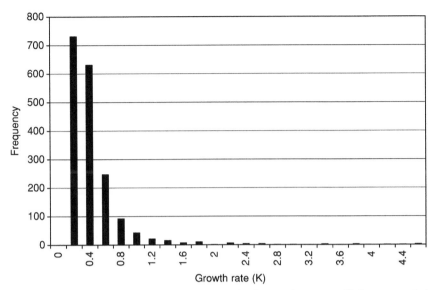

Fig. 3.4 The frequency distribution of the von Bertalanffy growth constant, K, for non-tropical fish species included in the FishBase database.

B. Do Tropical Fishes Grow Faster than Fishes in Colder Regions?

Because they are cold blooded and low temperature can clearly limit the growth of fish, it is likely that fishes in tropical regions have the potential to grow faster than those in temperate and polar regions. Tropical waters such as those in the Amazon can be nutrient-limited and very species-rich, suggesting that the production available to individual species may constrain growth. A simple comparison of the growth constants of tropical and non-tropical fishes in the FishBase database clearly shows that tropical fishes do have, on average, faster growth (Table 3.1, see also Chapter 5, this volume).

However, this may be an artefact of the relationship between K and L_∞ (Figure 3.6) if tropical fishes on average attain a smaller maximum size. In some habitats such as tropical freshwaters, there are clearly a far larger number of small species than would be observed in a comparable temperate water body. A typical example is the small stream habitat. In Amazonian forest streams, a well-defined submerged leaf-litter community comprises 20–30 small fish species (Henderson and Walker, 1986), many of which reach sexual maturity at a body mass smaller than any species found in a North European or North American temperate stream where salmonids are the dominant group. To allow for this possibility, a two-way analysis of variance was undertaken with L_∞ classified into ten groups each, covering a

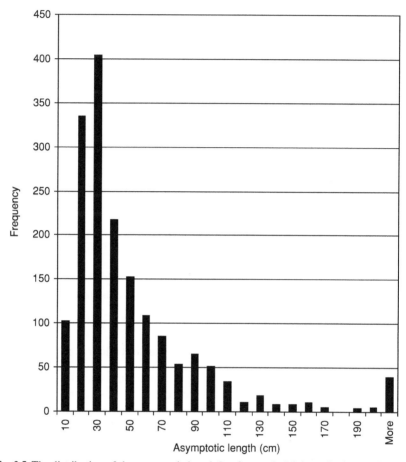

Fig. 3.5 The distribution of the asymptotic length L_∞ for tropical fish species in the FishBase database.

10 cm range and each population classified as either tropical or non-tropical. Even allowing for the fact that K declines with increasing L_∞, tropical fishes had a significantly higher average growth constant K ($p < 0.001$, see Table 3.2).

While the above analysis clearly demonstrates the general tendency of tropical fishes to reach their asymptotic, maximum length faster than temperate fish this conclusion may be biased by particular groups of tropical or cold-water fish, which are only found in one region of the world. It is therefore instructive to look in more detail at particular families that are well represented in both temperate and tropical waters. One of the most

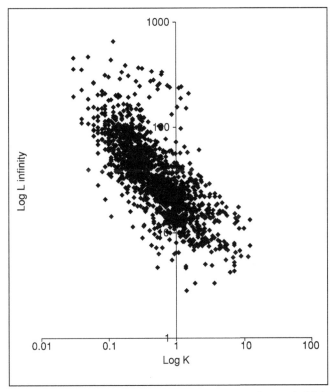

Fig. 3.6 The relationship between the growth constant, K and L_∞, for tropical fish population included in the FishBase database.

widespread families in both fresh and marine waters is the Clupeid or herring family. A Kruskal–Wallis one-way analysis of variance comparing the magnitude of K for tropical and non-tropical populations (Table 3.3) gave a test statistic of $H = 207.788$ with 1 degree of freedom showing that the K value of tropical clupeids was highly significantly larger ($p = <0.001$). As with fish in general, the median L_∞ of non-tropical clupeids is greater than that of tropical species but even allowing for this difference the median K value of tropical species is significantly greater. This result is consistent with the conclusions of Milton *et al.* (1993), who concluded that the tropical clupeids, which they studied, lived shorter lives but had a higher growth rate than temperate clupeids. Another widely distributed family, the anchovies, also shows a highly significantly greater median growth constant for tropical species ($H = 33.641$ with 1 degree of freedom, $p = <0.001$; see Table 3.4).

Table 3.1

Mean, Median and 25 and 75 Percentile Values for the Growth Constant K for Tropical and Non-tropical Fish

Locality	Mean of K	Median of K	25%	75%
Tropical K	0.78	0.530	0.240	1.030
Temperate K	0.4	0.250	0.150	0.410

Table 3.2

Results of a Two-way Analysis of Variance Investigating the Effects of Both L_∞ and Tropical and Non-tropical Locations on the Magnitude of the Average von Bertalanffy Growth Constant, K

Source of variation	DF	SS	MS	F	P
L_∞	9	767.460	85.273	178.123	<0.001
Tropical location	1	86.558	86.558	180.806	<0.001
Interaction	9	155.115	17.235	36.001	<0.001
Residual	3516	1683.222	0.479		
Total	3535	2968.193	0.840		

Table 3.3

Results of a Kruskal–Wallis Test for Significant Difference in the Growth Constant K for Tropical and Non-tropical Clupeid Fish

Locality	N	Median K	25%	75%
Tropical	188	1.050	0.690	2.015
Non-tropical	278	0.385	0.290	0.530

Table 3.4

Results of a Kruskal–Wallis Test for Significant Difference in the Growth Constant K for Tropical and Non-tropical Anchovies

Locality	N	Median K	25%	75%
Non-tropical	81	0.580	0.328	1.063
Tropical	135	1.080	0.813	1.715

While there is clear evidence that tropical fishes generally grow towards their maximum length faster, the minimum growth constants observed in the tropics are similar in value to those observed in Arctic and Antarctic waters. There are certainly tropical species and habitats where rapid growth does not, or possibly cannot, occur. While tropical fishes can grow towards their adult size faster than colder water fish it is clear that during their summer growing season many temperate fishes can grow exceedingly fast. It is likely that it is the length of the growing season rather than the maximum daily growth rate that is allowing tropical fishes to often grow more rapidly.

III. THE VARIATION BETWEEN HABITATS AND POPULATIONS

The growth rates achieved by a species can vary considerably between habitats. Lowe-McConnell (1987) makes the point that growth rates in lakes are often higher than in river habitats and concludes that lakes frequently offer better feeding grounds. The amount of between-habitat variability in growth is well demonstrated by highly studied commercial fish. Table 3.5 shows the variation in K and L_∞ for *Tilapia rendalli* in Africa. The range in K from 0.13 to 3.79 is dramatic, but is also shown by other species.

Table 3.5
Von Bertalanffy Growth Parameters for Different Populations
of *Tilapia rendalli*

Locality	K	L_∞
Uganda	3.79	13.1
Zambia	3.19	13.8
Zambia	1.1	20.8
Zambia	0.75	29.1
Madagascar	0.67	22.5
Zambia	0.62	21.7
Madagascar	0.53	21.8
Madagascar	0.53	27.2
Madagascar	0.52	30.1
Madagascar	0.5	24.9
Zambia	0.48	24.3
Zambia	0.47	27.8
Zambia	0.46	26.3
Madagascar	0.32	24.4
South Africa	0.31	26.5
South Africa	0.23	33.1
Zambia	0.19	33.9
South Africa	0.18	40.2
South Africa	0.16	41.1
Zimbabwe	0.14	40
Zimbabwe	0.14	48.5
Zambia	0.13	39.9

Data extracted from FishBase.

IV. SEASONALITY IN GROWTH

Given the lack of a well-defined winter, it might be assumed that the growth of tropical fishes lacks a seasonal pattern. As with the consideration of growth rates above, to compare the seasonality of tropical and non-tropical species we need a mathematical description of seasonal growth (see Chapter 4, this volume).

If length is plotted at time intervals of less than one year then it is frequently observed that growth varies with the seasons so that there are periods of the year when almost no growth occurs. This growth pattern is particularly pronounced in species living close to the northern or southern limit of their range. For example, Figure 3.7 shows the growth of the sole, *Solea solea*, a marine flatfish that has a geographical range extending from North Africa and the Mediterranean to British waters in the North. The graph shows that in British waters there is an extended period from autumn to mid spring when no growth occurs. This cessation of growth is almost certainly related to water temperature as in warmer years the growth season is extended.

Seasonality is not only a feature of temperate waters. Seasonality can be particularly pronounced in tropical freshwaters where there is a wet and dry season. In many tropical floodplains the seasonal variation in rainfall results

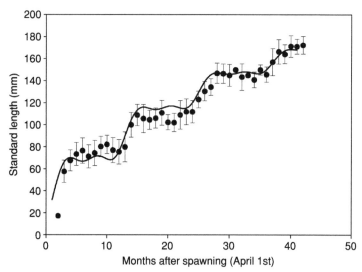

Fig. 3.7 The growth over the first two years of life of the sole *Solea solea* in British waters (From Henderson and Seaby, 2005).

in dramatic changes in water depth and the inundation of huge areas of forest. This change in habitat availability can impose a strict seasonality on reproduction and growth. In tropical seas there are often clear seasonal differences in climate with dramatic seasonal differences in the frequency of hurricanes and other tropical storms. Thus, there is often some degree of seasonality in shallow and coastal marine habitats. Even the deep-water habitats, where physical conditions of temperature and pressure and water chemistry are almost as constant as they can be on earth, there are still seasonal inputs from the surface waters above as is shown by studies in the Sargasso Sea (Deuser and Ross, 1980; Sayles *et al.*, 1994).

While no simple equation can entirely satisfactorily describe this seasonal pattern, one of the most popular is the seasonally adjusted von Bertalanffy which at least has the advantage of comparative simplicity and is an extension of the widely used non-seasonal growth model which was used above. The first published version of the VBGF with seasonality was Ursin (1963a,b) and improvements and methods for fitting were developed by Cloern and Nichols (1978), Pauly and Gaschütz (1979), Appeldoorn (1987), Somer (1988) and Soriano and Pauly (1989).

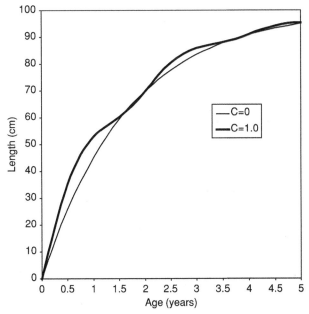

Fig. 3.8 An example of a seasonally varying von Bertalanffy growth equation. The curves with C = 0 and C = 1 are shown.

The equation for the growth model is:

$$L_t = L_\infty\{1-\exp-[K(t-t_0) + S(t)-S(t_0)]\}$$

where L_∞, K and t_0 are defined as in the standard VBGF and

$$S(t) = (CK/2\pi)\,\mathrm{Sin}\pi\,(t-t_s)$$

and

$$S(t_0) = (CK/2\pi)\,\mathrm{Sin}\pi\,(t_0-t_s).$$

The parameter C is the oscillation amplitude parameter ($0 < = C$). If $C = 0$ the model reverts to the standard von Bertalanffy equation, if $C = 1$ then growth is zero at one point in the annual cycle (Figure 3.8). The parameter t_s is the starting point of the oscillation (as fraction of year; $0 < = t_s < 1$) and the point of slowest growth occurs at $t_s + 0.5$.

Pauly and Ingles (1981) and Longhurst and Pauly (1987) have shown that even tropical fishes not living in clearly seasonal environments such as floodplains also show seasonal patterns of growth. Seasonal temperature differences as small as $2\,°C$ are sufficient to induce changes in the growth rate. The amplitude of the seasonal oscillations have been related to the seasonal difference in temperature so that for changes of greater than $10\,°C$

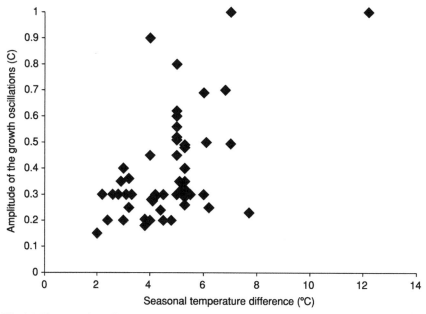

Fig. 3.9 The annual amplitude in the growth oscillations in relation to the seasonal temperature difference. (Data taken from the FishBase website and authored by D. Pauly.)

the value of C is approximately 1.0 – essentially there is a winter season during which growth becomes zero. The above seasonally adjusted growth curve works well for many tropical species but is not particularly useful for many temperate regions with extended winters because the equation is not able to model adequately situations with extended winter periods of zero growth (see Figure 3.7). Indeed, it is not unusual in temperate regions to get the illusion of growth during the winter months because smaller individuals consume their fat reserves and die of starvation first, resulting in an increase in mean population size (Henderson et al.,1988). This effect may have led to an exaggeration of the growth of some temperate populations.

The POPGROWTH table of FishBase includes most of the estimates of C so far published for fish, along with matching estimates of the summer–winter temperature difference (ΔT; difference of mean monthly values, in °C). These are plotted in Figure 3.9, which indicates that some seasonal variation in growth even occurs with a seasonal temperature difference of only 2 °C.

REFERENCES

Appeldoorn, R. S. (1987). Modification of a seasonally oscillating growth function for use with mark–recapture data. J. Cons. Int. L'Explor. Mer. **43**, 194–198.

Bertalanffy, L. von, and Müller, I. (1943). Untersuchungen über die Gesetzlichkeit des Wachstums. VIII. Die Abhängigkeit des Stoffwechsels von der Korpergrösse und der Zusammenhang von Stoffwechseltypen und Wachstumstypen. Rev. Biol. **35**, 48–95.

Cloern, J. E., and Nichols, F. H. (1978). A von Bertalanffy growth model with a seasonally varying coefficient. J. Fish. Res. Board Can. **35**, 1479–1482.

Deuser, W. G., and Ross, E. H. (1980). Seasonal change in the flux of organic carbon to the deep Sargasso Sea. Nature **283**, 364–365.

Edwards, R. R. C. (1985). Growth rates of Lutjanidae (snappers) in tropical Australian waters. J. Fish Biol. **26**, 1–4.

Fabre, N. N., and Saint-Paul, U. (1998). Annulus formation on scales and seasonal growth of the Central Amazonian anostomid Schizodon fasciatus. J. Fish Biol. **53**, 1–11.

Henderson, P. A., Bamber, R. N., and Turnpenny, A. W. T. (1988). Size-selective over wintering mortality in the sand smelt, Atherina boyeri Risso, and its role in population regulation. J. Fish Biol. **33**, 221–233.

Henderson, P. A., and Seaby, R. M. (2005). The role of climate in determining the temporal variation in abundance, recruitment and growth of sole (L) in the Bristol Channel. J. Mar. Biol. Ass. UK. **85**, 197–204.

Henderson, P. A., and Walker, I. (1986). On the leaf-litter community of the Amazonian blackwater stream Tarumazinho. J. Trop. Ecol. **2**, 1–17.

Jepsen, D. B., Winemiller, K. O., Taphorn, D. C., and Rodriguez-Olarte, D. (1999). Age structure and growth of peacock cichlids from rivers and reservoirs of Venezuela. J. Fish Biol. **55**, 433–450.

Kimura, S. (1995). Growth of the clupeid fishes, Stolothrissa tanganicae and Limnothrissa miodon, in the Zambian waters of Lake Tanganyika. J. Fish Biol. **47**, 569–575.

Longhurst, A. R., and Pauly, D. (1987). "Ecology of Tropical Oceans." Academic Press, San Diego, CA.

Lowe-McConnell, R. H. (1987). "Ecological Studies in Tropical Fish Communities." Cambridge University Press, Cambridge.

Milton, D. A., Blaber, S. J. M., and Rawlinson, N. J. F. (1993). Age and growth of three species of clupeids from Kiribati, tropical central south pacific. *J. Fish Biol.* **43**, 89–108.

Pauly, D., and Gaschütz, G. (1979). A simple method for fitting oscillating length growth data, with a program for pocket calculators. *I.C.E.S. CM 1979/6:24. Demersal Fish Cttee.*

Pauly, D., and Ingles, J. (1981). Aspects of the growth and natural mortality of exploited coral reef fishes, pp. 89–98. *In* "The Reef and Man. Proceedings of the Fourth International Coral Reef Symposium" (Gomez, E. D., Birkeland, C. E., Buddemeyer, R. W., Johannes, R. E., Marsh, J. A., and Tsuda, R. T., Eds.), Vol. 1. Marine Science Center, University of the Philippines, Quezon City.

Queiroz, H. L. (2000). Natural history and conservation of pirarucu, *Arapaima gigas*, in Amazonian *várzea*: Red giants in muddy waters. PhD thesis, University of St Andrews, Scotland.

Sayles, F. L., Martin, W. R., and Deuser, W. G. (1994). The response of benthic oxygen demand to particulate organic carbon supply in the deep sea near Bermuda. *Nature* **371**, 686–689.

Somer, I. F. (1988). On a seasonally oscillating growth function. *Fishbyte* **6**, 8–11.

Soriano, M., and Pauly, D. (1989). A method for estimating the parameters of a seasonally oscillating growth curve from growth increments data. *Fishbyte* **7**, 18–21.

Ursin, E. (1963a). On the incorporation of temperature in the von Bertalanffy growth equation. *Medd. Danm. Fisk. Havunders. N.S.* **4**, 1–16.

Ursin, E. (1963b). On the seasonal variation of growth rate and growth parameters in Norway pout (*Gadus esmarki*) in Skagerrak. *Medd. Danm. Fisk. Havunders. N.S.* **4**, 17–29.

4

BIOLOGICAL RHYTHMS

GILSON LUIZ VOLPATO
ELEONORA TRAJANO

I. INTRODUCTION

Biological rhythms are one of the most intriguing and exciting research fields in biology. Aquatic organisms are not exceptions; many studies have been devoted to different kinds of rhythms these organisms show. Despite the immense literature on this topic, little is found regarding fishes – and even less regarding the tropical ones. Even so, the studies mainly have pertained to the classical day–night cycles.

This chapter deals with regular fluctuations in the processes of tropical fishes whether governed by internal timing systems or not (emphasis, however, is on the former). The importance of the day–night cycles is also analyzed. Processes such as social organization (dominance rank, shoaling, schooling, and communication), migration and reproduction are analyzed

The Physiology of Tropical Fishes: Volume 21
FISH PHYSIOLOGY

concerning their relation with rhythmicity, irrespective of being controlled by endogenous timing systems. This chapter summarizes some of the important literature in this area, including practical considerations involving fishery and environmental transformations by dams used for producing electricity.

Most examples we have used concern freshwater fishes, but these biological models generate general concepts, most of which also are valid for marine fishes. In some cases, even correlates with animal groups other than fish were necessary for a broader explanation. Moreover, cave fishes are included showing some contrasts with species living in direct contact with the light–dark cycle and other drastic environmental fluctuations. In such a context, the evolution of the endogenously determined biological rhythms is also discussed.

In contrast with mammals, that have been intensively studied in the past (although centered on a few species, mostly rodents), relatively few chronobiological studies have been carried out on fishes, and also focusing on a limited number of species, especially from temperate regions. There is a large gap between naturalistic and laboratory studies, both quantitative, the great majority of publications on fishes refer to laboratory studies, and are qualitative, in terms of kind of data.

The simplified conditions observed in the laboratory, where variables can be precisely controlled, allow for an accurate investigation of the properties of biological clocks, revealing the main characteristics of time-control mechanisms in different taxa, including the exogenous or endogenous nature of biological rhythms, number and anatomical location of oscillators, mechanisms of entrainment, coupling between oscillators and in relation to their associated functions, hierarchy of *zeitgebers* (the German word for *time-givers*) acting over different species, masking, and so on. However, because chronobiological studies are by definition time-consuming, and possibly also because awareness of its relevance and knowledge of its methodology is not widespread among ichthyologists, relatively few species have been investigated in detail. This is especially true for teleost fishes, a most diversified group of vertebrates with more than 47 000 species around the world (Nelson, 1994), for which only a few dozens have been studied in the laboratory. In many cases, these are fishes from commercial stocks, bred in captivity for many generations, as the intensively studied goldfish, *Carassius auratus*, originated from warm temperate Asia, and the zebrafish, *Danio rerio*, from tropical Asia, or purchased in local markets, with no identifiable origin. Even in the case of wild caught experimental animals, few publications provide a precise locality of origin, and few data on the natural conditions are informed. In some cases, not even a detailed description of experimental conditions is given. Geographic and methodological differences, not always apparent in publications, may account for discrepancies in the results obtained by different authors for the same species.

On the other hand, naturalistic studies, that can provide the necessary input of information on fishes in their natural habitat, are still limited in number and there is a gap in quality of data hampering the connection with laboratory data – in the much more complex natural conditions, data on fish activity rhythms are still rather simple, partly due to methodological limitations related to the inherent difficulties of fieldwork. So far, most field studies carried out in tropical regions are based on direct observations during snorkeling or scuba diving. These studies focus mainly on fish species living in clear, transparent waters, such as tropical reefs and clear-water streams and lakes. Direct observation requires artificial illumination during the scotophase of the daily cycle. Due to the high light sensitivity of many fishes, which may react even to dim red light, this may interfere with their normal rhythms. In the laboratory, pulses of light have a significant effect on rhythmicity of fish activity, and the observation that some fishes do not visibly react when illuminated in the field is not a sure evidence that their rhythms have not been disrupted. More sophisticated techniques, such as electronic tags allowing for continuous individual recording and avoiding the bias of direct observation, are very recent and mostly restricted to medium to large-sized species.

An alternative method for chronobiological field studies consists of series of collections made at different times of the daily or annual cycle, with the collected individuals being counted and released (see, for instance, Naruse and Oishi, 1996). This method relies on the assumptions that the more active is an individual, the higher is its probability of being caught, thus a higher number of captures correspond to activity peaks in the population, and that the capture and release of a particular fish does not affect its probability of subsequent capture. Although clearly prone to bias, this method produces a gross estimation of locomotor rhythms in the studied populations. Likewise, feeding rhythms have been studied on the basis of analysis of stomach contents of fish collected at different times of the daily cycle, with focus on repletion degree, type of prey and digestion level.

In conclusion, field and laboratory work, each with its advantages and flaws, are equally important, complementary sources of knowledge about fish rhythmicity. Naturalistic data on fish behavior and habitat conditions are needed to understand the ecological significance of the variability in rhythms among species and individuals (Naruse and Oishi, 1996) and to help in the planning of laboratory studies which, by their turn, should orient further field studies. The integration of these two sets of data is necessary for consistent interpretations of results, from which ecophysiological and evolutionary chronobiological hypotheses can emerge.

II. BASIC CONCEPTS IN CHRONOBIOLOGY

The tropical area approximates the area between the Tropics of Capricorn and Cancer, including parts of Australia, India, Africa, Central and South America, as well as the totality of Indonesia and New Guinea, among other places. Thus, the tropical part of the Atlantic, Pacific, and Indian Oceans are included. In contrast with the temperate zone, these habitats are characterized by less drastic fluctuations in climate and photoperiod. Instead, other environmental cycles not necessarily related with temperate or tropical zones may be quite strong, like lunar phases, tidal movements, and dark–light phases of the day. However, the most specific environmental change in the tropical areas, mainly close to the Equator, is the distinction between wet and dry seasons. This offers interesting cases for discussing the biological cycles, mainly those rhythms endogenously controlled by biological clocks.

For many centuries, biological rhythms have been known empirically in living organisms, but the scientific study of these rhythms arose at the beginning of the eighteenth century, which may be accepted as the birth of chronobiology (Menna-Barreto, 1999; this author also provides extensive literature on basic concepts of biological rhythms). Nowadays, chronobiology is one of the most rapidly emerging areas in biological sciences, which is a consequence of both the widespread distribution of biological rhythms in the living organisms and the fact that these rhythms affect almost all organisms' activities. As a consequence, experimental designs in most areas of biological sciences must consider "time as variable."

Rhythmicity of environmental factors imposes on the animals a particular challenge: how to anticipate these "revisable" cyclical events to better cope with them. Thus, such anticipatory mechanisms may be an inherent trait of the organisms. In fact, cyclical environmental events, such as conditions of dark and light, temperature fluctuations, climate seasonal changes, lunar phases, tidal movements, rainy periods, etc., are conditions presumably prior to the origin of life on earth. Therefore, organic evolution was deeply affected by these cyclical environmental fluctuations, and thus biological mechanisms to anticipate them are expected to occur.

Responses to cyclical changes of the environment are of two types: (a) the passive ones, which are direct consequences of these environmental factors acting upon the organisms (producing *masking* effects, in chronobiological terms); or (b) the active ones, where the rhythm is intrinsic to the organism, going along with an environmental cycle from which modulates but not causes the rhythm (an *entraining* effect).

Biological significance also may be attributed to the organisms' rhythms passively imposed by environmental cycles (*masking* effect). Such a direct association provides the animals with an efficient biological condition to better utilize environmental resources. For instance, spawning concentrated in warmer months for those fishes reproducing more than once a year, like the Cichlidae Nile tilapia, is associated with food-supply availability and better temperature for development of offspring. Moreover, the seasonal metabolic changes detected in some fish may be also a direct (passive) response to water temperature (Wilhelm Filho *et al.*, 2001) rather than represent any endogenous rhythm. Also, most of the Amazon fishes adjust red cell concentration of ATP and GTP in response to natural oscillations of oxygen availability (Val, 1993). However, responses actively governed by self-sustained biological timing systems (sleeping and waking, motor-activity rhythms, etc.) are the primary cases discussed here.

A. Biological Rhythms Controlled by Timing Systems

Biological rhythms are synchronized by environmental cycles. For instance, swimming, feeding, hormone release (Bromage *et al.*, 2001), sleep (Kavanau, 1998, 2001), social interaction (Nejdi *et al.*, 1996), and learning (Reebs, 1996) are some of the activities associated with dark and light periods of the day. Other well-described connections are activities associated with tidal movements, temperature changes, rainy periods, and seasonal photoperiod fluctuations.

A first hypothesis to explain such an association between the biological cycle and the corresponding environmental one states that these biological rhythms were passive responses to the cyclic environment. This "exogenous-clock hypothesis" offers a simpler explanation, as expected from Parcimonian's Law of Science. However, studies from the twentieth century have undoubtedly shown a better explanation based on the "endogenous-clock hypothesis." This discussion, however, was not easily concluded, as will be shown later. To better understand the point, an explanation of the basic characteristics of biological rhythms is provided. Basic concepts in chronobiology are extensively described in Schwassmann (1971), Hill (1976), Aschoff (1981b), Brady (1987), Aschoff (1990), Ali (1992), Marques and Menna-Barreto (1999), and Menna-Barreto (1999).

A cyclical activity may be represented by a sine curve, with high and low activity periods in a time scale, as illustrated in Figure 4.1. The cyclical activity is associated with the environmental cycle (Figure 4.1A) and in this case, these cycles are in *phase* with each other. In Figure 4.1B, the cycles are out of *phase*. The *phase* of a cycle is represented by the Greek letter φ. When the biological cycle gets ahead of the referential environmental cycle, φ is negative; when behind the referential cycle, it is positive. The *period* of a cycle (represented by the Greek letter τ) indicates the distance between two

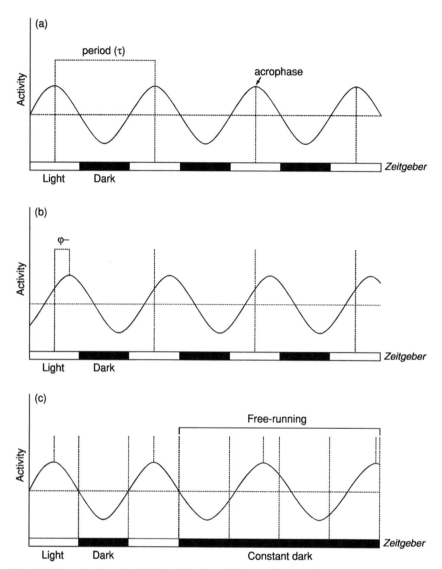

Fig. 4.1 Schematic view of a biological rhythm with the main properties mentioned in the text.

points of corresponding symmetrical positions immediately to subsequent each other (Figure 4.1A). Indeed, the shorter the *period*, the more frequent the cycles in a same interval of time, thus indicating higher *frequency*. Thus, *frequency* is the inverse of a *period*. If a cycle has *period* of 24 hours, the corresponding *frequency* is the inverse of it, that is, 1 cycle/24 hours.

Although synchronization between biological and environmental rhythms has been recognized for a long time, the behavior of such biological rhythms in the absence of the environmental cycles has clarified many aspects in the discussion on the exogenous or endogenous nature of the organisms' timing systems. Classical studies have shown in fish and many other organisms that most of the biological rhythms are sustained even when the coupled environmental cycle is abolished. In such a condition, these biological rhythms are said to be *free-running* and the *period* of the biological rhythm is not exactly equal to that of the environmental one (Figure 4.1C). For instance, a daily-activity rhythm may have a *period* of 24 hours while the animal *free-running* rhythm may have a shorter or longer *period*. As these rhythms (animal and environment) have similar but not identical periods, the biological rhythm is called circarhythm (*circa* = Latin "about, around"). Thus, they are classified as circadian (*ca.* 24 h), circatidal (*ca.* 12 h), circalunar (*ca.* 28 days), etc. The circadian rhythms are the most studied and best understood and are usually the background for theories on biological rhythms and used for classification of the rhythms. Thus, rhythms with *periods* shorter than 24 h are named *ultradians* (higher *frequencies*), while those longer than 24 h are *infradians* (lower *frequencies*).

Although *free-running* attracts attention to endogenous pacemakers, defenders of the "exogenous-clock hypothesis" argued that such a response is expected because the real clock was not the environmental cue selected and controlled; but instead it is some geophysical force which continuously cycles. However, another question for the defenders of the "exogenous-clock hypothesis" is how to explain individual variability if the pacemaker is exogenous and should affect equally all the individuals in the same area and time. In this controversy, well described by Hill (1976), the "exogenous-driver" sympathizers explain by postulating that the biochemical or biophysical structures of the timer can be affected by individual conditions, thus reflecting small differences among individuals. While the exogenous clock is precise, manifestation of overt processes (activity, color change etc.) is not precisely linked to this exogenous clock. They make an analogy with a mechanical watch, in which the hands are not firmly attached to the watch driver mechanism. Thus, the watch could work precisely, since the hands are free to turn somewhat more slowly or rapidly. In biological terms, the overt processes could progressively lag behind or get ahead of the exogenous clock; so, they show a circarhythm instead of the exact rhythm of the exogenous clock.

This controversy endured for years, and followers of the endogenous-clock hypothesis responded to these criticisms. Transference of cyclic rhythms by means of blood transference from one individual to another reinforced their hypothesis. Transplantation experiments have shown synchronization of a transplanted suprachiasmatic nucleus (SCN) with the host SCN in hamsters (see review by van Esseveldt *et al.*, 2000), strongly suggesting an endogenous rhythm. Moreover, better understanding of the genetic and molecular basis of the timing system has occurred since the 1980s and clearly refused the exogenous hypothesis.

While a biological rhythm is controlled endogenously, these rhythms maintain a narrow connection with the environmental cues, and are not determined by them. The environmental cues are called *zeitgebers* and they only modulate this rhythmic activity intrinsic to the organism (Figure 4.1). Only a flash of light is sufficient to maintain their connection to each other.

As the biological rhythms do not show exactly the same period of the environmental cycle, the *zeitgeber* contributes to the synchronization of these rhythms. When the endogenous rhythm shows a shorter period as compared with the environmental cycle, the correction to put these cycles in *phase* is due more to delay in the beginning of the cycle (e.g., beginning of the increased activity) than to accelerated advancement at the end of the endogenous cycle (end of the higher intensity activity). Otherwise, when the period of the endogenous cycle is longer than that of the environment, the adjustment is achieved mainly by increased advances at the end of the cycle. This kind of adjustment to phase the cycle is almost universal among the species, from unicellular to multicellular organisms, either diurnal or nocturnal (Marques and Menna-Barreto, 1999). A biological rhythm is thus entrained by environmental cues (*zeitgebers*, also named *phasing factor*), and this process is called *entrainment*. This is a very important characteristic because the biological rhythm can maintain its endogenous independence from the environment, but still have a connection with the environment.

Another interesting property of the *free-running* periodicities is the relative independence from environmental temperature. Effects of temperature on biological processes are easily understood in terms of Q_{10}, a coefficient that shows how much the intensity of a process increases or decreases under a 10 °C variation in the temperature. Thus, a $Q_{10} = 2$ means that if the temperature increases (or decreases) 10 °C the metabolism will double twice (or decrease by half). By using frequency values, it is possible to compute a Q_{10} for a rhythm. When a temperature increase shortens the frequency of a rhythm, a Q_{10} higher than 1.0 is expected; if the temperature decreases, this value will be below 1.0. However, the Q_{10} of the biological rhythms controlled by endogenous timing systems is very close to 1.0 (usually from 0.8 to 1.2) (Hill, 1976); that is, it is almost unchanged despite temperature fluctuations. This independence from the temperature is of biological significance, since temperature usually changes drastically over one day or longer periods of time (months, seasons, years, etc.), but the animals are able to maintain their normal cyclic processes.

This independence of the biological rhythms from environmental temperature is relative. These rhythms can be entrained by temperature cues, which is more pronounced in ectothermic than endothermic animals (Schwassmann, 1971). Furthermore, there is considerable evidence that temperature affects maturation in several fishes, including tropical and sub-tropical species (see Bromage *et al.*, 2001). In such cases temperature has a direct effect on gonads. Obviously, these are not effects on the biological timing system, but rather on activities controlled by them.

B. Organization and Genetics Basis of the Timing System

A biological rhythm controlled by an endogenous timing system depends on intrinsic oscillators (cell-autonomous timekeeper – pacemakers) generating self-sustained rhythmicity. They build a multioscillatory system, elegantly demonstrated by Moore-Ede *et al.* (1976). These oscillators are controlled by external cues coming from input pathways (sequence of events transducing and conducting the external environmental information to the oscillators). The rhythmicity of the oscillator is conducted by output pathways to the biological rhythm(s) it controls. This forms a hierarchical organization, which includes internal feedback mechanisms. The oscillator (autonomous timekeeper) imposes rhythm to passive structures or functions, which fluctuate accordingly.

The pineal gland is one of the structures involved in biological rhythms (Matty, 1985). In the excellent review by Bromage *et al.* (2001), the self-sustained activity of this gland in fishes is shown. These authors report *in vitro* studies showing free-running activity of the pineal gland releasing melatonin. Accordingly, tissues in the light–dark (LD) scheme increase melatonin release in the dark; and this cycle is maintained in the DD condition, but showing a period different from that exhibited in LD. In fact, the pineal gland of some fish oscillates in culture while in other species it does not (Underwood, 1990). In lampreys, there is little doubt that circadian oscillators regulate the rhythmic production of melatonin, which imposes a circadian cycle upon the locomotor activity (Menaker *et al.*, 1997). In pike, a clock controls melatonin synthesis of two enzymes (tryptophan hydroxylase – the first enzyme of melatonin synthesis; N-acetyltransferase – the penultimate enzyme of this chain), but not in trout, suggesting that, in trout, the single circadian system regulating the expressions of these two enzymes has been disrupted (Coon *et al.*, 1998).

Existence of endogenous rhythms in unicellular organisms (and also in the oscillator cells) supports that timekeeping is part of the cellular machinery. Recent developments in genetics regarding biological timing systems have clarified how the genes are involved in measuring time. Despite the very specific denominations that have emerged in this molecular genetics area, the general term *clock gene* has been used to refer to genes that encode any element of the oscillator system (maybe *timing-system gene* is a better choice).

Notwithstanding that the genetic basis of the biological timing system is logically expected, only in 1971 were Konopka and Benzer the first to describe sufficient evidence of circadian clock mutants by chemical mutagenesis in *Drosophila*. The period of the circadian rhythm of these mutants was longer, or shorter, than 24 hours, or even arrhythmic. These different expressions of the rhythms were all at the same genetic locus, termed *period* (*per*). Moreover, Ralph and Menaker (1988) accidentally discovered the circadian *tau* mutation in the hamster, producing very short cycles of 20 hours for homozygous animals. Other studies have shown that mutation changes circadian patterns of hormones secretion (Lucas *et al.*, 1999); even an independent circadian oscillator was found as a driver of melatonin rhythms in the retina (Tosini and Menaker, 1996). This mutation also disrupts the seasonal reproductive and endocrine responses to day-length variations (Stirland *et al.*, 1996). Nowadays, the molecular basis of *entrainment* and *free-running* have been shown (Young, 2000), although mainly using insects and mammals. Circadian regulation of gene expression has been shown in unicellular organisms, vertebrates and invertebrates. A gene-biochemical mechanism to mammal circadian clock was proposed by Carter and Murphy (1996). Accordingly, an autoregulatory loop including mRNA, protein synthesis and modified protein is the base of the "clock gene": time information is derived from duration of the loop mechanisms.

III. ACTIVITY RHYTHMS

Circadian rhythms are directly involved in the temporal and spatial organization of individuals and communities, and in the prediction of, and response to repetitive events (Boujard and Leatherland, 1992). Circadian activity patterns are among the most evident and conspicuous rhythms detectable in animals. Thus it is not surprising that they constitute the main object of so many chronobiological studies, including naturalistic and experimental works, carried out in the field and in laboratory. Daily activity rhythms are externally expressed as locomotor patterns associated to alternating activity and resting phases, habitat exploration, feeding, and intra- and interspecific interactions (schooling, agonistic behavior, defense of territory, mating, predator–prey interactions). In the ecological context, the importance of species-specific temporal patterns of activity seems obvious: they represent an adjustment of functions such as physiological states, locomotor activities and developmental steps to temporal changes in the environment; and concentrate or displace interactions among individuals (reproduction, competition, predator–prey interactions) (Lamprecht and Weber, 1992). Interspecific differences in activity phasing constitute an important factor involved in organization of fish communities, allowing the ecological separation of fish assemblages in accordance with the phase of the daily cycle when each species concentrate most of its activities (exploratory behavior, feeding, social interactions etc.).

In general, studies on activity rhythms in fish focus on a single or distantly related species, in many cases selected more for convenience than due to a specific chronobiological advantage. Thus, it is questionable whether the studied species constitute the best model systems for chronobiological investigation (Spieler, 1992). As an exception, one can cite the work by Erckens and Martin (1982a,b), who comparatively studied the epigean Mexican tetra characin, *Astyanax mexicanus*, and its cave derivative, "*A. antrobius*" (see below).

The lack of comparative data on closely related species and of reliable data on the natural conditions acting over the studied populations greatly limits interpretations on the ecophysiological and evolutionary meaning of rhythms found in laboratory. First, because, without a proper knowledge of the population habitat, one cannot be sure if such rhythms are not an artifact produced by a completely artificial laboratory situation, which would not express in the natural habitat. Secondly, because, without applying the comparative method, it is not possible to distinguish between the historical (genealogical) and the ecological factors involved in the expression of the studied rhythms.

It is well known that light–dark (LD) cycles are the main *zeitgeber* for animals (Aschoff, 1981a). Circadian rhythms have been reported for a wide range of behavioral and physiological variables in fish (Boujard and Leatherland, 1992). Herein, we will focus on activity rhythms, with emphasis on circadian ones. Activity *sensu lato* encompasses a set of behaviors: locomotor/swimming (spontaneous movements through the available space, either by swimming or displacing on the bottom, unrelated to feeding), feeding, reproduction, social interactions. Locomotor and feeding activities are hardly distinguishable in the field, but laboratory studies have shown that they represent the expression of distinct rhythms, rather loosely coupled in fish. These are the dominant activities in quantitative terms, and understandably have received the greater deal of attention by chronobiologists.

Fishes are classified as "diurnal", "nocturnal" or "crepuscular" (and their mixed types) (Iigo and Tabata, 1996), in accordance with the crepuscular phase of the daily cycle when they concentrate most of their activities (exploratory behavior, feeding, social interactions). Eriksson (1978) and

Sánchez-Vázquez *et al.* (1996) defined as nocturnal, patterns where respectively more than 67 or 65% of total activity occurs during the dark phase of a cycle, as diurnal, those with less than 33 or 35% activity during this phase, and as indifferent, those falling between these values.

Classifying fish species according to their predominant activity phase is the first step and, in many cases, the main goal of field, naturalistic chronobiological studies. When compared to laboratory works, field studies are generally poorly quantified, and usually no precise figure of individual variation is given. Nevertheless, they provide relevant information in an ecological context, revealing the main tendencies at the population level. Among tropical fishes, several publications include data on activity rhythms for stream and coastal species based on naturalistic studies (see Lowe-McConnell, 1964, 1987, for a review; for Brazilian fishes, see also Sazima and Machado, 1990, Sazima *et al.*, 2000a, b, among others).

Laboratory studies reveal a high flexibility of activity patterns in fishes, in contrast with mammals, which seem to present a relatively rigid internal control of activity. Variations in phasing of activity have been recorded not only among closely related species (within a genus, for instance), but also within a species and even in the same individual studied at different moments. Different activity patterns reported for the same species may be an artifact resulting from differences in recording techniques and data analysis in the case of independent studies, but may also reflect geographic, habitat, and seasonal changes in activity.

Ability for dual phasing, i.e., a shifting from a "diurnal" to a "nocturnal" behavior and vice-versa, was demonstrated for several temperate species (e.g., *Salmo* spp., *Lota lota*, *Cottus* spp., *Ictalurus nebulosus*), which show a seasonal inversion of daily activity patterns. As a matter of fact, seasonal changes in circadian variables are well known. For instance, under laboratory conditions, the brown trout, *Salmo trutta*, and the Atlantic salmon, *S. salar*, are predominantly diurnal during the summer, mostly nocturnal during the winter, and crepuscular or indifferent in transient periods (spring and autumn) (Eriksson, 1978). Likewise, *Cottus poecilopus* and *C. gobio* are active during the day in summer and during the night in winter. Conversely, *Lota lota* is night-active during the summer and day-active in the winter (Boujard and Leatherland, 1992). The tropical medaka, *Oryzias latipes*, is day-active at the surface layer, but night-active near the bottom during the winter, shifting to a diurnal activity at any layer in the summer (Naruse and Oishi, 1996).

The brown bullhead (*Ictalurus nebulosus*), a typical nocturnal catfish, tends to become diurnal when light intensity during the light phase of a 12:12 hour LD cycle is very low (around 1 lux) (Eriksson, 1978). However, both "nocturnal" and "diurnal" individuals show a predominant crepuscular feeding activity. A dual and independent phasing of locomotor and feeding rhythms was reported for the goldfish, *Carassius auratus*. Most individuals studied by Sánchez-Vázquez *et al.* (1996) tended to be day-active, but some displayed a nocturnal locomotor activity; however, some day-active fish displayed night feeding and vice-versa, and changes in feeding schedule inverted the activity patterns in some individuals. Iigo and Tabata (1996) also observed a high individual variability in *C. auratus* under LD cycles: the majority of the studied goldfish were active during the photophase, but some were active in the scotophase and others both in the photo- and scotophases. In addition, some individuals spontaneously switched activity patterns, demonstrating a high flexibility in the species. Likewise, another intensively studied laboratory model, the originally tropical zebrafish (*Danio rerio*), showed a considerable individual variation in phase, period and amplitude of activity rhythms under all experimental conditions tested (LD, DD, and LL), and a higher proportion of animals expressed significant rhythms at 21 °C than at other temperatures; most individuals were day-active under dim LD cycles, and no spontaneous switching between activity patterns was observed (Hurd *et al.*, 1998). A similar variation in the phasing of locomotor activity was demonstrated for other species, such as the loach, *Misgurus anguillicaudatus*, the salmon, *Oncorhynchus gorbuscha*, and the medaka, *Oryzias latipes* (Iigo and Tabata, 1996). Sex and age-related differences in activity phasing were reported for several species (Naruse and Oishi, 1996), and evidence for a significant interaction between temperature and sex was found for zebrafish (Hurd *et al.*, 1998).

A dual phasing ability, characteristic of a highly adaptable circadian system, appears to be a common fish trait, especially for temperate species, as an adaptation to accentuated seasonal changes in photoperiod, temperature and food availability. A dual phasing capacity seems not to be adaptive in tropical regions, where seasonal changes are not as accentuated as in the temperate zone. Nevertheless, in tropical areas with well-defined rainy cycles, there may be important annual fluctuations in food availability, both in quantitative and qualitative terms. In large tropical muddy rivers, such as the white water rivers in the Amazon basin, a light threshold as that observed for *I. nebulosus* (1 lux) is attained at depths of a few meters. Thus, many fishes in these large tropical rivers live temporarily or permanently under that threshold and, depending on the depth and the fish light sensitivity, even under free-running conditions (permanent darkness). The dual phasing ability is an interesting and exciting field for future investigation in tropical fishes.

It is well established that, besides light, feeding may entrain rhythms, as demonstrated for several fish species. In addition to generalized locomotor rhythms, specific behaviors may be entrained by feeding, including phototactic and agonistic behaviors, as well as some physiological variables, such as the levels of circulating cortisol. On the other hand, within a species, not all rhythms are entrainable by meal-feeding: for instance, in the medaka, *Oryzias latipes*, feeding entrains agonistic, but not reproductive behavior. There are also reports in which locomotor rhythms were not entrained by feeding schedules (Spieler, 1992).

A most characteristic and clearly adaptive component of circadian systems is the food anticipatory activity (FAA), i.e., a pronounced increase in activity beginning several hours before mealtime. Food-anticipatory activity rhythms exhibit the same oscillatory properties to those of light-entrainable rhythms. Scheduled feeding may act as a potent *zeitgeber* capable of inducing FAA. In the greenback flounder, *Rhombosolea tapirina*, both meal size and duration were involved in the development of FAA, indicating that the fish was capable of evaluating the energetic and temporal impacts of a single daily meal. FAA may persist (residual oscillations) for a number of days (3 in the greenback flounder, bluegill, largemouth bass and lesser sandeel, 3 to10 in the goldfish) during food deprivation, providing evidence that FAA is mediated by an endogenous food-entrainable circadian oscillator (Purser and Chen, 2001).

Behavioral differences between fish under differing feeding schedules may be interpreted in an adaptational, ecological context. From an adaptive point of view, the advantage of being prepared to feed when food is regularly available is self-evident (Boujard and Leatherland, 1992). When fish are given access to sufficient food and allowed adequate time to feed, there would be no immediate need to anticipate mealtime. However, when food access time and/or meal size is restricted, synchronization of feeding activity ensures that the feeding window is not missed, and may also maximize food utilization through the preparation of the digestive system for when food is available (Purser and Chen, 2001). This is a common situation in the natural habitat, due to cyclical decreases in food availability or to narrowing of food niches in order to reduce competition.

A relative independence between locomotor and feeding patterns was observed in several fish species (and in some birds and mammals as well), suggesting the existence of two independent, loosely coupled timing mechanisms, with the participation of independent oscillators, anatomically and functionally distinct from each other (Spieler, 1992; Sánchez-Vázquez *et al.*, 1996; Purser and Chen, 2001), the light-entrainable and the food-entrainable oscillators (LEO and FEO, respectively).

Many studied organisms, from protists to plants and animals, provide strong evidence of a more or less tight control of activity patterns by internal clocks. There is evidence for endogenous circadian rhythms in fishes, including activity patterns, growth of scales and otoliths, and vision (Boujard and

Leatherland, 1992). Nevertheless, because free-running circadian rhythms were not detected for some species, the question about endogenous *versus* exogenous control of activity in fishes is still matter for controversy.

Compared to tetrapods, especially mammals, free-running rhythms in fishes are usually more unstable. Many authors found circadian oscillations (mostly for locomotor activity) under constant conditions for a variety of freshwater and marine teleosts, e.g, *Carassius auratus*, and *Zacco temmincki* (Cyprinidae), *Nemacheilus barbatulus* (Balitoridae), *Catostomus commersoni* (Catostomidae), *Silurus asotus* (Siluridae), *Plecoglossus altivelis* (Plecoglossidae), *Lota lota* (Lotidae), *Fundulus heteroclitus* (Cyprinodontidae), *Solea vulgaris* (Soleidae), and, among tropical species, *Danio rerio* (Cyprinidae), *Astyanax mexicanus* (Characidae), *Pimelodella transitoria* and *Taunayia* sp. (Pimelodidae Heptapterinae), and *Halichoeres chrysus* (Labridae) (Boujard and Leatherland, 1992; Sánchez-Vázquez et al., 1996; Hurd et al., 1998; Trajano and Menna-Barreto, 1995, 2000; Gerkema et al., 2000).

In general, the ratio of individuals with free-running circadian rhythms and the signal energy of the oscillations are generally lower than in mammals. In *D. rerio*, up to 73% of fish expressed free-running circadian rhythms under DD, the remaining ones showed unstable rhythms or were arrhythmic (Hurd et al., 1998), and approximately 75% of catfishes, *Silurus asotus*, displayed circadian rhythms under different intensities of constant light (Tabata, 1992). Moreover, several fish species kept in constant darkness lose circadian activity patterns within some weeks (Tabata, 1992; Gerkema et al., 2000). Free-running rhythms could not be detected for *Alosa sapidissima* (Clupeidae), *Arius felis* (Ariidae), *Ictalurus punctatus* (Ictaluridae), and *Salmo trutta* (Salmonidae) (Boujard and Leatherland, 1992; Iigo and Tabata, 1996). In contrast, the studied Brazilian catfishes, *Pimelodella transitoria*, exhibited free-running circadian rhythms after living several months in constant darkness; likewise, several specimens of the troglobitic (cave-restricted) catfishes, *P. kronei*, that evolved under permanent darkness for generations, also showed significant free-running circadian locomotor rhythms (Trajano and Menna-Barreto, 1995), and a precise entrainment of locomotor activity to LD conditions was observed for the day-active yellow wrasse, *H. chrysus*, that presented the anticipatory behavior characteristic of endogenously controlled circadian systems. Possible differences in stability of circadian systems between tropical and temperate species deserve future investigation.

There is evidence that the teleost circadian system encompasses multiple self-sustained oscillators and that at least two organs, the pineal organ and the retina, contain oscillators. Photoreceptors in the retina, pineal organ, and deep in the brain would be involved in photosignal transduction to establish circadian rhythms in fish (Iigo and Tabata, 1996). Each of these organs alone may be involved in the entrainment of locomotor activity in certain fish species. Experiments with *Silurus asotus* indicate that these photoreceptor organs have different functional roles in circadian organization. The lateral eyes would be involved mainly under relatively intense light conditions, whereas the pineal organ is probably involved mainly in DD, and both would be involved in circadian organization under dim light conditions, when light information from eyes and pineal organ is integrated (Tabata, 1992).

Data from laboratory studies suggest that fish oscillators are loosely coupled to each other. A weak coupling among these oscillators or a lack of entraining signals from specific *zeitgebers* that typically synchronize these oscillators with one another may explain the variability observed in fish activity rhythms, including the apparent arrhythmicity in some animals. Data for zebrafishes suggest that temperature could be one of the environmental conditions affecting such coupling (Hurd *et al.*, 1998). The strength of coupling between oscillators and with overt rhythms, such as locomotor activity, may vary inter- and intra-individually according to internal, physiological, and external, environmental conditions, resulting in the plasticity of the teleostean circadian system. This hypothesis is supported by the seasonal changes in locomotor patterns observed for temperate species, indicating that such flexibility is a strategy, particularly important for ectothermic animals, to survive in ever-changing environments (Iigo and Tabata, 1996).

A multioscillator system of temporal integration, encompassing light-entrainable and food-entrainable oscillators, may be regarded as adaptive, because it might not be advantageous to entrain all circadian systems to the same *zeitgeber*. A circadian pre-feeding activity allows making a maximum benefit of a cyclic food resource, but this will not necessarily change the circadian organization for other kinds of behavior or physiological functions not directly involved in digestion (Boujard and Leatherland, 1992). Circadian systems are involved not only in locomotor and feeding activities, but also in reproductive rhythms. It would certainly be non-adaptive to starve in the presence of a phase-shifted food resource, but it could prove equally non-adaptive to align all rhythms to a new feeding time and thereby to spawn at a non-optimal season or time of day (Spieler, 1992). Flexibility in phasing and a certain degree of independence between feeding and locomotor rhythms could be seen as an adaptive response of fishes to a relatively stable aquatic environment but subject to periodic changes in some biotic factors (Sánchez-Vázquez *et al.*, 1996).

Locomotor activity phasing (diurnal versus nocturnal behavior) is frequently regarded as a taxonomy-related feature. Indeed, field studies point to general tendencies within families or genera. This is expected in view of similarities in ecology (e.g., modes of feeding, social behavior) and anatomy (feeding apparatus related to type of food items, sensorial systems adapted to specific light conditions) resulting from common genealogy.

Among tropical freshwater fishes, siluriforms and gymnotiforms are generally described as nocturnal or crepuscular (a generalization supported by studies in laboratory), as well as the marine holocentrids, scorpaenids, serranids, apogonids, priacanthids, and lutjanids. On the other hand, the freshwater characiforms and cichlids are predominantly diurnal, like the majority of reef fishes, such as chaetodontids, pomacentrids, labrids, acanthurids, balistids, tetraodontids, and diodontids, and predators such as synodontids, aulostomids, fistulariids, belonids, and sphyraenids (Lowe-McConnell, 1964, 1987). Nevertheless, many exceptions have been found, well illustrating the ecological plasticity characteristic of teleost fishes, and caution must underline generalizations.

This plasticity seems to vary according to the taxa. Cypriniforms (barbs, carps, loaches), many of which known as diurnal, visually oriented fish, seem to be particularly flexible, as shown by the individual variability and ability for dual-phasing documented for the goldfish and the loach, *Misgurnus anguillicaudatus*, and the existence of several troglobitic cypriniforms (Romero and Paulson, 2001), actually or probably derived from nocturnal epigean species. There are exceptions even among the more homogeneously night-active, chemically oriented siluriforms. For instance, diurnal species have been reported among the generally nocturnal *Trichomycterus* catfishes (Trichomycteridae) and *Ancistrus* armored catfishes (Loricariidae) (Buck and Sazima, 1995; Casatti and Castro, 1998). Nocturnal feeding appears uncommon for small loricarioids. However, Sazima *et al.* (2000a) documented a crepuscular and night foraging in the minute *Scoloplax empousa* (Scoloplacidae), which may search for prey visually, since the authors observed eye movements (in aquarium) and head orientation (in the field).

Intra-group variation in apparently homogeneous taxa was also observed for the carnivorous piranhas (Serrasalmidae), reported as predominantly day-active. Among three species studied in the field by Sazima and Machado (1990), larger individuals of *Serrasalmus marginatus* and *S. spilopleura* extended their feeding activity to early night time, and medium-sized to large individuals of *Pygocentrus nattereri* used to forage mainly at dawn and night up to around 22:00 h.

The coastal gobiids, many of which live in tide pools, are generally referred to as predominantly diurnal. However, Thetmeyer (1997), comparatively studying two species belonging to different genera, *Gobiusculus flavescens* and *Pomatoschistus minutus*, found daily rhythms in both locomotor activity and oxygen consumption, but the phasing was different according to the species: whereas *G. flavescens* was most active during the light phases,

P. minutus presented activity peaks during the dark phases. In these fishes, activity seems to be closely coupled to factors such as foraging (prey types, bottom *versus* off-bottom feeding), population density and presence of predators, which could provoke a switch in phasing.

Due to the moon- and starlight, airglow (light originated in the high atmosphere and associated with photochemical reaction of gases caused by solar radiation), and bioluminescence, there is no complete darkness in shallow water habitats during the night. Since many fishes present a high photosensitivity, at least some of them being able to perceive light at intensities as low as 0.01 lux or less (Eriksson, 1978), the amount of light available during the night would allow for visual orientation and feeding in fishes living in shallow coastal habitats, ponds, clear-water streams, and in superficial layers of larger water bodies (Thetmeyer, 1997). Therefore, a phylogenetic or individual dual-phasing is not surprising for visually oriented fishes from these habitats.

Day/night differences are accentuated for fishes from shallow aquatic habitats and/or in transparent waters, but not so for fishes living in murky, turbid waters of large rivers, near the bottom of deep lakes and seas. The relative weak expression of the circadian system in many fishes may be a response to relatively small differences in irradiance levels between light and dark in marine and deep freshwater habitats (Gerkema *et al.*, 2000). The highly flexible circadian system of teleosts permits the adjustment to non-daily changes in light levels to which fishes are subject as a consequence of movements between water layers, short- to medium-term changes in turbidity due to rains, and so on.

In a symbiotic relationship such as cleaning activity (removal of parasites, diseased or injured tissue, and mucus from the body of other fish), that has been studied in detail in some reef fish communities, the activity of cleaner and clients may act as mutual *zeitgebers*. The barber goby, *Elacatinus figaro*, starts the cleaning activity at dawn and ends it shortly before nightfall; diurnal fish clients are cleaned mainly in mid-afternoon, whereas nocturnal species are mostly cleaned close to twilight periods. The two peaks of cleaning activity of *E. figaro*, in early morning and middle afternoon, would correspond respectively to periods in which nocturnal and diurnal client species had already taken a meal and thus had time to seek the cleaning stations (Sazima *et al.*, 2000b).

Behaviors other than the locomotor/swimming and the feeding ones have been shown to express circadian rhythms. Such is the case with the air-gulping behavior, typical of fishes with facultative or obligatory aerial respiration as an adaptation to an oxygen-poor aquatic environment. Periodical (either daily or seasonal) lowering in oxygen concentrations is observed in habitats as diverse as tide pools and tropical slow-moving shallow freshwaters, as well as streams subject to temporary droughts leaving isolated pools where fish can survive. Rhythmicity of air-gulping behavior, measured as surfacing activity (frequency of intermittent excursions to the air-water interface to gulp air), was studied, among others, in the facultative air-breather Indian catfishes *Heteropneustes fossilis* (Heteropneustidae) and *Clarias batrachus* (Clariidae) (Maheshwari *et al.*, 1999). In addition to 24 h rhythms, circannual rhythms were detected in surfacing activity of both species, with peaks included in the pre-spawning (time of gonadal development) – spawning (reproductive activity) period. A negative correlation was obtained between annual curves of air-gulping activity and dissolved oxygen of water (DO), corroborating the notion that oxygen concentration in the habitat is an important factor modulating air-breathing frequency in fish.

At a daily scale, usually variation in DO is not important, and yet air-breathers may present circadian rhythms of air-gulping behavior (possibly endogenous). *Heteropneustes fossilis* and *C. batrachus* are, as many catfishes, night-active and probably an increased frequency of aerial respiration during the night is correlated with the higher oxygen demand for locomotor activity. On the other hand, over a seasonal scale air-breathing activity of these catfishes increases during a complex situation characterized by a phase of initial oxygen depletion accompanied by raised temperatures (pre-monsoon) and another phase of energy demanding reproductive activities in a post-monsoon environment saturated with oxygen (Maheshwari *et al.*, 1999). In conclusion, rhythms in air-gulping behavior would represent an adaptation to a cyclical availability of oxygen in relation to the species requirements, either due to a decrease of DO in the habitat (drought, temperature raise – typical of tropical regions), or to an increase in the species demands (enhancement of locomotor activity, reproduction, etc.).

Evidence of circadian rhythmicity of electric discharge was found for the tropical electrogenic fish, *Eigenmannia virescens* (Deng and Tseng, 2000). Circadian rhythms were detected both under LD (12:12 h) and DD conditions and the circadian oscillator seems to be temperature-compensated, providing evidence for an endogenous control of circadian rhythmicity of electric discharge. Unexpectedly for a night-active fish, peaks of electric discharge, that are used for orientation and feeding in electrogenic fishes, were observed in the middle of the subjective day. This is another interesting field open for future chronobiological research in tropical fishes.

The existence of daily rhythms in phototactic behavior has long been reported for fishes. For instance, Davis (1962) noticed that the light-shock reaction of bluegills immediately following a sudden (and random) exposure to bright light decreased in duration along the dark phase of the 24 h LD cycle. Studies on tropical troglobitic fishes (see below) also provided evidence for day–night differences in the phototactic behavior. Phreatobic clariid cat-fishes, *Uegitglanis zammaranoi*, showed a slightly stronger photonegative behavior in the light phase than in the dark phase during choice-chambers experiments on light reaction (Ercolini and Berti, 1977), the opposite being observed for the cyprinid *Barbopsis devecchii* (Ercolini and Berti, 1978). According to Pradhan *et al.* (1989), the Indian cave balitorid, *Nemacheilus evezardi*, presents a significant circadian rhythm in its phototactic behavior, which may be synchronized by meal scheduling. However, the methodology used in these studies does not allow distinguishing between a masking effect of exposure to LD cycles over the phototactic behavior or a true endogenous circadian rhythm in this function.

Tidal rhythms, among non-circadian short-term oscillations, have been consistently found in coastal fishes, and a number of species inhabiting intertidal zones exhibit circatidal activity rhythms under constant condi-tions. Population differences related to geographic differences in tidal re-gimes were described for juvenile gobiids, *Chasmichthys gulosus*, studied in laboratory under constant light intensity and water temperature (Sawara, 1992): individuals from a rocky shore with a rhythmic tidal pattern and large tidal range exhibited a semicircadian activity rhythm (period around 12 h), but no such a rhythm was detected for fish from a rocky shore with an irregular tidal pattern and small tide range.

Several *zeitgebers* have been shown to be effective for circatidal rhythms in different organisms to varying extents: cycles of change in light intensity, inundation, mechanical agitation, temperature, salinity and hydrostatic pressure caused by tide flow. However, it seems that the most reliable among these *zeitgebers* is hydrostatic pressure, because it is less affected by weather conditions and seasonal changes (Northcott *et al.*, 1991a). A circatidal phase–response curve was demonstrated for the rock-pool blennie, *Lypophrys pholis*, providing evidence of a circatidal endogenous oscillator because it cycles once every 12.5 h. On the other hand, no circadian component was detected in *Lypophrys* tidal rhythm (Northcott *et al.*, 1991b).

Circadian time-control in fish seems to be more tightly, precisely con-trolled in the natural habitat than in laboratory settings. If classifying a particular fish species as diurnal or nocturnal on the basis of laboratory studies may be quite hazardous in view of the great variation observed for the same species, it is more straightforward and less controversial in the field. Although frequently labile under laboratory conditions, activity rhythms are probably strictly entrained in the natural habitat by the multitude of *zeitge-bers* that interact with each other and the organism, producing the overt rhythms observed in the field. A more strict and stable species-specific behavior is expected in nature in view of the importance of the temporal organization in the structuration of fish communities. For instance, in the presence of potential competitors in the natural habitat, fishes may adjust their behavior in order to reduce the utilization of similar resources.

The role of temporal ecological separation allowing the coexistence of closely related species is well illustrated in the case of the fish community of Lake Kiwu, Central Africa, where 13 species of the *Haplochromis* (Cichlidae) are found. Ulyel *et al.* (1991) comparatively studied four among these species, basing the study on the analysis of stomach contents of specimens captured at different times of the daily cycle. The studied species are predominantly diurnal, becoming active at sunrise until a few hours after sunset, but differ not only in the number (one or two) and time of activity peaks, but also in the kind of prey items taken along the day, possibly as a consequence of temporal differences in prey activity. This is considered an example of ecological strategy to reduce interspecific contact and competition, especially important for closely related, generalized feeders found in sintopy, illustrating the importance of spatial and temporal exploitation and differential use of resources in the organization of fish communities.

In conclusion, the expression of behavior in the natural habitat probably reflects the relative contributions of exogenous factors (e.g., food availability, presence of potential competitors, predators and mates), that can mask or abolish rhythms, and endogenous influences (internal clocks) (Burrows and Gibson, 1995). This led to the statement that biological rhythms would find their full expression only under suitable environmental conditions (Gerkema *et al.*, 2000). On the other hand, an intrinsic highly flexible and adaptable circadian system provides material for differential selection, and is possibly one of the factors at the base of the great diversification observed in fishes (for diversity of tropical fishes, see Chapter 2, this volume).

IV. SOCIAL ORGANIZATION

A. Dominance Rank

Establishment of social hierarchies among individuals from a group is an important aspect in the lives of many organisms. The resulting social status, initially established by overt confrontations, is then maintained by less aggressive encounters, mostly characterized by displays, which avoid serious damage between conspecifics (Haller and Wittenberger, 1988).

Despite this significance, social order imposes on the organism's energy costs, and many studies have reported biochemical, physiological, and behavioral consequences of social stress to the subordinate fish in tropical and non-tropical species (Ejike and Schreck, 1980; Schreck, 1981; Haller and Wittenberger, 1988; Volpato *et al.*, 1989; Zayan, 1991; Fernandes and

Volpato, 1993; Haller, 1994; Volpato and Fernandes, 1994; Alvarenga and Volpato, 1995). This stress state is characterized by an increased metabolism in the subordinates. However, even the dominants have their metabolism increased as a consequence of hierarchical fights. That is, social rank in a group is an adaptation which is maintained by energy cost for every fish, mainly to the subordinates. This cost may result in distress, since growth, reproduction, and the immune system may be impaired or even suppressed (Moberg, 1999).

During this association between metabolism and social rank, Alvarenga and Volpato (1995) showed that energy cost in the tropical fish Nile tilapia is mostly a consequence of the previous hierarchical history than of the social rank. They show that metabolism of dominant and subordinate fish is positively correlated with fighting and other aggressive displays, and that each pair of fish could exhibit a very different profile of aggressive interaction. That is, a subordinate fish in a low aggressive group may have lower metabolism than both a size- and sex-matched dominant in an aggressive group.

Regarding social order and biological rhythm, a cyclically changing dominance–subordinance relationship over time has not been described. However, social interaction may have different intensities not only as a result of time of grouping, but also as a consequence of daily change in individual activity. The more active the fish, the more likely they will encounter each other – and fights occur. Such a circadian rhythm of this activity is one of the factors which affects evaluations about social interactions in a group and must be considered carefully when studying hierarchical fishes. Moreover, seasonal changes in aggression are also found, but this could be mostly a passive consequence of hormonal rhythm (Matty, 1985). For instance, testosterone is a reproductive hormone released with a seasonal rhythm (Crim, 1982; Matty, 1985), which increases aggression in territorial fishes (Munro and Pitcher, 1985). Thus, before attributing rhythmic changes of aggressive interactions in a group to any timing system, these two main reasons, activity and hormonal fluctuations, should be weighed.

B. Shoaling

Shoaling refers to a group of fish which maintain social relationship among themselves (Pitcher, 1986). They can or cannot exhibit schooling behavior. Such grouping has important effects on individual fish lives. Biological adaptive value of shoaling is related with the intrinsic advantages of shoals by decreasing risk of predation, improving feeding and homing. Advantages of shoaling in fishes are summarized by Pitcher (1986).

Tropical fishes, such as some Cichlidae and Characidae, exhibit shoals which are supposed to bring the advantages pointed to above. The Nile tilapia, *Oreochromis niloticus*, exhibits a certain degree of social behavior mainly during larval and young juvenile phases, and also during the mouth-brood care (McBay, 1961). Barki and Volpato (1998) examined the social behavior of this species looking for social learning effects. They observed naïve normal fish kept with dorsal-finless mutants (associated with the corresponding control groups) from the early stage of free-swimming up to 2 month and found that dorsal display is a characteristic affected by social learning. These fish have disrupted part of their normal dorsal fin display (an important aggressive display for maintaining social rank and territory and to avoid predators), thus showing another advantage of forming shoals instead of living alone. Moreover, Helfman *et al.* (1982) showed that young grunts joining a shoal might even learn migration routes by following the larger fish.

Species of the neotropical freshwater Callichthyidae (Teleostei: Siluriformes) show different degrees of shoaling behavior. Paxton (1997) studied shoaling behavior of two *Corydoras* species, *C. pygmaeus* and *C. ambiacus*, representing open-water and benthic fishes, respectively. Both species are crepuscular, although the former displayed more activity during the day. However, whether such rhythms are passively controlled by light or actively endogenous remains uncertain. Paxton (1997) found that *C. ambiacus* shoaled more than *C. pygmaeus* and that both shoaled more during daylight than twilight. Paxton (1997) attributes vision as a factor more important for shoaling than olfaction in these species, as earlier shown in other catfishes (Browen, 1931, cited in Paxton, 1997). This is very interesting because these are catfishes, a group in which olfaction is well-recognized as the most important sensory modality (Liley, 1982; Pfeiffer, 1982; Giaquinto and Volpato, 2001). This means that the relative importance of each sensory modality depends more on specific selective pressures than on generalizations about a whole group.

The relative stability of a shoal depends on behavioral synchronicity of each fish in the group. While individual behavior could disrupt such order, individual synchronization promoted by timing systems entrained by the same environmental cue is of indubitable importance to stabilize the group, which is one of the most remarkable contributions of chronobiology to studies on social behavior.

Inter-species synchronic behavior as described for the gobiid fishes and burrowing alpheid shrimps should also be considered. Their activity rhythms (inside and outside the burrow) are relatively synchronized with each other (Karplus, 1987), so that they can optimize the time spent in such an association. The fish stay inside the burrow with the opening closed (without external light cues), so they start activity guided mainly by endogenous rhythm. Such a synchronicity of activities among species (even considering strong taxonomic differences – shrimps and fishes) shows important roles of intra-species aggregations from a chronobiological point of view.

C. Schooling

Schooling indicates a highly organized group of fish where their relative positions contribute to the whole movement of the school. It is a leaderless social grouping based on mutual attraction of conspecifics of similar age and size engaged in the same activities at a given time. Schooling is an adaptive characteristic of some fishes providing advantages which are discussed by Pitcher (1986) relying on optimized foraging, hydrodynamic benefits, social facilitation for growth, synchronized cooperation which confuses predators, and increased vigilance. Moreover, many tropical fishes migrate long distance before reproduction (Winemiller and Jepsen, 1998; Silvano and Begossi, 2001), usually forming schools. Here the emphasis is on the relation of schooling with chronobiology.

Indeed, schooling may occur during some periods of the year, thus representing a seasonal or circannual rhythm. While schooling may change over the day, being reduced (or absent) at night in diurnal species, in some circumstances the school maintains continuous swimming for several days (in long distance migrations). At this condition, what should be the rest–wake cycle of these fish? Do they sleep during these long distance travels? This is an intriguing area of study for understanding the function of sleep, where fish biology contributes significantly. As the behavioral consequences of sleep deprivation are intensively destructive to the organism, how are these animals adapted to cope with this condition of continuous swimming?

As pointed out by Stopa and Hoshino (1999), rebound sleep (sleep compensation after a period of deprivation), as shown in some fishes (Tobler and Bórbely, 1985), reveals the importance of sleep among numerous and distantly phylogenetically related species. That is, sleep is very important for the homeostatic equilibrium of the fish so that the assumption that some fish species do not sleep may be part of a misinterpretation of observations.

Kavanau (1998), however, stresses that in some fish a clear pattern of motionless behavior has not been found, thus suggesting absence of sleep. However, this author argues that the "sleep function" is still preserved. Sleep or restful waking are states that enable the central nervous system to process complex information (mainly visual) acquired during previous activities. Such cerebral states refresh neural circuits related to these previous experiences. Frequently used circuits maintain experiential and also inherited memories. Spontaneous oscillatory activities also refresh neural circuits (Kavanau, 1998). Thus, circadian cyclic changes in fish activity (sleep–awake; rest–activity; restful waking–active waking) have been associated with needs for memory circuitry refreshment. The resultant question is how should these cycles behave in those species or conditions in which sleep, rest or restful waking are not necessary?

One possibility is that, in these conditions, the fish engages in unihemispheric sleep. That is, sleep with one brain hemisphere at a time. This has been described in dolphins, which close the lid of one eye while the brain hemisphere, usually of the opposite side, sleeps (their optic nerves cross completely in the optic chiasma, thus inverting the side of the nervous control below this level) (Cloutier and Ahlberg, 1996). Fishes, however, are very unlikely to show unihemispheric sleep during continuous swimming, because ocular obstruction (by either eyeball rotation in sharks or decreasing size of the pupils in some teleosts) was never reported when these fish were engaged in continuous swimming (Kavanau, 1998).

Perhaps another behavioral pattern replaces the function of sleep or rest in these animals. While schooling, the fish are in conditions more or less similar to those of sleep or restful waking, which are favorable for refreshing memory circuits (minimal interference from sensory processing). According to Kavanau (1998), the basic function of schooling is related to facilitation of brain activities: while schooling, less sensory processing is required. Fishes at inner positions of schools do not have to "listen", "smell", "taste" or process complex visual information; they need only to stay aware of their position relative to the nearest neighbors, obtained by a very important contribution from the lateral line (Kavanau, 1998). Similarly, during migration some birds do not need unihemispheric sleep. While flying thousands of kilometers for many days, they have little need for visual inputs because there is no detailed information to be seen and they fly most of the time in dim light or darkness (terrestrial or celestial visual cues employed do not require detailed visual processing) (Kavanau, 1998).

In short, schooling provides sufficiently low stimulatory inputs so that, in this condition, the fish can refresh memory while swimming. This idea enlarges the concept of sleep to a more general state with the same biological function, the refreshment of neural circuitry.

D. Communication

Environmental cues serve as external synchronizers for animal rhythms. Tidal movements (for coastal fishes) and day–light cycles are the most obvious *zeitgebers* regulating rhythm phases. Despite that, intraspecific communication may also help intra-group synchronization. In gregarious fish, Jordão and Volpato (2000) showed that grouping is also modulated by chemical cues. They found that while facing a predator, the Amazonian fish pacu, *Piaractus mesopotamicus*, releases chemicals that disperse conspecifics. On the other hand, when facing a sympatric non-predator heterospecific, these pacus chemically attracted the conspecifics. At night, schooling may be weak in species where visual communication is essential, and another way of communication is important. While sound may be used by the fish to keep shoal at night, predators may intercept the sounds, thus minimizing any anti-predator advantages provided by shoaling (Hawkins, 1986). But chemical communication provides these fish with sufficient information to avoid group disruption and safely warrants schooling at night (Liley, 1982). In fact, "schooling substance" has been shown in some fish species, such as the minnow *Phoxinus phoxinus* and the catfish eel *Plotosus anguillaris* (Matty, 1985). Moreover, individual recognition by chemical cues, also used to identify hierarchical status as described in the Nile tilapia (Giaquinto and Volpato, 1997), is another important phenomenon governed by chemicals that may play a role in grouping. As grouping may appear during some periods of the year in some species (for reproduction, for example), or even some period of the day, its periodical aspect is in accordance with cyclical release of grouping chemicals. In fact, reproductive pheromones attracting males or females are released in a certain season, thus providing a chemical attractant controlling a cyclical event. These chemical rhythms are governed by internal timing systems, while the grouping rhythm is a passive consequence of such a chemical variation.

Other sensory modalities are also important to maintain grouping in conditions where light is poor or even absent. Electrolocation is of special interest. The Gymnotiform fish are South American teleosts living in turbid and murky conditions, such as "white" water in the Upper Amazon. They are also found in "black-water" rivers in South America. At the confluence of Rio Negro and Rio Branco, electric fish were caught in water 5–10 m deep (Bullock, 1969). In this environment, vertical movements of a few meters expose the fish to major differences in light intensity.

Zupanc *et al.* (2001) determined that *Apteronotus leptorhynchus* changes electric organ discharge (EOD) in response to light intensity and has minor influence from endogenous rhythm. Despite the very low variation of the EOD, it can be classified in two categories: "chirps", which are complex modulations of frequency and amplitude lasting between 10 to several hundred milliseconds; and "GFRs" (gradual frequency rises), characterized by a relatively fast rise in EOD frequency followed by a slow decline in the baseline value (it lasts from some 100 ms to more than 1 min). Zupanc *et al.* (2001) found that "chirping" is predominant at night, when these fish also show higher locomotor activity. GFRs, however, are more frequent in the light period. However, these rhythms are not endogenously generated; instead, these authors showed a clear dependence on external light intensity. "Chirps" may function as an advertisement signal during agonistic interactions between two fish, thus decreasing interference of any neighboring conspecific (Engler et al., 2000). Zupanc et al. (2001) suggested an antipredator function for "chirps", as this EOD pattern was more frequent when the fish were more active. Furthermore, considering the habitat characteristics of the rivers in which these fish live, these authors conclude that this dynamic environment (vertical movements drastically changing external light intensity) favored the control of spontaneous EOD modulations and locomotor activity by environmental light, rather than an endogenous timing system.

Stopa and Hoshino (1999) described behavioral sleep in the neotropical electric fish *Gymnotus carapo*. This species showed a particular sleep posture, relative immobility, increased sensory thresholds, and reversibility of this state, which are the most accepted behavioral criteria for inference of sleep occurrence. While sleeping, however, these fish still maintain EOD, a response independent from those controlling behavioral sleep, thus neglecting the hypothesis that concomitant patterns of EOD and behavioral sleep occur in this species. However, these authors argue that maintaining EOD during behavioral sleep is also of adaptive value because it may help detection of potential cannibals. That is, while exhibiting a normal awake–asleep rhythm, they still maintain sensory channels to continuously inspect the surrounding environment.

Other ways of communication may also contribute to grouping activities in fish. Behavioral studies reveal that very specific internal conditions of the fish may be subtly communicated to each other, although the specific sensory modality involved is not known. In the tropical Nile tilapia, Volpato et al. (1989) described that two fish of the same size were isolated from each other and maintained the same frequency of opercular movement. After a partition of the aquarium was gently lifted assuring pairing, one of these fish almost doubled this frequency while the other remained unchanged. This difference was maintained even though no confrontation could be detected between them. Some minutes later, fights started and the fish with the higher ventilatory frequency was the subordinate of the pair. These authors interpreted this as a neurovegetative anticipatory response of the subordinate fish to an eminent social stressor. This clearly indicates that subtle communication occurred between these fish before fighting. Chemicals and/or sound may have played a role in this case.

Sound emission in fish has been well reported (Bone and Marshall, 1982; Hawkins, 1986). It is produced in several ways (grinding the teeth, rasping spines and fin rays, burping, farting or gulping air, or by swimbladder mechanisms). Sound is usually related to social interaction, being associated with aggressive species (Hawkins, 1986; Ladich, 1989). The relation of this behavior to chronobiology, however, is more scarce in the literature.

Circadian variation of sound emission is described in some fish (Brawn, 1961; Takemura, 1984; Nakazato and Takemura, 1987). *Abudefduf luridus* is a fish occurring in shallow waters of the west African coast which shows circadian rhythm of acoustic emission (Santiago and Castro, 1997). According to these authors, the frequency of sound emission of this species is increased at sunrise and sunset, and drastically reduced at noon. Higher periods of sound emission in fish have been associated with feeding activities (Miyagawa and Takemura, 1986; Nakazato and Takemura, 1987), but the biological significance of such an association is still not clear.

V. REPRODUCTION

According to Schwassmann (1971), reproductive rhythms entrained by environmental cues have biological significance because: (a) reproduction occurs in favorable seasons, so that offspring are most likely to survive; and (b) it assures sexual maturity at the same time for both sexes in brief-spawning species (for example, some catfishes and characids), which might not be important to prolonged-breeding species (for example, Cichlidae). In temperate regions, such environmental cues are mainly photoperiod and temperature. For the tropical fishes, however, the *zeitgebers* are quite different.

In central tropical areas, "cold" and "warm" seasons are replaced by "dryer" and "wetter" seasons. Even so, the "dry" period is not necessarily dry, but represents a considerable lack of precipitation. Most of the tropical fishes, mainly those concentrated in the area between the latitudes 10° North and 10° South, spawn very soon after a good rain during the reproductive period (Bone and Marshall, 1982). However, some fishermen have failed to facilitate spawning by artificial rain in tanks of migratory fishes (such as the catfish pintado, *Pseudoplatystoma coruscans*, and the Characidae pacu, *Piaractus mesopotamicus*). Like this example, many doubts are still present regarding environmental cues eliciting reproduction in tropical fishes. Although several candidate entrainment cues (rains, atmospheric pressure, lunar phase, water transparency, etc.) are currently recognized by some fishermen and some scientists, scientific support remains weak.

For a full acceptance of a circannual cycle as a consequence of an endogenous rhythm entrained by *zeitgebers*, five conditions must be attained, as admitted by Bromage *et al.* (2001): (a) the rhythm must be represented over more than one year (preferably several cycles); (b) under *free-running*, the *period* must approximate to a year, but be significantly longer or shorter than a year; (c) the cycle should be *entrained* by environmental cues; (d) temperature is not expected to affect the rhythm; and (e) the rhythm should display a phase–response curve (the phase of the rhythm exposed to the photoperiod).

Unfortunately, most of the earlier studies on fish do not fit such conditions for an endogenously-based rhythm, because only part of these "criteria" have been investigated in each case. Sundararaj *et al.* (1982) is one of the few to show evidence that females of the tropical catfish *Heteropneustes fossilis* have an endogenous self-sustained reproductive annual cycle. In rainbow trout, a non-tropical fish, Randall *et al.* (1999; cited by Bromage *et al.*, 2001) has shown seasonal phase–response curve in the spawning time, corroborating participation of endogenous timing systems in this circannual process.

Although identification of the environmental cues is quite difficult to be achieved, the circannual rhythm is still clear. In the tropical freshwater Asian catfish *Clarias macrocephalus*, Tan-Fermin *et al.* (1997) studied interactive effects between period of the year and hormonal induction of reproduction. They tested luteinizing hormone-releasing analog (LHRHa) in combination with pimozide on initial egg size, ovulation rate, egg production, fertilization, hatching and survival rates of the larvae upon yolk resorption. This study was also characterized by using fish from a same batch and age group. The spawning was induced during the off-season (February), before (May), at the peak (August), and the end (November) of the natural spawning period. This experimental design has the size of the fish (increased during the year) as an additional factor acting together with the period of the year, thus permitting misinterpretation of the data. However, the results did not correlate with female size, thus validating the conclusions in terms of effect of the period of the year. These authors found that all the parameters investigated were high at the pre- and peak breeding months (May and August) and lower during the off-season; some were increased only in November. This clearly shows a circannual responsiveness fluctuation in reproduction in this species, while the specific cues *entraining* such a cycle is still not known.

In tropical Brazilian species, the situation is not so different. The periods of reproduction are known for several species, as listed in Table 4.1. Fishermen know, however, that these periods may change somewhat according to the climate history in the previous months. It is observed in practical activities that warmer years shorten the reproductive cycle, thus postulating a clear environmental modulation in a presumably self-sustained rhythm.

In mammals which breed only in certain seasons, the photoperiod is the main cue. It is perceived visually and conducted via a neural pathway to the

Table 4.1

Months of Reproduction for Some Freshwater Migratory Fish Species in Central and
Southeastern Brazil

Family	Species	Months of reproduction in nature
Prochilodontidae	*Prochilodus* spp.	Nov–Apr
Anastomidae	*Leporinus* sp.	Oct–Feb
Pimelodidae	*Pseudoplatystoma corruscans*	Dec–Mar
	Pseudoplatystoma fasciatum	Dec–Mar
Characidae	*Brycon orbygnianus*	Nov–Jan
	Brycon cephallus	Nov–Jan
	Salminus maxillosus	Dec–Jan
	Piaractus mesopotamicus	Oct–Feb

The seasons are concentrated in the following months: Jun–Sep (winter), Sep–Dec (Spring), Dec–Mar (Summer), and Mar–Jun (Fall).

pineal gland in the brain. This gland secretes melatonin in a daily rhythm controlling reproduction. A similar melatonin rhythm occurs in fishes, which is also under photoperiod control. This establishes a link between melatonin and time of reproduction.

Synchronization of spawning with lunar cycles is also a matter of interest. It is common in many marine teleosts, especially in the tropical regions (Schwassman, 1971; Johannes, 1978; Taylor, 1984; Rahman *et al.*, 2000). The seagrass rabbitfish *Siganus canaliculatus* lives in different regions of the Pacific Ocean and shows a circannual reproductive cycle. During the reproduction period, however, the gonadosomatic index and serum vitellogenin levels showed peaks at around the time of new moon and the waning moon, respectively. Rahman *et al.* (2000) reported rhythmical changes in reproductive hormones and suggested that lunar periodicity is the main factor synchronizing testicular activity. Hoque *et al.* (1999) have also shown synchronization between lunar cycle and reproduction (gonadosomatic index and serum vitellogenin with peaks at around the new moon and the waning moon, respectively) in marine fish.

Annual rhythm of spawning in fish demands the availability of market-sized fish for certain times of year. Techniques of manipulating the photoperiod, rains, temperature and hormones have been used by a number of fishermen to determine the desirable times for supplying young fish to industry. As more understanding is available about how timing systems control reproduction, better ways of commercially raising fish will be developed.

The most important Brazilian fishes for food are those which migrate several hundred or thousand kilometers before reproduction, such as curimbata (*Prochilodus scrofa*), pacu (*Piaractus mesopotamicus*), tambaqui (*Colossoma mcropomum*), dourado (*Saliminus maxilosus*), matrinxã (*Brycon cephalus*), piracanjuva (*Brycon orbygnianus*), piau (*Leporinus fasciatus*), and the catfishes pintado (*Pseudoplathystoma coruscans*), cachara (*Pseudoplatystoma fasciatum*), jaú (*Pauliceia luetkeni*), and jurupensen (*Sorubim lima*). All these fishes are of great economic interest in this country because each may reach about 1 kg in one year and tens of kilos over years.

Species that depend on long distance migration before reproduction in nature represent more difficulty in terms of induced spawning in captivity because migration is crucial for gonad development. Thus, artificial induction of spawning in such tropical fishes is necessary. The method for inducing spawning by pituitary hormone treatment was developed in Brazil in the early 1930s by Rodolph von Ihering in the northeastern region close to the Equator. Nowadays this method using extract of carp pituitary gland to induce spawning is widespread among Brazilian fisheries and also globally. It consists of a double injection of this extract (prepared in saline) in females, with an interval of about 6 to 12 h between them, and only one dose in the males (coincident with the second dose for the females). Some hours after the second hormonal dose, the fish usually are handled (the abdomen compressed) for inducing egg or sperm releases and fecundation occurs in a small receptacle before hydration of the eggs.

An intriguing fact about hormonally induced spawning of these tropical species is how to determine the time to start extrusion (pressuring the abdomen to force the fish to release the gametes). Fishermen calculate the so-called "degree-hour", which is the sum of the mean water temperature of each hour after injection of the second dose of the pituitary extract. Table 4.2 shows the "degree-hour" expected for each species, as depicted by Ceccarelli *et al.* (2000). After this sum reaches the expected range of "degree-hour", extrusion is tried usually every 30 min until successful gamete release. Usually good results are obtained in the first or second trial, which shows the adequacy of this method. This clearly shows a temperature effect, rather than a time dependence, for the effectiveness of hormonally-induced spawning.

Some researchers believe that calculations of "degree-hour" for the period of the year (or some months) previous to the reproductive season is the best indicator of the period these tropical species will reproduce. This idea agrees with the variation in the reproductive period described in Table 4.1. In fact, data from this table sum observations of the same species in different regions in Brazil, which represents variations of about 25° in latitude. Thus, daily variations of temperature may range from a relatively wide variation in the southern region to only a few degrees over a day in the northern and northeastern regions (closer to the Equator).

Table 4.2

"Degree-hour" for Extrusion of Gametes during Artificially Induced
Reproduction in Some Brazilian Freshwater Species

Family	Species	°C	Degree-hour[a]
Prochilodontidae	*Prochilodus* spp.	23–25	190–240
Anastomidae	*Leporinus* sp.	23–25	210–220
Pimelodidae	*Pseudoplatystoma corruscans*	~ 24	~ 255
	Pseudoplatystoma fasciatum	~ 24	~ 255
Characidae	*Brycon orbygnianus*	~ 24	140–160
	Brycon cephallus	23–25	150–160
	Salminus maxillosus	23–25	130–150
	Piaractus mesopotamicus	~ 25	240–320
	Colossoma macropomum	~ 27	~ 290

"Degree-hour" is the sum of temperature values for each hour in a period. In this table, "degree-hour" refers to the period between the final dose of carp pituitary extract and time elapsed for extrusion of the gametes by gently compressing the fish's abdomen. Adapted from Woynarovich and Horváth (1980), Sallum and Cantelmo (1999), and Ceccarelli *et al.* (2000).

[a]These values are affected by fish size and treatment conditions. By increasing hormonal doses or frequency of doses, the "degree-hour" is decreased. In warmer water the "degree-hour" is also decreased. Shortening the time elapsed between the first and the second dose of the hormone decreases "degree-hour".

During artificial induction of reproduction in fisheries, after the second dose of the pituitary extract in the females (first in the males), both sexes are grouped in the same tank. This practice improves the success of reproduction. During male–female courtship, several kinds of stimuli are involved. Chemicals (usually pheromones), sounds, and visual displays are the most common, but their relative importance depends on the species considered. For instance, curimbata emits sound (audible even out of the water) soon before the emergence of spawning (Ceccarelli *et al.*, 2000). Motor displays for reproduction (mainly chasing and sideways tail beating) are the behavioral patterns most widespread among fishes. Other important cues for inducing or facilitating reproduction in fishes are the pheromones, or other kind of chemicals released by the consort (Liley, 1982). These innate mechanisms are complementary conditions that enable reproduction, thus aggregating the whole history of the fish, characterized by supposedly self-sustained oscillators and environmental cues that modulate their reproductively ability.

Handling stressor conditions of the fish during hormonal induction sessions has provided good results. Quiet environment, darker and better water quality, careful handling of each fish are among the most frequent form of care. Environmental color manipulation is an incipient alternative which may give good conditions for better fish reproduction. Volpato (2000) tested the effect of environmental color on the period between the first and the second injection of pituitary extract in the Amazonian matrinxã, *Brycon cephalus*. After matching the fish for similar external indicators of gonadal development (abdomen size and facility to release gametes after lightly compressing the abdomen), half of them were maintained in a tank covered with green cellophane and the remainder under white light. This author found spawning was successfully induced in 8 out of 9 females under the green color and only 4 out of 9 in the control females. Males released sperm more abundantly under green color than the controls. Volpato *et al.* (2004) showed that Nile tilapia reproduction was more frequent and intense in the presence of blue light ($\sim$100–120 Lux). These results are very suggestive that environmental color modulates reproduction in fish. Reinforcing this idea, Volpato and Barreto (2001) showed in Nile tilapia that blue environmental color abolishes the characteristic cortisol increase in response to a stressor, an effect not related to light intensity. While reproductive circannual rhythm may be controlled by internal time systems, short-term adjustments are provided by environmental factors setting ahead – or back – the exact time of spawning.

Regarding non-annual reproductive fishes, the Cichlidae merit special interest. The Nile tilapia, *Oreochromis niloticus*, is known by its high reproductive capacity. After hatching, the larvae grow rapidly and reach maturity within 3 months. Despite that, spawning pauses by the wintertime (Rothbard, 1979). More recently, Gonçalves-de-Freitas and Nishida (1998) described spawning of this species concentrated during the afternoon, but a clear circadian endogenous rhythmicity was not tested. In two other Cichlidae, the convict *Cichlasoma nigrofasciatum* and the rainbow *Herotilapia multispinosa*, Reebs and Colgan (1991) found increased fanning activity at night, but were not able to show a self-sustained nature of this rhythm. During the night, the dissolved oxygen is decreased in the water mainly because of respiration and lack of photosynthesis by aquatic plants (Reebs *et al.*, 1984), so that increased fanning at night may represent, at least, an adjustment increasing conditions for egg respiration.

VI. MIGRATION

Some species of fish make an exhausting yearly journey from downstream back upstream where they were born. Once these so-called anadromic fishes reach their destinations, they are able to reproduce. Conversely, catadromic fishes migrate downstream. Biological significance of migration, however, is not exclusively related with reproduction induction. Ramirez-Gil *et al.* (1998) studied the Amazonian Pimelodidae *Callophysus macropterus*, a migratory fish, and suggest that migration helps genetic flow, thus explaining the genetic similarity between individuals of this species caught from two distant places.

Annual reproductive migrations, however, are restricted to certain seasons and months (Table 4.1, section V, shows these periods for some Brazilian species). The ecological significance of such a time concentration is evident as it assures a better period for feeding and developing larvae and fries, but also guarantees that both sexes attain maturity at the same time (Schwassmann, 1971).

How do these fish know it is time to migrate; that is, what are the environmental cues entraining migration? What are the cues driving the fish to the right place? These are fundamental questions about migration still unsolved for most migratory species. Studies on salmon and other temperate fishes have indicated that the *entrainment* of this circannual cycle is determined by photoperiod or the daily amount of light. However, some controversy still exists, because in other species (including the tropical ones) *free-running* of these rhythms was not tested (Bromage *et al.*, 2001). The environmental factors driving a salmon to its home have also been the subject of several pages in migration writings.

Female rainbow trout maintained under a constant schedule of light and dark, constant temperature, and constant feeding rate for 4 to 5 years spawn on a cycle ranging from 11 to 15 months. This shows a rhythm of spawning although external time cues (such as longer days or warmer weather) were absent, which indicates that such a rhythm is under the control of endogenous rhythms. When artificial light was used, thus expanding or compressing the daylight period (compared with the natural seasonal cycle), time of spawning was delayed or advanced, respectively. A similar control is supposed to exist in tropical fishes, although the environmental cues may be quite different from those for temperate species.

While many studies are dedicated to temperate fish migration, very little is known about the tropical ones. The importance of chemical cues directing salmon back home, postulated by Buckland in 1880 (according to Hara, 1986), is almost certain (Hara, 1970, 1986; Cooper and Hirsch, 1982), while the factors conducting *Piaractus mesopotamicus* (pacu), *Pseudoplatystoma coruscans* (pintado), *Prochilodus scrofa* (curimbatá), *Salminus maxilosus* (dourado), and many other South American migrating fishes are still inferences from studies on salmon species.

As the tropical rivers close to the Equator are under quite constant photoperiod and temperature throughout the year, other cues are more likely to govern migration in this region. Precipitation changes are very marked in these places, like the Amazon region, and are supposed to be involved. Water-current-directed swimming is also expected because much of the migration is spent in the same direction in the same river. Winemiller and Jepsen (1998) summarize important literature on effects of season on migration in tropical rivers emphasizing the effects on food webs.

While migrating, the fish are under extreme conditions and thus several morphological, biochemical and physiological changes may occur (Farrell *et al.*, 1991; Leonard and McCormick, 1999). As migration is a rhythmic activity (usually annual), these other changes should not be attributed to any timing system, but to a passive consequence of a biological timing system controlling migration; and this behavior imposes such biological modifications.

River damming for electricity is a widespread problem for migrating fishes, usually preventing upstream migration (Pringle, 1997; Fièvet *et al.*, 2001a, b). Alternative methods have employed three main techniques to overcome this barrier: stairs, elevators, and replacement stocks. The stairs are constructed with large steps (about 8×5 m), each about 0.8 m high. An entire stair may reach 500 m or more length. Water flows from the upper reservoir down the steps to the river. This water flow is sufficiently intense to attract the fish in the lowest level (river), which are in migratory behavior, swimming upstream. These fish jump when they find the first step and then successively upward to the end of the stairs. Another technique uses elevators. Near the river surface, the fish are attracted to the door of the elevator (a large tank) by water flow from strong submerged pumps. As they reach the elevator, the door is closed and the entire tank is moved up to the reservoir.

Although these two methods are still effective to transport the fish and thus solve the problem imposed by the river damming, some consequences on the fish population may occur. The stair technique will allow only some fish to arrive upstream, a real artificial selective pressure. What is the profile of the fish downstairs? Is it similar to those that reach upstairs and thus can continue the migratory journey? In western São Paulo, Brazil, a research project sponsored by the Energy Company of São Paulo (CESP) has shown that the fish that are able to reach upstairs are statistically longer and heavier than those downstairs (Volpato, unpublished). Such a size selection, however, is unlike to occur when fish are transported by the elevator, but this needs be tested.

The third way to manage this problem of blocking upstream migration is required by Brazilian law and consists of culturing migratory species of the respective rivers and releasing the juveniles (total length about 10 cm) above the dam. The major concern in this method is to be certain that the released fish are still growing up and reproducing. Addressing this question, 300 000 fries of pacu (*Piaractus mesopotamicus*) marked with oxytetracycline were released in the reservoir of the Jupiá electricity company, in western São

Paulo state in Brazil (CESP and Volpato, unpublished). After 7 months, capture of this species was started to evaluate the proportion of marked and non-marked captured fish. Initial results showed that some marked fish could be recaptured. As this reservoir is about $544 \, \text{km}^2$, inferences about the whole population are still premature because capture was concentrated partially in some areas of this large reservoir and the mortality rate is unknown. Pacu is also a schooling fish, thus making this inference more difficult. Nonetheless, the marked fish captured confirm that the released fish are able to survive in the natural environment. This method to overcome the problems imposed by the dams in upstream migration must be carefully considered because the genetic structure of the cultured population must be accurately inspected to avoid drastic changes in the natural population.

VII. EVOLUTION OF CIRCADIAN RHYTHMICITY AND CAVE FISHES

The subterranean, or hypogean, realm comprises the network of interconnected subsoil spaces, with variable sizes from microvoids to large spaces accessible to humans (caves), filled with water or air, and which may develop in different kinds of rocks, but mainly in karst areas with outcrops of soluble rocks, specially limestones. This domain of the biosphere comprises habitats varying from interconnected crevices, fissures and caves, lava tubes, interstitial habitats, phreatic, alluvial and perched aquifers, among others (Juberthie, 2001). It is noteworthy that "cave" corresponds to an anthropocentric concept, linked to human size and locomotor ability – as a matter of fact, caves in general are inserted into a continuum of smaller and larger spaces where many smaller vertebrates and invertebrates may freely transit.

Due to the limited contact with the surface (epigean) environment, where the great majority of species live, hypogean habitats are generally characterized by permanent darkness, and thus absence of photoperiods, which is their most relevant feature from the biological point of view and more readily associated to caves. Likewise, these habitats are characterized by the tendency towards environmental stability. As a consequence of the insulating properties of soil and subsoil, daily variations in temperature are minimized in subterranean cavities. Therefore, air and water temperatures in such spaces tend to equal the mean annual temperature in the epigean environment, unless the existence, number and position of large openings allow for important exchanges with the surface. Due to the thermal inertia of water, epigean streams sinking into the ground are another source of daily variations in subterranean temperatures. Thus, important *zeitgebers* for epigean species, especially light–dark cycles and, less frequently, daily cycles of temperature, are absent in subterranean habitats.

On the other hand, many subterranean ecosystems are subject to more or less pronounced seasonality due to rain cycles in the surface. Cyclical increases in food availability – organic matter washed into caves and temperature fluctuations induced by floods – are more important for aquatic organisms, and are potential *zeitgebers* for subterranean organisms in the annual or semiannual range. Moreover, daily variations in food availability may also occur as a consequence of cyclical guano deposition by animals regularly found in hypogean habitats (trogloxenes, see below) and which leave caves every day to forage, such as bats and some echolocating birds. This is a possible *zeitgeber* operating in the circadian range.

Organisms regularly found in the subterranean environment, i.e., organisms that are not there by accident and for which this is part of (or all) the habitat naturally occupied, are classified into: **trogloxenes**, organisms habitually found in caves but which must return periodically to the surface in order to complete their life cycle; **troglophiles**, facultative subterranean species, able to complete their life cycle both in hypogean and in epigean habitats; and **troglobites**, species restricted to subterranean habitats (Holsinger and Culver, 1988). As a consequence of genetic isolation, troglobites may develop a series of autapomorphies (exclusive character states) related to the hypogean life (troglomorphisms). Most troglomorphisms are related to the absence of light: structures and behaviors that become functionless under this condition (e.g., visual organs, melanic pigments) may regress due to accumulation of neutral mutations, selection for energy economy or pleiotropic effects (Culver and Wilkens, 2001). Thus, troglobites in general may be distinguished by some degree of eyes and pigmentation reduction.

Daily patterns of activity are one of the most evident chronobiological behavioral characteristics of many organisms, from protists to animals and plants, along with seasonal reproductive patterns. Thus, it is not surprising that they constitute a main point of debate, raising many questions regarding their evolutionary origin and function in living organisms, especially in the case of circadian rhythms (there is a general agreement that reproductive cycles, when they occur, are a response to the seasonal availability of nutrients necessary for the extra costs of reproduction, although the alternative hypothesis of anticipation of favorable surviving conditions for the offspring should also be considered).

Basically, there are two main evolutionary hypotheses regarding the factors involved in the appearance and maintenance of circadian rhythms – those of internal versus external, ecological selection. According to the first one, circadian rhythms are important to help assure an adequate sequence of metabolic reactions, which could then be distributed to different phases of an oscillation (maintenance of the Internal Temporal Organization). Ecological factors include the advantage of adapting functions (locomotion, feeding, etc.) to daily changes in the environment; concentrating or dislocating interactions among individuals (mating, competition, predation); allowing for the measurement of the length of the day, necessary to adjust functions such as seasonal reproduction (Lamprecht and Weber, 1992).

Troglobitic species, especially those evolving in very constant environments, provide good opportunities to test these hypotheses. In many cases, such species evolved during generations in the absence of 24-hour *zeitgebers*, and it is predicted that, if external factors provide the main selective forces for circadian rhythms, these could regress to some degree in troglobites, as observed for eyes, melanic pigmentation, and other characteristics related to light. On the other hand, if internal order were of prime importance for the maintenance of circadian rhythms, these would not be lost in troglobites.

Several studies pointed to the loss or weakening of circadian components of activity in troglobites as diverse as beetles, crustaceans, and fishes, suggesting that such rhythmicity is not necessary for the maintenance of internal temporal order (Lamprecht and Weber, 1992). However, different methods of data analysis were used in these studies, including visual examination of actograms, which may not be reliable enough to detect oscillations. In order to support such generalization, more studies are needed encompassing a larger number of unrelated troglobitic species, and fishes are a good material for this in view of their size and relatively easy maintenance in laboratory.

More than 100 species of troglobitic fishes are known, from all continents except Europe, and also in many islands (e.g., Cuba, Madagascar). The highest species richness is observed in China, Mexico, Brazil, and Southeast Asia. Most are siluriforms (several families in the Americas, Africa, a few in Asia) or cypriniforms, mostly cyprinids or balitorids (throughout tropical Asia, some cyprinids in Africa). Hence, these groups of freshwater fishes present a high potential for easy adaptation to subterranean life. On the other hand, characiforms, another important group of neotropical freshwater fishes, are in general poorly represented among the troglobitic fauna. Likewise, cichlids, another important group of freshwater fishes, have no troglobitic derivatives. Those are good examples of low potential for subterranean life (Trajano, 2001).

Preadaptation to hypogean life is clear for siluriforms, which are generally nocturnal, chemo-oriented fishes, mostly omnivores or generalist carnivores, but not so for cypriniforms which, like characiforms, include many diurnal, visually oriented species. For the latter, preadaptations for subterranean life must be sought among the closest epigean relatives, which could have retained the preadaptive character states shown by the troglobite's ancestors.

It is worthy noting that these species present different degrees of troglo-morphism, the more evident being the variable degrees of reduction of eyes and pigmentation, suggestive of different times of isolation in the subterra-nean habitat. Some species present only a slight reduction of eyes and pigmentation in relation to their epigean relatives, others exhibit a consider-able individual variability, with populations encompassing from individuals with reduced but still visible eyes and pigmentation to those externally anophthalmic and depigmented; yet others are homogeneously anophthal-mic and depigmented. Because such regressions seem to be mostly gradual, it is assumed that the degree of regressive troglomorphisms provides a crude measurement of time in isolation. Thus, variable populations are considered recent troglobites and those with advanced regressive character states ob-served throughout the population would be ancient troglobites. However, ability to perceive and react to light was demonstrated for many species, including even some of most advanced troglobites, apparently through extra-ocular, extra-pineal receptors (Langecker, 1992).

In spite of the relative richness, especially in tropical areas, and scientific interest for chronobiological studies, few troglobitic fishes have been inves-tigated in detail with a focus on rhythmicity: the blind Mexican tetras, genus *Astyanax* (Characiformes: Characidae); *Nemacheilus evezardi* (= *Oreonectes evezardi*) (Cypriniformes: Balitoridae), from India; and catfish species (Siluriformes), from Brazil.

Mexican tetra characins, with 29 populations showing different degrees of troglomorphism in caves from the Huastecan Province, constitute a rare example of troglobitic characiform fishes. An explanation for this relies on the unusual features of their putative epigean ancestor, *Astyanax mexicanus* (= *A. fasciatus*), which, unlike most *Astyanax* species, would present pre-adaptations to the cave life such as crepuscular activity, ability to feed in darkness, and a chemically stimulated spawning behavior (Wilkens, 1988). The Mexican cave tetras are by far the most intensively studied troglobitic fishes, with hundreds of publications (almost 200 up to the mid-70s, Mitchell *et al.*, 1977). Because these cave populations may introgress with epigean, eyed, and pigmented *Astyanax mexicanus*, producing fertile hybrids, empha-sis has been given to genetic studies in the laboratory, including behavioral aspects. Nevertheless, not much has been done in the chronobiological field, with relatively few publications on circadian rhythmicity of locomotor ac-tivity (Thinès *et al.*, 1965; Erckens and Weber, 1976; Thinès and Weyers, 1978; Erckens and Martin, 1982a, b; Cordiner and Morgan, 1987).

These studies, performed under *free-running* conditions (DD) and 24-h light–dark cycles (LD) of different phase lengths, allow for a comparison between eyed, epigean fishes (*Astyanax mexicanus*) and two (probably) different troglobitic populations, a specialized, totally anophthalmic one,

"*A. antrobius*" from El Pachon cave, and another probably from La Chica cave (fish from commercial stocks). The latter, "*A. jordani*," is characterized by the presence of intermediate fenotypes as regards to development of eyes and pigmentation, and it is considered a hybrid population resulting from the introgression of epigean genotypes (epigean fishes entering the cave periodically) into an already established troglobitic population. The great majority of fishes in commercial stocks came from this population.

In *Astyanax mexicanus*, free-running activity rhythms were detected, as expected for an epigean species. All applied LD cycles (12:12 h, 6:6 h, 4:4 h, 16:8 h etc.) acted as *zeitgebers*, entraining the locomotor activity, with no need of a swing-in time to become entrained when starting an LD. Furthermore, in nearly all applied LDs a non-synchronized circadian rhythm was observed in addition to the dominant entrained frequency. Residual oscillations (post-oscillations) were observed after the transition from LD to DD, during one or a few cycles. These results suggest the existence of an endogenous circadian oscillator, whose effects are overlapped under forcing conditions (masking), but which becomes obvious in free-running conditions. The passive system has a nearly unlimited range of response (Erckens and Martin, 1982a). On the other hand, in the phylogenetically old "*A. antrobius*" (the specific name has been questioned), although activity was entrained by all applied LDs, the signal energies were lower than in the tests with epigean fishes, the rhythms of total activity disappeared immediately after the transition from LD to DD (no residual oscillations), and in no LD with a period frequency differing from 24 h a circadian rhythm could be observed in addition to the entrained frequency. The activity responses to changing environmental conditions were not as uniformly quick as in *A. mexicanicus*, but the system hardly needed a swing-in time to be synchronized by the imposed LD. The authors concluded that the internal clock of "*A. antrobius*" was simplified in relation to its epigean ancestor: the passive system has developed into an extremely passive one, incapable of synchronizing, thus the circadian oscillator was subject to regression, but it was not completely lost (Erckens and Martin, 1982b).

"*Astyanax jordani*" from La Chica seems to be intermediate also in this aspect, because one or two residual oscillations were observed after a transition from LD (12:12) to DD (Erckens and Weber, 1976). The persistence of a circadian clock in this cave fish was pointed out by Cordiner and Morgan (1987), who also studied fish from a commercial stock. These authors recorded free-running circadian activity rhythms (not shown by Erckens and Weber, 1976), often masked by apparently random, infradian oscillations.

It is interesting to note that circadian differences between surface and bottom activity were observed for both epigean and cave fishes. In both

A. mexicanus and "*A. antrobius*" the maximum surface activity was observed during the dark phases of a LD cycle, and the bottom activity in the light phases. Moreover, in the latter *free-running* rhythms were detected in surface activity after the LD (12:12) to DD transition, but not for bottom and total activity (Erckens and Martin, 1982a,b). Cordiner and Morgan (1987) observed that cave fish probably from La Chica spend less time in the upper level of the tank during the light phase than in the dark, when locomotor activity is more evenly distributed.

Four Brazilian catfish species had their locomotor rhythmicity investigated: the trichomycterid *Trichomycterus itacarambiensis*, and the heptapterines *Pimelodella kronei* (see details of the head of this species in Figure 4.2), *Taunayia* sp. and an undescribed species of a new genus from Chapada Diamantina, NE Brazil (Trajano and Menna-Barreto, 1995, 1996, 2000). The former two present a considerable degree of individual variation in development of eyes and pigmentation (including one-third of the population as true albinos in the latter – see Trajano and Pinna, 1996), suggestive of a shorter time of isolation in the subterranean habitat (recent troglobites), whereas the two latter are homogeneously anophthalmic and depigmented, being considered ancient troglobites. All these troglobitic species belong

Fig. 4.2 Head of the blind catfish *Pimelodella kronei*, a troglobitic heptapterid from caves in the Ribeira Valley karst area, southeastern Brazil. (Photo: José Sabino.)

to typically nocturnal taxa and, when known and available, the epigean sister-species were also studied for comparison.

A high degree of individual variation concerning the presence of *free-running* circadian rhythms and the frequency and periodicity of other rhythms was reported for these Brazilian catfishes. This is in accordance with the variability observed for other characters. The less specialized troglobite *P. kronei* presented the higher proportion of specimens showing significant circadian rhythms (7 out of 9 studied catfishes) as well as the higher number of ultradian and infradian rhythms, superimposed to the circadian ones. This species was followed by *T. itacarambiensis*, apparently also a recent troglobite, and the new heptapterid from Chapada Diamantina, a morphologically specialized troglobite, each one with three out of six studied catfishes showing significant circadian rhythms, including both pigmented and albino specimens in *T. itacarambiensis* (Trajano and Menna-Barreto, 1995, 1996). However, the average number of rhythms with other periodicities was higher in the latter, as expected based on its cave-related morphology. The other highly morphologically specialized troglobite, *Taunayia* sp., presented the weaker rhythmicity among these species: none of the three studied specimens exhibited free-running circadian rhythms, and two were completely arrhythmic, not even showing ultradian rhythms, the only case among the studied Brazilian catfishes. On the other hand, all the studied epigean catfishes, *P. transitoria* (seven specimens) and *Taunayia bifasciata* (two specimens), exhibited strong, significant *free-running* circadian components of locomotor activity. As in cave *Astyanax*, there is evidence of differences in temporal patterns of surface and bottom activity in *Taunayia* sp.: under *free-running* conditions, one studied specimen exhibited a significant circadian rhythm in surface activity, but not in bottom and total activity (Trajano and Menna-Barreto, 2000).

The individual variability observed in the *free-running* circadian rhythmicity of *P. kronei*, *T. itacarambiensis*, and the new heptapterid from Chapada Diamantina may be due to one or more of the following factors: (1) relatively short time in isolation in the subterranean habitat, insufficiently long to genetically fix modifications in time-control mechanisms throughout the populations (for *P. kronei* and *T. itacarambiensis*); (2) *zeitgebers* are nowadays acting over part of the studied populations (temperature cycles for *P. kronei*, LD cycles for the new heptapterid from Chapada Diamantina); (3) circadian rhythms are selected for among catfishes that live near the entrances, in order to prevent those individuals which leave the cave during the night to be overtaken outside by the daylight, when they would be more vulnerable to predators and climatic fluctuations (for *P. kronei* and, possibly, *T. itacarambiensis*) (Trajano and Menna-Barreto, 1995, 1996).

Taunayia sp. and the new heptapterid from Chapada Diamantina were also studied under LD cycles, 12:12 h (Trajano and Menna-Barreto, 2000; Trajano *et al.*, 2001). In both species, activity was entrained by these cycles, but no residual oscillations were observed, indicating a possible masking effect (Figure 4.3).

The comparison between troglobitic fishes and their epigean close relatives (*Astyanax antrobius* × *A. mexicanus*; *Pimelodella kronei* × *P. transitoria*; *Taunayia* sp. × *T. bifasciata*), analyzed under the same conditions and using the same protocols, provides good evidence for the hypothesis of an evolutionary regression of time-control mechanisms in troglobitic species, either affecting the oscillator(s) itself (themselves) or due to an uncoupling between the oscillators and at least one of their related functions – in this case, the locomotor activity. Regression of retina and, at least for some of the studied species, possibly also of the photoreceptors of the pineal organ, where circadian oscillators of fishes would be located (see above), may be involved in the disorganization of the circadian system verified in several troglobitic species.

Therefore, data on cave fishes favor the notion of external, ecological factors as the main factors stabilizing selection for circadian rhythms. The progressive reduction of locomotor rhythmicity in different catfish species in parallel with the reduction of eyes, pigmentation, and other characters indicates that similar processes may be involved in such regression. As in

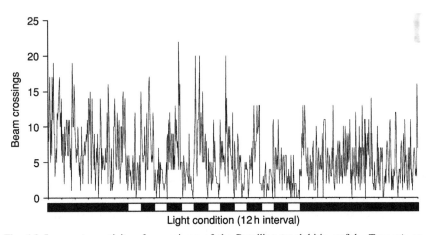

Fig. 4.3 Locomotor activity of a specimen of the Brazilian troglobitic catfish, *Taunayia* sp. (Siluriformes: Heptapteridae), recorded over 14 consecutive days under DD (3 days), LD (7 days), and DD (4 days). Activity = IR beam crossings totaled every 30 minutes from 6 photocells. (From Trajano and Menna-Barreto, 2000, reproduced with permission.)

the case with eyes, melanic pigmentation, and other light-related characters that become functionally neutral in the perpetually dark subterranean environment, the regression of time-control mechanisms of locomotor activity (and possibly other circadian functions as well) evidenced for several troglobitic species may be due to the accumulation of deleterious mutations (Culver and Wilkens, 2001) affecting those mechanisms. It is hypothesized that in the absence of stabilizing selection eliminating such mutations in subterranean animals as it does for epigean species, the ecologically selected circadian rhythms present in the epigean ancestors may be lost by troglobites.

No particular ultradian or infradian rhythms were consistently found throughout the studied Brazilian specimens. These fishes showed a great variation concerning the number and periodicity of non-circadian components of locomotor activity, both in epigean and troglobitic specimens. This makes the biological significance of such rhythms difficult to interpret.

The Indian balitorid *Nemacheilus evezardi* was subject to several studies on circadian and circannual rhythmicity, with a focus on locomotor activity, air-gulping behavior, and phototactic behavior (Pradhan *et al.*, 1989; Biswas *et al.*, 1990a, b; Biswas, 1991). In these publications, the name *N. evezardi* is used for both epigean and subterranean populations, although the latter present clear-cut specialization to the hypogean life (e.g., reduction of eyes and pigmentation, air-gulping behavior) not present in the epigean ones, which *per se* would justify the status of a separate species. The epigean *N. evezardi* is basically a bottom-dweller that shows schooling behavior (not observed in the cave loaches, that are not gregarious – Pradhan *et al.*, 1989). The epigean loaches are dusk-active, with maximum locomotor activity at the early part of the dark phase, concealing themselves under stones during most of the day (Biswas, 1991). This is another example of epigean fish preadapted to the subterranean life.

Differences in habitat account for the occurrence of air-gulping behavior (surfacing activity related to extra oxygen uptake) only in hypogean *N. evezardi*. In contrast with epigean populations that live in well-oxygenated hill streams, the cave population inhabits small pools subject to lowered oxygen concentrations during the dry season. Significant circadian and circannual rhythms were detected for the air-gulping activity under free-running conditions (Biswas *et al.*, 1990a). Semicircannual or circannual modulation of the circadian rhythms in surfacing activity may be a response to the rain regime leading to seasonal fluctuations in water oxygen concentrations. However, the authors could not identify *zeitgebers* synchronizing the surfacing activity inside the cave at the daily scale. Considering that surfacing is probably an expression of general locomotor activity, and that the hypogean *N. evezardi* is clearly a recent, little modified troglobite, it is possible that circadian locomotor rhythms recorded for the cave loaches represent a relictual, plesiomorphic trait retained from their epigean ancestors.

Biswas *et al.* (1991b) studied the locomotor activity and surfacing frequency during the pre-spawning, spawning and post-spawning phases, comparing epigean (kept under natural LD cycles) and hypogean individuals (kept under DD). Significant rhythms were detected for total activity in both epigean and hypogean loaches during the pre-spawning phase, but only for the epigean fishes during the spawning and post-spawning phases. Significant circadian rhythms of surfacing behavior related to air-gulping were observed during the pre-spawning and spawning phases only. In hypogean loaches, the levels of total and surfacing activity were lower in the pre-spawning and spawning phases, suddenly increasing during the post-spawning phase, whereas no change in total activity was noticed for the epigean loaches. These results demonstrate the influence of the reproductive condition over the expression of circadian rhythmicity, which differs among cave and epigean populations.

Finally, a probable example of behavioral trait in troglobitic fishes subject to the influence of *zeitgebers* other than LD cycles (daily temperature cycles, in the case) is provided by the armored catfish, *Ancistrus cryptophthalmus* (Loricariidae), from the State of Goiás, Central Brazil. Part of the population found in Angélica Cave lives in the aphotic zone, but not far from the cave sinkhole (input of an epigean river). We observed a daily variation in the number of catfishes exposed on the rocky substrate, which is higher during the morning and conspicuously decreases in the afternoon, possibly as a response to an increase of 1 °C in water temperature observed along a period of 4 hours (from 10:00 to 14:00 h) (E. Bessa and E. Trajano, personal observation). This may be a consequence of a daily fluctuation in hiding habits, which would be synchronized by the 24-hour cycle of environmental temperature.

VIII. FUTURE DIRECTIONS

Despite the extensive literature on fish biological rhythms, tropical fishes have been much less studied on this aspect. The very expressive number of tropical fish species provides a useful biological material for testing hypotheses on the nature, mechanisms, and significance of the internal synchronizers. Moreover, as the tropical area provides some less evident *zeitgebers*, the passive rhythms (masking effect) are also of great importance. Specially for fish of economic interest, comprehension of the environmental factors modulating reproduction is still necessary and provides a developing area of study. In this respect, factors of migration acting on gonadal development and effects of environmental colors on behavior and physiology of tropical fishes still deserve much more attention.

The boom of molecular biology and genetics has impressed many biologists, and thus chronobiologists. Indeed, this is an exciting area providing very specific information mainly on the genetic control of biological timing systems. Indeed, a precise understanding of the molecular mechanisms controlling the internal synchronizers may be a key-point for developing technologies based on biological rhythms. However, for a complete understanding of the global phenomenon of the biological rhythms a holistic view cannot be overlooked, and thus ecological and behavioral studies are still of great interest.

ACKNOWLEDGMENTS

We are grateful to Dr. Luiz Menna Barreto, from the Grupo Interdisciplinar de Desenvolvimento e Ritmos Biológicos – ICB/USP, for the critical reading of the manuscript. The authors are partially supported by Conselho Nacional de Desenvolvimento Científico e Tecnológico (CNPq; G.L. Volpato No. 300644-86-8; E. Trajano No. 306066-88-2). The first author's research is also partially supported by Companhia Energética de São Paulo – CESP. Permission to publish Figure 4.3, extracted from Trajano and Menna-Barreto (2000), was granted by the copyright holder, Swets & Zeitlinger Publisher.

REFERENCES

Ali, M. A. (1992)."Rhythms in Fishes." Plenum Press, New York.
Alvarenga, C. M. D., and Volpato, G. L. (1995). Agonistic profile and metabolism in alevins of the Nile tilapia. *Physiol. Behav.* **57,** 75–80.
Aschoff, J. (1981a). Freerunning and entrained circadian rhythms. *In* "Handbook of Behavioral Neurobiology; Biological Rhythms" (Aschoff, J., Ed.), Vol. 4, pp. 81–93. Plenum Press, New York.
Aschoff, J. (1981b)."Handbook of Behavioral Neurobiology; Biological Rhythms," Vol. 4. Plenum Press, New York.
Aschoff, J. (1990). From temperature regulation to rhythm research. *Chronobiol. Int.* **7,** 179–186.
Barki, A., and Volpato, G. L. (1998). Early social environment and the fighting behaviour of young *Oreochromis niloticus* (Pisces, Cichlidae). *Behaviour* **135,** 913–929.
Biswas, J. (1991). Annual modulation of diel activity rhythm of the dusk active loach *Nemacheilus evezardi*(Day): A correlation between day length and circadian parameters. *Proc. Indian Natl. Sci. Acad.* B57 **5,** 339–346.
Biswas, J., Pati, A. K., and Pradhan, R. K. (1990a). Circadian and circannual rhythms in air gulping behaviour of cave fish. *J. Interdiscipl. Cycle Res.* **21,** 257–268.
Biswas, J., Pati, A. K., Pradhan, R. K., and Kanoje, R. S. (1990b). Comparative aspects of reproductive phase dependent adjustments in behavioural circadian rhythms of epigean and hypogean fish. *Comp. Physiol. Ecol.* **15,** 134–139.
Bone, Q., and Marshall, N. B. (1982)."Biology of Fishes." Chapman & Hall, New York.
Boujard, T., and Leatherland, J. F. (1992). Circadian rhythms and feeding time in fishes. *Environm. Biol. Fish.* **35,** 109–131.

Brady, J. (1987). Circadian rhythms: Endogenous or exogenous. *J. Comp. Physiol.* **161A**, 711–714.

Brawn, V. M. (1961). Sound production by the cod (*Gadus callarias* L.). *Behaviour* **18**, 239–255.

Bromage, N., Porter, M., and Randall, C. (2001). The environmental regulation of maturation in farmed finfish with special reference to the hole of photoperiod and melatonin. *Aquaculture* **197**, 63–98.

Buck, S. M. C., and Sazima, I. (1995). An assemblage of mailed catfishes (Loricariidae) in southeastern Brazil: Distribution, activity, and feeding. *Ichthyol. Explor. Freshwaters* **6**, 325–332.

Bullock, T. H. (1969). Species differences in effect of electroreceptor input on electric organ pacemakers and other aspects of behavior in electric fish. *Brain Behav. Evolut.* **2**, 85–118.

Burrows, M. T., and Gibson, R. N. (1995). The effects of food, predation risk and endogenous rhythmicity on the behaviour of juvenile plaice, *Pleuronectes platessa* L. *Anim. Behav.* **50**, 41–52.

Carter, D. A., and Murphy, D. (1996). Circadian rhythms and autoregulatory transcription loops – going round in circles? *Mol Cell. Endocrinol.* **124**, 1–5.

Casatti, L., and Castro, R. M. C. (1998). A fish community of the São Francisco River headwaters riffles, southeastern Brazil. *Ichthyol. Explor. Freshwaters* **9**, 229–242.

Ceccarelli, P. S., Senhorini, J. A., and Volpato, G. L. (2000)."Dicas em Piscicultura." Santana, Botucatu.

Cloutier, R., and Ahlberg, P. E. (1996). Morphology, characters, and the interrelationships of basal sarcopterygians. *In* "Interrelationships of Fishes" (Stiassny, M. L. J., Parenti, L. R., and Johnson, C. D., Eds.), pp. 445–479. Academic Press, New York.

Coon, S. L., Bégay, V., Falcón, J., and Klein, D. C. (1998). Expression of melatonin synthesis genes is controlled by a circadian clock in the pike pineal organ but not in the trout. *Biol. Cell* **90**, 399–405.

Cooper, J. C., and Hirsch, P. J. (1982). The role of chemoreception in salmonid homing. *In* "Chemoreception in Fishes" (Hara, T. J., Ed.), pp. 343–362. Elsevier, Amsterdam.

Cordiner, S., and Morgan, E. (1987). An endogenous circadian rhythm in the swimming activity of the blind Mexican cave fish. *In* "Chronobiology and Chronomedicine" (Hildebrandt, G., Moog, R., and Raschke, F., Eds.), pp. 177–181. Verlag Peter Lang, Frankfurt.

Crim, L. W. (1982). Environmental modulation of annual and daily rhythms associated with reproduction in teleost fishes. *Can. J. Fish. Aquat. Sci.* **39**, 17–21.

Culver, D. C., and Wilkens, H. (2001). Critical review of the relevant theories of the evolution of subterranean animals. *In* "Ecosystems of the World 30. Subterranean Ecosystems" (Wilkens, H., Culver, D. C., and Humphreys, W. F., Eds.), pp. 381–398. Elsevier, Amsterdam.

Davis, R. E. (1962). Daily rhythm in the reaction of fish to light. *Science* **137**, 430–432.

Deng, T.-S., and Tseng, T.-C. (2000). Evidence of circadian rhythm of electric discharge in *Eigenmannia virescens* system. *Chronobiol. Int.* **17**, 43–48.

Ejike, C. B., and Schreck, C. B. (1980). Stress and social hierarchy rank in coho salmon. *T. Am. Fish. Soc.* **109**, 423–426.

Engler, C., Fogarty, C. M., Banks, J. R., and Zupanc, G. K. H. (2000). Spontaneous modulations of the electric organ discharge in the weakly electric fish, *Apteronotus leptorhynchus*: A biophysical and behavioral analysis. *J. Comp. Physiol. A* **186**, 645–660.

Erckens, W., and Martin, W. (1982a). Exogenous and endogenous control of swimming activity in *Astyanax mexicanus* (Characidae, Pisces) by direct light response and by a circadian oscillator. I. Analysis of the time-control systems of an epigean river population. *Z. Naturforsch.* **37c**, 1253–1265.

Erckens, W., and Martin, W. (1982b). Exogenous and endogenous control of swimming activity in *Astyanax mexicanus* (Characidae, Pisces) by direct light response and by a circadian oscillator. II. Features of time-controlled behaviour of a cave population and their comparison to a epigean ancestral form. *Z. Naturforsch.* **37 c,** 1266–1273.

Erckens, W., and Weber, F. (1976). Rudiments of an ability for time measurement in the cavernicolous fish *Anoptichthys jordani* Hubbs and Innes (Pisces, Characidae). *Experientia* **32,** 1297–1299.

Ercolini, A., and Berti, R. (1977). Morphology and response to light of *Uegitglanis zammaranoi* Gianferrari, anophthalmic phreatic fish from Somalia. *Monit. Zool. Ital.* Suppl. **9/8,** 183–199.

Ercolini, A., and Berti, R. (1978). Morphology and response to light of *Barbopsis devecchii* Di Caporiacco (Cyprinidae), microphthalmic phreatic fish from Somalia. *Monit. Zool. Ital.* Suppl. **10/15,** 299–314.

Eriksson, L.-O. (1978). Nocturnalism versus diurnalism – dualism within fish individuals. *In* "Rhythmic Activity of Fishes" (Thorpe, J. E., Ed.), pp. 69–89. Academic Press, New York.

van Esseveldt, K. E., Lehman, M. N., and Boer, G. J. (2000). The suprachiasmatic nucleus and the circadian time-keeping system revisited. *Brain Res. Rev.* **33,** 34–77.

Farrell, A. P., Johansen, J. A., and Suarez, R. K. (1991). Effects of exercise-training on cardiac performance and muscle enzymes in rainbow trout, *Oncorhynchus mykiss. Fish Physiol. Biochem.* **9,** 303–312.

Fernandes, M. O., and Volpato, G. L. (1993). Heterogeneous growth in the Nile tilapia: Social stress and carbohydrate metabolism. *Physiol. Behav.* **54,** 319–323.

Fièvet, E., Dolédec, S., and Lim, P. (2001a). Distribution of migratory fishes and shrimps along multivariate gradients in tropical island streams. *J. Fish Biol.* **59,** 390–402.

Fièvet, E., Tito de Morais, L., Tito de Morais, A., Monti, D., and Tachet, H. (2001b). Impacts of an irrigation and hydroelectric scheme in a stream with a high rate of diadromy: Can downstream alterations affect upstream faunal assemblages? *Arch. Hydrobiol.* **151,** 405–425.

Gerkema, M. P., Videler, J. J., Wiljes, J. de, van Lavieren, H., Gerritsen, H., and Karel, M. (2000). Photic entrainment of circadian activity patterns in the tropical labrid fish *Halichoeres chrysus. Chronobiol. Int.* **17,** 613–622.

Giaquinto, P. C., and Volpato, G. L. (1997). Chemical communication, aggression, and conspecific recognition in the fish Nile tilapia. *Physiol. Behav.* **62,** 1333–1338.

Giaquinto, P. C., and Volpato, G. L. (2001). Hunger suppresses the onset and the freezing component of the antipredator response to conspecific skin extract in pintado catfish. *Behaviour* **138,** 1205–1214.

Gonçalves-de-Freitas, E., and Nishida, S. M. (1998). Sneaking behavior of the Nile tilapia. *Bol. Téc. CEPTA* **11,** 71–79.

Haller, J. (1994). Biochemical costs of a three day long cohabitation in dominant and submissive male *Betta splendens. Aggressive Behav.* **20,** 369–378.

Haller, J., and Wittenberger, C. (1988). Biochemical energetics of hierarchy formation in *Betta splendens. Physiol. Behav.* **43,** 447–450.

Hara, T. J. (1970). An electrophysiological basis for olfactory discrimination in homing salmon; a review. *J. Fish. Res. Board Can.* **27,** 565–586.

Hara, T. J. (1986). Role of olfaction in fish behaviour. *In* "The Behaviour of Teleost Fishes" (Pitcher, T. J., Ed.), pp. 152–176. Croom Helm, London and Sydney.

Hawkins, A. D. (1986). Underwater sound and fish behaviour. *In* "The Behaviour of Teleost Fishes" (Pitcher, T. J., Ed.), pp. 114–151. Croom Helm, London and Sydney.

Helfman, G. S., Meyer, J. L., and McFarland, W. N. (1982). The ontogeny of twilight migration patterns in grunts (Pisces: Haemulidae). *Anim. Behav.* **30,** 317–326.

Hill, R. W. (1976)."Comparative Physiology of Animals; an Environmental Approach." HarperCollins, New York.

Holsinger, J. R., and Culver, D. C. (1988). The invertebrate cave fauna of Virginia and a part of Eastern Tennessee: Zoogeography and ecology. *Brimleyana* **14,** 1–162.

Hoque, M. M., Takemura, A., Matsuyama, M., Matsuura, S., and Takano, K. (1999). Lunar spawning in *Siganus canaliculatus*. *J. Fish Biol.* **55**, 1213–1222.

Hurd, M. W., Debruyne, J., Straume, M., and Cahill, G. (1998). Circadian rhythms of locomotor activity in zebrafish. *Physiol. Behav.* **65**, 465–472.

Iigo, M., and Tabata, M. (1996). Circadian rhythms of locomotor activity in the goldfish *Carassius auratus*. *Physiol. Behav.* **60**, 775–781.

Johannes, R. E. (1978). Reproductive strategies of coastal marine fishes in the tropics. *Environ. Biol. Fish.* **3**, 65–84.

Jordão, L. C., and Volpato, G. L. (2000). Chemical transfer of warning information in non-injured fish. *Behaviour* **137**, 681–690.

Juberthie, C. (2001). The diversity of the karstic and pseudokarstic hypogean habitats in the world. *In* "Ecosystems of the World 30. Subterranean Ecosystems" (Wilkens, H., Culver, D. C., and Humphreys, W. F., Eds.), pp. 17–39. Elsevier, Amsterdam.

Karplus, I. (1987). The association between gobiid fishes and burrowing alpheid shrimps. *Oceanogr. Mar. Biol. Ann. Rev.* **25**, 507–562.

Kavanau, J. L. (1998). Vertebrates that never sleep: Implications for sleep's basic function. *Brain Res. Bull* **46**, 269–279.

Kavanau, J. L. (2001). Brain-processing limitations and selective pressures for sleep, fish schooling and avian flocking. *Anim. Behav.* **62**, 1219–1224.

Konopka, R. J., and Benzer, S. (1971). Clock mutants of *Drosophila melanogaster*. *Proc. Natl Acad. Sci.* **68**, 2112–2116.

Ladich, F. (1989). Sound production by the river bullhead, *Cottus gobio* L. (Cottidae, Teleostei). *J. Fish Biol.* **35**, 531–538.

Lamprecht, G., and Weber, F. (1992). Spontaneous locomotion behaviour in cavernicolous animals: The regression of the endogenous circadian system. *In* "The Natural History of Biospeleology" (Camacho, A. I., Ed.), pp. 225–262. Monografias del Museo Nacional de Ciencias Naturales, Madrid.

Langecker, T. G. (1992). Light sensitivity of cave vertebrates. Behavioral and morphological aspects. *In* "The Natural History of Biospeleology" (Camacho, A. I., Ed.), pp. 295–326. Monografias del Museo Nacional de Ciencias Naturales, Madrid.

Leonard, J. B. K., and McCormick, D. (1999). Changes in haematology during upstream migration in American shad. *J. Fish Biol.* **54**, 1218–1230.

Liley, N. R. (1982). Chemical communication in fish. *Can. J. Fish. Aquat. Sci.* **39**, 22–35.

Lowe-McConnell, R. H. (1964). The fishes of the Rupununi savanna district of British Guiana, South America. Part 1. Ecological groupings of fish species and effects of the seasonal cycle on the fish. *J. Linn. Soc. (Zool.)* **45**, 103–144.

Lowe-McConnell, R. H. (1987)."Ecological Studies in Tropical Fish Communities." Cambridge University Press, Cambridge.

Lucas, R. J., Stirland, J. A., Darraw, J. M., Menaker, M., and Loudon, A. S. I. (1999). Free running circadian rhythms of melatonin, luteinizing hormone, and cortisol in Syrian hamsters bearing the circadian *tau* mutation. *Endocrinology* **140**, 758–764.

Maheshwari, R., Pati, A. K., and Gupta, S. (1999). Annual variation in air-gulping behaviour of two Indian siluroids, *Heteropneustes fossilis* and *Clarias batrachus*. *Ind. J. Anim. Sci.* **69**, 66–72.

Marques, N., and Menna-Barreto, L. (1999)."Introdução ao Estudo da Cronobiologia." EDUSP and Ícone Editora, São Paulo.

Matty, A. J. (1985)."Fish Endocrinology." Croom Helm, London and Sydney.

McBay, L. G. (1961). The biology of *Tilapia nilotica* Linnaeus. *Proc. Annu. Conf. Southeast Assoc. Game Fish Comm.* **15**, 208–218.

Menaker, M., Moreira, L. F., and Tosini, G. (1997). Evolution of circadian organization in vertebrates. *Braz. J. Med. Biol. Res.* **30**, 305–313.

Menna-Barreto, L. (1999). Human chronobiology. *ARBS Annu. Rev. Biomed. Sci.* **1**, 103–131.
Mitchell, R. W., Russell, W. H., and Elliott, W. R. (1977). Mexican eyeless characin fishes, genus *Astyanax*: Environment, distribution, and evolution. *Spec. Publ. Mus. Texas Tech. Univ.* **12**, 1–89.
Miyagawa, N. B., and Takemura, A. (1986). Acoustical behaviour of scorpaenoid fish *Sebasticus mamoratus. Bull. Japn Soc. Sci. Fish.* **52**, 411–415.
Moberg, G. P. (1999). When does stress become distress? *Lab. Anim.* **28**, 22–26.
Moore-ede, M. C., Schmelzer, W. S., Kass, D. A., and Herd, J. A. (1976). Internal organization of the circadian timing system in multicellular animals. *Fed. Proc.* **35**, 2333–2338.
Munro, A. D., and Pitcher, T. J. (1985). Steroid hormones and agonistic behavior in a cichlid teleost, *Aequidens pulcher. Horm. Behav.* **19**, 353–371.
Nakazato, M., and Takemura, A. (1987). Acoustical behavior of Japanese parrot fish *Oplenathus fasciatus. Nippon Suisan Gakkaishi* **53**, 967–973.
Naruse, M., and Oishi, T. (1996). Annual and daily activity rhythms of loaches in an irrigation creek and ditches around paddy fields. *Environm. Biol. Fish.* **47**, 93–99.
Nejdi, A., Guastavino, J. M., and Lalonde, R. (1996). Effects of the light–dark cycle on a water tank social interaction test in mice. *Physiol. Behav.* **59**, 45–47.
Nelson, J. S. (1994)."Fishes of the World." John Wiley and Sons, NewYork.
Northcott, S. J., Gibson, R. N., and Morgan, E. (1991a). Phase responsiveness of the activity rhythm of *Lipophrys pholis* (L.) (Teleostei) to a hydrostatic pressure pulse. *J. Exp. Mar. Biol. Ecol.* **148**, 47–57.
Northcott, S. J., Gibson, R. N., and Morgan, E. (1991b). The effect of tidal cycles of hydrostatic pressure on the activity of *Lipophrys pholis* (L.) (Teleostei). *J. Exp. Mar. Biol. Ecol.* **148**, 35–45.
Paxton, C. G. M. (1997). Shoaling and activity levels in *Corydoras. J. Fish Biol.* **51**, 496–502.
Pfeiffer, W. (1982). Chemical signals in communication. *In* "Chemoreception in Fishes" (Hara, T. J., Ed.), pp. 306–326. Elsevier, Amsterdam.
Pitcher, T. J. (1986). Functions of shoaling behaviour in teleosts. *In* "The Behaviour of Teleost Fishes" (Pitcher, T. J., Ed.), pp. 294–337. Croom Helm, London and Sydney.
Pradhan, R. K., Pati, A. K., and Agarwal, S. M. (1989). Meal scheduling modulation of circadian rhythm of phototactic behaviour in cave dwelling fish. *Chronobiol. Int.* **6**, 245–249.
Pringle, C. M. (1997). Exploring how disturbance is transmitted upstream: Going against the flow. *J. North Am. Benthol. Soc.* **16**, 425–438.
Purser, G. J., and Chen, W.-M. (2001). The effect of meal size and meal duration on food anticipatory activity in greenback flounder. *J. Fish Biol.* **58**, 188–200.
Rahman, M. S., Takemura, A., and Takano, K. (2000). Lunar synchronization of testicular development and plasma steroid hormone profiles in the golden rabbitfish. *J. Fish Biol.* **57**, 1065–1074.
Ralph, M. R., and Menaker, M. (1988). A mutation of the circadian system in golden hamster. *Science* **241**, 1225–1227.
Ramirez-Gil, H., Feldberg, E., Almeida-Val, V. M. F., and Val, A. L. (1998). Karyological, biochemical, and physiological aspects of *Callophysus macropterus* (Siluriformes, Pimelodidae) from the Solimões and Negro rivers (Central Amazon). *Braz. J. Med. Biol. Res.* **31**, 1449–1458.
Reebs, S. G. (1996). Time–place learning in golden shiners (Pisces: Cyprinidae). *Behav. Process* **36**, 253–262.
Reebs, S. G., and Colgan, P. W. (1991). Nocturnal care of eggs and circadian rhythms of fanning activity in two normally diurnal cichlid fishes, *Cichlasoma nigrofasciatum* and *Hetotilapia multispinosa. Anim. Behav.* **41**, 303–311.
Reebs, S. G., Whoriskey, F. G., and FitzGerald, G. J. (1984). Diel patterns of fanning activity, egg respiration, and the nocturnal behavior of male three-spined sticklebacks, *Gasterosteus aculeatus* L. (*f. trachurus*). *Can. J. Zool.* **62**, 329–334.

Romero, A., and Paulson, K. M. (2001). It's a wonderful hypogean life: A guide to the troglomorphic fishes of the world. *In* "The Biology of Hypogean Fishes" (Romero, A., Ed.), pp. 13–41. Kluwer Academic, Dordrecht.

Rothbard, S. (1979). Observations on the reproductive behavior of *Tilapia zillii* and several *Sarotherodon* spp. under aquarium conditions. *Bamidgeh* **31**, 35–43.

Sallum, W. B., and Cantelmo, A. O. (1999). Cultivo e Reprodução das Principais Espécies de Peixes UFLA/FAEPE, Lavras.

Sánchez-Vázquez, F. J., Madrid, J. A., Iigo, M., and Tabata, M. (1996). Demand feeding and locomotor circadian rhythms in the goldfish, *Carassius auratus*: Dual and independent phasing. *Physiol. Behav.* **60**, 665–674.

Santiago, J. A., and Castro, J. J. (1997). Acoustic behavior of *Abudefduf luridus*. *J. Fish Biol.* **51**, 952–959.

Sawara, Y. (1992). Differences in the activity rhythms of juvenile gobiids fish, *Chasmichthys gulosus*, from different tidal localities. *Japan. J. Ichthyol.* **39**, 201–209.

Sazima, I., and Machado, F. A. (1990). Underwater observations of piranhas in western Brazil. *Environm. Biol. Fish.* **28**, 17–31.

Sazima, I., Machado, F. A., and Zuanon, J. (2000a). Natural history of *Scoloplax empousa* (Scoloplacidae), a minute spiny catfish from the Pantanal wetlands in western Brazil. *Ichthyol. Expl. Freshwaters* **11**, 89–95.

Sazima, I., Sazima, C., Francini-Filho, R., and Moura, R. L. (2000b). Daily cleaning activity and diversity of clients of the barber goby, *Elacatinus figaro*, on rocky reefs in southeastern Brazil. *Environm. Biol. Fish.* **59**, 69–77.

Schreck, C. B. (1981). Stress and compensation in teleostean fishes: Response to social and physical factors. *In* "Stress and Fish" (Pickering, A. D., Ed.), pp. 295–321. Academic Press, London.

Schwassmann, H. O. (1971). Biological rhythms. *In* "Fish Physiology" (Hoar, W. S., and Randall, D. J., Eds.), pp. 371–428. Academic Press, London.

Silvano, R. A. M., and Begossi, A. (2001). Seasonal dynamics of fishery at the Piracicaba river (Brazil). *Fish. Res.* **51**, 69–86.

Spieler, R. E. (1992). Feeding-entrained circadian rhythms in fishes. *In* "Rhythms in Fishes" (Ali, M. A., Ed.), pp. 137–147. Plenum Press, NewYork.

Stirland, J. A., Mohammad, Y. N., and Loudon, A. S. I. (1996). A mutation of the circadian timing system (*tau* gene) in the seasonally breeding Syrian hamster alters the reproductive response to photoperiod change. *Proc. R. Soc. Lond. B Biol.* **263**, 345–350.

Stopa, R. M., and Hoshino, K. (1999). Electrolocation-communication discharges of the fish *Gymnotus carapo* L. (Gymnotidae: Gymnotiformes) during behavioral sleep. *Braz. J. Med. Biol. Res.* **32**, 1223–1228.

Sundararaj, B., Vasal, S., and Halberg, F. (1982). Circannual rhythmic ovarian recrudescence in the catfish *Heteropneustes fossilis* (Bloch). *Adv. Biosci.* **41**, 319–337.

Tabata, M. (1992). Photoreceptor organs and circadian locomotor activity in fishes. *In* "Rhythms in Fishes" (Ali, M. A., Ed.), pp. 223–234. Plenum Press, New York.

Takemura, A. (1984). Acoustical behaviour of the freshwater goby *Odontobutis obscura*. *Bull. Jap. Soc. Scient. Fish.* **50**, 561–564.

Tan-Fermin, J. D., Pagador, R. R., and Chavez, R. C. (1997). LHRHa and pimozine-induced spawning of Asian catfish *Clarias macrocephalus* (Gunther) at different times during an annual reproductive cycle. *Aquaculture* **148**, 323–331.

Taylor, M. H. (1984). Lunar synchronization of fish reproduction. *T. Am. Fish. Soc.* **133**, 484–493.

Thetmeyer, H. (1997). Diel rhythms of swimming activity and oxygen consumption in *Gobiusculus flavescens* (Fabricius) and *Pomatoschistus minutus* (Pallas) (Teleostei: Gobiidae). *J. Exp. Mar. Biol. Ecol.* **218**, 187–198.

Thinès, G., and Weyers, M. (1978). Résponses locomotrices du poisson cavernicole *Astyanax mexicanus* (Pisces, Characidae) à des signaux périodiques et apériodiques de lumière et de température. *Int. J. Speleol.* **10,** 35–55.

Thinès, G., Wolff, F., Boucquey, C., and Soffie, M. (1965). Etude comparative de l'activité du poisson cavernicole *Anoptichthys antrobius* Alvarez et son ancêtre épigé *Astyanax mexicanus* Philipe. *Ann. Soc. R. Zool. Belg.* **96,** 61–115.

Tobler, I., and Bórbely, A. A. (1985). Effect of rest deprivation on motor activity in fish. *J. Comp. Physiol.* **157,** 817–822.

Tosini, G., and Menaker, M. (1996). Circadian rhythms in cultured mammalian retina. *Science* **272,** 419–421.

Trajano, E. (2001). Ecology of subterranean fishes: An overview. *Environ. Biol. Fish.* **62,** 133–160.

Trajano, E., Duarte, L., and Menna-Barreto, L. (2001). Subterranean organisms as models for chronobiological studies: The case of troglobitic fishes. Abstracts of the 25th Conference of the International Society for Chronobiology, p. 48. Antalya.

Trajano, E., and Menna-Barreto, L. (1995). Locomotor activity pattern of Brazilian cave catfishes under constant darkness (Siluriformes, Pimelodidae). *Biol. Rhythm Res.* **26,** 341–353.

Trajano, E., and Menna-Barreto, L. (1996). Free-running locomotor activity rhythms in cave-dwelling catfishes, *Trichomycterus* sp., from Brazil (Teleostei, Siluriformes). *Biol. Rhythm Res.* **27,** 329–335.

Trajano, E., and Menna-Barreto, L. (2000). Locomotor activity rhythms in cave catfishes, genus *Taunayia*, from eastern Brazil (Teleostei: Siluriformes: Heptapterinae). *Biol. Rhythm Res.* **31,** 469–480.

Trajano, E., and Pinna, M. C. C. (1996). A new cave species of *Trichomycterus* from eastern Brazil (Siluriformes, Trichomycteridae). *Rev. Franç. d'Aquariol* **23,** 85–90.

Ulyel, A.-P., Ollevier, F., Ceusters, R., and Thys van den Audenaerde, D. (1991). Food and feeding habits of *Haplochromis* (Teleostei: Cichlidae) from Lake Kivu (Central Africa). II. Daily feeding periodicity and dietary changes of some *Haplochromis* species under natural conditions. *Belg. J. Zool.* **121,** 93–112.

Underwood, H. (1990). The pineal and melatonin: Regulators of circadian function in lower vertebrates. *Experientia* **46,** 120–128.

Val, A. L. (1993). Adaptation of fishes to extreme conditions in fresh water. *In* "Vertebrate Gas Transfer Cascade: Adaptations to Environment and Mode of Life" (Bicudo, J. E., Ed.), pp. 43–53. CRC Press, Boca Raton.

Volpato, G. L. (2000). Aggression among farmed fish. *In* "Aqua 2000: Responsible Aquaculture in the New Millenium" (Flos, R., and Creswell, L., Eds.), *European Aquaculture Society Special Publication* **28,** 803.

Volpato, G. L., and Barreto, R. E. (2001). Environmental blue light prevents stress in the fish Nile tilapia. *Braz. J. Med. Biol. Res.* **34,** 1041–1045.

Volpato, G. L., and Fernandes, M. O. (1994). Social control of growth in fish. *Braz. J. Med. Biol. Res.* **27,** 797–810.

Volpato, G. L., Duarte, C. R. A., and Luchiari, A. C. (2004). Environmental color affects Nile tilapia reproduction. *Braz. J. Med. Biol. Res.* **37,** 479–483.

Volpato, G. L., Frioli, P. M. A., and Carrieri, M. P. (1989). Heterogeneous growth in fishes: Some new data in the Nile tilapia *Oreochromis niloticus* and a general view about the causal mechanisms. *Bol. Fisiol. Animal* **13,** 7–22.

Wilhelm Filho, D., Torres, M. A., Tribess, T. B., Pedrosa, R. C., and Soares, C. H. I. (2001). Influence of season and pollution on the antioxidant defenses of the cichlid fish acará (*Geophagus brasiliensis*). *Braz. J. Med. Biol. Res.* **34,** 719–726.

Wilkens, H. (1988). Evolution and genetics of epigean and cave *Astyanax fasciatus* (Characidae, Pisces). *Evol. Biol.* **23,** 271–367.

Winemiller, K. O., and Jepsen, D. B. (1998). Effects of seasonality and fish movement on tropical river food webs. *J. Fish Biol.* **53,** 267–296.

Woynarovich, E., and Horváth, L. (1980). The artificial propagation of warm-water finfishes – a manual for extension. *FAO Fish. Tech.* **201,** 1–183.

Young, M. W. (2000). Life's 24-hour clock: Molecular control of circadian rhythms in animal cells. *Trends Biochem. Sci.* **25,** 601–606.

Zayan, R. (1991). The specificity of social stress. *Behav. Process.* **25,** 81–93.

Zupanc, M. M., Engler, G., Midson, A., Oxberry, H., Hurst, L. A., Symon, M. R., and Zupanc, G. K. H. (2001). Light–dark-controlled changes in modulations of the electric organ discharge in the teleost *Apteronotus leptorhynchus. Anim. Behav.* **62,** 1119–1128.

5

FEEDING PLASTICITY AND NUTRITIONAL PHYSIOLOGY IN TROPICAL FISHES

KONRAD DABROWSKI
MARIA CELIA PORTELLA

I. FOOD AND FEEDING

A. Feeding Behavior, Territoriality, Group Foraging, Food Preferences, and Quality

The feeding process is composed of nine stereotyped movement patterns (particulate intake, gulping, rinsing, spitting, selective retention of

The Physiology of Tropical Fishes: Volume 21
FISH PHYSIOLOGY

food, transport, crushing, grinding, and deglutition). The sequence and frequency of these movements are adjusted to the type, size, and texture of food (Sibbing, 1988). Better understandings of food intake and mechanisms of food processing reveal intraspecies plasticity and interspecies trophic interactions. As a consequence, this knowledge is essential to manage multispecies communities and maximize productivity of polyculture systems.

In common carp (*Cyprinus carpio*), an omnivorous species successfully established as a member of fish fauna on all continents, the complexity of food intake and processing requires consideration of (1) the dimensions of the mouth opening, (2) the protrusion of the upper jaw, (3) the shape of the pharyngeal cavity, (4) the palatal and postlingual organs, (5) the branchial sieve, (6) the pharyngeal masticatory apparatus, (7) the distribution of taste buds, and (8) the mucus cells and muscle fibers along the oropharyngeal surface (Sibbing, 1988). The question is how does an omnivore achieve an apparently highly selective efficiency between food and non-food materials? Sibbing (1988) hypothesized that selective retention is due to efficiently expelling small waste material while retaining food between the pharyngeal roof and floor, where taste buds are at a density of 820 per mm^2, the highest reported in any fish species. This anatomical adaptation in carp may have resulted in the loss of pharyngeal mastication efficiency, the principal apparatus in herbivorous fish and pharyngeal mollusk crushers. Therefore, the common carp is very limited in processing elongated vegetable material and handling other fish as prey. In grass carp (*Ctenopharyngodon idella*), using pharyngeal teeth to masticate plants, time spent on foraging (assessing and taking food) and chewing (orally transporting and masticating food) varied between 16 and 56% and 13 and 56%, respectively, depending on the plant (*Elodea, Lemna, Typha*, among others) offered and ingested (Vincent and Sibbing, 1992). Interestingly, all plants were chewed upon at the same frequency of jaw movement independent of their toughness or unpleasant taste. Plant material is subjected to an antero-posterior stroke of teeth, then pulled apart laterally and sheared between the teeth. Damage, measured as area with plant cells broken open, frequently reaches 40% in particles less than $1\,mm^2$, but falls quickly to less than 10% with larger particles.

Two tropical characids, tambaqui (*Colossoma macropomum*) and pirapitinga (*C. bidens*), possess oral teeth characterized by broad, multicusped molariform and incisive features (Goulding, 1980). The pirapitinga premaxillary teeth, forming a triangular shape, are a possible adaptation to the more diverse diet than that of the tambaqui, a seed and fruit eater during the wet season in the Amazon. Juvenile Characiformes from Central Amazon switched from mostly cladoceran (zooplankton) diet at sizes 10–30 mm to predominantly filamentous algae and/or wild rice seeds at 30–50 mm body length (Araujo-Lima *et al.*, 1986). The adult tambaqui crushes seeds before ingesting them, whereas another characid fish, the matrinchã (*Brycon* sp.), can remove the kernels from the shells and ingests only the more nutritious part.

In comparison to temperate herbivorous fishes (Horn, 1989), tropical rocky shore fishes are rarely studied (Ferreira *et al.*, 1998). The major finding in South American shore fishes is not unexpected; larger sized fish had an increased number of algae bites, ingestion rate, and gut fullness. Interestingly, the authors found that fish fill their gut 2.5–3 times a day despite dramatically different food processing strategies. The scarid fish, with a pharyngeal meal to grind the algae to minute particles, has the fastest food evacuation rate, whereas *Acanthurus bahianus*, likely harboring symbiotic bacteria (Clements and Choat, 1995), had much slower and seasonally variable gut evacuation rates. Feeding strategy related to sex in scarid fish (males are solitary feeders while females feed in groups of 3–8) was not addressed.

Molluscivorous cichlids from Lake Victoria, Africa, appear to crush small bivalves and swallow them without shell ejection, but rinse and eject shell fragments after crushing gastropods (Hoogerhoud, 1987). The type of mollusks ingested has a profound effect on fish vertical movements as the rate of shell ingestion results in a change of swimbladder volume. Consequently, in order to maintain neutral buoyancy, Lake Victoria cichlids can only swim from the depth of 5 m to the surface with empty guts. After a meal, depending on the ejection rate of shells of consumed mollusks, they can swim vertically only 0.4–2 m to the surface. In other words, space partitioning in the abdominal cavity between gut contents and swimbladder affects fish behavior, response to food, avoidance of predators and "socializing" with conspecifics. Meyer (1989) showed that in a molluscivorous cichlid from Lake Jiloa, Nicaragua, *Cichlasoma citrinellum*, a bimodal distribution of pharyngeal jaw structure determined a preference for soft- or hard-shell prey. Two forms with morphological specialization coexisted as seasonality and annual variation in prey abundance prevented the competitive extirpation of either form.

Gianquinto and Volpato (2001) demonstrated that under non-hazardous conditions feed-deprived or well-fed South American catfish, *Pseudoplatystoma coruscans*, responded similarly to offered food, whereas exposure to skin extracts (alarm substances) significantly increased the latency to food response in fed individuals. In other words, hunger suppressed alarm signals and feed-deprived catfish were willing to risk predation. The aquatic chemical signaling in fishes in tropical systems has to be reliable, stable in variable environmental conditions and energetically inexpensive to produce. Several species of tropical characins use purine-N-oxides, strongly polarized molecules (such as hypoxanthine-3-N-oxide), byproducts of the purine degradation pathway, as their alarm pheromones (Brown *et al.*, 2001). Their functions in feeding behavior of conspecifics and sympatric heterospecifics need to be explored. For instance, red bellied piranha, *Pygocentrus nattereri*, make routinely non-random schooling decisions avoiding intraspecific aggression and cannibalism (Magurran and Queiroz, 2003). In fact, smaller fish that occupied the outer zone of the school showed a greater motivation to feed, whereas large fish in the central, safer zone of the school were slower to attack.

Food preferences are established in the process of adaptive radiation and colonization of different habitats. Changes in morphology, physiology, and behavior evolve in response to trophic differences. Central African lakes and their cichlid fauna are a unique example of how adaptive radiation is linked to the ability to exploit different sources of food. Sturmbauer *et al.* (1992) provides an example of such specialization in the detritivorous and microalgivorous cichlid from Lake Tanganyika, *Petrochromis orthognathus*. This cichlid, with a particularly long intestine (6–10-fold body length) ingests diatoms and is "equipped" with a marker enzyme, laminarinase, which is able to digest the major polysaccharide in diatoms but is absent in *Chlorophyta* and *Cyaonophyta*. Fish are the only vertebrates capable of producing endogenous laminarinase.

The quality of food in fish feeding on animal-, plant- or detritus-based diets in a tropical reservoir was addressed by De Silva *et al.* (1984). They provided proximate analyses of stomach contents in *Sarotherodon mossambicus* in several reservoirs in Sri Lanka. Average protein, lipid, and carbohydrate concentrations were 18.5–35.1, 5.9–9.8 and 11.6–34.7%, respectively. These values showed relatively high variations considering that contributions of animal, plant or detritus in the diet may have reached in particular reservoirs (populations) 60.3, 94.4 or 88.4%, respectively. The authors pointed out the enormous plasticity of species that can be detritivorous or carnivorous in adjacent reservoirs and that utilization of nutrients (digestibility) does not change significantly unless detritus exceeds 70%. This, however, is not exactly direct evidence for the high growth rate, efficiency of the trophic food chain, or the "unprecedented success" of the species. Cichlid fishes, with their enormous feeding plasticity, still require a balanced diet and, as reported by Hassan and Edwards (1992), differences in crude fiber concentrations (6.9 versus 11.7%) between two species of duckweed resulted in significant differences in growth of Nile tilapia. More importantly, feeding on duckweed, *Lemna* or *Spirodella*, alone led to a growth rate decrease or sight loss after 8–10 weeks. Feeding exclusively on one species of plant may result in fish mortality. Parrotfish (*Sparisoma radians*) died more rapidly when fed only seagrass *Penicillus pyriformis* (high $CaCO_3$

concentration) than a starved group (Lobel and Ogden, 1981). Parrotfish from the US Virgin Islands died or showed high mortality when fed solely five different seagrasses, otherwise abundant in its diet. The most nutritiously valuable plant, eaten most frequently in the natural habitat, when offered without accompanying epiphytes (algae) also led to over 60% mortality. Clearly, energetic value, abundance of seagrasses, and predator preferences are not synonymous with the highest quality food for herbivorous fishes.

Appler (1985) compared utilization of the green filamentous alga *Hydrodictyon reticulatum* in diets of two cichlids, *Oreochromis niloticus* and *Tilapia zilli*. When 17% of animal protein (fish meal) was replaced with algae protein, fish weight increased by 5% in *T. zilli* but decreased by 10% in *O. niloticus*. However, 50% animal protein replacement decreased growth in both species by approximately 50%. Interestingly, the "herbivorous nature" of *T. zilli* can be associated with a positive effect on growth only when a small proportion of the diet consists of algal material. Appler and Jauncey (1983) also concluded that replacement of fish meal protein by filamentous green alga (*Cladophora glomerata*) decreased growth rate of Nile tilapia juveniles by half, 3.1 and 1.85% per day, respectively. Bitterlich (1985b) compared the quality of ingested phytoplankton and detritus in two stomachless cyprinids and the stomach possessing *Oreochromis mossambicus* in a Sri Lankan reservoir using *Scenedesmus* cell numbers in the foregut and hindgut as a marker. She concluded that tilapia feeding predominantly on *Diatomophyceae* utilized more than 90% of available nutrients whereas stomachless cyprinids were unable to efficiently digest algae and were able to extract only 25–40% of their nutrients. However, the conclusion that detrital material provides high nutrient value for stomach-possessing fish is questionable; if for no other reason, lipid degradation will result in loss of essential fatty acids. Harvey and Macko (1997) analyzed microbially mediated lipid degradation in a marine diatom and cyanobacterium in oxic and anoxic conditions. They concluded that even at 19 °C, total fatty acid methyl esters dropped rapidly within 3 days and decreased to 10% (cyanobacterium) or 0% (diatoms) within 20 days from oxic decay. In anoxic conditions, concentrations of polyunsaturated fatty acids during phytoplankton lipid degradation decreased to trace amounts within 20 days. In tropical systems, likely under anoxic conditions, major lipid components in detritus will be lost.

B. Circadian Rhythms

Tropical ecosystems are extremely variable in respect to transparency and visibility and these conditions impose different types of interactions between prey and predator. Photoinhibition may also occur to prevent photosynthesis whereas in most scenarios increased photosynthesis results in accumulation of nutrients and, inversely, a degradation of reserves during the night. The assumption is frequently made that variation in the content of nutrients will correlate, induce, and structure the feeding periodicity of fish.

Feeding periodicity has been observed in marine fishes feeding on algae at the time of peak algal energy due to early afternoon photosynthesis. However, simple chemical analysis for protein or carbohydrates did not sufficiently explain the diel pattern in fish feeding (Zoufal and Taborsky, 1991). Evidence was provided that feeding periodicity is not endogenous and was established via natural selection with 10–20 generations. Zemke-White *et al.* (2002) analyzed other components of this fish–algae interaction and documented that the feeding pattern of the herbivorous *Stegastes nigricans* from the Great Barrier Reef in Australia matched an increase in the nutritional component of algae, floridoside (major sugar alcohol compound found in rodophyte algae, *Gracilaria, Acanthophora*). The authors examined alpha-galactosidase activity in fish intestines and provided evidence that indeed this nutrient increased by 51–82% in the mid-afternoon, significantly affecting feeding efficiency.

However, if foraging efficiency and food quality can be extended beyond algae feeders to fishes utilizing animal and other mixed diets it may have implications for foraging models in the wild and controlled fish culture conditions. The correspondence between meal timing, proximate composition of the meal and fish activity (metabolic rhythms) is the basis of the hypothesis to explain differences in 35–50% better growth of *Piaractus brachypomus* fed at night compared to fish fed during the day (Baras *et al.*, 1996). Fish fed during the night were expected to save energy by limiting their activity whereas the experiments demonstrated the opposite results. Fish fed at night were more active at night and did not differ in activity during the day compared to fish fed during light hours (Baras, 2000). The authors speculated that fish fed at night grew faster because they showed less agonistic interactions and likely deposited dietary protein more efficiently. Tilapia, *Oreochromis aureus*, categorized as a food-conformer, adapted their feeding behavior to the time of meal provision in the previous day and did not show differences in growth rate dependent on meal timing.

An equatorial stream illustrates the interaction related to a diel periodicity between fish fauna and their potential invertebrate prey that may be overlooked in other aquatic ecosystems. In water with a mean monthly temperature of 24–25 °C and variation of 1 °C, the mean night:day drift ratio in the number of aquatic invertebrates was 10 (Jacobsen and Bojsen, 2002). The night:day ratio of invertebrates correlated only with the richness of loricariids catfish in the stream. Although the authors were not able to state conclusively what caused this correlation, their a priori expectations were not confirmed. One possible explanation of drift periodicity is that catfishes are benthic, night feeders, and cause a physical disturbance in the stream.

C. Feeding Migrations and Reproduction

De Godoy (1959) described downstream migrations of curimbata (*Prochilodus scrofa*) from the Upper Mogi Guassu River to feeding grounds in the River Grande, South America, over a distance of 500 km. This species, and the closely related *P. platensis*, are iliophagous fish, feeding as adults on the organic mud formed by the partial breakdown of macrophytes (Bayley, 1973). Upstream migration of the same distance with no suitable food sources (Bayley confirmed empty stomachs in *P. platensis*) forces the fish to use body reserves. Calculated growth of *P. platensis* suggests a total length of 31.2 cm in 2-year-old fish entering maturity (males) and an increase of 3–5 cm body length in the following 4–5 years. This clearly illustrates a direct impact of long migrations, and a more speculative effect of the nutritiously poor diet of iliophagous characins. The Amazonian tambaqui (*Colossoma macropomum*) remains in flooded forests for 4–7 months annually and feeds largely on fruits and seeds (Goulding and Carvalho, 1982). Fish stores in the body following this diet consist of up to 10% visceral fat reserves and large fat reserves in parts of all tissues (muscle). Subsequent to the flood season, tambaqui spends the low water period in large rivers. The upstream migration begins 1 to 2 months before floodplain inundation and results in spawning in nutrient-poor, but turbid, whitewater rivers and tributaries. The distance of the feeding migration from whitewater to floodplains can extend for more than 200 km. The annual floods of central Amazonian rivers occur with regularity and have resulted in the development of complex relationships between fishes and fruit/seed producing plants (Araujo-Lima and Goulding, 1998). Characins are principally seed-destructive predators whereas the large doradid catfish, *Litodoras dorsalis*, described by Goulding and Carvalho (1982) as a consumer of "the root and leaves of some aquatic plants," turned out to be the major seed dispenser, being unable to masticate and destroy them (Kubitzki and Ziburski, 1994). The latter authors investigated the distance reached in a single dispersal event. A daily distance of 20–30 km upstream migration would account for a long range if seeds remain in characin guts up to one week. An Amazonian catfish, *Auchenipterichthys longimanus*, reaching 25 cm in length, contained seeds of 20 plant species, both buoyant and sinking, and some fruits as large as 20% of fish length (Mannheimer *et al.*, 2003). As the most abundant fish species in the mining silt degraded area of the lake, this species is likely a contributor to the massive regeneration of "igapo" (blackwater rivers) forests by migrations and seed dispersal.

Horn (1997) analyzed the dispersal of seeds of *Ficus glabrata*, a major canopy-forming riparian tree in Central America tropical rain forests, by *Brycon guatemalensis*. This riverine fish, 29–46 cm in length, eats fruits as a major part of its diet. The seeds were defecated in 15–33 hours and were viable. The time to first germination was somewhat longer but growth of the plant from seeds passed through the fish gut was faster. Contrary to the annual regularity of flooding in the Amazon, the neotropical rain forests are characterized by 15–30 unpredictable floods per year. These frequent inundations allow the riverine fish migrations and access to terrestrial ecosystems. As a consequence, fruits such as figs become an important part of their diets and fish moving upstream during flooding disperse the seeds (Banack *et al.*, 2002). As deforestation continues in the tropics, dependence on fish-tree interactions increases even more.

II. MORPHOLOGY AND PHYSIOLOGY OF DIGESTIVE TRACT

A. "Metamorphosis" of Digestive Tract in Larval–Juvenile Transition

The ontogenetic changes in digestive tract development during the larval–juvenile transition can be categorized into three types: (1) stomachless fish with an increase in complexity of the coiling pattern (cyprinids); (2) "stomachless" larvae which develop a stomach structure after ingestion of food (coregonids, silurids, serrasalmids) (Segner *et al.*, 1993); and (3) alevin and/or juvenile stages of fish capable of ingesting the first food when the stomach is present as a distinguished feature (salmonids, cichlids) (Dabrowski, 1984, 1986a; Stroband and Dabrowski, 1981).

Morphological features of the digestive system are of great consequence in respect to the type of diet larval/juvenile fish are able to utilize, especially at the highest growth rates during early ontogenetic development (50% per day in larval common carp, *Cyprinus carpio* – Bryant and Matty, 1981; 30–50% per day in *Clarias gariepinus* larvae – Terjesen *et al.*, 1997). Cichlids are exceptional as their digestive gastrointestinal tract appears to be completely formed with a functional stomach and an elongated intestine prior to the use of yolk sac reserves (Figure 5.1). Unlike most other teleosts, cichlid juveniles pass through an extended period of "mixed" feeding of endogenous (yolk sac) and exogenous feeding. This modulation shifts the focus to maternal–offspring nutrient transfer in juveniles rather than a sole dependence on external food intake and its quality (nutrient presence and availability) for larval fish. Juvenile, first feeding Nile tilapia, for instance,

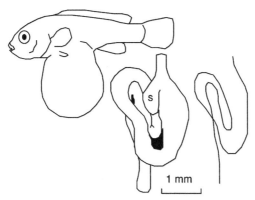

Fig. 5.1 Digestive tract of yolk sac larvae of *Petrochromis polyodon*. Note the large yolk sac, presence of the stomach, and complex intestinal coiling pattern at the beginning of exogenous feeding. (From Yamaoka, 1985, reproduced by permission of the author and the Linnean Society of London.)

were able to grow on phytoplankton (*Nannochloris, Chlorella, Scenedesmus*) provided during the first several weeks of life (Pantastico *et al.*, 1982), although the small larval tilapia density and rear exchange of water may have resulted in production of protozoans as a supplementary food. In a similarly designed study with larval milkfish (*Chanos chanos*) reared at low density (2 fish/l) on freshwater algae diets, Pantastico *et al.* (1986) came to the conclusion that a mix of *Oscillatoria* and *Chrococcus* supported the best growth of this species. However, the growth rate was only from 6 to 16 mg in 40 days. Control groups with zooplanktonic food need to be introduced in this type of experiment in order to make a conclusion about the efficiency of the larval digestive tract to utilize algae as a sole food source. Rotifers and small zooplankton are lysed within 20 min at 20 °C by gut fluids of planktivorous cyprinids and slight mechanical triturating render these organisms unrecognizable, a mass of detritus material (Bitterlich and Gnaiger, 1984). Most phytoplankton cells remain undigested in the gut. Segner *et al.* (1987) demonstrated that in *Chlorella* algae-fed milkfish larvae, histopathology in the intestine was dissimilar to starvation conditions, with the enterocytes containing bizarre-shaped nuclei, enlarged and branched mitochondria and intracellular vacuolization. The conclusion was that *Chlorella* acts as an additional stress in the larval milkfish digestive tract.

Most neotropical fish larvae hatch with relatively small endogenous reserves and exogenous feeding starts after a few days when the digestive tract is not fully differentiated into the gastrointestinal system. The beginning of exogenous feeding is one of the most critical periods for tropical fish

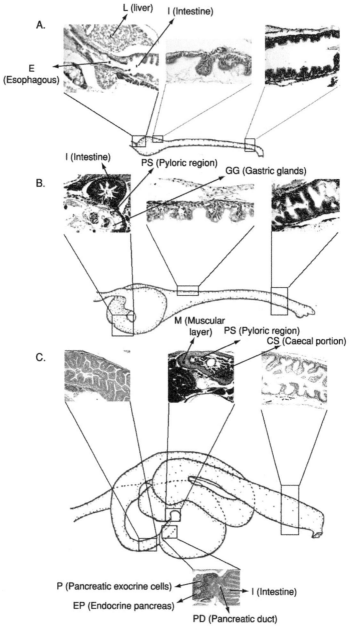

Fig. 5.2 Digestive tract changes in ontogenesis of pacu (*Piaractus mesopotamicus*). **A**, **B**, and **C** refer to fish sizes of 6, 12, and 22 min in length, respectively. The digestive tract morphology drawing is from Yamanaka, N., Fisheries Institute, Sao Paulo (1988). **A.** Longitudinal sections through (A1) the liver (L), esophagus (E) and (A2) the anterior and (A3) posterior intestine (I). The

survival, as it is in temperate climate fish, but higher water temperatures and the consequent increased metabolic rate accelerate the possibility of starvation. In general, at hatching, the gut of tropical fish larvae presents as an "apparently" undifferentiated straight tube lying over the yolk sac, with mouth and anus closed. During the endotrophic phase morphological and physiological changes must occur to allow larvae to search for, ingest and digest food.

The major alterations in the digestive tract during the early larval phase are related to the gut, pancreas, and liver development. These structures are present in pacu (*Piaractus mesopotamicus*) larvae (Figure 5.2) and in bared surubim (*Pseudoplatystoma fasciatum*) larvae sampled soon after hatching (Portella and Flores-Quintana, 2003a). The stomach structure is missing at the time of the first feeding, but the intestine is differentiated into three segments as described earlier in African catfish (*Clarias lazera*) (Stroband and Kroon, 1981). The gut is comprised of a single layer of columnar epithelium. Three segments with distinguished cell morphology are also present in pacu larvae. Differences in the apical border of microvilli, the presence (first segment) or absence (second and third segments) of large lipid droplets, and the presence (second segment) or absence (first and third segments) of pinocytotic vesicles are evident. Similar ultrastructural changes in larval–juvenile fish intestine were observed in other teleosts (Albertini-Berhaut, 1988).

In surubim larvae sampled 1 day after hatching (~3.86 mm, standard length, SL), groups of undifferentiated round cells, the precursors of liver and pancreas, were observed between the anterior part of the digestive tract and the yolk sac. In 2-day-old larvae, these two structures were well defined and two portions of exocrine pancreas were observed, one close to the liver and another above the yolk sac. Pancreatic cells were basophilic with many acidophilic zymogen granules, which increased in quantity with development (Portella and Flores-Quintana, 2003a). In pacu (*P. mesopotamicus*) larvae

---

intestine (I) in the "presumptive stomach" area (A1) presented as mucosal folds lined with a single-layer of columnar cells. The mid-intestine section shows few mucous cells. The posterior intestine contains flattened folds and numerous supranuclear vacuoles in the enterocytes. **B.** (B1) Pyloric region (PS) and part of corpus region of the stomach with gastric glands (GG). The muscular layer near the pyloric sphincter of the intestine (I) shows mucosal folds, submucosa, and *tunica muscularis*. (B2) The anterior intestine shows numerous mucous goblet cells. **C.** (C1) The anterior intestine shows developed folds, a single layer of epithelium with columnar absorptive enterocytes and goblet cells. (C2) The pyloric region (PS) of the stomach is non-glandular with a cuboidal single cell epithelium and a well developed muscular (M) layer. At the corpus portion (CS), gastric glands can be recognized. Goblet (C3) The posterior intestine shows smaller and flattened mucosal folds with only few mucous cells. (C4) Pancreatic exocrine cells (P) show acinar arrangement, basophilic cytoplasm, and zymogen granules. The endocrine pancreas (EP) is also present. The pancreatic duct (PD) opens into the pyloric area of the intestine.

hepatic tissue was well organized at 2 days after hatching (DAH) but pancreatic cells seemed to be more dispersed and formed a lobe-like structure only in 4-day-old larvae (~5.5 mm, total length, TL) (Tesser, 2002). However, Peña *et al.* (2003) provided evidence of the liver and pancreas as small patches of undifferentiated cells in newly hatched spotted sand bass, *Paralabrax maculatofasciatus*, larvae, a similar structure as described in African catfish (*Clarias gariepinus*) larvae (Verreth *et al.*, 1992). The yolk sac was completely depleted at the third and fifth days in surubim (5.65 mm TL) and pacu larvae (5.5–6 mm TL), respectively, when reared at 28 °C and at the third day in spotted sand bass at 25 °C. The differentiation of gut segments started in 3-day-old *P. fasciatum* (Portella and Flores-Quintana, 2003a) and in *P. maculatofasciatus*. The buccopharynx and the incipient esophagus of both species exhibited squamous epithelium, the last section before the stomach with some folds and discernible connective tissue. The esophagus in bared surubim had some mucous, PAS-positive cells that were absent in 3-day-old spotted sand bass. In the bared surubim a presumptive stomach was observed as a single layer of cubic cells with a short PAS-negative brush border, but the brush border was also absent in the cubic cells of spotted sand bass (Peña *et al.*, 2003).

The bared surubim intestine showed a columnar epithelium with some folds and mucous goblet cells moderately PAS-positive. The enterocytes showed a developed brush border. The same characteristics were described in 6-day-old *P. maculofasciatum* larvae (Pena *et al.*, 2003). The liver vacuolization increased as well as the PAS-positive reaction in bared surubim and spotted sand bass, indicating glycogen storage and liver functionality (Bouhic and Gabaudan, 1992; Peña *et al.*, 2003). Pancreatic cells observed in bared surubim showed an acinar arrangement and exhibited basophilic cytoplasm, round nuclei and a large number of acidophilic zymogen granules. The development of the surubim digestive tract continued with an increase in the number of mucous cells in the esophagus, folds in the intestines, and the intensity of PAS reaction in the hepatocytes and intestinal mucous cells, which also presented an AB moderately positive reaction.

Stomach differentiation is considered an important event during the gastrointestinal tract development of fish (Govoni *et al.*, 1986), mainly from a perspective of increased adaptation to a variable diet and ability to digest complex proteins (Grabner and Hofer, 1989). In 5-day-old (18 mm total length) alligator gar (*Atractosteus spatula*) juveniles, the stomach was formed and pepsin activity (pH 2.0) was present despite simultaneous utilization of yolk sac reserves (Mendoza *et al.*, 2002). In surubim (*P. fasciatum*) the first gastric glands were observed at 10 days after hatching (11.3 mm SL) when fish had already increased body weight several times (Portella and Flores-Quintana, 2003a), whereas in pacu (*P. mesopotamicus*) they were observed at 7–10.3 mm TL, in *P. maculatofasciatus* at 16 days after hatching (Tesser, 2002; Peña *et al.*, 2003), and in *Mugil platanus* at 38 days (Galvão *et al.*, 1997a). The stomach was developed much earlier in *C. gariepinus* reared at 27.5 °C: at 4 days after initiation of feeding or 6 days after hatching, in fish of 12.1 mm TL (Verreth *et al.*, 1992). In *Clarias lazera* the first gastric glands appeared on day 4, but a functional stomach, based on the observation of exocytosis of secretory granules, was present in juveniles of about 11 mm TL, approximately 12 days after fertilization at 23–24 °C (Stroband and Kroon, 1981). However, if the criterion of stomach function is the acidic pH required for pepsinogen activation and optimum activity of pepsin, then the juvenile African catfish has to be larger than 11.5 mm to meet this requirement. Surprisingly, in a cichlid fish, the ornamental discus, *Symphysodon aequifasciata*, the stomach was detected as early as 10 days after hatching, but pepsin-like activity was only significant several days after (Chong *et al.*, 2002a). This species and a Central American cichlid, *Cichlasoma citrinellum*, utilize very special food at an early stage, the mucus of its parents, although larvae of both species separated from their parents and reared on live food also survived (Schutz and Barlow, 1997).

In 12-day-old (15.2 mm SL) bared surubim larvae, the stomach was well developed and showed glandular and non-glandular regions and a well-differentiated *tunica muscularis*. In the medium intestine, mucosal folds were well developed, the mucous cells were abundant, and the absorptive enterocytes exhibited supranuclear vacuoles under the brush border in the second segment. From this stage on, the modifications observed in surubim were only in regard to the development of the size (hypertrophy) and complexity (hyperplasia) of the structures (Portella and Flores-Quintana, 2003a). Peña *et al.* (2003) also reported an antero-medium glandular and posterior non-glandular regions corresponding to the corpus and pyloric parts of the stomach in spotted sand bass at 16 days after hatching. The authors mentioned the diminished size and number of supranuclear vacuoles in the spotted sand bass anterior intestine after stomach development. In African catfish, 24 hours after first feeding, vacuoles containing absorbed lipid (assumed by the authors based on the holes in the structure caused by routine procedures used) were observed in the first part of the intestine. In the posterior half of the intestine, the size of these vacuoles decreased significantly (Verreth *et al.*, 1992). However, following horseradish peroxidase ingestion, presented to larval catfish "encapsulated" in *Artemia* nauplii, the second segment cells were positively stained for this enzyme activity (Stroband and Kroon, 1981). This confirmed the ability of African catfish to absorb protein macromolecules by endocytosis and intracellular digestion. In warmwater common carp (*Cyprinus carpio*), Fishelson and Becker (2001) described the development of the liver and pancreas in embryos at the free tail stage and 3 days after fertilization when these tissues were detected for the first time. The organ was composed of cellular buds of the embryonic mid-gut, enveloped by the coelomic mesothelium. Two days after hatching or 9 days after fertilization (5.6 mm TL), this primordium is divided, making the separation between liver and pancreas. In larger juveniles the hepato-pancreas is formed. Pancreatic islets were also observed in pacu liver juveniles (T. Ostaszewska, personal communication). Exocrine pancreatic tissues were also found in this species around the vein of the portal-hepatic system, in the mesenteric adipose tissues and in the liver and spleen (Ferraz de Lima *et al.*, 1991). The major concentration of pancreatic tissue in pacu adults was found around pyloric ceca, but it was not a completely discrete organ.

B. Stomach, Intestine, Rectum, and Diet–Morphology Relationships

The strategy of expanding intestinal absorptive surface is taking place in both cold-water and tropical fishes. In the cold-water salmonid rainbow trout, which possesses on average 70 pyloric ceca, the total length of the pyloric ceca is 6-fold larger than the total intestine. The serosal surface of pyloric ceca represents a 2-fold larger surface area than the whole intestine (Bergot *et al.*, 1975). Several tropical characid fishes also possess numerous pyloric ceca. However, detailed analyses of their absorptive surface and function are lacking. Frierson and Foltz (1992) provided analysis of intestinal surface area in *O. aureus* and *Tilapia zilli*, classified by the authors as predominantly feeding on detritus and macrophytes, respectively. The relative intestinal lengths in these two species were not different and amounted to approximately 3.5 and 7 for fish of 100 and 200 mm, respectively (see Figure 5.3). The major difference was in the intestinal diameter (nearly 4.5-fold larger in *T. zilli*), probably an adaptation to macrophytes in the diet, of which large pieces were ingested. However, when intestinal folds and microvilli dimensions were combined in calculating the surface area, it came rather close for both species (1819 and 1504 cm^2 for a standardized fish size of 145 mm TL, *O. aureus* and *T. zilli*, respectively). Since microvilli accounted for 90% of the absorptive surface in tilapia, a mucosal surface can be calculated. A comparison to rainbow trout of 206 mm (Bergot *et al.*, 1975) should demonstrate a major taxonomic and/or diet–morphology related difference. However, the total mucosal surface of the trout intestine was 132 cm^2 (pyloric region 91.7 cm^2) and only marginally smaller than in tilapia. Most certainly this area requires further studies.

In a series of descriptions of functional anatomy and morphometry of the intestine of two South American freshwater species, piracanjuba *Brycon orbignyanus* and piau *Leporinus fridericci*, of two different size classes (total range 14–29 cm), Seixas-Filho *et al.* (2000a, b) have described the similarities

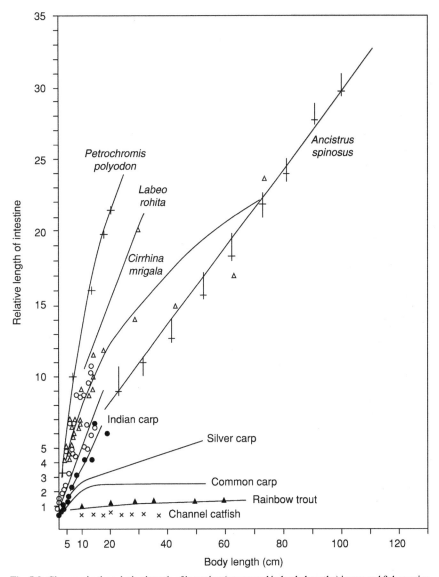

Fig. 5.3 Changes in the relative length of intestine (expressed in body lengths) in several fish species (modified from Dabrowski, 1993 and Kafuku, 1975). Data for *Petrochromis polyodon* (Yamaoka, 1985) were recalculated assuming a coefficient of 2 for "intersecting point" measurements. (Data for *Ancistrus spinosus* are from Kramer and Bryant, 1995a; data for *Labeo rohita* are from Girgis, 1952.)

of a pattern of two intestinal loops. The gut of these two species can be divided into anterior, medium, and posterior intestines with a noticeable absence of the ileo-rectal valve. Pyloric ceca were observed in the first loop of the anterior intestine. In piau they were less numerous (8–13) than in piracanjuba (42–93). The relative intestinal length (see Figure 5.3 to relate to other species) varied between 1.17 and 1.03 in piracanjuba and from 1.09 to 1.1 body length in piau. The authors concluded that these values are compatible with the omnivorous feeding behavior of those two species. Seixas-Filho *et al.* (2001) divided intestine of spotted surubim, *Pseudoplatystoma coruscans*, into medial and rectal parts due to the presence of the ileo-rectal valve and intestinal valve invaginations between these segments. Based on the final rings of the medium intestine, the authors considered these adaptations as signaling omnivory or preferential carnivory.

Albrecht *et al.* (2001) reported the anatomical features and histology of the digestive tract in two related species, *Leporinus friderici* and *L. taeniofasciatus*. The number of pyloric ceca found was 12 and 10 and the relative intestine length was 1.25 and 1.14 body length, respectively. The size range of sampled fish (9.8–48 cm) may have contributed to an elongated intestine. The values found agree with the classification of both species as omnivores. The most remarkable difference was the presence of a sphincter between the cardiac and pyloric portions of *L. taeniofasciatus* stomach. This may suggest that this species feeds at a constant rate, the food being digested partially in the anterior chamber and when it fills up; digestion is completed in the pyloric portion.

A generalization predicts that diets composed of voluminous food, algae, and detritus should result in elongation of the intestine. Hofer (1988) analyzed eight species of cyprinids in a Sri Lankan lake with relative intestinal lengths ranging from 1.4 to 6.1 body length and concluded that mucosal surface expressed per body length was 2-fold larger in omnivorous species than in phytoplanktivores. Clearly, different strategies are at play. Intestinal length was positively allometric with slopes of the log of intestine length against log of body length from 1.09 to 2.11 for 21 species representing four orders in a Panama stream community (Kramer and Bryant, 1995a). The relative gut length of *Ancistrus spinosus* (Loricariidae) increased from 4.6 at 10 cm to 34 at 120 cm standard body length (Figure 5.3). This is the longest recorded measurement in the literature. Carnivorous fish feeding on scales and smaller fish most consistently had the lowest allometric coefficient of intestinal growth (Kramer and Bryant, 1995b).

One strategy in order to compensate for the metabolic demand of a continuously high growth rate and the switch to more voluminous food (lower nutrient concentration) is an allometric increase in absorptive surface area of the intestine, as seen in cichlids (Yamaoka, 1985). However, ontogenetic isometric growth of the intestine may be accompanied by allometric changes in mucosal surface, intestinal diameter, and thickness of the intestinal wall (Hofer, 1989). In warmwater cyprinids, the gut passage time will be substantially different depending on the "gut storage capacity" that may vary between 2 and 10% among species. Based on the multispecies analysis of relative intestinal length that included ontogenetic changes for the entire size range of the species, the transition from animal to plant food as the principal component can be modeled as a descending linear regression that explains 83% of the variation (Piet, 1998). The author concluded that in fish from a Sri Lankan lowland reservoir the morphological changes during ontogeny explained dietary changes in cyprinids, clariids, and cichlids of the tropical reservoir community. Albertini-Berhaut (1987) demonstrated a linear increase of the intestinal length of four species of Mediterranean Mugilidae (genus *Mugil* and *Liza*). Juveniles of 50–100 mm body length increased the relative length of the intestine from 2- to 4-fold body length, which corresponded to a change in diet from predominantly animal (crustacean) to diatomids and algae.

The alimentary tract of teleosts has attracted considerable interest because of its diversity of form related to diet (Albrecht *et al.*, 2001). However, knowledge about the neotropical ichthyofauna, the most morphologically diverse of any epicontinental fish fauna in the world (Vari and Malabarba, 1998), is still largely unexplored and descriptive information is lacking for many South American fish (Albrecht *et al.*, 2001). Delariva and Agostinho (2001) studied six subtropical loricariids in the Paraná River and found in all species the same intestinal arrangement, characterized by a network of loops in a nearly horizontal plane within the ventral region of the abdominal cavity. However, the relative intestinal length was different among the studied species. *Rhinelepis aspera*, which feeds on fine-grained detritus, possesses a thin stomach wall and a long intestine, while species such as *Megalancistrus aculeatus* and *Hypostomus microstomus* that scrap the substrate and feed on coarser material with a high incidence of animal prey showed a well-developed stomach and shorter intestine. This systematic relatedness turned out to be an excellent choice for examining diet–morphology relationships. As Pouilly *et al.* (2003) demonstrated, taxonomic relatedness is likely to strengthen conclusions that can be drawn based on examining phylogenetically diverse groups of fishes.

The gut morphology in relation to diet was investigated in three species of glassy perchlets (*Ambassius products, A. natalensis*, and *A. gymnocephalus*) sampled in different estuaries in the Indian Ocean off the Natal Coast, South Africa (Martin and Blaber, 1984). The distensible stomach that terminates in a well-defined muscular constriction (sphincter) and a low relative gut length suggest predatory and carnivorous feeding habits in all three species.

However, *A. products* and *A. natalensis* collected in the Mdloti Estuary exhibited lower values of relative gut lengths than the specimens sampled from other habitats. The authors argued that in the Mdloti Estuary food is less abundant, which could have an effect on the intestine length.

The histology of the stomach in juvenile *Pseudoplatystoma coruscans* in relation to its feeding habit was studied with light microscopy by Souza *et al.* (2001a). They described a revetment of a single layer of columnar muco-secretory epithelial cells. The *lamina propria*, formed of loose connective tissue, is glandular at the cardiac and non-glandular at the pyloric region. Glands are a simple tubular type, non-ramified or slightly ramified. The secretory cells present in these glands are responsible for both the production of HCl and pepsinogen, and received the denomination of oxyntopeptic cells (Souza *et al.*, 2001b). In the adult *Piaractus mesopotamicus* stomach, very similar oxyntopeptic cells were detected with transmission electron microscopy (Aires *et al.*, 1999).

Columnar epithelium was also the revetment of *L. fridericci* and *L. taeniofasciatus* stomachs, which occurs as glandular in the cardiac portion and non-glandular in the pyloric region, therefore showing functional differences (Albrecht *et al.*, 2001). The same authors described the pyloric cecal mucosa having similar histological structure as the intestine, formed of a single-layer of epithelium of columnar absorptive cells with evident brush borders and two types of goblet cells. This fact suggests increasing of the absorptive zone. Rodlet cells were found only in *L. fridericci*, but they seem to play no role in digestion, although appearing in the gastrointestinal system.

The histological features of the tilapia *Oreochromis niloticus* digestive system have been investigated and some discrepancies were found among the structures (Al-Hussaini and Kholy, 1953; Caceci *et al.*, 1997; Smith *et al.*, 2000). Recently, Morrison and Wright (1999) carried out a study aiming to resolve these discrepancies; however, neither the size of fish examined was mentioned, nor was nutritional history. In the stomach, from the entry of the esophagus, across the anterior part of the stomach to the pyloric valve, a region with large tubular glands consisting of mucous cells forms a bypass, circumventing the sack-like portion of the stomach. This region contains striated muscle and may therefore be a means of disposing unwanted material, either by regurgitating it, or by passing it rapidly along to the intestine. The apical cytoplasm of the columnar cells in the pyloric and sac-like regions of the stomach exhibits a PAS-positive reaction. In the cardiac portion, typical gastric glands were observed, although a well-developed region of large goblet cells remained at the neck of the gastric glands, probably secreting substances with the function of protecting the mucosa from the very acidic contents of the stomach. The *muscularis* of the pyloric region extends only a short distance into the main portion of the stomach and consists of an inner circular and an outer longitudinal layer of smooth muscle. In the transitional region, from tubular to gastric glands, the smooth muscle wall is thicker. In the intestinal epithelium, consisting of columnar epithelium with small goblet cells, the structure is similar in size to the small mucous cells found in the esophagus. Also similar histochemical reactions were present in the large mucous cells in the tubular glands. The presence of the ileo-rectal valve separates the intestine and rectum.

C. Digestive Mechanisms

Based on the importance of the mucus in the gut for digestion and absorption processes, a hypothesis was proposed to explain the herbivory in snub-nosed garfish *Arrhamphus sclerolepis krefftii*, a Hemiramphidae fish, Beloniformes (Tibbetts, 1997). This species lack many of the adaptations of the digestive tract (acidic stomach, long intestine, pyloric ceca) that could increase digestion and absorption efficiencies. The garfish has a very short and straight gut, about 0.5 body length, which would indicate a carnivorous habit (Al Hussaini, 1946). However, Tibbetts (1997) argues that the relative gut length criterion is not valid to predict garfish trophic habits as herbivorous hemiramphids contradict what would be inferred based on their morphological features. In addition, *A. sclerolepis krefftii* have a short gut passage time and large intestinal diameter, which would also lead to the assumption of low assimilation efficiency. However, the presence of abundant mucous cells in the digestive tract, especially in the pharyngeal and esophageal regions, and a coat of mucus around the gut contents led the author to investigate the histochemistry of the mucous cells along the digestive tract and the importance of mucus in garfish digestion. The pharyngeal mill is the sole macerative process in hemiramphids and Tibbetts (1997) suggests that the particulate and dissolved nutrients released from plant cells come in close contact with acidic glycoproteins (AGP) produced by the mucous cells in the pharynx and esophagus. The highly viscous mucus absorbs water and water-soluble nutrients and becomes a gel that mixes with food during pharyngeal transport, forming a cylinder of food. In the intestine, the digestive and absorptive processes act upon nutrients held within the mucous matrix. Tibbetts (1997) proposed that mucus also plays a role in the enhancement of food utilization, in addition to trapping and aggregating particles and lubricating the digestive tract.

D. Digestive Enzymes

The chronological development of larval fish digestive systems and the related enzymes have been under intense investigation during the past 20 years, particularly larvae of marine fishes with potential for aquaculture.

However, much less work was focused on the early ontogeny of digestive enzymes in tropical fish larvae.

Several authors emphasized the importance of live organisms as a first food, suggesting that larvae could utilize the enzymes in the food to improve the process of digestion until the digestive tract becomes completely differentiated and developed (Dabrowski and Glogowski, 1977; Lauff and Hofer, 1984; Munilla-Moran et al., 1990; Kolkovski et al., 1993; Galvão et al., 1997b). Kolkovski et al. (1993) found 30% assimilation and 200% growth increases in gilthead sea bream (Sparus aurata) larvae, a warmwater marine fish, fed enzyme-supplemented diets. In contrast, Kolkovski (2001) found no effects of enzyme supplementation for juvenile sea bass (D. labrax). New studies argue that the contribution from zooplanktonic exogenous enzymes in the Japanese sardine Sardinops melanoticus (Kurokawa et al., 1998) and D. labrax (Cahu and Zambonino-Infante, 1995) larval gut was not significant. Therefore, the effect of exogenous enzymes on the digestive process of larval fish is not uniformly accepted. Kolkovski (2001) stated that enzymes of live food origin contribute to larval digestion and assimilation, but their contribution may be in functions other than direct enzyme activity in hydrolysis of food. Another hypothesis regarding the possibility of the products of prey autolysis stimulating secretion of pancreatic trypsinogen and/or activating endogenous zymogens (Dabrowski, 1984; Person Le-Ruyet et al., 1993) requires further study. Garcia-Ortega et al. (2000) used decapsulated cysts of dormant Artemia in a feeding experiment with larval African catfish (Clarias) and concluded that there is no effect of "live organisms" on contribution of proteolytic enzymes in comparison to activity of enzymes in "homogenized digestive tracts" of fish. The difference among fish species is certainly a plausible explanation. However, the methods that this conclusion is based upon must be first validated. Pan et al. (1991) found that Artemia nauplii autolysis is likely due to cathepsins present in these organisms, only marginally recognized in biochemical assays because of substrate affinity differences compared to serine endoproteases. Applebaum et al. (2001) determined the optimum pH (7.8) and temperature (50 °C) for chymotrypsin activity in larval red drum. The authors stressed, however, that proteolytic activity monitored with biochemical assays (synthetic substrate) may have resulted in part from other proteases (cathepsins). Zymograms on day 1 post hatching showed an absence of bands in the area of proteins with caseinolytic activity and molecular weight (28 kD) corresponding to serine endoproteases.

According to Cahu and Zambonino-Infante (2001), fish larvae do not lack digestive enzymes, and the onset of digestive functions follows a sequential chronology during the morphological and physiological development of fish larvae. Indeed, most studied larvae that lack a stomach at the beginning of exogenous feeding possess alkaline proteases at the time of mouth opening, certainly prior to the first feeding (*Paralichthys olivaceus*, Srivastava *et al.*, 2002; *Sparus aurata*, Moyano *et al.*, 1996; *Solea senegalensis*, Ribeiro *et al.*, 1999; *Morone saxatilis*, Baragi and Lovell, 1986; *Lates calcarifer*, Walford and Lam, 1993; *Dicentrarchus labrax*, Cahu and Zambonino-Infante, 1994). Verri *et al.* (2003) forced zebrafish (*Danio rerio*) larvae of 4–10 days (after fertilization) to ingest particles from a solution of 0.04% m-cresol purple to measure intestinal pH *in vivo* (7.5). The microclimate of the epithelial brush border membrane in the larval fish lumen is critical for peptide transport.

Srivastava *et al.* (2002) demonstrated, through whole mount *in situ* hybridization, expression of the mRNA precursors of trypsin, chymotrypsin, lipase, elastase, carboxypeptidase A and B in the flounder *P. olivaceus* pancreas at first feeding (3 days after fertilization). *In vitro* assays revealed that some proteins (thyroglobulin, albumin, and lactate dehydrogenase) are rapidly cleaved to polypeptides whereas other proteins (ferritin or catalase) were resistant to hydrolysis. Based on literature data, Srivastava *et al.* (2002) summarized the larval fish digestion processes at early stages as follows: the food particle is engulfed and arrives at the intestinal lumen without any pre-digestion; in the intestinal lumen the pancreatic proteases cleave proteins in the food to amino acids and polypeptides; further, as a result of the action of the intestinal epithelium aminopeptidases, the polypeptides are digested to amino acids and smaller peptides; monomeric amino acids are absorbed by the enterocytes while the remaining peptides are taken up by pinocytosis in the epithelium of the second segment of the intestine. Without a stomach, protein digestion occurs first in the larval intestine, where the pH remains alkaline until the development of gastric glands and secretion of HCl that reduces pH to acidic (Walford and Lam, 1993; Stroband and Kroon, 1981). Yamada *et al.* (1993) purified a proteolytic enzyme from Nile tilapia stomach that had optimum activity at pH 3.5 and 50 °C. This is a considerably different pH than in most other studies with fish pepsins where maximum activities for crude stomach extracts were between 1.8 and 2.5. Tilapia stomach protease was inhibited by pepstatin and classified as aspartate protease.

In tropical catfish, surubim *Pseudoplatystoma fasciatum* (1-2 DAH, ~3.86 mm SL) pancreatic protease activities were already detectable at the time of endogenous feeding (Portella *et al.*, 2004) when zymogen granules were also observed on histological sections of the larval pancreas (Portella and Flores-Quintana, 2003a). Trypsin and chymotrypsin activities increased soon after the beginning of exogenous feeding (3 DAH, 5.65 mm SL). Later increases in activities of both endopeptidases were observed after 30 DAH (26.8 mm SL). Pepsin-like activity increased at 10 DAH (11.3 mm SL) and coincided with the appearance of the first gastric glands in the stomach and the decrease in pancreatic alkaline protease activities. Amylase activity increased from 6 DAH on.

A similar tendency of a decrease in trypsin-like activity after the stomach became functional was found in *Lates calcarifer* larvae (Walford and Lam, 1993), but the synchronization found in appearance and increase of peptic activity of surubim gastric glands was less evident (Portella *et al.*, 2004). The stomach was well differentiated on day 13 but pepsin-like activity rose only on day 17, when the pH at the presumptive stomach region became more acidic. In a tropical cichlid, discus *Symphysodon aequifasciata* juveniles, Chong *et al.* (2002a) observed trypsin activity since hatching, increasing after exogenous feeding and having a peak at 10 DAH. At 20 DAH the trypsin-like activity increased again. The chymotrypsin-like activity was very low at the time of first feeding and increased at 15 DAH. While pancreatic enzymes showed considerable activity in discus early larvae, pepsin-like activity was very low until 25 DAH but significantly increased only at 30 DAH. In discus juveniles, despite the appearance of the stomach on day 10 after hatching, pepsin activity was observed several days later. In the euryhaline species, *Mugil platanus*, trypsin activity was detectable at the first feeding while chymotrypsin and pepsin activities were not found up to 29 days after hatching (Galvão *et al.*, 1997b).

In a marine teleost, red drum *Sciaenops ocellatus* from the warm coastal waters of Texas (27.6 °C), trypsin, amylase, and lipase activities were measurable at hatching and reached highest activity at 3 DAH, before first feeding (Lazo *et al.*, 2000). The activities then decreased in the following days and subsequently increased again 10 DAH. In contrast to their initial hypothesis, the authors concluded that the dietary regime (live or mixed live and artificial food), the presence of live organisms, and their possible influence on the analyzed enzymes were not significant during the early development of larval red drum. The authors concluded that "diet type does not appear to be a controlling factor regulating trypsin, lipase and amylase activity in the first feeding" larval fish. This conclusion strongly contrasts to what is known about digestive processes in other animals. However, that conclusion may not be accurate, first, because a "snapshot" activity measurement does not account for the dynamics of digestive processes (synthesis, secretion, activation of zymogens, reabsorption and degradation) (see Rothman *et al.*, 2002); second, no distinction was made for zymogen and active enzyme activity (see Hjelmeland *et al.*, 1988); and third, no attempt was made to quantify protease and other cytoplasmic inhibitors. Lazo *et al.* (2000) addressed the question of activities of enzymes from non-digestive tissues, and reported that trypsin-like activity amounted to only between 2 and 7% of that in the digestive tract. However, this approach does not account for protease inhibitors present in the whole body or "eviscerated whole body."

In contrast to these findings, Peres *et al.* (1998) concluded that when the protein intake of *Artemia* was differentiated in the two parallel dietary groups (satiation and 1/8 of satiation level), specific activity of trypsin was significantly higher in the "satiation" group. The proteolytic enzyme response highly correlates with the observed growth rate. However, when 20–40-day-old juvenile sea bass (*Dicentrarchus labrax*) were offered diets containing 29.2 or 59.9% protein (fish meal-based), there was no significant difference in trypsin activity. Further analysis of experimental design indicates that the result was not surprising if considered in conjunction with the fact that there was no growth of fish during the first 10 days of the experiment (signifying no feed intake, or no feed utilization) and negligible weight gain in both treatments in the following 10-day period. It may be then speculated that despite different "concentrations" of protein in the diets, the absolute amount ingested was the same and resulted in the same growth rates. In other words, it would be more appropriate to measure protein intake and correlate these levels with enzyme activity response.

In the euryhaline species *Mugil platanus*, trypsin, chymotrypsin, and pepsin activities were present in 1-year-old juveniles (Galvão *et al.*, 1997b). Albertini-Berhaut *et al.* (1979) found that gastric specific activity measured at pH 2.2 (expressed per g of stomach or soluble extractable protein) in three species of *Mugil*, *M. auratus*, *M. capito*, and *M. saliens*, decreased exponentially with size between 15 (larvae) and 135 (juvenile) millimeter standard length. The authors associated these declines in pepsin activity with dramatic changes in diets of mullets where animal food is exclusively present in size classes of 10–30 mm, mixed food in 30–55 mm, and from 55 mm on, benthic diatomids and multicellular algae predominate. Pyloric ceca of adult *Mugil cephalus* (size not given) were used to purify (92-fold) trypsin and characterize its optimum pH (8.0) and temperature (55 °C). Thermostability of mugil trypsin was much lower than that of bovine trypsin. It was lost above 75 °C, when bovine trypsin appeared to be resistant to degradation (Guizani *et al.*, 1991). Synthetic serine protease inhibitor (SBTI) resulted in 93% inhibition of trypsin in mullet compared to 61–83% inhibition in crude enzyme preparations of sea bream (Diaz *et al.*, 1997).

Intestinal enzymes associated with the brush border of enterocytes were present in some fish species at first feeding, but not in others. Segner *et al.* (1989b) did not observe aminopeptidase activity in larvae of fresh-coldwater whitefish *Coregonus lavaretus* while this activity was present in marine sea bream *Sparus auratus* (Moyano *et al.*, 1996) and turbot *Scophthalmus maximus* (Couisin *et al.*, 1987). Kurokawa *et al.* (1998), using immunohistochemical assay, showed that synthesis of the brush border aminopeptidase in the epithelial cells of Japanese flounder (*P. olivaceus*) begins before hatching and is completed before first feeding. Mullet, *Mugil platanus*, larvae exhibited carboxypeptidase A and B activities since the beginning of exogenous feeding (Galvão *et al.*, 1997b). Cahu and Zambonino-Infante (2001) showed the high leucine–alanine peptidase activity in young sea bass that decreased by day 25. The same pattern was reported by Ribeiro *et al.* (1999) in sole, *S. senegalensis*, larvae. The decrease of cytosolic enzyme activity is accompanied by an increase of the activity of the brush border enzymes, such as alkaline phosphatase (Cahu and Zambonino-Infante, 1997; Ribeiro *et al.*, 1999). These changes are characteristic of the final differentiation of enterocytes in developing animals (Arellano *et al.*, 2001).

The presence of food in the gut induces the process of synthesis of pancreatic enzymes and starvation turns it off. Chakrabarti and Sharma (1997) documented that after 4 days of feeding cyprinid *Catla catla* larvae with zooplankton, withdrawal of food resulted in a decline of proteolytic activity to less than half, indicating an effective mechanism preventing digestive enzyme loss.

Information about digestive enzymes in tropical juvenile and adult fish is scarce and in some cases contradictory. In juvenile discus *S. aequifasciata*, an acidic protease in the stomach region with an optimum pH of 2–3, and an alkaline protease in the intestinal region have been shown, with two optimum alkaline pH ranges (8–9 and 12–13), suggesting the presence of two groups of alkaline proteases in the intestinal lumen of this species (Chong *et al.*, 2002b). Specific biochemical analysis revealed the presence of trypsin and chymotrypsin, as well as metallo-proteases and non-trypsin/chymotrypsin serine proteases. These characteristics led the authors to suggest a protein digestion model similar to that found in other fish species with endoprotease hydrolysis followed by the release of individual amino acids by exoproteases. Yamada *et al.* (1991) characterized serine protease activity isolated from Nile tilapia intestines as having maximum activity at 55 °C and optimum pH at 8.5–9.

Activities of trypsin, chymotrypsin, amylase, and lipase were determined in liver, pancreas, two portions of the medium intestine, and rectum of 1-year-old surubim, *P. fasciatum*, fed with artificial diets (Portella *et al.*, 2002). The highest enzymatic activity was observed in the compact pancreas. Chymotrypsin and lipase were observed in the lumen of all analyzed segments. High levels of trypsin activity showed that this enzyme is of predominant importance for this species. Pepsin-like activity was found in stomachs of juvenile surubim. A similar trend in the digestive enzyme activities was reported by Uys and Hecht (1987) in an African catfish, *Clarias gariepinus*.

However, Olatunde and Ogunbiyi (1977) reported higher pepsin activity than trypsin-like activity in three tropical catfishes of the family Schilbeidae (*Physailia pellucida, Eutropius niloticus,* and *Schilbe mystus*). Seixas-Filho *et al.* (2000c) argued that trypsin activity in the intestinal lumen corresponds to feeding habits of fish, however; specific activity differed in two omnivorous fishes, *Brycon orbignyanus* and *Leporinus fridericci*, by 10-fold and a carnivorous fish, *Pseudoplatystoma coruscans* had an intermediate trypsin activity.

Sabapathy and Teo (1993) reported that trypsin-like activity was restricted to the intestine and pyloric ceca of *Lates calcarifer*, a carnivorous species, while in the herbivorous rabbitfish, *Siganus canaliculatus*, the activity appeared over the entire digestive tract. On a per weight basis of digestive tract tissue, proteolytic activity in the stomach was 20 times higher in the carnivore, whereas activity of trypsin-like enzymes was 20 times higher in herbivorous rabbitfish. Again, these conclusions, although interesting, remain largely speculative, if not misleading, as no data on the nutritional status was provided. Pepsin-like activity in rabbitfish was almost 10-fold higher in the esophagus than in the stomach and this finding needs re-examination, as most likely, cross-contamination occurred. A significant quantity of trypsin-like activity was also found in the esophagus and stomach, which seems to suggest some *postmortem* changes. The authors did not analyze esophagus in these species for the presence of "gastric" secretory glands that may have been involved in production of pepsinogen so this result remains to be re-analyzed. In *Sarotherodon mossambicus* the total proteolytic activity in the lumen drastically decreased along the digestive tract (Hofer and Schiemer, 1981). In carnivorous species, proteolytic activities are reported to be much higher than those of other enzymes (Hofer and Schiemer, 1981; Kuzmina, 1996; Hidalgo *et al.*, 1999), although a clear-cut comparison (for instance, expression per unit of reaction product, e.g. glucose, amino acid) is missing.

Analyzing activities of enzymes solely secreted into the gastrointestinal lumen rather than in homogenates of tissues and presenting results grouped into classes of activities in an ascending manner for the tropical matrinxã, *Brycon melanopterus*, Reimer (1982) showed the enormous individual variation of activity within a particular treatment. No data were provided, however, on food acceptance and, as a consequence, the intake of diets with different protein levels (11, 28, and 57%) may have differed considerably as absolute protein intake per fish may overshadow the effect of "percent of protein" in the diet. The author suggested positive responses of trypsin and lipase activities to increased levels of their respective substrates in diets.

In the juvenile tropical fish surubim, the pancreatic proteolytic activity was represented by higher activity of trypsin than chymotrypsin (Portella *et al.*, 2002). However, this comparison does not include correction for affinities to specific substrates. The same preponderance of trypsin over chymotrypsin was found in *Silurus glanis* (Jónás *et al.*, 1983) and *Clarias batrachus* (Mukhopadhyay *et al.*, 1977). However, in silver and common carp (*Hypophthalmichthys molitrix* and *Cyprinus carpio*, respectively) the chymotrypsin activity was nearly four times that of trypsin (Jónás *et al.*, 1983). According to Zendzian and Barnard (from Buddington and Doroshev, 1986), the existence of interspecific variations of the trypsin: chymotrypsin ratio has been reported in several vertebrate species. Garcia-Carreno *et al.* (2002) examined proteolytic activity in tissues of *Brycon orbignyanus*, a fish genus occurring in the Amazon, Paraná and Uruguay Rivers) following washing contents of the stomach and intestine and homogenization in water. The optimum pH for pepsin (hemoglobin substrate) was 2.5 and activity declined to zero at pH 4. The first contradiction in this work was the fact that as the authors provided evidence for pepsin being unstable at pH 5 as they used water (pH 6.5) to extract the enzyme from the tissue. Intestinal alkaline protease activities reached a peak at pH 10–10.5 in *Brycon* and the authors offered an explanation of trypsin-like activity (based on the response to TLCK specific inhibitor) in the intestine as being a result of a reabsorption process of pancreatic enzymes.

Sabapathy and Teo (1995) characterized activities of herbivorous, warm-water rabbitfish proteolytic enzymes trypsin-, chymotrypsin-like, and leucine aminopeptidase as having optimum temperatures at 55, 30, and 60 °C respectively, when incubated at the optimum pH 8–9. The optimum pH for trypsin in marine herbivore corresponded to that found in phytoplanktoni-vorous cyprinids (Bitterlich, 1985a). In tropical tambaqui, optimum temperatures for pepsin-like activity and alkaline protease activity were 35 and 65 °C, respectively, and these values corresponded to the thermal stability of these proteases; for instance, proteases isolated from pyloric ceca in tambaqui remained active following a 90 min incubation at 55 °C (De Souza *et al.*, 2000).

Pancreatic enzymes were also detected in the liver of surubim. In some species, the pancreas can be diffused in the abdominal cavity (Fange and Grove, 1979) and histological studies revealed that besides the compact pancreas (Portella and Flores-Quintana, 2003b) the pancreatic tissues also infiltrated into the liver of *P. fasciatum* (unpublished data). The exocrine pancreatic system was also observed in a closely related species *P. coruscans* (Souza *et al.*, 2001b; Seixas-Filho *et al.*, 2001), with endocrine cells embedded in the intestinal tissues. Benitez and Tiro (1982) prepared enzyme extracts from nine distinct regions of the milkfish digestive tract including the esophagus, stomach, intestine, pancreas, and liver. Trypsin- and chymo-trypsin-like activities were determined with synthetic substrates and total proteases with casein. After using a series of buffers, the optimum pH was found to be between 9.5 and 10, with maximum activity occurring at 60 °C. The authors suggested that the spiral folds and numerous mucous glands in the milkfish esophagus need to be considered as a site of caseinolytic activity. However, the lack of measurable pepsin activity in the stomach of milkfish contradicts the findings of Lobel (1981), who indicated that milkfish feeding on green algae had one of the lowest pH values recorded in fish (1.9). Lobel (1981) concluded that without grinding food, the thin-walled stomachs of fish such as Pomacanthidae and Pomacentridae are capable of forming a pH of 3.4 (2.4–4.2) that can be as effective as trituration in releasing algal cell contents. Indeed, Lobel (1981) also confirmed that in *Mugil cephalus* and *Crenimugil crenilabrus*, the pH of stomachs were in the range of 7.2. Some more light is shed on the effect of algal food on a pH in the stomach by data collected in milkfish (*Chanos chanos*). Milkfish sampled in ponds rich in benthic unicellular algae had alkaline pH in the stomach (7.8), while fish reared in ponds with dominant food being filamentous green algae (*Chaetomorpha*) exhibited an acidic pH (4.28–4.62) (Chiu and Benitez, 1981).

Many plant ingredients in fish diets contain antiproteases. El-Sayed *et al.* (2000) replaced fish meal with different sources of soybean in Nile tilapia and found that growth rate declined in all treatments. *In vitro* tests of protease inhibition or the degree of dietary protein hydrolysis did not correspond to the biological value of soybean protein estimated in a growth trial.

Amylase activity was observed in the pancreas, liver, and anterior intestine of *P. fasciatum* (Portella *et al.*, 2002), suggesting that this species can digest carbohydrates and use plant food in their natural diets. The possibility of pancreatic tissue infiltration in the liver of surubim might suggest the origin of amylolytic activity in this organ. Other Siluriformes such as *Clarias batrachus* (Mukhopadhyay, 1977), *C. gariepinus* (Uys *et al.*, 1987) and *Schilbe mystus* (Olatunde and Ogunbiyi, 1977) also have high amylase activity. Das and Tripathi (1991) found higher amylase activity in the hepatopancreas of grass carp, *Ctenopharyngodon idella*, than in the intestine, whereas in milkfish amylase was the major carbohydrase (Chiu and Benitez, 1981). Hidalgo *et al.* (1999) stated that the amylolytic activity is a more reliable indicator of the nutritional habits (carnivorous or herbivorous) than proteolytic activity. For instance, their findings indicated that proteolytic activity of *Cyprinus carpio* and tench *Tinca tinca* represented 99.8% and 69.8% of the total proteolytic activity encountered in rainbow trout *Oncorhynchus mykiss*. In contrast, the amylase activity in trout represented only 0.72% of the carp activity.

Lipase activity was found in the pancreas, liver, intestine, and rectum (Portella *et al.*, 2002) of *P. fasciatum* 1-year-old juveniles. Higher activities were reported in the pancreas and the distal part of the medium intestine than in the rectum. Borlongan (1990) also demonstrated lipase activity in all segments of *Chanos chanos* intestines, with major activity in the anterior intestine, pancreas, and pyloric ceca. Das and Tripathi (1991) reported lipase activity in the intestine and hepatopancreas of *C. idella* but Olatunde and Ogunbiyi (1977) did not find lipase activity in the digestive tract of several catfish species, *Physailia pellucida, Eutropius niloticus,* and *Schilbe mystus.* The most intense lipase activity ever reported was in the brush border of the first two intestinal segments of Nile tilapia juveniles (10–12 months old) (Tengjaroenkul *et al.*, 2000).

In the stomach of surubim juveniles considerable lipase activity was found (Portella and Pizauro, unpublished data), and these results differ from the findings of Koven *et al.* (1997) that reported lower lipase activity in the stomach and anterior intestine in comparison to midintestine and rectum of turbot, *Scophthalmus maximus.* Two optimal pH ranges were found for lipase activity in milkfish *C. chanos,* one slightly acidic (6.8–6.4) and the other at alkaline pH (8.0–8.6). These results indicate the presence of intestinal and pancreatic lipases and the physiological versatility of milkfish in respect to lipid digestion (Borlongan, 1990).

The presence and role of endogenous cellulase in tropical fish digestive tracts are somewhat contradictory although most authors associate this activity with symbiotic bacteria harbored in fish intestines. In *Clarias batrachus* cellulase activity was detected in assays that used microcrystalline cellulose as substrate (Mukhopadhyay, 1977). However, this substrate and Na-carboxymethyl cellulose were not hydrolyzed by extracts of different segments of milkfish digestive tract (Chiu and Benitez, 1981). Das and Tripathi (1991) used an antibiotic (tetracycline) to separate bacterial and endogenous cellulase in the digestive tract of grass carp; a significant decline of cellulase activity being observed following the treatment. A more recent study by Saha and Ray (1998), which also used tetracycline, concluded that cellulase activity in the cyprinid fish rohu (*Labeo rohita*) is largely from intestinal bacteria. Prejs and Blaszczyk (1977) have found that the cellulase activity of intestinal contents depends on the types of plants or plant detritus ingested, although there was no relationship between cellulase activity and the concentration of cellulose in the diet.

E. Intestinal Nutrient Trafficking: Protein, Peptide, Amino Acid, Sugar and Vitamin Absorption in Fish

Intestinal transepithelial transport of nutrients reflects a general tendency for fish species to consume diets containing either more carbohydrates (herbivores and omnivores) or more protein/amino acids (carnivores). Seasonality, feeding migrations, and ontogeny result in dietary modulations that are reflected in the intestinal nutrient transport/uptake as well. The mechanistic basis for changes in nutrient absorption can be analyzed using brush border membrane vesicles (BBMV), intact intestinal tissue preparations (*in vitro*), and the "whole animal" approach where nutrients acquired are measured along the digestive tract. In Mozambique tilapia (*O. mossambicus*) the first approach was used to measure uptake of an amino acid (proline) and glucose when fish were fed for 4 weeks on a diet with either 60% or 17% carbohydrates (Titus *et al.*, 1991). The authors concluded that an increase of carbohydrates in the diet of this species, maintained at the lower end of acceptable water temperatures for growth (24–25 °C), resulted in a higher maximum uptake rate of glucose. What the authors did not mention was that changing the carbohydrate source (plant meals) also changed protein quantity, 65 and 4% fish meal in low and high carbohydrate diets, respectively. There was no significant effect of protein concentration in tilapia diets on Na-mediated uptake of L-proline (dispensable amino acid). In two out of four experiments, K_m (Michaelis constant) values for proline uptake were 2.5 to 3-fold higher in BBMV preparations from fish fed a high protein level diet that would suggest increased transporter-substrate interaction, and/or affinity. In other words, amino acid traffic is modulated by dietary protein level, but this interpretation certainly goes beyond the authors' intent. Uptake of indispensable phenylalanine in a similar preparation from tilapia was characterized by a 10-fold higher affinity (K_m values) than that for proline (Reshkin and Ahearn, 1991). This work is also significant because it demonstrates for the first time in tropical fish that dipeptides are absorbed by different transporters than amino acids and are characterized by much higher affinity (9.8 and 0.74 mM in the case of Phe-Gly and Phe, respectively). The dipeptide-containing essential phenylalanine was 95% hydrolyzed intravesicularly within 10 seconds. This evidence of intercellular hydrolysis in intestinal epithelial cells added a new dimension to our understanding of hydrolysis by the brush border membrane (Figure 5.4). Further studies by Thamotharan *et al.* (1996a) addressed the mechanism of dipeptide uptake in tilapia brush border membrane vesicles. Results suggested that dipeptides are transported via mucosal surfaces by Na-independent, proton gradient dependent, saturable (high affinity), and unsaturable (low affinity, 1–10 mM) mechanisms. Interestingly, when BBMV preparations were preloaded with a suite of different dipeptides (some containing indispensable amino acids; Gly-Leu, Gly-Phe), uptake of Gly-Sar (nonhydrolyzable dipeptide) was not significantly affected, neither inhibited nor trans-stimulated. The authors concluded that only some dipeptides shared common transporters, whereas others are relatively specific. Recently, Verri *et al.* (2003) reported that the peptide transporter of the mammalian PEPT1-type is abundantly expressed in larval zebrafish (*Danio rerio*) prior to digestive tract

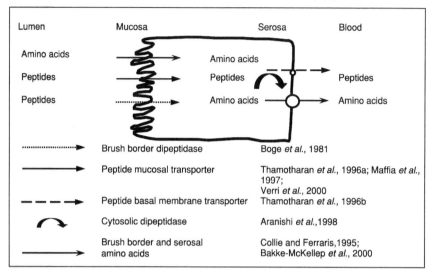

Fig. 5.4 Dietary protein, peptide, and free amino acid absorption and/or hydrolysis and absorption in teleost intestinal enterocytes.

differentiation, 4 days after fertilization (28 °C) in the proximal intestine. It suggests that the stomachless larval teleost is completely adapted to a high capacity transport system of dipeptides at alkaline pH at the time of first exogenous feeding. To complete the picture, Thamotharan et al. (1996b) examined the basolateral (serosal) transporters of dipeptides in tilapia and concluded that they are distinctly different from mucosal transporters. The former are characterized by inhibition of Gly-Sar transport by several other dipeptides and are evidently shared transporters. Intestinal aminopeptidases isolated from tilapia (*O. niloticus*) were characterized by optimum temperatures at 40–50 °C and significant specificity (affinity) toward substrates (K_m varied between 0.1 (Ala-pNA) and 2.0 mM (Val-pNA) (Taniguchi and Takano, 2002).

In warmwater common carp, intestinal cytoplasmic dipeptidase was characterized as most active at pH 9 and temperature 60 °C against the substrate L-leucine-glycine (Aranishi et al., 1998). The specificity of this enzyme that targets X-glycine and L-leucine-X peptides has been established, although within this group of peptides the efficiency against L-Lys-Gly and L-Met-Gly were 1 and 5%, respectively, in comparison to L-Leu-Gly (100%). There are many cytoplasmic dipeptidases in epithelial cells of the intestine, however, as demonstrated in crude extracts that had much broader activity against multiple peptides. For example, L-Lys-Gly and L-Met-Gly hydrolysis rates amounted to 8 and 123% of the major (control) substrate

(L-Leu-Gly). Some substrates were completely unhydrolyzed, such as Gly-L-Tyr or L-Pro-Gly.

These findings point out that transepithelial transport of intact dipeptides may be as important in fish as observed in mammals and of greater quantitative significance than transport of free amino acids (Figure 5.4). Intact peptide transport in concert with the hydrolytic capacity of intestinal mucosa, present as brush border dipeptidases early on in tilapia larvae (Tengjaroenkul *et al.*, 2000), may explain the nature of absorption and utilization of a mixture of dietary dipeptides in the fish digestive tract. Dabrowski *et al.* (2003a) demonstrated that a dipeptide-based diet resulted in growth of cold water salmonid fish when a free amino acid mixture-based diet failed and resulted in weight loss of first feeding rainbow trout alevins. Administration of dipeptide-based diets to larvae of tropical fishes needs to be tested.

Peptic digestion in the fish stomach does lead to a very limited release of free amino acids, however, the most significant part of protein hydrolyzates in the rainbow trout stomach are in the form of small peptides of 300–1700 Daltons (MW) (Grabner and Hofer, 1989). In both warmwater common carp and cold-water rainbow trout a high molar concentration of free amino acids is maintained until the posterior intestinal lumen, although in the stomachless carp it was higher (543 mM) compared to trout (147 mM) (Dabrowski, 1986b). This phenomenon in concert with a high proportion as peptides (50% of total protein amino acids; less than 10 kD peptides), points to a major role of protein hydrolyzates as securing continuous, stable rates of amino acid provisions to sites of protein synthesis. Partial hydrolysis allows the maintenance of fractions that show gastrin and cholecystokinin immunoreactivity and may result in stimulation of exocrine secretion and acceleration of digestion (Cancre *et al.*, 1999).

After reviewing the mechanisms of nitrogen compound uptake at the digestive tract level, we may turn to the speculative accounts presented by Bowen *et al.* (1984), who claimed that amino acids, detected as organic nitrogen in detritus "are part of non-protein nitrogen in some unknown form." However, in another work, the same author provided quantitative data as mg/g of total amino acids (Bowen, 1984) in comparison to mg/100 mg of amino acids (Bowen, 1980). Based on the Bowen (1980) data, concentrations of 197 (protein) and 674 (non-protein) mg amino acid per 1 g of ash in the tilapia stomach would correspond to 1.97 and 6.74 mg amino acids per 100 mg of diet. This concentration is probably characteristic for detrital material but extremely low in comparison to any plant or animal material (20–60 mg amino acids per 100 mg dry weight). During phytoplankton blooms, the dissolved amino acids combined, were composed of at least 60% of small 1000 Dalton peptides and significant amounts of free amino acids and were used for bacterial production (transferred to dissolved protein amino acids) (Coffin, 1989; Rosenstock and Simon, 2001). There was no indication of a measurable amount of some "unknown" amino acid fraction other than peptides that should be detected by standard ninhydrin methods. In effect, Mambrini and Kaushik (1994) documented that when 25 or 50% of protein in Nile tilapia diets were substituted with a mixture of six dispensable amino acids and were fed to fish, 10 and 50% growth depressions were found, respectively. This was accompanied by decreases in ammonia excretion rates and almost doubled urea excretion rates. The authors suggested that an excess of dispensable amino acids reduced protein synthesis rates in tilapia and free amino acids entered intermediary metabolism, resulting in deamination and enhanced urea synthesis. To extend these results to the ecological context, we may argue that protein sources other than "detrital aggregates" (highly deficient in indispensable methionine that would almost eliminate weight gains) must have been the source of essential amino acids in tilapia food. An incidental sampling may have missed periods of abundant, protein-rich dietary sources. In other words, exclusive detritivory in fish can hardly support rapid growth based on nutritional requirements as we know them.

Intestinal absorption of vitamins has only been characterized in warmwater catfish in respect to riboflavin, biotin, nicotinamide, folic acid (Casirola et al., 1995), and ascorbic acid (Buddington et al., 1993). In catfish, in contrast to riboflavin and biotin, which had a saturable mechanism of absorption, nicotinamide and folic acid absorptions were not inhibited by their increased concentrations and appear to be transferred by simple diffusion. Riboflavin and biotin transfer is mediated by specific carrier(s) on intestinal cells in catfish. Ascorbic acid is translocated via the high affinity, Na-dependent transporter and negligible amounts are absorbed via passive influx (Maffia et al., 1993). Buddington et al. (1993) calculated that the capacity to absorb reduced ascorbate in the catfish intestine, measured at relatively low temperatures of $20\,^\circ C$, still exceeded estimated daily requirements by 3 orders of magnitude.

Vilella et al. (1989), in studies on myoinositol transport in tilapia brush border and basolateral membrane vesicles, demonstrated that the intestinal brush border of herbivorous tilapia exhibited a significantly higher apparent binding capacity for myoinositol and a lower apparent maximal uptake rate than did the intestine of the carnivorous eel (*Anguilla anguilla*). However, no such adaptation is present at the epithelial basolateral membrane.

F. Gut Microflora and Symbiotic Organisms

The bacterial flora is always present in the fish environment in association with the trophic chain of microalgae and zooplankton or formulated diets. Bacteria may be ingested by rotifers or other zooplankton. Bacteria may be releasing nutrients to be utilized by zooplankton. Alternatively, bacteria can be directly ingested by larval (Nicolas *et al.*, 1989) and adult fish (Beveridge *et al.*, 1989; Rahmatullah and Beveridge, 1993). Nicolas *et al.* (1989) indicated that none of the bacteria in algae culture was subsequently isolated from rotifers, and most bacteria from rotifers were not found in fish larvae, suggesting that bacteria very selectively choose their biotope. Exoenzymatic activities of bacteria released in fish digestive tracts may be one of many attributes in which the bacterial flora enhances food utilization in the fish intestine. Hansen and Olafsen (1999) demonstrated endocytosis of bacteria in the posterior intestine (second segment) of herring larvae, but no similar accounts were described in tropical fish.

The involvement of intestinal bacterial microflora in vitamin (and most likely other essential nutrients) synthesis in warmwater and tropical fishes seems to be quantitatively more important than in cool- and cold-water fishes (see Nutrient requirements: Vitamins). However, as Sugita *et al.* (1992) pointed out, the net dietary requirements for essential nutrients may be the difference in the metabolic capacity of vitamin-producing (anterior intestine) and vitamin-consuming (posterior intestine) microflora. Consequently, the authors suggest that in the case of dietary biotin, coolwater ayu, *Plecoglossus altivelis*, with a large population density of biotin-producing bacteria, does not require a dietary vitamin source. The warmwater common carp and goldfish, with dominant biotin-consuming microflora in their intestine, are completely dependent on dietary biotin needs.

Both free-living and particle-bound bacteria are ingested by warmwater fishes and their concentrations in the water correlate with the number of colonies found in the stomach (Beveridge *et al.*, 1989). Based on counting the colonies of viable bacteria grown from a series of dilutions of intestinal fluids, juveniles of four cyprinids were associated with the ability of active ingestion of *Chromobacterium violaceum*, but the mechanism of ingestion by drinking has been ruled out (Rahmutullah and Beveridge, 1993). However, because the bacteria generation time is shorter than 1 hour, the processes of bacterial replication and cell disintegration (lysis) were not accounted for in the experiments, which lasted up to 4 hours. In similar conditions ingestion rates of free-living bacteria by *O. niloticus* were 4–5 orders of magnitude higher than in cyprinids.

Adult surgeonfish (*Acanthurus nigrofuscus*), herbivorous fish in the Red Sea, have very few small pyloric cecae, but the relative intestine length is from 2.1 to 3.9 body lengths depending on the season (Montgomery and Pollak, 1988a). Despite this, the gut of this species harbors symbiotic megabacteria *Epulopiscium fishelsoni* (Montgomery and Pollak, 1988b), "big bacteria" (Schultz and Jorgensen, 2001) that are 70–200 μm long. In surgeonfish collected at night (non-feeding), gastric pH dropped to as low as 2.4 and 2.9 in the gastric and pyloric regions suggesting a significant impact of ingested algae on the neutralization of the intestinal lumen. The maximum density of symbionts was found at 40–60% of gut length (Fishelson *et al.*, 1985), corresponding to a significant change in the lumen pH from 7.5 to 6.5. The pH rose then again to 7.7 as the number of megabacteria declined by 3 orders of magnitude. Collectively, bacterial symbionts exert strong, seasonally and diurnally variable influences on the digestive physiology of herbivorous tropical surgeonfish (Fishelson *et al.*, 1987), but the mechanisms in which they may help the process of algal food digestion and nutrient accretion were not addressed. There were other investigators working with marine warmwater kyphosid fishes (*Kyphosus cornelli* and *K. sydneyanus*) at 22.9 °C, who realized that gut resident microflora harbored in specialized parts of the digestive tract, were capable of producing volatile fatty acids (Rimmer and Wiebe, 1987). This was the first report of fermentative digestion of algal carbohydrates in fish. The unique features of these fishes include an elongated digestive tract (relative intestine length 3.3–5.8 body lengths), blind digestive sacs (cecae) with adjacent valves, diverse microflora and large ciliates. Clements and Choat (1995) provided a comprehensive account of 32 species representing 5 families, including surgeonfishes and kyphosids, that were capable of producing volatile fatty acids through fermentation in the posterior intestine, where concentrations of acetate ranged from 3 to 40 mM. Substantial amounts of acetate and other short chain fatty acids in the blood of these fishes suggest that fermentation of algal polysaccharides, and possibly proteins, is an important source of energy. Therefore, Fishelson *et al.*'s (1985) discovery of megabacteria in the gut of surgeonfish led to the explanation of their role in digestion. There is now conclusive evidence that fermentation is a quite common process contributing to the nutrition of many marine, predominantly herbivorous, fishes.

III. NUTRIENT REQUIREMENTS

A. Protein Quantity and Quality

Despite the fact that there is a measurable requirement for essential amino acids in fish rather than protein, the discussion frequently drifts toward protein needs and consequently numerous controversies may arise. Bowen (1987) very strongly argued, based on comparison of 13 teleost fishes and higher vertebrates (birds and mammals), that based on weight or growth achieved per weight of protein ingested, there is no reason to suggest differences in protein requirements between fishes and terrestrial homeotherms. Although protein requirements for maximum growth determined when provided *ad libitum* or at restricted ration is considerably different because protein is used for different physiological purposes, requirements will differ considerably (by 2 orders of magnitude when expressed per unit weight) depending on fish size and water temperatures. These factors were not considered in the above comparison. Clearly, intake of diet in fish is not related to concentration of protein but rather to daily protein intake that linearly correlates with specific growth rate (Tacon and Cowey, 1985). There is an opinion that water temperature does not affect dietary protein requirements, although the evidence in some cases is not convincing (Hidalgo and Alliot, 1988). Studies with typical tropical fish are, however, missing. The problem is also confounded by the fact that lower protein concentrations in feeds (plant, detritus, bacterial) are associated with lower biological value of those materials in comparison to animal protein. Bowen (1980, 1984) argued that "detrital non-protein amino acids" may explain the rapid growth of *Sarotherodon mossambicus* (Lake Valencia, Venezuela) and *Prochilodus platensis* (Rio de la Plata, Argentina), two species in distinctly different environments, where microorganisms and detritus in their diet are a common denominator. The first question that arises is the amount and quality of the "non-protein" source of amino acids (Dabrowski, 1982, 1986a). The difference in expression of amino acid composition in stomach content differs by an order of magnitude. The interpretation in terms of essential amino acid concentrations is misleading as methionine is highly deficient in detritus in comparison to requirements (see also Table 5.1).

Table 5.1
Amino Acid Requirements in Cold-water and Tropical Fishes (Expressed as % Dietary Protein)

	Pacific salmon[1]	Tilapia[2]	Cyprinid[3]	Milkfish[4]
Water temperature	14–16 °C	27 °C	27 °C	28 °C
Aminoacid				
Arginine	6.0	4.20	4.80	5.2
Histidine	1.6	1.72*	2.45*	2.0*
Isoleucine	2.4	3.11*	2.35	4.0*
Leucine	3.8	3.39	3.70	5.1*
Lysine	4.8	5.12*	6.23*	4.0
Methionine	3.0	3.21*	3.55*	3.2*
Phenylalanine	6.3	5.59	3.70	5.2
Threonine	3.0	3.75*	4.95*	4.5*
Valine	3.0	2.80	3.55*	3.6*
Tryptophan	0.7	1.00*	0.93*	0.6

*Indicates values higher than in salmon.

Sources: [1]Akiyama et al., 1985; [2]Santiago and Lovell, 1988; [3]*Catla catla*; Ravi and Devaraj, 1991; [4]Borlongan and Coloso, 1993.

The most researched among tropical fish, in respect to nutrient requirements, is the cichlid Nile tilapia, although it is debatable to what extent it is a "representative" species for tropical fishes. De Silva *et al.* (1989) summarized data on protein requirements in four species of cichlids analyzed in a range of water temperatures (23–31 °C) and sizes (0.8 mg to 70 g) and calculated that maximum growth was supported by a diet containing 34% protein. Wang *et al.* (1985) made their conclusion in regard to the optimum level of protein based on better performance of Nile tilapia with 30 rather than 40% protein, with fish increasing weight only 2- to 2.5-fold. This was hardly a sufficient weight gain to make requirement estimation. Santiago *et al.* (1982) demonstrated that diets for Nile tilapia juveniles with excess of 63% of fish meal, depressed fish growth.

In pursuit of more accurate protein requirements for juvenile (0.8 g) and young (40 g individual weight) Nile tilapia, Siddiqui *et al.* (1988) suggested respective optimum levels of 40 and 30% protein. However, this study suffered from the excessive use of fish meal and high levels of ash (14.8%) in the "high protein" diets that had a detrimental effect on growth rates. Kaushik *et al.* (1995) presented in an elegant manner probably the best option to formulate the diets for protein need estimation in Nile tilapia. The authors used a constant proportion of fish meal:soybean meal (animal: plant) protein proportion (1:3) and demonstrated that weight gains continued to increase up to 38.5% protein in the diet. Furthermore, Kaushik *et al.* (1995) were able to confirm the earlier finding with common carp juveniles that the endogenous nitrogen (ammonia) excretion in fasted tilapia were significantly smaller than in fish fed up to 16% protein in their diets. This can be explained that at low protein intake, dietary energy sources (carbohydrates and lipids) have a sparing effect on endogenous protein use. This has far reaching implications for situations in the wild where many species of fish on low protein (herbivorous) diets can save body proteins and improve condition factor.

Further increases in percentage of fish meal up to 83% with high ash concentrations in tilapia diets seemed to further depress growth rates and led to an unrealistically low "optimum" protein requirement in tilapia hybrids in seawater (Shiau and Huang, 1989) and *Cichlasoma synspilum* (Olvera-Novoa *et al.*, 1996). The realization of a limited biological value of processed fish meal in diets for cichlids can be traced back to the work of Kesamaru and Miyazono (1978) who demonstrated a higher value of wheat germ protein than fish meal protein. This controversy is somewhat clarified when semi-purified, casein–gelatin-based diets were used in a series of studies to determine optimum protein levels for Nile tilapia reproductive efficiency, and offspring production and quality (Gunasekera *et al.*, 1995, 1996). Higher growth, earlier maturation, and high fertilization rate of eggs and hatching rates of larvae were associated with females fed diets containing 35–40% protein. A similar conclusion was reached in experiments with a fast growing and early maturing (4 months of age) tropical dwarf gourami (*Colisa lalia*) reared at 27 °C (Shim *et al.*, 1989). Maximum growth was achieved with diets containing 45% protein, whereas females with the largest ovaries were found in the 35% protein group. The highest quality of eggs (hatchability 94.1%) occurred in females fed a 45% protein diet in comparison to fish fed diets for over 20 weeks with low protein concentrations (5–15%; hatching rates 23.7–77.3%, respectively). It is evident that low quality food (low protein content) will predispose fish to reproductive failures.

An Amazonian frugivorous fish, such as tambaqui (*Colossoma macropomum*), utilizes seeds and fruits in inundated forests, an abundant food base, however, with relatively poor protein concentrations of 4.1 to 21.3% (Roubach and Saint-Paul, 1994). A protein-sparing effect may be at play as some fruit-bearing trees in the tropical forests provide fruits with extremely high lipid concentrations (e.g. 65% in dry matter; *Caryocar villosum*) (Marx *et al.*, 1997). However, in the Roubach and Saint-Paul experiment, the growth rate of fish was related to protein concentrations in diets and in general did not exceed 1.3% per day. In optimum conditions of 29.1 °C and fed a 48% protein diet, juvenile tambaqui of 1.5 g gained weight at 4.6% per day, whereas juveniles of 30 g grew best fed a 24:22 ratio of fish meal/soybean meal, 40% protein diet at the daily rate of 1.7% ($55 \, g \, kg^{-0.8}$; van der Meer *et al.*, 1995). Interestingly, fish of initial weight of 30 and 96 g achieved the highest growth rates at 40% protein in the diet, although maximum growth rates were significantly smaller, 22.7 and 15.3 $g \, kg^{-0.8}$, respectively. These results were dramatically different from the estimates of Vidal Junior *et al.* (1998), who concluded that tambaqui in the size range of 37–240 g individual weight had maximum gain on feeds containing only 21% protein. This discrepancy is puzzling as diet formulations were very similar. Experiments with tambaqui in tanks, however, tend to underestimate the growth potential for this species in comparison to growth in ponds (Melard *et al.*, 1993).

Tambaqui are known to consume equal amounts of fruits/seeds and zooplankton as adults (Goulding and Carvalho, 1982). When subjected to a purified protein, casein-based diet at 25 °C, juveniles gained weight at the highest rate when provided with a diet containing 47.7% protein (Hernandez *et al.*, 1995). This may be related to the juvenile size of fish (8.4 g) used in this experiment, when adaptation of the digestive tract to a voluminous, low nutrient/energy food, is not yet present. In juvenile milkfish of 2.8 g body weight, the protein requirement was estimated to be 43% (Coloso *et al.*, 1988) when raised at a water temperature of 25–29 °C and salinity 28–34 ppt. However, during the whole trial fish increased weight by only 139% and the diet was supplemented with a mixture of essential free amino acids up to 32%

of total protein. This may lead to an underestimated requirement if availability of raw proteins in the practical diets or natural foods decreases. Characin fish from the Paraná and Uruguay Rivers, *Brycon orbignyanus*, fed semi-purified diets based on casein and gelatin, grew best on a diet containing 29% protein, but the feed conversion ratio was best when protein level was raised to 36% (Carmo e Sa and Fracalossi, 2002). In warmwater cyprinid grass carp juveniles of 0.2 g weight maintained at 23 °C, a maximum growth rate was achieved with 43–52% dietary protein (Dabrowski, 1977). It needs to be emphasized that no growth depression was observed in both studies on *Brycon* and *Ctenopharyngodon* at high protein concentrations (see Shearer, 2000) when purified proteins were used in contrast to fish meal-based diets. It is an important consideration because it links artifacts in protein requirement (inaccurate term) determination with indispensable amino acids requirements. In other words, there is no need of protein as a dietary component because amino acids are the compounds essential for growth. For instance, Ravi and Devaraj (1991) noticed significant growth depression in the tropical cyprinid *Catla catla*, when purified amino acid mixture diets were fed with supplements of several amino acids (phenylalanine, threonine, tryptophan, and valine) above an apparent optimum level for growth (see also Table 5.1). These results may be related to a feeding rate in these experiments, which was set at 10% fish body weight per day whereas gains amounted to only 2.6–3.5%/day. The decrease in weight gain may have been due to feed waste.

Teleosts require 10 indispensable amino acids in food and tropical fishes are no exception (Table 5.1). It appears that requirements for most amino acids are higher in tropical fishes than in cold-water salmon, with a marked difference in arginine and phenylalanine. As alluded to earlier (Dabrowski and Guderley, 2002), data for juvenile tilapia (15–87 mg individual weight) amino acid requirements were collected over an extended period of time with body weight gains 15–79 times the initial weights (Santiago and Lovell, 1988). For comparison, data for *Catla catla* represent only a 1.6- to 2-fold weight increase and fish grew much slower (Ravi and Devaraj, 1991). Evidently free amino acid mixture diets are a major difficulty in arriving at acceptable growth rates in fish, including the tropical cyprinid, *Labeo rohita* (Khan and Jafri, 1993). It remains to be verified if weight gains of 20–25% over a period of 6 weeks can provide meaningful estimates of quantitative amino acid requirements. In contrast, in juvenile Nile tilapia, utilization of diets which contained up to 82% protein as crystalline amino acids was excellent and resulted in weight gains of 1672–7902% over the same period of time (Santiago and Lovell, 1988).

In warmwater common carp at 25 °C, requirements for indispensable amino acids were determined at a maximum growth rate of 1.5–3.5% per day

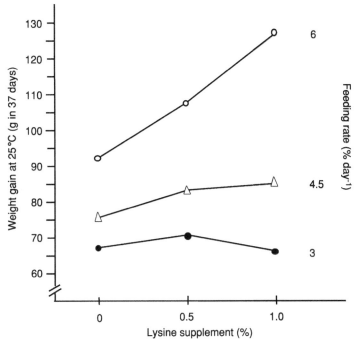

Fig. 5.5 Effect of lysine supplementation in common carp (125 g initial weight). (Based on data presented by Viola and Arieli, 1989.)

(Nose, 1979). As demonstrated earlier (Dabrowski, 1986a), these estimates were made for fish frequently performing at 10 times lower than maximum weight gains. This must lead to somewhat skewed results as illustrated in the case of common carp fed diets with two levels of supplemented lysine and three feeding rates (Figure 5.5; Viola and Arieli, 1989). The lysine requirement for maximum growth would have been higher than 2.6% in accordance with Figure 5.5 in comparison to an earlier estimate at a suboptimal growth rate of 2.2%, according to Nose (1979). Amino acid requirements in tropical fishes are clearly not sufficiently studied. They are not determined at optimum water temperatures for growth or with optimized diet formulations.

B. Lipids and Fatty Acids

In most cichlid fishes examined, the best growth was achieved within the first two weeks of feeding on diets supplemented with 10% lipids and growth enhancement was 150% over the lipid-free diet (Chou and Shiau, 1996). In subtropical catfish from the Paraná River basin, *Pseudoplatystoma coruscans*, the best performance of fish maintained at 26.5 °C was on diets containing 18% lipid (Martino *et al.*, 2003). This study was followed by work where lipids of profoundly different fatty acid profiles (pig lard and/or squid liver oil) were compared either as a single source or as mixture (Martino *et al.*, 2003). No differences were found as fish increased weight almost 10-fold. The proportion of polyunsaturated fatty acids (PUFA) (20:5n3 and 22:6n3) in fish carcasses differed among treatments, reflecting the nature of saturated fats in pig lard supplemented feeds. Similarly, Maia *et al.* (1995) found no PUFA in the neutral lipid fraction extracted from muscle of pond-raised pacu (*P. mesopotamicus*), most likely reflecting a plant ingredients-based diet. Viegas and Guzman (1998) compared gradual substitution of deodorized soybean oil (DSO) by crude palm oil (CPO) in diets of tambaqui and although weight gains of juveniles (14 g initial weight) were somewhat erratic, the best gain of 10-fold was achieved with 6% CPO (11.6% total dietary lipids). However, CPO contains considerable amount of carotenoids (500–700 mg/kg) and tocopherols (560–1000 mg/kg) that are thermally destroyed during refining, bleaching, and deodorization (Edem, 2002), so a comparison to DSO may also suggest the importance of carotenoids and vitamins E in tambaqui diets. An unfavorable profile of essential n3 and n6 fatty acids in CPO (10.5% unsaturates) compared to DSO (61% unsaturates) had lowered linoleate and linolenate in tambaqui.

Lipids are the source of essential linoleic and linolenic fatty acids that are common to fish and all other vertebrates. The ability of tropical fish to elongate and desaturate fatty acids is the metabolic feature that may differ among species and consequently may create limitations on growth. The differences in fatty acid requirements and optimum lipid levels seem, however, to occur between carnivores and herbivores, cold and tropical species, more decisively than between freshwater and marine fishes. The need for linolenic acid (18:3n3) or its derivatives (PUFA), is so profound among salmonids and most marine fishes that a need for linoleic (18:2n6) was frequently questioned or overlooked.

To the contrary, the quantitative linoleic acid requirements in Nile and Zillii's tilapia for maximum growth were estimated to be 0.5 and 1% of diets, respectively, met also by arachidonic acid (20:4n6), whereas the need for linolenic fatty acid tends to be neglected. When Nile tilapia were fed an experimental diet containing 1% linoleate as the only PUFA, carbon from radiolabelled linolenic acid was identified in all n3 series of 20 and 22 carbon fatty acids, mostly in phospholipids (Olsen *et al.*, 1990). Tilapia fed a diet containing a high level of n3 PUFAs exhibited a much lower rate of incorporation of radiolabelled carbon from linoleate and linolenate precursors into PUFAs. The authors concluded that tilapia desaturases and elongases are perfectly capable of converting linoleate to arachidonate as the predominant end product and docosahexaenate as the favored product of delta-4 desaturase. The conversion of linoleate/linolenate to longer chain PUFAs is suppressed by an elevated level of dietary 22:6n3. This phenomenon may be responsible for the discrepancy in demonstrating the dietary essentiality of both the n3 and n6 series in cichlid fishes. A supplement of cod liver oil, for instance, containing a high level of n3, resulted in the highest growth of tilapia in comparison to other oils of plant origin (corn or soybean); however, it coincided with the poorest reproductive performance measured by spawning frequency and number of offspring produced per female (Santiago and Reyes, 1993).

In studies with common carp larvae, Radunz-Neto *et al.* (1996) used a control diet with 2% phospholipid supplementation instead of a lipid-free diet. This source of lipids alone provided 0.192 and 0.014% of n6 and n3, respectively, on a dry diet basis, and was responsible for an increase in body weight of fish over 50-fold. This result illustrates how difficult it is to conclusively demonstrate the requirements of essential fatty acids in fish when yolk lipid reserves and traces in the purified diets must be accounted for. However, Radunz-Neto *et al.* (1996) were able to show that a supplement of 0.25% linoleate over the base level in a phospholipid containing diet improved growth by 27% in carp within 21 days of rearing. Deficiency symptoms became apparent within the next 5 days, when growth depression reached 67% in carp juveniles fed a linoleate/linolenate-deficient diet. Takeuchi (1996) confirmed the requirements of both n6 and n3 in common carp and (herbivorous as adults) grass carp and added new information regarding the pathology observed in the latter species when fed a diet without n6 and n3, but supplemented with methyl laurate (C 12:0). Grass carp showed in 85% of the population vertebral column abnormalities, such as lordosis, never before associated with lipid deficiency, although Meske and Pfeffer (1978) described similar pathologies in grass carp juveniles fed diets predominantly composed of plant (algae) proteins. Additional experiments further elucidated that vitamin E (tocopherol) deficiency was a contributing factor. The observation of increased deposition of Mead fatty acid (20:3n9) in polar lipids of the hepatopancreas in cyprinids as a marker of abnormal lipid metabolism is of great value to lipid physiology in warmwater and tropical fishes. In larvae of the subtropical fish curimbata (*Prochilodus scrofa*), enrichment of rotifers with cod oil-derived PUFAs did not result in appearance of these fatty acids in the fish body (Portella *et al.*, 2000). To the contrary, juvenile *Tilapia zilli* fed for 21 days with rotifers reared with microalgae (containing 20:5n3) had body lipids dominated by 22:6n3 (10.7–14.3%) (Işik *et al.*, 1999). Unfortunately, the authors did not present data of arachidonate (20:4n6) concentrations in the tilapia body and the fate of linoleate, one of two major fatty acids in rotifer lipids, remained unknown. In four species of tropical aquarium fishes at the larval or early juvenile stage (as soon as *Artemia* cysts were acceptable), offered either freshwater-cladoceran *Moina* or brine shrimp decapsulated cysts, fatty acids in fish bodies reflected precisely the dietary sources of lipids (Lim *et al.*, 2002). For instance, linolenate, which was the predominant PUFA in *Artemia* and 10-fold higher than in *Moina*, accumulated in the fish body in the same proportion (10:1). All four species synthesized 22:6n3; however, differences among fish species were significant.

Although most freshwater fish adhere to the established requirements of n3 and n6 as described in cichlids, Henderson *et al.* (1996) were the first to address this question in regard to tropical Serrasalmid fish, the herbivorous *Mylossoma aureum* and carnivorous red piranha *Serrasalmus nattereri*. The authors indicated that the fatty acid composition of these two species was notably influenced by the pattern of fatty acids available in the plant- (n6/n3 ratio, 34.7) or animal- (n6/n3 ratio, 4.4-6.2) based diets. Despite these ratios in the diets (dominated by linoleate), herbivorous *Mylossoma* is characterized by high activities of delta-6, delta-5, and delta-4 desaturases and the ability to convert linolenate to 22:6n3 (4.9% in total brain lipids). Linoleate is efficiently converted to arachidonic acid (5.4% in liver) in herbivorous piranha, and is fairly high in comparison to carnivorous piranha (7.6% in liver). The loss of delta-5 desaturase activity and inability to utilize C18 unsaturated precursors in marine carnivorous fish evidently does not apply to this tropical carnivore. It must be noted, however, that some of the features of fatty acid composition that characterized the herbivorous and carnivorous tropical species at 25 °C (actual water temperature for 9 months of experimental feeding) may change dramatically when temperatures will approach those characteristic for tropical waters. Craig *et al.* (1995) have shown in marine warmwater fishes that lowering water temperature over a 6-week period from 26 °C to chronic lethal temperatures (3–9 °C) resulted in an increase of highly unsaturated fatty acids in polar lipids. This was particularly evident in fish fed initially on plant, low n3 diets (corn oil). As desaturases and elongases in fish increase in activity due to lowered environmental temperatures (Hager and Hazel, 1985), it remains to be examined how an increase to 32–35 °C would affect lipid metabolism in piranhas compared to studies at 25 °C. In Nile tilapia reared at three temperatures, 15, 20, and 25 °C, the ratio of n3/n6 was highest at 15 °C; however, the prediction of the trend in the range of 30–35 °C (optimum for growth) would be speculative (Tadesse *et al.*, 2003). The proportion of 22:6n3 decreased in muscle lipids of tilapia, whereas arachidonate, linoleate, and linolenate showed the opposite trends with increasing water temperature.

In zooplanktivorous catfish from the Amazon River, mapara, *Hypophthalmus* sp., the composition of fatty acids in muscle total lipids included both C18 precursors of essential n3 and n6 and 2.4% of 22:6n3 and 2.5% of 20:4n6; their quantity was not greatly impacted by the dry or wet period of the year (Inhamuns and Franco, 2001).

C. Vitamins

Warmwater fishes, and tropical fishes in particular, may have a sufficient biomass of microorganisms in the digestive tract to provide a significant amount of water-soluble vitamins as an exogenous source (Burtle and Lovell, 1989; Limsuwan and Lovell, 1981). In the case of Nile tilapia juveniles kept at 28 °C, intestinal synthesis of cobalamine increased the level of this vitamin in feces over 100-fold in comparison to the level in food (Lovell and Limsuwan, 1982). Sugita *et al.* (1990) identified bacteria (*Aeromonas, Pseudomonas*) in intestinal contents of tilapia with the ability to synthesize vitamin B_{12} and argued that because tilapia harbor anaerobes well, it gives this species an advantage over channel catfish. The estimated synthesis of cobalamine expressed per unit body weight in tilapia juveniles (7.1 g) was nearly 10-fold higher than in channel catfish when diets supplemented with cobalt were given. The use of the antibiotic succinylsulfatiasole suppressed cobalamine synthesis in the intestine of tilapia. It may be speculated that in larger cichlids with an elongated intestine (Figure 5.3) the importance of bacterial synthesis of vitamins will be even more enhanced. Both qualitative and quantitative vitamin requirements of tropical fishes determined thus far (Table 5.2) are similar to those established in salmonids, common carp, or channel catfish. Kato *et al.* (1994) reported the qualitative requirement for water-soluble vitamins in a marine fish, the tiger puffer, maintained at 22–28.5 °C. The need for inositol, folate, and ascorbic acid were demonstrated only after 7–12 weeks of feeding, whereas growth depression due to missing choline or nicotinic acid was dramatic after only 2–3 weeks.

Burtle and Lovell (1989) suggested that *de novo* synthesis of inositol takes place in the liver and intestine of channel catfish and no dietary need was reported. However, earlier findings suggested that the common carp intestinal flora may provide sufficient amounts of inositol as well. To the contrary, Meyer-Burgdorff *et al.* (1986) were able to demonstrate hemorrhages, skin lesions, and fin erosion in common carp within 3–5 weeks on a myoinositol-free diet. A decrease of food intake followed the observed pathologies. The optimum requirement for common carp was estimated to be 1200 mg myoinositol/kg diet. This finding is significant because myoinositol, being a structural component of cell membranes as phosphatidylinositol, is critical in animals experiencing changes in water temperatures.

Vitamin C has been the most extensively studied vitamin in tropical teleost fishes and, as suggested earlier, ascorbic acid was confirmed to be

Table 5.2
Vitamin Requirements in Cichlids and Other Tropical Fish

Vitamins	Species	Concentration (mg/kg)	Deficiency signs	Reference
Pantothenic acid	*O. aureus*	10	Anorexia, fin and tail hemorrhages, sluggishness, clubed gills, intracellular proliferative lesions	Soliman and Wilson, 1992a
Riboflavin	*O. aureus*	6	Short body dwarfism, lethargy, lens cataracts, fin erosion, anemia	Soliman and Wilson, 1992b
	O. mossambicus x *O. niloticus*	5	Anorexia, lens cataracts, short body dwarfism	Lim *et al.*, 1993
Tocopherol	*O. niloticus*	50–100	Low hepatosomatic index, swollen, pale liver (low water temperature, 20 °C)	Satoh *et al.*, 1987
Ascorbic acid	*O. aureus*	50	Scoliosis, hemorrhages of fins, mouth, and swim bladder. Shortening and thickening of gill lamellae, irregular gill chondrocytes	Stickney *et al.*, 1984
	O. niloticus x *O. aureus*	20	Loss of pigmentation, necrosis of fins, hemorrhages all over the body	Shiau and Hsu, 1995
	Cichlasoma urophthalmus	40	Inflammatory response, spongiosis in epidermis, muscle inflammation, gill thickening, edema, hyperplasia, shrinkage of acinar cells in pancreas	Chavez de Matinez, 1990
				Chavez de Martinez and Richards, 1991
	Astronotus ocellatus	25	Deformed opercula and jaws, hemorrhage in the eyes, fins, lordosis	Fracalossi *et al.*, 1998
	Piaractus mesopotamicus	50	Hyperplasia, hypertrophy of gill filaments, twisted gill lamellas, inflammatory infiltrates at the end of gill filaments	Martins, 1995
	Clarias gariepinus	46	Broken-skull, hemorrhages, and dorsal fin erosion	Eya, 1996
	Pterophylum scalare	120	Not observed	Blom *et al.*, 2000

Nutrient	Species	Requirement	Deficiency signs	Reference
Choline	O. niloticus	3000	Reduced growth below and above requirement, anorexia, lens cataracts, short body dwarfism	Kasper et al., 2000
	O. niloticus x O. aureus	1000	Decreased lipid level in liver	Shiau and Lo, 2000
Pyridoxine	O. niloticus x O. aureus	4 and 15 at 28% and 36% protein, respectively	Anorexia, ataxia, edema, mortality in 3 weeks; anemia in hyperdose	Shiau and Hsieh, 1997
Provitamin A (astaxanthin)	O. niloticus	71–132	Pathological structural alterations in hepatocytes	Segner et al., 1989b
Niacin	O. niloticus x O. aureus	26, diet with 38% glucose; 120, diet with 38% dextrin	Hemorrhages and lesions in skin and fins, deformed snout, exophthalmia, gill filaments edema, fatty infiltration in liver	Shiau and Suen, 1992
Cholecalciferol (D3)	O. niloticus x O. aureus	375 IU/kg	Lower hemoglobin, hepatosomatic index, and alkaline phosphatase in plasma (28, 12 and 55% less than in control)	Shiau and Hwang, 1993
Biotin		0.1	Not observed	Shiau and Chin, 1999
	Clarias batrachus	1.0*	Anorexia, dark skin, convulsion	Shaik Mohamed et al., 2000

*Differs from that suggested by the authors (2.5 mg/kg).

essential in the Amazon River teleosts irrespective of their feeding habits (Fracalossi *et al.*, 2001). As the Amazon floodplain forests are inundated from January to May each year, fruits and nuts are the most important components of frugivorous fish diets (Goulding, 1980; Araujo-Lima and Goulding, 1998). The nutritional composition of fruits, such as cashew apple (*Anacardium occidentale*) and camu-camu (*Myrciaria dubia*), includes high levels of ascorbic acid, 400–518 mg/100 g and 1570 mg/100 g, respectively (Egbekun and Otiri, 1999; Justi *et al.*, 2000). If these values are combined with ascorbic acid concentrations frequently encountered in many species of microalgae (130–300 mg/100 g; Brown *et al.*, 1999; Brown and Hohmann, 2002), it may be concluded that tropical herbivorous and omnivorous fish are provided with amounts 100 times their requirements for optimum growth (Table 5.2).

Ascorbate is synthesized by a representative of Chondrichthyes in the Amazon, the freshwater stingray, and by lungfishes (Dipnoi) (Fracalossi *et al.*, 2001). Some authors report the presence of gulonolactone oxidase, the enzyme responsible for the final step of ascorbic acid synthesis, activity in the hepatopancreas of common carp (Sato *et al.*, 1978) and kidney of tilapia (Soliman *et al.*, 1985). However, these results have been discounted based on the inaccuracies of the methods used (see Moreau and Dabrowski, 2001). Similarly, some studies that have addressed the requirements of Nile tilapia used the lowest level of ascorbic acid supplementation of 500 mg/kg (estimated to provide 146 mg/kg based on retention) (Soliman *et al.*, 1994). These authors estimated the vitamin C requirements in Nile tilapia to be 420 mg/kg of dry diet which is 10-fold higher than in most other studies on cichlids (Table 5.2) and other fish. The reason for this overestimation can be attributed to the use of the colorimetric method for ascorbate analysis without correction for interfering substances (see Moreau and Dabrowski, 2001).

Shiau and Hsu (1995) argue that "ascorbyl monophosphate and ascorbyl sulfate have similar antiscorbutic activity as a vitamin C source for tilapia". However, an inspection of the relationships between the equivalent amounts of the two esters in diets and responses in liver concentrations of ascorbate suggest 20–40% lower levels in fish fed the ascorbyl sulfate form. Therefore, contrary to the authors' conclusion, tilapias do not differ in their inferior ability to utilize ascorbyl sulfate as demonstrated in cyprinids and salmonids. Shiau and Hsu (2002) investigated the vitamin C/vitamin E interaction in hybrid tilapia and opted to examine two levels of ascorbate supplementation rather than a more typical two-factorial design with a group receiving "no vitamin C" and "no vitamin E". After only 8 weeks, weight gains were significantly lower in treatments with no vitamin E and "optimum vitamin C", whereas vitamin C supplemented at a 3-fold higher level prevented growth depression. This is an interesting finding taking into account that no differences in tissue concentrations of tocopherol between two "no vitamin E" groups were shown. It seems apparent that what was observed was not a "vitamin E sparing effect" by ascorbate but ascorbic acid-reduced toxicity of accumulating oxygen free radicals in the liver (measured as hepatic thiobarbituric acid-reactive substances, TBARS).

Soliman et al. (1986) demonstrated significantly reduced hatching rates (54% versus 89% in control) and an increased percentage of newly hatched fish with severe spinal deformities (57% versus 1.28% in control) in O. mossambicus when parent fish were fed a diet devoid of ascorbic acid for 21 weeks. As growth in this species proceeds from embryo to the juvenile stage based on utilization of yolk reserves, spinal deformities and malformations were the result of endogenous feeding.

Some comment is also necessary regarding the ascorbate requirement of oscars (Astronotus ocellatus) (Table 5.2), "sufficient to prevent growth reduction and vitamin C deficiency signs" (Fracallosi et al., 1998). In this experiment, fish receiving a vitamin C-devoid diet only doubled their weight over a period of 26 weeks, which was accompanied by very low concentrations of ascorbate in the liver ($6 \mu g/g$). Concentrations below $20 \mu g/g$ are considered to signify vitamin C deficiency in salmonids (Matusiewicz et al., 1994). Therefore, studies with oscars growing at much faster rates should address more precisely the requirement for this species.

Lim et al. (2002) claimed that an increased provision of ascorbic acid in the diet of guppies (Poecilia reticulata) increased resistance of fish to osmotic (35 ppt) stress. However, the data presented is suspect considering no ascorbic acid was found in control Artemia and levels in enriched shrimp were 10-fold lower than normally encountered in zooplanktonic organisms.

Hybrid tilapia (O. niloticus x O. aureus) utilized food best and had the highest protein deposition efficiency at 50 and 75 mg alpha-tocopherol/kg when fed diets containing 5 and 12% lipids, respectively (Shiau and Shiau, 2001). Supplemented oils (maize and cod liver) were tocopherol-stripped; however, no data were provided on actual levels of vitamin E in the diets, nor were deficiency signs reported. Baker and Davies (1997) found no effect on growth of Clarias gariepinus fed with a fish meal-based (60%) diet at 27 °C supplemented with no or 5–100 mg/kg alpha-tocopherol. In this study fish increased body weight 12-fold. The requirement was estimated to be 35 mg/kg based on tissue concentration of tocopherol and the lipid peroxide value in the liver.

Segner et al. (1989a) demonstrated a requirement of astaxanthin for growth in tropical fish (Table 5.2). Kodric-Brown (1989) has further shown that guppy males fed a diet supplemented with 25 mg/kg each of astaxanthin and canthaxantin were preferred by females and had a higher mating success than their siblings raised on a diet without provitamin A supplementation. Therefore, a dietary nutrient can have an impact on the flow of genetic information in populations of tropical fish.

Pyridoxine supplementation of 100 mg/kg resulted in weight gain decline, severe anemia, and a hematocrit of 5% in comparison to 18–21% in other treatments in fish fed low protein (28%) diets (Shiau and Hsieh, 1997). Anemia was also reported in channel catfish fed a high pyridoxine-containing diet (Andrews and Murai, 1979). In cold-water fishes, anemia is associated with low pyridoxine diets. In juvenile Indian catfish (*Heteropneustes fossilis*) offered diets with pyridoxine ranging from 0 to 27.2 mg/kg, Shaik Mohamed (2001) reported anorexia, lethargy, pale body color, and mortality when the diet was deficient in pyridoxine. However, the overall growth was extremely slow; weight gains of only 105–215% over 15 weeks of feeding were obtained. The estimated requirement, 3.2 mg/kg, may differ for maximum growth of fish.

Weight gains of tilapia hybrids on glucose- (38%) supplemented diets were only half in comparison to fish fed dextrin-supplemented diets, whereas the niacin level had much less impact on the food conversion coefficient (Shiau and Suen, 1992). Therefore, the differences in niacin requirements, ranging from 20 to 120 mg/kg, are related to overall metabolic differences as a result of different food intakes. Consequently, nutrient requirements "determined under suboptimal growth conditions" (Shiau and Suen, 1992) can be misleading.

It took 16 weeks and more than a 15-fold body weight increase to observe growth depression in tilapia fed a diet devoid of vitamin D_3 (Shiau and Hwang, 1993). This is probably the reason that this study stands alone among unsuccessful attempts to show vitamin D needs in warmwater fishes. The growth depression was accompanied by a significantly lower condition factor and bone Ca concentrations, whereas blood plasma Ca and P were not different among treatments (see also Table 5.2). However, the requirement level based on data presented is only an approximation because of the variation in weight gains among all vitamin D-supplemented groups, which was +15% of the mean. O'Connell and Gatlin (1994) concluded that dietary vitamin D_3 was not required for growth or to utilize dietary Ca for mineralization in *O. aureus*. This contradiction may have arisen from the fact that blue tilapia grew at a significantly slower rate (25-fold in 24 weeks) than tilapia hybrids in Shiau and Hwang's studies and did not reach the limiting "dilution" of body reserve vitamin D to demonstrate deficiency. In earlier studies, Ashok et al. (1998) suggested that the warmwater cyprinid *Labeo rohita* does not require vitamin D as an essential nutrient. However, growth rates and final weights of fish kept in the dark and fed a diet devoid of vitamin D_3 for six months were not given, nor was the level of vitamin D_3 in the liver analyzed prior to the experimental treatments. If accepted, these results are inconclusive at best, based on previous work that suggests a body weight gain of 15–20-fold is needed to demonstrate deficiency in lipid-soluble vitamins. Fish fed a diet devoid of vitamin D_3 had either undetectable (dark conditions) or significantly decreased (light conditions) levels in liver tissue in comparison to control treatments. It is somewhat unfounded to conclude that vitamin D "is not an essential nutrient" (Ashok *et al.*, 1998) in freshwater fish.

The same group of researchers attempted to extend this conclusion of vitamin D inessentiality to a cichlid fish, *O. mossambicus*, based on the lack of an apparent role of vitamin D_3 and its hydroxylated derivatives in changes of blood Ca and P levels, intestinal Ca absorption, or gill Ca binding protein activity (Rao and Raghuramulu, 1999). However, these conclusions were made based on fish sacrificed and sampled 3 days after an intraperitoneal injection of vitamin D3. There was no evidence that control fish were depleted of this vitamin, so in effect the authors experimented with tilapia that were administered an additional dose of vitamin D and the result was "no response". It was Wendelaar Bonga *et al.* (1983) who demonstrated in the past in *O. mossambicus* that 1, 25 dihydroxy-vitamin D has an antagonistic effect on acellular bone in tilapia and injections of this derivative led to a decrease of Ca and P in bones (demineralization). Therefore, vitamin D_3 and its hydroxylated derivatives may have completely different physiological roles in fish.

D. Minerals

Fish can absorb some minerals from the aquatic environment, although the dietary essentiality of phosphorus, magnesium, iron, copper, manganese, zinc, selenium, and iodine have been documented in many freshwater and marine fishes. Quantitative requirements for dietary calcium and potassium are somewhat elusive in fish as it depends on concentrations in the water and absorption through either the gills and skin in freshwater, or the gastrointestinal tract (drinking) in marine environments. Robinson *et al.* (1987) reared *O. aureus* in Ca-free water and using a purified, casein-based diet determined that 0.8% Ca and 0.5% P were required for optimum growth. Differences in growth between 0.17 and 0.7% dietary Ca groups resulted in 727 and 1112% gain, respectively, but no impact on body composition was observed. In *O. niloticus*, Watanabe *et al.* (1980) added Na-monophosphate to an already phosphorus-rich diet (1.4% P) and did not observe any impact on growth. No abnormalities in tilapia skeletal structures were observed when fed diets containing fish meal-derived

P and Ca, leading to the conclusion that tilapia, in comparison to stomachless fishes, have a higher capacity to utilize more complex phosphate sources.

Interestingly, in *O. mossambicus* maintained in freshwater on low or high magnesium (Mg)-diets, no significant differences in either extra-intestinal intake of Mg, or growth suppression following 3 weeks of feeding were observed (Van der Velden *et al.*, 1991). However, neither growth nor food utilization data were provided. When *O. niloticus* was fed Mg-devoid diets for 10 weeks, growth depression was highly significant in comparison to diets supplemented with 0.6–0.77% Mg. Dabrowska *et al.* (1989a) concluded that *O. niloticus* requires 0.5–0.7% Mg when fed high protein diets. However, Mg-oxide and Mg-sulfate were 15 and 28% less efficacious, respectively, than Mg-acetate in supporting growth of tilapia. This suggests that organic Mg sources are characterized by significantly higher bioavailability in tilapia. Mg-sulfate and -oxide supplements resulted in almost double the amounts of Ca and P in tilapia body in comparison to a group fed a Mg-acetate supplemented diet (Dabrowska *et al.*, 1989b). These interactions seem to be very important for fish growth, mineralization, and possible endocrine regulation, but are frequently overlooked when feeding and analyzing diet compositions.

E. Carbohydrates and Cellulose

Fish are generally glucose-intolerant and signs of dietary carbohydrate "overdose" include high levels of glucose in blood plasma and hyperinsulinemia (Moon, 2001). Surprisingly, glucose transporters that allow glucose to pass through tissue membranes in a Na-independent fashion, called GLUT 1-4, were not detectable in skeletal muscle of tilapia (Wright *et al.*, 1998). Expression of GLUT transporters is regulated by insulin; however, in fish, in contrast to mammals, the number of insulin receptors is far less than the number of IGF-1 (insulin-like growth factor-1) receptors (Navarro *et al.*, 1999). Therefore it should be less surprising that hybrid tilapia (*O. niloticus x O. aurea*) utilized a glucose-containing diet (34%) much more poorly than a diet containing equivalent amount of starch (Shiau and Liang, 1995). The apparent absorption of glucose and starch, measured with an indirect marker method, was high, 92.6 and 92.9%, respectively, but the authors failed to identify the cause of 3-fold weight gain differences between the two experimental groups. This is particularly intriguing in light of an earlier study by Anderson *et al.* (1984) in which utilization of glucose was compared to sucrose, dextrin, and starch in diets fed to juvenile Nile tilapia of 2 g individual weight for 63 days. It was found that an increase of carbohydrates from 10 to 40% of the diet improved growth rates significantly, and there were no differences in growth of fish between carbohydrate sources provided at equivalent levels of 10, 25, or 40%. In the case of sucrose, dextrin, and starch, an increase in the percentage of the diet resulted in significant improvement of dietary protein utilization, i.e., a protein energy-sparing effect by carbohydrates. An increase of glucose in the diet of tilapia did not show this effect. Therefore, it is indeed urgently necessary to explain how the proportion and type of carbohydrates in the diet of tropical fishes will effect protein utilization and retention in the body. The mechanism of absorption and high plasma levels of glucose following dietary availability of glucose may involve disturbances in amino acids and/or water-soluble vitamin absorption (GLUT transporters are shared by ascorbic acid; Vera et al., 1993) and need to be addressed in herbivorous/frugivorous tropical fish.

A level of 40% starch in the diet of tilapia improved fish performance according to one account. Kihara and Sakata (1997) argued that part of this enhanced utilization may be explained by fermentation processes in the intestine and production of easily absorbable short chain fatty acids such as acetate, propionate, and butyrate. Thus, microbial activity in cichlid fishes and possibly many other species of fish, can contribute to utilization of otherwise indigestible fractions of some dietary carbohydrates, although microbial degradation of cellulose was also occurring in the gut. Wang et al. (1985) concluded that 20% cellulose had a negative impact on Nile tilapia growth, although a high lipid content in the diet (15%) may have decreased feed intake in fish (the authors used an ad libitum feeding method) and resulted in the same "net" growth depression effect. Dioundick and Stom (1990) concluded with authority that O. mossambicus maintained at 29 °C on diets with 10% cellulose had depressed growth in comparison to an optimum level of 5%. However, an inspection of the data shows that there was no significant difference in the final weight of fish. The evidence provided by Anderson et al. (1984) seems to be the most compelling thus far. Only cellulose levels above 10% appeared to have a negative effect on diet utilization in cichlids. Based on evidence provided by Anderson et al. (1984) and others, suggesting a linear decrease in growth of tilapia on diets supplemented with increasing cellulose contents (10–40%), results of optimization of protein/energy ratios in O. aureus (Winfree and Stickney, 1981) fed diets supplemented with 19–46% cellulose (or 16.8–35.4% estimated fiber) should be regarded as highly unsatisfactory. Indeed, most likely due to a cellulose depression of nutrient availability, diets with 34–56% protein (casein/albumin) supported rather low growth rates of tilapia and poor feed efficiency was obtained (feed/gain, 1.9–4.2). Diets containing more than 10% cellulose are not desirable for tropical cichlids.

IV. ENVIRONMENTAL CONDITIONS AND FISH FORAGING IMPACT ON ECOSYSTEM

A. Effect of Seasons and Extreme Environments

Oreochromis alcalicus grahami, a fish adapted to alkaline hot-springs, where temperatures are in excess of 42 °C, frequently become hypoxic, adopting ureotelism and gulping air oxygen as a means of living at a pH of 9.98 and high ammonia concentrations (Randall *et al.*, 1989; Franklin *et al.*, 1995). Synthesis of urea creates conditions of compartmentalized arginine synthesis, a metabolic characteristic absent in most teleosts (see Chapter 8). Coincidently, alkaliphilic cyoanobacteria, mainly represented by *Spirulina,* occur in Lake Magadi (Dubinin *et al.*, 1995) and may constitute a significant proportion of their diet (Walsh, P., personal communication). *Spirulina* protein contains almost twice as much arginine as fish protein, however, the overall value of cyanophitic protein in the diet of tilapia, *Oreochromis mossambicus*, was extremely low (Olvera-Novoa *et al.*, 1998). Several plausible explanations can be proposed: (1) the small size of fish used in testing *Spirulina* protein prevented better utilization for growth, (2) major differences occur between the digestive physiologies of *O. mossambicus* and *O. alcalicus* which are decisive in respect to protein utilization efficiency, or (3) dietary arginine is rapidly metabolized in a system with a high affinity for arginine as a substrate.

In the aquatic environment, oxygen limitation by ambient variation and the capacity of diffusion through the gills sets the margin of embryonic development (Dabrowski *et al.*, 2003b), feed intake and, consequently, growth. Van Dam and Pauly (1995) argued that amino acid and lipid oxidation account for at least 90–95% of total oxygen demand in actively feeding fish, whereas at the maximum feeding rate biosynthesis costs absorb 45% of total energy. Simulations thus far have not included tropical fishes with unlimited access to atmospheric oxygen (*Lepisosteus, Arapaima*) or conditions of oxygen supersaturation.

B. Impact of Herbivores and Piscivores

Overall productivity in aquatic ecosystems is regulated by nutrient availability to primary producers. In other words, an increase in upper trophic levels of consumers will cascade down and alter lower trophic levels. In the case of tilapia feeding on phytoplankton and zooplankton, such as *Tilapia galilea*, community-level effects have been clearly demonstrated as zooplankton and large dinoflagellates (*Peridinium* sp.) declined and nanoplankton (smaller than 10 μm) reached its highest abundance at an intermediate fish density (Drenner et al., 1987). As expected, removal of primary producers (Peridinium) resulted in chlorophyll concentration decreases and nanoplankton density decreases in treatments with the highest fish density, although those fish were severely undernourished and lost weight. The implications from this study seem to be that in order to achieve an increase of system productivity, fish stock enhancement methods need to be introduced. However, responses to changing density of herbivorous (or omnivorous) consumers in tropical ecosystems can frequently be counterintuitive. Diana et al. (1991) demonstrated that a 3-fold increase of Nile tilapia density in fertile ponds resulted in a similar biomass yield but a decreased growth rate of adult fish. This suggested that even at the lowest fish density, the carrying capacity (productivity) of these ecosystems was reached. However, despite the limitation imposed on tilapia growth at the higher fish density, no response was measurable in zooplankton or phytoplankton productivity. Thus "top-down" control of these ponds in the tropics had an unpredicted result.

In freshwater ecosystems, the Amazon floodplains can be considered as extreme environments where the macrophytes and trees of the flooded forests contribute 65 and 28% of the net primary production, respectively (Melack et al., 1999). Therefore, intuitively, algae cannot support secondary consumers alone and nutrients must enter the food chain indirectly through a microbial loop. However, as demonstrated by Leite et al. (2002) in the case of larvae of eight species of fishes from the Amazon, microscopic algae were likely the main plants that contributed to larval fish production. Lewis et al. (2001) documented that in the Orinoco River, macrophytes and litterfall from the floodplain forests composed 98% of the potentially available carbon. $\delta^{15}N$ trophic changes in respect to algal C source suggested that fishes (18 species, 50% biomass) were predominantly carnivorous (trophic level 2.8, where 3.0 is primary carnivore) and only 20% (of fish production) can be directly related to algal consumption. The authors argue that the reliance of fishes in floodplains of the Orinoco River on algal rather than vascular plant carbon is probably due to the higher nutritional value of the former, a conclusion that is "hard to digest" (see section on food preferences). If the explanation is the "nutritional superiority" of algal material over vascular plant detritus transferred in the food chain, then this evolutionary incentive is hard to reject. However, such an account of trophic dynamics of tropical rivers puts into question earlier concepts of dominance of herbivores and detritivores (consumers of aquatic macrophytes and derived detritus) in trophic networks (Winemiller, 1990).

Cyanobacteria are present in nature as mixtures of strains of varying toxicity and therefore the impact of phytoplanktivores on these primary producers will be quite complex. Keshavanath et al. (1994) found in Nile tilapia a linear decrease in the grazing rate of cyanobacterium *Microcystic aeruginosa* with an increase in the proportion (over 25%) of the toxic strain in the population. However, the rate of filtration and intake correlated with cell surface properties and the rate of particle binding to secreted mucus rather than an extracellularly released microcystin, earlier deemed more important (Beveridge *et al.*, 1993).

Nile tilapia effectively use "pump filtering", an intermediary process between random filtering and particulate feeding. The mucus produced on the gill rakers and the pharyngeal jaws increases the efficiency of the entrapment mechanism. This last mechanism seemed to make a difference between the ability of *O. niloticus* and its sister species, *O. esculentus* to feed on small green algae and colonial cyanobacteria (Batjakas *et al.*, 1997) and consequently survive in Lake Victoria. As anthropogenic changes of the limnology of the lake took place so did the dominance of the phytoplankton community. In changing environments a more adaptable species with a wider dietary breadth (and specialized morphological structures) has eliminated a very specialized feeder.

Fast-growing freshwater shrimp (*Cardina nilotica*) are the major component of the diet of juvenile Nile perch (*Lates niloticus*) in the Lake Victoria littoral zone, although δ^{13}C studies have indicated that some other benthic organisms, possibly chironomids, may also contribute. Interestingly, δ^{15}N studies have indicated that copepods and cladocerans are not significant prey in the diet of Nile perch. A pelagic indigenous cyprinid, *Rastrineobola argentea*, is an important food source to adult Nile perch. Stable isotope data in Lake Victoria have provided quantitative evidence of two food chains that have implications for Nile perch fisheries. A significant decline of Nile perch stocks over 1999–2001 was noticed based on acoustic surveys (Getabu *et al.*, 2003). Oxygen depletion (1.2 ± 0.7 mg/l) at depths >40 m contributed to a decline in Haplochromine cichlids in the past whereas, at the current level, it seems to some extent to disfavor large Nile perch which are less tolerant to low-oxygen levels. Balirwa *et al.* (2003) conclude that the resurgence of a native zooplanktivore, *R. argentea,* and some Haplochromines does not suggest that a new food web structure will be less dynamic. It is anticipated that some changes in water turbidity may already have caused hybridization among Haplochromines and resurgent populations already represent genetically mosaic stocks.

ACKNOWLEDGMENTS

We wish to thank Michael Penn, DVM, for his gracious revision of the manuscript. This work was a component of the Pond Dynamics/Aquaculture Collaborative Research Support Program (PD/A CRSP) supported by the US Agency for International Development, Grant No. RD010A-12 (KD) and by the FAPESP grant (00/07314-0) (MCP).

REFERENCES

Aires, E. D., Dias, E., and Orsi, A. M. (1999). Ultrastructural features of the glandular region of the stomach of *Piaractus mesopotamicus* (Holmberg, 1887) with emphasis on the oxyntopeptic cell. *J. Submicr. Cytol. Path.* **31,** 287–293.

Akiyama, T., Arai, S., Murai, T., and Nose, T. (1985). Threonine, histidine and lysine requirements of chum salmon fry. *B. Jpn Soc. Sci. Fish.* **51,** 635–639.

Albertini-Berhaut, J. (1987). L-intestin chez Mugilidae (poisons; Teleosteens) a differentes etapes de leur croissance. I. Aspects morphologiques et histologiques. *J. Appl. Ichthyol.* **3,** 1–12.

Albertini-Berhaut, J. (1988). L'Intestin chez les mugilidae (Poissons Teleosteens) à differentes étapes de leur croissance. II. Aspectes ultrastructuraux et cytophysiologiques. *J. Appl. Ichthyol.* **4,** 65–78.

Albertini-Berhaut, J., Alliot, E., and Raphel, D. (1979). Evolution des activites proteolytiques digestives chez les jeunes Mugilidae. *Biochem. Syst. Ecol.* **7,** 317–321.

Albrecht, M. P., Ferreira, M. F. N., and Caramaschi, E. P. (2001). Anatomical features and histology of the digestive tract of two related neotropical omnivorous fishes (Characiformes; Anostomidae). *J. Fish Biol.* **58,** 419–430.

Al-Hussaini, A. H. (1946). The anatomy and histology of the alimentary tract of the bottom-feeder *Mulloides auriflamma. J. Morphol.* **78,** 121–153.

Al-Hussaini, A. H., and Kholy, A. A. (1953). On the functional morphology of the alimentary tract of some omnivorous fish. *P. Egyptian Acad. Sci.* **4,** 17–39.

Anderson, J., Jackson, A. J., Matty, A. J., and Capper, B. S. (1984). Effects of dietary carbohydrate and fibre on the tilapia *Oreochromis niloticus* (Linn.). *Aquaculture* **37,** 303–314.

Andrews, J. W., and Murai, T. (1979). Pyridoxine requirements of channel catfish. *J. Nutr.* **109,** 533–537.

Applebaum, S. L., Perez, R., Lazo, J. P., and Holt, G. J. (2001). Characterization of chymotrypsin activity during early ontogeny of larval red drum (*Sciaenops ocellatus*). *Fish Physiol. Biochem.* **25,** 291–300.

Appler, H. N. (1985). Evaluation of *Hydrodictyon reticulatum* as protein source in feeds for *Oreochromis (Tilapia) niloticus* and *Tilapia zillii. J. Fish Biol.* **27,** 327–334.

Appler, H. N., and Jauncey, K. (1983). The utilization of a filamentous green alga (*Cladophora glomerata* (L) Kutzin) as a protein source in pelleted feeds for *Sarotherodon* (Tilapia) *niloticus* fingerlings. *Aquaculture* **30,** 21–30.

Aranishi, F., Watanabe, T., Osatomi, K., Cao, M., Hara, K., and Ishihara, T. (1998). Purification and characterization of thermostable depeptidase from carp intestine. *J. Mar. Biotechnol.* **6,** 116–123.

Araujo-Lima, C., and Goulding, M. (1998). So Fruitful a Fish. *In* "Ecology, Conservation, and Aquaculture of the Amazon's Tambaqui." "Biology and Resource Management in the Tropics Series" (Balick, M. J., Anderson, A. B., and Redford, K. H., Eds.). Columbia University Press, New York.

Araujo-Lima, C. A. R. M., Portugal, L. P. S., and Ferreira, E. G. (1986). Fish-macrophyte relationship in the Anavilhanas Archipelago, a black water system in the Central Amazon. *J. Fish. Biol.* **29,** 1–11.

Arellano, J. M., Storch, V., and Sarasquete, C. (2001). A histological and histochemical study of the oesophagus and oesogaster of the Senegal sole, *Solea senegalensis. Eur. J. Histochem.* **45,** 279–294.

Ashok, A., Rao, D. S., and Raghuramulu, N. (1998). Vitamin D is not an essential nutrient for Rora (*Labeo rohita*) as a representative of freshwater fish. *J. Nutr. Sci. Vitaminol.* **44,** 195–205.

Baker, R. T. M., and Davies, S. J. (1997). The quantitative requirement for α-tocopherol by juvenile African catfish, *Clarias gariepinus* Burchell. *Anim. Sci.* **65**, 135–142.

Bakke-McKellep, A. M., Nordrum, S., Krogdahl, A., and Buddington, R. K. (2000). Absorption of glucose, amino acids, and dipeptides by the intestines of Atlantic salmon (*Salmo salar* L.). *Fish Physiol. Biochem.* **22**, 33–44.

Balirwa, J. S., Chapman, C. A., Chapman, L. J., Cowx, I. G., Geheb, K., Kaufman, L., Lowe-McConnell, R. H., Seehausen, O., Wanink, J. H., Welcomme, R. L., and Witte, F. (2003). Biodiversity and fishery sustainability in the Lake Victoria basin: An unexpected marriage. *Bioscience* **53**, 703–715.

Banack, S. A., Horn, M. H., and Gawlicka, A. (2002). Disperser- vs. establishment-limited distribution of a riparian fi tree (*Ficus insipida*) in a Costa Rican tropical rain forest. *Biotropica* **34**, 232–243.

Baragi, V., and Lovell, R. T. (1986). Digestive enzyme activities in stripped bass from first feeding through larval development. *Trans. Am. Fish. Soc.* **115**, 478–481.

Baras, E. (2000). Day–night alternation prevails over food availability in synchronizing the activity of *Piaractus brachypomus* (Characidae). *Aquat. Living Resour.* **13**, 115–120.

Baras, E., Melard, C., Grignard, J. C., and Thoreau, X. (1996). Comparison of food conversion by pirapatinga *Piaractus brachypomus* under different feeding times. *Progr. Fish Cult.* **58**, 59–61.

Batjakas, I. E., Edgar, R. K., and Kaufman, L. S. (1997). Comparative feeding efficiency of indigenous and introduced phytoplanktivores from Lake Victoria: Experimental studies on *Oreochromis esculentus* and *Oreochromis niloticus*. *Hydrobiologia* **347**, 75–82.

Bayley, P. B. (1973). Studies on the migratory characin, *Prochilodus platensis* Holmberg 1889, (Pisces, Characoidei) in the River Pilcomayo, South America. *J. Fish Biol.* **5**, 25–40.

Benitez, L. V., and Tiro, L. B. (1982). Studies on the digestive proteases of the milkfish *Chanos chanos*. *Mar. Biol.* **71**, 309–315.

Bergot, P., Solari, A., and Luquet, P. (1975). Comparaison des surfaces absorbantes des CÆCA pyloriques et de l'intestin chez la truite arc-en-ciel (*Salmo gairdneri* Rich.). *Ann. Hydrobiol.* **6**, 27–43.

Beveridge, M. C. M., Baird, D. J., Rahmatullah, S. M., Lawton, L. A., Beattie, K. A., and Codd, G. A. (1993). Grazing rates on toxic and non-toxic strains of cyanobacteria by *Hypophthalmichthys molitrix* and *Oreochromis niloticus*. *J. Fish Biol.* **43**, 901–907.

Beveridge, M. C. M., Begum, M., Frerichs, G. N., and Millar, S. (1989). The ingestion of bacteria in suspension by the tilapia *Oreochromis niloticus*. *Aquaculture* **81**, 373–378.

Bitterlich, G. (1985). Digestive enzyme pattern of two stomachless filter feeders, silver carp, *Hypophthalmichthys molitrix* Val. bighead carp, *Aristichthys nobilis* Rich. *J. Fish Biol.* **27**, 103–112.

Bitterlich, G. (1985). The nutrition and stomachless phytoplanktivorous fish in comparison with *Tilapia*. *Hydrobiologia* **121**, 173–179.

Bitterlich, G., and Gnaiger, E. (1984). Phytoplanktivorous or omnivorous fish? Digestibility of zooplankton by silvercarp, *Hypophthalmichthys molitrix* (Val.). *Aquaculture* **40**, 261–263.

Blom, J. H., Dabrowski, K., and Ebeling, J. (2000). Vitamin C requirements of the angelfish *Pterophylum scalare*. *J. World Aquacult. Soc.* **32**, 115–118.

Boge, G., Rigal, A., and Peres, G. (1981). Rates of *in vivo* intestinal absorption of glycine and glycylglycine by rainbow trout (*Salmo gairdneri* R.). *Comp. Biochem. Physiol.* **69A**, 455–459.

Borlongan, I. G. (1990). Studies on the digestive lipases of milkfish, *Chanos chanos*. *Aquaculture* **89**, 315–325.

Borlongan, E. G., and Coloso, R. M. (1993). Requirements of juvenile milkfish (*Chanos chanos* Forsskal) for essential amino acids. *J. Nutr.* **123**, 125–132.

Bouhic, M., and Gabaudan, J. (1992). Histological study of the organogenesis of the digestive system and swim bladder of the Dove sole *Solea solea*. *Aquaculture* **102**, 373–396.

Bowen, S. H. (1980). Detrital nonprotein amino acids are the key to rapid growth of *Tilapia* in Lake Valencia, Venezuela. *Science* **207**, 1216–1218.

Bowen, S. H. (1984). Detrital amino acids and the growth of *Sarotherodon mossambicus* – a reply to Dabrowski. *Acta Hydroch. Hydrob.* **12**, 55–59.

Bowen, S. H. (1987). Dietary protein requirements of fishes – a reassessment. *Can. J. Fish. Aquat. Sci.* **44**, 1995–2001.

Bowen, S. H., Bonetto, A. A., and Ahlgren, M. O. (1984). Microorganisms and detritus in the diet of a typical neotropical riverine detritivore, *Prochilodus plantesis* (Pisces: Prochilodontidae). *Limnol. Oceanogr.* **29**, 1120–1122.

Brown, G. E., Adrian, J. C., Jr, Kaufman, I. H., Erickson, J. L., and Gershaneck, D. (2001). Responses to nitrogen-oxides by Characiforme fishes suggest evolutionary conservation in Ostariophysan alarm pheromones. *In* "Chemical Signals in Vertebrates 9" (Marchlewska-Koj, A., Lepri, J. L., and Muller-Schwarze, D., Eds.). Kluwer Academic/Plenum Publishers, New York.

Brown, M. R., Mular, M., Miller, I., Farmer, C., and Trenerry, C. (1999). The vitamin content of microalgae used in aquaculture. *J. Appl. Phycol.* **11**, 247–255.

Brown, M. R., and Hohmann, S. (2002). Effects of irradiance and growth phase on the ascorbic acid content of *Isochrysis* sp. T.ISO (Prymnesiophyta). *J. Appl. Phycol.* **14**, 211–214.

Bryant, P. L., and Matty, A. J. (1981). Adaptation of carp (*Cyprinus carpio*) larvae to artificial diets. 1. Optimum feeding rate and adaptation age for a commercial diet. *Aquaculture* **23**, 275–286.

Buddington, R. K., and Doroshev, S. I. (1986). Development of digestive secretion in white sturgeon juveniles. *Comp. Biochem. Physiol.* **83A**, 233–238.

Buddington, R. K., Puchal, A. A., Houpe, K. L., and Diehl III, W. J. (1993). Hydrolysis and absorption of two monophosphate derivatives of ascorbic acid by channel catfish *Ictalurus punctatus* intestine. *Aquaculture* **114**, 317–326.

Burtle, G. J., and Lovell, R. T. (1989). Lack of response of channel catfish (*Ictalurus punctatus*) to dietary myoinositol. *Can. J. Fish. Aquat. Sci.* **46**, 218–222.

Caceci, T., El-Habback, H. A., Smith, S. A., and Smith, B. J. (1997). The stomach of *Oreochromis niloticus* has three regions. *J. Fish Biol.* **50**, 939–952.

Cahu, C., and Zambonino-Infante, J. L. (1994). Early weaning of sea bass (*Dicentrarchus labrax*) larvae with a compound diet: effect on digestive enzymes. *Comp. Biochem. Physiol.* **109A**, 213–222.

Cahu, C., and Zambonino-Infante, J. L. (1995). Effect of molecular form of dietary nitrogen supply in sea bass larvae: Response of pancreatic enzymes and intestinal peptidase. *Fish Physiol. Biochem.* **14**, 209–214.

Cahu, C., and Zambonino-Infante, J. L. (1997). Is the digestive capacity of marine fish larvae sufficient for compound diet feeding? *Aquacult. Int.* **5**, 151–160.

Cahu, C., and Zambonino-Infante, J. L. (2001). Substitution of live food by formulated diets in marine fish larvae. *Aquaculture* **200**, 161–180.

Cancre, I., Ravallec, R., Van Wormhoudt, A., Stenberg, E., Gildberg, A., and Le Gal, Y. (1999). Secretagogues and growth factors in fish and crustacean protein hydrolysates. *Mar. Biotechnol.* **1**, 489–494.

Carmo e Sa, M. V., and Fracalossi, D. M. (2002). Dietary protein requirement and energy to protein ratio for piracanjuba (*Brycon orbignyanus*) fingerlings. *Rev. Bras. Zootecn.* **31**, 1–10.

Casirola, D. M., Vinnakota, R. R., and Ferraris, R. P. (1995). Intestinal absorption of water-soluble vitamins in channel catfish (*Ictalurus punctatus*). *Am. Physiol. Soc.* R490–R496.

Chakrabarti, R., and Sharma, J. (1997). Ontogenic changes of amylase and proteolytic enzyme activities of Indian major carp, *Catla catla* (Ham.) in relation to natural diet. *Indian. J. Anim. Sci.* **67**, 932–934.

Chavez de Martinez, M. C. (1990). Vitamin C requirement of the Mexican native cichlid *Cichlasoma urophthalmus* (Günther). *Aquaculture* **86**, 409–416.

Chavez de Martinez, M. C., and Richards, R. H. (1991). Histopathology of vitamin C deficiency in a cichlid, *Cichlasoma urophthalmus* (Günther). *J. Fish Dis.* **14**, 507–519.

Chiu, Y. N., and Benitez, L. V. (1981). Studies on the carbohydrases in the digestive tract of the milkfish *Chanos chanos*. *Mar. Biol.* **61**, 247–254.

Chong, A. S. C., Hashim, R., Chow-Yang, L., and Ali, A. B. (2002a). Characterization of protease activity in developing discus *Symphysodon aequifasciata* larva. *Aquac. Res.* **33**, 663–672.

Chong, A. S. C., Hashim, R., Chow-Yang, L., and Ali, A. B. (2002b). Partial characterization and activities of proteases from the digestive tract of discus fish, *Symphysodon aequifasciata*. *Aquaculture* **203**, 321–333.

Chou, B.-S., and Shiau, S.-Y. (1996). Optimal dietary lipid level for growth of juvenile hybrid tilapia, *Oreochromis niloticus* x *Oreochromis aureus*. *Aquaculture* **143**, 185–195.

Clements, K. D., and Choat, J. H. (1995). Fermentation in tropical marine herbivorous fishes. *Physiol. Zool.* **68**, 355–378.

Coffin, R. B. (1989). Bacterial uptake of dissolved free and combined amino acids in estuarine waters. *Limnol. Oceanogr.* **34**, 531–542.

Collie, N. L., and Ferraris, R. P. (1995). Nutrient fluxes and regulation in fish intestine. *In* "Metabolic Biochemistry" (Hochachka, P. W., and Mommsen, T. P., Eds.), pp. 221–239. Elsevier Science, Amsterdam.

Coloso, R. M., Benitez, L. V., and Tiro, L. B. (1988). The effect of dietary protein-energy levels on growth and metabolism of milkfish *(Chanos Chanos Forsskal)*. *Comp. Biochem. Physiol.* **89A**, 11–17.

Couisin, J. C. B., Baudin-Laurencin, F., and Gabaudan, J. (1987). Ontogeny of enzymatic activities in fed and fasting turbot, *Scophthalmus maximus* L. *J. Fish Biol.* **30**, 15–33.

Craig, S. R., Neill, W. H., and Gatlin III, D. M. (1995). Effects of dietary lipid and environmental salinity on growth, body composition, and cold tolerance of juvenile red drum (*Sciaenops ocellatus*). *Fish Physiol. Biochem.* **14**, 49–61.

Dabrowska, H., Günther, K-D., and Meyer-Burgdorff, K. (1989a). Availability of various magnesium compounds to tilapia (*Oreochromis niloticus*). *Aquaculture* **76**, 269–276.

Dabrowska, H., Meyer-Burgdorff, K., and Günther, K.-D. (1989b). Interaction between dietaryprotein and magnesium level in tilapia (*Oreochromis niloticus*). *Aquaculture* **76**, 277–291.

Dabrowski, K. (1977). Protein requirements of grass carp fry (*Ctenopharyngodon idella* Val.). *Aquaculture* **12**, 63–73.

Dabrowski, K. (1982). *Tilapia* in Lakes and aquaculture – ecological and nutritional approach. *Acta Hydroch. Hydrobiol.* **10**, 265–271.

Dabrowski, K. (1984). The feeding of fish larvae: Present "state of art" and perspectives. *Reprod. Nutr. Dev.* **24**, 807–833.

Dabrowski, K. (1986a). Ontogenetical aspects of nutritional requirements in fish. *Comp. Biochem. Physiol.* **85A**, 639–655.

Dabrowski, K. (1986b). Protein digestion and amino acid absorption along the intestine of the common carp (*Cyprinus carpio* L.), a stomachless fish : An *in vivo* study. *Reprod. Nutr. Dev.* **26**, 755–766.

Dabrowski, K. (1993). Ecophysiological adaptations exist in nutrient requirements of fish: True or false? *Comp. Biochem. Physiol.* **104A**, 579–584.

Dabrowski, K., and Guderley, H. (2002). Intermediary metabolism. *In* "Fish Nutrition" (Halver, J. E., and Hardy, R., Eds.), pp. 309–365. Academic Press, New York.

Dabrowski, K., and Glogowski, J. (1977). The role of exogenic proteolytic enzymes in digestion processes in fish. *Hydrobiologia* **54**, 129–134.

Dabrowski, K., Lee, K.-J., and Rinchard, J. (2003). The smallest vertebrate, teleost fish, can utilize synthetic dipeptide-based diets. *J. Nutr.* **133,** 4225–4229.

Dabrowski, K., Rinchard, J., Ottobre, J. S., Alcantara, F., Padilla, P., Ciereszko, A., de Jesus, M. J., and Kohler, C. C. (2003). Effects of oxygen saturation in water on reproductive performances of pacu *Piaractus brachypomus*. *J. World Aquacult. Soc.* **34,** 441–449.

Das, K. M., and Tripathi, S. D. (1991). Studies on the digestive enzymes of grass carp, *Ctenopharyngodon idella* (Val.). *Aquaculture* **92,** 21–32.

De Godoy, P. (1959). Age, growth, sexual maturity, behaviour, migration, tagging and transplantation of the curimbata (*Prochilodus scrofa* Steindachner, 1881) of the Mogi Guassu River, Sao Paulo State, Brasil. *Anais. Acad. Brasil. Cienc.* **31,** 447–477.

Delariva, R. L., and Agostinho, A. A. (2001). Relationship between morphology and diets of six neotropical loricariids. *J. Fish Biol.* **58,** 832–847.

De Silva, S. S., Gunasekera, B. M., and Atapattu, D. (1989). The dietary protein requirements of young tilapia and an evaluation of the least cost dietary protein levels. *Aquaculture* **80,** 271–284.

De Silva, S. S., Perera, M. K., and Maitipe, P. (1984). The composition, nutritional status and digestibility of the diets of *Sarotherodon mossambicus* from 9 man-made lakes in Sri Lanka. *Environ. Biol. Fish.* **11,** 205–219.

De Souza, R., Ferreira dos Cantos, J., Da Silva Lino, M. A., Almeida Vieira, V. L., and Bezerra Carvalho, L., Jr (2000). Characterization of stomach and pyloric caeca proteinases of tambaqui (*Colossoma macropomum*). *J. Food Biochem.* **24,** 189–199.

Diana, J. S., Dettweiler, D. J., and Lin, C. K. (1991). Effect of Nile tilapia (*Oreochromis niloticus*) on the ecosystem of aquaculture ponds, and its significance to the trophic cascade hypothesis. *Can. J. Fish. Aquat. Sci.* **48,** 183–190.

Diaz, M., Moyano, F. J., Garcia-Carreño, F. L., Alarcón, F. J., and Sarasquete, M. C. (1997). Substrate-SDS-PAGE determination of protease activity through larval development in sea bream. *Aquacult. Int.* **5,** 461–471.

Dioundick, O. B., and Stom, D. I. (1990). Effects of dietary α-cellulose levels on the juvenile tilapia, *Oreochromis mossambicus* (Peters). *Aquaculture* **91,** 311–315.

Drenner, R. W., Vinyard, G. L., Hambright, K. D., and Gophen, M. (1987). Particle ingestion by *Tilapia galilaea* is not affected by removal of gill rakers and microbranchiospines. *Trans. Am. Fish. Soc.* **116,** 272–276.

Dubinin, A. V., Gerasimenko, L. M., and Zavarzin, G. A. (1995). Ecophysiology and species diversity of cynaobacteria from Lake Magadi. *Microbiology* **64,** 717–721.

Edem, D. O. (2002). Palm oil: Biochemical, physiological, nutritional, haematological, and toxicological aspects: A review. *Plant Food Hum. Nutr.* **57,** 319–341.

Egbekun, M. K., and Otiri, A. O. (1999). Changes in ascorbic acid contents in oranges and cashew apples with maturity. *Ecol. Food Nutr.* **38,** 275–284.

El-Sayed, A.-F. M., Nmartinez, I., and Moyano, F. J. (2000). Assessment of the effect of plant inhibitors on digestive proteases of Nile tilapia using *in vitro* assays. *Aquacult. Int.* **8,** 403–415.

Eya, J. C. (1996). "Broken-skull disease" in African catfish *Clarias gariepinus* is related to a dietary deficiency of ascorbic acid. *J. World Aquacult. Soc.* **27,** 493–498.

Fange, R., and Grove, D. (1979). Digestion. *In* "Fish Physiology" (Hoar, W. S., Randall, D. J., and Brett, J. R., Eds.). Academic Press, New York.

Ferraz de Lima, J. S., Ferraz de Lima, C. L. B., Krieger-Axxolini, J. H., and Boschero, A. C. (1991). Topography of the pancreatic region of the Pacu, *Piaractus mesopotamicus* Holmberg, 1887. *Bull. Tec. CEPTA, Pirassununga.* **4,** 47–56.

Ferreira, C. E. L., Peret, A. C., and Coutinho, R. (1998). Seasonal grazing rates and food processing by tropical herbivorous fishes. *J. Fish Biol.* **53A,** 222–235.

Fishelson, L., and Becker, K. (2001). Development and aging of the liver and pancreas in the domestic carp, *Cyprinus carpio*: From embryogenesis to 15-year-old fish. *Environ. Biol. Fish.* **61,** 85–97.

Fishelson, L., Montgomery, L. W., and Myrberg, A. H., Jr. (1987). Biology of surgeonfish *Acanthurus nigrofuscus* with emphasis on change over in diet and annual gonadal cycles. *Mar. Ecol. Prog. Ser.* **39,** 37–47.

Fishelson, L., Montgomery, W. L., and Myrberg, A. A., Jr. (1985). A unique symbiosis in the gut of tropical herbivorous surgeonfish (Acanthuridae: Teleostei) from the Red Sea. *Science* **229,** 49–51.

Fracalossi, D. M., Allen, M. E., Yuyama, L. K., and Oftedal, O. T. (2001). Ascorbic acid biosynthesis in Amazonian fishes. *Aquaculture* **192,** 321–332.

Fracalossi, D. M., Allen, M. E., Nichols, D. K., and Oftedal, O. T. (1998). Oscars, *Astronotus ocellatus*, have a dietary requirement for vitamin C. *J. Nutr.* **128,** 1745–1751.

Franklin, C. E., Johnston, I. A., Crockford, T., and Kamunde, C. (1995). Scaling of oxygen consumption of Lake Magadi tilapia, a fish living at 37 °C. *J. Fish Biol.* **46,** 829–834.

Frierson, E. W., and Foltz, J. W. (1992). Comparison and estimation of absorptive intestinal surface areas in two species of cichlid fish. *Trans. Am. Fish. Soc.* **121,** 517–523.

Galvão, M. N. S., Fenerich-Verani, N., Yamanaka, N., and Oliveira, I. (1997a). Histologia do sistema digestivo de tainha *Mugil platanus* Gunther, 1880 (Ostheichthyes, Mugilidae) durante as fases larval e juvenil. *Bol. Inst. de Pesca* **34,** 91–100.

Galvão, M. S. M., Yamanaka, N., Fenerich-Verani, N., and Pimentel, C. M. M. (1997b). Estudos preliminares sobre enzimas digestivas proteolíticas da tainha (*Mugil platanus*) Günther 1880 (Osteichthyes, Mugilidae) durante as fases larval e juvenil. *Bol. Inst. Pesca* **24,** 101–110.

Garcia-Carreno, F. L., Albuquerque-Cavalcanti, C., Toro, M. A. N., and Zaniboni-Filho, E. (2002). Digestive proteinases of *Brycon orbignyanus* (Characidae, Teleostei): Characteristics and effects of protein quality. *Comp. Biochem. Physiol. Part B* **132,** 343–352.

Garcia-Ortega, A., Verreth, J., and Segner, H. (2000). Post-prandial protease activity in the digestive tract of African catfish *Clarias gariepinus* larvae fed decapsulated cysts of *Artemia. Fish Physiol. Biochem.* **22,** 237–244.

Getabu, A., Tumwebaze, R., and MacLennan, D. N. (2003). Spatial distribution and temporal changes in the fish populations of Lake Victoria. *Aquat Living Resour.* **16,** 159–165.

Gianquinto, P. C., and Volpato, G. L. (2001). Hunger suppresses the onset and the freezing component of the antipredator response to conspecific skin extract in pintado catfish. *Behaviour* **138,** 1205–1214.

Girgis, S. (1952). On the anatomy and histology of the alimentary tract of an herbivorous bottom-feeding cyprinoid fish, *Labeo horie* (Cuvier). *J. Morphol.* **90,** 317–362.

Goulding, M. (1980). "The Fishes and The Forest. Explorations in Amazonian Natural History." University of California Press, Berkeley, Los Angeles and London.

Goulding, M., and Carvalho, M. L. (1982). Life history and management of the tambaqui (*Colossoma macropomum*, Characidae): An important Amazonian food fish. *Revta bras. Zool., S. Paulo.* **1,** 107–133.

Govoni, J. J., Boehlert, G. W., and Watanabe, Y. (1986). The physiology of digestion in fish larvae. *Environ Biol. Fish.* **16,** 59–77.

Grabner, M., and Hofer, R. (1989). Stomach digestion and its effect upon protein hydrolysis in the intestine of rainbow trout (*Salmo gairdneri* Richardson). *Comp. Biochem. Physiol.* **92A(1),** 81–83.

Guizani, N., Rolle, R. S., Barshall, M. R., and Wei, C. I. (1991). Isolation, purification and characterization of a trypsin from the pyloric caeca of mullet (*Mugil cephalus*). *Comp. Biochem. Physiol.* **98B,** 517–521.

Gunasekera, R. M., Shim, K. F., and Lam, T. J. (1995). Effect of dietary protein level on puberty, oocyte growth and egg chemical composition in the tilapia, *Oreochromis niloticus* (L.). *Aquculture* **134**, 169–183.

Gunasekera, R. M., Shim, K. F., and Lam, T. J. (1996). Influence of protein content of broodstock diets on larval quality and performance in Nile tilapia, *Oreochromis noloticus* (L.). *Aquaculture* **146**, 245–259.

Hager, A. F., and Hazel, J. R. (1985). Changes in desaturase activity and the fatty acid composition of microsomal membranes from liver tissue of thermally-acclimating rainbow trout. *J. Comp. Physiol. B* **156**, 35–42.

Hansen, G. H., and Olafsen, J. A. (1999). Bacterial interactions in early life stages of marine cold water fish. *Microbial Ecol.* **38**, 1–26.

Harvey, H. R., and Macko, S. A. (1997). Kinetics of phytoplankton decay during stimulated sedimentation: changes in lipids under oxic and anoxic conditions. *Org. Geochem.* **27**, 129–140.

Hassan, M. S., and Edwards, P. (1992). Evaluation of duckweed (*Lemna perpusilla* and *Spirodela polyrrhiza*) as feed for Nile tilapia (*Oreochromis niloticus*). *Aquaculture* **104**, 315–326.

Henderson, R. J., Tillmanns, M. M., and Sargent, J. R. (1996). The lipid composition of two species of Serrasalmid fish in relation to dietary polyunsaturated fatty acids. *J. Fish Biol.* **48**, 522–538.

Hernandez, M., Takeuchi, T., and Watanabe, T. (1995). Effect of dietary energy sources on the utilization of protein by *Colossma macropomum* fingerlings. *Fish. Sci.* **61**, 507–511.

Hidalgo, F., and Alliot, E. (1988). Influence of water temperature on protein requirement and protein utilization in juvenile sea bass, *Dicentrarchus labrax*. *Aquaculture* **72**, 115–129.

Hidalgo, M. C., Urea, E., and Sanz, A. (1999). Comparative study of digestive enzymes in fish with different nutritional habits. Proteolytic and amylase activities. *Aquaculture* **170**, 267–283.

Hjelmeland, K., Pedersen, B. H., and Nilssen, E. M. (1988). Trypsin content in intestines of herring larvae, *Clupea harengus*, ingesting inert polystyrene spheres or live crustacea prey. *Mar. Biol.* **98**, 331–335.

Hofer, R. (1988). Morphological adaptations of the digestive tract of tropical cyprinids and cichlids to diet. *J. Fish Biol.* **33**, 399–408.

Hofer, R. (1989). Digestion. *In* "Cyprinid Fishes, Systematics, Biology and Exploitation" (Winfield, I. J., and Nelson, J. S., Eds.). Chapman & Hall, London.

Hofer, R., and Schiemer, F. (1981). Proteolytic activity in the digestive tract of several species of fish with different feeding habitats. *Oecologia* **48**, 342–345.

Hoogerhoud, R. J. C. (1987). The adverse effects of shell ingestion for molluscivorous cichlids, a constructional morphological approach. Netherlands. *J. Zool.* **37**, 277–300.

Horn, M. H. (1989). Biology of marine herbivorous fishes. *Oceanogr. Mar. Biol.* **27**, 167–272.

Horn, M. H. (1997). Evidence for dispersal of fig seeds by the fruit-eating characid fish *Brycon guatemalensis* Regan in a Costa Rican tropical rain forest. *Oecologia* **109**, 259–264.

Inhamuns, A. J., and Franco, M. R. B. (2001). Composition of total, neutral, and phospholipids in mapara (*Hypopthalmus* sp.) from the Brazilian Amazonian area. *J. Agr. Food Chem.* **49**, 4859–4863.

Işik, O., Sarihan, E., Kuşvuran, E., Gül, Ö., and Erbatur, O. (1999). Comparison of the fatty acid composition of the freshwater fish larvae *Tilapia zillii*, the rotifer *Brachionus calyciflorus*, and the microalgae *Scenedesmus abundans*, *Monoraphidium minitum* and *Chlorella vugaris* in the algae-rotifer-fish larvae food chains. *Aquaculture* **174**, 299–311.

Jacobsen, D., and Bojsen, B. (2002). Macroinvertebrate drift in Amazon streams in relation to riparian forest cover and fish fauna. *Arch. Hydrobiol.* **155**, 177–197.

Jónás, E., Rágyanszky, M., Oláh, J., and Boross, L. (1983). Proteolytic digestive enzymes of carnivorous (*Silurus glanis* L.), herbivorous (*Hypophthalmichthys molitrix* Val.) and omnivorous (*Cyprinus carpio* L.) fishes. *Aquaculture* **30**, 145–154.

Justi, K. C., Visentainer, J. V., Evelázio de Souza, N., and Matusushita, M. (2000). Nutritional composition and vitamin C stability in stored camu-camu (*Myrciaria dubia*) pulp. *Arch. Latinam. Nutr.* **50**, 405–408.

Kafuku, T. (1975). An ontogenetical study of intestinal coiling pattern on Indian major carps. *Bull. Freshwater Fish. Res. Lab.* **27**, 1–19.

Kasper, C. S., White, M. R., and Brown, P. B. (2000). Choline is required by tilapia when methionine is not in excess. *J. Nutr.* **130**, 238–242.

Kato, K., Ishibashi, Y., Murata, O., Nasu, T., Ikeda, S., and Kumai, H. (1994). Qualitative water-soluble vitamin requirements of Tiger Puffer. *Fisheries Sci.* **60**, 589–596.

Kaushik, S. J., Doudet, T., Medale, F., Aguirre, P., and Blanc, D. (1995). Protein and energy needs for maintenance and growth of Nile tilapia (*Oreochromis niloticus*). *J. Appl. Ichthyol.* **11**, 290–296.

Kesamaru, K., and Miyazono, I. (1978). Studies on the nutrition of *Tilapia nilotica*. II. The nutritive values of diets containing various dietary proteins. *Bull. Fac. Agric. Miyazaki Univ.* **25**, 351–359.

Keshavanath, P., Beveridge, M. C. M., Baird, D. J., Lawton, L. A., Nimmo, A., and Codd, G. A. (1994). The functional grazing response of a phytoplanktivorous fish *Oreochromis niloticus* to mixtures of toxic and non-toxic strains of the cyanobacterium *Microcystis aeruginosa*. *J. Fish Biol.* **45**, 123–129.

Khan, M. A., and Jafri, A. K. (1993). Quantitative dietary requirement for some indispensable amino acids in the Indian major carp, *Labeo Rohita* (Hamilton) fingerling. *J. Aquacult. Trop.* **8**, 67–80.

Kihara, M., and Sakata, T. (1997). Fermentation of dietary carbohydrates to short-chain fatty acids by gut microbes and its influence on intestinal morphology of a detritivorous teleost tilapia (*Oreochromis niloticus*). *Comp. Biochem. Physiol.* **118A**, 1201–1207.

Kodric-Brown, A. (1989). Dietary carotenoids and male mating success in the guppy: And environmental component to female choice. *Behav. Ecol. Sociobiol.* **25**, 393–401.

Kolkovski, S. (2001). Digestive enzymes in fish larvae and juveniles: implications and applications to formulated diets. *Aquaculture* **200**, 181–201.

Kolkovski, S., Tandler, A., Kissil, G. W., and Gertler, A. (1993). The effect of dietary exogenous digestive enzymes on ingestion, assimilation, growth and survival of gilthead seabream (*Sparus aurata*, Sparidae, Linnaeus) larvae. *Fish Physiol. Biochem.* **12**, 203–209.

Koven, W. M., Henderson, R. J., and Sargent, J. R. (1997). Lipid digestion in turbot (*Scophthalmus maximus*): *In-vivo* and *in-vitro* studies of the lipolytic activity in various segments of the digestive tract. *Aquaculture* **151**, 155–171.

Kramer, D. L., and Bryant, M. J. (1995a). Intestine length in the fishes of a tropical stream: 1. Ontogenetic allometry. *Environ. Biol. Fish* **42**, 115–127.

Kramer, D. L., and Bryant, M. J. (1995b). Intestine length in the fishes of a tropical stream: 2. Relashionship to diet – the long and short of a convoluted issue. *Environ Biol. Fish* **42**, 129–141.

Kubitzki, K., and Ziburski, A. (1994). Seed dispersal in flood plain forests of Amazonia. *Biotropica* **26**, 30–43.

Kurokawa, T., Shiraishi, M., and Suzuki, T. (1998). Quantification of exogenous protease derived from zooplankton in the intestine of Japanese sardine *Sardinops melanoticus* larvae. *Aquaculture* **161**, 491–499.

Kuzmina, V. V. (1996). Influence of age on digestive enzyme activity in some freshwater teleosts. *Aquaculture* **148**, 25–37.

Lauff, M., and Hofer, R. (1984). Proteolytic enzymes in fish development and the importance of dietary enzymes. *Aquaculture* **37**, 335–346.

Lazo, J. P., Holt, G. J., and Arnold, C. R. (2000). Ontogeny of pancreatic enzymes in larval red drum *Scianops ocellatus*. *Aquacult. Nutr.* **6**, 183–192.

Leite, R. G., Araujo-Lima, C. A. R. M., Victoria, R. L., and Martinelli, L. A. (2002). Stable isotope analysis of energy sources for larvae of eight fish species from the Amazon floodplain. *Ecol. Freshw. Fish* **11**, 56–63.

Lewis, W. M., Jr., Hamilton, S. K., Rodriguez, M. A., Saunders, III, J. F., and Lasi, M. A. (2001). Food web analysis of the Orinoco floodplain based on production estimates and stable isotope data. *J. N. Am. Benthol. Soc.* **20**, 241–254.

Lim, C., Leamaster, B., and Brock, J. A. (1993). Riboflavin requirement of fingerling red hybrid tilapia grown in seawater. *J. World Aquacult. Soc.* **24**(4), 451–458.

Lim, L. C., Cho, Y. L., Dhert, P., Wong, C. C., Nelis, H., and Sorgeloos, P. (2002). Use of decapsulated *Artemia* cysts in ornamental fish culture. *Aquac. Res.* **33**, 575–589.

Limsuwan, T., and Lovell, R. T. (1981). Intestinal synthesis and absorption of vitamin B-12 in channel catfish. *J. Nutr.* **111**, 2125–2132.

Lobel, P. S. (1981). The trophic biology of herbivorous reef fishes: Alimentary pH and digestive capabilities. *J. Fish Biol.* **19**, 365–397.

Lobel, P. S., and Ogden, J. C. (1981). Foraging by the herbivorous parrotfish *Sparisoma radians*. *Mar. Biol.* **64**, 173–183.

Lovell, R. T., and Limsuwan, T. (1982). Intestinal synthesis and dietary nonessentiality of vitamin B_{12} for *Tilapia nilotica. Trans. Am. Fish Soc.* **111**, 485–490.

Maffia, M., Ahearn, G. A., Vilella, S., Zonno, V., and Storelli, C. (1993). Ascorbic acid transport by intestinal brush-border membrane vesicles of the teleost *Anguilla anguilla. Am. J. Physiol. Reg. 1.* **264**, R1248–R1253.

Maffia, M., Verri, T., Danieli, A., Thamotharan, M., Pastore, M., Ahearn, G. A., and Storelli, C. (1997). H^+-glycyl-L-proline cotransport in brush-border membrane vesicles of eel (*Anguilla anguilla*) intestine. *Am. J. Physiol. Reg. 1.* **272**, R217–R225.

Magurran, A. E., and Queiroz, H. L. (2003). Partner choice in piranha shoals. *Behaviour* **140**, 289–299.

Maia, E. L., Rodriguez-Amaya, D. B., and Hotta, L. K. (1995). Fatty acid composition of the total, neutral and phospholipids of pond-raised Brazilian *Piaractus mesopotamicus. Int. J. Food Sci. Tech.* **30**, 591–597.

Mambrini, M., and Kaushik, S. J. (1994). Partial replacement of dietary protein nitrogen with dispensable amino acids in diets of Nile tilapia, *Oreochromis niloticus. Comp. Biochem. Physiol.* **109A**, 469–477.

Mannheimer, S., Bevilacqua, G., Caramaschi, E. P., and Scarano, F. R. (2003). Evidence for seed dispersal by the catfish *Auchenipterichthys longimanus* in an Amazonian lake. *J. Tropic. Ecol.* **19**, 215–218.

Martin, T. J., and Blaber, S. J. M. (1984). Morphology and histology of the alimentary tracts of Ambassidae (Cuvier) (Teleostei) in relation to feeding. *J. Morphol.* **182**, 295–305.

Martino, R. C., Trugo, L. C., Cyrino, J. E. P., and Portz, L. (2003). Use of white fat as a replacement for squid liver oil in practical diets for surubim *Pseudoplatystoma coruscans. J. World Aquacult. Soc.* **34**, 192–202.

Martins, M. L. (1995). Effect of ascorbic acid deficiency on the growth, gill filament lesions and behavior of pacu fry (*Piaractus mesopotamicus* Holmberg, 1887). *Braz. J. Med. Biol. Res.* **28**, 563–568.

Marx, F., Andrade, E. H. A., and Maia, J. G. (1997). Chemical composition of the fruit pulp of *Caryocar villosum. Z. Lebensm Unters For.* **204**, 442–444.

Matusiewicz, M., Dabrowski, K., Volker, L., and Matusiewicz, K. (1994). Regulation of saturation and depletion of ascorbic acid in rainbow trout. *J. Nutr. Biochem.* **5**, 204–211.

Melack, J. M., Forsberg, B. R., Victoria, R. L., and Richey, J. E. (1999). Biogeochemistry of Amazon floodplain lakes and associated wetlands. *In* "The Biogeochemistry of the Amazon

Basin and Its Role in a Changing World" (McClain, M., Victoria, R., and Richey, J., Eds.). Oxford University Press, Oxford.

Melard, Ch., Orozco, J. J., Uran, L. A., and Ducarme, Ch. (1993). Comparative growth rate and production of *Colossoma macropomum* and *Piaractus brachypomus* (*Colossoma bidens*) in tanks and cages using intensive rearing conditions. *In* "Production, Environment and Quality" (Barnabe, G., and Kestemont, P., Eds.). *European Aquacult. Soc. Spec. Publ.* **18**, 433–442.

Mendoza, R., Aguilera, C., Rodriquez, G., González, M., and Castro, R. (2002). Morphophysiological studies on alligator gar (*Atractosteus spatula*) larval development as a basis for their culture and repopulation of their natural habitats. *Rev. Fish Biol. Fisher.* **12**, 133–142.

Meske, Ch., and Pfeffer, E. (1978). Growth experiments with carp and grass carp. *Arch. Hydrobiol. Beih.* **11**, 98–107.

Meyer, A. (1989). Cost of morphological specialization: Feeding performance of the two morphs in the tropically polymorphic cichlid fish, *Cichlasoma citrinellum*. *Oecologia* **80**, 431–436.

Meyer-Burgdorff, von K.-H., Becker, K., and Günther, K.-D. (1986). M-Inosit: Mangelerscheinungen and bedarf beim wachsender spiegelkarpfen (*Cyprinus carpio* L.). *J. Anim. Physiol. An. N.* **56**, 232–241.

Montgomery, W. L., and Pollak, P. E. (1988a). *Epulopiscium fishelsoni* N. G., N. Sp., a protist of uncertain taxonomic affinities from the gut of an herbivorous reef fish. *J. Protozool.* **35**, 565–569.

Montgomery, W. L., and Pollak, P. E. (1988b). Gut anatomy and pH in a Red Sea surgeonfish, *Acanthurus nigrofuscus*. *Mar. Ecol. Prog. Ser.* **44**, 7–13.

Moon, T. W. (2001). Glucose intolerance in teleost fish: Fact or fiction? *Comp. Biochem. Physiol. Part B* **129**, 243–249.

Moreau, R., and Dabrowski, K. (2001). Gulonolactone oxidase presence in fishes: activity and significance. *In* "Ascorbic Acid in Aquatic Organisms" (Dabrowski, K., Ed.), pp. 13–31. CRC Press, Boca Raton, FL.

Morrison, C. M., and Wright, J. R., Jr (1999). A study of the histology of the digestive tract of the Nile tilapia. *J. Fish Biol.* **54**, 597–606.

Moyano, F. J., Diaz, M., Alarcón, F. J., and Serasquete, M. C. (1996). Characterization of digestive enzyme activity during larval development of gilthead seabream *Sparus aurata*. *Fish Physiol. Biochem.* **15**, 121–130.

Mukhopadhyay, P. (1977). Studies on the enzymatic activities related to varied pattern of diets in the air-breathing catfish, *Clarias batrachus* (Linn.). *Hydrobiologia* **52**, 235–237.

Munilla-Moran, R., Stark, J. R., and Barbour, A. (1990). The role of exogenous enzymes in digestion in cultured turbot larvae, *Scophthalmus maximus* (L.). *Aquaculture* **88**, 337–350.

Navarro, I., Leibush, B. N., Moon, T. W., Plisetskaya, E. M., Banos, N., Mendez, E., Planas, J. V., and Gutierrez, J. (1999). Insulin, insulin-like growth factor-I (IGF-1) and glucagons: The evolution of their receptors. *Comp. Biochem. Physiol.* **122B**, 137–153.

Nicolas, J. L., Bobic, E., and Ansquer, D. (1989). Bacterial flora associated with a trophic chain consisting of microalgae, rotifers and turbot larvae: influence of bacteria on larval survival. *Aquaculture* **83**, 237–248.

Nose, T. (1979). *In* "Summary report on the requirements of essential amino acids for carp Proceedings of World Symposium on Finfish Nutrition and Fishfeed Technology," 20–23 June, 1978, pp. 145–156. Heenemann, Berlin.

O'Connell, J. P., and Gatlin, III, D. M. (1994). Effects of dietary calcium and vitamin D_3 on weight gain and mineral composition of the blue tilapia (*Oreochromis aureus*) in low-calcium water. *Aquaculture* **125**, 107–117.

Olatunde, A. A., and Ogunbiyi, O. A. (1977). Digestive enzymes in the alimentary canal of three tropical catfish. *Hydrobiologia* **56**(1), 21–24.

Olsen, R. E., Henderson, R. J., and McAndrew, B. J. (1990). The conversion of linoleic acid and linolenic acid to longer chain polyunsaturated fatty acids by *Tilapia (Oreochromis) nilotica in vivo. Fish Physiol. Biochem.* **8**(3), 261–270.

Olvera-Novoa, M. A., Dominguez-Cen, L. J., Olivera-Castillo, L., and Martinez-Palacios, C. A. (1998). Effect of the use of the microalga *Spirulina maxima* as fish meal replacement in diets for tilapia, *Oreochromis mossambicus* (Peters), fry. *Aquacult. Res.* **29**, 709–715.

Olvera-Novoa, M. A., Gasca-Leyva, E., and Martinez-Palacios, C. A. (1996). The dietary protein requirements of *Cichlasoma synspilum* Hubbs, 1935 (Pisces: Cichlidae) fry. *Aquacult. Res.* **27**, 167–173.

Ostaszewska, T. (personal communication).

Pan, B. S., Lan, C. C., and Hung, T. Y. (1991). Changes in composition and proteolytic enzyme activities of *Artemia* during early development. *Comp. Biochem. Physiol.* **100A**, 725–730.

Pantastico, J. B., Baldia, J. P., and Reyes, D. M., Jr. (1986). Feed preference of milkfish (*Chanos chanos* Forsskal) fry given different algal species as natural feed. *Aquaculture* **56**, 169–178.

Pantastico, J. B., Espegadera, C., and Reyes, D. (1982). Fry-to-fingerling production of tilapia nilotica in aquaria using phytoplankton as natural feed. *Philipp. J. Biol.* **11**(2–3), 245–254.

Peña, R., Dumas, S., Villalejo-Fuerte, M., and Ortiz-Galindo, J. L. (2003). Ontogenetic development of the digestive tract in reared spotted sand bass *Paralabrax maculofasciatus* larvae. *Aquaculture* **219**, 633–644.

Peres, A., Zambonino Infante, J. L., and Cahu, C. (1998). Dietary regulation of activities and mRNA levels of trypsin and amylase in sea bass (*Dicentrarchus labrax*) larvae. *Fish Physiol. Biochem.* **19**, 145–152.

Person-Le Ruyet, J., Alexandre, J. C., Thébaud, L., and Mugnier, C. (1993). Marine fish larvae feeding: Formulated diets or live prey? *J. World Aquacult. Soc.* **24**(2), 211–224.

Piet, G. J. (1998). Ecomorphology of size-structured tropical freshwater fish community. *Environ. Biol. Fish* **51**, 67–86.

Portella, M. C., and Flores-Quintana, C. (2003a). Histology and histochemistry of the digestive system development of *Pseudoplatystoma fasciatum* larvae. World Aquaculture 2003, Book of Abstracts, p. 590. Salvador, Bahia, Brazil.

Portella, M. C., and Flores-Quintana, C. (2003b). Histological analysis of juvenile *Pseudoplatystoma fasciatum* digestive system. World Aquaculture 2003, Book of Abstracts, p. 591. Salvador, Bahia, Brazil.

Portella, M. C., Pizauro, J. M., Tesser, M. B., and Carneiro, D. J. (2002). Determination of enzymatic activity in different segments of the digestive system of *Pseudoplatystoma fasciatum*. World Aquaculture 2002 Book of Abstracts, p. 615. Beijing, China.

Portella, M. C., Pizauro, J. M., Tesser, M. B., Jomori, R. K., and Carneiro, D. J. (2004). Digestive enzymes activity during the early development of surubim *Pseudoplatystoma fasciatum*. World Aquaculture 2004 Book of Abstracts. Honolulu, Hawaii.

Portella, M. C., Verani, J. R., Tercio, J., Ferreira, B., and Carneiro, D. J. (2000). Use of live and artificial diets enriched with several fatty acid sources to feed *Prochilodus scrofa* larvae and fingerlings, 1. Effects on body composition. *J. Aquacult. Trop.* **15**, 45–58.

Pouilly, M., Lino, F., Bretenous, J.-G., and Rosales, C. (2003). Dietary-morphological relationships in a fish assemblage of the Bolivian Amazonian floodplain. *J. Fish Biol.* **62**, 1137–1158.

Prejs, A., and Blaszczyk, M. (1977). Relationships between food and cellulase activity in freshwater fishes. *J. Fish Biol.* **11**, 447–452.

Radunz-Neto, J., Corraze, G., Bergot, P., and Kaushik, S. J. (1996). Estimation of essential fatty acid requirements of common carp larvae using semi-purified artificial diets. *Arch. Anim. Nutr.* **49**, 41–48.

Rahmatullah, S. M., and Beveridge, M. C. M. (1993). Ingestion of bacteria in suspension Indian major carps (*Catla catla, Lubeo rohita*) and Chinese carps (*Hypophthalmichthys molitrix, Aristichthys nobilis*). *Hydrobiologia* **264**, 79–84.

Randall, D. J., Wood, C. M., Perry, S. F., Bergman, H., Maloiy, G. M. O., Mommsen, T. P., and Wright, P. A. (1989). Urea excretion as a strategy for survival in a fish living in a very alkaline environment. *Nature* **337**, 165–166.

Rao, D. S., and Raghuramulu, N. (1999). Vitamin D$_3$ and its metabolites have no role in calcium and phosphorus metabolism in *Tilapia mossambica. J. Nutr. Sci. Vitaminol.* **45**, 9–19.

Ravi, J., and Devaraj, K. V. (1991). Quantitative essential amino acid requirements for growth of catla, *Catla catla* (Hamilton). *Aquaculture* **96**, 281–291.

Reimer, G. (1982). The influence of diet on the digestive enzymes of the Amazon fish Matrinchã, *Brycon* cf. *melanopterus. J. Fish Biol.* **21**, 637–642.

Reshkin, S. J., and Ahearn, G. A. (1991). Intestinal glycyl-L-phenylalanine and L-phenylalanine transport in a euryhaline teleost. *Am. J. Physiol.* **29**, R563–R569.

Ribeiro, L., Sarasquete, C., and Dinis, M. (1999). Histological and histochemical development of the digestive system of *Solea senegalensis* (Kaup,1858) larvae. *Aquaculture* **171**, 293–308.

Rimmer, D. W., and Wiebe, W. J. (1987). Fermentative microbial digestion in herbivorous fishes. *J. Fish Biol.* **31**, 229–236.

Robinson, E. H., LaBomascus, D., Brown, P. B., and Linton, T. L. (1987). Dietary calcium and phosphorus requirements of *Oreochromis aureus* reared in calcium-free water. *Aquaculture* **64**, 267–276.

Rosenstock, B., and Simon, M. (2001). Sources and sinks of dissolved free amino acids and protein in a large and deep mesotrophic lake. *Limnol. Oceanogr.* **46**(3), 644–654.

Rothman, S., Liebow, C., and Isenman, L. (2002). Conservation of digestive enzymes. *Physiol. Rev.* **82**, 1–18.

Roubach, R., and Saint-Paul, U. (1994). Use of fruits and seeds from Amazonian inundated forest in feeding trials with *Colossoma macropomum* (Cuvier, 1818) (Pisces, Characidae). *J. Appl. Ichthyol.* **10**, 134–140.

Sabapathy, U., and Teo, L. H. (1993). A quantitative study of some digestive enzymes in the rabbitfish, *Siganus canaliculatus* and the sea bass, *Lates calcarifer. J. Fish. Biol.* **42**, 595–602.

Sabapathy, U., and Teo, L-H. (1995). Some properties of the intestinal proteases of the rabbitfish, *Siganus canaliculatus* (Park). *Fish Physiol. Biochem.* **14**(3), 215–221.

Saha, A. K., and Ray, A. K. (1998). Cellulase activity in rohu fingerlings. *Aquacult. Int.* **6**, 281–291.

Santiago, C. B., and Lovell, R. T. (1988). Amino acid requirements for growth of Nile tilapia. *J. Nutr.* **118**, 1540–1546.

Santiago, C. B., and Reyes, O. S. (1993). Effects of dietary lipid source on reproductive performance and tissue lipid levels of Nile tilapia *Oreochromis niloticus* (Linnaeus) broodstock. *J. Appl. Ichthyol.* **9**, 33–40.

Santiago, C. B., Bañez-Aldaba, M., and Laron, M. A. (1982). Dietary crude protein requirement of *Tilapia nilotica* fry. *Kalikasan, Philipp. J. Biol.* **11**, 255–265.

Sato, M., Yoshinaka, R., and Yamamoto, Y. (1978). Nonessentiality of ascorbic acid in the diet of carp. *B. Jpn Soc. Sci. Fish* **49**, 1151–1156.

Satoh, S., Takeuchi, T., and Watanabe, T. (1987). Requirement of *Tilapia* for α-tocopherol. *Nippon Suisan Gakk.* **53**, 119–124.

Schultz, H. N., and Jorgensen, B. B. (2001). Big bacteria. *Annu. Rev. Microbiol.* **55**, 105–137.

Schutz, M., and Barlow, G. W. (1997). Young of the Midas cichlid get biologically active nutrients by eating mucus from the surface of their parents. *Fish Physiol. Biochem.* **16**, 11–18.

Segner, H., Arend, P., Von Poeppinghausen, K., and Schmidt, H. (1989a). The effect of feed astaxanthin to *Oreochromis niloticus* and *Colisa labiosa* on the histology of the liver. *Aquaculture* **79**, 381–390.

Segner, H., Rosch, R., Schmidt, H., and von Poeppinghausen, K. J. (1989b). Digestive enzymes in larval *Coregonus lavaretus* L. *J. Fish Biol.* **35**, 249–263.

Segner, H., Burkhardt, P., Avila, E. M., Storch, V., and Juario, J. V. (1987). Effects of *Chlorella*-feeding on larval milkfish, *Chanos chanos*, as evidenced by histological monitoring. *Aquaculture* **67**, 113–116.

Segner, H., Rösch, R., Verreth, J., and Witt, U. (1993). Larval nutritional physiology: Studies with *Clarias gariepinus*, *Coregonus lavaretus* and *Scophtalmus maximus*. *J. World Aquacult. Soc.* **24**, 121–134.

Seixas Filho, J. T., Brás, J. M., Gomide, A. T. M., Oliveira, M. G. A., and Donzele, J. L. (2000a). Anatomia funcional e morfometria dos intestinos e dos cecos pilóricos do Teleostei (pisces) de água doce *Brycon orbignyanus* (Valenciennes, 1849). *Rev. Bras. Zootecn.* **29**, 313–324.

Seixas Filho, J. T., Brás, J. M., Gomide, A. T. M., Oliveira, M. G. A., and Donzele, J. L. (2000b). Anatomia funcional e morfometria dos intestinos e dos cecos pilóricos do Teleostei (pisces) de água doce piau (*Leporinus friderici*, Bloch, 1794). *Rev. Bras. Zootecn.* **29**, 2181–2192.

Seixas-Filho, J. T., Almeida-Oliveira, M. G., Donzele, J. L., Mendonca-Gomide, A. T., and Menin, E. (2000c). Trypsin acitivity in the chime of three tropical teleost freshwater fish. *Rev. Bras. Zootecn.* **29**, 2172–2180.

Seixas-Filho, J. T., Moura-Bras, J., Mendonca-Gomide, A. T., Almeida-Oliveira, M. G., Lopes-Donzele, J., and Menin, E. (2001). Functional anatomy and morphometry of the intestine of fresh water teleoste (Pisces) de agua doce surubim (*Pseudoplatystoma coruscans* -Agassiz, 1829). *Rev. Bras. Zootecn.* **30**, 1670–1680.

Shaik Mohamed, J. (2001). Dietary pyridoxine requirement of the Indian catfish, *Heteropneustes fossilis*. *Aquaculture* **194**, 327–335.

Shaik Mohamed, J., Ravisankar, B., and Ibrahim, A. (2000). Quantifying the dietary biotin requirement of the catfish, *Clarias batrachus*. *Aquacult. Int.* **8**, 9–18.

Shearer, K. D. (2000). Experimental design, statistical analysis and modelling of dietary nutrient requirement studies for fish: A critical review. *Aquaculture* **6**, 91–102.

Shiau, S.-Y., and Chin, Y.-H. (1999). Estimation of the dietary biotin requirement of juvenile hybrid tilapia, *Oreochromis niloticus* x *O. aureus*. *Aquaculture* **170**, 71–78.

Shiau, S.-Y., and Hsieh, H.-L. (1997). Vitamin B_6 requirements of tilapia *Oreochromis niloticus* x *O. aureus* fed two dietary protein concentrations. *Fisheries Sci.* **63**, 1002–1007.

Shiau, S.-Y., and Hsu, T.-S. (1995). L-ascorbyl-2-sulfate has equal antiscorbutic activity as L-ascorbyl-2-monophosphate for tilapia, *Oreochromis niloticus* x *O. aureus*. *Aquaculture* **133**, 147–157.

Shiau, S.-Y., and Hsu, T.-S. (2002). Vitamin E sparing effect by dietary vitqamin C in juvenile hybrid tilapia, *Oreochromis niloticus* x *O. aureus*. *Aquaculture* **210**, 335–342.

Shiau, S., and Huang, S. (1989). Optimal dietary protein level for hybrid tilapia (*Oreochromis niloticus* x *O. aureus*) reared in seawater. *Aquaculture* **81**, 119–127.

Shiau, S.-Y., and Hwang, J.-Y. (1993). Vitamin D. requirements of juvenile hybrid tilapia *Oreochromis niloticus* x *O. aureus*. *Nippon Suisan Gakk.* **59**, 553–558.

Shiau, S.-Y., and Liang, H.-S. (1995). Carbohydrate utilization and digestibility by tilapia, *Oreochromis niloticus* x *O. aureus*, are affected by chromic oxide inclusion in the diet. *J. Nutr.* **125**, 976–982.

Shiau, S.-Y., and Lo, P.-S. (2000). Dietary choline requirements of juvenile hybrid tilapia, *Oreochromis niloticus* x *O. aureus*. *J. Nutr.* **130**, 100–103.

Shiau, S. Y., and Shiau, L. F. (2001). Re-evaluation of the vitamin E requirements of juvenile tilapia (*Oreochromis niloticus x O. aureus*). *Anim. Sci.* **72,** 529–534.

Shiau, S.-Y., and Suen, G.-S. (1992). Estimation of the niacin requirements for tilapia fed diets containing glucose or dextrin. *J. Nutr.* **122,** 2030–2036.

Shim, K. F., Landesman, L., and Lam, T. J. (1989). Effect of dietary protein on growth, ovarian development and fecundity in the dwarf gourami, *Colisa Lalia* (Hamilton). *J. Aquacult. Trop.* **4,** 111–123.

Sibbing, F. A. (1988). Specializations and limitations in the utilization of food resources by the carp, *Cyprinus carpio*: A study of oral food processing. *Environ. Biol. Fish* **22,** 161–178.

Siddiqui, A. Q., Howlader, M. S., and Adam, A. A. (1988). Effects of dietary protein levels on growth, feed conversión and protein utilization in fry and young tilapia, *Oreochromis niloticus* (1988). *Aquaculture* **70,** 63–72.

Smith, B. L., Smith, S. A., and Laurance, T. A. (2000). Gross morphology and topography of the adult intestinal tract of the tilapia fish. *Oreochronis niloticus* L. *Cells, Tissues, Organs* **166,** 294–303.

Soliman, A. K., Jauncey, K., and Roberts, R. J. (1985). Qualitative and quantitative identification of L-gulonolactone oxidase activity in some teleosts. *Aquacult. Fish Manag.* **1,** 249–256.

Soliman, A. K., and Wilson, R. P. (1992a). Water-soluble vitamin requirements of tilapia, 1. Pantothenic acid requirement of blue tilapia, *Oreochromis aureus*. *Aquaculture* **104,** 121–126.

Soliman, A. K., and Wilson, R. P. (1992b). Water-soluble vitamin requirements of tilapia, 2. Riboflavin requirement of blue tilapia, *Oreochromis aureus*. *Aquaculture* **104,** 309–314.

Soliman, A. K., Jauncey, K., and Roberts, R. J. (1986). The effect of dietary ascorbic acid supplementation on hatchability, survival rate and fry performance in *Oreochromis mossambicus* (Peters). *Aquaculture* **59,** 197–208.

Soliman, A. K., Jauncey, K., and Roberts, R. J. (1994). Water-soluble vitamin requirements of tilapia: Ascorbic acid (vitamin C) requirement of Nile tilapia, *Oreochromis niloticus* (L.). *Aquacult. Fish Manag.* **25,** 269–278.

Souza, S., Menin, E., Juarez Lopez, D., and Fonseca, C. (2001a). Histologia do estômago de alevinos de surubim e sua relação com o hábito alimentar. Anais da 38 Reunião Anual da Sociedade Brasileira de Zootecnia. Piracicaba, SP, Brazil, 1435–1436.

Souza, S., Menin, E., Fonseca, C., and Juarez Lopez, D. (2001b). Sistema endócrino difuso enteropancreático em alevinos de surubim e sua potencialidade para o controle das secreções digestivas. Anais da 38 Reunião Anual da Sociedade Brasileira de Zootecnia. Piracicaba, SP, Brazil, 1437–1438.

Srivastava, A. S., Kurokawa, T., and Suzuki, T. (2002). mRNA expression of pancreatic enzymes precursors and estimation of protein digestibility in first feeding larvae of the Japanese flounder, *Paralichthys olivaceus*. *Comp. Biochem. Physiol.* **132A,** 629–635.

Stickney, R. R., McGeachin, R. B., Lewis, D. H., Marks, J., Riggs, A., Sis, R. F., Robinson, E. H., and Wurts, W. (1984). Response of *Tilapia aurea* to dietary vitamin C. *J. World Maricult. Soc.* **15,** 179–185.

Stroband, H. W. J., and Dabrowski, K. (1981). Morphological and physiological aspects of the digestive system and feeding in fresh water fish larvae. *In* "La Nutrition des Poissons" (Fontaine, M., Ed.), pp. 355–376. Paris, Actes du Colloque CNERNA Paris, May 1979.

Stroband, H. W. J., and Kroon, A. G. (1981). The development of the stomach in *Clarias lazera* and the intestinal absorption of protein macromolecules. *Cell Tissue Res.* **215,** 397–415.

Sturmbauer, C., Mark, W., and Dallinger, R. (1992). Ecophysiology of Aufwuchs-eating cichlids in Lake Tanganyika: Niche separation by trophic specialization. *Environ. Biol. Fish* **35,** 283–290.

Sugita, H., Miyajima, C., and Deguchi, Y. (1990). The vitamin B12-producing ability of intestinal bacteria isolated from tilapia and channel catfish. *Nippon Suisan Gakk.* **56,** 701.

Sugita, H., Takahashi, J., and Deguchi, Y. (1992). Production and consumption of biotin by the intestinal microflora of cultured freshwater fishes. *Biosci. Biotech. Bioch.* **56,** 1678–1679.

Tacon, A. G. J., and Cowey, C. B. (1985). Protein and amino acid requirements. *In* "Fish Energetics. New Perspectives" (Tyler, P., and Calow, P., Eds.). Johns Hopkins University Press, Baltimore, MD.

Tadesse, Z., Boberg, M., Sonesten, L., and Ahlgren, G. (2003). Effects of algal diets and temperature on the growth and fatty acid content of the cichlid fish *Oreochromis niloticus* L. – a laboratory study. *Aquatic Ecol.* **37**, 169–182.

Takeuchi, T. (1996). Essential fatty acid requirements in carp. *Arch. Anim. Nutr.* **49**, 23–32.

Taniguchi, A. Y., and Takano, K. (2002). Purification and properties of aminopeptidase from *Tilapia* intestine – digestive enzyme of *Tilapia*-IX-. *Nippon Suisan Gakk.* **68**, 382–388.

Tengjaroenkul, B., Smith, B. J., Caceci, T., and Smith, S. A. (2000). Distribution of intestinal enzyme activities along the intestinal tract of cultured Nile tilapia, *Oreochromis niloticus* L. *Aquaculture* **182**, 317–327.

Terjesen, B. F., Verreth, J., and Fyhn, H. J. (1997). Urea and ammonia excretion by embryos and larvae of the African catfish *Clarias gariepinus* (Burchell, 1822). *Fish Physiol. Biochem.* **16**, 311–321.

Tesser, M. B. (2002). Desenvolvimento do trato digestório e crescimento de larvas de pacu, *Piaractus mesopotamicus* (Holmberg, 1887) em sistemas de co-alimentação com náuplios de *Artemia* e dieta microencapsulada. Centro de Aqüicultura da Universidade Estadual Paulista. Master Dissertation.

Thamotharan, M., Gomme, J., Zonno, V., Maffia, M., Storelli, C., and Ahearn, G. A. (1996a). Electrogenic, proton-coupled, intestinal dipeptide transport in herbivorous and carnivorus teleosts. *Am. J. Physiol.* **39**, R939–R947.

Thamotharan, M., Zonno, V., Storelli, C., and Ahearn, G. A. (1996b). Basolateral dipeptide transport by the intestine of the teleost *Oreochromis mossambicus*. *Am. J. Physiol.* **39**, R948–R954.

Tibbetts, I. (1997). The distribution and function of mucous cells and their secretions in the alimentary tract of *Arrhamphus sclerolepis* Krefftri. *J. Fish Biol.* **50**, 809–820.

Titus, E., Karasov, W. H., and Ahearn, G. A. (1991). Dietary modulation of intestinal nutrient transport in the teleost fish tilapia. *Am. J. Physiol.* **30**, R1568–R1574.

Uys, W., and Hecht, T. (1987). Assays on the digestive enzymes of sharptooth catfish, *Clarias gariepinus* (Pisces: Clariidae). *Aquaculture* **63**, 301–313.

Uys, W., Hecht, T., and Walters, M. (1987). Change in digestive enzyme activities of *Clarias gariepinus* (Pisces: Clariidae) after feeding. *Aquaculture* **63**, 243–250.

Van Dam, A. A., and Pauly, D. (1995). Simulation of the effects of oxygen on food consumption and growth of Nile tilapia, *Oreochromis niloticus* (L.). *Aquacult. Res.* **26**, 427–440.

Van der Meer, M. B., Machiels, M. A. M., and Verdegem, M. C. J. (1995). The effect of dietary protein level on growth, protein utilization and body composition of *Colossoma macropomum* (Cuvier). *Aquacult. Res.* **26**, 901–909.

Van der Velden, J. A., Kolar, Z. I., and Flik, G. (1991). Intake of magnesium from water by freshwater tilapia fed on a low-Mg diet. *Comp. Biochem. Physiol.* **99A**(1/2), 103–105.

Vari, R. P., and Malabarba, L. R. (1998). Neotropical/Icthyology: an overview. *In* "Phylogeny and Classification of Neotropical Fishes" (Malabarba, L. R., Reis, R. E., Vari, R. P., Lucena, Z. M. S., and Lucena, C. A. S., Eds.), pp. 7–11. EDIPUCRS, Porto Alegre, Brazil.

Vera, J. C., Rivas, C. I., Fischbarg, J., and Golde, D. W. (1993). Mammalian facilitative hexose transporters mediate the transport of dehydroascorbic acid. *Nature* **364**, 79–82.

Verreth, J. A. J., Torreele, E., Spazier, E., and Sluiszen, A. V.der. (1992). The development of a functional digestive system in the African catfish *Clarias gariepinus* (Burchell). *J. World Aquacult. Soc.* **23**, 286–298.

Verri, T., Maffia, M., Danieli, A., Herget, M., Wenzel, U., Daniel, H., and Storelli, C. (2000). Characterisation of the H^+/peptide cotransporter of eel intestinal brush-border membranes. *J. Exp. Biol.* **203**, 2991–3001.

Verri, T., Kottra, G., Romano, A., Tiso, N., Peric, M., Maffia, M., Broll, M., Argenton, F., Daniel, Hannedore, and Storelli, C. (2003). Molecular and functional characterisation of the zebrafish (*Danio rerio*) PEPT1-type peptide transporter. *FEBS Lett.* **549**, 115–122.

Vidal Junior, M. V., Donzele, J. L., Silva Camargo, A. C., Andrade, D. R., and Santos, L. C. (1998). Levels of crude protein for tambaqui (*Colossoma macropomum*), in the phase of 30 to 250 grams 1. The tambaquis performance. *Rev. Bras. Zootecn.* **27**(3), 421–426.

Viegas, E. M. M., and Guzman, E. C. (1998). Effect of sources and levels of dietary lipids on growth, body composition, and fatty acids of the tambaqui (*Colossoma macropomum*). *World Aquacult.* 66–70.

Vilella, S., Reshkin, S. J., Storelli, C., and Ahearn, G. A. (1989). Brush-border inositol transport by intestines of carnivorous and herbivorous teleosts. *Am. J. Physiol.-Gastr. L.* **256**, G501–G508.

Vincent, J. F. V., and Sibbing, F. A. (1992). How the grass carp (*Ctenopharyngodon idella*) chooses and chews its food – some clues. *J. Zool.* **226**, 435–444.

Viola, S., and Arieli, Y. (1989). Changes in the lysine requirement of carp (*Cyprinus Carpio*) as a function of growth rate and temperature. Part I. Juvenile fishes in cages. *Isr. J. Aquacult.-Bamid.* **41**, 147–158.

Walford, J., and Lam, T. (1993). Development of digestive tract and proteolytic enzyme activity in sea bass (*Lates calcarifer*) larvae and juveniles. *Aquaculture* **109**, 187–205.

Walsh, P. (Personal communication).

Wang, K., Takeuchi, T., and Watanabe, T. (1985). Effect of dietary protein levels on growth of *Tilapia nilotica*. *B. Jpn. Soc. Sci. Fish.* **51**, 133–140.

Watanabe, T., Takeuchi, T., Murakami, A., and Ogino, C. (1980). The availability to *Tilapia nilotica* of phosphorus in white fish meal. *B. Jpn. Soc. Sci. Fish* **46**, 897–899.

Wendelaar Bonga, S. E., Lammers, P. I., and vander Meij, J. C. A. (1983). Effects of 1,25- and 24,25-dihydroxyvitamin D_3 on bone formation in the cichlid teleost *Sarotherodon mossambicus*. *Cell Tissue Res.* **228**, 117–126.

Winemiller, K. O. (1990). Spatial and temporal variation in tropical fish trophic networks. *Ecol. Monogr.* **60**, 331–367.

Winfree, R. A., and Stickney, R. R. (1981). Effects of dietary protein and energy on growth, feed conversion efficiency and body composition of *Tilapia aurea*. *J. Nutr.* **111**(6), 1001–1011.

Wright, J. R., Jr, O'Hali, W., Yang, H., Han, X.-X., and Bonen, A. (1998). GLUT-4 deficiency and severe peripheral resistance to insulin in the teleost fish tilapia. *Gen. Comp. Endocrinol.* **111**, 20–27.

Yamada, A., Takano, K., and Kamoi, I. (1991). Purification and properties of proteases from tilapia intestine. *Nippon Suisan Gakk.* **57**, 1551–1557.

Yamada, A., Takano, K., and Kamoi, I. (1993). Purification and properties of protease from tilapia stomach. *Nippon Suisan Gakk.* **59**, 1903–1908.

Yamanaka, N. (1988). Descrição, desenvolvimento e alimentação de larvas e pré juvenis de pacu *Piaractus mesopotamicus* (Holmberg, 1887) (Teleostei, Characidae), mantidos em confinamento. Tese de Doutoramento PhD thesis, Instituto de Biociências. Universidade de Saõ Paulo, Brazil.

Yamaoka, K. (1985). Intestinal coiling pattern in the epilithic algal-feeding cichlids (Pisces, Teleostei) of Lake Tanganyika, and its phylogenetic significance. *Zool. J. Linn. Soc.-Lond.* **84**, 235–261.

Zemke-White, W. L., Choat, J. H., and Clements, K. D. (2002). A reevaluation of the diel feeding hypothesis for marine herbivorous fishes. *Mar. Biol.* **141**, 571–579.

Zoufal, R., and Taborsky, M. (1991). Fish foraging periodicity correlates with daily changes of diet quality. *Mar. Biol.* **108**, 193–196.

6

THE CARDIORESPIRATORY SYSTEM IN TROPICAL FISHES: STRUCTURE, FUNCTION, AND CONTROL

STEPHEN G. REID
LENA SUNDIN
WILLIAM K. MILSOM

I. Introduction
II. Respiratory Strategies
III. Respiratory Organs
    A. Water Breathing
    B. Air Breathing
IV. Ventilatory Mechanisms (Pumps)
V. Circulatory Patterns
VI. Cardiac Pumps
VII. Cardiorespiratory Control
    A. Water-Breathing Fishes
    B. Air-Breathing Fishes

I. INTRODUCTION

Earlier chapters in this book have outlined the diverse nature (spatial and temporal) of the tropical aquatic environment and the adaptive radiation that it has given rise to in tropical fishes. The high temperatures of these waters, often accompanied by hypoxia and hypercarbia/acidosis, have also given rise to a tremendous adaptive radiation in cardiorespiratory strategies designed to enhance survival under these conditions. The subject of this chapter is the structure, function, and control of the respiratory and circulatory systems in these fishes. Unfortunately, space limitations do not allow a comprehensive discussion of all aspects of this topic. Fortunately, the myriad of adaptations seen in structure and function have been the subject of several excellent, recent reviews to which the reader is referred for more detail (Randall *et al.*, 1981; Val and Almeida-Val, 1995, 1999; Val *et al.*,

The Physiology of Tropical Fishes: Volume 21
FISH PHYSIOLOGY

1996; Graham, 1997; Maina, 2003). Instead, the emphasis of this chapter will be placed on recent advances in our understanding of the control of cardio-respiratory processes in these fish with a brief review of structure and function designed to place discussion of control mechanisms in perspective.

II. RESPIRATORY STRATEGIES

The great majority of tropical fishes continue to breathe water like their temperate relatives. For the most part they have developed strategies (be-havioral, morphological, anatomical, physiological, and biochemical), either to avoid low oxygen conditions, increase oxygen transfer from the environ-ment to the tissues, reduce oxygen demands, or some combination of these. Thus, many species of tropical fish have evolved no special mechanisms for dealing with harsh conditions such as hypoxia/anoxia but constantly sense and monitor environmental conditions and migrate to better areas. These migrations are usually short, moving between stagnant areas and areas with higher water flow (Junk *et al.*, 1983; Wootton, 1990). Other species do not leave their habitat when environmental oxygen levels fall but simply increase oxygen extraction and/or reduce oxygen demands through a host of physio-logical and biochemical adjustments. These strategies are similar to those exhibited by fish from temperate climes. The mechanisms involved include regulation of different hemoglobin fractions, adjustment of intra-erythrocyt-ic levels of organophosphates, changes in hematocrit/[hemoglobin] and metabolic suppression; almost all under catecholaminergic control (Milligan and Wood, 1987; Perry and Kinkead, 1989; Nikinmaa, 1990; Randall, 1990; Val *et al.*, 1992; Almeida-Val and Val, 1993; see Chapter 7, this volume). These are slow processes however (Hochachka and Somero, 1984; Wooton, 1990), that do not protect fish from the sudden rapid changes in O_2 avail-ability that can occur, such as when temperature drops induce turnover of the water column leading to the rapid replacement of O_2-rich surface waters with anoxic water from the bottom of the water column (Val and Almeida-Val, 1995).

To deal with these more severe conditions, many water-breathing species of tropical fish have evolved adaptations to enhance skimming of the O_2-rich surface layers of the water. This behavior is observed in many unrelated species indicating the convergent nature of the behavior. While some species have no special adaptations for performing aquatic surface respira-tion (ASR), others develop a swollen lower lip that acts as a funnel to direct the surface water across the gills. This is found in such Brazilian fish as tambaqui (*Colossoma macropomum*), pacu (*Piaractus mesopotamicus* and *brachypomum*), and various species of *Brycon* (*erythropterum, cephalus*) and *Mylossoma* (*duriventris, aureus*), as well as in such African fishes as the Lake Magadi tilapia (*Oreochromis alcalicus*) and the cyprinid, *Barbus neumayer* (Braum and Junk, 1982; Kramer and McClure, 1982; Val and Almeida-Val, 1995; Olowo and Chapman, 1996).

Perhaps most notable amongst the respiratory adaptations of tropical fishes for dealing with hypoxic/anoxic waters is the use of air breathing. For some species, this is a facultative event that occurs only when water oxygen levels are low, while for others it is an obligatory behavior and these species rely primarily, if not exclusively, on O_2 taken from the air. While some species of fish are clearly facultative air-breathers and others obligate air-breathers, there are many species that utilize both strategies, either as a function of developmental age, or environmental conditions. Thus, many species of *Anabis, Clarius, Heteropneustes*, and *Arapaima* begin life as gill-breathers and slowly progress through stages as facultative, and then obligate air-breathers as they mature (Johansen *et al.*, 1970; Rahn *et al.*, 1971; Singh and Hughes, 1971; Stevens and Holeton, 1978). *Piabucina* is a facultative air-breather under normoxic conditions but an obligate air-breather under hypoxic conditions (Graham, 1997). The gar, *Lepisosteus*, is a facultative air-breather at low temperatures but becomes an obligate air-breather when oxygen uptake increases at higher temperatures (Rahn *et al.*, 1971). In all cases, however, obligatory air-breathers remain bi-modal breathers to some extent and, while they may be obliged to breathe air for oxygen uptake, they always remain obligatory water-breathers for CO_2 excretion and pH regulation.

The adaptations associated with air breathing in fishes are diverse and intriguing. These are briefly described in the following section.

III. RESPIRATORY ORGANS

A. Water Breathing

The primary adaptations seen in the respiratory organs of water-breathers living in oxygen-poor waters are associated with gill diffusing capacity. Here we see both interspecies and intraspecies adaptations.

The diffusing capacity of any species can be increased by alterations in the number of gill arches, the length and number of gill filaments on each arch, the spacing of the lamellae along the filament, the surface area of individual gill lamellae, the thickness of the water/blood interface, and the resistance to water flow through the gill sieve (Hughes, 1984). Changes in any or all of these variables occur as a function of life style and habitat throughout all taxonomic groups of fishes (see Chapter 7).

The diffusing capacity of any individual can also be changed in a number of ways. These include increasing the number of lamellae perfused at any one time (and hence the functional area available for gas transfer; Booth, 1978), redirecting blood through sections of lamellae exposed to gill water flow, and reducing lymphatic space (Randall *et al.*, 1981). All result in a reduction in diffusion distance between blood and water and an increase in the surface area across which gas exchange occurs.

B. Air Breathing

The alternate strategy employed by a small, but notable, fraction of tropical fishes is air breathing. Utilizing air as a source of oxygen provides fish with relative independence from the fluctuations in dissolved oxygen associated with some tropical waters. Although the number of air-breathing species is small when compared with other fish species that share the same habitats, it is greatest in fish from the tropics and is correlated with the incidence of waters that are naturally low in oxygen. The number of air-breathing fish species living outside the tropics in normoxic waters is proportionately reduced (Carter and Beadle, 1931; Beebe, 1945; Packard, 1974; Kramer and Graham, 1976; Junk *et al.*, 1983). The diversity of sites and surfaces that are utilized for gas transfer from air to blood, in fish, is remarkable. While a few species do utilize their gills for gas exchange in air, this is a rare occurrence and most air-breathing fishes utilize other surfaces.

Graham (1997) put forward a simplified classification scheme for structures utilized by fish for aerial gas exchange (air-breathing organs; ABO). He suggests that "even though air-breathing has evolved numerous times and independently, the location of aerial exchange sites has remained largely under the conservative influence of structures predisposed for air gulping and sites in the body where gas storage and the requisite vascularization could be developed." This scheme divides structures into three groups: (1) those associated with the skin, (2) structures associated with organs in the head region or along the digestive tract, and (3) the lungs and respiratory gas bladders (Figure 6.1a).

1. CUTANEOUS GAS EXCHANGE

Many fish that spend time out of water (amphibious fish) do use their skin for aerial gas transfer and, although subject to uncontrolled water loss and limited as an organ for oxygen uptake, the skin is adequate for CO_2 excretion (Graham, 1997). Most air-breathing fishes, however, remain in water (aquatic air-breathers) and the gills and/or skin become the major site of CO_2 excretion (into water) while other specialized exchange surfaces

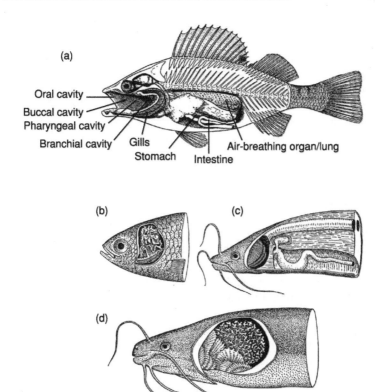

Fig. 6.1 (a) A schematic diagram of a mid-sagittal section of a fish indicating various areas where adaptations for air breathing are known to occur. (b–d) illustrate accessory air-breathing organs: in (b) the climbing perch (*Anabas, testudinosus*), (c) the Indian catfish (*Heteropneustes fossilis*) and (d) the African catfish (*Clarius lazera*). (From Greenwood, 1961; reproduced with permission.)

become the major site of oxygen uptake from air with normally little involvement in CO_2 excretion.

2. STRUCTURES ASSOCIATED WITH THE HEAD REGION OR THE DIGESTIVE TRACT

(a) Gills are generally ill suited for air breathing since the moist lamellae stick together in air, due to surface tension, and collapse without the buoyant support of water. Modifications to the gills for aerial gas exchange include increased structural support (cartilaginous rods within the lamellae or "cytoplasmic stiffening material" in the pillar cells), fusion of secondary lamellae, widely spaced lamellae, and thickening and mucus-sequestering in secondary lamellae (Graham, 1973). The net result is that fish that utilize their gills for

gas exchange in both air and water have a reduced gill surface area, usually about one-half that of their non-air-breathing relatives (see Table 6.1 on p. 263; Chapter 7, this volume; Fernandes *et al.*, 1994; Graham, 1997).

(b) Buccal, pharyngeal, branchial and opercular surfaces have all been reported to display specialized respiratory epithelia for aerial gas exchange. These have been reported for at least 16 genera of air-breathing fishes. Most of these fish are either amphibious (spend time out of water) or hold air in their mouths while air breathing. Modifications for air breathing range from increased vascularization, to elaborations of the epithelial surface through expanded diverticulae or pouches (Figure 6.1b–d).

Several groups [*Channa* (3 species), *Monopterus* and most *Synbranchids*] utilize chambers in the roof of the pharynx above the gills and adjacent to the skull for ABOs. The respiratory epithelium that lines these chambers takes the form of vascular rosettes consisting of numerous vascular papillae bulging into the lumen of the ABO in wave-like patterns that are believed to increase blood–air contact (Figure 6.1d; Munshi *et al.*, 1994; Graham, 1997).

Other groups of fish air breathe using structures derived from their gills, branchial chambers or both. These include the *Clariidae, Heteropneustidae* and the *Anabantoidei*. In all cases, the ABO consists of a suprabranchial chamber containing arborescent organs and gill fans with a vascular epithelial lining. The labyrinth apparatus (an intricately laminated bony element) of the *Anabantoidei* is amongst the most elaborate (Figure 6.1b, c; Munshi, 1961; Peters, 1978; Graham *et al.*, 1995).

(c) Parts of the digestive tract, including the esophagus (*Dallia pectoralis, Blennius pholis*), pneumatic duct (*Anguilla*), stomach (*Loricariids* and *Trichomycterids*), and intestine (*Cobitids* and *Calichthyids*) (Graham, 1997) have also been shown to play roles as ABOs in various species.

3. LUNGS AND RESPIRATORY GAS BLADDERS

Finally, at least 47 species from 24 genera of bony fish are known to breathe air using a lung or a respiratory gas bladder. While several different sets of criteria have been used to classify ABOs as lungs or gas bladders, the scheme put forward by Graham (1997) is perhaps the most thorough and explicit. By this scheme, gas bladders have an embryonic origin from the side or dorsal aspect of the alimentary canal, are not paired, do not always have a glottis (and may or may not retain an open pneumatic duct) and, in most cases, receive blood in parallel with the systemic circulation and lack a specialized pulmonary circulation (Figure 6.2a, c, e). Lungs, on the other hand, have an embryonic origin from the ventral wall of the alimentary canal, are paired, possess a valvular glottis in the floor of the alimentary canal and have a proper pulmonary circulation in which efferent vessels return blood directly to the heart (rather than to the vena cava) (Figure 6.2b, d, f).

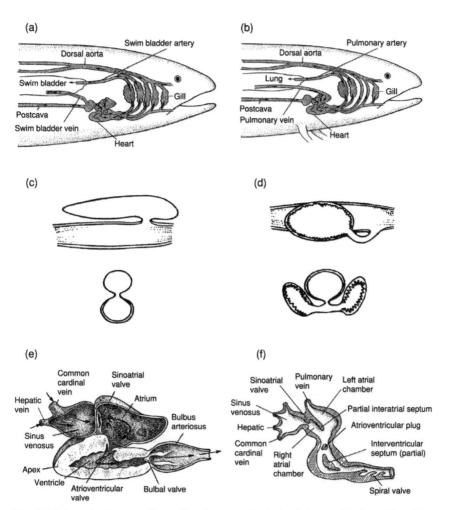

Fig. 6.2 Schematic diagrams illustrating the generalized circulation to (a) the air-breathing organ of a teleost fish and (b) the lung of a lungfish (see text for details). Panels (c) and (d) illustrate the relation of the air-breathing organ and lung to the esophagus as seen from the side and in cross-section. Panels (e) (teleost) and (f) (lungfish) illustrate the general structure of the heart of the two groups and the differences in venous return to the heart from the air-breathing organ. (Modified from Kardong, 2002.)

By this scheme, lungs are possessed only by the lungfishes (*Neoceratodus, Lepidosiren* and *Protopterus*) and the polypterids (*Polypterus* and *Erpetoichthys*). Gas bladders are found in both *Amia* and the gars and are scattered throughout the teleosts. Among these fishes, respiratory gas bladders differ greatly in complexity (Graham, 1997).

For species possessing well-developed ABOs there are conflicting functional requirements placed on the design of their gills. This arises since more O_2-rich blood draining the ABO returns to the heart and must then pass through the gills before entering the systemic circulation. In the process, the potential exists for significant loss of O_2 to hypoxic water during transit through the gills (Randall *et al.*, 1981). As a result, many of these fish exhibit a reduction in functional gill surface area (see Chapter 7). This may be in the form of reductions in the number of gill arches, the number of secondary lamellae, secondary lamellar thickening, or presence of gill vascular shunts. The extent to which any or all of these occurs is generally a function of the dependence of the species on air breathing.

It is interesting to note that both species that use their gills for air breathing, and those that use other structures, have greatly reduced gill surface areas; one to prevent collapse of the filaments and to enhance O_2 uptake and one to prevent O_2 loss.

IV. VENTILATORY MECHANISMS (PUMPS)

No matter what the gas exchange organ, water or air must move actively across the respiratory surfaces to increase the rate of diffusion. Invariably this requires muscular action. In the case of cutaneous exchange, this may involve general body movements or more specialized movements. For instance, in the newly hatched larvae of the Asian teleost, *Monopterus albus*, the large and heavily vascularized pectoral fins drive a stream of water backwards across the surface of the larva and its yolk sac. Since blood in superficial skin vessels flows forward, this establishes a countercurrent exchange between blood and water (Liem, 1988).

In most fishes, the buccal and opercular cavities form dual pumps on either side of the gill curtain. Both cavities are expanded simultaneously by muscular action creating a suction that closes the operculae and draws water in through the mouth. Both cavities are then compressed by muscular action while the mouth closes, forcing water over the gill curtain and out through the operculae. Because of a slight difference in pressure between buccal and opercular cavities, water flows almost continuously across the gills in one direction (Hughes, 1984).

This mechanism remains the same for species that employ aquatic surface respiration to irrigate the gills with the more oxygen-rich surface film. Some of these fishes, however, posses a flap-like valve in the mouth that closes to prevent water from refluxing during the buccal compression phase of the ventilatory cycle while the mouth remains agape for skimming surface water (*Colossoma;* Sundin *et al.*, 2000). This flap appears to be composed of thin epithelial sheets, which extend from the margins of the upper and lower jaw and act like a pocket valve. These flaps collapse against the roof and floor of the mouth during the negative pressure expansion phase of the buccal cycle but fill with water and close sealing the entrance to the mouth during the positive pressure compression phase. As such, they prevent reflux of water back through the open mouth allowing the fish to efficiently ventilate the gills while still maintaining the mouth gape.

This system is only slightly modified in air-breathing fishes. Now, however, ventilation is usually produced by the buccal pump exclusively. In actinopterygian fishes this occurs in four phases while in sarcopterygian fishes it occurs in two phases. In the former case, initial buccal expansion occurs with the mouth closed and draws air from the ABO into the buccal cavity. This may be assisted by elastic recoil of the ABO as well as compression of muscles in the wall of the ABO. Hydrostatic pressure gradients in submerged fish may also assist in this air movement. This air is then expelled during buccal compression through the mouth or operculae. A second buccal expansion now draws in fresh air through the open mouth and the subsequent buccal compression, which takes place with the operculae and mouth closed, forces this air into the air-breathing organ (Figure 6.3b) (Liem, 1988; Brainerd, 1994). In sarcopterygian fishes, an initial buccal expansion phase draws previously inspired air from the air-breathing organ and fresh air from the environment into the buccal cavity simultaneously. Again, lung emptying is due to a combination of elastic recoil, contraction of muscles within the lung wall, hydrostatic forces and the negative pressure created by buccal expansion. In the next step, buccal compression, in series with jaw closure and sealing of the operculae, forces mixed air into the lungs with any excess being expelled through the mouth, operculae or nares (Figure 6.3a) (McMahon, 1969; Brainerd, 1994).

There are, of course, exceptions to this general trend. In the jeju, a freshwater Amazonian fish that uses a modified swim bladder as an air-breathing organ, the gas bladder is subdivided into an anterior and posterior chamber by a muscular sphincter. As the jeju breaks the water's surface, the fresh air gulped into the buccal cavity is forced along the pneumatic duct and preferentially enters the anterior chamber of the ABO. The sphincter then closes and spent air in the posterior chamber exits into the buccal cavity and out under the operculum. Finally the sphincter opens and the muscular walls of the anterior chamber contract forcing the fresh air into the vascularized posterior chamber (Figure 6.3c) (Randall *et al.*, 1981). Given the variety of structures that have evolved associated with the buccal, pharyngeal, branchial, and opercular surfaces, it is not surprising that other exceptions to

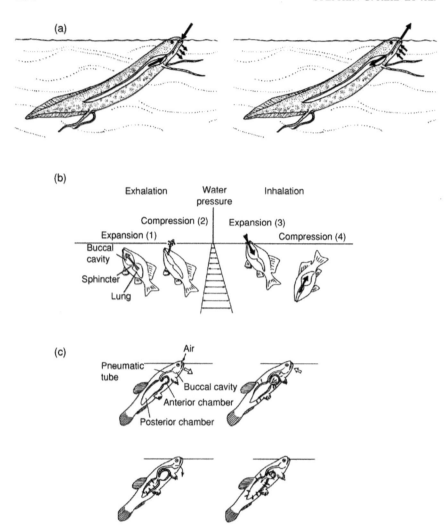

Fig. 6.3 Schematic diagrams illustrating (a) the two-stroke buccal pump found in sarcopterygian fishes, (b) the four-stroke pump found in most actinopterygian fishes, and (c) the modified pumping mechanism used by jeju. See text for details on all pumping mechanisms. (From Randall *et al.*, 1981 and Kardong, 2002; reproduced with permission.)

the general trend have evolved to ventilate these structures. A full description of these is beyond the scope of this chapter but can be found in the authoritative work of Graham (1997).

Another notable exception to this general trend is found in the polypterids (*Polypterus* and *Erpetoichthys*) in which elastic recoil from emptying of the lungs leads to aspiration breathing. Exhalation in these fishes is driven by contraction of the lung wall, which also deforms the body wall. When the muscles subsequently relax, a negative, recoil pressure is created within the lungs, enhanced by the ganoid scale-reinforced skin and body wall, which serves to re-inflate the lungs (Purser, 1926; Brainerd et al., 1989). Claims of suctional filling, of lungs by estivating *Protopterus* (Lomholt et al., 1975) and of ABOs by *Arapaima* (Farrell and Randall, 1978), have not been substantiated (DeLaney and Fishman, 1977; Greenwood and Liem, 1984).

V. CIRCULATORY PATTERNS

The basic circulatory pattern of water-breathing fishes is a single, serial pattern in which blood passes only once through the heart during each complete circuit. With this design, blood moves from the heart to the gills to the systemic tissues and back to the heart, and hence must pass through at least two capillary beds before it returns to the heart. Not surprisingly, the independent origin of the tremendous variety of air-breathing structures has given rise to a tremendous variety of circulatory modifications. As pointed out by Graham (1997), most of these modifications are designed to alleviate three basic problems that all arise as a result of the basic single circulation pattern. In most fish with supplementary air-breathing organs, the oxygenated blood leaving these organs enters the general venous circulation (Figures 6.4, 6.5). Thus, the first problem that arises is venous admixture stemming from the mixing of oxygenated and deoxygenated blood. The second problem is that this mixed blood will normally re-enter the gills where there is the potential for oxygen to be lost from the blood to the water during periods of air breathing in fish in oxygen-poor water. While one solution to both of these problems would be to place all air-breathing organs in series between the gills and the systemic circulation, this would require that blood traverse three capillary beds before returning to the heart. Raising pressure sufficiently to counter the flow resistance that this would create would lead to problems of its own. Presumably this is why blood from all ABOs invariably returns directly to the venous circulation, bringing us back to problem number one. A full description of the modifications that are seen to alleviate these problems is beyond the scope of this chapter but they include modifications in the pattern of afferent arterial supply (blood may be delivered to ABOs either from the ventral aorta directly, from the afferent or efferent branchial arteries or from the post-branchial dorsal aorta) or efferent venous return. Graham (1997) describes eight basic patterns of afferent and efferent circulation to ABOs and points out that because of the limitations that arise from the problems just described, in all cases, there is a need to regulate blood

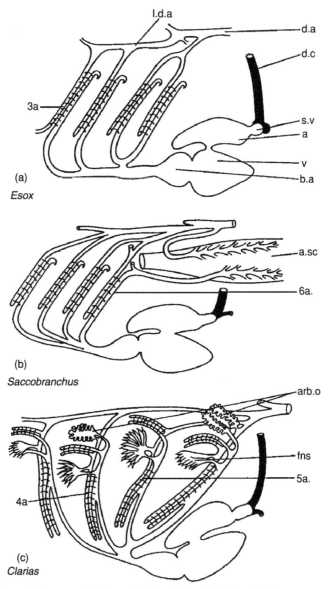

Fig. 6.4 Schematic diagrams of the branchial circulation of fish in which the oxygenated blood is returned to the dorsal aorta in (a) an unmodified water-breathing fish, *Esox*, and in (b–c) two water-breathing fishes (b, *Saccobranchus* and c, *Clarias*). Abbreviations: *a*, atrium; *arb.o*, arborescent organs; *a.sc*, air sac; *b.a*, bulbus arteriosus; *d.a*, dorsal aorta; *d.c*, ductus cuvieri; *fns*, fans; *l.d.a*, lateral dorsal aorta; *s.v*, sinus venosus; *v*, ventricle; 3a, 4a, 5a, 6a, third, fourth, fifth and sixth aortic arches. (From Satchell, 1976; reproduced with permission.)

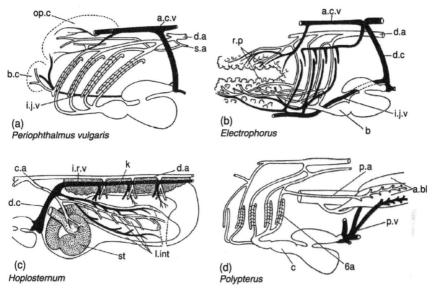

Fig. 6.5 Schematic diagrams of the branchial circulation of fish in which oxygenated blood is returned to a central vein. Abbreviations: *a.bl*, air bladder; *a.c.v*, anterior cardinal vein; *b*, bulbus arteriosus; *b.c*, buccal cavity; *c*, conus arteriosus; *c.a*, coeliac artery; *i.j.v*, internal jugular vein; *i.r.v*, inter renal vein; *k*, kidney; *l.int*, internal loops of intestine; *op.c*, opercular cavity; *p.a*, pulmonary artery; *p.v*, pulmonary vein; *r.p*, respiratory papillae; *s.a*, subclavian artery; *st*, stomach. Other abbreviations are the same as in Figure 6.4. (From Satchell, 1976; reproduced with permission.)

flow to the ABO, gills, and systemic circulation to enhance gas exchange and minimize these problems. Mechanisms that favor shunting of blood to the gills during water breathing, and away from the gills and to the ABO during air breathing, have been described for most species of air-breathing actinopterygian fish (see Graham, 1997 for review).

The lungfishes are a notable exception to this trend. In these fish, blood leaving the lungs returns directly to the heart via a separate pulmonary vein (Figure 6.2b). Phylogenetically, separate left and right atria appear first in lungfishes, establishing a separate pulmonary circuit from the lungs. As described below, despite an anatomically incomplete internal septation of their heart, blood entering the heart from the pulmonary and systemic circulations does not tend to mix and as the oxygenated and deoxygenated blood exit from the heart, they enter different sets of aortic arches. In the lungfishes, as in other bony fishes, the first pharyngeal slit is reduced and has no respiratory function. In the Australian lungfish (*Neoceratodus*), the remaining five pharyngeal slits open to fully functional gills supplied by four aortic arches. In the African lungfish (*Protopterus*), the functional gills are reduced further. The third and fourth gills are absent entirely but their aortic arches persist. In all lungfishes, the efferent vessel of the most posterior aortic arch gives rise to the pulmonary artery but maintains its connection to the dorsal aorta via a short *ductus arteriosus* (Figure 6.2b). Oxygenated blood returning to the heart from the lungs is shunted straight through gill arches III and IV, which lack gills, and flows to systemic tissues directly. Venous blood returning from the body is shunted through the posterior arches (V and VI) and then diverted to the lungs. In addition to the preferential shunting of oxygenated blood to the anterior arches, secondary mechanisms exist which involve shunts at the base of the gill capillaries, as well as at the *ductus arteriosus*, that redistribute blood flow to favour water or air breathing (Johansen, 1970; Delaney *et al.*, 1974).

VI. CARDIAC PUMPS

The hearts of bony fishes consist of four basic chambers, the sinus arteriosus, atrium, ventricle, and conus arteriosus, with one-way valves between compartments (Figure 6.2e). Like the other chambers, the muscular conus arteriosus contracts, acting as an auxiliary pump to help maintain blood flow into the ventral aorta after the onset of ventricular relaxation. In teleosts, this fourth chamber is an elastic, non-contractile, bulbus arteriosus which acts as a passive elastic reservoir to maintain blood flow into the ventral aorta during ventricular relaxation. The lungfish heart is modified from this basic plan. The sinus venosus still receives blood returning from the systemic circulation. The single atrium, however, is partially subdivided internally by an inter-atrial septum that produces right and left atrial chambers. In the Australian lungfish (*Neoceratodus*) the pulmonary veins, returning blood from the lungs, empty into the sinus venosus as does blood returning from the body. In the African and South American lungfishes (*Protopterus* and *Lepidosiren*) the pulmonary veins empty directly into the left atrial chamber (Figure 6.2f). In place of an atrioventricular valve, these fish have an atrioventricular plug, and the ventricle is also partially divided internally by an interventricular septum. Within the lungfishes, the greatest degree of internal subdivision of both the atrium and ventricle is seen in the South American lungfish and the least is seen in the Australian lungfish. Alignment of the interventricular septum, the atrioventricular plug, and the interatrial septum establishes internal channels through the heart that partially separate blood returning from the body and lungs. Within the conus arteriosus is a spiral valve that also aids in separating the two blood streams (Figure 6.2f). Oxygenated blood returning from the lungs enters the left channel while deoxygenated blood returning from the body enters the right channel and, as the two streams of blood exit from the conus arteriosus, they enter different sets of aortic arches (as described above; Kardong, 2002).

VII. CARDIORESPIRATORY CONTROL

Breathing is produced as a conditional rhythm by a central respiratory rhythm generator located within the brainstem. The respiratory rhythm generator operates at a sub-threshold level, requiring some external (biasing) input for its output to rise above a threshold and be expressed. This, in turn, leads to activity in the respiratory motor neurons/nerves that drive the respiratory muscles and produce breathing (Richter, 1982; Ballintijn and Juch, 1984; Feldman et al., 1990). Some of the more common "biasing" inputs that modulate the respiratory rhythm originate from chemoreceptors that sense oxygen (O_2) and carbon dioxide (CO_2)/pH levels in water and blood, mechanoreceptors that monitor stretch or displacement of the respiratory organ, and higher brain centers which, in mammals at least, allow for emotions and sleep state to influence breathing.

As described already, the hypoxic and anoxic conditions that frequently occur in the tropics have given rise to a remarkable array of respiratory organs with specialized circulatory pathways, and the exploitation of both water and air as respiratory media. Ventilation and perfusion of these organs must be tightly controlled, and it is not clear to what degree the respiratory control systems differ amongst species with varying respiratory strategies and whether or not respiratory control differs in tropical, compared to temperate, polar, etc., fish. On the one hand, given the fundamental importance of breathing, it is reasonable to assume that the systems underlying the control of breathing are similar not only amongst various fish species in different regions of the world but also amongst the vertebrates as a whole. On the other hand, given the numerous types of respiratory organs and strategies found in tropical fishes, it is also reasonable to assume that a number of differences in respiratory control mechanisms will exist, which reflect this diversity as well as the evolutionary transition from aquatic to terrestrial life (Ultsch, 1996).

In fish, respiratory-related afferent input arises from peripheral chemoreceptors and mechanoreceptors located on the gill arches and in the orobranchial cavity (Burleson et al., 1992; Perry and Gilmour, 2002; Sundin and Nilsson, 2002). Additionally, other inputs that may be described as generalized afferent inputs, rather than specific respiratory-related feedback, can also influence breathing; particularly the breathing pattern (Reid et al., 2003). Arguably the most important external factor controlling breathing in both tropical and non-tropical fish is the level of O_2, and to a lesser degree the levels of pH/CO_2, within the environment (water and/or air) and subsequently within the arterial blood and/or cerebral spinal fluid (CSF). Given this, the majority of the discussion of cardiorespiratory control systems in this chapter will focus on the role of chemoreceptors in tropical fishes. A number of reviews have summarized the current state of knowledge regarding the role of O_2 and pH/CO_2 chemoreceptors in the cardiorespiratory control of fish in general (Smatresk, 1988; Perry and Wood, 1989; Burleson *et al.*, 1992; Milsom, 1995, 1996, 2002; Graham, 1997; Gilmour, 2001; Remmers *et al.*, 2001; Perry and Gilmour, 2002). While the majority of research to date has examined cardiorespiratory control mechanisms in temperate species, a growing number of studies have focused on the cardiorespiratory-related function of chemoreceptors in tropical fishes. It is clear from the data that the relative role of different groups of chemoreceptors in cardiorespiratory control is highly variable amongst species. For the most part, this variability has prevented the development of an all-encompassing, and teleologically satisfying, model that explains the chemoreceptor-mediated control of breathing in fish and the evolution of these systems.

Milsom (1996) identified a number of questions regarding the role of chemoreceptors in cardiorespiratory control in tropical fish that remain, for the most part, unanswered. These include: (1) Do fish possess central O_2 chemoreceptors? (2) Do all fish possess both water-sensing (externally oriented) and blood-sensing (internally oriented) O_2 chemoreceptors? If so, do they exist on all gill arches? (3) Do branchial (gill) O_2 chemoreceptors always elicit both cardiovascular and ventilatory responses? (4) Are the afferent fibers arising from gill O_2 chemoreceptors carried in different nerves? (5) Where else do O_2 chemoreceptors exist in fish? (6) How is the input of O_2 chemoreceptors transformed in O_2-conformers and in bi-modal (water and air) breathers? (7) Do fish possess pH/CO_2 chemoreceptors? If so, where are they located? Until the answers to these, and other, questions become clear, it is unlikely that a single model of respiratory control in fish, tropical or otherwise, will emerge that can satisfactorily explain all of the data from numerous species.

While the discussion in this chapter focuses on a limited number of studies of control systems in tropical fishes, examples from temperate species, particularly air-breathing fishes (see review by Graham, 1997), are included, where appropriate. While recognizing the variability that exists amongst fish species and that, currently, no one model can adequately explain cardiorespiratory control in these animals, this chapter will highlight as many generalizations as possible in order to illustrate the most important aspects of cardiorespiratory control in tropical fishes. Furthermore, the discussion will focus on those mechanisms that regulate the cardiorespiratory changes that occur during challenges such as hypoxia (low environmental O_2) and hypercarbia (high environmental CO_2).

A. Water-Breathing Fishes

1. PERIPHERAL CHEMORECEPTORS IN WATER-BREATHING FISHES

The primary sites of peripheral O_2 and CO_2/pH chemoreceptors in tropical, and temperate, fishes are the gills and oro-branchial cavity. Peripheral chemoreceptors in tropical fishes are either internally oriented and monitor O_2, CO_2 or pH levels in the blood, or externally oriented and monitor these variables in the water (Smatresk, 1988; Burleson et al., 1992; Sundin et al., 1999, 2000; Rantin et al., 2002; Milsom et al., 2002; Florindo et al., 2002; Reid et al., 2000, 2003). Chemoreceptors on the gill arches are innervated by branches of the ninth (glossopharyngeal) and/or tenth (vagus) cranial nerves (cn) while those in the oro-branchial cavity are innervated by branches of the fifth (trigeminal) and/or seventh (facial) cranial nerves (Figure 6.6) (Milsom et al., 2002).

Fish gills, and air-breathing organs, contain chromaffin cell-like neuroendocrine (neuroepithelial) cells (NEC) derived from the sympathetic-adrenal lineage (see reviews by Zaccone et al., 1997; Sundin and Nilsson, 2002). It is likely that these cells, or at least a population of them, are the peripheral chemoreceptors and are analogous to the O_2-sensing glomus cells in the mammalian/avian carotid body and amphibian/reptilian carotid labyrinth and aortic arch. Indeed, recent studies have demonstrated that NEC from the gills of two species of temperate fish, the channel catfish (*Ictalurus punctatus*) and the zebrafish (*Danio reiro*), exhibit O_2-sensitive K^+ currents (Burleson, 2002; Jonz and Nurse, 2002), that are the hallmark of mammalian O_2-sensitive cells that trigger respiratory reflexes during hypoxia. The afferent fibers from the peripheral chemoreceptor cells report to sensory nuclei within the brainstem. Reviews by Taylor et al. (1999, 2001) provide a detailed account of the neuroanatomical basis of the central control of cardiorespiratory function in vertebrates, including that in water- and air-breathing fishes.

The location (i.e., gills or oro-branchial cavity), distribution (i.e., which gill arches), and stimulus modality (i.e., O_2, CO_2 or pH) of the peripheral chemoreceptors have been described in a number of temperate fish species (see Burleson et al., 1992; Perry and Gilmour, 2002 for reviews). Based on these studies, current dogma suggests that populations of both blood-sensing (internal) and water-sensing (external) chemoreceptors regulate the increase in breathing that occurs during environmental hypoxia, while, as described in the following section, water-sensing chemoreceptors trigger the

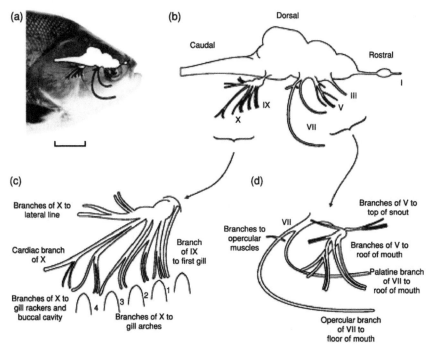

Fig. 6.6 A schematic diagram showing the cranial nerve roots that innervate the gill and oro-branchial chemoreceptors in the tambaqui. (a) The location of the nerves relative to the external anatomy of the fish. (b) An enlargement showing the origin of the various cranial nerve roots. (c, d) Details of the branches of cranial nerves V, VII, IX, and X. (From Milsom *et al.*, 2002; reproduced with permission.)

cardiovascular responses to hypoxia. The chemoreceptor control of the hypercarbic ventilatory response is more complicated and is discussed below (see review by Gilmour, 2001).

2. Hypoxic Ventilatory Responses

The hypoxic ventilatory response in tropical fishes consists of an increase in breathing frequency and breath amplitude (see Milsom, 1996 for review). The overall magnitude of the hypoxic ventilatory response is influenced by factors such as hypoxia tolerance (e.g., Rantin *et al.*, 1992, 1993), gill surface area (Fernandes *et al.*, 1994; Severi *et al.*, 1997), body weight (Kalinin *et al.*, 1993), activity level (De Salvo Souza *et al.*, 2001) and environmental temperature (Fernandes and Rantin, 1989). Although most species increase breathing initially during hypoxia, in order to maintain O_2 uptake (i.e., they are O_2 regulators), in many species, a critical O_2 threshold is reached beyond which the partial pressure of O_2 in the water (P_wO_2) is not sufficient to maintain O_2 uptake, regardless of the level of ventilation. At this point, the fish become O_2 conformers and blood O_2 levels and metabolic rate are allowed to fall.

Denervation experiments, in which various branches of cnV and cnVII to the oro-branchial cavity as well as cnIX and cnX to the gills were cut (Figure 6.6), in combination with injections of the chemoreceptor stimulant, sodium cyanide (NaCN), into the branchial circulation and/or inspired water, have revealed the distribution and stimulus modality of respiratory-related peripheral chemoreceptors in two species of tropical fish. In traira, the increase in breathing frequency during hypoxia was triggered primarily by external O_2 receptors located on all gill arches, innervated by cnIX and cnX, while the increase in breathing amplitude arose from extra-branchial O_2 chemoreceptors (Sundin et al., 1999). In the tambaqui, the frequency component of the hypoxic ventilatory response was triggered by internal and external O_2 receptors on all gill arches while the increase in breath amplitude arose from extra-branchial chemoreceptors (Sundin et al., 2000). Milsom et al. (2002) demonstrated, in tambaqui, that these extra-branchial receptors are located in the oro-branchial cavity and are innervated by the trigeminal (cnV) and facial (cnVII) nerves.

Sundin et al. (1999) also reported evidence for two populations of water-sensing O_2 chemoreceptors on the first gill arch of traira that inhibit ventilation during hypoxia. One group of chemoreceptors, innervated by the IXth cranial nerve, inhibits the increase in breathing amplitude during hypoxia while the second population is innervated by the pre-trematic branch of the Xth cranial nerve and inhibits the increase in breathing frequency. Burleson and Smatresk (1990) observed a similar phenomenon in a temperate species of fish, the channel catfish.

In the studies of traira, breathing frequency and breath amplitude began to increase when the water PO_2 reached 80 and 60 Torr, respectively, while in tambaqui both frequency and amplitude began to increase when the water PO_2 was approximately 120 Torr. In a recent study on hypoxic traira (Perry et al., 2004), exposure to a P_wO_2 of 60–80 Torr caused arterial PO_2 (PaO_2) levels to fall to 40–60 Torr. Based on in vivo O_2 equilibrium curves obtained in that study, the p50 for traira is approximately 8.5 Torr. Given this, exposure of traira to hypoxic water with a PO_2 of 60–80 Torr would be unlikely to change O_2-hemoglobin saturation from normoxic levels. A recent study on tambaqui (Brauner et al., 2001) reported a p50 value of 2.4 Torr suggesting a high degree of hypoxia tolerance in this species. Clearly, both traira and tambaqui begin to increase ventilation during hypoxia when PaO_2 is falling yet arterial O_2 content remains elevated. The implication of these data is that the increase in ventilation, mediated by internally oriented chemoreceptors, is triggered by a decrease in the arterial PO_2 rather than a decrease in arterial O_2 content.

3. HYPOXIC CARDIOVASCULAR RESPONSES

As the primary function of the cardiovascular system is to convey substances to and from cells, it is crucial for these fish to regulate the cardiovascular system to minimize energy expenditure and allow optimal oxygen uptake at the respiratory organs. The hemodynamic alterations that occur in the gills of water-breathing fishes during hypoxic exposure are designed to optimize gas exchange and they are dependent on flow and pressure changes caused by modified cardiac performance and adjustments of systemic and branchial vascular resistances.

(a) Systemic responses: A common response to hypoxia in fish is a slowing of the heart, a hypoxic bradycardia. The heart rate decrease can either be reflexively through cholinergic, vagal cardio-inhibitory fibers (Smith and Jones, 1978; Fritsche and Nilsson, 1989; Burleson and Smatresk, 1990) or, if the ensuing hypoxemia is severe enough, through a direct effect on cardiac myocytes (Rantin *et al.*, 1993).

In hypoxia-tolerant species like traira, *Hoplias malabaricus*, slow graded hypoxia does not produce a sustained bradycardia until aquatic oxygen tensions fall below 20 mmHg, while the closely related but less hypoxia tolerant *Hoplias lacerdae* develops a bradycardia at 35 mmHg (Rantin *et al.*, 1993). A bradycardia appears already at a PWO_2 of 70 mmHg in the hypoxia-intolerant species, dourado (*Salminus maxillosus*; De Salvo Souza *et al.*, 2001). The aquatic oxygen tension inducing the sustained fall in heart rate corresponds to the critical PO_2 at which metabolic rate also begins to fall and it appears that the difference in sensitivity of the heart muscle towards hypoxia corresponds to each species choice of habitat and life style (Rantin *et al.*, 1993; De Salvo Souza *et al.*, 2001). In contrast, if hypoxia is introduced rapidly, a reflex fall in heart rate appears at a PO_2 of approximately 100 mmHg in traira (Sundin *et al.*, 1999) and tambaqui, *Colossoma macropomum* (Sundin *et al.*, 2000). Selective denervation of branchial branches in traira and tambaqui demonstrate that the bradycardia elicited by rapidly induced hypoxia is a reflex mediated by branchial oxygen receptors innervated by cranial nerves IX and X (Sundin *et al.*, 1999; Sundin *et al.*, 2000, Figure 6.7). With the exception of traira, these receptors (in all six species investigated to date) are externally oriented and monitor the respiratory water (Milsom *et al.*, 1999; Sundin *et al.*, 1999). However, most studies did not attempt to clarify (using atropine or denervation) whether the graded, hypoxia-induced bradycardia involved oxygen receptors or was a result of direct cardiac hypoxemia.

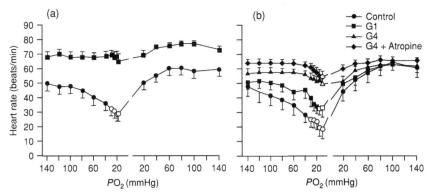

Fig. 6.7 Heart rate responses to rapid hypoxia and recovery in (a) traira, *Hoplias Malabaricus* (b) and Tambaqui, *Colossoma macropomum*. Circles indicate intact fish (control) with no gill denervation. Squares indicate fish in which cnIX and the pre-trematic branch of cnX to the first gill arch were sectioned (G1). Triangles indicate fish with cnIX and branchial branches of cnX to all four gill arches sectioned (G4). Diamonds indicate G4 fish pre-treated with atropine (G4 + atropine). Note that the hypoxic bradycardia in traira is eliminated after denervation of only the first gill arch (G1) while it remains in tambaqui until all gills (G4) are denervated. Atropine was used to verify that the remaining decrease in heart rate in G4 tambaqui was a direct effect of hypoxemia on cardiac myocytes and not related to neurotransmission. Open symbols are significantly different from starting values. (Adapted from Sundin *et al.*, 1999, 2000.)

Although the adaptive value of the hypoxic bradycardia has yet to be established, the prolonged cardiac filling time may, apart from reducing cardiac energy expense, lower diastolic pressure in the ventral aorta and allow more efficient O_2 extraction by the myocardium itself (Farrell *et al.*, 1989). In this regard it is interesting to note that tambaqui, instantaneously exposed to hypoxic conditions slightly below their critical PO_2 tension of 22 mmHg (20 mmHg for 6 hours), display a bradycardia that is only maintained for 60 min before the heart rate returns to pre-hypoxic values (Rantin and Kalinin, 1994). This implies that the heart of this species can function well below its critical oxygen tension and only temporarily needs the possible advantages of a heart rate reduction. Equally intriguing is the finding that fish embryos and early larvae, even during severe hypoxia, do not develop a bradycardia (Holeton, 1971; Barrionuevo and Burggren, 1999).

Accompanying the hypoxic bradycardia is an elevation of the systemic vascular resistance. Depending on the magnitude of the peripheral vasoconstriction, this may result in an increase in blood pressure, maintenance of a constant blood pressure (as in the tambaqui; Sundin *et al.*, 2000) or a slight hypotension (as in the traira; Sundin *et al.*, 1999). The reflex elevation of the systemic resistance appears to be adrenergic (Fritsche and Nilsson, 1990;

Wood and Shelton, 1980) and the chemoreceptor control of the vascular reflexes involves extra-branchial sites.

(b) Branchial responses: While there is little data describing the branchial vascular responses of tropical fishes to environmental hypoxia, it is unlikely that they deviate to any great extent from the responses observed in species from temperate waters. In fish in general, the changes in vasomotor tone of the branchial vasculature that occur in response to hypoxia act primarily to enlarge the respiratory surface area and enhance oxygen uptake. Hypoxia increases the branchial vascular resistance in rainbow trout *in vivo* (Holeton and Randall, 1967; Sundin and Nilsson, 1997), as well as in the isolated perfused heads of rainbow trout and Atlantic cod (Ristori and Laurent, 1977; Pettersson and Johansen, 1982). Visual observations of the branchial microvasculature *in vivo* reveal that hypoxia produces a reflex cholinergic constriction of the proximal efferent filamental arteries in trout (Sundin and Nilsson, 1997), increasing the perfusion pressure, which in turn facilitates lamellar recruitment (Booth, 1978) and an even lamellar sheet flow (Farrell *et al.*, 1980). If the reflex bradycardia is compensated by a concomitant rise in stroke volume (Fritsche and Nilsson, 1989; Sundin, 1995; Wood and Shelton, 1980), the ensuing rise in pulse pressure will also recruit previously unperfused lamellae (Farrell *et al.*, 1979). The reduced heart frequency may also enhance oxygen uptake by increasing the equilibrium time of blood in the secondary lamella (Holeton and Randall, 1967). The sum of these hemodynamic alterations should enhance oxygen uptake across the gills.

Aquatic hypoxia is also a potent stimulus for the release of circulating catecholamines. Adrenaline administration both *in vitro* and *in vivo* enhances oxygen uptake by the gills (Perry and Reid, 1992; for reviews see Randall and Perry, 1992; Reid *et al.*, 1998). This involves both an α-adrenoceptor mediated constriction distal to, and a β-adrenoceptor mediated dilation proximal to, the secondary lamellae. These actions are believed to increase the respiratory surface area of the secondary lamellae and thereby augment gas exchange (see Nilsson, 1984; Nilsson and Sundin, 1998; Sundin and Nilsson, 2002 for review).

In addition, it has been suggested that contraction of both smooth and striated interbranchial muscles in the loricariid fish, *Hypostomus plecostomus*, could affect the arterial system in the interconnecting septal area between filaments, which could alter the perfusion pressure at the tips of the filaments and aid in lamellar recruitment (Fernandes and Perna, 1995).

The complex microvasculature in the gills also permits the shunting of blood from either the afferent or the efferent branchial arteries down into the venous system in the center of the filaments. Shunting blood in this manner from the efferent branchial arteries, as has been observed in the hypoxia-intolerant trout (Sundin and Nilsson, 1997), could provide the energy-demanding chloride cells located adjacent to the central venous system (CVS) with oxygen and nutrients. Shunting blood from the afferent branchial arteries, as has been observed in the hypoxia tolerant epaulette shark, *Hemiscyllium ocellatum* (living on sometimes very hypoxic reef plateaus in northeast Australia; K-O. Stens-Lökken, L. Sundin, G. Renshaw, and G. E. Nilsson, unpublished observations; see Chapter 12, this volume), would allow "venous" blood coming from the heart to bypass the respiratory units of the secondary lamellae. This may prevent oxygen loss from the blood to the water during conditions of extreme hypoxia and may be particularly important in air-breathing fishes (see below).

4. HYPERCARBIC VENTILATORY RESPONSES

As with environmental hypoxia, exposure to aquatic hypercarbia (elevated water PCO_2; P_WCO_2) also elicits an increase in breathing in most water-breathing fishes (see reviews by Gilmour, 2001; Remmers *et al.*, 2001; Milsom, 2002). Recent data from tropical and temperate species indicate that environmental CO_2 can be a direct respiratory stimulus in fish rather than acting indirectly via effects on blood O_2 content mediated by Bohr and Root Effects (Perry and Wood, 1989; see below).

Denervation studies demonstrate that the chemoreceptors which trigger the increase in breathing frequency during hypercarbia in traira are located on all of the gill arches. Extra-branchial (i.e., not on the gills) chemoreceptors are also involved in the increase in breathing amplitude (Reid *et al.*, 2000). In the tambaqui, the increase in ventilation rate during hypercarbia is mediated by chemoreceptors on all gill arches while ventilatory amplitude does not increase (Sundin *et al.*, 2000). Interestingly, decerebration restored part of the increase in breathing following branchial denervation in tambaqui suggesting the presence of olfactory chemoreceptors that inhibit breathing during exposure to CO_2 (Milsom *et al.*, 2002). CO_2 chemoreceptors that inhibit breathing have also been reported in the South American lungfish (*Lepidosiren paradoxa*; Sanchez and Glass, 2001) and in other vertebrates (see review by Coates, 2001).

In both traira and tambaqui, injections of HCl into either the inspired water or branchial circulation did not elicit any respiratory, or cardiovascular, responses. This suggests that the chemoreceptors that elicit cardiorespiratory reflexes during hypercarbia are sensitive to CO_2, in the water or blood, rather than changes in pH. This does not, however, preclude a change in intracellular pH within the chemoreceptor cell from being the ultimate stimulus for chemoreceptor activation. Indeed, this interpretation is consistent with current cellular mechanisms of pH/CO_2 sensing within the mammalian carotid body (Gonzalez *et al.*, 1994).

In a series of recent experiments on tambaqui (Perry *et al.*, 2004), blood gas levels were measured using an extracorporeal blood loop in experiments designed to determine whether the ventilatory responses to CO_2 were triggered by internally (blood) or externally (water) oriented receptors. An intra-arterial injection of a carbonic anhydrase inhibitor, acetazolamide, caused significant retention of CO_2 in the arterial blood with no concomitant increase in ventilation. On the other hand, exposure to aquatic hypercarbia (5% CO_2 equilibrated water) caused a significant increase in breathing. Furthermore, if the hypercarbic water was rapidly replaced by air-equilibrated normocarbic water, breathing immediately returned to resting levels despite the fact that arterial PCO_2 remained elevated for some time. The data indicate that the ventilatory response to CO_2 in tambaqui is mediated by external (water-sensing) CO_2 chemoreceptors rather than internal, blood-sensing receptors. Additionally, attempts to alter ventilation by superfusing the brain with acidic, alkaline, hypoxic, hyperoxic, and hypercarbic saline were without effect suggesting the absence of central chemoreceptors in this species (Milsom *et al.*, 2002).

Until recently, the increase in breathing during hypercarbia in fish was though to be triggered by the diminished arterial O_2 content arising from Bohr and Root effects on O_2-hemoglobin binding (Perry and Wood, 1989). The data from the studies on traira and tambaqui suggest the presence of specific pH/CO_2 chemoreceptors that can trigger an increase in breathing, as well as cardiovascular reflexes, during hypercarbia independently of changes in blood O_2 content (see Gilmour, 2001 for review). This is consistent with recent work on temperate fish such as the rainbow trout and Pacific salmon (McKendry and Perry, 2001; McKendry *et al.*, 2001).

5. HYPERCARBIC CARDIOVASCULAR RESPONSES

(a) Systemic responses: The common response to hypercarbia (external hypercapnia) in teleost fish (traira, Reid *et al.*, 2000; tambaqui, Sundin *et al.*, 2000; trout, Perry *et al.*, 1999) and elasmobranchs (Kent and Peirce II, 1978; McKendry *et al.*, 2001) is a bradycardia. This is superficially similar to the response to hypoxia, although a comparison of the effects of selective denervation of branchial nerves in traira and tambaqui reveal some interesting differences. For instance, the receptors triggering the hypercarbic bradycardia in tambaqui are located on the first arch, while those triggering the hypoxic bradycardia are probably located on all gill arches (Sundin *et al.*, 2000). By contrast, in traira the hypercarbic bradycardia is mediated by receptors most likely located on all gill arches (Reid *et al.*, 2000), while the hypoxic bradycardia is elicited by receptors on only the first gill arch (Sundin *et al.*, 1999). The channel catfish shows a profound bradycardia on exposure

to hypoxia yet lacks any cardiovascular responses to hypercapnic acidosis (Burleson and Smatresk, 2000). The sum of these results tends to suggest that the receptors involved in mediating the hypoxic and hypercarbic responses are different.

Elevated levels of ambient CO_2 increase blood pressure in tambaqui (Sundin *et al.*, 2000) and trout (Perry *et al.*, 1999), but decrease blood pressure in traira (Reid *et al.*, 2000) and dogfish (McKendry *et al.*, 2001). The hypertension in trout is a result of an α-adrenoceptor-mediated vaso-constriction (Perry *et al.*, 1999). The mild hypotension in traira might arise from active inhibition of systemic vasoconstriction, arising from the stimulation of receptors or the first gill arch (Reid *et al.*, 2000). This is consistent with the observation that the first gill arch is a site of inhibitory signals that dampen the hypoxic respiratory response in this species during environmental hypoxia (Sundin *et al.*, 1999). The vascular changes in dogfish appear to be of cholinergic origin (McKendry *et al.*, 2001).

The physiological significance of the cardiovascular responses to hypercarbia in fish is not clear, although they are believed to lead to enhanced gas transfer (see Perry *et al.*, 1999).

(b) Branchial responses: Hypercarbia increases the gill resistance in dogfish (Kent and Peirce II, 1978), but has no effect on branchial resistance in trout (Perry *et al.*, 1999).

6. BREATHING PATTERN FORMATION IN WATER-BREATHERS

It has recently been shown that there is a breathing pattern *continuum* in tambaqui ranging from regular continuous breathing, to frequency cycling (continuous breathing that occurs in alternating fast and slow cycles), to classical episodic breathing (where the breaths are separated by periods of no active ventilation), and finally to periods of no breathing or apnea (Reid *et al.*, 2003). Figure 6.8 illustrates examples of frequency cycling and episodic breathing in the tambaqui. The position of the breathing pattern on the *continuum* at any given time is influenced by chemoreceptor (O_2 and pH/CO_2) drive and possibly by input from gill mechanoreceptors and non-respiratory related afferent vagal traffic. Additionally, there appear to be O_2 and CO_2/pH chemoreceptors that regulate breathing pattern independent of the overall level of ventilation (i.e., independent of changes in breathing frequency or amplitude).

The five bearded rockling (*Ciliata mustela*), a temperate, intertidal fish commonly found on the coast of Europe in the North Sea, possesses a novel chemosensory system that modulates both the overall level and pattern of breathing. The anterior dorsal fin of this fish consists of a fringe of small rays that contain approximately five million secondary sensory cells called

(a) Continuous breathing

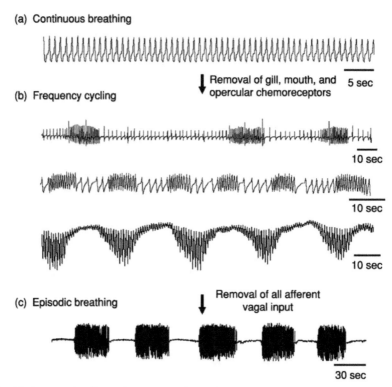

(b) Frequency cycling

Removal of gill, mouth, and 5 sec
opercular chemoreceptors

10 sec

10 sec

10 sec

(c) Episodic breathing

Removal of all afferent
vagal input

30 sec

Fig. 6.8 An example of breathing patterns observed in a decerebrate, spinalized and artificially-ventilated tambaqui. (a) Continuous breathing; (b) three examples of frequency cycling; (c) episodic breathing. (Modified from Reid *et al.*, 2003.)

solitary chemosensory cells (SCC; Kotrschal *et al.*, 1998). These cells sense, amongst other things, body mucus of other fish. Upon stimulation, these receptors trigger a cessation of breathing followed by an ataxic breathing pattern with an overall reduction in respiratory rate (Kotrschal *et al.*, 1993). To date, these cells have been identified in lampreys, sea robins, and the rockling. Whether the SCC play a role in the control of breathing in many fish species remains unknown. It is possible that the non-specific afferent vagal traffic that modifies the breathing pattern in tambaqui (Reid *et al.*, 2003) arises from some external factor similar to the SCC.

While the role of mechanoreceptors in respiratory control in temperate fishes has been examined in numerous studies (see De Graff and Ballintijn, 1987; De Graff *et al.*, 1987; Burleson *et al.*, 1992 for references), no such studies have been performed on tropical, water-breathing fishes.

7. AQUATIC SURFACE RESPIRATION

Under conditions of moderate to severe environmental hypoxia, some species of water-breathing tropical, and temperate, fishes perform aquatic surface respiration (ASR). This involves skimming the well-oxygenated surface layer of water across the gills (Kramer and McClure, 1982). While some species have no special adaptations for performing ASR, others (e.g., Amazonian fish such as the tambaqui, *Colossoma macropomum* and the silver mylossoma, *Mylossoma duriventris*; African fish such as the Lake Magadi tilapia, *Oreochromis alcalicus*, and the African cyprinid, *Barbus neumayeri*) develop a swollen lower lip that acts as a funnel to direct the surface water across the gills. The O_2 threshold for ASR is lower in *Barbus* living in dense, O_2-poor, papyrus swamps compared to fish of the same species living in well-oxygenated streams (Olowo and Chapman, 1996). In addition to ASR, the Lake Magadi tilapia will also breathe air under severely hypoxic conditions (Narahara *et al.*, 1996).

Some fishes, such as Australian gobies (Gee and Gee, 1991) and the New Zealand black mudfish (*Neochanna diversus* Stokell; McPhail, 1999) utilize an air bubble held within the buccal cavity to facilitate ASR. This "buccal bubble" performs two primary respiratory functions. First, it increases the oxygenation of water immediately prior to the gills (Burggren, 1982; Gee and Gee, 1991; Thompson and Withers, 2002). Second, it plays a hydrostatic role by providing lift to the head, helping to position it at the water surface to facilitate effective skimming. Gee and Gee (1995) have demonstrated the role of the buccal air bubble in providing lift to the head and body in several species of Australian gobies (see Figure 6.9 taken from Gee and Gee, 1995). Furthermore, these authors speculate that both ASR and the buoyancy regulation provided by the buccal air bubble during ASR were necessary steps in the evolution of air breathing in these fish.

Rantin *et al.* (1998) reported that the pacu (*Piaractus mesopotamicus*) began to perform ASR when the water PO_2 was lowered to 34 Torr and that denying fish access to the water surface led to a significant hypoxic bradycardia which otherwise did not occur. Rantin and Kalinin (1996) demonstrated that tambaqui began to perform ASR when the water PO_2 fell to 50 Torr. According to Kramer and McClure (1982), 29 of 31 species of tropical fish they examined spent 50–90% of the time performing ASR when the water PO_2 was lowered to 8–24 Torr. These figures are comparable to those reported by Gee *et al.* (1978), who observed ASR in 22 of 26 temperate species.

Intra-arterial adrenaline injections do not influence lip swelling in tambaqui (Moura, 1994). Sundin *et al.* (2000) demonstrated that complete

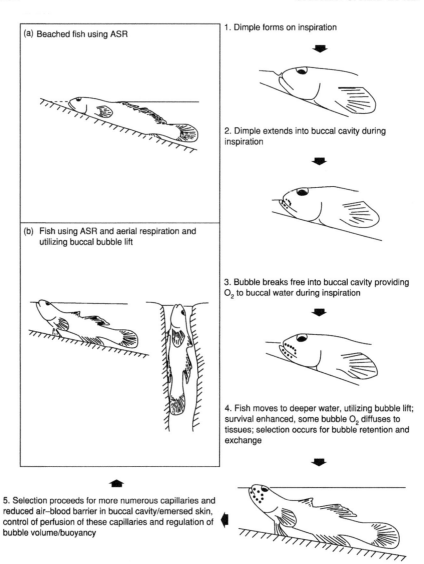

Fig. 6.9 The role of a buccal air bubble in providing hydrostatic lift during ASR or aerial respiration in a goby. (From Gee and Gee, *J. Exp. Biol.* 1995. **198**, 79–89. Reproduced with permission from the Company of Biologists Ltd.)

denervation of tambaqui gills attenuated, but did not prevent, inferior lip swelling nor did it prevent the fish from performing ASR (Sundin *et al.*, 2000; Rantin *et al.*, 2002). Recently, Florindo *et al.* (2002) observed that denervation of cranial nerves V and VII, to the oro-branchial cavity, did not affect the development of the swollen inferior lip during severe hypoxia in tambaqui. On the other hand, denervation of cranial nerve V alone prevented the fish from performing ASR. This indicates that ASR is dependent upon the stimulation of O_2 chemoreceptors in the oro-branchial cavity, innervated by cnV but that lip swelling is not. Although the data of Sundin et al. (2000) indicate that stimulation of gill O_2 receptors, innervated by cranial nerves IX and X, is partially responsible for initiating lip swelling, other factors (i.e., endocrine or paracrine) excluding chemoreceptor input from cnV or cnVII, must also be involved.

8. DO CIRCULATING CATECHOLAMINES INFLUENCE BREATHING IN TROPICAL FISHES?

While the role of catecholamines in the control of breathing in fish has, at times, been controversial (e.g., Randall and Taylor, 1991), the sum of the evidence to date suggests that circulating catecholamines do not exert an excitatory influence on breathing (Perry et al., 1992). Milsom et al. (2002) administered intra-arterial injections of adrenaline in tambaqui and observed a dose-dependent decrease in breathing frequency, amplitude, and total ventilation. Pre-treatment with sotalol (a β-adrenoceptor antagonist) abolished the adrenaline-induced decrease in breathing frequency but not the decrease in breath amplitude. Sotalol treatment also altered the frequency component of the hypoxic ventilatory response, as breathing frequency reached a peak value at a lower water PO_2, compared to the non-sotalol-treated fish, and actually fell during severe hypoxia (Milsom et al., 2002). While the data do not support an excitatory role for circulating catecholamines in the hypoxic ventilatory response, it is not clear where the injected catecholamines are exerting their effects. Superfusion of the tambaqui brain with adrenaline, with and without concomitant sotalol superfusion, caused mixed effects on breathing in a decerebrate, spinalized, and artificially ventilated preparation (Reid, Sundin, Rantin and Milsom, unpublished observations).

9. SUMMARY: WATER-BREATHING FISHES

The general scheme of respiratory control in water-breathing tropical fishes appears to be similar to that in water-breathing temperate fishes. Our studies on traira and tambaqui have not produced a unifying model of respiratory control in tropical fishes but rather have revealed an unexpected degree of complexity. While this has led to a certain degree of frustration, it is not surprising that an assortment of receptor actions and configurations exist in such a tremendous diversity of species.

B. Air-Breathing Fishes

While this section will, again, focus primarily on responses to chemore-ceptor stimuli, these are not the only factors that play key roles in cardiore-spiratory control in tropical air-breathing species. Undoubtedly the daily and seasonal variations in aquatic O_2 levels found in the tropics have contributed to the diversity of air-breathing strategies and the evolution of air breathing in tropical fishes. However, it would seem unlikely that sea-sonal environmental rhythms, *per se*, would not also exert a significant role in ventilatory control and the responses to respiratory challenges indepen-dent of changes in aquatic O_2 levels. Accordingly, Maheshwari *et al.* (1999) have reported that in *Heteropneustes fossilis* and *Clarias batrachus*, main-tained under constant conditions (including $P_{W}O_2$) in the laboratory for 15 months, air-breathing frequency increased during the summer months (which are normally characterized by a decrease in environmental O_2 levels) and during times that correspond with periods of intense reproductive activity (times when water O_2 levels are normally saturated). These results indicate that endogenous, seasonal rhythms can affect air breathing in tropical fish, independent of aquatic O_2 levels. Environmental disasters, such as crude oil spills, can also influence air breathing, as Brauner *et al.* (1999) have demonstrated in *Hoplosternum littorale,* where exposure to the water-soluble fraction of crude oil led to an increase in air breathing (see review by Val and Almeida-Val, 1999). Behavioral factors can also influence air breathing. For instance, since forays to the water's surface increase the risk of predation, visual and auditory cues have been shown to exert a role in ventilatory control; even overriding powerful chemoreceptor input (Kramer and Graham, 1976; Smith and Kramer, 1986). This aside, oxygen availabili-ty remains the key factor regulating air breathing in most fish. Oxygen availability has also been correlated, not only with respiratory strategies, but also with the ability of 64 species of gymnotiforms from the Tefe region of the upper Amazon basin to produce various forms of electrical signals (Crampton, 1998).

Ventilatory control mechanisms in air-breathing fishes are believed to be more complex than the primarily O_2-driven system of respiratory control seen in exclusively water-breathing fishes (see review by Smatresk, 1988). Three groups of questions arise in relation to the predominant issues surrounding ventilatory control in air-breathing fishes. First, is air breathing a reflex response driven exclusively by peripheral chemoreceptors? If not, does an "air-breathing rhythm" arise centrally or via a combination of central mechanisms modified by afferent peripheral input? Second, if air breathing is driven by peripheral chemoreceptors, what is the role, if any, of externally oriented chemoreceptors monitoring aerial versus aquatic gas levels (O_2 and CO_2) in initiating air breathing? Third, is air breathing initiated by internally oriented chemoreceptors monitoring blood gas levels and, if so, do respiratory-related central pH/CO_2 chemoreceptors exist in these fish?

Air breathing in fish is periodic. In other words, the air-breathing organ is not ventilated continuously. Air breathing in amphibians (e.g., the bullfrog, *Rana catesbeiana*), is also periodic, and like fish, the *Rana* tadpole exhibits both gill and lung ventilation, while the adult of this species exhibits both lung ventilation and buccal oscillations. The latter are the developmental equivalent of gill ventilation. Recently, several studies have shown that there are separate central rhythm generators (CRG) that control buccal (gill) ventilation versus lung ventilation in the bullfrog (Torgerson et al., 2001; Wilson et al., 2002). While the data on the central control of breathing in fish, including tropical species, is sparse, it is highly likely that separate central rhythm-generating mechanisms for air breathing exist in fish. Pack et al. (1992) demonstrated that lung mechanoreceptors had virtually no influence on gill ventilation in *Protopterus* but did influence air breathing, lending support to the idea that gill and lung ventilation are independently generated centrally.

It is also highly likely that both the air- and water-breathing control systems influence one another. Thus, removal of the left air-breathing organ (arborescent organ) from the walking catfish (*Clarias batrachus*) causes an increase in water-breathing frequency in fish denied access to air but not in those fish with access to air (Gupta and Pati, 1998). The mechanics of air breathing in lungfish such as *Protopterus* also suggest that there must be significant coordination between a buccal/gill CRG and a lung CRG. In *Protopterus*, the process of air breathing utilizes a two-stroke pump (see above). This involves surfacing, expanding the buccal cavity while opening the glottis and exhaling, and then submerging with air trapped in the buccal cavity. Only following submersion do the buccal muscles pump the air into the lungs (Bishop and Foxon, 1968; McMahon, 1969; DeLaney et al., 1977). Unlike the situation seen when this animal is water-breathing, central output is now also required to open the glottal sphincter during exhalation, and to hold the glottis open during buccal compression (lung inflation). Interactions between a buccal and lung CRG have recently been described in the bullfrog (Wilson et al., 2002) and it has been suggested (Perry et al., 2001) that the neural circuitry underlying lung ventilation evolved prior to the emergence of air breathing and the specialization of the air-breathing organ for respiratory gas exchange.

Finally, it must be mentioned that while air-breathing organs may play key roles in oxygen extraction for these fish, the skin and gills remain the key sites of CO_2 exchange. As a consequence, the respiratory quotient (RQ), which is the ratio of CO_2 production/O_2 consumption measured at the lungs will be less than 1 while that for the skin and/or gills will be greater than 1. This has implications for both O_2 and CO_2 exchange processes as well as for acid-base balance. It has been argued that in fish, the pattern of CO_2 excretion is designed to retain control of HCO_3^- movement across the gills and that carbon dioxide loss via an air-breathing organ would reduce the capacity of the animal to regulate pH in this manner (Randall *et al.*, 1981). These are the topics of the following chapter.

1. HYPOXIC VENTILATORY RESPONSES

(a) Facultative air-breathers: From a control systems perspective, facultative air-breathing fishes are likely to be similar to obligate water-breathing fishes whereas obligate air-breathers are likely to possess many of the control mechanisms common to terrestrial air-breathing vertebrates such as amphibians. Facultative air-breathing fishes breathe water under normoxic conditions and continue to do so during mild to moderate hypoxia. Ultimately, however, a critical O_2 threshold is reached at which point O_2 extraction from the water is no longer sufficient to sustain a normal level of O_2 uptake; the fish then begins to breathe air. The stimulus to begin air ventilation likely arises from gill and/or oro-branchial chemoreceptors with the potential for involvement of both water-sensing and blood-sensing chemoreceptors in various loci.

Graham (1997) provides tables of the aquatic PO_2 threshold for air breathing in a number of air-breathing species. The jeju (*Hoplerythrinus unitaeniatus*) is a facultative air-breathing fish that utilizes a modified swim bladder as an air-breathing organ (Kramer, 1977). According to Farrell and Randall (1978), this fish often breathes air in episodes, taking approximately three breaths every three minutes even under normal conditions. In a recent study (Lopes *et al.*, 2002) on the responses of jeju to aquatic hypoxia, forays to the surface began when the water PO_2 was lowered to approximately 40 mmHg. Selective denervation of either cranial nerve IX alone or both cranial nerve IX and the pre-trematic branch of X to the first gill arch did not alter the frequency of trips to the surface or the duration of the periods spent air breathing. However, when the branches of cnIX and cnX to all gill arches were cut, air breathing was abolished entirely. These data suggest that the trigger for air breathing in jeju arises from O_2 chemoreceptors on the gill arches. What the data do not say is whether air breathing was triggered by water-sensing or blood-sensing chemoreceptors. This was also the conclusion from one study on the temperate facultative air-breathing *Amia calva* (McKenzie *et al.*, 1991) but not from another (Hedrick and Jones, 1993) where branchial denervation depressed but did not eliminate the hypoxic air-breathing response.

For the majority of air-breathing fishes, gill ventilation increases leading up to the initiation of air breathing and is subsequently inhibited after an air breath. This pattern has been observed in *Protopterus, Lepidosiren, Erpetoichthys, Leoiisosteus, Amia, Ancistrus, Gymnotus, Anabas,* and *Piabucina* (see Smatresk, 1988; Graham, 1997 for reviews). As Graham (1997) points out, because the blood returning from the air-breathing organ in most of these fishes returns to the systemic venous circulation and goes immediately to the gills, the inhibition of branchial ventilation immediately following an air breath will serve to reduce the possibility of losing oxygen obtained from the ABO to the water in the gills.

(b) Obligate air-breathers: Given that obligate air-breathing fishes depend on aerial respiration for O_2 uptake, it would be reasonable to assume that water-sensing (external) O_2 chemoreceptors have taken on a reduced role in respiratory control compared with water-breathing or facultative air-breathing species. Aquatic hypoxia has little or no effect on gill ventilation in many obligate air-breathing fishes (e.g., the African lungfish (*Protopterus*), the South American lungfish (*Lepidosiren*), the electric eel (*Electrophorus*), the Asian Swamp eel (*Monopterus*), the Australian lungfish (*Neoceratodus*) and species of *Synbranchus*) while in other species (e.g., the African reed fish (*Erpetoichtys*), the weatherfish (*Misgurnis*) and armoured catfishes such as *Hopolosternum, Brochis* and *Piabucina*) gill ventilation is reduced in aquatic hypoxia. Still other species of obligate air-breathers increase gill ventilation during aquatic hypoxia (e.g., juvenile *Protopterus*, the bichir (*Polypterus*), and catfishes such as *Heteropneustes* and *Clarius*). Given that aquatic hypoxia can modify gill ventilation in some obligate air-breathing tropical fishes, and assuming that aquatic hypoxia is having a minimal effect on blood O_2 levels, it is evident that water-sensing O_2 chemoreceptors are present in some of these fishes. However, it is unclear why activation of external O_2 receptors should modify breathing in some obligate air-breathers. The presence and role of external O_2 chemoreceptors in obligate air-breathing fishes likely represents an intermediate stage in the evolution of air breathing where gill ventilation still affects blood O_2 levels to some small degree.

What is the trigger for air breathing in obligate air-breathing fishes? In *Protopterus*, branchial denervation abolished the increase in gill ventilation and attenuated the increase in air breathing (Lahiri *et al.*, 1970). This led to the suggestion that air breathing in fish is driven by peripheral O_2 chemoreflexes rather than by central mechanisms of respiratory rhythm and pattern generation (see below). Although external O_2 chemoreceptors can alter gill ventilation in obligate air-breathers (see above), it is reasonable to assume that air breathing is triggered by a reduction in blood, rather than water, O_2 levels that activate internal O_2 chemoreceptors in the periphery (i.e., the gills and/or oro-branchial cavity). In a recent study, Sanchez *et al.* (2001b) exposed the South American lungfish, *Lepidosiren paradoxa*, to aquatic and aerial hypoxia either separately or in combination. While aquatic hypoxia had no effect on ventilation (frequency or tidal volume), aerial hypoxia and a combination of aerial and aquatic hypoxia caused a significant increase in air breathing mediated by an increase in breathing frequency with no change in tidal volume. These results are in agreement with previous studies in which aquatic hypoxia did not stimulate lung ventilation in *Neoceratodus* (Johansen *et al.*, 1967), *Protopterus* (Johansen and Lenfant, 1968) or *Electrophorus* (Johansen *et al.*, 1968), but in opposition to the results of a study on *Hypostomus* (Graham and Baird, 1982). Other studies have also documented an increase in breathing during aerial hypoxia in lungfish (e.g., Jesse *et al.*, 1967; Fritsche *et al.*, 1993).

Given that gill ventilation does not participate in O_2 uptake in *Lepidosiren*, it is unlikely that the aquatic hypoxia imposed by Sanchez *et al.* (2001b) had any effect on blood O_2 status, suggesting that the ventilatory response to aerial hypoxia in *Lepidosiren* arises from stimulation of blood-sensing O_2 chemoreceptors. Indeed, the sum of these studies would suggest that changes in blood, rather than water, O_2 levels are the stimulus for air breathing in obligate air-breathing fishes. Furthermore, a reduction in arterial O_2 content (via anemia) did not lead to an increase in breathing during aerial normoxia, suggesting that the stimulus modality that triggers air breathing is a decrease in PaO_2 rather than a decrease in blood oxygen content. Such an interpretation is consistent with the cellular models of O_2-sensing in mammals (e.g., Gonzalez *et al.*, 1994) and the results of studies on water-breathing fishes (see above).

2. HYPERCARBIC VENTILATORY RESPONSES

(a) Peripheral chemoreceptors: The effects of aquatic hypercarbia on air breathing are variable. For example, *Protopterus* (Johansen and Lenfant, 1968), *Neoceratodus* (Johansen *et al.*, 1967) and *Channa argus* (the Northern Snakehead; Glass *et al.*, 1986) increase ventilation in response to aquatic hypercarbia while elevated levels of aquatic CO_2 have little effect on air breathing in *Electrophorus* (Johansen *et al.*, 1968) and *Misgurnus anguillicaudatus* (the Oriental Weatherloach; McMahon and Burggren, 1987). Given that air-breathing freshwater fish rely predominantly on the water as a medium for CO_2 excretion, it is curious that aquatic hypercarbia triggers an increase in air breathing, which is predominately used for O_2 uptake, rather than water breathing, which is predominantly used for CO_2 excretion. Note, however, that during aestivation (see Harder *et al.*, 1999), while encased in a dry subterranean cocoon, lungfishes utilize their lungs for both O_2 and CO_2 exchange, with no contribution from the gills. Under these

conditions, breathing becomes episodic with clusters of breaths separated by periods of apnea (DeLaney and Fishman, 1977).

Recently, Sanchez and Glass (2001) examined the effects of aquatic versus aerial hypercarbia on lung ventilation in the South American lungfish, *Lepidosiren paradoxa*. While increasing the water PCO_2 from 10 to 35 Torr did not affect lung ventilation, once the water PCO_2 reached 55 Torr there was a significant increase in breathing mediated by an increase in breathing frequency. On the other hand, aerial hypercapnia ($P_ICO_2 = 55$ Torr) in combination with aquatic normocapnia had no effect on breathing. These results illustrate several important points and highlight the potential complexities of respiratory control in air-breathing fish. At first glance it may appear teleologically unsatisfying that air-breathing fishes would increase ventilation in response to aquatic hypercarbia but not aerial hypercarbia. However, these authors also document the presence of CO_2 receptors in the airway of *Lepidosiren* that inhibit breathing. Amphibians such as the bullfrog also possess upper airway (olfactory) CO_2 receptors that, when stimulated, inhibit breathing (see review by Coates, 2001). Although the gills of *Lepidosiren* are reduced, they are still a significant site of CO_2 exchange with approximately 70% of CO_2 excretion occurring across the gills (Johansen, 1970). As such, aquatic hypercarbia is likely to lead to an increase in arterial PCO_2 which appears to trigger ventilation by activating internally oriented CO_2 receptors. Although aerial hypercarbia would also lead to an elevation of arterial PCO_2 and should trigger the same internal CO_2 receptors, the stimulatory effects of activating these receptors may have been countered by the inhibitory effects of the airway CO_2 receptors. Such an interpretation would explain why, in an air-breathing fish, aquatic hypercarbia elicited an increase in air breathing while aerial hypercarbia did not. These data do not exclude the possibility that water-sensing pH/CO_2 chemoreceptors are also involved in triggering the initiation of air breathing in response to elevated water PCO_2 levels. The potential location of internally oriented CO_2 receptors is discussed below.

(b) Lung (airway) receptors: In addition to the CO_2-sensitive airway receptors described by Sanchez and Glass (2001), air-breathing fishes also possess lung (air-breathing organ; ABO) mechanoreceptors. In terrestrial vertebrates, lung (pulmonary) stretch receptors (PSR) monitor the overall degree of lung inflation (tonic component of slowly adapting receptor discharge) as well as the phasic inflation and deflation that occur on a breath-by-breath basis (phasic component of slowly adapting receptor discharge). Activation of PSR during inspiration acts to terminate a breath thereby functioning as an inspiratory off-switch (the Hering–Breuer reflex; see Milsom, 1990 for review). In the dipnoi lungfish, lung deflation elicits a

lung breath (Smith, 1931) and increases breathing frequency (Delaney *et al.*, 1974), suggesting the presence of a Hering–Breuer-like reflex in these animals. Although there are few studies on lung mechanoreceptors in tropical fishes, it appears that they play a similar role to those in higher vertebrates (DeLaney *et al.*, 1983; Pack *et al.*, 1990, 1992), as shown for the air-breathing temperate fish, the spotted gar (*Lepisosteus oculatus*; Smatresk and Azizi, 1987).

Working with isolated lung preparations, Delaney *et al.* (1983) demonstrated the presence of rapidly and slowly adapting mechanoreceptors in both *Lepidosiren* and *Protopterus*. These receptors increased their rate of discharge with progressive lung inflation and responded to changes in both lung volume and the rate of lung inflation. These responses were similar, but not identical, to those exhibited by stretch receptors from facultative air-breathers (*Lepisosteus*; Smatresk and Azizi, 1987; *Amia*, Milsom and Jones, 1985) and amphibians (e.g., McKean, 1969; Milsom and Jones, 1977). Like amphibian PSR, the lungfish mechanoreceptors studied by Delaney *et al.* (1983) exhibited CO_2 sensitivity with their firing rate decreasing as CO_2 levels in the lung increased.

Pack *et al.* (1990) observed, in the African lungfish (*Protopterus*), that an increase in intrapulmonary pressure prolonged the interval between breaths and that inflating the lungs in the early phase of the interbreath interval had a greater effect on the duration of that interbreath interval compared to inflating the lungs in the later phase of this period. Working with decerebrate and spinalized *Protopterus*, Pack *et al.* (1992) demonstrated that lung inflation led to a decrease in breath duration which was not altered by the composition of the gas used to ventilate the lungs (air, O_2 or N_2) but was abolished by vagotomy. Note that PSR report to the brain via the pulmonary branch of the vagus nerve. A similar relationship between tidal volume and breath duration, as well as the effects of vagotomy, is seen in mammals (Milsom, 1990).

The results of Pack *et al.* (1992) suggest that the neural circuitry responsible for breath timing is well developed in lungfishes and is similar to the control system that functions in mammals. This observation is consistent with the idea advanced by Perry *et al.* (2001) that the neural mechanisms responsible for the genesis of air breathing were well developed prior to the emergence of air breathing and the use of air-breathing organs, including lungs, in respiratory-related gas exchange. Furthermore, given that mammals breathe using an aspiration pump while lower vertebrates, including lungfishes (DeLaney and Fishman, 1977), breathe with a buccal force pump, Pack *et al.* (1992) also suggested that the neural circuitry that controls the timing of breaths must be distinct from the circuits that shape the pattern of respiratory motor output.

The sum of these studies suggest that, like other air-breathing vertebrates, lungfishes (and presumably, or possibly, other air-breathing fishes) possess mechanoreceptors in their air-breathing organ that sense the rate of lung inflation and/or lung volume and function to terminate inspiration.

(c) Central chemoreceptors: The transition from aquatic to terrestrial life meant that air, rather than water, became the primary respiratory medium (Ultsch, 1996). This led to an elevation of arterial HCO_3^-/CO_2 levels in air-breathers and a significantly greater CO_2 drive to breathe compared with water-breathing animals. Although peripheral chemoreceptors in terrestrial animals do contribute to the hypercarbic ventilatory response, the primary, steady-state CO_2 drive to breathe arises from central chemoreceptors, located on the ventrolateral surface of the medulla, that sense changes in pH/CO_2 within the cerebral spinal fluid (see reviews by Smatresk, 1990; Ballantyne and Scheid, 2001). Until recently, dogma suggested that fish do not posses central respiratory chemoreceptors. Recent evidence, however, is beginning to challenge this view, at least for air-breathing fishes (Wilson *et al.*, 2000; Sanchez *et al.*, 2001a; see reviews by Gilmour, 2001; Remmers *et al.*, 2001; Milsom, 2002).

Gill denervation experiments performed on a variety of species have led to the conclusion that gill denervation can have highly varied effects on different components of the ventilatory response (frequency versus amplitude) under different conditions and that there are receptors outside the gills, at least in some fishes, that can give rise to ventilatory responses to changes in CO_2/H^+. The strongest evidence to suggest that the extra-branchial receptors in water-breathing fishes might be central CO_2/H^+ chemoreceptors comes from brain perfusion studies in tench (Hughes and Shelton, 1962) and lamprey (Rovainen, 1977), but this evidence is far from compelling (see Milsom, 2002 for review). Other attempts to demonstrate a role for central chemoreceptors in regulating water breathing in skate, amia, trout, and tambaqui, have not been successful (Graham *et al.*, 1990; Wood *et al.*, 1990; Hedrick *et al.*, 1991; Burleson *et al.*, 1992; Sundin *et al.*, 2000; Milsom *et al.*, 2002). All of these data indicate that the normal hypercarbic response of these species is to changes in the CO_2 of the external environment and provide no evidence for the existence of central chemoreceptors.

Several recent reports suggest that in some species of air-breathing fish, central CO_2/H^+ chemoreceptors may contribute to these responses. These studies employed *in vitro* brainstem–spinal cord preparations of the holostean fish, the gar (*Lepisosteus osseus*) and the teleost fish, the siamese fighting fish (*Beta splendens*). Superfusion of isolated brainstems from these species with high CO_2/low pH solutions did lead to an increase in motor output, believed to represent lung breathing, but had no effect on fictive gill ventilation (Wilson *et al.*, 2000). The interpretation of this data is not straightforward (see review by Milsom, 2002). In another holostean fish, the bowfin (*Amia calva*), superfusion of the brain stem of intact fish with acidotic and alkalotic solutions had no effect on ventilation (Hedrick *et al.*, 1991). On the other hand, it has also been reported that the South American lungfish (a Sarcopterygian fish belonging to the lineage giving rise to higher vertebrates) does respond well to perfusion of the IVth cerebral ventricle with mock CSF of differing pH (Sanchez *et al.*, 2001a).

Based on this sparse data, several interpretations are possible. If central chemoreceptors are present in the lamprey, the data would suggest that central CO_2 receptors predate the origin of the Agnatha and Elasmobranchs, and perhaps have been secondarily lost in some species. If they are only present in some fishes that exhibit various forms of air breathing, then the possibility exists that they have arisen multiple times, in association with the evolution of air breathing. Finally, until it can be shown that the responses of some Actinopterygian fishes to changes in CSF pH are physiologically relevant, the possibility remains that central CO_2 receptors are only present in the true lungfish and arose only once in the line giving rise to Sarcopterygian fishes, amphibians, and terrestrial vertebrates.

3. Air Breathing in Marine and Intertidal Fishes

Marine air-breathing fishes rely on cutaneous respiration and modified gills for aerial respiration and lack the specialized air-breathing organs found in many freshwater species (Johansen, 1970). As such they are bimodal gill-breathers like those species of freshwater fish that also utilize gills for air breathing. The major difference between these two groups is that the amphibious fishes often leave relatively well-oxygenated water to make excursions onto land. In comparison to species that utilize accessory structures for air breathing, even less is known about the respiratory control mechanisms of amphibious fishes (marine and freshwater) that use their gills in both media.

Many intertidal fishes, such as the *Blenniidae* and *Cottidae*, are facultative air-breathers, which, like their temperate relatives, have specific behavioral and physiological attributes that permit aerial respiration and survival out of water (Martin, 1996; see Chapters 11 and 12). Many are capable of exchanging both O_2 and CO_2 in air (e.g., Martin, 1993, 1995; Bridges, 1988) with aerial CO_2 excretion being dependent upon the activity of carbonic anhydrase (Pelster *et al.*, 1988).

Mudskippers (*Gobiidae*: *Oxudercinae*) are intertidal fishes which have adapted to an amphibious life style, spending a significant amount of time exposed to air on the intertidal mudflats throughout the Indo-Pacific Ocean and West Africa (Clayton, 1993; see Chapter 11). They can maintain relatively constant rates of O_2 uptake and CO_2 excretion when exposed to air (e.g., Steeger and Bridges, 1995; see review by Graham, 1997). Ishimatsu

Table 6.1
Ratios of gill area (cm^2) to body mass (g) and total body surface area (cm^2) of
some air-breathing and non-air-breathing fish species

Species	Gill area/body mass	Gill area/body area
Amphibious air-breathers		
Periophthalmus cantonensis[1]	1.24	0.38
P. koelreuteri[5]		0.46
P. dipus[5]		0.35
P. chrysospilos[5]		0.34–0.36
P. vulgaris[5]		0.27–0.32
P. schlosseri[9]		0.20–0.50
P. chrysophilos[9]		0.25–0.3
Boleophthalmus chinensis[1]	0.94	0.56
B. viridis[5]		0.72
B. boddarti[5]		0.68–0.83
B. boddarti[6]		0.52
B. boddarti[9]		0.65–0.75
Aquatic air-breathers		
Heteropneustes fossilis[2]	0.32	0.34
Anabas testudineus[3]	0.39	0.40
Channa argus[1]	0.85	0.38
Clarias batrachus[7]	0.83	0.48
C. mossambicus[8]	0.17	
Non-air-breathers		
Gobius jozo[5]		1.00
G. auratus[5]		1.17
G. caninus[5]		1.40
Lophius piscatorius[4]	1.96	2.99
Anguilla japonica[1]	3.32	1.45
Tautoga onitus[4]	3.92	4.35
Cyprinus carpio[1]	4.16	1.74
Carassius auratus[1]	4.49	2.91
Stenotomus chrysops[4]	5.06	4.78
Sarda sarda[4]	5.95	11.55
Mugil cephalus[4]	9.54	6.54
Scomber scombrus[4]	11.58	8.38
Brevoortia tyrannus[4]	17.73	18.28
Gymnosarda alleterata[4]	19.39	48.54

References: [1]Tamura and Moriyama, 1976; [2]Hughes et al., 1974; [3]Hughes et al., 1973; [4]Gray, 1954; [5]Schöttle, 1931; [6]Biswas et al., 1981; [7]Munshi, 1985; [8]Maina and Maloiy, 1986; [9]Low et al., 1990.

Source: Graham, 1997.

et al. (1999) subjected an Asian mudskipper (*Periophthalmodon schosseri*) to three hours of aerial exposure, aquatic hypoxia (P_WO_2 <7 Torr) and forced submergence in normoxic water. While aerial exposure and aquatic hypoxia were without effect on blood gas levels, forced submergence caused a 50% reduction in blood O_2 content and hemoglobin-O_2 saturation, indicating that this species is poorly adapted for aquatic O_2 extraction. This is consistent with the study of Takeda *et al.* (1999), who report that this species can quickly repay an O_2 debt while breathing air but not while breathing water. Aguilar *et al.* (2000) exposed *P. schosseri* to both aquatic and aerial hypoxia. While aquatic hypoxia did not affect ventilation, exposure to an inspired aerial PO_2 of approximately 37 Torr caused an increase in overall breathing that was mediated by both an increase in frequency and tidal volume. This response is similar to that described above for lungfish. Aerial hypercapnia also caused an increase in breathing mediated exclusively by an increase in breathing frequency with no change in tidal volume. The sum of the data suggest that the O_2 chemoreceptors mediating the increase in breathing during hypoxia in this species of mudskipper are likely to be internally oriented (blood-sensing) but do not allow for speculation on the site of CO_2-sensing (Ishimatsu *et al.*, 1999; Aguilar *et al.*, 2000). Further studies on marine and intertidal species will be required before data from these groups of fishes can be incorporated into attempts at modeling respiratory control systems in tropical fishes.

4. CARDIOVASCULAR RESPONSES

Since hypoxia and hypercarbia serve to increase the ventilation of air-breathing organs or accessory structures in air-breathing fishes, it is extremely difficult to separate out the effects of hypoxia and hypercarbia on the cardiovascular system of these fishes under such conditions, from the effects of air-breathing *per se*. Clearly there must be sensory and motor integration between gills, air-breathing organs and other accessory structures used for gas exchange with cardiac output and blood flow distribution to enhance bimodal gas exchange in these fishes.

(a) Cardiac responses: In most air-breathing fishes, air breathing is associated with a tachycardia, increased blood flow to the ABO and reduced gill ventilation (see Graham, 1997 for review). Interestingly, as noted by Graham (1997), the tachycardia is least in the lungfishes and *Lepisosteus*. These fish all lack adrenergic (sympathetic) cardiac innervation and it would appear that the small tachycardia they exhibit is due to the release of a low resting vagal (parasympathetic) tone. It has been shown that externally oriented branchial chemoreceptors and ABO mechanoreceptors contribute to this tachycardia in various species (McKenzie *et al.*, 1991; Roberts and Graham, 1985; Graham *et al.*, 1995).

(b) Vasomotor responses: In all studies to date, air-breathing initiates increases in blood flow to the ABO which gradually subside over the subsequent breath hold as O_2 levels in the ABO decline. This blood flow

redistribution is under both cholinergic and adrenergic control. In *Protopterus*, acetylcholine constricts the pulmonary artery but dilates the ductus arteriosus shunting blood away from the lungs and to the systemic circulation during periods between air breaths (Johansen and Reite, 1967; Johansen *et al.*, 1968; Laurent *et al.*, 1978). In *Hoplerythrinus*, cholinergic stimulation of the gills resulted in selective perfusion of the posterior gill arches favouring increased flow to the coeliac artery and the ABO (Smith and Gannon, 1978).

5. SUMMARY: AIR-BREATHING FISHES

Studies of cardiorespiratory control in air-breathing fishes have produced a plethora of intriguing suggestions that bear further study. These include implications about the evolution of multiple central rhythm generators for breathing, the evolution of central chemoreceptors for sensing CO_2/pH, the evolution of partially divided hearts, central cardiac shunts, and the switch from a hypoxic bradycardia to a hypoxic tachycardia. Ongoing and future studies on a wide range of tropical fishes ranging from hypoxia-tolerant and hypoxia-intolerant water-breathers, to fishes that perform aquatic surface respiration, facultative air-breathers, and obligate air-breathers should, with luck, ultimately lead to an encompassing model of cardiorespiratory control in fish.

REFERENCES

Aguilar, N. M., Ishimatsu, A., Ogawa, K., and Huat, K. K. (2000). Aerial ventilatory responses of the mudskipper, *Periophthalmodon schlosseri*, to altered aerial and aquatic respiratory gas concentrations. *Comp. Biochem. Physiol. A* **127**, 285–292.

Almeida-Val, V. H. M., and Val, A. L. (1993). Evolutionary trends of LDH isozymes in fishes. *Comp. Biochem. Physiol.* **105B**, 21–28.

Ballantyne, D., and Scheid, P. (2001). Central chemosensitivity of respiration: A brief overview. *Respir. Physiol.* **129**, 5–12.

Ballintijn, C. M., and Juch, P. J. W. (1984). Interaction of respiration with coughing, feeding, vision and oculomotor control in fish. *Brain. Behav. Evol.* **25**, 99–108.

Barrionuevo, W. R., and Burggren, W. W. (1999). O_2 consumption and heart rate in developing zebrafish (*Danio rerio*): Influence of temperature and ambient O_2. *Am. J. Physiol.* **276**, R505–R513.

Beebe, W. (1945). Vertebrate fauna of a tropical dry season mud-hole. *Zoologica* **30**, 81–87.

Bishop, I. R., and Foxon, G. E. H. (1968). The mechanism of breathing in the South American lungfish, *Lepidosiren paradoxa*: A radiological study. *J. Zool. Lond.* **154**, 263–271.

Biswas, N., Ojha, J., and Munshi, J. S. D. (1981). Morphometrics of the respiratory organs of an estuarine goby, *Boleophthalmus boddaerti*. *Japan. J. Ichthyol.* **27**, 316–326.

Booth, J. H. (1978). The distribution of blood flow in the gills of fish: Application of a new technique to rainbow trout (*Salmo gairdneri*). *J. Exp. Biol.* **73**, 119–129.

Brainerd, E. L. (1994). The evolution of lung-gill bimodal breathing and the homology of vertebrate respiratory pumps. *Am. Zool.* **34**, 289–299.

Brainerd, E. L., Liem, K. F., and Samper, C. T. (1989). Air ventilation by recoil aspiration in polypterid fishes. *Science* **246**, 1593–1595.

Braum, E., and Junk, W. A. (1982). Morphological adaptation of two Amazonian characoids (Pisces) for surviving in oxygen deficient waters. *Int. Rev. Gesamten. Hydrobiol.* **67**, 869–886.

Brauner, C. J., Ballantyne, C. L., Vijayan, M. M., and Val, A. L. (1999). Crude oil exposure affects air-breathing frequency, blood phosphate levels and ion regulation in an air-breathing teleost fish, *Hoplosternum littorale. Comp. Biochem. Physiol. C.* **123**(2), 127–134.

Brauner, C. J., Wang, T., Val, A. L., and Jensen, F. B. (2001). Non-linear release of Bohr protons with hemoglobin-oxygenation in the blood of two teleost fishes; Carp (*Cyprinus carpio*) and tambaqui (*Colossoma macropomum*). *Fish Physiol. Biochem.* **24**, 97–104.

Bridges, C. R. (1988). Respiratory adaptations in intertidal fish. *Am. Zool.* **28**, 79–96.

Burggren, W. W. (1982). "Air gulping" improves blood oxygen transport during aquatic hypoxia in the goldfish *Carassius auratus. Physiol. Zool.* **55**, 327–334.

Burleson, M. L. (2002). Oxygen-sensitive chemoreceptors in fish: Where are they? How do they work? *In* "Cardiorespiratory Responses to Oxygen and Carbon Dioxide in Fish," pp. 25–27. International Congress of the Biology of Fish, Extended abstract.

Burleson, M. L., and Smatresk, N. J. (1990). Effects of sectioning cranial nerves IX and X on cardiovascular and ventilatory responses to hypoxia and NaCN in channel catfish. *J. Exp. Biol.* **154**, 407–420.

Burleson, M. L., and Smatresk, N. J. (2000). Branchial chemoreceptors mediate ventilatory responses to hypercapnic acidosis in channel catfish. *Comp. Biochem. Physiol. A.* **125**, 403–414.

Burleson, M. L., Smatresk, N. J., and Milsom, W. K. (1992). Afferent inputs associated with cardioventilatory control in fish. *In* "Fish Physiology" (Randall, D. J., and Farrell, A. P., Eds.), Vol XIIB, pp. 389–426. Academic Press, New York.

Carter, G. S., and Beadle, L. C. (1931). Notes on the habits and development of *Lepidosiren paradoxa. J. Linn. Soc. Lond. Zool.* **37**, 197–203.

Clayton, D. A. (1993). Mudskippers. *Oceanogr. Mar. Biol. Annu. Rev.* **31**, 507–577.

Coates, E. L. (2001). Olfactory CO_2 chemoreceptors. *Respir. Physiol.* **129**, 219–229.

Crampton, W. G. R. (1998). Effects of anoxia on the distribution, respiratory strategies and electric signal diversity of gymnotiforms fishes. *J. Fish Biol.* **53**, 307–330.

De Graff, P. J., and Ballintijn, C. M. (1987). Mechanoreceptor activity in the gills of the carp. II. Gill arch proprioceptors. *Respir. Physiol.* **69**(2), 183–194.

De Graff, P. J., Ballintijn, C. M., and Maes, F. W. (1987). Mechanoreceptor activity in the gills of the carp. I. Gill filament and gill raker mechanoreceptors. *Respir. Physiol.* **69**, 173–182.

DeLaney, R. G., and Fishman, A. P. (1977). Analysis of lung ventilation in the aestivating lungfish *Protopterus aethiopicus. Am. J. Physiol.* **233**(5), R181–R187.

DeLaney, R. G., Lahiri, S., and Fishman, A. P. (1974). Aestivation of the African lungfish, *Protopterus aethiopicus*: Cardiovascular and respiratory functions. *J. Exp. Biol.* **61**, 111–128.

DeLaney, R. G., Lahiri, S., Hamilton, R., and Fishman, A. P. (1977). Acid-base balance and plasma composition in the aestivating lungfish (*Protopterus*). *Am. J. Physiol. Regul. Integr. Comp. Physiol.* **232**, R10–R17.

DeLaney, R. G., Laurent, P., Galante, R., Pack, A. I., and Fishman, A. P. (1983). Pulmonary mechanoreceptors in the dipnoi lungfish *Protopterus* and *Lepidosiren. Am. J. Physiol.* **244**(3), R418–R428.

De Salvo Souza, R. H., Soncini, R., Glass, M. L., Sanches, J. R., and Rantin, F. T. (2001). Ventilation, gill perfusion and blood gases in dourado, *Salminus maxillosus* Valenciennes (*Teleosti, Characidae*), exposed to graded hypoxia. *J. Comp. Physiol. B.* **171**, 483–489.

Farrell, A. P., and Randall, D. J. (1978). Air-breathing mechanics in two Amazonian teleosts, *Arapaima gigas* and *Hoplyerythrinus unitaeniatus*. *Can. J. Zool.* **56,** 939–945.

Farrell, A. P., Daxboeck, C., and Randall, D. J. (1979). The effect of input pressure and flow on the pattern and resistance to flow in the isolated perfused gill of a teleost fish. *J. Comp. Physiol.* **133,** 233–240.

Farrell, A. P., Sobin, S. S., Randall, D. J., and Crosby, S. (1980). Intralamellar blood flow patterns in fish gills. *Am. J. Physiol.* **8,** R428–R436.

Farrell, A. P., Small, S., and Graham, M. S. (1989). Effect of heart rate and hypoxia on the performance of a perfused trout heart. *Can. J. Zool.* **67,** 274–280.

Feldman, J. L., Smith, J. C., Ellenberger, H. H., Connelly, C. A., Lui, G. S., Greer, J. G., Lindsay, A. D., and Otto, M. R. (1990). Neurogenesis of respiratory rhythm and pattern: Emerging concepts. *Am. J. Physiol.* **259**(5, Pt. 2), R879–886.

Fernandes, M. N., and Perna, S. A. (1995). Internal morphology of the gill of a loricariid fish, *Hypostomus plecostomus*: Arterio-arterial vasculature and muscle organization. *Can. J. Zool.* **73,** 2259–2265.

Fernandes, M. N., and Rantin, F. T. (1989). Respiratory responses of *Oreochromis niloticus* (Pisces, Cichlidae) to environmental hypoxia under different thermal conditions. *J. Fish Biol.* **35,** 509–519.

Fernandes, M. N., Rantin, F. T., Kalinin, A. L., and Moron, S. E. (1994). Comparative study of gill dimensions of three erythrinid species in relation to their respiratory function. *Can. J. Zool.* **72,** 160–165.

Florindo, L. H., Kalinin, A. L., Reid, S. G., Milsom, W. K., and Rantin, F. T. (2002). The role of orobranchial chemoreceptors on the control of aquatic surface respiration in tambaqui, *Colossoma macropomum*. In "Cardiorespiratory Responses to Oxygen and Carbon Dioxide in Fish." International Congress on the Biology of Fish, pp. 51–67.

Fritsche, R., and Nilsson, S. (1989). Cardiovascular responses to hypoxia in the Atlantic cod, *Gadus morhua*. *Exp. Biol.* **48,** 153–160.

Fritsche, R., and Nilsson, S. (1990). Autonomic nervous control of blood pressure and heart rate during hypoxia in the cod, *Gadus morhua*. *J. Comp. Physiol.* **160B,** 287–292.

Fritsche, R., Axelsson, M., Franklin, C. E., Grigg, G. G., Holmgren, S., and Nilsson, S. (1993). Respiratory and cardiovascular responses to hypoxia in the Australian lungfish. *Respir. Physiol.* **94,** 173–187.

Gee, J. H., and Gee, P. A. (1991). Reactions of gobioid fishes to hypoxia: Buoyancy and aquatic surface respiration. *Copeia.* **1991,** 17–28.

Gee, J. H., and Gee, P. A. (1995). Aquatic surface respiration, buoyancy control and the evolution of air-breathing in gobies (*Gobiidae*; Pisces). *J. Exp. Biol.* **198,** 79–89.

Gee, J. H., Tallman, R. F., and Smart, H. J. (1978). Reactions of some great plains fishes to progressive hypoxia. *Can. J. Zool.* **56,** 1962–1966.

Gilmour, K. M. (2001). Review. The CO_2/pH ventilatory drive in fish. *Comp. Biochem. Physiol. A.* **130,** 219–240.

Glass, M. L., Ishimatsu, A., and Johansen, K. (1986). Responses of the aerial ventilation to hypoxia and hypercapnia in *Channa argus*, an air-breathing fish. *J. Comp. Physiol.* **156,** 165–174.

Gonzalez, C., Almarez, L., Obeso, A., and Rigual, R. (1994). Carotid body chemoreceptors: From natural stimuli to sensory discharge. *Physiol. Rev.* **74,** 829–898.

Graham, J. B. (1973). Terrestrial life of the amphibious fish *Mnierpes macrocephalus*. *Mar. Biol.* **23,** 83–91.

Graham, J. B. (1997). "Air-Breathing Fishes: Evolution, Diversity and Adaptation." Academic Press, San Diego and London.

Graham, J. B., and Baird, T. (1982). The transition to air breathing in fishes. I. Environmental effects on the facultative air breathing of *Ancistrus chagresi* and *Hypostomus plecostomus* (*Loricariidae*). *J. Exp. Biol.* **96**, 53–67.

Graham, J. B., Lai, N. C., Chiller, D., and Roberts, J. L. (1995). The transition to air-breathing in fishes: V. Comparative aspects of cardiorespiratory regulation in *Synbranchus marmoratus* and *Monopterus albus* (Synbranchidae). *J. Exp. Biol.* **198**, 1455–1467.

Graham, M. S., Turner, J. D., and Wood, C. M. (1990). Control of ventilation in the hypercapnic skate *Raja ocellata*: I. Blood and extradural fluid. *Respir. Physiol.* **80**(2, 3), 259–277.

Gray, J. E. (1954). Comparative study of the gill area of marine fishes. *Biol. Bull.* **107**, 219–225.

Greenwood, P. H. (1961). A revision of the genus *Dinotopterus* BLGR (Pisces, *Clariidae*) with notes on the comparative anatomy of the suprabranchial organs in the *Clariidae*. *Bull. Br. Mus.* **7**, 215–241.

Greenwood, P. H., and Liem, K. F. (1984). Aspiratory respiration in *Arapaima gigas* (*Teleostei, Osteoglossomorpha*): A reappraisal. *J. Zool., London* **203**, 411–425.

Gupta, S., and Pati, A. K. (1998). Effects of surfacing prevention and unilateral removal of air-breathing organs on opercular frequency in an Indian air-breathing catfish, *Clarias batrachus*. *Indian J. Anim. Sci.* **68**(9), 997–1000.

Harder, V., Souza, R. H. S., Severi, W., Rantin, F. T., and Bridges, C. R. (1999). The South American lungfish – adaptations to an extreme environment. *In* "Biology of Tropical Fishes" (Val, A. L., and Almeida-Val, V. M. F., Eds.), Chapter 7, pp. 87–98. INPA, Manaus.

Hedrick, M. S., and Jones, D. R. (1993). The effects of altered aquatic and aerial respiratory gas concentrations on air-breathing patterns in a primitive fish (*Amia calva*). *J. Exp. Biol.* **181**, 81–94.

Hedrick, M. S., Burleson, M. L., Jones, D. R., and Milsom, W. K. (1991). An examination of central chemosensitivity in an air-breathing fish (*Amia calva*). *J. Exp. Biol.* **155**, 165–174.

Hochachka, P. W., and Somero, G. N. (1984). "Biochemical Adaptation." Princeton University Press, Princeton, NJ.

Holeton, G. F. (1971). Respiratory and circulatory responses of rainbow trout larvae to carbon monoxide and to hypoxia. *J. Exp. Biol.* **55**(3), 683–694.

Holeton, G. F., and Randall, D. J. (1967). Changes in blood pressure in the rainbow trout during hypoxia. *J. Exp. Biol.* **46**, 297–305.

Hughes, G. M. (1984). Measurement of gill area in fishes: Practices and problems. *J. Mar. Biol. Assn. U.K.* **64**, 637–655.

Hughes, G. M., and Shelton, G. (1962). Respiratory mechanisms and their nervous control in fish. *Adv. Comp. Physiol. Biochem.* **1**, 275–364.

Hughes, G. M., Dube, S. C., and Munshi, J. S. D. (1973). Surface area of the respiratory organs of the climbing perch, *Anabas testudineus* (Pisces: *Anabantidae*). *J. Zool. London* **170**, 227–243.

Hughes, G. M., Singh, B. R., Guha, G., Dube, S. C., and Munshi, J. S. D. (1974). Respiratory surface areas of an air-breathing siluroid fish *Saccobranchus* (= *Heteropneustes*) *fossilis* in relation to body size. *J. Zool. London* **172**, 215–232.

Ishimatsu, A., Aguilar, N. M., Ogawa, K., Hishida, Y., Takeda, T., Oikawa, S., Kanda, T., and Huat, K. K. (1999). Arterial blood gas levels and cardiovascular function during varying environmental conditions in a mudskipper, *Periphthalmodon Schlosseri*. *J. Exp. Biol.* **202**, 1753–1762.

Jesse, M. J., Shrub, C., and Fishman, A. P. (1967). Lung and gill ventilation of the African lungfish. *Respir. Physiol.* **3**, 267–287.

Johansen, K. (1970). Air-breathing in fishes. *In* "Fish Physiology" (Hoar, W. S., and Randall, D. J., Eds.), Vol. IV, pp. 361–411. Academic Press, New York.

Johansen, K., and Lenfant, C. (1968). Respiration in the African lungfish, *Protopterus aethiopicus*. II Control of breathing. *J. Exp. Biol.* **49,** 453–468.

Johansen, K., and Reite, O. B. (1967). Effects of acetylcholine and biogenic amines on pulmonary smooth muscle in the African lungfish, *Protoperus aethiopicus*. *Acta Physiol. Scand.* **71,** 248–252.

Johansen, K., Lenfant, C., and Grigg, G. C. (1967). Respiratory control in the lungfish, *Neoceratodus forsteri* (Krefft). *Comp. Biochem. Physiol.* **20,** 835–854.

Johansen, K., Lenfant, C., and Schmidt-Nielsen, K. (1968). Gas exchange and control of breathing in the electric eel, *Electrophorus electricus*. *Z. Vergl. Physiol.* **61,** 137–163.

Johansen, K., Hansen, D., and Lenfant, C. (1970). Respiration in a primitive air breather, *Amia calva*. *Respir. Physiol.* **9,** 162–174.

Jonz, M. G., and Nurse, C. A. (2002). A potential role for neuroepithelial cells of the gill in O_2 sensing. *In* "Cardiorespiratory Responses to Oxygen and Carbon Dioxide in Fish." International Congress on the Biology of Fish. Extended abstract, pp. 29–33.

Junk, W. J., Soares, G. M., and Carvalho, F. M. (1983). Distribution of fish species in a lake of the Amazon river floodplain near Manaus (Lago Camaleão), with special reference to extreme oxygen conditions. *Amazioniana* **7**(4), 397–431.

Kalinin, A. L., Rantin, F. T., and Glass, M. L. (1993). Dependence on body size of respiratory function in *Hoplias malabaricus* (*Teleosti, Erythrinidae*) during graded hypoxia. *Fish Physiol. Biochem.* **12**(1), 47–51.

Kardong, K. V. (2002). "Vertebrates: Comparative Anatomy, Function, Evolution," 3rd edn. McGraw-Hill, New York.

Kent, B., and Peirce II, E. C. (1978). Cardiovascular responses to changes in blood gases in dogfish shark, *Squalus acanthias*. *Comp. Biochem. Physiol.* **60C,** 37–44.

Kotrschal, K., Peters, R., and Atema, J. (1993). Sampling and behavioral evidence for mucus detection in a unique chemosensory organ – the anterior dorsal fin in rocklings (*Ciliata mustela, Gadidae, Teleostei*). *Zoologische Jahrbucher-Abteilung fur Allgemeine zoologie und physiologie der tiere* **97**(1), 47–67.

Kotrschal, K., Royer, S., and Kinnamon, J. C. (1998). High voltage electron microscopy and 3-D reconstruction of solitary chemosensory cells in the anterior dorsal fin of the gadid fish *Ciliata mustela* (Teleostei). *J. Struct. Biol.* **124,** 59–69.

Kramer, D. L. (1977). Ventilation of the respiratory gas bladder in *Hoplyerythrinus unitaeniatus* (Pisces, *Characoidei, Erythrinidae*). *Can. J. Zool.* **56,** 931–938.

Kramer, D. L., and Graham, J. B. (1976). Synchronous air breathing, a social component of respiration in fishes. *Copeia* **1976,** 689–697.

Kramer, D. L., and McClure, M. (1982). Aquatic surface respiration, a widespread adaptation to hypoxia in tropical freshwater fishes. *Environ. Biol. Fish* **7,** 47–55.

Lahiri, S., Szidon, J. P., and Fishman, A. P. (1970). Potential respiratory and circulatory adjustments to hypoxia in the African lungfish. *Ann. Rev. Physiol.* **29,** 1141–1148.

Laurent, P., Delaney, R. G., and Fishman, A. P. (1978). The vasculature of the gills in the aquatic and aestivating lungfish (*Protopterus aethiopicus*). *J. Morph.* **156**(2), 173–208.

Liem, K. F. (1988). Form and function of lungs: The evolution of air breathing mechanisms. *Am. Zool.* **28,** 739–759.

Lomholt, J. P., Johansen, K., and Maloiy, G. M. O. (1975). Is the aestivating lungfish the first vertebrate with sectional breathing? *Nature* **257,** 787–788.

Lopes, J. M., Boijink, C. L., Kalinin, A. L., Reid, S. G., Perry, S. F., Gilmour, K. M., Milsom, W. K., and Rantin, F. T. (2002). Abstract. Cardiovascular and respiratory reflexes in the air-breathing fish, jeju (*Hoplerythrinus unitaeniatus*): O_2 Chemoresponses. *In* "Cardiorespiratory Responses to Oxygen and Carbon Dioxide in Fish." Proceedings from the International Congress on the Biology of Tropical Fishes, pp. 69–74.

Low, W. P., Ip, Y. K., and Lane, D. J. W. (1990). A comparative study of the gill morphometry in the mudskippers – *Periophthalmus chrysospilos, Boleophthalmus boddaerti* and *Periophthalmadon schlosseri. Zool. Sci.* **7,** 29–38.

Maheshwari, R., Pati, A. K., and Gupta, S. (1999). Annual variation in air-breathing behavior in two Indian siluroids, *Heteropneustes fossilis* and *Clarias batrachus. Indian J. Anim. Sci.* **69** (1), 66–72.

Maina, J. N. (2003). Functional morphology of the vertebrate respiratory systems. *In* "Biological Systems in Vertebrates" (Dutta, H. M., and Kline, D. W., Eds.), Vol. 1. Science Publishers, Enfield, NH.

Maina, J. N., and Maloiy, G. M. O. (1986). The morphology of the respiratory organs of the African air-breathing catfish (*Clarius mossambicus*): A light, electron and scanning microscopic study, with morphometric observations. *J. Zool. Lond.* **209A,** 421–445.

Martin, K. L. M. (1993). Aerial release of CO_2 and respiratory exchange ratio in intertidal fishes out of water. *Environ. Biol. Fishes* **37,** 189–196.

Martin, K. L. M. (1995). Time and tide wait for no fishes: Intertidal fishes out of water. *Environ. Biol. Fishes* **44,** 165–181.

Martin, K. L. M. (1996). An ecological gradient in air-breathing ability among marine cottid fishes. *Physiol. Zool.* **69**(5), 1096–1113.

McKean, T. A. (1969). A linear approximation of the transfer function of pulmonary mechanoreceptors of the frog. *J. Appl. Physiol.* **27,** 775–781.

McKendry, J. E., and Perry, S. F. (2001). Cardiovascular effects of hypercapnia in rainbow trout (*Oncorhynchus mykiss*): A role for externally-oriented receptors. *J. Exp. Biol.* **204,** 115–125.

McKendry, J. E., Milsom, W. K., and Perry, S. F. (2001). Branchial CO_2 receptors and cardiorespiratory adjustments during hypercarbia in Pacific spiny dogfish (*Squalus acanthias*). *J. Exp. Biol.* **204,** 1519–1527.

McKenzie, D. J., Burleson, M. L., and Randall, D. J. (1991). The effects of branchial denervation and pseudobranch ablation on cardioventilatory control in an air-breathing fish. *J. Exp. Biol.* **161,** 347–365.

McMahon, B. R. (1969). A functional analysis of the aquatic and aerial respiratory movements of an African lungfish, *Protopterus aethiopicus*, with reference to the evolution of the lung ventilation mechanism in vertebrates. *J. Exp. Biol.* **51,** 407–430.

McMahon, B. R., and Burggren, W. W. (1987). Respiratory physiology of intestinal air-breathing in the teleost fish *Misgurnus anguillicaudatus. J. Exp. Biol.* **133,** 371–393.

McPhail, J. D. (1999). A fish out of water: Observations on the ability of the black mudfish, *Neochanna diversus*, to withstand hypoxic water and drought. *NZ. J. Mar. Freshwater Res.* **33**(3), 417–424.

Milligan, C. L., and Wood, C. M. (1987). Regulation of blood oxygen transport and red cell pHi after exhaustive activity in rainbow trout (*Salmo gairdneri*) and starry flounder (*Platichthys stellatus*). *J. Exp. Biol.* **133,** 263–282.

Milsom, W. K. (1990). Mechanoreceptor modulation of endogenous respiratory rhythm in vertebrates. *Am. J. Physiol.* **259**(5, Pt. 2), R898–R910.

Milsom, W. K. (1995). The role of CO_2/pH chemoreceptors in ventilatory control. *Braz. J. Med. Biol. Res.* **11–12,** 1147–1160.

Milsom, W. K. (1996). Control of breathing in fish: Role of chemoreceptors. *In* "Physiology and Biochemistry of the Fishes of the Amazon" (Val, A. L., and Almeida-Val, V. M. F., Eds.), pp. 359–377. INPA, Manaus.

Milsom, W. K. (2002). Phylogeny of CO_2/pH chemoreception in vertebrates. *Respir. Physiol. Neurobiol.* **131**, 1–2:29–41.

Milsom, W. K., and Jones, D. R. (1977). Carbon dioxide sensitivity of pulmonary receptors in the frog. *Experientia* **33**, 1167–1168.

Milsom, W. K., and Jones, D. R. (1985). Characteristics of mechanoreceptors in the air-breathing organ of the holostean fish, *Amia calva. J. Exp. Biol.* **117**, 389–399.

Milsom, W. K., Reid, S. G., Rantin, F. T., and Sundin, L. (2002). Extrabranchial chemoreceptors involved in respiratory reflexes in the neotropical fish; *Colossoma macropomum* (The Tambaqui). *J. Exp. Biol.* **205**, 1765–1774.

Milsom, W. K., Sundin, L., Reid, S. G., Kalinin, A. L., and Rantin, F. T. (1999). Chemoreceptor control of cardiovascular reflexes. *In* "Biology of Tropical Fishes" (Val, A. L., and Almedia-Val, V. M. F., Eds.), Vol. 29, pp. 363–374. INPA, Manaus.

Moura, M. A. F. (1994). Efeito da anemia, do exercício físico e da adrenalina sobre o baço e eritrócitos de *Colossoma macropomum* (Pisces). MSC thesis, PPG INPA/FUA.

Munshi, J. S. D. (1961). The accessory respiratory organs of *Clarias batrachus* (Linn.). *J. Morphol.* **109**, 115–139.

Munshi, J. S. D. (1985). The structure, function and evolution of the accessory respiratory organs of air-breathing fishes of India. *Fortsch. Zool.* **30**, 353–366.

Munshi, J. S. D., Roy, P. K., Ghosh, T. K., and Olson, K. R. (1994). Cephalic circulation in the air-breathing snakehead fish, *Channa punctata, C. gachua,* and *C. marulius* (*Ophiocephalidae, Ophiocephaliformes*). *Anat. Rec.* **238**, 77–91.

Narahara, A., Bergman, H. L., Laurent, P., Maina, J. N., Walsh, P. J., and Wood, C. M. (1996). Respiratory physiology of the Lake Magadi Tilapia (*Oreochromis alcalicus* grahami), a fish adapted to a hot, alkaline and frequently hypoxic environment. *Physiol. Zool.* **69**(5), 1114–1136.

Nikinmaa, M. (1990). "Vertebrate Red Blood Cells. Adaptations of Function to Respiratory Requirements." Springer, Berlin, Heidelberg, and New York.

Nilsson, S. (1984). Innervation and pharmacology of the gills. *In* "Fish Physiology" (Hoar, W. S., and Randall, D. J., Eds.), Vol. XA, pp. 185–227. Academic Press, Orlando, FL.

Nilsson, S., and Sundin, L. (1998). Gill blood flow control. *Comp. Biochem. Physiol.* **119A**, 137–147.

Olowo, J. P., and Chapman, L. J. (1996). Papyrus swamps and variation in the respiratory behavior of the African fish *Barbus neumayeri. African J. Ecol.* **34**(2), 211–222.

Pack, A. I., Galante, R. J., and Fishman, A. P. (1990). Control of interbreath interval in the African lungfish. *Am. J. Physiol.* **259**(1), R139–R146.

Pack, A. I., Galante, R. J., and Fishman, A. P. (1992). Role of lung inflation in control of air breath duration in African lungfish (*Protopterus annectens*). *Am. J. Physiol.* **262**(5), R879–R884.

Packard, G. C. (1974). The evolution of air-breathing in Paleozoic gnathostome fishes. *Evolution* **28**, 320–325.

Pelster, B., Bridges, C. R., and Grieshaber, M. K. (1988). Physiological adaptations of the intertidal rockpool teleost *Blennius pholis L.,* to aerial exposure. *Respir. Physiol.* **71**, 355–374.

Perry, S. F., and Gilmour, K. M. (2002). Sensing and transfer of respiratory gases at the fish gill. *J. Exp. Zool.* **293**(3), 249–263.

Perry, S. F., and Kinkead, R. (1989). The role of catecholamines in regulating arterial oxygen content during acute hypercapnic acidosis in rainbow trout (*Salmo gairdneri*). *Resp. Physiol.* **77**, 365–377.

Perry, S. F., and Reid, S. D. (1992). Relationship between blood O_2 content and catecholamine levels during hypoxia in rainbow trout and American eel. *Am. J. Physiol.* **32**, R240–R249.

Perry, S. F., and Wood, C. M. (1989). Control and coordination of gas transfer in fishes. *Can. J. Zool.* **67,** 2961–2970.

Perry, S. F., Fritsche, R., Hoagland, T. M., Duff, D. W., and Olson, K. R. (1999). The control of blood pressure during external hypercapnia in the rainbow trout (*Oncorhynchus mykiss*). *J. Exp. Biol.* **202,** 2177–2190.

Perry, S. F., Kinkead, R., and Fritsche, R. (1992). Are circulating catecholamines involved in the control of breathing by fishes? *Rev. Fish Biol. Fisheries* **2,** 65–83.

Perry, S. F., Reid, S. G., Gilmour, K. M., *et al* (2004). A comparison in three tropical teleosts exposed to acute hypoxia. *Am. J. Physiol.* **287,** R188–R197.

Perry, S. F., Wilson, R. J. A., Straus, C., Harris, M. B., and Remmers, J. E. (2001). Which came first, the lung or the breath? *Comp. Biochem. Physiol.* **129A,** 37–47.

Peters, H. M. (1978). On the mechanism of air ventilation in anabantoids (*Pisces: Teleostei*). *Zoomorphologie* **89,** 93–123.

Pettersson, K., and Johansen, K. (1982). Hypoxic vasoconstriction and the effects of adrenaline on gas exchange efficiency in fish gills. *J. Exp. Biol.* **97,** 263–272.

Purser, G. L. (1926). *Calamoichthys calabaricus* (Smith. J. A). Part I. The alimentary and respiratory systems. *Trans. R. Soc., Edinb.* **54,** 767–784.

Rahn, H., Rahn, K. B., Howell, B. J., Gans, C., and Tenney, S. M. (1971). Air breathing of the garfish (*Lepisosteus osseus*). *Respir. Physiol.* **11,** 285–307.

Randall, D. J. (1990). Control and co-ordination of gas exchange in water breathers. *In* "Advances in Comparative and Environmental Physiology. Vertebrate Gas Exchange: From Environmental to Cell" (Boutilier, R. G., Ed.), pp. 253–278. Springer, Berlin, Heidelberg, and New York.

Randall, D. J., and Perry, S. F. (1992). Catecholamines. *In* "Fish Physiology" (Hoar, W. S., and Randall, D. J., Eds.), Vol. XIIB, pp. 255–301. Academic Press, Orlando, FL.

Randall, D. J., and Taylor, E. W. (1991). Control of breathing in fishes: Evidence for a role of catecholamines. *Rev. Fish Biol. Fisheries* **1,** 139–157.

Randall, D. J., Burggren, W. W., Farrell, A. P., and Haswell, M. S. (1981). "The Evolution of Air-Breathing Vertebrates." Cambridge University Press, Cambridge.

Rantin, F. T., and Kalinin, A. L. (1996). Cardiorespiratory function and aquatic surface respiration in *Colossoma macropomum* exposed to graded and acute hypoxia. *In* "Physiology and Biochemistry of the Fishes of the Amazon" (Val, A. L., and Almeida-Val, V. M. F., Eds.), pp. 169–180. INPA, Manaus.

Rantin, F. T., Florindo, L. H., Kalinin, A. L., Reid, S. G., and Milsom, W. K. (2002). Cardiorespiratory responses to O$_2$ in water-breathing fish. *In* "Cardiorespiratory Responses to Oxygen and Carbon Dioxide in Fish." International Congress on the Biology of Fish, pp. 35–50.

Rantin, F. T., Glass, M. L., Kalinin, A. L., Verzola, M. M., and Fernandes, M. N. (1993). Cardio-respiratory responses in two ecologically distinct erythrinids (*Hoplias malabaricus* and *Hoplias lacerdae*) exposed to graded environmental hypoxia. *Environ. Biol. Fishes* **36,** 93–97.

Rantin, F. T., Guerra, C. D. R., Kalinin, A. L., and Glass, M. L. (1998). The influence of aquatic surface respiration (ASR) on cardio-respiratory function of the serrasalmid fish *Piaractus mesopotamicus. Comp. Biochem. Physiol.* **119A,** 991–997.

Rantin, F. T., Kalinin, A. L., Glass, M. L., and Fernandez, M. N. (1992). Respiratory responses to hypoxia in relation to mode of life of two erythrinid species (*Hoplias malabaricus* and *Hoplias lacerdae*). *J. Fish Biol.* **41,** 805–812.

Reid, S. G., Bernier, N. J., and Perry, S. F. (1998). The adrenergic stress response in fish: Control of catecholamine storage and release. *Comp. Biochem. Physiol.* **120C,** 1–27.

Reid, S. G., Sundin, L., Kalinin, A. L., Rantin, F. T., and Milsom, W. K. (2000). Cardiovascular and respiratory reflexes in the tropical fish, traira (*Hoplias malabaricus*): CO_2/pH chemoresponses. *Respir. Physiol.* **120**, 47–59.

Reid, S. G., Sundin, L., Florindo, L. H., Rantin, F. T., and Milsom, W. K. (2003). The effects of afferent input on the breathing pattern *continuum* in the tambaqui. *Respir. Physiol. Neurobiol.* **36**, 39–53.

Remmers, J. E., Torgerson, C., Harris, M. B., Perry, S. F., Vasilakos, K., and Wilson, R. J. A. (2001). Evolution of central respiratory chemoreception: A new twist on an old story. *Respir. Physiol.* **129**, 211–217.

Richter, D. E. (1982). Generation and maintenance of the respiratory rhythm. *J. Exp. Biol.* **100**, 93–107.

Ristori, M. T., and Laurent, P. (1977). Action de l'hypoxie sur le système vasculaire branchial de la tête perfusée de truite. *C. R. Acad. Sci., Ser. Biologiques* **171**, 809–813.

Roberts, J. L., and Graham, J. B. (1985). Adjustments of cardiac rate to changes in respiratory gases by a bimodal breather, the Panamanian swamp eel, *Synbranchus marmoratus*. *Am. Zool.* **25**, 51A.

Rovainen, C. M. (1977). Neural control of ventilation in the lamprey. *Fed. Proc.* **36**(10), 2386–2389.

Sanchez, A. P., and Glass, M. L. (2001). Effects of environmental hypercapnia on pulmonary ventilation of the South American lungfish. *J. Fish Biol.* **58**, 1181–1189.

Sanchez, A. P., Hoffmann, A., Rantin, F. T., and Glass, M. L. (2001a). Relationship between cerebro-spinal fluid pH and pulmonary ventilation of the South American lungfish, *Lepidosiren paradoxa* (Fitz.). *J. Exp. Zool.* **290**, 421–425.

Sanchez, A., Soncini, R., Wang, T., Koldkjaer, P., Taylor, E. W., and Glass, M. L. (2001b). The differential cardio-respiratory responses to ambient hypoxia and systemic hypoxaemia in the South American lungfish, *Lepidosiren paradoxa*. *Comp. Biochem. Physiol.* **130A**, 677–687.

Satchell, G. H. (1976). The circulatory system of air-breathing fish. *In* "Respiration of Amphibious Vertebrates" (Hughes, G. M., Ed.), pp. 105–123. Academic Press, London.

Schöttle, E. (1931). Morphologie und physiologie der atmung bei wasser-, schlamm-, und landlebenden Gobiiformes. *Z. Wiss. Zool.* **140**, 1–114.

Severi, W., Rantin, F. T., and Fernandez, M. N. (1997). Respiratory gill surface of the serrasalmid fish, *Piaractus mesopotamicus*. *J. Fish Biol.* **50**, 127–136.

Singh, B. N., and Hughes, G. M. (1971). Respiration of an air-breathing catfish *Clarias batrachus* (Linn.). *J. Exp. Biol.* **55**, 421–434.

Smatresk, N. J. (1988). Control of the respiratory mode in air-breathing fishes. *Can. J. Zool.* **66**, 144–151.

Smatresk, N. J. (1990). Chemoreceptor modulation of endogenous respiratory rhythms in vertebrates. *Am. J. Physiol.* **259**(5, Pt. 2), R887–R897.

Smatresk, N. J., and Azizi, S. Q. (1987). Characteristics of lung mechanoreceptors in spotted gar, *Lepisosteus oculatus*. *Am. J. Physiol.* **252**(6), R1066–R1072.

Smith, H. W. (1931). Observations on the African lungfish, *Protopterus aethiopicus*, and on evolution from water to land environments. *Ecology* **12**, 164–181.

Smith, D. G., and Gannon, B. J. (1978). Selective control of branchial arch perfusion in an air-breathing Amazonian fish *Hoplerythrinus unitaeniatus*. *Can. J. Zool.* **56**, 959–964.

Smith, F. M., and Jones, D. R. (1978). Localization of receptors causing hypoxic bradycardia in trout (*Salmo gairdneri*). *Can. J. Zool.* **56**(6), 1260–1265.

Smith, R. S., and Kramer, D. L. (1986). The effect of apparent predation risk on the respiratory behavior of the Florida gar (*Lepisosteus platyrhincus*). *Can. J. Zool.* **64**, 2133–2136.

Steeger, H.-U., and Bridges, C. R. (1995). A method for long term measurement of respiration in intertidal fishes during simulated intertidal conditions. *J. Fish Biol.* **47**, 308–320.

Stevens, E. D., and Holeton, G. F. (1978). The partitioning of oxygen uptake from air and from water by the large obligate air-breathing teleost pirarucu (*Arapaima gigas*). *Can. J. Zool.* **56**, 974–976.

Sundin, L. (1995). Responses of the branchial circulation to hypoxia in the Atlantic cod, *Gadus morhua. Am. J. Physiol.* **268**, R771–R778.

Sundin, L., and Nilsson, G. E. (1997). Neurochemical mechanisms behind gill microcirculatory responses to hypoxia in trout: *In vivo* microscopy study. *Am. J. Physiol.* **272**, R576–R585.

Sundin, L., and Nilsson, S. (2002). Branchial innervation. *J. Exp. Zool.* **293**, 232–248.

Sundin, L., Reid, S. G., Kalinin, A. L., Rantin, F. T., and Milsom, W. K. (1999). Cardiovascular and respiratory reflexes in the tropical fish, traira (*Hoplias malabaricus*): O_2 chemoresponses. *Respir. Physiol.* **116**, 181–199.

Sundin, L., Reid, S. G., Rantin, F. T., and Milsom, W. K. (2000). Branchial receptors and cardiovascular reflexes in a neotropical fish, the tambaqui (*Colossoma macropomum*). *J. Exp. Biol.* **203**, 1225–1239.

Takeda, T., Ishimatsu, A., Oikawa, S., Kanda, T., Hishida, Y., and Khoo, K. H. (1999). Mudskipper *periophthalmodon schlosseri* can repay oxygen debts in air but not in water. *J. Exp. Zool.* **284**(3), 265–270.

Tamura, O., and Moriyama, T. (1976). On the morphological feature of the gill of amphibious and air-breathing fishes. *Bull. Fac. Fish. Nagaski Univ.* **41**, 1–8.

Taylor, E. W., Al-Ghamdi, M. S., Ihmied, I. H., Wang, T., and Abe, A. S. (2001). The neuroanatomical basis of central control of cardiorespiratory interactions in vertebrates. *Exp. Physiol.* **86**(6), 771–776.

Taylor, E. W., Jordan, D., and Coote, J. H. (1999). Central control of the cardiovascular and respiratory systems and their interactions in vertebrates. *Physiol. Rev.* **79**(3), 855–916.

Thompson, G. G., and Withers, P. C. (2002). Aerial and aquatic respiration of the Australian desert goby *Chlamydogobius eremius. Comp. Biochem. Physiol.* **131A**, 871–879.

Torgerson, C. S., Gdovin, M. J., and Remmers, J. E. (2001). Sites of respiratory rhythmogenesis during development in the tadpole. *Am. J. Physiol.* **280**(4), R913–R920.

Ultsch, G. R. (1996). Gas exchange, hypercarbia and acid-base balance, paleocology, and the evolutionary transition from water-breathing to air-breathing among vertebrates. *Palaeogeol. Palaeoclim. Palaeoecol.* **123**, 1–27.

Val, A. L., and Almeida-Val, V. M. F. (1995). "Fishes of the Amazon and Their Environment." Springer-Verlag, Berlin.

Val, A. L., and Almeida-Val, A. M. F. (1999). Effects of crude oil on respiratory aspects of some fish species of the Amazon. *In* "Biology of Tropical Fishes" (Val, A. L., and Almeida-Val, V. M. F., Eds.), Chapter 22, pp. 277–291. INPA, Manaus.

Val, A. L., Affonso, E. G., and Almeida-Val, V. M. F. (1992). Adaptive features of Amazon fishes: Blood characteristics of Curimatã (*Prochilodus cf nigricans*, Osteichthyes). *Physiol. Zool.* **65**(4), 832–843.

Val, A. L., Almeida-Val, V. M. F., and Randall, D. J. (1996). "Physiology and Biochemistry of the Fishes of the Amazon." INPA, Manaus.

Wilson, R. J. A., Harris, M. B., Remmers, J. E., and Perry, S. F. (2000). Evolution of air-breathing and central CO_2/pH respiratory chemosensitivity: New insights from an old fish. *J. Exp.Biol.* **203**, 3505–3512.

Wilson, R. J. A., Vasilakos, K., Harris, M. B., Straus, C., and Remmers, J. E. (2002). Evidence that ventilatory rhythmogenesis in the frog involves two distinct neuronal oscillators. *J. Physiol. (London)* **540**(2), 557–570.

Wood, C. M., and Shelton, G. (1980). The reflex control of heart rate and cardiac output in the rainbow trout: Interactive influences of hypoxia, hemorrhage, and systemic vasomotor tone. *J. Exp. Biol.* **87,** 271–284.

Wood, C. M., Turner, J. D., Munger, R. S., and Graham, M. S. (1990). Control of ventilation in the hypercapnic skate *Raja ocellata*: II. Cerebrospinal fluid and intracellular pH in the brain and other tissues. *Respir. Physiol.* **80**(2, 3), 279–298.

Wootton, R. J. (1990). "Ecology of Teleost Fishes." Chapman & Hall, New York.

Zaccone, G., Fasulo, S., Ainis, L., and Licata, A. (1997). Paraneurons in the gills and airways of fishes. *Micro. Res. Tech.* **37,** 4–12.

7

OXYGEN TRANSFER

COLIN J. BRAUNER
ADALBERTO L. VAL

I. Introduction
II. Oxygen and the Evolution of Air Breathing
III. Gas Exchange Organs: Diversity in Structure and Function
   A. Water-breathers
   B. Air-breathers
IV. Transport of Oxygen
   A. Whole Blood
   B. Hemoglobin
   C. Erythrocyte Function
V. Environmental Effects on Oxygen Transport
   A. Hypoxia
   B. Hyperoxia
   C. Hypercapnia
   D. Water Level
   E. Temperature
   F. Exercise
   G. Anemia
VI. Contaminant Effects on Oxygen Transport
VII. Concluding Remarks

I. INTRODUCTION

The conservative number of living vertebrate species in the world is estimated to be 50 000. Estimates for the number of fish species range from 20 000 to 30 000, almost half the total number of vertebrates (Lauder and Liem, 1983; Nelson, 1984; Val and Almeida-Val, 1995; Castro and Menezes, 1998). According to Moyle and Cech (1996), 58% of teleosts are marine, 41% are freshwater, and only 1% migrate between both habitats. Tropical fishes constitute almost 75% of the total number of fish species and, despite their dominance among fishes, far less is known about the

The Physiology of Tropical Fishes: Volume 21
FISH PHYSIOLOGY

physiology of tropical relative to temperate fishes. Tropical fishes exhibit an enormous diversity at all levels of biological organization, from morphology to behavior, and from coloration patterns to their physiological ability to acclimate/adapt to challenging environmental conditions. Clearly, tropical fishes constitute a unique group of vertebrates. This chapter will focus predominantly on what is known about oxygen transport in tropical fishes, with emphasis upon freshwater fishes of the Amazon and India, for which the greatest amount is known. When possible and as needed, reference to other tropical and neotropical fishes will be made. This chapter will discuss the basic aspects of O_2 transport, including differences between water- and air-breathing fishes, and how changes in environmental variables found within tropical systems affect O_2 transport and are compensated for by tropical fishes.

II. OXYGEN AND THE EVOLUTION OF AIR BREATHING

The content of O_2 in air is much higher than that in water for a given PO_2, the exact value of which varies with temperature. In distilled water, the ratio of air to water O_2 content is about 20:1 at $0\,°C$, 30:1 at $20\,°C$ and 38:1 at $40\,°C$ (see Dejours, 1988). Thus, to achieve a given O_2 extraction from the ventilated medium at a constant ΔPO_2 between inspired and expired media, ventilation volume of water would have to be 20- to 40-fold that of air. In fish, this value is lower due to the counter-current design of the gill, which permits a greater ΔPO_2 across the gills than occurs across the gas exchange organ in air-breathers. A further compounding factor, however, is that water has a viscosity about 60-fold that of air; thus respiring an aquatic medium is costly relative to air.

Diffusion of O_2 and CO_2 across a gas exchange organ is passive, driven only by the respective partial pressure differences. The rate of diffusion across the gills is governed by Fick's Law of Diffusion:

$$R = \frac{DxAx\Delta p}{d}$$

which relates the rate of diffusion (R) with the respective gas diffusion constant (D), area over which the diffusion occurs (A), the difference in gas partial pressure between blood and water (Δp) and distance across which diffusion occurs (d). During evolution, gas diffusion has been optimized across the gills by an increase in surface area (A), decrease in diffusion distance (d), and increase in gas concentration difference (Δp). This is well documented in both temperate and tropical fishes.

Freshwater tropical environments experience severe hypoxia on a daily and seasonal basis which, given the constraints of breathing an aquatic medium, pose a serious challenge. Fish residing in these waters must either deal with the hypoxic environment directly, by relying upon behavioral, morphological or biochemical/physiological adjustments, or indirectly, by breathing air.

Air breathing appeared early in the evolution of tropical fishes, possibly as a response to low dissolved O_2, as atmospheric O_2 was below that of present levels (see Dudley, 1998). The first group of air-breathers, the lung-fishes, appeared early in the Devonian, and extant species consist of: the South American lungfish, *Lepidosiren paradoxa*, the Australian lungfish, *Neoceratodus forsteri*, and African species belonging to genus *Protopterus*. They were followed by a myriad of fish able to breathe air, using an enormous diversity of structures for gas exchange (Table 7.1). Air breathing has been hypothesized to have evolved independently over 68 times (Graham, 1997).

Two types of tropical air-breathing fishes can be distinguished, amphibious and aquatic. Among the amphibious group are animals able to breathe air during periods they are out of water, a situation faced during dry periods and during overland locomotion in search for new water bodies; species of the genus *Hoplosternum* are included in this group. Among aquatic air-breathers, fish are defined as facultative or continuous (obligate) air-breathers and a gradient of efficiency in extracting oxygen from water is observed. Facultative air-breathers usually breathe air when water becomes

Table 7.1

Air-breathing Organs in Selected Fish Families of the Amazon. The Families are Organized from Generalized to Specialized According to Nelson (1984)

Family	Air-breathing organs				
	Lung	Swim bladder	Skin	Stomach intestine	Pharyngeal branchial and mouth diverticula
Lepidosirenidae	✓				
Arapaimidae	✓				
Erythrinidae	✓	✓			
Doradidae			✓		
Callichthyidae			✓		
Loricariidae			✓		
Rhamphychthyidae					✓
Electrophoridae				✓	
Synbranchidae				✓	

hypoxic or oxygen demand increases. This group includes species of several families: Loricariidae, Erythrinidae, Doradidae, Ramphychthidae, Synbran- chidae, Callichthyidae (Val and Almeida-Val, 1995; Brauner and Val, 1996; Graham, 1997). The continuous air-breathers, or obligate air-breathers, breathe air constantly, regardless of the oxygen content of the water, and include the lungfishes, as well as species such as *Arapaima gigas* and *Elec- trophorus electricus*.

III. GAS EXCHANGE ORGANS: DIVERSITY IN STRUCTURE AND FUNCTION

In water-breathing fishes, the gills are the primary site for gas exchange; although, in many cases a significant proportion of oxygen uptake can occur across the skin. All air-breathing fishes possess gills. However, total gill surface area may be greatly reduced and oxygen uptake is often facilitated (and in some cases dominated) by the air-breathing organ (ABO), which can take a variety of forms.

A. Water-breathers

In general, gill morphology and total gill surface area of water-breathing tropical fishes is similar to that of well-studied temperate fishes (see Fernandes, 1996; see Chapter 6). Although there is great diversity in the internal and external morphology of teleost gills, the general design consists of four gill arches, each containing two sets of filaments. Each filament contains a row of equally spaced lamellae on each side, and the tips of the lamellae from neighboring filaments of different gill arches form a fine sieve to maximize water contact with the lamellae. Blood flow through the lamel- lae is regulated by the parasympathetic and sympathetic nervous system, as well as by humoral catecholamines (Sundin, 1999), and is directed counter- current to water flow to optimize gas transfer. Filaments are supported by cartilaginous rods and are moved by adductor and abductor muscles during the respiratory cycle (see Hughes, 1984 and Laurent, 1984 for reviews).

Among water-breathing fishes, there appears to be a positive relationship between total gill surface area and aerobic metabolic demand, where more active fish species, such as tuna, have high gill surface areas (Hughes, 1972; Brill, 1996), and sluggish fish tend to have low gill surface areas (Hughes, 1966). In addition to oxygen requirements, oxygen availability may also influence gill surface area. For example, *Hoplias malabaricus*, a water- breathing fish that can reside in very hypoxic waters, has a gill surface area almost 3-fold that of a closely related species, *Hoplias lacerdae*, which is found exclusively in well-oxygenated waters (Fernandes *et al.*, 1994) and has a gill surface area similar to that of trout (see Fernandes, 1996). The increase in gill surface area in *H. malabaricus* is predominantly due to an increase in individual lamellar surface area, achieved through an increase in filament length, and thus, total number of lamellae (Fernandes *et al.*, 1994). The increased gill surface area is correlated with a relatively lower rate of gill ventilation volume over a range of inspired water PO_2's (Kalinin *et al.*, 1996) as well as with a lower critical PO_2 (Rantin *et al.*, 1992) in *H. malabaricus* relative to *H. lacerdae*. Neither of these fish is very active and both have metabolic rates that are low relative to other tropical fishes at similar temperatures (Cameron and Wood, 1978; Rantin *et al.*, 1992). Thus, the greater total gill surface area in *H. malabaricus* relative to *H. lacerdae* is most likely related to life in an environment that is more prone to hypoxia.

B. Air-breathers

Facultative air-breathers have reduced filaments but possess lamellae that are still functional for gas exchange and acid-base balance. For example, in *Hoplerythrinus unitaeniatus* the lamellar surface area is reduced relative to its closely related water-breathing species, *H. malabaricus*. Despite the reduction in surface area, the gills of *H. unitaeniatus* are still sufficient to satisfy the metabolic oxygen demand in normoxic waters. However, this appears to be associated with a higher gill ventilation and lower water oxygen extraction to maintain the same metabolic rate observed in *H. malabaricus* (Mattais *et al.*, 1996). Facultative air-breathers, however, rely on some degree of aerial respiration during exposure to hypoxia. The ABO may take on a variety of forms and be as simple as modification to the epithelium and structure of the buccal, pharyngeal, esophageal or opercular chambers, as well as modifications to the gills, skin, stomach or intestine. More complex modifications to the pneumatic duct and gas bladder are also observed (see Graham, 1997 for a review).

Obligate air-breathers, by definition, satisfy at least a part of their metabolic oxygen demand in normoxia through aerial respiration. In general, they have greatly reduced total gill surface area relative to body mass when compared to water-breathers and facultative air-breathers, and this pattern is well described (see Graham, 1997 for a review). What has received much less attention is how the gill surface area and gill morphology change during development in air-breathers. The air-breathing teleost, *Arapaima gigas*, provides an interesting model to investigate this. Shortly following hatch *A. gigas* begins to breathe air, and by the time they reach 10 g (about 1 month), they drown within about 20 minutes without access to air (Brauner and Val, personal observations). By the time they reach 0.6–1 kg (4–5

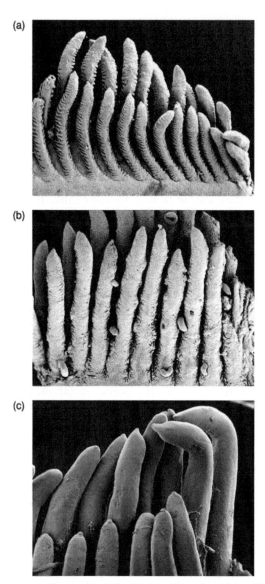

Fig. 7.1 Scanning electron micrographs of the gills from three sizes of the obligate air-breathing teleost *Arapaima gigas*; (a) 10g; (b) 100g; and (c) 1kg body mass (scale bar corresponds to 500μm). (From Brauner *et al.*, 2004a; reproduced with permission.)

months) they have become slightly more dependent upon aerial respiration and drown within about 10 minutes without access to air (Brauner and Val, personal observations), securing approximately 80% of their oxygen uptake from air (Stevens and Holeton, 1978; Brauner and Val, 1996). Over this relatively short duration, there are dramatic changes in gill morphology. In 10 g animals, the lamellae are stubby, but well formed, and mitochondria-rich cells are found at the base of the lamellae as observed in water-breathing fishes (Figures 7.1, 7.2). As *A. gigas* grow, the space between the lamellae fill

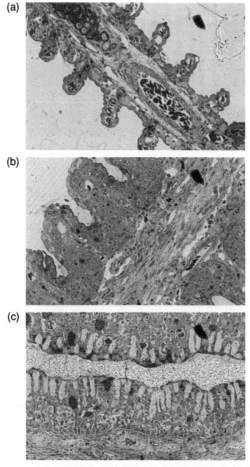

Fig. 7.2 Light micrographs of the gills from three different sizes of the obligative air-breathing teleost *Arapaima gigas*; (a) 10 g; (b) 100 g; and (c) 1 kg body mass (MR indicates mitochondria-rich cells). (From Brauner *et al.*, 2004a reproduced with permission.)

with developing cells, including mitochondria-rich cells, and the lamellae gradually disappear to the point where lamellae are not visible on the filaments by scanning electron microscopy. In association with the dramatic reduction in gill surface area (Figure 7.1), there is a large increase in the diffusion distance between water and blood (Figure 7.2). For a given distance along the filament, there is a doubling of mitochondria-rich cells (Brauner *et al.*, 2004a). Thus, it appears that during development, the gill of *A. gigas* is converted from a structure similar to a typical water-breathing gill, to a low surface area, high diffusion distance organ packed with mitochondria-rich cells. This surface is still responsible for the majority of CO_2 excretion (Randall *et al.*, 1978; Brauner and Val, 1996), but it may play a greater role in ion regulation or acid-base balance than that observed in water-breathing fishes (Brauner, Matey *et al.*, 2004).

IV. TRANSPORT OF OXYGEN

Oxygen uptake from the environment is dependent upon ventilation of the respiratory medium, perfusion of the gill or air-breathing organ, and diffusion of oxygen across the respiratory epithelium; described in a number of reviews (Randall and Daxboeck, 1984; Dejours, 1988; Graham, 1997). Once within the blood, the majority of oxygen is transported bound to hemoglobin (Hb) encapsulated within the red blood cells. The nature of O_2 transport and delivery is determined by the characteristics of Hb, in addition to the environment provided and regulated by the red blood cell.

A. Whole Blood

1. BLOOD–OXYGEN AFFINITY

The most commonly used parameter to describe O_2 transport characteristics of whole blood is P_{50}, which refers to the PO_2 at which blood is 50% saturated with oxygen. The lower the P_{50}, the greater the affinity of blood for O_2, and the more effectively O_2 can be removed from the water. However, a higher P_{50} elevates the PO_2 at which oxygen is offloaded to the tissues, which is beneficial to O_2 delivery. The P_{50} of whole blood *in vivo* presumably represents a compromise between these conflicting pressures. The P_{50} value in whole blood is a function of many interacting variables including concentration of Hb within the red cell, intrinsic affinity of the Hb for O_2, sensitivity of the Hb to ligands (particularly Cl^-, ATP and GTP), and their relative concentrations within the red cell, and temperature among others (see Nikinmaa, 1990 for a review). Furthermore, the P_{50} values can be altered during exposure to different environmental conditions (as discussed below). Despite the great number of interacting factors and plasticity associated with environmental acclimation/adaptation, P_{50} values in whole blood of fishes caught in the wild are a valuable index of O_2 transport characteristics of whole blood, particularly with regard to hypoxia tolerance.

Among tropical fishes, there is a great diversity in whole blood P_{50} values. Johansen and Lenfant (1972) were the first to propose that air-breathing fishes exhibit higher P_{50}'s than water-breathing fishes, presumably to facilitate oxygen delivery to the tissues given the reduced limitations to oxygen uptake associated with aerial respiration. However, in a more exhaustive investigation of whole blood affinity among 40 Genera of Amazonian fishes (Powers et al., 1979), this relationship was not supported based upon mode of respiration, and the best predictor of blood P_{50} was related to the rate of water flow in which the fish resided. They found that fish residing in "rapid" flowing waters, which tend to be more oxygenated, had blood P_{50} values approximately 50% greater than those from "slow" flowing waters that tend to have more variable oxygen levels, and are often hypoxic. This survey, however, represents a tremendous phylogenetic diversity of species and when closely related water- and air-breathing species are investigated (i.e. *H. malabaricus* and *H. unitaeniatus* or *A. gigas* and *Osteoglossum bicirrhosum*; Johansen et al., 1978a, b) air-breathers possess lower affinity blood than water-breathers. The generality of the relationship between P_{50} and mode of respiration remains controversial (Graham, 1997), but there appears to be a relationship between whole blood P_{50} and environmental water flow rate, and more likely, water oxygen levels.

2. HEMATOCRIT

In addition to blood–oxygen affinity, a major factor influencing O_2 delivery is the total hemoglobin content of blood. While an increase in temperature can result in an increase in hematocrit (Hct) associated within temperature acclimation within a species, Hct and blood Hb values taken from 25 species of Amazonian fishes, permitted to recover following capture in the wild, do not appear to be higher than those of temperate fishes. Hematocrit values in Amazonian fishes ranged from 21 to 35.5%, with blood Hb concentrations of 0.82 to 1.65 mM, respectively. In general, the more active the fish, the higher the Hct and Hb levels (Marcon et al., 1999). There also appears to be a higher Hct and total blood Hb in air-breathing relative to water-breathing fishes (Val and Almeida-Val, 1995).

3. BOHR EFFECT

The Bohr effect describes the change in Hb–O_2 affinity associated with a change in pH of the blood (Riggs, 1988; Jensen et al., 1998). The Bohr effect is generally thought to be important for enhancing oxygen delivery to the tissues, as CO_2 diffuses from the tissues into the blood, which results in a reduction in blood pH. The greater the magnitude of the Bohr effect, the greater the potential for oxygen delivery to the tissues for a given reduction in blood pH. Protons produced upon CO_2 dehydration, are bound to Hb as O_2 is off-loaded to the tissues due to the Haldane effect (Christiansen *et al.*, 1914), which is the reciprocal of the Bohr effect (Wyman, 1973). Thus, an optimal value exists for the Bohr coefficient ($\Delta\log P_{50}/\Delta pH$) to maximize O_2 delivery to the tissues under steady state conditions in the face of proton binding associated with the Haldane effect. This optimal value has been calculated to be between -0.35 and -0.5 (Lapennas, 1983).

In the most comprehensive analysis of Bohr coefficients measured in whole blood of Amazonian fishes, including 34 species, spanning 32 genera and 18 Families, there is a rather broad distribution of Bohr coefficients ranging from -0.1 to -0.79 (Figure 7.3). However, when a mean Bohr coefficient is calculated for each of these groupings they are: species, -0.38 $\pm$ 0.03, genus, -0.39 ± 0.3 and Family, -0.42 ± 0.04. When an average Bohr coefficient is calculated for the five Orders present in this group, the mean value is -0.39 ± 0.06. Thus, despite the great range of Bohr coefficients observed among Amazonian fishes, the mean value, regardless of the level of phylogenetic grouping, falls within the optimal range for maximizing oxygen delivery that was calculated by Lapennas (1983). These data indicate that the Bohr coefficient appears to have been optimized for O_2 delivery in

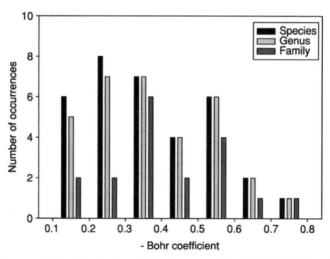

Fig. 7.3 The magnitude of the Bohr coefficients measured in whole blood from 34 species of Amazonian fishes, and the number of occurrences of these values based upon grouping of observations by species, genus or Family. (Data from Powers *et al.*, 1979.)

these species. Because many temperate teleost fishes have Bohr coefficients that greatly exceed -0.35 to -0.5, it has been concluded that those teleost fish Hbs may be more optimized for CO_2 than O_2 transport (Jensen, 1989, 1991). Whether there truly is a difference in this regard between temperate and tropical fishes remains to be determined. Given the prevalence of hypoxia in tropical systems, optimization of the Bohr coefficient for oxygen delivery is not unreasonable.

4. ROOT EFFECT

The Root effect is defined as a reduction in oxygen-carrying capacity of the blood, at atmospheric O_2 levels, when blood pH is reduced (Root, 1931). The Root effect is only found in teleost fishes (with the exception of *Amia calva*, Weber *et al.*, 1976) and at the level of Hb, the Root effect is thought to be an exaggerated Bohr effect (Brittain, 1987). The complete molecular basis for the Root effect remains unresolved despite great effort (Mylvaganam *et al.*, 1996; Fago *et al.*, 1997). Physiologically, however, the Root effect has very different implications for gas transport than the Bohr effect. The Root effect in conjunction with a rete (a structure capable of creating a large localized acidosis within capillaries), drives O_2 from Hb to the eye or swimbladder (Pelster and Weber, 1991). Thus, the Hb characteristics and rete system in teleost fish form the basis of an O_2 multiplication system that is unparalleled in the animal kingdom and capable of generating O_2 tensions over 20 times that found in arterial blood (Fairbanks *et al.*, 1969). This system permits fish to regulate swimbladder volume, and thus maintain neutral buoyancy as fish sojourn to different depths, and is one of the predominant traits responsible for the explosive radiation of teleosts that exist today (Moyle and Cech, 1996).

It was research on 56 genera of Amazonian fishes, however, that first indicated that the Root effect is best correlated with the presence of the choroid rete rather than the swimbladder (Farmer *et al.*, 1979). The largest Root effects, however, are observed in those species possessing both retia (Val and Almeida-Val, 1995). There does not appear to be a relationship between the presence of the Root effect and number of Hb fractions, level of activity, hypoxia tolerance, trophic levels or habitat preference (Val and Almeida-Val, 1995). In general, the Root effect is absent or very small in the Gymotoidei and the Siluroidei, as well as in most air-breathing fishes (Farmer *et al.*, 1979). In stripped hemolyzates, however, a pronounced Root effect has been demonstrated in at least two air-breathers, *A. gigas* and *H. unitaenaitus*. Because air-breathing fishes typically have higher blood PCO_2 and lower blood pH than water-breathers, it was proposed as early as 1931 that air-breathing fishes would possess Hbs relatively insensitive to pH (Carter, 1931), and thus would not be expected to possess a Root effect.

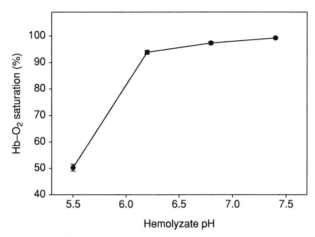

Fig. 7.4 The magnitude of the Root effect in hemolyzates from *Arapaima gigas*. Hb–O_2 saturation at each pH was measured spectrophotometrically according to Farmer *et al.* (1979). ($n = 6$, Brauner and Val, unpublished data.)

For the most part, this is the case. For *A. gigas* and *H. unitaenaitus*, the presence of the Root effect was confirmed in hemolyzates at pH = 5.5 (Farmer *et al.*, 1979). In another analysis, however, it is apparent that the Root effect in hemolyzates of *A. gigas*, is not observed until pH is reduced below pH 6.2 (Figure 7.4) which is unlikely to occur *in vivo* (with the exception of in a structure like a rete) where resting intracellular red cell pH is 7.22 ± 0.02 (Brauner and Val, unpublished data). Thus, it may not be so surprising that air-breathing fish possess a Root effect, provided it does not impair O_2 uptake at the gills or ABO *in vivo*. It should be noted that the presence of the Root effect in these air-breathers was confirmed in the absence of organic phosphates which tend to potentiate the Root effect and increase the pH of onset (Pelster and Weber, 1990).

5. NON-LINEAR BOHR EFFECT

In several species of teleost fishes, the magnitude of the Bohr effect is non-linearly distributed over the oxygen equilibrium curve (Jensen, 1986; Brauner *et al.*, 1996; Lowe *et al.*, 1998), where almost the entire effect exists between 50 and 100% Hb–O_2 saturation. The non-linear Bohr effect has large implications for O_2 and CO_2 transport, as well as for acid-base regulation, depending upon the region of the OEC being used for gas exchange (Brauner and Randall, 1998). The non-linear Bohr effect has been proposed to be a general feature of teleost fish Hbs that possess a Root effect (Brauner and Jensen, 1999) and has been found to exist in the blood of tambaqui, *C. macropomum* (Brauner *et al.*, 2001), and tilapia from Lake Magadi

(Narahara *et al.*, 1996), the only tropical fishes investigated to date. Given the tremendous diversity in Hb characteristics observed in tropical fishes, this group provides an important opportunity to gain further insight into the ubiquity, functional significance, and molecular basis for the non-linear Bohr effect in fish (Brauner and Jensen, 1999).

B. Hemoglobin

The majority of tropical fishes possess multiple Hbs, similar to that observed in temperate fishes (Figure 7.5). In a survey of 77 genera of teleost fishes, only 8% possessed a single component. Among the superorders Ostariophysi (including Characoidei, Gymnotoidei and Siluroidei) and Acanthopterygii, the average number of Hb isoforms was 3.3 ± 0.15 and 6.7 ± 0.38, respectively (Fyhn *et al.*, 1979). There were no obvious correlations between fish behavior or habitat preference and the number of Hb isoforms. The reason that so many fish possess multiple Hbs is at least partly related to gene duplication and polyploidy that is widespread among fishes; however, it remains controversial whether Hb multiplicity is adaptive.

Different isoforms do have different characteristics. Those with a net negative charge (anodic components) tend to have a low oxygen affinity, large Bohr/Haldane effects and often a Root effect. Those with a net positive

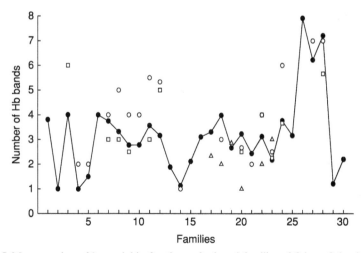

Fig. 7.5 Mean number of hemoglobin fractions of selected families of fishes of the Amazon basin. The families are organized along the abscissa from the generalized to the specialized. Observe that multiple Hb occur in almost all fish families. Different symbols refer to fish species analyzed by different authors and using different hemoglobin separation methods. (This figure appeared in *Fish of the Amazon and their Environment*, by A. L. Val and V. M. F. Almeida-Val and is reproduced here with kind permission of Springer Science and Business Media.)

charge (cathodic components), tend to exhibit high oxygen affinity and small, non-existent Bohr/Haldane effects (Weber and Jensen, 1988; Weber, 1990). It has been proposed that these differences in Hb properties in fish that possess multiple Hbs permit a division of labour with respect to oxygen transport, where the cathodic component would still bind O_2 in the face of a severe acidosis (Weber, 1990). This strategy would only require two isoforms while some fish, such as *Hassar* sp. of the Amazon, exhibit up to 12 (Fyhn *et al.*, 1979)! There is evidence for switching of different isoforms according to season, stage of development, and dissolved oxygen levels (see Val and Almeida-Val, 1995) and it may be that the high levels of Hb heterogeneity observed in Amazonian fishes is related to environmental variability (Fyhn *et al.*, 1979; Riggs, 1979).

Another hypothesis is that Hb multiplicity may reduce the chances of Hb crystallization within the red blood cell. Hemoglobin exists at the limit of its solubility in the red blood cell and deoxy-Hb is less soluble than the oxy form. As tropical fishes routinely experience pronounced hypoxia or even anoxia, deoxygenated red blood cells may run the risk of Hb crystallization. According to the phase rule, saturated protein solutions of different components contain more protein than those with a single component. The presence of multiple Hb isoforms may reduce the chances of Hb crystallization without affecting red blood cell Hb concentration, or even increasing total Hb concentration. Whether a relationship exists between red blood cell Hb concentration and the number of Hb isoforms remains to be investigated.

C. Erythrocyte Function

Encapsulation of Hb within erythrocytes allows for fine tuning of Hb function through regulation of the intracellular environment. The two factors that have the greatest influence on many fish Hbs are organic phosphate levels and pH.

1. ORGANIC PHOSPHATE LEVELS

The majority of fish Hbs are sensitive to organic phosphates (Weber and Jensen, 1988) which act as negative allosteric modifiers of Hb–O_2 affinity. In general, organic phosphates bind to specific amino acid residues at the central cavity between the two β-chains, stabilizing the T-state (Nikinmaa, 1990; Jensen *et al.*, 1998). Thus, whole blood P_{50} is determined in part by the type and/or absolute concentration of organic phosphate present within the red cell. While the predominant organic phosphates tend to be ATP and GTP (both referred to as NTP) in a great number and diversity of fish species (see Val, 2000 for a review), ATP often occurs in greater abundance than GTP within fish red cells. Because GTP is able to form one more hydrogen bond with the globin chain residues than ATP, it is often a more potent allosteric effector than ATP (Val, 1995; Marcon *et al.*, 1999). This is not always the case and in some tropical fishes the effect of ATP is equal to or greater than that of GTP, the basis for which is not understood (see Val, 2000). The Hbs of some tropical fishes also respond to other phosphates such as 2,3 DPG, as observed in the armored catfishes *Pterygoplichthys* spp. (Isaacks *et al.*, 1978) and *Hoplosternum littorale* (Affonso, 1990; Val, 1993). In very rare instances, IPP (inositol pentaphospate, see Val, 2000 for a review) has been found in red blood cells. Inositol pentaphosphate is found widely in birds and reptiles, however, to date it has only been found in three species of fish, *A. gigas* (Isaacks *et al.*, 1977), and two species of elasmobranchs (Borgese and Nagel, 1978). Another ester, inositol diphosphate (IP2), has been found in two species of air-breathing fishes (*Liposarcus pardalis* and *Protopterus aethiopicus*, Bartlett, 1978; Isaaks *et al.*, 1978), however, the effect of IP2 on Hb–O_2 affinity has not yet been characterized.

In addition to interspecific differences in the concentration and type of organic phosphates found in fish erythrocytes, there are instances where ontogenetic differences are observed. One of the most interesting examples is exhibited by *A. gigas*, where large differences in both the type and concentration of organic phosphate differ with development. As fry, *A. gigas* are not yet air-breathers and possess high levels of ATP and GTP. As they grow and become more dependent upon aerial respiration, erythrocyte IPP levels increase, and ATP and GTP levels decrease (Val *et al.*, 1992). Because both ATP and GTP are less potent allosteric effectors than IPP, this may ensure that whole blood O_2 affinity is high when fish are water-breathers, but lower when the fish are extracting the majority of O_2 from the air.

2. RED CELL pH REGULATION AND THE β-ADRENERGIC RESPONSE

The Hb of many teleosts exhibit a Root effect (Farmer *et al.*, 1979; Ingerman and Terwilliger, 1982; Brittain, 1987; Cossins and Kilbey, 1991; Pelster and Weber, 1991; Val *et al.*, 1998), where O_2 carrying capacity of the blood is reduced as intracellular pH of the red cell decreases (see Nikinmaa, 1990). During exposure to hypoxia or following exhaustive exercise, fish may experience a generalized acidosis which could dramatically reduce O_2-carrying capacity of the blood via the Root effect. To secure O_2 uptake under these conditions, many teleost red blood cells possess an adrenergic mechanism through which red cell pH can be regulated in the face of an extracellular acidosis (Nikinmaa, 1990). Catecholamines, specifically adrenaline and noradrenaline, are released into the blood when arterial blood saturation falls below 50% saturation (Perry and Reid, 1992), and they stimulate the β-adrenergic receptors of the red cell. This activates red cell Na^+/H^+ exchange, which initiates a cascade of events ultimately leading to an increase in RBC volume, [Na$^+$] and pH, and a reduction in NTP levels (see Nikinmaa, 1990 for a review). While the mechanism through which these processes occur is well described in temperate fish, whether tropical fish release catecholamines, and the ubiquity of the red cell adrenergic response among tropical fishes is not well studied.

In the only study where catecholamine release during exposure to environmental hypoxia has been measured in neotropical fish, two of three species (jeju *H. unitaeniatus*, and traira, *H. malabaricus*) released catecholamines at arterial PO_2 levels corresponding to approximately 50% Hb–O_2 saturation. A third species, pacu (*Piaractus mesopotamicus*) did not release appreciable levels of catecholamines (Reid *et al.*, 2002; Perry *et al.*, 2004). In all cases, catecholamine release was far less than that observed in the hypoxia-intolerant rainbow trout.

A recent study on Amazonian fishes indicates that the RBC adrenergic response *in vitro* is present in two of four species of Characiformes. The response was present in tambaqui (*Colossoma macropomum*) and jaraqui (*Semaprochilodus insignis*) but absent in the black piranha (*Serrasalmus rhombeus*) and the aracu (*Leporinus fasciatus*). In two species of siluriformes (piranambu (*Pinirampus pirinampu*) and the acari-bodo (*Pterygoplichthys multiradiatus*)) the response was also absent (Val *et al.*, 1998). In another study (Brauner *et al.*, 2000), the presence of the RBC adrenergic response in tambaqui was confirmed, and interestingly, the response was present in pacu that do not appear to release catecholamines during exposure to hypoxia (Reid *et al.*, 2002). Adrenergic activation of red cell Na$^+$/H$^+$ exchange was not apparent in the red bellied piranha (consistent with that observed in the black piranha, Val *et al.*, 1998) or in the osteoglossids (*O. biccirhosum* and *A. gigas*). The only fish species that exhibited a reduction in red cell NTP levels associated with adrenergic stimulation was pacu, however, the degree of change was modest (<10%), compared with that observed in salmonids. The lack of adrenergically mediated changes in red cell NTP levels under a similar experimental protocol have also been reported for other Amazonian fishes, including characins and catfishes (Val *et al.*, 1998).

V. ENVIRONMENTAL EFFECTS ON OXYGEN TRANSPORT

Tropical climates by definition exist near the Equator and are generally characterized as warm and thermostable with a constant photoperiod (see Chapter 1, this volume). Some tropical climates have considerable annual rainfall, resulting in extensive rainforests such as those found in the Amazon region of South America, the Congo region of Central Africa, and the Indonesian islands of Asia. Most of these rainforests experience some degree of annual fluctuation in precipitation which can result in seasons that are defined by rainfall (i.e. flood *vs* dry season) rather than photoperiod and/or temperature. This large annual variation in precipitation imposes dramatic effects upon the aquatic and, to a lesser extent, marine environments in which a tremendous diversity of fishes can be found.

Within the tropical freshwater systems, the dry season can result in a reduction of river water level by 10 m or more (Junk *et al.*, 1989), greatly curtailing the flow in rivers and isolating large and small bodies of water. Due to the extensive algal and plant growth under these conditions, large daily oscillations in water O_2 and CO_2 tension are routinely observed. Environmental changes in O_2 (hypoxia, hypercapnia), CO_2, temperature, water level, as well as in conditions that result in exercise or anemia, all affect O_2 transport, that in some cases can be compensated for in tropical fishes as described below.

A. Hypoxia

One of the characteristic features of tropical waters, especially in the Amazon, is pronounced bouts of hypoxia, to which fishes have adapted the ability to maintain metabolism as aerobic as possible under the conditions. Adaptations that are seen in tropical fishes cover a broad range including behavioral, morphological, physiological, and biochemical. The first line of defense during exposure to hypoxia is behavioral, more specifically, to avoid hypoxia if possible. When hypoxia cannot be avoided, morphological, cardiorespiratory, and hematological adjustments are routinely observed.

During exposure to hypoxia, there are at least two behavioral changes among tropical fish: a change in position within the water column and lateral migration. Changes in position within the water column have been reported for several fish species and have been described as adaptive convergence, as they occur in distantly related fish species. When exposed to hypoxia, fishes such as the serrasalmids (*Colossoma* and *Mylossoma*), the bryconins (*Brycon* and *Hyphesssobrycon*), some cichlids (*Astronotus* and *Herus*), and the freshwater stingray (*Potamotrygon*) all move to the upper region of the water column, close to the air–water interface, where more dissolved oxygen is available. While this behavioral response aids in securing O_2 uptake during aquatic hypoxia, it also increases susceptibility to predation (Graham, 1997). In some species, movement to the air–water interface is accompanied by color changes to reduce predation risks. Control of this color change is unknown, but is presumably under similar control as other physiological and biochemical adjustments to hypoxia (Hochachka, 1996; Hochachka *et al.*, 1997; Ratcliffe *et al.*, 1998).

In many tropical water bodies, sunset is accompanied by a decrease in dissolved oxygen due to a reduction in plant/algal photosynthesis and an increase in respiration (Junk et al., 1983; Val et al., 1999). Water oxygen levels can be completely depleted within 2 hours of sunset (Val and Almeida-Val, 1995; Val and Antunes de Moura, unpublished data). Not surprisingly, dusk serves as a cue for some fish species to migrate back to main river channels, or to open water, where dissolved oxygen is more stable. This movement is known as lateral migration and has been described for several fish species in the Amazon (Goulding, 1980; Lowe McConnell, 1987), the Pantanal (Antunes de Moura, 2000), and in Africa (Bénech and Quensière, 1982).

If hypoxia cannot be avoided, one of the most noticeable adaptations other than air breathing, is seen in *C. macropomum*, where the lower lip of the fish becomes greatly enlarged, and acts as a broad surface used to facilitate skimming of the water at the air–water interface. This mechanism can increase arterial blood content by about 30%, at a water PO_2 of 35 mmHg, relative to those fish denied access to the surface (Val, 1995). Although not as noticeable as in *Colossoma*, surface skimming using expanded lips has been observed in several other fish groups of the Amazon, such as *Mylossoma*, *Brycon*, and *Triportheus*, and this is also thought to be an example of adaptive convergence.

In fish incapable of this rather unique morphological adaptation, hypoxia generally results in a reduction in spontaneous activity and metabolic oxygen consumption (Brauner et al., 1995; Almeida-Val et al., 2000), an increase in gill ventilation rate (Rantin et al., 1992) and bradycardia (Rantin et al., 1995), as observed in temperate fishes (Randall, 1982).

In addition to cardiorespiratory adjustments to hypoxia, an increase in Hct, associated with adrenergically mediated red cell swelling and release from the spleen (Moura, 1994), and an increase in blood–oxygen affinity, due to reduction in red cell NTP:Hb ratio, is often observed in water-breathing fishes (see Val and Almeida-Val, 1995, and Val, 2000 for a review). Interestingly, the NTP:Hb ratio for 25 species of fresh-water Amazonian fishes is about half that found in marine fishes, which may be associated with greater levels of hypoxia routinely encountered by Amazonian fishes (Marcon et al., 1999). In air-breathing fishes, the magnitude of the increase in blood–oxygen affinity appears to be associated with the relative importance of air breathing to securing O_2 transport. In the lungfish, there is a 14% increase in affinity associated with moderate aquatic hypoxia (PO_2 of 60 mmHg) despite an increase in air-breathing frequency (Kind et al., 2002). A further reduction in water oxygen levels to 40 mmHg did not elicit additional effects on blood–oxygen affinity. In the facultative air-breathing catfish *Hypostomus* sp., exposure to extreme hypoxia (PO_2 of 20–25 mmHg) results in a 30% increase in blood–oxygen affinity, due to reductions in intracellular GTP levels.

B. Hyperoxia

Hyperoxia is a common occurrence during the day in natural aquatic systems dense in vegetation and algae, due to the high levels of photosynthesis. Dense vegetation is characteristic of many tropical environments, and water oxygen levels can reach 250% of atmospheric levels (Kramer *et al.*, 1978). Except for the swimbladder and retina of fish that routinely experience extremely high PO_2's, hyperoxia is deleterious to tissues (Pelster, 2001). Hyperoxia can induce oxidative cell damage in the gills within 6 hours of exposure to 200% of atmospheric oxygen levels (Liepelt *et al.*, 1995), which could affect O_2 transport. Whether fish that live in waters that routinely become hyperoxic have an elevated capacity to deal with oxidative cell damage is not known.

Changes in erythocyte phosphate levels in tropical fish exposed to hyperoxia yield conflicting data. In the cichlid, *Astronotus ocellatus*, exposure to a PO_2 of 360 mmHg for 15 days had no effect upon erythrocyte NTP levels, while under similar conditions, a reduction in NTP levels has been observed for *Colossoma* (Marcon and Val, 1996). A decrease in erythrocyte ATP levels has also been described for temperate fish *Pleuronectes platessa* exposed to a PO_2 of 300 mmHg (Wood *et al.*, 1975). In the tropical piranha, *Serrasalmus rhombeus*, exposed to hyperoxia (300 mmHg) for 6 hours, a 2-fold increase in GTP levels have been observed despite no changes in ATP levels (A. L. Val, unpublished data). While an increase in erythrocyte GTP may reduce blood affinity and thus minimize tissue damage during hyperoxia, changes in erythrocyte NTP levels during hyperoxia among studies are very variable and it is difficult to generalize how NTP levels are regulated during exposure to hyperoxia.

C. Hypercapnia

Environmental hypercapnia commonly occurs in tropical freshwater systems, particularly in areas covered with dense mats of vegetation, where CO_2 tensions can rise to as high as 60 mmHg (Heisler, 1982). In addition to the direct effects of CO_2 on ventilation rate (see Chapter 6), short-term exposure to hypercapnia influences O_2 transport at the level of Hb. In many vertebrates, CO_2 binds to the terminal amine groups of Hb, stabilizing the T-state, and reducing Hb–O_2 affinity. In fish, the terminal amino groups of the α-subunits are acetylated, and unavailable to bind CO_2 (Farmer *et al.*, 1979; Riggs, 1979), and β-subunits preferentially bind organic phosphates over CO_2 (Gillen and Riggs, 1973; Weber and Lykke-boe, 1978). Consequently, there is little direct effect of CO_2 on $Hb-O_2$ transport. Through the effects of pH, however, hypercapnia reduces $Hb-O_2$ affinity via the Bohr effect, and O_2-carrying capacity of the blood via the Root effect. The degree to which O_2 transport is affected by hyper-capnia depends upon the magnitude of the acidosis and the magnitude of the Bohr and Root effects. As discussed above, many teleost fishes have the ability to regulate red cell pH_i in the face of an extracellular acidosis associated with elevations in circulating catecholamines. In trout, catecho-lamines are released during exposure to hypercapnia which at least partially protects $Hb-O_2$ transport under these conditions (Perry et al., 1989). Only one study has investigated whether catecholamines are released in tropical fishes subjected to hypoxia (Reid et al., 2002), and no similar studies have been conducted during exposure to hypercapnia. Thus, the effect of a short-term hypercapnic acidosis on O_2 transport in tropical fish in vivo, remains to be investigated.

The acidosis associated with longer-term hypercapnia is compensated within 24–72 hours by elevation in plasma HCO_3^- levels associated with Cl^-/HCO_3^- exchange or Na^+/H^+ exchange predominantly at the gills (Heisler, 1993; Larsen and Jensen, 1997). Once the acidosis has been com-pensated, there will be relatively minor effects on O_2 transport. Acid-base regulation during hypercapnia is greatly influenced by water ionic composi-tion, where water Cl^-, HCO_3^-, and Ca^{2+} all increase the rate and degree of pH compensation in rainbow trout (Larsen and Jensen, 1997). Many tropical freshwater systems, such as the Amazon, are characteristically dilute in ions, with water total conductivity in the Rio Negro as low as $9\,\mu S/cm$, and water Cl^-, Na^+, and Ca^{2+} of 50, 17, and $5\,\mu M$, respectively (see Val and Almeida-Val, 1995 for a review). Consequently, acid-base regulation during hypercapnia in these ion-poor waters may be impeded. In the air-breathing catfish *Liposarcus pardalis*, this certainly appears to be the case. Acid-base regulation during exposure to water hypercapnia of 7–40 mmHg, resulted in a very limited ability to acid-base regulate, with blood pH falling from 7.90 to 6.99, with little change over the following 4 days (Brauner et al., 2004b).

Liposarcus pardalis lacks the ability to regulate red cell pH adrenergically (Brauner et al., 2004b), however, its Hb is relatively insensitive to changes in pH (Bossa et al., 1982), preventing an impairment to O_2 transport. If the ability to acid-base regulate during hypercapnia is limited in all fishes living in dilute waters, the effects of hypercapnia on O_2 transport may be greater than in fishes living in water with higher ionic contents; this remains to be investigated.

D. Water Level

In regions like the Amazon, one of the largest seasonal changes is the result of an annual flood cycle. Annual changes in river water level can result in a difference of up to 10 m (Junk, 1979) and is thought to be one of the most important environmental cues for organisms living in that system (Junk et al., 1989).

In several fish species (i.e. *C. macropomum, Mylossoma duriventris* and *Pterygoplichthys multiradiatus*), the relative concentrations of the three-five Hb fractions differ between low and high water levels (Val, 1986; Val and Almeida-Val, 1995). Changes in Hct and total blood Hb levels also correlate with the water levels (Val and Almeida-Val, 1995). In *Mylossoma duriventris* (collected from the *várzea* lakes of Marchantaria Island, Amazon River) erythrocyte ATP and GTP levels were highest during the low water season when dissolved oxygen was high. This resulted in differences in whole blood oxygen affinity between the flood and dry season (Monteiro et al., 1987).

During the low water season, many fish are trapped in receding water bodies along the *várzea* of the Amazon that is deeply hypoxic and warm (Junk et al., 1983; Val and Almeida-Val, 1995; Val et al., 1999). Animals exposed to such stressful environmental situations, may display hematocrit values significantly higher than those observed for recovered animals of the same species or collected from nearby flowing rivers.

In many cases, the change in a given physiological/biological parameter is indirectly associated with water level, and directly associated with changes in other specific environmental variables such as dissolved oxygen, temperature, and hydrogen sulfide. The greatest direct effect of a change in water level is hydrostatic pressure. While marine fishes and mammals do exhibit anatomical, biochemical, and physiological adaptations to withstand high hydrostatic pressure (up to 200 atmospheres) (Castellini et al., 2002), there do not seem to be any direct effects of small changes in hydrostatic pressure in shallow freshwater tropical environments.

E. Temperature

Increasing temperature lowers $Hb-O_2$ affinity (Nikinmaa, 1990), which must be compensated for when fish simultaneously face a decrease in water oxygen solubility and an increase in oxygen demand. In *C. macropomum*, $Hb-O_2$ affinity at pH 7.1 is reduced 3-fold as the temperature increases from 19 to 29 °C (Val, 1986). Above 30 °C, *C. macropomum* reduces oxygen consumption, reflecting an overall decrease in metabolic rate (Saint-Paul, 1983). An increase in acclimation temperature (20, 25, and 30 °C) in the facultative air-breather *Hypostomus regani* results in a progressive increase in Hct from 21 to 26% with no change in red cell volume (Fernandes et al., 1999). These data indicate that the elevation in temperature leads to an increase in erythropoiesis (or at least splenic release of red blood cells), as has been observed in rainbow trout (Tun and Houston, 1986) and goldfish (Houston and Murad, 1992). An increase in oxygen-carrying capacity of the blood during exposure to elevated temperature is presumably beneficial for ensuring adequate oxygen delivery at a time when metabolic rate becomes elevated, and it has been proposed that changes in hemoglobin content of the blood may be more important than changes in P_{50} in securing oxygen transport under a number of environmental conditions (Brauner and Wang, 1997).

As the temperature increases, *H. littorale* increases air-breathing frequency and this is accompanied with an increase in red blood cell 2,3 DPG levels, which may favour oxygen unloading to tissues. No change in ATP and GTP occurs under these conditions (Val and Almeida-Val, 1995).

F. Exercise

Swimming activity, to feed, migrate, escape predators, or spawn, greatly increases oxygen demand (Brett, 1972). Fish exposed to these conditions adjust several blood parameters to increase oxygen transfer to tissues. Migrating prochilodontids *Semaprochilodus insignis* and *Semaparochilodus taeniurus* (known in the Amazon as jaraqui), exhibit decreased levels of ATP and GTP in the red blood cells, and increased circulating red blood cells and hematocrit, compared to non-migrating animals (Val et al., 1986). In contrast, short-term burst swimming in the ornamental fish species *Pterophylum scalare* and *Symphysodum aequifasciata* resulted in increased Hct, but no changes in NTP levels (Val et al., 1994b). These data suggest that adjustments induced by short-term burst activity occur slowly compared to those observed in animals acutely exposed to hypoxia.

G. Anemia

Anemia, like environmental hypoxia, results in internal hypoxia but due a reduction in blood oxygen-carrying capacity. Anemia elicits different physiological responses in fish than does environmental hypoxia, as the animal must improve oxygen unloading to the tissues rather than securing oxygen uptake at the gills (Val et al., 1994a; Brauner and Wang, 1977). In the fish species analyzed to date, a significant increase in NTP:Hb ratio is observed during progressive anemia for both temperate and tropical fish species (Val, 2000), possibly representing a mechanism to compensate for the decreased oxygen transport capacity. Functional studies need to be conducted to confirm this.

VI. CONTAMINANT EFFECTS ON OXYGEN TRANSPORT

While there are a number of anthropogenic sources for environmental contamination, crude oil toxicity has become a primary concern in areas such as the Amazon due to recently discovered large reserves. Crude oil is a very complex mixture of thousands of large and short chain hydrocarbons. Short chain hydrocarbons are the most toxic and make up what is called the water-soluble fraction (WSF); however, they tend to be relatively short-lived within the aquatic environment because the majority are aromatic (Neff, 1979). Long chain hydrocarbons are much less toxic but persist in the environment, posing a physical threat by creating a viscous barrier at the air–water interface. In many systems, the WSF is of greatest concern because the majority of fishes are water-breathers that do not access the air–water interface. In the Amazon and other tropical areas that routinely experience hypoxia, the air–water interface is heavily utilized by air-breathing fishes, as well as surface skimming fishes. Thus, both the WSF and long chain hydrocarbons are of concern. In general, exposure of fish to the WSF results in variable responses ranging from no observed effects, to disruption of ion regulation and impairment of gill Na^+,K^+-ATPase activity (Boese *et al.*, 1982). In *H. littorale*, exposure to WSF resulted in an increase in air-breathing frequency, and ingestion of crude oil resulted in some disruption in ion regulation, as well as hypoxemia, as indicated by a reduction in whole blood ATP:Hb and GTP:Hb ratios (Brauner *et al.*, 1999). Contact with oil at the air–water interface during exposure to hypoxia has been demonstrated to reduce blood-oxygen content relative to hypoxia in the absence of oil, in the facultative air-breathing fishes, *H. littorale* and *L. pardalis*, and in the surface skimmer *C. macropomum* (Val and Almeida-Val, 1999). Exposure to the WSF results in an elevation in Hct and blood Hb levels in *H. littorale*, however, this is counteracted by the appearance of methemoglobin levels reaching 50% within 24 hours. In *C. macropomum*, methemoglobin levels reached over 70% within 72 hours' exposure to crude oil (Val and Almeida-Val, 1999), indicating that oxygen transfer must be severely limited.

VII. CONCLUDING REMARKS

A great deal is known about O_2 transport in fish and how it is affected by changes in environmental conditions; however, most of what is known comes from a few model temperate fish species. Almost half the vertebrates on the planet are fish, 75% of which live in the highly diverse niches of the tropics. While there are many similarities in O_2 transport between the few temperate and tropical fishes studied to date, a tremendous potential exists to discover novel mechanisms and variations on established mechanisms related to O_2 transport in tropical fishes, and fishes in general.

ACKNOWLEDGMENTS

Supported by an NSERC Discovery grant to CJB and CNPq Brazil research grant to ALV. We thank Brian Sardella and Jodie Rummer for helpful comments on the text.

REFERENCES

Affonso, E. G. (1990). Estudo sazonal de caracteristicas respiratórias do sangue de *Hoplosternum littorale* (Siluriformes, Callichthyidae) da ilha da Marchantaria, Amazonas. MSc thesis, PPG INPA/FUA, Manaus.

Almeida-Val, V. M. F., Val, A. L., Duncan, W. P., Souza, F. C. A., Paula-Silva, M. N., and Land, S. (2000). Scaling effects on hypoxia tolerance in the Amazon fish *Astromotus ocellatus* (Perciformes, Cichlidae): Contribution of tissue enzyme levels. *Comp. Biochem. Physiol.* **125B**, 221–226.

Antunes de Moura, N. (2000). Influência de fatores físico-químicos e recursos alimentares na migração lateral de peixes no lago Chocororé, Pantanal de Barão de Melgaço, estado de Mato Grosso. *In* "Programa de Pós Graduação em Biologia Tropical e Recursos Naturais." p. 88. INPA/UFAM, Manaus, AM.

Bartlett, G. R. (1978). Phosphates in red cells of two lungfish: The South American, *Lepidosiren paradoxa*, and the African, *Protopterus aethiopicus. Can. J. Zool.* **56**, 882–886.

Bénech, V., and Quensière, J. (1982). Migrations de poisson vers le lac Tchad à la dècrue de la plaine inondèe du Nord Cameroum. *Revue du Hydrobiologie Tropicale* **15**, 253–270.

Boese, B. L., Johnson, V. G., Chapman, D. E., and Ridlington, J. W. (1982). Effects of petroleum refinery wastewater exposure on gill ATPase and selected blood parameters in the pacific staghorn sculpin (*Leptocottus armatus*). *Comp. Biochem. Physiol.* **71**, 63–67.

Borgese, T. A., and Nagel, R. L. (1978). Inositol pentaphosphate in fish red blood cells. *J. Exp. Zool.* **205**, 133–140.

Bossa, F., Savi, M. R., Barra, D., and Brunori, M. (1982). Structural comparison of the haemoglobin components of the armoured catfish *Pterygoplichthys pardalis. Biochemical J.* **205**, 39–42.

Brauner, C. J., and Jensen, F. B. (1999). O_2 and CO_2 Exchange in Fish: The nonlinear release of Bohr/Haldane protons with oxygenation. *In* "Biology of Tropical Fishes" (Val, A. L., and Almeida-Val, V. M. F., Eds.), pp. 393–400. INPA, Manaus.

Brauner, C. J., and Randall, D. (1998). The linkage between the oxygen and carbon dioxide transport. *In* "Fish Physiology" (Perry, S. F., and Tufts, B. L., Eds.), Volume 17, pp. 283–319. Academic Press, New York.

Brauner, C. J., and Val, A. L. (1996). The interaction between O_2 and CO_2 exchange in the obligate air breather, *Arapaima gigas*, and the facultative air breather *Liposarcus pardalis. In* "Physiology and Biochemistry of the Fishes of the Amazon" (Val, A. L., Almeida-Val, V. M. F., and Randall, D. J., Eds.), pp. 101–110. INPA, Manaus.

Brauner, C. J., and Wang, T. (1997). The optimal oxygen equilibria curve: A comparison between environmental hypoxia and anemia. *Am. Zool.* **37**, 101–108.

Brauner, C. J., Ballantyne, C. L., Randall, D. J., and Val, A. L. (1995). Air breathing in the armoured catfish (*Hoplosternum littorale*) as an adaptation to hypoxic, acidic, and hydrogen sulphide rich waters. *Can. J. Zool.* **73**, 739–744.

Brauner, C. J., Ballantyne, C. L., Vijayan, M. M., and Val, A. L. (1999). Crude oil exposure affects air-breathing frequency, blood phosphate levels and ion regulation in an air-breathing teleost fish, *Hoplosternum littorale*. *Comp. Biochem. Physiol. C:* **123**, 127–134.

Brauner, C. J., Gilmour, K. M., and Perry, S. F. (1996). Effect of haemoglobin oxygenation on Bohr proton release and CO_2 excretion in the rainbow trout. *Resp. Physiol.* **106**, 65–70.

Brauner, C. J., Matey, V., Wilson, J. M., Bernier, N. J., and Val, A. L. (2004a). Transition in organ function during the evolution of air-breathing; insights from *Arapaima gigas*, an obligate air-breathing teleost from the Amazon. *J. Exp. Biol.* **207**, 1433–1438.

Brauner, C. J., Vijayan, M. M., and Val, A. L. (2000). Organic phosphate, pH and ion regulation in normoxic and hypoxic red blood cells of Amazonian fish following adrenergic stimulation. *In* "International Congress on the Biology of Fish. Surviving Extreme Physiological and Environmental Conditions" (Val, A. L., Wilson, R., and MacKinlay, D., Eds.), pp. 9–11. American Fisheries Society.

Brauner, C. J., Wang, T., Val, A. L., and Jensen, F. B. (2001). Non-linear release of Bohr protons with haemoglobin-oxygenation in the blood of two teleost fishes; carp (*Cyprinus carpio*) and tambaqui (*Colossoma macropomum*). *Fish Physiol. Biochem.* **24**, 97–104.

Brauner, C. J., Wang, T., Wang, Y., Richards, J. G., Gonzalez, R. J., Bernier, N. J., Xi, W., Patrick, M., and Val, A. L. (2004b). Limited extracellular but complete intracellular acid-base regulation during short term environmental hypercapnia in the armoured catfish, Liposarcus pardalis. *J. Exp. Biol.* **207**, 3381–3390.

Brett, J. R. (1972). The metabolic demand for oxygen in fish, particularly salmonids, and a comparison with other vertebrates. *Resp. Physiol.* **14**, 151–170.

Brill, R. W. (1996). Selective advantages conferred by the high performance physiology of tunas, billfishes, and dolphin fish. *Comp. Biochem. Physiol.* **133A**, 3–15.

Brittain, T. (1987). The Root effect. *Comp. Biochem. Physiol.* **86B**, 473–481.

Cameron, J. N., and Wood, C. M. (1978). Renal function and acid-base regulation in two Amazonian erythrinid fishes: *Hoplias malabaricus*, a facultative air breather. *Can. J. Zool.* **56**, 917–930.

Carter, G. S. (1931). Aquatic and aerial respiration in animals. *Biol. Rev.* **6**, 1–35.

Castellini, M. A., Rivera, P. M., and Castellini, J. M. (2002). Biochemical aspects of pressure tolerance in marine mammals. *Comp. Biochem. Physiol.* **133A**, 893–899.

Castro, R. M. C., and Menezes, N. A. (1998). Estudo diagnóstico da diversidade de peixes do estado de São Paulo. *In* "Biodiversidade do Estado de São Paulo, Brasil," Vol. 6, "Vertebrados" (Castro, R. C. M., Ed.), pp. 1–13. FAPESP, São Paulo.

Christiansen, J., Douglas, C. G., and Haldane, J. S. (1914). The absorption and dissociation of carbon dioxide by human blood. *J. Physiol. Lond.* **48**, 244–277.

Cossins, A. R., and Kilbey, R. V. (1991). Adrenergic responses and the Root effect in erythrocytes of freshwater fish. *J. Fish Biol.* **38**, 421–429.

Dejours, P. (1988). "Respiration in Water and Air: Adaptations, Regulation, Evolution." Elsevier, Amsterdam.

Dudley, R. (1998). Atmospheric oxygen, giant paleozoic insects and the evolution of aerial locomotor performance. *J. Exp. Biol.* **201**, 1043–1050.

Fago, A., Bendixen, E., Malte, H., and Weber, R. E. (1997). The anodic hemoglobin of *Anguilla anguilla*. Molecular basis for allosteric effects in a Root-effect hemoglobin. *J. Biol. Chem.* **272**, 15628–15635.

Fairbanks, M. B., Hoffert, J. R., and Fromm, P. O. (1969). The dependence of the oxygen-concentrating mechanism of the teleost eye (*Salmo gairdneri*) on the enzyme carbonic anhydrase. *J. Gen. Physiol.* **54**, 203–211.

Farmer, M., Fyhn, H. J., Fyhn, U. E. H., and Noble, R. W. (1979). Occurrence of Root effect hemoglobins in Amazonian fishes. *Comp. Biochem. Physiol.* **62**, 115–124.

Fernandes, M. (1996). Morpho-functional adaptations of gills in tropical fish. *In* "Physiology and Biochemistry of the Fishes of the Amazon" (Val, A. L., Almeida-Val, V. M. F., and Randall, D. J., Eds.), pp. 181–190. INPA, Manaus.

Fernandes, M., Rantin, F., Kalinin, A., and Moron, S. (1994). Comparative study of gill dimensions of three erythrinid species in relation to their respiratory function. *Can. J. Zool.* **72**, 160–165.

Fernandes, M., Sanches, J., Matsuzaki, M., Panepucci, L., and Rantin, F. T. (1999). Aquatic respiration in facultative air-breathing fish: Effects of temperature and hypoxia. *In* "Biology of Tropical Fishes" (Val, A. L., and Almeida-Val, V. M. F., Eds.), pp. 341–350. INPA, Manaus.

Fyhn, E. H., Fyhn, H. J., Davis, B. J., Powers, D. A., Fink, W. L., and Garlick, R. L. (1979). Hemoglobin heterogeneity in Amazonian fishes. *Comp. Biochem. Physiol.* **62**, 39–66.

Gillen, R. G., and Riggs, A. (1973). Structure and function of the isolated hemoglobins of the American eel (*Anguilla rostrata*). *J. Biol. Chem.* **248**, 1961–1969.

Goulding, M. (1980). "The Fishes and the Forest. Explorations in Amazonian Natural History." University of California Press, Los Angeles.

Graham, J. B. (1997). "Air-Breathing Fishes. Evolution, Diversity and Adaptation." Academic Press, San Diego.

Heisler, N. (1982). Intracellular and extracellular acid-base regulation in the tropical fresh-water teleost fish *Synbranchus marmoratus* in response to the transition from water breathing to air breathing. *J. Exp. Biol.* **99**, 9–28.

Heisler, N. (1993). Acid-base regulation. *In* "The Physiology of Fishes" (Evans, D. H., Ed.), pp. 343–378. CRC Press, Boca Raton, FL.

Hochachka, P. W. (1996). Oxygen sensing and metabolic regulation: Short, intermediate and long term roles. *In* "Physiology and Biochemistry of the Fishes of the Amazon" (Val, A. L., Almeida-Val, V. M. F., and Randall, D. J., Eds.), pp. 233–256. INPA, Manaus.

Hochachka, P. W., Land, S. C., and Buck, L. T. (1997). Oxygen sensing and signal transduction in metabolic defense against hypoxia: Lessons from vertebrate facultative anaerobes. *Comp. Biochem. Physiol.* **118A**, 23–29.

Houston, A. H., and Murad, A. (1992). Erythrodynamics in goldfish, Carassius auratus L.: Temperature effects. *Physiol. Zool.* **65**, 55–76.

Hughes, G. M. (1966). The dimensions of fish gills in relation to their function. *J. Exp. Biol.* **45**, 177–195.

Hughes, G. M. (1972). Morphometrics of fish gills. *Respir. Physiol.* **14**, 1–25.

Hughes, G. M. (1984). General anatomy of the gills. *In* "Fish Physiology" (Hoar, W. S., and Randall, D. J., Eds.), Vol. XA, pp. 1–63. Academic Press, New York.

Ingerman, R. L., and Terwilliger, R. C. (1982). Presence and possible function of Root effect hemoglobins in fishes lacking functional swim bladders. *J. Exp. Zool.* **220**, 171–177.

Isaacks, R. E., Kim, D. H., and Harkness, D. R. (1978). Inositol diphosphate in erythrocytes of the lungfish, *Lepidosiren paradoxa*, and 2,3 diphosphoglycerate in erythrocytes of the armored catfish, *Pterygoplichthys* sp. *Can. J. Zool.* **56**, 1014–1016.

Isaacks, R. E., Kim, H. D., Bartlett, G. R., and Harkness, D. R. (1977). Inositol pentaphosphate in erythrocytes of a fresh water fish, pirarucu (*Arapaima gigas*). *Life Sciences* **20**, 987–990.

Jensen, F. B. (1986). Pronounced influence of Hb–O_2 saturation on red cell pH in tench blood *in vivo* and *in vitro*. *J. Exp. Zool.* **238**, 119–124.

Jensen, F. B. (1989). Hydrogen ion equilibria in fish haemoglobins. *J. Exp. Biol.* **143**, 225–234.

Jensen, F. B. (1991). Multiple strategies in oxygen and carbon dioxide transport by haemoglobin. *In* "Physiological Strategies for Gas Exchange and Metabolism" (Woakes, A. J., Greishaber, M. K., and Bridges, C. R., Eds.), pp. 55–78. Cambridge University Press, Cambridge.

Jensen, F. B., Fago, A., and Weber, R. E. (1998). Hemoglobin structure and function. *In* "Fish Physiology" (Perry, S. F., and Tufts, B. L., Eds.), Vol. 17, pp. 1–32. Academic Press, New York.

Johansen, K., and Lenfant, C. (1972). A comparative approach to the adaptability of O_2-Hb affinity. *In* "Oxygen Affinity of Hemoglobin and Red Cell Acid Base Status" (Rorth, M., and Astrup, P., Eds.), pp. 750–783. Munksgaard, Copenhagen.

Johansen, K., Mangum, C. P., and Lykkeboe, G. (1978a). Respiratory properties of the blood of amazon fishes. *Can. J. Zool.* **56**, 891–897.

Johansen, K., Mangum, C. P., and Weber, R. E. (1978b). Reduced blood O_2 affinity with air breathing in osteoglossid fishes. *Can. J. Zool.* **56**, 891–897.

Junk, W. (1979). Macrófitas aquáticas nas várzeas da Amazônia e possibilidades de seu uso na agropecuária. INPA, Manaus.

Junk, W. J., Bayley, P. B., and Sparks, R. E. (1989). The flood pulse concept in river-floodplain systems. *In* "Proceedings of the International Large River Symposium" (Dodge, D. P., Ed.), Vol. 106, pp. 110–127. *Can. Spec. Publ. Fish. Aquat. Sci.*, Canada.

Junk, W. J., Soares, M. G., and Carvalho, F. M. (1983). Distribution of fish species in a lake of the Amazon river floodplain near Manaus (lago Camaleao), with special reference to extreme oxygen conditions. *Amazoniana* **7**, 397–431.

Kalinin, A., Rantin, F., Fernandes, M., and Glass, M. L. (1996). Ventilatory flow relative to intrabuccal and intraopercular volumes in two ecologically distinct erythrinids (*Hoplias malabaricus* and *Hoplias lacerdae*) exposed to normoxia and graded hypoxia. *In* "Physiology and Biochemistry of the Fishes of the Amazon" (Val, A. L., Almeida-Val, V. M. F., and Randall, D. J., Eds.), pp. 191–202. INPA, Manaus.

Kind, P., Grigg, G., and Booth, D. (2002). Physiological responses to prolonged aquatic hypoxia in the Queensland lungfish *Neoceratodus forsteri*. *Resp. Physiol.* **132**, 179–190.

Kramer, D. L., Lindsey, C. C., Moodie, G. E. E., and Stevens, E. D. (1978). The fishes and the aquatic environment of the central Amazon basin, with particular reference to respiratory patterns. *Can. J. Zool.* **56**, 717–729.

Lapennas, G. N. (1983). The magnitude of the Bohr coefficient: Optimal for oxygen delivery. *Resp. Physiol.* **54**, 161–172.

Larsen, B. K., and Jensen, F. B. (1997). Influence of ionic composition on acid-base regulation in rainbow trout (*Oncorhynchus mykiss*) exposed to environmental hypercapnia. *Fish Physiol. Biochem.* **16**, 157–170.

Lauder, G. V., and Liem, K. F. (1983). The evolution and interrelationships of the Actinopterygian fishes *Bull. Mus. Comp. Zool.* **150**, 95–197.

Laurent, P. (1984). Gill internal morphology. *In* "Fish Physiology" (Hoar, W. S., and Randall, D., Eds.), Vol. XA, pp. 73–172. Academic Press, New York.

Liepelt, A., Karbe, L., and Westendorf, J. (1995). Induction of DNA strand breaks in rainbow trout *Oncorhynchus mykiss* under hypoxic and hyperoxic conditions. *Aquatic Toxicology* **33**, 177–181.

Lowe, T. E., Brill, R. W., and Cousins, K. L. (1998). Responses of the red blood cells from two high-energy-demand teleosts, yellowfin tuna (*Thunnus albacares*) and skipjack tuna (*Katsuwonus pelamis*), to catecholamines. *J. Comp. Physiol. B* **168**, 405–418.

Lowe McConnell, R. H. (1987). "Ecological Studies in Tropical Fish Communities." Cambridge University Press, Cambridge.

Marcon, J. L., and Val, A. L. (1996). Intraerythrocytic phosphates in *Colossoma macropomum* and *Astronotus ocellatus* (Pisces) of the Amazon. *In* "International Congress of the Biology of Fishes. The Physiology of Tropical Fish" (Val, A. L., Randall, D. J., and MacKinlay, D., Eds.), pp. 101–107. American Fisheries Society, San Francisco.

Marcon, J., Chagas, E., Kavassaki, J., and Val, A. L. (1999). Intraerythrocyte phosphates in 25 fish species of the Amazon: GTP as a key factor in the regulation of Hb–O_2 affinity. *In* "Biology of Tropical Fishes" (Val, A. L., and Almeida-Val, V. M. F., Eds.), pp. 229–240. INPA, Manaus.

Mattias, A. T., Moron, S. E., and Fernandes, M. (1996). Aquatic respiration during hypoxia of the facultative air-breathing *Hoplerythrinus unitaeniatus*. A comparison with the water-breathing *Hoplias malabaricus*. *In* "Physiology and Biochemistry of the Fishes of the Amazon" (Val, A. L., Almeida-Val, V. M. F., and Randall, D. J., Eds.), pp. 203–211. INPA, Manaus.

Monteiro, P. J. C., Val, A. L., and Almeida-Val, V. M. F. (1987). Biological aspects of Amazonian fishes. Hemoglobin, hematology, intraerythrocytic phosphates and whole blood Bohr effect of *Mylossoma duriventris*. *Can. J. Zool.* **65,** 1805–1811.

Moura, M. A. F. (1994). Efeito da anemia, do exercício físico e da adrenalina sobre o baço e eritrócitos de *Colossoma macropomum* (Pisces). PPG INPA/FUA.

Moyle, P. B., and Cech, J. J. Jr (1996). *Fishes: An Introduction to Ichthyology.* Prentice Hall, Englewood Cliffs, NJ.

Mylvaganam, S. E., Bonaventura, C., Bonaventura, J., and Getzoff, E. D. (1996). Structural basis for the Root effect in haemoglobin. *Nature* **3,** 275–283.

Narahara, A., Bergman, H. L., Laurent, P., Maina, J. N., Walsh, P. J., and Wood, C. M. (1996). Respiratory physiology of the Lake Magadi tilapia (*Oreochromis alcalicus grahami*), a fish adapted to a hot, alkaline, and frequently hypoxic environment. *Physiol. Zool.* **69,** 1114–1136.

Neff, J. M. (1979). "Polycyclic Aromatic Hydrocarbons in the Aquatic Environment: Sources, Fates and Biological Effects." Applied Science Publishers Ltd, Essex.

Nelson, J. S. (1984). "Fishes of the World." John Wiley & Sons, New York.

Nikinmaa, M. (1990). "Vertebrate Red Blood Cells. Adaptations of Function to Respiratory Requirements." Springer-Verlag, Heidelberg.

Pelster, B. (2001). The generation of hyperbaric oxygen tensions in fish. *News in Physiological Science* **16,** 287–291.

Pelster, B., and Weber, R. E. (1990). Influence of organic phosphates on the root effect of multiple fish haemoglobins. *J. Exp. Biol.* **149,** 425–437.

Pelster, B., and Weber, R. E. (1991). The physiology of the Root effect. *In* "Advances in Comparative and Environmental Physiology," Vol. 8, pp. 51–77. Springer-Verlag, Berlin.

Perry, S. F., and Reid, S. D. (1992). Relationship between blood O_2-content and catecholamine levels during hypoxia in rainbow trout and American eel. *Am. J. Physiol.* **263,** R240–R249.

Perry, S. F., Kinkead, R., Gallaugher, P., and Randall, D. J. (1989). Evidence that hypoxemia promotes catecholamine release during hypercapnic acidosis in rainbow trout (*Salmo gairdneri*). *Resp. Physiol.* **77,** 351–364.

Perry, S. F., Reid, S. D., Gilmour, K. M., Boijink, C. L., Lopes, J. M., Milsom, W. K., and Rantin, F. T. (2004). A comparison of adrenergic stress responses in three tropical teleosts exposed to acute hypoxia. *Am. J. Physiol. Regul. Integr. Comp. Physiol.* **287,** R188–R197.

Powers, D. A., Fyhn, H. J., Fyhn, U. E. H., Martin, J. P., Garlick, R. L., and Wood, S. C. (1979). A comparative study of the oxygen equilibria of blood from 40 genera of amazonian fishes. *Comp. Biochem. Physiol.* **62,** 67–85.

Randall, D. J. (1982). The control of respiration and circulation in fish during exercise and hypoxia. *J. Exp. Biol.* **100**, 275–288.

Randall, D. J., and Daxboeck, C. (1984). Oxygen and carbon dioxide transfer across fish gills. *In* "Fish Physiology" (Hoar, W. S., and Randall, D. J., Eds.), Vol. XA, pp. 263–307. Academic Press, New York.

Randall, D. J., Farrell, A. P., and Haswell, M. S. (1978). Carbon dioxide excretion in the pirarucu (*Arapaima gigas*), an obligate air-breathing fish. *Can. J. Zool.* **56**, 977–982.

Rantin, F., Kalinin, A., Glass, M. L., and Fernandes, M. (1992). Respiratory responses to hypoxia in relation to mode of life of two erythrinid species (*Hoplias malabaricus and Hoplias lacerdae*). *J. Fish Biol.* **41**, 805–812.

Rantin, F., Kalinin, A., Guerra, C., Maricondi-Massari, M., and Verzola, R. (1995). Electron-cardiographic characterization of myocardial function in normoxic and hypoxic teleosts. *Braz. J. Med. Biol. Res.* **28**, 1277–1289.

Ratcliffe, P. J., O'Rourke, J. F., Maxwell, P. H., and Pugh, C. W. (1998). Oxygen sensing, hypoxia-inducible factor-1 and the regulation of mammalian gene expression. *J. Exp. Biol.* **201**, 1153–1162.

Reid, S., Gilmour, K. M., Perry, S. F., and Rantin, T. (2002). Hypoxia-induced catecholamine secretion in jeju, traira and pacu. *In* " Proceedings of the International Congress on the Biology of Fish. Tropical Fishes: News and Reviews" (Val, A., Brauner, C., and McKinlay, D., Eds.), pp. 1–3. American Fisheries Society.

Riggs, A. (1979). Studies of the hemoglobins of Amazonian fishes: An overview. *Comp. Biochem. Physiol.* **62A**, 257–272.

Riggs, A. F. (1988). The Bohr effect. *Ann. Rev. Physiol.* **50**, 181–204.

Root, R. W. (1931). The respiratory function of the blood of marine fishes. *Biol. Bull. Mar. Biol. Lab. Woods Hole* **61**, 427–456.

Saint-Paul, U. (1983). Investigation on the respiration of the neotropical fish, *Colossoma macropomum* (Serrasalmidae). The influence of weight and temperature on the routine oxygen consumption. *Amazoniana* **VII**, 433–443.

Stevens, E. D., and Holeton, G. F. (1978). The partitioning of oxygen uptake from air and from water by the large obligate air-breathing teleost pirarucu (*Arapaima gigas*). *Can. J. Zool.* **56**, 974–976.

Sundin, L. (1999). Hypoxia and blood flow control in fish gills. *In* "Biology of Tropical Fishes" (Val, A. L., and Almeida-Val, V. M. F., Eds.), pp. 353–362. INPA, Manaus.

Tun, N., and Houston, H. (1986). Temperature, oxygen, photoperiod, and the hemoglobin system of the rainbow trout (*Salmo gairdneri*). *Can. J. Zool.* **64**, 1883–1888.

Val, A. L. (1986). Hemoglobinas de Colossoma macropomum Cuvier, 1818 (Characoidei, Pisces): Aspectos adaptativos (Ilha da Marchantaria, Manaus, AM). Instituto Nacional de Pesquisas da Amazonia e Universidade do Amazonas, Manaus, AM.

Val, A. L. (1993). Adaptations of fish to extreme conditions in fresh waters. *In* "The Vertebrate Gas Transport Cascade: Adaptations to Environment and Mode of Life" (Bicudo, J. E. P. W., Ed.), pp. 43–53. CRC Press, Boca Raton, FL.

Val, A. L. (1995). Oxygen transfer in fish: Morphological and molecular adjustments. *Braz. J. Med. Biol. Res.* **28**, 1119–1127.

Val, A. L. (2000). Organic phosphates in the red blood cells of fish. *Comp. Biochem. Physiol.* **125A**, 417–435.

Val, A. L., and Almeida-Val, V. M. F. (1995). "Fishes of the Amazon and their Environments. Physiological and Biochemical Features." Springer Verlag, Heidelberg.

Val, A. L., and Almeida-Val, V. M. F. (1999). Effects of crude oil on respiratory aspects of some fish species of the Amazon. *In* "Biology of Tropical Fishes" (Val, A. L., and Almeida-Val, V. M. F., Eds.), pp. 277–291. INPA, Manaus.

Val, A. L., Affonso, E. G., Souza, R. H. S., Almeida-Val, V. M. F., and Moura, M. A. F. (1992). Inositol pentaphosphate in the erythrocytes of an Amazonian fish, the pirarucu (*Arapaima gigas*). *Can. J. Zool.* **70**, 852–855.

Val, A. L., Marcon, J. L., Costa, O. T. F., Barcellos, F. M., Maco Garcia, J. T., and Almeida-Val, V. M. F. (1999). Fishes of the Amazon: Surviving environmental changes. *In* "Ichthyology. Recent Research Advances" (Saksena, D. N., Ed.), pp. 389–402. Science Publishers Inc., Enfield, NH.

Val, A. L., Mazur, C. F., Salvo-Souza, R. H., and Iwama, G. (1994a). Effects of experimental anaemia on intra-erythrocytic phosphate levels in rainbow trout, *Oncorhynchus mykiss*. *J. Fish Biol.* **45**, 269–279.

Val, A. L., de Menezes, G. C., and Wood, C. M. (1998). Red blood cell adrenergic responses in Amazonian teleosts. *J. Fish Biol.* **52**, 83–93.

Val, A. L., Moraes, G., Barcelos, J. F., Roubach, R., Yossa-Perdomo, M. I., and Almeida-Val, V. M. F. (1994b). Effects of extreme environmental conditions on respiratory parameters of the ornamental fish *Pterophylum scalare* and *Symphysodon aequifasciata*. *In* "Physiology and Biochemistry of the Fish of the Amazon" (Val, A. L., Randall, D. J., and Almeida-Val, V. M. F., Eds.), p. 64. INPA, Manaus, Amazon.

Val, A. L., Schwantes, A. R., and Almeida-Val, V. M. F. (1986). Biological aspects of Amazonian fishes. VI. Hemoglobins and whole blood properties of *Semaprochilodus* species (Prochilodontidae) at two phases of migration. *Comp. Biochem. Physiol.* **83B**, 659–667.

Weber, R. E. (1990). Functional significance and structural basis of multiple hemoglobins with special reference to ectothermic vertebrates. *In* "Comparative Physiology; Animal Nutrition and Transport Processes; 2. Transport, Respiration and Excretion: Comparative and Environmental Aspects. II. Blood Oxygen Transport: Adjustment to Physiological and Environmental Conditions" (Truchot, J. P., and Lahlou, B., Eds.), Vol. 6, pp. 58–75. Karger, Basel.

Weber, R. E., and Jensen, F. B. (1988). Functional adaptations in hemoglobins from ectothermic vertebrates. *Ann. Rev. Physiol.* **50**, 161–179.

Weber, R. E., and Lykkeboe, G. (1978). Respiratory adaptations in carp blood: Influences of hypoxia, red cell organic phosphates, divalent cations and CO_2 on hemoglobin-oxygen affinity. *J. Comp. Physiol.* **128**, 127–137.

Weber, G., Sullivan, G. V., Bonaventura, J., and Bonaventura, C. (1976). The hemoglobin system of the primitive fish, *Amia Calva*: Isolation and functional characterization of the individual hemoglobin components. *Biochim. Biophys. Acta* **434**, 18–31.

Wood, S. C., Johansen, K., and Weber, R. E. (1975). Effects of ambient Po_2 on hemoglobin-oxygen affinity and red cell ATP concentrations in a benthic fish, *Pleuronectes platessa*. *Resp. Physiol.* **25**, 259–267.

Wyman, J. (1973). Linked functions and reciprocal effects in haemoglobin: A second look. *Adv. Prot. Chem.* **19**, 223–228.

8

NITROGEN EXCRETION AND DEFENSE AGAINST AMMONIA TOXICITY

SHIT F. CHEW

JONATHAN M. WILSON

YUEN K. IP

DAVID J. RANDALL

I. INTRODUCTION

Ammonia is mainly produced in fish through catabolism of amino acids. Ingested proteins are hydrolyzed producing amino acids, the catabolism of which leads to the release of ammonia. Ammonia production takes place

The Physiology of Tropical Fishes: Volume 21
FISH PHYSIOLOGY

mainly in the liver in fish (Pequin and Serfaty, 1963) but other tissues are also capable of doing so (Walton and Cowey, 1977). Ammonia is produced by the transamination of amino acids followed by the deamination of glutamate and/or by the deamination of adenylates in fish muscle during severe exercise (Driedzic and Hochachka, 1976). Most aquatic animals keep body ammonia levels low by simply excreting excess ammonia produced through digestion and metabolism.

Tropical regions are characterized by high temperatures of ~30 °C. The dietary protein requirement of fish increases at higher temperatures, possibly due to increased oxidation of amino acids (Delong *et al.*, 1958). Wood (2001) reviewed the effects of temperature on nitrogen metabolism and concluded that as temperature increased an increasing percentage of aerobic metabolism was fueled by oxidation of protein. That is, ammonia production increases with increasing temperature due to both increased metabolic rate and a switch to greater protein utilization. However, this conclusion was based on studies of temperate fishes and, therefore, may not be applicable to tropical fishes. Tropical fishes show a marked increase in food intake and ammonia excretion with increasing temperature (Leung *et al.*, 1999a) but the effect on protein metabolism is not known. Although intensive fish farming is dominated by production from Southeast Asia, much less is known of these species compared with that for temperate species, especially salmonids. In making comparisons it is important to note that tropical and subtropical farmed fishes are fed a high protein diet of trash fish, rather than pellets, and there is a much higher nitrogen loss to the environment compared with that reported in temperate regions (Leung *et al.*, 1999b).

In aqueous solution ammonia has two species, NH_3 NH_4^+; the equilibrium reaction can be written as $NH_3 + H_3O^+ \Leftrightarrow NH_4^+ + H_2O$. In this article, NH_3 represents unionized molecular ammonia, NH_4^+ represents ammonium ion, and ammonia refers to both NH_3 and NH_4^+. Total ammonia (TAmm) is the sum of $[NH_3] + [NH_4^+]$ and the pK of this NH_3/NH_4^+ reaction is around 9.0 to 9.5. Thus, the total ammonia in solution will have two components – the gas NH_3 and the cation NH_4^+; the separate fractions of each can be calculated from the Henderson–Hasselbalch equation if the pH and appropriate pK are known: $NH_4^+ = TAmm/\{1 + antilog(pH - pK)\}$ $= TAmm - NH_3$. Biological membranes are not very permeable to NH_4^+, but are more permeable to NH_3. Thus, in most cases, ammonia crosses membranes as NH_3.

Ammonia is usually excreted as NH_3 across the body surface of fish, usually the gills, into the surrounding water (Figure 8.1A). Under acidic environmental conditions, NH_3 diffusing across the gills is converted to NH_4^+ and trapped in the water (Figure 8.1A). Thus, acidic conditions in the environment augment ammonia excretion. The diffusion of NH_3 into the environment is inhibited if the pH of the environment is high, and ammonia accumulates in the body. Excretion is also reduced if the fish moves out of water. Finally, decomposition of organic matter or the use of fertilizers will increase ammonia levels in the water and this will result in elevated ammonia levels in the fish. The partition of ammonia between the water and the fish will be largely determined by the pH of the two systems.

Oxygen solubility in water decreases with increasing temperature; so the oxygen content of water in tropical regions is less than in more temperate waters. Both photosynthetic and respiration rates tend to be higher at higher temperatures so both oxygen production and consumption are elevated in tropical waters. At night, when oxygen production is reduced but respiration rates remain high, aquatic hypoxia is common. Thus, the tropical aquatic environment is warm and often hypoxic. Because aquatic hypoxia is a frequent event in tropical waters, air breathing is a common occurrence. With

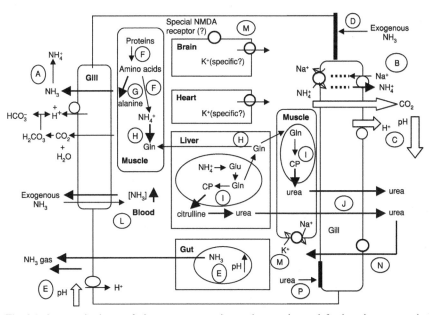

Fig. 8.1 A general scheme of nitrogenous excretion and strategies to defend against ammonia toxicity in fish. **A**, NH_3 excretion and NH_3 trapping; **B**, active transport of NH_4^+; **C**, environmental acidification; **D**, decreased membrane permeability to NH_3; **E**, NH_3 volatilization; **F**, reduction in proteolysis and amino acid catabolism; **G**, partial amino acid catabolism and alanine formation; **H**, glutamine synthesis; **I**, ureogenesis; **J**, urea excretion; **L**, ammonia build-up in the blood; **M**, extreme ammonia tolerance at the cellular level with special NMDA receptor and/or K^+ specific channel; **N**, active absorption of urea; **P**, decreased membrane permeability to urea.

the development of air-breathing capabilities, some tropical fishes (e.g. mudskippers, catfishes, etc.) are equipped to leave water and make short excursions on land. Some (e.g. the weatherloach and the sweep eel) even burrow into semi-solid mud during drought. Consequently, many tropical species have evolved mechanisms to deal with increasing body ammonia loads resulting from reductions in ammonia excretion associated with aerial exposure. Similar mechanisms are used to avoid ammonia toxicity due to increases in ammonia production associated with feeding and increases in tissue ammonia due to elevated environmental ammonia levels. These mechanisms include actively excreting NH_4^+ in water, excreting NH_3 into air by volatilization, reducing ammonia production, and converting ammonia to less toxic compounds such as glutamine or urea. It appears that some tropical fishes have reduced skin permeability to NH_3 to avoid a back flux of NH_3 from the environment and several species acidify the environment in order to trap ammonia in the water as NH_4^+. All of these mechanisms can be found in various species of tropical fishes, the responses of which are many and varied, determined by the behavior of the fish species and the nature of the environment in which they live.

II. EFFECTS OF ENVIRONMENTAL pH OR TEMPERATURE ON AMMONIA TOXICITY

In many studies and reviews, ammonia toxicity is expressed as $[NH_3]$ in water. This is because the NH_3 gradient is an important determinant of the rate of ammonia uptake, especially in freshwater animals. Water pH has a marked effect on the $[NH_3]/[NH_4^+]$ equilibrium and, as a result, the $[NH_3]$ in water. The acute and chronic toxicity of ammonia, expressed as TAmm in the water, increases with pH; very low levels of TAmm are toxic if water pH exceeds 9.5 (Ip, Chew *et al.*, 2001a). Many fishes cannot survive in extremely alkaline waters because of their inability to excrete ammonia at adequate rates, even if the water is essentially ammonia free. Ammonia toxicity, expressed as $[NH_3]$ in the water, however, increases with decreasing pH, presumably because of the additional toxic effects of the increasing $[NH_4^+]$ in the water.

Thurston *et al.* (1981) observed a decrease in toxicity of $[NH_3]$ in water with increasing temperature. Compiled and normalized data on acute toxicity in various species of fishes expressed as TAmm indicate that the effect of increased temperature on toxicity is minimal between 3 and 30 °C in freshwater systems (USA Environmental Protection Agency, 1998). Thus, many of the data collected at lower temperatures in temperate waters may not be applicable to tropical fishes at temperatures around 30 °C.

III. MECHANISM OF AMMONIA TOXICITY

A. General Effects

Ammonia accumulated in the body is toxic (reviewed by Ip, Chew et al., 2001a). Rainbow trout, *Oncorhynchus mykiss*, becomes hyper-excitable when exposed to water with $3 \mu g$ NH_3 ml^{-1} $(176.5 \mu mol l^{-1})$. Any disturbance to the tank or movement above the tank visible to the fish results in disoriented escape attempts, which send the fish crashing into the sides of the tank (Olson and Fromm, 1971). At $36 \mu mol l^{-1}$ NH_3, the rainbow trout develops convulsions and dies (Arillo et al., 1981). Elevated water ammonia levels reduce feeding and inhibit growth (Hampson, 1976; Alderson, 1979; Dabrowska and Wlasow, 1986). Swimming ability is reduced in coho salmon (Randall and Wicks, 2000), probably because increased NH_4^+ levels in the fish cause muscle depolarization (Taylor, 2000).

Chronic environmental ammonia exposure results in gill hyperplasia (Burrows, 1964; Reichenbach-Klinke, 1967; Smart, 1976; Thurston et al., 1978), and changes in mucus production, growth, and stamina (Lang et al., 1987). The gill lamellae of rainbow trout exposed to $5 \mu g$ NH_3 ml^{-1} $(294.1 \mu mol l^{-1})$ become shorter and thicker with bulbous ends. Many filaments show limited hyperplasia, with cells containing large vacuoles with proteins (Olson and Fromm, 1971). Trout exposed to ammonia show increased rates of diuresis with an increase in ammonia concentration (Lloyd and Orr, 1969).

Exposure of rainbow trout to ammonia causes changes in the percentage of blood cells, resulting in an increase in the number of erythroblasts (Dabrowska and Wlasow, 1986). At the same time, the total number of leukocytes decreases significantly, leading to an increase in the red:white cell ratio. Results obtained from coho salmon also indicate that exposure of the fish to ammonia results in an increase in the percentage of circulating immature erythrocytes. At $0.72 mg$ NH_3-N l^{-1} $(42.3 \mu mol l^{-1})$, hemoglobin and hematocrit values are reduced to levels that are indicative of anemia (Buckley et al., 1979).

Chronic exposure of fish to ammonia also leads to an increase in the level of the stress hormone, cortisol, rendering them more susceptible to stress and diseases as the immune system is suppressed by the corticosteroid (Mommsen and Walsh, 1992). When the common carp, *Cyprinus carpio*, is exposed to varying concentrations of ammonia, the levels of catecholamines increase along with the ammonia concentration.

Ammonia acts on the central nervous system of vertebrates, including fish, causing hyperventilation (Hillaby and Randall, 1979; McKenzie et al., 1993), hyper-excitability, convulsions, coma, and finally death. NH_3 can cross the blood–brain barrier in mammals (Sears *et al.*, 1985) and high ammonia levels modify many aspects of the blood–brain barrier (Cooper and Plum, 1987). In addition, NH_4^+ can substitute for K^+ (Binstock and Lecar, 1969) affecting the membrane potential.

B. Effects on Branchial Ionic Transport

In freshwater teleost fishes, Na^+ and Cl^- are actively taken up to compensate for passive ion losses. The uptake of these ions is coupled (directly or indirectly) to respective H^+ and HCO_3^- exchange, however, the exact mechanisms of these exchanges are still the subject of debate. The currently accepted mechanisms for Na^+ uptake are directly coupled Na^+/H^+ exchange facilitated by an Na^+/H^+ antiporter (NHE) or indirect coupling of Na^+ uptake via an epithelial Na^+ channel (ENaC) driven by a vacuolar type proton ATPase (vH^+-ATPase). The latter mechanism is more widely accepted (Lin and Randall, 1995; Marshall, 2002; Chapter 9, this volume). Cl^- uptake is generally accepted to be facilitated by a Cl^-/HCO_3^- anion exchanger (AE) which involves a directly coupled exchange (Wilson *et al.*, 2000a). Cytoplasmic carbonic anhydrase is important in providing an intracellular pool of H^+ and HCO_3^- from CO_2 hydration for the vH^+-ATPase/NHE and AE, respectively. These exchange processes are also important in acid-base regulation.

Marine fishes engage in active ion elimination to compensate for the passive gain of ions. The mechanism of secondary active transport of Cl^-, which has been described in the gills of fish, is the same as in other NaCl-secreting organs (seabird salt gland, shark rectal gland). Cl^- is taken up into specialized chloride-secreting cell via a basolateral $Na^+:K^+:2Cl^-$ cotransporter (NKCC) driven by the Na^+ gradient across the basolateral membrane maintained by Na^+, K^+-ATPase. A basolateral K^+ channel recycles the K^+. The intracellular Cl^- exits the cell apically via a CFTR Cl^- channel down its electrochemical gradient (Marshall, 2002; Chapter 9, this volume). Na^+ efflux takes a paracellular path between leaky tight junctions found between chloride cells and neighboring accessory cells (Sardet *et al.*, 1979). Na^+/H^+ and Cl^-/HCO_3^- exchange processes are also important in marine fishes for acid-base regulation (Claiborne *et al.*, 2002).

NH_4^+ can be transported by a number of ion transport proteins mentioned above. The evidence comes largely from *in vitro* studies of homologous mammalian ion transport proteins and the list includes Na^+, K^+-ATPase, NKCC, NHE, and ENaC (Kinsella and Aronson, 1981; Knepper *et al.*, 1989; Wall, 1996; Nakhoul *et al.*, 2001a). NH_4^+ competes for transport sites with either H^+ or K^+ and thus might adversely affect the regulation of these ions and others indirectly. Conversely, the interaction of

NH_4^+ with these ion transport proteins may have beneficial ends as will be discussed in the section on active NH_4^+ excretion (section IV.A). So far in fish, there is only evidence for the Na^+, K^+-ATPase being a potential target of ammonia toxicity, although this effect may be species-specific (Mallery, 1983; Randall et al., 1999; Salama et al., 1999).

Despite knowing the potential adverse effects of ammonia on ionic transport, the only direct evidence for deleterious effects of environmental ammonia exposure on ionic transport are from studies on the temperate rainbow trout (O. mykiss) in vivo (Twitchen and Eddy, 1994; Wilson et al., 1994) or using an in vitro perfused head preparation (Avella and Bornancin, 1989), and from in vivo studies on the goldfish (Carassius auratus; Maetz and García Romeu, 1964; Maetz, 1973). In the studies by Avella and Bornancin (1989; $1 \, mmol \, l^{-1}$ NH_4Cl), Wilson et al. (1994; $1 \, mmol \, l^{-1}$ NH_4Cl) and Maetz and García Romeu (1964; $<0.2 \, mmol \, l^{-1}$ $(NH_4)_2SO_4$), acute exposure to high environmental ammonia concentrations (HEA) results in an inhibition of Na^+ influx.

There are two potential explanations of this data based on available information of the ion transport processes in freshwater fishes. The addition of ammonia to the medium would generally result in an acute intracellular alkalinization of the epithelial cells as a result of NH_3 entry by diffusion and its consequent binding with an intracellular H^+ resulting in an increase in intracellular pH (pH_i). This increase in pH_i would decrease the availability of H^+ for the vH^+-ATPase that drives Na^+ uptake via ENaC. Notably, higher pH_i within the physiological range has been shown to decrease V-ATPase activity in isolated vesicle preparations; however, pH does not appear to be the principal regulator of activity in intact cells (Gluck et al., 1992). The pH_i perturbation could also potentially have adverse affects on the performance of other ion transport proteins by shifting the pH_i away from their optimal pH. Alternatively, NH_4^+ may interfere directly with Na^+ uptake via the ENaC. In Xenopus oocytes expressing mouse ENaC, ammonia has been shown to inhibit Na^+ transport, which is not due to the effects of environmental NH_4Cl on pH_i (Nakhoul et al., 2001a). It has been suggested that permeability and selectivity of ENaC to NH_4^+ may play a role in that study.

In the goldfish (C. auratus), Maetz and García Romeu (1964) were unable to find a consistent effect of HEA exposure on Cl^- uptake. This indicates that the deleterious effect is specific to Na^+ uptake – specialized Na^+ uptake cells (PVCs) or mechanisms of uptake (NHE, ENaC) – and not general to the epithelium and all ion-uptake mechanisms. There is also some evidence in the channel catfish (Ictalurus punctatus) that HEA does not affect Cl^- homeostasis (plasma and tissue Cl^- levels) while under the same conditions plasma Na^+ levels are perturbed (Tomasso et al., 1980).

    In the study by Wilson *et al.* (1994) on the trout, Na^+ uptake recovered within hours of the initial exposure to HEA and was significantly elevated after 24 hours. Twitchen and Eddy (1994) also observed an increase in Na^+ uptake after approximately 24 hours of HEA exposure. These results indicate that the deleterious effects on pH_i and/or ENaC are compensated for. They also suggest that during prolonged exposure to HEA, direct Na^+/NH_4^+ exchange (NHE) may come into play; however, the rates of increased Na^+ uptake can only partially account for NH_4^+ excretion based on a 1:1 exchange mechanism (Wilson *et al.*, 1994). This increase may only represent a compensation for the interruption in Na^+ uptake.

    In the study by Twitchen and Eddy (1994) on juvenile rainbow trout, no deleterious effect of HEA exposure (up to $28.2\,\mu mol\,l^{-1}$ NH_3-N or $5.2\,mmol\,l^{-1}$ TAmm-N) was seen on Na^+ uptake; however, Na^+ efflux was stimulated by ammonia levels greater than $6.4\,\mu mol\,l^{-1}$ NH_3-N ($1.2\,mmol\,l^{-1}$ TAmm-N). This increase in efflux is likely to have occurred through an increased Na^+ permeability of the gills (Gonzalez and McDonald, 1994), which is also observed with a base load (Goss and Wood, 1990). The mechanism of the permeability change in the gill is not well understood but could be related to a modulation of the paracellular pathway (Madara, 1998). An increase in Na^+ efflux may be directed at compensating the effect of alkalinization from NH_3 entry by decreasing the strong ion difference and consequently pH_i (Stewart, 1983; Goss and Wood, 1990). Exposure to HEA can also cause diuresis (Lloyd and Orr, 1969), and increased gill ventilation and perfusion (Smart, 1978), which would also contribute to Na^+ losses, in fishes.

    In addition to these possible direct effects of ammonia on ion transport proteins, the increase in intracellular ammonia may also disrupt the supply of ATP used to drive ion transport (vH^+-ATPase, Na^+, K^+-ATPase). In tilapia (tropical *Oreochromis mossambicus*), Begum (1987) has shown that 2-d exposure to $1\,mmol\,l^{-1}$ NH_4Cl disrupts gill glucose metabolism and reduces oxidative capacity (decreases in succinate dehydrogenase and cytochrome c oxidase activities). Correlated with these findings is the observation that ammonia at these sub-lethal concentrations causes damage to the mitochondria of branchial and skin epithelial cells (temperate *Pungitius pungitius*, Matei, 1983, 1984; tropical *Nibea japonica*, Guillén *et al.*, 1994). Similar findings are also found in other organs (liver and brain; Arillo *et al.*, 1981).

    HEA is also known to predispose the gill to histopathological changes that will potentially disrupt ion transport (Smart, 1976; Daoust and Ferguson, 1984). The disruptions of epithelial integrity will have adverse consequences for ion transport and other cellular processes. HEA alone causes the proliferation of branchial mucous cells, which will affect diffusion distances across the gill (Ferguson *et al.*, 1992).

C. Effects on Blood pH and Plasma Ionic Concentration

During exposure to HEA, blood pH is expected to increase as the weak base NH_3 diffuses from the water into the fish. At typical plasma pH levels, NH_3 would bind an H^+ resulting in the pH increase. Avella and Bornancin (1989) were able to clearly demonstrate this using the freshwater trout perfused head preparation during exposure to $1 \, mmol \, l^{-1} \, NH_4Cl$. The concentration of ammonia in the perfusate also increased during HEA exposure. Cameron and Heisler (1983) were also able to produce a blood alkalosis *in vivo* in freshwater rainbow trout, by using a combination of high environmental pH (pH_{env}) and HEA. However, in the majority of *in vivo* studies, the results have been far from consistent since non-statistically significant and temporal increases in plasma pH usually result from similar exposure conditions (*O. mykiss*, Smart, 1978; Wilson and Taylor, 1992; Wilson *et al.*, 1994; temperate *Myoxocephalus octodecimspinosus*, Claiborne and Evans, 1988; tropical *Periophthalmodon schlosseri*, Ip, Randall *et al.*, 2004).

In marine fishes, the leaky paracellular junctions represent another potential path for ammonia entry as NH_4^+. This, however, would not be expected to result in a plasma pH change as was concluded from studies on the sculpin (*M. octodecimspinosus*, Claiborne and Evans, 1988). In the tropical brackish water mudskipper *P. schlosseri*, HEA (75 mM NH_4Cl at pH 7.0) also has no effect on blood pH; however, blood ammonia does increase (Ip, Randall *et al.*, 2004). In the tropical freshwater weatherloach (*Misgurnus anguillicaudatus*), blood pH is unaffected by 48 hours' exposure to 30 mM TAmm-N (pH 7.2; Tsui *et al.*, 2002). In the same study, 48 hours of aerial exposure, which results in a 4-fold build-up of endogenous ammonia (Chew *et al.*, 2001), also causes blood alkalosis. Chronic sub-lethal exposure of the tropical freshwater tilapia (*Tilapia zilli*) to ammonia results in blood alkalosis (64–194 $\mu mol \, l^{-1} \, NH_3$-N; El-Shafey, 1998). Some caution should be taken in interpreting the results from these studies on tropical fishes because blood samples were not taken in cannulated fishes. Blood pH values are affected by anesthesia, fish handling, and taking blood by puncture, however, since these fish tend to be small ($\ll 100 \, g$) such procedures are unavoidable (see Wood, 1993).

In freshwater fishes, Na^+ losses during exposure to HEA have been reported in catfish (75 $\mu mol \, l^{-1} \, NH_3$-N or 1.7 $mmol \, l^{-1}$ TAmm-N; Tomasso *et al.*, 1980) and trout fry (24 h 36 $\mu mol \, l^{-1} \, NH_3$-N, 15.8 $mmol \, l^{-1}$ TAmm-N; Paley *et al.*, 1993) and juveniles (24 h 6.5–28.2 $\mu mol \, l^{-1} \, NH_3$-N, 1.2–5.2 $mmol \, l^{-1}$ TAmm, Twitchen and Eddy, 1994). However, the use of lower environmental ammonia levels in these studies had no effect on plasma and body Na^+ and Cl^- levels. In two additional studies on the rainbow trout, plasma ammonia levels were also unperturbed (0.5 $mmol \, l^{-1}$ TAmm-N pH 8.25 for up to 4 days, Vedel *et al.*, 1998; 1 $mmol \, l^{-1}$ TAmm-N pH 8 for 24 h,

Wilson and Taylor, 1992). Similar observations were made on seawater and brackish water (33% seawater) adapted trout in the latter study.

In seawater acclimated Atlantic salmon (temperate *Salmo salar*), acute exposure to HEA (18.2 μmol l^{-1} NH$_3$-N; 1.8 mmol l^{-1} TAmm-N) results in large increases in plasma osmolality and ion levels prior to death; however, sub-lethal HEA (1.5 μmol l^{-1} NH$_3$-N; 0.17 mmol l^{-1} TAmm-N for 2 weeks) has no effect on plasma ions (Knoph and Thorud, 1996). In adult turbot (temperate *Scophthalmus maximus*) exposed chronically to 6.4–57 μmol l^{-1} NH$_3$-N, plasma osmolality and Na$^+$ decreases (Person Le Ruyet *et al.*, 1997).

With the exception of the turbot, data on temperate teleosts, both freshwater and marine, indicate that at higher (lethal) environmental ammonia concentrations, iono-regulatory impairment or failure results; while at lower (sub-lethal) concentrations, fishes are able to adequately regulate their plasma ions.

For the mudskipper *P. schlosseri*, plasma [Na$^+$] (and [Cl$^-$]) increases during exposure to HEA (6 days in 8 and 100 mmol l^{-1} TAmm-N; 36 and 446 μmol l^{-1} NH$_3$-N, respectively). Despite the fact that these environmental ammonia concentrations are markedly higher than the lethal dose for temperate teleosts, they are still sub-lethal to the mudskipper (Peng *et al.*, 1998; Randall *et al.*, 1999). Conflicting interpretations of the mudskipper data exist since this species is capable of active ammonia excretion, which is thought to involve a Na$^+$/NH$_4^+$ exchange process. The increase in plasma [Na$^+$] might simply reflect an increase in Na$^+$/NH$_4^+$ exchange. This view is taken up in detail in the section on active NH$_4^+$ excretion (section IV.A). In a study by Buckley *et al.* (1979), chronic acclimation of the coho salmon (*Oncorhynchus kisutch*) to HEA (1.1–19.4 μmol l^{-1} NH$_3$-N; 0.2–3.4 mmol l^{-1} TAmm-N) in freshwater did not change plasma ammonia concentration but did result in an increase in plasma [Na$^+$]. This presumably is a result of the induction of some sort of NH$_4^+$/Na$^+$ exchange mechanism (non-obligatory NH$_4^+$/Na$^+$ exchange or Na$^+$ uptake coupled to H$^+$ excretion) (see Salama *et al.*, 1999).

The effects of HEA exposure on plasma ion levels, flux rates, and pH are lacking in other "tropical fishes" but we would expect that the majority respond to HEA in a similar manner as their temperate relatives rather than the giant mudskipper *P. schlosseri. Periophthalmodon schlosseri* is exceptional in its ability to eliminate ammonia against large inwardly directed gradients and, thus, is unlikely to be a typical "tropical fish."

D. Effects at the Cellular and Sub-cellular Levels

In mice, high ammonia levels in the brain induce an increase in extracellular glutamate, due to increased neuronal release, decreased re-uptake, or both (Hilgier *et al.*, 1991; Bosman *et al.*, 1992; Rao *et al.*, 1992; Schmidt *et al.*, 1993; Felipo *et al.*, 1994). It has been proposed that ammonia toxicity is mediated by excessive activation of NMDA-type glutamate receptors in the brain (Marcaida *et al.*, 1992), leading to cerebral ATP depletion (Marcaida *et al.*, 1992; Felipo *et al.*, 1994) and increases in intracellular Ca^{2+}, with subsequent increases in extracellular K^+ and cell death. L-Carnitine prevents acute ammonia toxicity because it prevents glutamate neurotoxicity by increasing the affinity of glutamate for the quisqualate type of glutamate receptors (Felipo *et al.*, 1994). Blocking NMDA receptors by antagonists such as MK-801 also significantly reduces ammonia toxicity in rats (Marcaida *et al.*, 1992; Hermenegildo *et al.*, 1996) and the weatherloach, *M. anguillicaudatus* (K. N. T. Tsui, D. J. Randall and Y. K. Ip, unpublished data). Thus, it has been suggested that elevated ammonia leads to increased glutamate levels and excessive activation of NMDA receptors. Hermenegildo *et al.* (2000), however, showed that activation of NMDA receptors preceded the increase in extracellular glutamate. Blocking NMDA receptors with MK-801 was also shown to prevent the increase in extracellular glutamate. It has been suggested much earlier that NH_4^+ can substitute for K^+ and affect the membrane potential in the squid giant axon (Binstock and Lecar, 1969). In addition, Beaumont *et al.* (2000) reported measured levels of depolarization of muscle fibers in trout with elevated levels of ammonia in their tissues (from $-87\,mV$ to $-52\,mV$) that matched the effect predicted on the basis of the measured gradient for NH_4^+ across the cell membranes. Thus, it is proposed that the excessive activation of NMDA receptors is due to neuronal depolarization caused by an increase in extracellular NH_4^+. Excessive activation of NMDA-type glutamate receptors results in an increase in intracellular Ca^{2+}, activating Ca^{2+}-dependent enzymes including protein kinases, phosphatases, and proteases. This would lead to alterations in MAP-2 and decreases in PKC-mediated phosphorylation and concomitant activation of Na^+, K^+-ATPase. Increased ATPase activity leads to consumption of larger amounts of ATP and could explain the depletion of cerebral ATP, which, in turn, reverses the Na^+-dependent glutamate uptake mechanism. The increase in extracellular glutamate is thus a consequence, and not a cause, of activation of NMDA receptors. Therefore, the primary cause of ammonia toxicity may be the depolarization effect of NH_4^+ on neurons, leading to excessive activation of NMDA receptors and subsequent death of the cell (Randall and Tsui, 2002).

Elevated ammonia levels are associated with increased glutamine levels in the brain in many vertebrates (Ip, Chew *et al.*, 2001a; Brusilow, 2002). Glutamine synthetase (GS) activity is increased during postprandial hyperammonia in trout (Wicks and Randall, 2002), reducing the magnitude of the ammonia pulse in the brain. It has also been suggested, however, that high glutamine levels can contribute to toxicity (Brusilow, 2002). It is proposed that increased glutamine production and accumulation causes increased astrocyte cell volume, leading to cellular dysfunction, brain edema, and death. L-methionine S-sulfoximine (MSO) inhibits GS, reduces edema, attenuates ammonia-induced increases in brain extracellular K^+, and ameliorates ammonia toxicity (Brusilow, 2002). Thus, glutamine formation may either exacerbate or ameliorate ammonia toxicity, depending on the site of formation and the species in question (section V.C).

Formation of glutamine requires the participation of the cytosolic and mitochondrial compartments. Glutamate formation usually involves glutamate dehydrogenase (GDH) in the mitochondrial matrix and GS in the cytosol. If a large amount of glutamate is made during hyperammonia in the fish, equally large amounts of α-ketoglutarate (α-KG) and NADH are needed. Consumption of α-KG pulls it away from the Krebs cycle, and oxidation of NADH disrupts redox balance. This can lead to a reduction in ATP production through the electron transport chain (Campbell, 1973). The efflux of a large amount of glutamate from the mitochondria would interfere with the movement of NADH across the inner mitochondrial membrane through the malate–aspartate shuttle. In addition, the increased ATP demand for the GS reaction may account for the decrease in cerebral ATP noted in ammonia-exposed rainbow trout (Smart, 1978; Arillo *et al.*, 1981; Mommsen and Walsh, 1992).

Thus, ammonia toxicity may be due to membrane depolarization and a rise in extracellular K^+ in the brain, exacerbated by NMDA receptor activation, glutamine-mediated astrocyte swelling, depletion of Krebs cycle intermediates and disruption of redox balance. These mechanisms are not mutually exclusive and could be additive in their effects. Drugs like MSO and MK-801 ameliorate but do not prevent ammonia toxicity.

IV. DEFENSE AGAINST AMMONIA TOXICITY AT THE BRANCHIAL AND EPITHELIAL SURFACES

A. Active NH_4^+ Transport

Active elimination of NH_4^+ is required to maintain nitrogen balance when the NH_3 partial pressure gradient (ΔP_{NH3}) and NH_4^+ electrochemical gradient are directed inward and therefore cannot contribute to net excretion (Figure 8.1B). HEA and/or alkaline pH as well as aerial exposure are conditions that may require active NH_4^+ excretion in order to maintain a positive nitrogen balance. Compared to detoxification and storage options such as urea synthesis, active NH_4^+ is an energetically more favorable alternative.

Synthesis of 1 mol of urea from 2 mol of NH_4^+-N requires 5 mol of ATP versus only 1 mol ATP for the removal of 2 mol of NH_4^+-N through the Na^+, K^+ (or NH_4^+)-ATPase (1 ATP: 2 K^+ or NH_4^+: 3 Na^+). There are a few tropical fishes (mudskippers, Randall et al., 1999; Chew, Wong et al., 2003; Ip, Randall et al., 2004; the African sharptooth catfish, Ip, Subaidah et al., 2004) which can excrete ammonia against large inward gradients, indicating active NH_4^+ excretion is being utilized. Out of these, the giant mudskipper *P. schlosseri* has been studied the most extensively.

1. MUDSKIPPERS

The giant mudskipper *P. schlosseri*, belonging to the Order Perciformes and Family Gobiidae, is amphibious and commonly found on muddy shores in estuaries and in the tidal zone of rivers in Singapore (Ip et al., 1990), Indonesia, New Guinea, India, Peninsular Malaysia, Sarawak, and Thailand (Murdy, 1989). It is carnivorous and can grow up to 27 cm in length. It is the only species of mudskipper that has not been found outside of the tropics. Similar to other mudskippers, *P. schlosseri* builds burrows in the mud, in which it lays and rears eggs during the breeding season.

Amongst the tropical fishes *P. schlosseri* is exceptional in its capability of maintaining its plasma ammonia concentration and excreting ammonia against large inward NH_3 and NH_4^+ gradients (up to at least 30 mmol l^{-1} TAmm-N at pH 7.2; Randall et al. 1999; Ip, Chew et al. 2001a). In more recent works using a shorter, 1 hour time course, Ip, Randall et al. (2004) showed that the net total ammonia flux (J_{AMM}) was actually markedly enhanced in the presence of HEA with a maximal response at 20 mmol l^{-1} TAmm-N pH 7 or 8. In order to clearly establish if active NH_4^+ transport is indeed taking place, alternative explanations need to be eliminated. Alternatives include, (a) diffusion trapping of NH_3 through boundary layer acidification ($NH_3 + H^+ \rightarrow NH_4^+$), and (b) paracellular diffusion of NH_4^+ down its electrochemical gradient via leaky tight junctions (see Ip, Chew et al., 2001a; Wilkie, 2002).

(a) NH_3 trapping can be discounted since elimination of the acidic conditions of the boundary layer by buffer addition (HEPES or TRIS) to the medium does not affect J_{AMM} (Wilson et al., 2000b; Chew, Hong et al., 2003). Also, the addition of the V-ATPase inhibitor bafilomycin A1, which significantly decreases the net acid flux (J_{ACID}), does not alter J_{AMM} (Ip, Randall et al., 2004).

(b) The paracellular efflux of NH_4^+ is another possible path for ammonia excretion; however, transepithelial potential (TEP) measurements do not indicate that it contributes to J_{AMM} during HEA (TEP = +10 mV; Randall et al., 1999). Also, although *P. schlosseri* has very high densities of mitochondria-rich cells (MRCs) in its gills, there are very few leaky tight junctions, which would typically be associated with Na^+ efflux path in hypo-osmoregulation (Wilson *et al.*, 1999). The remaining viable explanation is active NH_4^+ elimination.

The mechanism of NH_4^+ elimination can be ascertained through the use of specific inhibitors of ion transport proteins. NH_4^+ can substitute for K^+ in a number of ion transport proteins owing to similarity in the size of these ions (hydration radius). It has been shown that the kidney Na^+, K^+-ATPase can also transport NH_4^+ into cells (see Wall, 1996). In *P. schlosseri*, Randall *et al.* (1999) showed that active J_{AMM} during HEA was sensitive to the Na^+, K^+-ATPase inhibitor ouabain and NH_4^+ could stimulate the ATPase at physiological K^+ concentrations. In addition, Mallery (1983) has shown that gill Na^+, K^+-ATPase in the semi-tropical toadfish, *Opsanus tau*, is stimulated by NH_4^+. This contrasts with the finding in temperate rainbow trout by Salama *et al.* (1999) that ouabain-sensitive ATPase activity was not stimulated with NH_4^+ in the presence of physiological levels of K^+.

The NHE also has high selectivity for NH_4^+ (Kinsella and Aronson, 1981). In *P. schlosseri*, inhibition of NHE by amiloride decreases J_{AMM} (Randall *et al.*, 1999). Wilson *et al.* (2000b) have immunolocalized NHE 2- and 3-like proteins to the apical crypts of branchial MRCs. As a corollary, plasma Na^+ level increases in response to HEA (Randall *et al.*, 1999) and increasing environmental Na^+ levels, while keeping osmolality constant, increases J_{AMM} (Kok, 2000). Frick and Wright (personal communication) found in the mangrove killifish, *Rivulus marmoratus*, that $5\,mmol\,l^{-1}$ NH_4Cl resulted in 100% mortality in freshwater-acclimated killifish although this dose was well below lethal concentration for those exposed to brackish water. This suggests a role of Na^+ in survival of hyperammonia stress in *R. marmoratus*. In freshwater species making use of active NH_4^+ excretion, an alternative strategy to the use of NHE must be used since Na^+ levels will be low in their environment.

Carbonic anhydrase reversibly catalyzes the CO_2 hydration and HCO_3^- dehydration reactions and has an important role in providing acid-base equivalents for ion exchange, notably Na^+/H^+ and Cl^-/HCO_3^- exchanges (Henry and Heming, 1998). In *P. schlosseri*, inhibition of CA with acetazolamide significantly inhibits J_{AMM} (Kok, 2000; Wilson *et al.*, 2000b). Carbonic anhydrase has been immunolocalized to the apical cytoplasmic region of gill mitochondria-rich cells (Wilson *et al.*, 2000b).

These enzymes are all present in the branchial mitochondria-rich cells of *P. schlosseri* as determined by immunohistochemistry (Wilson *et al.*, 2000b), which strongly indicates the gill as the site of active NH_4^+ elimination. The active J_{AMM} has also been shown to occur predominantly through the head region (Ip, Randall *et al.*, 2004). The gills of *P. schlosseri* are highly modified for air breathing and its secondary lamellar have a high density of mitochondria-rich cells (Low et al., 1988, 1990; Wilson et al., 1999).

Wilson et al. (2000b) have suggested that the cystic fibrosis transmembrane receptor (CFTR)-like anion channel immunolocalized to the apical crypt of MRCs may be involved in direct HCO_3^- efflux to account for the lack of correlation of net acid excretion (J_{ACID}) and J_{AMM}, and the example of the role of the CFTR in HCO_3^- excretion in the duodenum has been cited (Hogan et al., 1997). However, recently it has been shown that the duodenal CFTR does not directly facilitate transmembrane HCO_3^- movement (Praetorius et al., 2002). Instead, it would appear that the CFTR has an indirect role in HCO_3^- excretion, cycling Cl^- while HCO_3^- efflux is mediated through a Cl^-/HCO_3^- anion exchanger (AE). In support, we have recently found that J_{AMM} is sensitive to SITS, a Cl^-/HCO_3^- exchange inhibitor of which the CFTR is insensitive (Kok, 2000). Thus, in the revised version of the P. schlosseri model, HCO_3^- efflux is mediated through an apical AE in exchange for Cl^- and the Cl^- cycles through the apical CFTR Cl^- channel.

A cohabitant of the mangrove mud flats (in Singapore and Peninsular Malaysia) with P. schlosseri is the Boddart's goggle-eyed mudskipper, Boelophthalmus boddaerti (Order: Perciformes; Family Gobiidae) which is herbivorous. It is often found on mudflats during low tides browsing on surface algae. Unlike P. schlosseri, B. boddaerti has wide distribution and can be found in the subtropical regions in Asia.

Boelophthalmus boddaerti is also capable of excreting ammonia against inward gradients although of a much smaller magnitude ($8 \, mmol \, l^{-1}$ TAmm-N pH 7; Kok, 2000; Chew, Hong et al., 2003). Ammonia excretion was also shown to be sensitive to the respective NHE and Na^+, K^+-ATPase inhibitors amiloride and ouabain, and stimulated by increased environmental Na^+ levels (Kok, 2000). Unfortunately, TEP measurements have not been made in B. boddaerti exposed to ammonia, although information is available on specimens exposed to salinity changes (Lee et al., 1991). So, an accurate assessment of NH_4^+ electrochemical gradients cannot be made. However, if we assume similar values as obtained from P. schlosseri, active NH_4^+ elimination seems likely in this species, although, unlike in P. schlosseri, active NH_4^+ elimination in B. boddaerti is not capable of maintaining J_{AMM} at $pH_{env} > 8$ or ammonia concentrations $>20 \, mmol \, l^{-1}$ (Kok, 2000; Chew, Hong et al., 2003).

Another potential mechanism that may play a role in ammonia excretion in fish is an NH_4^+ transporter. In plants and micro-organisms NH_4^+ is transported across cell membranes via NH_4^+ transporters (AMT; Howitt and Udvardi, 2000). Since NH_4^+ is an important source of nitrogen for plants this mechanism has been well studied. In the roots and leaves of higher plants a high affinity uniport mechanism of NH_4^+ uptake has been described.

NH_4^+ uptake is driven by the proton motive force which is generated by the plasma membrane P-type H^+-ATPase. In animals, the Rh type B glycoprotein, which shows some homology with the AMT, has been suggested to be a putative ammonia transporter in mammals (Heitman and Agre, 2000; Liu *et al.*, 2001). Importantly, it has recently been shown that Rh proteins can mediate NH_4^+ transport (Marini *et al.*, 2000; Westhoff *et al.*, 2002). This presents a very interesting future area of exploration in the study of ammonia excretion in fish.

B. Environmental Acidification

In water-breathing fishes at circum-neutral and alkaline pH_{env}, the water passing over the gills is acidified by the hydration of excreted CO_2, giving rise to $HCO_3^- + H^+$, and to a lesser extent by direct H^+ excretion (Wright *et al.*, 1989). The hydration reaction is very rapid and does not necessarily require extracellular CA (Henry and Heming, 1998). Since the gill functions as a counter-current exchanger and ventilation volumes are generally quite large, the acidification of expired water on ammonia speciation will only be relevant in the boundary layer close to the surface of the epithelium. This boundary layer acidification has been shown to be important in shifting the ammonia equilibrium toward NH_4^+ formation thereby lowering NH_3 to maintain the transbranchial ΔP_{NH3} gradient (Wright *et al.*, 1989; Wilson *et al.*, 1994). Generally, the transbranchial ΔP_{NH3} gradient can account for most of the J_{AMM} (Cameron and Heisler, 1983).

For fishes living in stagnant water of a finite volume (puddle, tidepool or water-filled burrow), it is possible for the acidifying effects of their excretions of CO_2 and H^+ to have a significant impact on pH_{env}. This lowering of pH_{env} has obvious advantages for dealing with the elevations in environmental ammonia levels which would also occur under such conditions by reducing the concentration of the more permeable species, NH_3 (Figure 8.1C).

1. MUDSKIPPERS

When the giant mudskipper *P. schlosseri* is kept in an artificial burrow, the ambient ammonia increases to $10 \, mmol \, l^{-1}$ within 5–6 days (Ip, Randall *et al.*, 2004). *Periophthalmodon schlosseri* is capable of active NH_4^+ excretion so it needs not worry about maintaining transbranchial ΔP_{NH3} gradients to maintain J_{AMM} (see section IV.A). However, NH_3 shows moderate membrane permeability and given suitable gradients will diffuse into the animal (Barzaghi, 2002). This is made obvious by the fact that only exposure to $30 \, mmol \, l^{-1}$ TAmm-N at high pH_{env} (pH 9) is lethal (Kok, 2000; Barzaghi, 2002). This mudskipper has been shown to be able to respond to HEA as well as high pH_{env} (which will increase fNH_3) by increasing both CO_2

excretion (M_{CO2}) and net acid excretion (J_{ACID}) (Kok, 2000; Ip, Randall *et al.*, 2004). Surprisingly, the large increases in M_{CO2} are not accompanied by complementary increases in oxygen uptake (M_{O2}). So, the respiratory exchange ratio (RER, $M_{CO2}:M_{O2}$) jumps from 1 (50% SW pH 7) to $\sim$10. High pH_{env} (pH 9) gives the greatest RER while HEA (50–75 mmol l^{-1} NH$_4$Cl pH 7 in 50% seawater) results in RER values around 6–7. A combination of high pH_{env} and HEA gives an additive effect (25–50 mmol l^{-1} NH$_4$Cl at pH 8 in 50% seawater). Radioisotope labeling studies indicate that glucose, Krebs cycle intermediates, and certain amino acids can act as substrates for the excess CO_2 excreted (Koh, 2000).

J_{ACID}, which is equal to the summation of titratable acid flux (J_{TA}) and J_{AMM}, is inhibited by acidic pH_{env} (pH 6) and maximally stimulated (4-fold) by alkaline pH_{env} up to pH 8.5. J_{ACID} is stimulated by HEA maximally at 20 mmol l^{-1} NH$_4$Cl at both pH 7 and 8. J_{AMM} is not affected by external pH (6–9); however, in the presence of HEA there is an unexpected significant increase in J_{AMM}. In *P. schlosseri* exposed to 25–75 mmol l^{-1} NH$_4$Cl, the blood ammonia level is unaffected. This would suggest that increasing M_{CO2} and J_{ACID} to lower environmental pH minimizes ammonia uptake by NH$_3$ diffusion (Randall *et al.*, 1999; Ip, Randall *et al.*, 2004).

In vivo inhibitor studies on *P. schlosseri* indicate that at least 50% of the J_{ACID} is sensitive to the specific V-type H$^+$-ATPase inhibitor bafilomycin A1 (Ip, Randall *et al.*, 2004). This H$^+$-ATPase has been immunolocalized to the apical crypt of branchial mitochondria-rich cells in this species (Wilson *et al.*, 2000b). The inhibition of J_{ACID} by bafilomycin has no effect on J_{AMM}, indicating that there is no dependence of ammonia excretion on acidification (as would be related to NH$_3$ trapping in the boundary layer). This is supported by the lack of an effect of boundary pH perturbation by buffers on J_{AMM} (Wilson *et al.* 2000b). This contrasts markedly with other teleostean fishes (trout) that depend on ΔP_{NH3} for ammonia excretion (Cameron and Heisler, 1983; Wilson *et al.*, 1994).

Both the CA inhibitor acetazolamide and the Cl$^-$/HCO$_3^-$ exchange inhibitor SITS significantly reduce J_{ACID} (Kok, 2000; Wilson *et al.*, 2000b). CA is found in the branchial MRCs and is likely important in catalyzing the hydration of CO_2 to provide an intracellular supply of H$^+$ and HCO$_3^-$ for the various ion transport proteins. Notably, the bafilomycin sensitive H$^+$-ATPase contributes significantly to J_{ACID}. Protons will also be consumed in NH$_3$ hydration which supplies NH$_4^+$ for an apical Na$^+$/NH$_4^+$ exchanger. The SITS-sensitive Cl$^-$/HCO$_3^-$ exchanger likely plays an important role in removing the accumulated HCO$_3^-$ since its build-up will affect the CO_2 hydration reaction and the supply of H$^+$. It is also possible that these inhibitors are affecting CO_2 delivery to the gills by knocking out erythrocyte CA and AE. Presumably, if this were the case then M_{CO2} would also be inhibited but data is lacking.

In summary, it would appear that these acidifying mechanisms are not necessary for maintaining J_{AMM} in *P. schlosseri* since the addition of buffers to the medium to abolish the boundary layer pH and proton pump inhibition does not affect J_{AMM} (Wilson *et al.*, 2000a). Rather, acidification serves to prevent the entry of NH_3 by reducing its concentration in the medium.

The mudskipper *B. boddaerti* is also capable of increasing the production and excretion of CO_2 in response to an alkaline pH_{env} and HEA (at constant pH) (Koh, 2000; Kok, 2000). An alkalinization of pH_{env} induces an increase in M_{CO2} resulting in an increase in the respiratory exchange ratio (RER) to ~ 10 (M_{CO2}/M_{O2}) since M_{O2} is largely unaffected by the increase in pH_{env} (pH 6–9). An increase in environmental ammonia also induces a similar response in this species (8 and 15 mmol l^{-1} NH_4Cl pH 7 in 50% seawater). In *B. boddaerti*, the excess CO_2 likely originates from the Krebs cycle although an increase in O_2 consumption does occur, indicating that O_2 is not utilized as the terminal electron acceptor in mitochondrial redox balance. It has also been demonstrated that NADH can be oxidized to NAD under anoxic condition although at a lower rate than under normonic conditions. These results suggest a unique NADH dehydrogenase that can regenerate NAD in the absence of O_2 in *B. boddaerti* (Koh, 2000).

Similar to *P. schlosseri*, *B. boddaerti* also has a branchial bafilomycin-sensitive ATPase and is capable of increasing J_{ACID} with increases in pH_{env}. (Y. K. Ip and S. F. Chew, unpublished data). However, unlike *P. schlosseri*, *B. boddaerti* is incapable of increasing J_{ACID} in response to HEA.

C. Lower Permeability of Epithelial Surfaces to NH_3

In order to properly address the topic of ammonia permeation through the plasma membrane of a cell we will briefly review the current state of the art in this area (see reviews by Boron *et al.*, 1994; Zeidel, 1996). NH_3 is believed to permeate the membrane by a solubility-diffusion mechanism involving its solvation in the lipid of the membrane, diffusion across the bilayer, and its re-emergence on the other side. The solubility of NH_3, as measured by the oil:water partition coefficient (~ 0.1), is not particularly high; however, NH_4^+ is much less permeable than NH_3 because of its charge and larger size. It is generally accepted that the diffusion of NH_3 across the bilayer is facilitated by "kinks" in the phospholipid acyl chains (decrease in packing). In this respect, the composition of the bilayer is thus a key determinant of membrane permeability. Lande *et al.* (1995) were able to modulate NH_3 permeability in artificial vesicle preparations through changes in lipid fluidity by way of membrane composition with higher permeability correlating positively with lipid fluidity. Generally, unsaturated hydrocarbon chains of fatty acids lead to higher fluidity and permeability whereas saturated fatty acids lead to lower permeability. Cholesterol at certain concentrations increases packing, reducing fatty acids mobility (Zeidel, 1996). In addition to the lipid composition, permeability is also determined by bilayer asymmetry, that is, the presence of intrinsic trans-membrane proteins and the outer mucin coat (Lande et al., 1994). It has been shown that changes in the composition of a single leaflet will affect permeability and that the outer or exofacial leaflet is the most restricted in its mobility and thus likely represents the diffusion barrier (Negrete et al., 1996; Hill et al., 1999).

Epithelial cells are polarized cells with apical and basolateral membrane domains which are kept separated by the tight junction complex. The composition, and thus the properties of these two domains, can be quite distinct and modified independently. An important point to consider regarding apical versus basolateral membrane permeabilities is the large difference in surface area. For a typical cell, the area of the apical membrane is one-fifteenth that of the basolateral membrane. So, even if the permeability properties of the membranes were identical, on a unit area basis, the apical membrane would provide the greater barrier to transepithelial diffusion because of its much smaller area. In a cell like the branchial chloride cell with its extensive basolateral tubular system the apical:basolateral surface area ratio would be much smaller.

Since NH_3 does have low lipid solubility, the presence of water-filled channels would be expected to greatly enhance NH_3 fluxes. There is now direct evidence that NH_3 can traverse the membrane through water channel proteins or aquaporins (AQP1, Nakhoul et al., 2001b). Aquaporins (AQP3) have been reported in fish gills (Cutler and Cramb, 2002; J. Fuentes, personal communication) although only in the basolateral membrane and intracellular vesicles (Lignot et al., 2002). In a recent review, Wilkie (2002) has given the AQP an apical localization, which may be suitable in aiding transepithelial NH_3 fluxes but would be disastrous for water fluxes because of the presence of large osmotic gradients across the gills of both marine and freshwater fishes. In light of the above discussion on the barrier function of apical versus the basolateral membrane, it is unlikely that branchial AQPs have a significant role in transepithelial NH_3 fluxes.

In addition to the transcellular pathway described above, paracellular permeability is also an important component of epithelial permeability. Tight junctions, which provide the gate function or barrier to the movement of molecules through the paracellular path, are not static features of the epithelium. They are dynamic and subject to physiological regulation and show ion selectivity (Madara, 1998; Yap *et al.*, 1998; Denker and Nigam, 1998). In fishes, the best example of the modulation of paracellular permeability is in response to an increase in environmental salinity to facilitate a paracellular Na^+ efflux down its electrochemical gradient (see Evans *et al.*, 1999; Chapter 9, this volume).

Membranes of several cell types facing the gastric and urinary tracts have been found to have relatively low permeability to NH_3. Kikeri *et al.* (1989) found that when ammonium was applied in the lumen of the medullary ascending limb of Henle of the mouse, the initial pH_i change was in the acid direction. This acid change could be blocked pharmacologically by the application of furosemide. NH_4^+ then had no effect on pH_i. Therefore, Kikeri *et al.* (1989) concluded that the membranes were relatively impermeable to NH_3. Despite a subsequent suggestion by Good (1994) that there could be rapid efflux of NH_3 through the basolateral membranes in these experiments, later works have indeed substantiated the existence of plasma membranes in the urinary tract of the rabbit (Yip and Kurtz, 1995) and other animal cell membranes with low NH_3 permeability. A particular elegant demonstration of such a low NH_3 permeability is that of Singh *et al.* (1995) on the luminal (apical) surface of colonic crypt cells of the rabbit. The apical membranes of bladder cells of the rabbit, which also have a low NH_3 permeability, have an unusual composition, 70–90% of the membrane area being occupied by paracrystalline arrays of proteins called uroplakins (Chang *et al.* 1994). It therefore seems likely that relatively high NH_3 permeability is a normal property of cell membranes that is only reduced when the phospholipid composition is altered and/or when lipids are replaced by proteins (Marcaggi and Coles, 2001).

1. MUDSKIPPERS

The giant mudskipper *P. schlosseri*, which has an exceptionally high tolerance to external ammonia (Ip *et al.*, 1993; Peng *et al.*, 1998), is capable of preventing an elevation in its plasma ammonia when confronted with large inward NH_3 and NH_4^+ gradients (Randall *et al.*, 1999). Besides being able to actively excrete NH_4^+ across its gills (see section IV.A), *P. schlosseri* apparently possesses a skin surface which has a low permeability to ammonia. Using an Ussing-type apparatus, Ip, Randall *et al.* (2004) were able to demonstrate that the skin of *P. schlosseri* had a lower permeability to ammonia than the abdominal skin of the frog *Rana catesbiana*. Since the skin *P. schlosseri* is highly vascularized for cutaneous respiration, a low ammonia permeability in the skin is essential to reduce the influx of NH_3 during HEA exposure, and to prevent a back flux of NH_3 to render the branchial active NH_4^+ transport process (see section IV.A) effective.

Although membrane fluidity in the skin of *P. schlosseri* has not been measured directly, the skin membranes are predicted to have a low fluidity based on measured cholesterol levels and phospholipid composition (Ip, Randall *et al.*, 2004). The cholesterol level in the skin of *P. schlosseri* is high (4.5 μmol g^{-1}). This gives rise to a very high cholesterol:phospholipid ratio of >6, compared to values of <1 in other teleosts which would be expected to have high NH$_3$ permeability (pNH$_3$) because they make use of ΔP_{NH_3} for J_{AMM} (Wood, 1993; Ip, Chew *et al.*, 2001a; Wilkie, 2002). The apical membranes of bladder cells of the rabbit, which have a low NH$_3$ permeability, contain mainly (70–90% of the membrane area) uroplakins, a para-crystalline protein (Chang *et al.*, 1994). This led Marcaggi and Coles (2001) to suggest that the relatively high NH$_3$ permeability of cell membranes is reduced only when the normal lipids are replaced with proteins. Results obtained from the skin of the giant mudskipper (Ip, Randall *et al.*, 2004) suggest that cholesterol may serve a similar function in lowering NH$_3$ permeability.

Phosphatidylcholine is known to stabilize lipid bilayer (Lande *et al.*, 1995) and is the most abundant phospholipid present (50%) in the skin of *P. schlosseri* (Ip, Randall *et al.*, 2004). Phosphatidylethanolamine, which is associated with decreased membrane order (i.e., higher membrane fluidity), is present in low amounts (<15%). The phosphatidylcholine:phosphatidyl-ethanolamine ratio, an indicator of membrane fluidity, is much higher in the skin of *P. schlosseri* (3.4) compared to values obtained in temperate teleosts (1.7; *Anguilla rostrata* and *O. mykiss*; Crockett, 1999). The fraction of saturated (52%) to unsaturated (28% mono- and 20% poly-unsaturated) fatty acids is high (Barzaghi, 2002; Ip, Randall *et al.*, 2004) in comparison with the skin of the trout *O. mykiss* (Ghioni *et al.*, 1997). This again would indicate low membrane fluidity in the skin of this mudskipper.

The cholesterol content in the skin of *P. schlosseri* increases significantly to 5.5 μmol g^{-1} following exposure to HEA (30 mmol l^{-1} NH$_4$Cl pH 7 in 50% seawater for 6 days) (Ip, Randall *et al.*, 2004). In addition, the levels of sphingomyelin, which is structurally rigid and hydrophobic, also increase in the skin of *P. schlosseri* following HEA acclimation (Ip, Randall *et al.*, 2004). Sphingomyelin has a similar effect as cholesterol on membrane fluidity (Lande *et al.*, 1995). These results indicate that *P. schlosseri* can further reduce membrane fluidity, and hence ammonia permeability, in its skin, after long period of exposure to HEA (Ip, Randall *et al.*, 2004; Figure 8.1D).

Gill ATPase levels in *P. schlosseri* are the highest recorded in fish (Randall *et al.*, 1999) and this may be a response to low membrane fluidity of its skin. In addition, *P. schlosseri* is restricted to tropical regions, possibly because the low fluidity of its skin restricts it to high temperatures (Ip, Randall *et al.*, 2004).

D. Volatilization of NH₃

In a number of terrestrial ammonotelic invertebrates, NH_3 volatilization contributes significantly to total ammonia elimination and also water conservation (e.g. Greenaway, 1991; Wright and O'Donnell, 1993). Amongst the vertebrates, however, terrestrial ammonotely is uncommon and examples of significant NH_3 volatilization are rare. That is not to say that ammonia volatilization does not occur in non-ammonotelic vertebrates, such as humans, only that its contribution to nitrogenous waste elimination is negligible (Jacquez *et al.*, 1959; Robin *et al.*, 1959).

In teleost fishes, the vast majority of which are ammonotelic, the first report of ammonia volatilization is in the temperate intertidal blenny, *Blennius pholis*, which can only account for 8% of the total ammonia excreted during emersion (Davenport and Sayer, 1986; 13 °C). In this and in the majority of cases, aerial exposure or emersion results in a drastic reduction in ammonia excretion because the loss of a ventilatory water flow deprives the fish of the positive diffusional gradients on which aquatic ammonia excretion is based (see Ip, Chew *et al.*, 2001a). However, there are a few studies on ammonotelic tropical fishes capable of enduring emersion, and volatilizing significant amounts of ammonia (Rozemeijer and Plaut, 1993; Frick and Wright, 2002b; Tsui *et al.*, 2002) (Figure 8.1E).

High temperatures and humidity, characteristic of the tropical climates and the experimental conditions noted above, are factors that increase the likelihood of the ammonia excreted into the film of water covering the body surface being volatilized in significant quantities. The most important factor is the effect of temperature on evaporation rate. Higher temperatures will also decrease the ammonia equilibrium constant (pK_{amm}), resulting in a higher fraction of NH_3 at a given pH [$fNH_3 = 1/(10^{pKamm-pH} + 1)$, where pK_{amm} equals to 9.4 and 9.0 at 20 and 30 °C, respectively].

It should be noted that while the tropical climate provides some prerequisites to enable significant ammonia volatilization, not all tropical fish species that can survive prolonged emersion make use of ammonia volatilization for nitrogen balance [<1% of terrestrial J_{AMM}: *P. schlosseri, B. boddaerti* (Y. K. Ip, J. M. Wilson and D. J. Randall, unpublished data) and *Bostrichthys sinensis* (Leong, 1999)].

1. WEATHERLOACH

The Oriental weatherloach *Misgurnus anguillicaudatus* (Order Cypriniformes; Family Cobitidae) inhabits rivers, lakes, and ponds, also in swamps and rice fields, and prefers muddy bottoms. It can be found from Myanmar to China (53° N to 27° S). During dry seasons, it often buries itself into the mud until it rains. It is an intestinal air-breather and is capable of volatilizing significant amounts of ammonia during aerial exposure (56% of terrestrial J_{AMM}, Tsui et al., 2002; 25 °C) although J_{AMM} is much lower (5–15%) during the terrestrial conditions compared to aquatic control rates (~750 μmol TAmm kg^{-1} h^{-1}; Chew et al., 2001; Tsui et al., 2002). The rate of ammonia volatilization increases with time of aerial exposure (1–4 days; Tsui et al., 2002). For *M. anguillicaudatus*, which is able to increase its skin and gut pH to 8.2, the fNH$_3$ would be 4.7% versus 13.7% at 20 and 30 °C, respectively. The significance of temperature is demonstrated in the finding by Tsui and co-workers (2002), in which the rate of ammonia volatilization doubled with a 10 °C difference in temperature (31% vs 57%). The non-volatilized component of total ammonia excretion is unaffected by temperature, indicating that this is not merely an effect of temperature on metabolic rate and hence endogenous ammonia production.

The elimination of ammonia by volatilization can be divided into a two-step process. First, the ammonia must be transported from the inside (plasma) to the outside (luminal fluid, and/or body surface film of water) of the animal across an epithelial surface. Secondly, the ammonia must be volatilized from the external fluid. In the terrestrial invertebrates, some detailed studies of mechanism have been made and are thus referred to for comparison.

(a) Transepithelial movement: The movement of ammonia across the epithelium may involve passive diffusion of NH$_3$ and/or NH$_4^+$ or active transport of NH$_4^+$. In the terrestrial isopod (*Porcellio scaber*), passive diffusion processes likely dominate (Wright and O'Donnell, 1993) while in the grapsid crab (*Geograpsus grayi*) active NH$_4^+$ excretion is important (Greenaway and Nakamura, 1991; Varley and Greenaway, 1994). The mechanism (s) involved in the transepithelial movement of ammonia into the boundary layer of water covering the weatherloach, mangrove killifish, and leaping blenny are not known. The skin represents an obvious site for excretion, as in the killifish and weatherloach it is well vascularized (Jakubowski, 1958; Grizzle and Thiyagarajah, 1987). In the weatherloach, the gut is also a likely site since it is involved in gas exchange (McMahon and Burggren, 1987) (Figure 8.1E).

In the weatherloach, we know that plasma total ammonia concentration increases to very high levels during HEA (4.2 mmol l^{-1}) and emersion (5.1 mmol l^{-1}), suggesting that excretion may be by diffusion of NH$_3$ using ΔP_{NH3} (Chew et al., 2001; Tsui et al., 2002). Emersion also results in an increase in blood pH which would further increase plasma NH$_3$. Alternatively, the increase in plasma ammonia concentration may simply be related to the inhibition in J_{AMM}. We can make some calculations which will give us an indication of whether passive NH$_3$ gradients are involved in this species using some parameters measured by Tsui et al. (2002). Aerial exposure will result in ammonia building up to 20 mmol l^{-1} in the small volume of water available to the animal (1 ml) within 24 h at 25 °C (calculated from Tsui *et al.*, 2002). It would be reasonable to assume that the concentration of ammonia in the boundary layer covering the skin would be at least as high. The pH of the boundary layer is 8.2 and thus the P_{NH3} would be 6.368 Pa (0.0477 torr). Even after 48 h aerial exposure the plasma P_{NH3} would only be 0.272 Pa (0.0020 torr) with a measured blood pH of 7.5, and TAmm $\sim$5.1 mmol l^{-1} (pK$_{amm}$ and αNH$_3$ calculated from Boutilier *et al.*, 1984). If these assumptions are correct the ΔP_{NH3} gradient would be directed inwards and thus it would not be possible for passive NH$_3$ diffusion to account for the excretion of ammonia into the boundary layer for volatilization. In the terrestrial isopod *P. scaber*, in which passive diffusion is thought to dominate, hemolymph ammonia concentrations can increase up to 100 mmol l^{-1} and an equilibration with external pleon fluid is found (Wright and O'Donnell, 1993).

In the terrestrial grapsid (*G. grayi*) and gecarcinid (*Gecarcoidea natalis*) crabs, the urine, which contains NaCl and low ammonia, is also passed over the gills and the NaCl reabsorbed and NH$_4^+$ excreted (Greenaway and Nakamura, 1991). In *G. grayi*, the exchange of Na$^+$ and Cl$^-$ for NH$_4^+$ and HCO$_3^-$, respectively, are sensitive to apically applied amiloride and SITS indicating NHE and AE are involved (Varley and Greenaway, 1994). The exchange of Na$^+$ for NH$_4^+$ by the NHE is important in increasing branchial fluid ammonia and the exchange of Cl$^-$ for HCO$_3^-$ in alkalinizing the branchial fluid to facilitate significant ammonia volatilization (82% of total nitrogenous waste excreted in *G. grayi*; Greenaway and Nakamura, 1991). Chloride-dependent alkalinization is also seen in the ocypodid (ghost) crab gill (*Ocypode quadrata*; de Vries and Wolcott, 1993).

(b) Alkalinization and volatilization: Alkalinization of the external ammonia-containing fluid (Figure 8.1E) will greatly enhance volatilization by increasing the fraction and hence concentration of the gaseous form of ammonia (fNH$_3$). The grapsid and ocypodid land crabs both make use of alkalinization to enhance volatilization but the terrestrial isopod does not (Wright and O'Donnell, 1993). This makes sense because in the latter case NH$_3$ diffusion is used for transepithelial ammonia movements and increasing the pH$_{env}$ will also result in back diffusion into the animal. Therefore, a prerequisite for effective enhancement of volatilization by alkalinization would be a low NH$_3$ epithelial permeability (see section IV.C).

In *M. anguillicaudatus* the surface pHs of both the skin and gut increase to >8 in response to aerial exposure, which correlate with increased ammonia volatilization (Tsui *et al.*, 2002). The weatherloach is an intestinal air-breather with a highly vascularized intestine containing intraepithelial capillaries (McMahon and Burggren, 1987). Air is gulped from the surface, swallowed, and passed through the digestive tract for O_2 extraction but would also provide a convenient sink for ammonia elimination through excretion and volatilization. The finding that the weatherloach is capable of volatilizing ammonia during HEA ($30\,\mathrm{mmol\,l}^{-1}$ NH_4Cl) exposure indicates that the gut is an important site since the skin route is lost under those conditions (Tsui *et al.*, 2002). Water conditions are not altered during HEA exposure to enhance volatilization from the medium, clearly indicating that the gut is the prime site for volatilization. Indeed, if the animal is denied access to air then ammonia volatilization does not occur (Tsui *et al.*, 2002). Air that is swallowed for O_2 uptake and CO_2 excretion is hence also used as a sink for excreted ammonia.

The volatilization of NH_3 from the boundary layer will result in a decrease in pH as each molecule of NH_3 volatilized leaves a proton behind. The effect would be to decrease the fraction of ammonia as NH_3 and thus the partial pressure gradient for volatilization. The mechanism of boundary alkalinization to enhance ammonia volatilization would thus also be required to maintain volatilization. If alkalinization was achieved through HCO_3^- excretion as in the land crabs, the buffering of the remaining proton would form CO_2 which could then also be volatilized.

The "either or" argument for passive NH_3 excretion and surface alkalinization and the extremely high ammonia tolerance ($100\,\mathrm{mmol\,l}^{-1}$; Y. K. Ip and S. F. Chew, unpublished data) of the weatherloach would suggest that NH_3 permeability of the epithelium is likely kept low, and that active excretion of NH_4^+ into the boundary layer is the dominant mechanism of transepithelial NH_4^+ movement. An increase in surface pH in response to HEA and terrestrial exposure would also indicate that acid trapping of NH_3 (facilitated NH_3 diffusion) is not an important mechanism for J_{AMM}. Thus in conjunction with a predicted low epithelial NH_3 permeability, alkalinization of the boundary layer would increase the P_{NH3} to facilitate volatilization without increasing a back flux into the animal.

Recently, an apical Cl^-/HCO_3^- exchanger has been identified in mouse intestine and stomach as the putative anion transporter 1 (PAT1 or SCL26A6; Wang *et al.*, 2002). PAT1 also has low expression in the colon in mice. This pattern of expression of PAT1 in the mouse gut correlates with the areas of alkalinization in the weatherloach gut and may represent the mechanism for gut alkalinization. In teleost fishes, there is accruing evidence for an apical Cl^-/HCO_3^- exchanger in the intestine involved primarily in HCO_3^- excretion (Wilson *et al.*, 2002).

At first it may seem odd to consider the stomach as a site for alkalinization in light of the significant HCl secretion associated with digestion. However, the gastric mucosa actually protects itself from the deleterious effects of high luminal fluid acidity (pH 1) by producing an adjacent thin HCO_3^- film and a thick mucus layer (Flemström and Isenberg, 2001). This gastric juxtamucosal pH is maintained by DIDS-sensitive HCO_3^- transport. The HCl originates from the parietal cells of the deep gastric glands (or pits), which are unlikely to secrete acid during non-feeding periods (Trischitta *et al.*, 1998). We can speculate that the Cl^-/HCO_3^- exchange mechanism used for ameliorating damage to the gastric and intestinal epithelia by neutralizing gastric acid are used in the loach for enhancing ammonia volatilization. Future work looking at PAT1 expression and the effects of pharmalogical inhibition in the loach gut may yield interesting results.

2. MANGROVE KILLIFISH AND LEAPING BLENNY

The mangrove killifish, *Rivulus marmoratus*, is distributed in North, Central and South America; from the eastern coasts of Florida in the USA to Cuba, Jamaica, Brazil, Mexico, and throughout the Caribbean. It inhabits shallow, mud-bottomed ditches, bays, salt marshes, and other brackish-water environments. Very often, it can be found in crab burrows. *Rivulus marmoratus* is capable of surviving out of water for one month and can excrete ammonia at 57% of aquatic rates ($1000\,\mu$mol TAmm-$N\,kg^{-1}\,h^{-1}$; Frick and Wright, 2002b; 28–30 °C). Frick and Wright (2002b) determined that ammonia volatilization accounted for 42% of the terrestrial J_{AMM} (23% aquatic rate) and remained quite constant during exposure (up to 11 days).

The leaping blenny *Alticus kirki* (Order: Perciformes; Family Blennidae) is found in the intertidal zone from Mozambique to India. It lives in moist shaded pockets of pitted limestone in the spray zone, leaping from hole to hole when disturbed. *Alticus kirki* can excrete its nitrogenous wastes (ammonia + urea) at 28% of its aquatic rate after 24 hours of emersion ($1140\ vs\ 327\,\mu$mol $N\,kg^{-1}\,h^{-1}$; Rozemeijer and Plaut, 1993; 25 °C). Unlike the weatherloach and mangrove killifish, greater than 50% of N-excretion in *A. kiki* is accounted for by urea compared to only 20% and 5% in *R. marmoratus* (Frick and Wright, 2002b) and *M. anguillicaudatus* (Chew *et al.*, 2001), respectively. In the leaping blenny, terrestrial J_{AMM} accounts for 45% of the total terrestrial N-excretion ($147\,\mu$mol $N\,kg^{-1}\,h^{-1}$) and the volatilized ammonia component ($93\,\mu$mol $N\,h^{-1}\,kg^{-1}$) was 28% of the total terrestrial N-excretion or 63% of the terrestrial J_{AMM}. Interestingly, urea volatilization accounted for 27% ($87\,\mu$mol $N\,h^{-1}\,kg^{-1}$) of the total terrestrial N-excretion (Rozemeijer and Plaut, 1993). Urea volatilization has not been measured in the other two species but would be expected to be smaller since terrestrial urea excretion rates remain low (Chew *et al.*, 2001; Frick and Wright, 2002b).

V. DEFENSE AGAINST AMMONIA TOXICITY AT THE CELLULAR AND SUB-CELLULAR LEVELS

A. Reduction in Proteolysis and/or Amino Acid Catabolism

In order to slow down the build-up of ammonia internally, fish can augment excretion by decreasing the rate of ammonia production through amino acid catabolism (Figure 8.1F). The steady-state concentrations of free amino acids (FAAs) in tissues are maintained by the balance between their rates of degradation and rates of production (through proteolysis or synthesis). Alteration of these two rates would lead to changes in concentrations of FAAs. Being able to alter these rates is a valuable strategy to a fish that has to endure short periods of water shortage, because it would slow down the build-up of endogenous ammonia. This strategy may also be essential to the survival of certain fish species which are naturally exposed to an ammonia-loading situation, especially after the ammonia in the body has built up to a sub-critical level.

1. MUDSKIPPERS

The ammonia and urea excretion rates of *P. schlosseri* and *B. boddaerti* decrease upon aerial exposure, with only very small amounts of ammonia accumulating in their tissues (Lim *et al.*, 2001). When exposed to terrestrial conditions under constant darkness, during which the mudskippers remain quiescent, total FAA (TFAA) levels in the liver and plasma of *P. schlosseri* decrease significantly (Ip, Chew *et al.*, 2001a; Lim *et al.*, 2001). In *B. boddaerti*, the decrease is observed in the muscle instead. Since experimental specimens were unfed, the rate of protein degradation should be higher than the rate of protein synthesis, leading to a net proteolysis. If the rate of proteolysis remains relatively constant and is unaffected by aerial exposure, there would be accumulation of FAAs, leading to an increase in the TFAA content. Therefore, the decreases in TFAA contents, as observed in some tissues of these two mudskippers, indicate that simultaneous decreases in rates of proteolysis and amino acid catabolism have occurred. In addition, the decrease in proteolytic rate is likely to be greater than the decrease in the rate of amino acid catabolism, leading to decreases in the steady-state concentrations of various FAAs and, consequently lowering the TFAA concentrations.

An analysis of the balance between nitrogenous excretion and nitrogenous accumulation in *P. schlosseri* and *B. boddaerti* supports the above conclusion (Lim *et al.*, 2001). However, it is obvious that the decrease in the rate of nitrogenous excretion is much greater than the reduction in the rate of ammonia production, leading to the accumulation of ammonia in the tissues and organs of these two mudskippers.

Reduction in rates of protein and amino acid catabolism constitutes an effective strategy to slow down the internal build-up of ammonia. However, it also prevents the utilization of amino acids as an energy source. Operating by itself, it may not be a useful mechanism for fish species (e.g. *P. schlosseri*, Kok *et al.*, 1998) that are usually active on land. To overcome this, *P. schlosseri* adopts the strategies of partial amino acid catabolism (see section V.B) in conjunction with a suppression of ammonia production in general. This allows the mudskipper to use protein and amino acids as energy sources to support locomotory activities on land without releasing ammonia.

2. WEATHERLOACH AND FOUR-EYED SLEEPER

The weatherloach, *M. anguillicaudatus,* can burrow into the mud during drought and survive therein for a long period. After exposure to the terrestrial condition for 24 hours, the deficit between the reduction in ammonia excretion and the increase in accumulation of N is $-99\,\mu$mol, and the deficit becomes greater with time, reaching -274 and $-558\,\mu$mol N by hour 48 and hour 72, respectively (Chew *et al.*, 2001). The amount of N accumulated appears to saturate after 48 hours of aerial exposure. Therefore, Chew *et al.* (2001) concluded that *M. anguillicaudatus* has high tolerance to aerial exposure partially because it is capable of suppressing proteolysis and amino acid catabolism.

The four-eyed sleeper, *Bostrichyths sinensis*, is an eleotrid belonging to the Order Perciformes, Class Actinopterygii, and Family Eleotridae. The sleepers are placed in a separate family from those in the Family Gobiidae (e.g. mudskippers) because they have separate pelvic fins, whereas the gobies possess fused, cup-like pelvic fins. *Bostrichyths sinensis* can be found in the Indo-Pacific, from India to Australia and Taiwan. In southern China and Hong Kong, *B. sinensis* is regarded highly as a food fish. It inhabits brackish water, seeking out crevices of the river mouths. In its natural environment, *B. sinensis* may be passively exposed to air by a receding tide, because it remains in its crevice above the water's edge.

An analysis of the nitrogenous balance sheet in a 150 *B. sinensis* reveals that it does not undergo a reduction in ammonia production during the first 24 hours of aerial exposure (Ip, Lim *et al.*, 2001). The reduction in ammonia excretion, which amounts to $370\,\mu$mol N, is completely accounted for by the accumulation of $441\,\mu$mol glutamine-N in the muscle. After 72 hours of aerial exposure, there is a much greater discrepancy between the reduction in nitrogenous excretion ($1110\,\mu$mol) and the retention of N ($595\,\mu$mol) in the body. Therefore, reductions in proteolysis and amino acid catabolism occur in *B. sinensis* during long periods of exposure to terrestrial conditions (Ip, Lim *et al.*, 2001).

3. SWAMP EEL

The swamp eel, *Monopterus albus*, is an obligatory air-breathing bony fish belonging to the Family Synbranchidae, Order Synbranchiformes, and Class Actinopterygii. It can be found in the tropics (34° N to 6° S) from India to Southern China, Malaysia and Indonesia. It has an anguilliform body reaching a maximal length of 100 cm at maturity. It has no scale and no pectoral and pelvic fins. Its dorsal, caudal, and anal fins are confluent and reduced to a skin fold. *Monopterus albus* lives in muddy ponds, swamps, canals, and rice fields, where it burrows in moist earth in the dry season, surviving for long periods without water. Thus, it encounters drought during summer and ammonia-loading during agricultural fertilization. While *M. albus* can tolerate the seasonal draining of rice paddies in the mud, it can survive only several days in the market, without water.

Without water to irrigate the gills or the body surface in the mud, *M. albus* has difficulties in excreting ammonia that is endogenously produced. Ammonia and glutamine, but not urea, accumulate in the tissues of specimens exposed to air (normoxia) for 6 days (see section V.C; Tay *et al.*, 2003), indicating that endogenous ammonia is detoxified to glutamine under such experimental conditions. In contrast, ammonia accumulation occurs only in the muscle of specimens buried in the mud for 6 days, with no increase in glutamine or glutamate contents in all the tissues and organs studied (Chew *et al.*, 2005). Similar results can be obtained with specimens buried in the mud for 40 days (Chew *et al.*, 2005). While staying inside the mud prevents excessive water loss from the surface of the skin through evaporation, the specimens are necessarily exposed to a lack of oxygen. Indeed, there are decreases in the energy charge and ATP content in the muscle of specimens buried in mud for 6 days (S. F. Chew and Y. K. Ip, unpublished data). Glutamine synthesis is energy-dependent, and is obviously discarded as the major strategy to ameliorate ammonia toxicity by *M. albus* when ATP production is at stake during hypoxic exposure in the mud. Instead, suppression of endogenous ammonia production is adopted as the major strategy to avoid ammonia intoxication. The reduction in ammonia production in *M. albus* in mud involves decreases in rates of proteolysis and amino acid catabolism. This may be the most effective strategy for a fish to defend against ammonia toxicity because it would impose no extra energy demand on the fish, as compared to glutamine or urea syntheses (see section V.C and V.D, respectively). At the same time, it slows down the build-up of internal ammonia concentrations as the fish entered into a stage of torpor to bear through this adverse period. Environmental hypoxia may be a more effective signal than increased internal ammonia concentration to initiate a suppression of ammonia production. This may explain why *M. albus* can estivate in the mud for >40 days but cannot survive in air for more than 10 days.

4. SNAKEHEAD AND MARBLE GOBY

Snakeheads are a group of fish belonging to the Class Actinopterygii, Order Perciformes, and Family Channidae. The small snakehead, *Channa asiatica,* is an obligatory air-breather found in Asia: Taiwan, Southern China, and Sri Lanka. *Channa asiatica* is a valued freshwater food fish in Southern China. There, people believe the consumption of this fish aids in the recovery of injuries. It is a predaceous fish that resides in slow-flowing streams and in crevices near riverbanks. In its natural habitat, it may encounter bouts of aerial exposure during the dry seasons. Unlike *P. schlosseri, C. asiatica* is incapable of using its pectoral fins to move actively on land, and does not exhibit any feeding, territorial or courtship behavior while out of water. When trapped in puddles of water, it will try to struggle back to water through eel-like body movements, which may last several minutes. After struggling for a while without any success in returning to water, it will turn motionless and remain quiescent, with only occasional eel-like movements, for an extended period. Unlike mudskippers (Ip, Chew et al., 2001b), the reduction in nitrogenous excretion in the snakehead *C. asiatica* during 48 hours of aerial exposure is completely balanced by nitrogenous accumulation in the tissues (Chew, Wong et al., 2003). Hence, it is unlikely that the rates of proteolysis and amino acid catabolism are reduced during aerial exposure, as in the case of mudskippers (Ip, Chew et al., 2001b). This implies that, with respect to nitrogen metabolism alone, *C. asiatica* cannot measure up to the mudskipper's capability of surviving on land.

The marble goby, *Oxyeleotris marmoratus*, is also a sleeper, belonging to the Family: Eleotridae. However, it grows to a much larger size (>1 kg) than *B. sinensis* and does not stay in crevices. It can be found in rivers, swamps, reservoirs, and canals in Asia, from Thailand to Indonesia. It is considered a delicacy over much of eastern Asia, and exported fish command a high price. *Oxyeleotris marmoratus* is a facultative air-breather capable of surviving under terrestrial conditions for up to 7 days. In its natural environment, it may encounter an absence of water during voluntary emergence or habitat desiccation. Usually, it remains inactive on land. When exposed to terrestrial conditions for 72 hours, it does not undergo a reduction in amino acid catabolism. Instead, it appears that protein and/or amino acid catabolism may increase during aerial exposure because glutamine accumulates to levels far in excess of that required to detoxify the amount of ammonia which cannot be excreted during that period (Jow et al., 1999). In this respect, its adaptation to aerial exposure seems inferior to that of *B. sinensis*. From here, it is obvious that the ability to reduce the rate of endogenous ammonia production during aerial exposure is not common among tropical teleost fishes.

5. LUNGFISHES

The Dipnoi are an archaic group of fishes belonging to the Class Sarcopterygii, and are characterized by the possession of a lung opening off the ventral side of the esophagus. Their gills are reduced and inadequate for respiration, and they are entirely dependent on aerial respiration. Though unable to move about to any great extent on land, lungfishes can live for an extended period out of water. Members of three different Families: Protopteridae, Lepidosirenidae, and Ceratodontidae, are found in Africa, South America, and Australia, respectively. The African lungfishes are usually found in marginal swamps and backwaters of rivers and lakes. During the dry season, they (*Protopterus aethiopicus* and *Protopterus annectens*) estivate in subterranean mud cocoons or in a layer of dried mucus on land (*Protoperus dolloi*) (Poll, 1961). They can exist in this state for over a year, although normally they estivate only from the end of one wet season to the start of the next. African lungfish have been kept alive in estivation in the laboratory for 3 years (Smith, 1930), and Smith calculated that they have the metabolic resources to survive for 5 years. They can drastically reduce the rate of ammonia production during estivation (Janssens and Cohen, 1968a, b). The South American lungfish lives in the Amazon River basin and Paraguay–Paraná River basin. It prefers stagnant waters where there is little current. During the dry period, it burrows into the mud, to a depth of about 30–50 cm, and seals off the entrance with clay, leaving two or three holes for respiration. Metabolism is reduced during this period of estivation. The Australian lungfish can be found in still or slow-flowing waters, usually in deep pools. During periods of drought, it can tolerate stagnant conditions by breathing air. However, it lacks the ability to survive dry spells by estivation.

The slender lungfish, *P. dolloi*, is a lungfish found in Central Africa in the lower and middle regions of the Congo River basin. *Protopterus dolloi* retains the capacity to estivate when out of water, although unlike other African lungfishes, it does not estivate in a cocoon. During the dry season, the males guard the eggs and larvae in the nest, which is built in the mud of the swamps, and the females can be found in the open water of the rivers. Hence, *P. dolloi* represents an ideal specimen for direct comparison with tropical teleosts, which are often exposed to terrestrial conditions. During aerial exposure, the ammonia excretion rate in *P. dolloi* decreases significantly to 8–16% of the submerged control, because of a lack of water to flush the branchial and cutaneous surfaces (Chew, Ong *et al.*, 2003). However, there are no significant increases in ammonia contents in the muscle, liver, brain or plasma exposed to air for 6 days. In addition, the rate of ammonia excretion of the experimental animal remains low and does not return to the control level during the subsequent 24-hour period of re-immersion. These results suggest that (1) endogenous ammonia production is drastically reduced and (2) endogenous ammonia is detoxified effectively into urea (see section V.D). Indeed, there are significant decreases in glutamate, glutamine, and lysine levels in the liver of fish exposed to air, which lead to a decrease in the TFAA content. This indirectly confirms that the specimen has reduced its rates of proteolysis and/or amino acid catabolism to suppress the production of ammonia. *Protopterus dolloi* also reduces ammonia production during 40 days of estivation in a mucus cocoon on land (Chew, Chan *et al.* 2004). In addition, Chew, Ho *et al.* (2004) obtained results suggesting that *P. dolloi* was capable of reducing ammonia production during ammonia loading.

B. Partial Amino Acid Catabolism and Alanine Formation

Certain amino acids (e.g. arginine, glutamine, histidine, and proline) can be converted to glutamate. Glutamate can undergo deamination catalyzed by glutamate dehydrogenase (GDH), producing NH_4^+ and α-KG (Campbell, 1991). The latter is then channeled into the Krebs cycle. Glutamate can also undergo transamination with pyruvate, catalyzed by alanine aminotransferase (ALT), producing α-KG without the release of ammonia (Ip, Chew *et al.*, 2001a, b; Chew, Wong *et al.*, 2003). If there were a continuous supply of pyruvate, transamination would facilitate the oxidation of carbon chains of some amino acids without polluting the internal environment with ammonia. For fishes that have difficulty in excreting endogenous ammonia, partial catabolism of certain amino acids, leading to the formation of alanine (Figure 8.1G), coupled with a reduction in amino acid catabolism (Figure 8.1F) in general, would be the most cost-effective way to minimize ammonia build-up in the body. It allows amino acids to be used as an energy source during adverse conditions without polluting the internal environment. However, it cannot be regarded as a strategy for ammonia detoxification, simply because ammonia is not released and then converted back to alanine. Available information indicates that it is a major strategy adopted mainly by fishes which are relatively active on land, but not by those that have to deal with ammonia loading situations in their natural habitats.

The presence of malic enzyme and ALT together in the tissues is a prerequisite to the fish's capability of undergoing partial amino acid catabolism. In this pathway, malate is channeled out of the Krebs cycle through malic enzyme to form pyruvate. Pyruvate then undergoes transamination through the ALT reaction to form alanine (Figure 8.2). For this mechanism to work, α-KG produced must be channeled into the Krebs cycle to maintain a supply of pyruvate for transamination so that no endogenous ammonia is released in the process (Chew, Wong et al., 2003). However, α-KG also acts as a substrate for the GDH reaction, in the aminating direction, which also happens to be present in the mitochondria. If α-KG were turned into glutamate, partial amino acid catabolism would not proceed. Consequently, ATP would not be produced, and endogenous ammonia would start to accumulate in the terrestrial conditions at a higher rate. Hence, either the aminating or deaminating reaction of GDH must be modified to reduce the carbon flux from α-KG to glutamate (Figure 8.2).

1. MUDSKIPPERS

Increases in the contents of certain essential amino acids (e.g. isoleucine, leucine, proline, serine, lysine, and valine) occur in the tissues of *P. schlosseri* exposed to terrestrial conditions under a dark–light regime (Ip, Chew et al., 2001b). Since specimens are not fed during the experiment, this may indicate the mobilization of amino acids through proteolysis under such experimental conditions. Simultaneously, alanine levels increase in the muscle, liver, and plasma, and TFAA concentrations increase significantly in the muscle and plasma (Ip et al., 1993; Ip, Chew et al., 2001b). Under normal

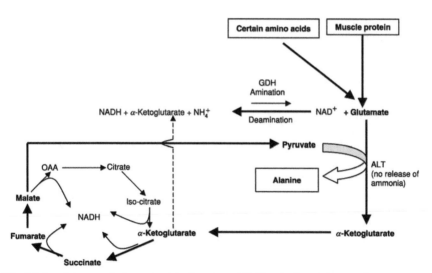

Fig. 8.2 Involvement of glutamate dehydrogenase (GDH), alanine aminotransferase (ALT), and Krebs cycle in the partial catabolism of amino acids leading to the formation of alanine without releasing ammonia.

circumstances, the carbon chain of an amino acid can be completely oxidized to CO_2 and water through the Krebs cycle and the electron transport chain, producing ATP and/or its equivalent. For *P. schlosseri* exposed to terrestrial conditions, the carbon chain may undergo partial oxidation only. α-Ketoglutarate can be metabolized through portions of the Krebs cycle to malate, which can then be turned into pyruvate in the presence of malic enzyme. This would cause a reduction in the efficiency of ATP production, because amino acids are not fully oxidized. However, since *P. schlosseri* has difficulty in excreting ammonia on land, partial amino acid catabolism leading to the formation of alanine would provide energy in the form of ATP without NH_4^+ release (Ip, Chew *et al.*, 2001b). This would allow the utilization of certain amino acids as energy sources and, at the same time, minimize ammonia accumulation.

When *P. schlosseri* is forced to exercise until exhaustion on land after 3 hours' exposure to terrestrial conditions, there is no change in the glycogen level despite an increase in the lactate content in the muscle. Muscle ammonia and alanine levels also increase, with ammonia accumulating to a level twice that of the fish exercised in water (Ip, Chew *et al.*, 2001b). The efficiency of alanine production through partial amino acid catabolism is apparently dependent on the period of aerial exposure. For *P. schlosseri* forced to exercise on land after exposure to terrestrial conditions for 24 hours, there is no change in glycogen level and no further accumulation of ammonia. However, muscle alanine content increases 4-fold, and the amount accumulated is much higher than the amount found in fish exercised in water. This adaptation reduces the dependence of *P. schlosseri* on carbohydrate, sparing the glycogen store, and can, therefore, sustain the higher metabolic rate on land (as observed by Kok *et al.*, 1998). Exercise in water leads to a decrease in glycogen content, and increases in the levels of lactate, ammonia, and alanine in the muscle of *P. schlosseri* (Ip, Chew *et al.*, 2001b). Thus, it can be concluded that both glycogen and amino acids are mobilized in this situation.

The absence of any significant accumulation of alanine, glutamine or glutamate in the muscle of *B. boddaerti* exposed to terrestrial conditions in a dark–light regime suggests that it does not rely on protein as an energy source during aerial exposure. Unlike *P. schlosseri, B. boddaerti* uses glycogen as a metabolic fuel to support activity on land. This strategy offers a limited amount of energy for a short period and, as a consequence, the muscle ATP content decreased to one-tenth of the control value after a short period of locomotory activity on land. Together with the high levels of ammonia accumulated in the muscle of the exercised specimens, these results suggest that it is highly unfavorable for *B. boddaerti* to stay away from water for long periods. In fact, *B. boddaerti* lives closer to the water's edge than *P. schlosseri*.

2. SNAKEHEAD

Alanine increases 4-fold, from 3.7 to 12.6 μmol g^{-1}, in the muscle of the small snakehead *C. asiatica* after 48 hours of aerial exposure (Chew, Wong *et al.*, 2003). The accumulated alanine accounts for 70% of the deficit in ammonia excretion during that period. This allows the utilization of certain amino acids as energy sources and, at the same time, minimizes ammonia accumulation. The ratio of equilibrium constant:mass action ratio indicates that there is a greater tendency for the ALT reaction to proceed in the direction of alanine formation in the muscle and liver of specimens kept in the submerged condition (S. F. Chew and Y. K. Ip, unpublished data). Such a tendency (a ratio of 13.22) is sustained, although at a smaller magnitude, in the liver of the specimens exposed to 48 hours of terrestrial conditions. As for the muscle, a ratio of 0.423 suggests that this experimental condition favors alanine degradation. The apparent discrepancy between the tendency of alanine degradation and the accumulation of alanine (12 μmol g^{-1}) in the muscle is attributable to the fact that metabolites are compartmentalized *in situ*, which is not considered in the calculation of the mass action ratio (S. F. Chew and Y. K. Ip, unpublished data). It may also indicate the possible presence of ALT isozymes in *C. asiatica*, with the one in the muscle favoring the formation of alanine.

In addition, aerial exposure significantly decreases the aminating activities of GDH from the muscle and liver of *C. asiatica* (Chew, Wong *et al.*, 2003). This can be seen as an important adaptation to facilitate the formation of alanine through partial amino acid catabolism (Figure 8.2) in order to reduce the rate of accumulation of endogenous ammonia in adverse conditions. However, it is not a good strategy to handle exogenous ammonia during ammonia-loading, when ammonia has to be detoxified through glutamate and glutamine formation (Figure 8.3) (Ip, Chew *et al.*, 2001a). On the contrary, Iwata *et al.* (1981) reported that the aminating GDH activities in the muscle and liver of the mudskipper *Periophthalmus modestus* (previously as *P. cantonensis*) increased significantly after aerial exposure. They (Iwata *et al.*, 1981) interpreted the results as ammonia (endogenous) being detoxified to amino acids through amination of α-KG. However, it would be a futile effort if GDH acts, on one hand, to release ammonia through transdeamination (Mommsen and Walsh, 1992) and, on the other hand, to detoxify the released ammonia back to glutamate simultaneously, as suggested by Iwata *et al.* (1981). If indeed alanine is formed through partial amino acid catabolism, the amino groups of various amino acids (e.g. glutamate, valine, and leucine) are not physically released, and then incorporate back into amino acids. Rather, it is formed through transamination. The glutamate that acts as a substrate may have its amino group originated

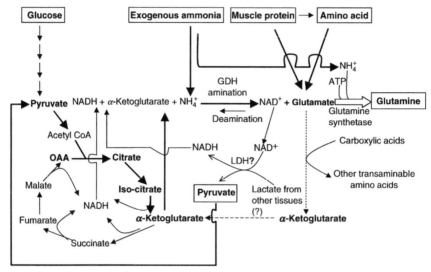

Fig. 8.3 Involvement of glutamate dehydrogenase (GDH), glutamine synthetase, and Krebs cycle in the detoxification of ammonia to glutamine; and possible solutions to the associated problems of redox balance and depletion of Krebs cycle intermediates.

from other amino acids (Ip, Chew *et al.*, 2001a, b) rather than from NH_4^+ in the GDH reaction (Chew, Wong *et al.*, 2003).

Unlike the giant mudskipper, *P. schlosseri* (Ip, Chew *et al.*, 2001b), *C. asiatica* is incapable of increasing the rate of partial amino acid catabolism to sustain locomotory activities on land, although it can use partial amino acid catabolism to reduce ammonia production during aerial exposure (Chew, Wong *et al.*, 2003). This difference may explain, at least partially, why *P. schlosseri* can maintain long periods of locomotory activity on land, but *C. asiatica* can only afford short bursts of muscular activity to try to get back to water under similar conditions.

C. Glutamine Synthesis and Possible Solutions for Redox Balance and Depletion of Krebs Cycle Intermediates

Glutamine plays a role in detoxifying exogenous and endogenous ammonia in fish, especially in the brain, during exposure to HEA (Levi *et al.*, 1974; Arillo *et al.*, 1981; Dabrowska and Wlasow, 1986; Mommsen and Walsh, 1992; Peng *et al.*, 1998) (Figure 8.1H). Ammonia toxicity can be avoided by converting ammonia to glutamine. Glutamine is produced from glutamate and NH_4^+, the reaction catalyzed by GS in the muscle and/or liver. Glutamate may in turn be produced from α-KG and NH_4^+, catalyzed by GDH, or α-KG and other amino acids catalyzed by various transaminases (Figure 8.3). In other words, formation of one mole of glutamine allows the uptake of two moles of ammonia (Campbell, 1973). In contrast to the production of alanine, the formation of glutamine is energetically intensive. One mole of ATP is required for the production of every amide group of glutamine via GS. If the reaction begins with ammonia and α-KG, every mole of ammonia detoxified would result in the hydrolysis of 2 mol of ATP-equivalent (Ip, Chew et al., 2001a). Hence, glutamine formation would be more effective than ureogenesis (2.5 mol ATP for every mole of nitrogen detoxified via carbamoyl phosphate synthetase III – CPS III) with respect to energy expenditure. More importantly, glutamine is stored within the body after being synthesized, which can be used for other anabolic processes (e.g., syntheses of purine, pyrimidine, mucopolysaccharides etc.) when the environmental conditions become more favorable. Urea, being a small and uncharged molecule permeable to biomembranes, can be excreted easily upon its synthesis and ureotely actually represents a loss of both nitrogen and carbon to the environment (see section V.D).

Whether the synthesis of glutamine begins with glutamate or α-KG can be related to the distribution of GS within the cell. In fish, GDH is a mitochondrial enzyme (Campbell, 1973). For the detoxification of endogenous ammonia, produced through transdeamination, to glutamine or urea (via CPS III; see section V.D), it would be essential for GS to be located in the mitochondria. However, mitochondrially located GS would render the detoxification of exogenous ammonia (e.g., in HEA) inefficient as NH_3 has to permeate through both the plasmalemma and the mitochondrial membranes to be accessed by the mitochondrial GS. NH_3 would bind with H^+ after entering the cytosol to form NH_4^+, and ammonia begins to exert its toxic effects. Hence, GS should be located in the cytosol if its function were to detoxify both exogenous and endogenous ammonia. In the case of endogenous ammonia detoxification, cytosolic GS would have to function in cooperation with GDH in the mitochondria through the glutamate transporter in the inner mitochondria.

Indeed, in the case of elasmobranchs that depend on urea synthesized through CPS III in the liver for osmoregulation, two GS isozymes exist separately in the brain and the liver, localized in the cytosol and the mitochondria, respectively (Smith et al., 1983). Ammonia asserts its toxic effects on the brain, and it is essential to have GS in the cytosol to protect the brain against ammonia toxicity (Korsgaard et al., 1995). High levels of GS activity are present in fish brains with a binding constant for ammonia in the micromolar range (Mommsen and Walsh, 1991; Peng et al., 1998; Ip, Chew et al., 2001b). As a result, the brain is often the organ undergoing the largest increases in glutamine content in fish exposed to ammonia.

Sleepers (marble goby and four-eyed sleeper) belonging to the Family Eleotridae and the swamp eel belonging to Family Synbranchidae are exceptional because they can detoxify endogenous ammonia to glutamine in non-cerebral tissues during aerial exposure (Ip, Chew *et al.*, 2001b; Tay *et al.*, 2003). Apparently, they can also detoxify both endogenous and exogenous ammonia to glutamine during ammonia-loading (Anderson *et al.*, 2002; Ip, Tay *et al.*, 2004). Since sleepers and the swamp eel remain quiescent on land and in the mud, respectively, the reduction in energy demand for muscular activity may provide them with the opportunity to exploit glutamine formation as a means to detoxify ammonia. The Gulf toadfish, *Opsanus beta*, is another unique example; it detoxifies endogenous ammonia to glutamine to suppress ammonia excretion during confinement stress (Walsh and Milligan, 1995).

1. SLEEPERS

The presence of GS activity in fish liver is variable (Campbell and Anderson, 1991). It is below the level of detection in the liver of non-ureosmotic fishes. GS activities in the muscle of fishes are small to insignificant (Mommsen and Walsh, 1992). In this regard the four-eyed sleeper, *B. sinensis*, is atypical because high activities of GS are detected in the stomach, intestine (foregut and hindgut), liver, and muscle of control specimens (Anderson *et al.*, 2002). This may partially explain its capability of survival during extended period of aerial exposure (up to 7 days) and tolerating HEA (>30 mmol l^{-1} NH_4Cl at pH 7).

The marble goby, *O. marmoratus*, is a facultative air-breather capable of surviving under terrestrial conditions for up to several days without any significant increase in the ammonia level in its muscle during aerial exposure (Jow *et al.*, 1999). The glutamine content in the muscle of *O. marmoratus* exposed to the terrestrial condition, however, increases 3-fold in 72 hours (Jow *et al.*, 1999). The hepatic glutamine content peaks at 24 hours of aerial exposure, while that in the muscle peaks only after 24 hours. This indicates that the glutamine formed in the liver is perhaps later shuttled to the muscle, which acts as a reservoir for glutamine accumulation. Consequently, the steady-state concentration of glutamine in the plasma is maintained.

Bostrichthys sinensis detoxifies endogenous ammonia to glutamine $(5.42 \mu mol\,g^{-1})$ during the first 24 hours of aerial exposure. Beyond hour 24, it resorts to reducing endogenous ammonia production, possibly through reductions in proteolysis and amino acid catabolism (see section IV.A), to avoid the build-up of ammonia to an unbearable state. Probably because of this, glutamine formation is adopted as a short-term strategy only, and the glutamate levels in the muscle and livers of specimens exposed to terrestrial conditions remain relatively unchanged (Ip, Chew *et al.*, 2001a). The glutamine content in the muscle of *B. sinensis* returns back to the control level after 48 hours of aerial exposure, indicating that the accumulated glutamine may be transformed into other nitrogenous compounds via anabolic pathways during long-term aerial exposure.

For *B. sinensis* exposed to HEA, the ammonia levels in the muscle and liver increase to $13\,\mu\text{mol g}^{-1}$ and $9\,\mu\text{mol g}^{-1}$, respectively, while that in the plasma reaches $2.8\,\mu\text{mol ml}^{-1}$ (Anderson *et al.*, 2002). This suggests that ammonia is accumulated in non-cerebral tissues. However, it is doubtful if the proposition of Korsgaard *et al.* (1995) that fishes simply accumulate ammonia and tolerate its toxicity can be applied here. Ammonia may have been sequestered in these non-cerebral tissues for detoxification purposes. For *B. sinensis* exposed to $15\,\text{mmol l}^{-1}$ NH_4Cl, the muscle glutamine level increases 3-fold to $8.1\,\mu\text{mol g}^{-1}$. The glutamine level in the liver increases from undetectable to $3.3\,\mu\text{mol g}^{-1}$. The muscle and liver of this fish are apparently capable of detoxifying ammonia to glutamine. In fact, this is the first non-ureosmotic teleost fish known to respond to environmental ammonia by increasing the expression of GS in non-cerebral tissues, that is the liver, muscle, and intestine (Anderson *et al.*, 2002). Previously, it was believed that only cerebral GS was inducible by sub-lethal concentrations of environmental ammonia (Korsgaard *et al.*, 1995; Peng *et al.*, 1998).

Hepatic GS of *B. sinensis* is present mainly in the cytosol (Anderson *et al.*, 2002). Hence, as in the case of the brain (Campbell and Anderson, 1991), the cytosolic compartmentalization of hepatic GS in this fish would eliminate the necessity for ammonia to penetrate the mitochondrial membranes before being detoxified, and ammonia is likely to be converted into the amide-N, and not the amino-N of glutamine. Glutamine synthetase in elasmobranchs and holocephalans are present as two isozymes. In the liver, where the enzyme is mitochondrial, a larger isozyme is present. In the brain, where the enzyme is cytosolic, a smaller isozyme occurs (Smith *et al.*, 1983; Ritter *et al.*, 1987). Recently, more than one GS gene have been isolated from the trout (Murray *et al.*, 2003) and the toadfish (Walsh *et al.*, 2003). Two GS genes (within the lineage containing the trout sequences) have also been reported for *B. sinensis*. Due to their sequence similarity these genes appear to be alleles of a single locus (Anderson *et al.*, 2002). However, it is uncertain if functional GS isozymes are present in *B. sinensis*. The fact that GS in various tissues responded to ammonia-loading differently (e.g. liver *vs* stomach) (Anderson *et al.*, 2002) indicates indirectly that either GS isozymes are present or different tissue-specific promoters are involved.

Formation of glutamine requires glutamate as a substrate. Glutamate is formed in the mitochondria matrix, and GDH is a mitochondrial enzyme. If a large amount of glutamate is to be made when ammonia builds up in the fish, equally large amounts of α-KG and NADH are needed. Consumption of α-KG would pull it away from the Krebs cycle, and oxidation of NADH would disrupt redox balance. It would also lead to a reduction in ATP production through the electron transport chain. So, there must be mechanisms, which specifically deal with redox balance and the supply of α-KG for glutamate production, yet to be discovered in the marble goby and the four-eyed sleeper. Incidentally, the mitochondria isolated from the liver of these two sleepers contain high levels of lactate dehydrogenase (K. N. T. Tsui, D. J. Randall, and Y. K. Ip, unpublished data). Lactate produced elsewhere (e.g., muscle) may act as a substrate for pyruvate and NADH production to supply α-ketoglutarate and replenish NADH for glutamate formation (Figure 8.3).

2. SWAMP EEL

Tay *et al.* (2003) discovered that the glutamine content in the muscle of *M. albus* increased 6- and 4.5-fold after exposure to terrestrial conditions for 72 and 144 hours, respectively, reaching respective absolute values of 10.11 and 7.62 μmol g^{-1}. In the liver, the increase in glutamine level is even more drastic, reaching 39-fold and 31-fold at 72 and 144 hours, respectively. As for the brain, glutamine increases from 1.57 μmol g^{-1} to 5.19 μmol g^{-1} after 24 hours of aerial exposure and stays relatively constant thereafter. The brain of *M. albus* has very high level of GS activity, probably the highest known in fish, which may contribute significantly to its high tolerance to aerial and HEA exposure (120 mmol l^{-1} NH$_4$Cl at pH 7 for at least 144 hours). The glutamine levels in the muscle, liver, and brain increase to 10.8, 17.01 and 9.42 μmol g^{-1}, respectively, after 144 hours of exposure to 75 mmol l^{-1} NH$_4$Cl at pH 7 (Ip, Tay *et al.*, 2004), which are greater than the corresponding values of specimens being exposed to terrestrial conditions for a similar period (Tay *et al.*, 2003). Glutamine synthetase activity increases significantly in the liver (2.8-fold) and gut (1.5-fold) of specimens being exposed to HEA for 144 hours. The liver is the main site of ammonia detoxification and the gut, as in the case of *B. sinensis* (Anderson *et al.*, 2002), appears to have functions beyond being a digestive/absorptive organ in *M. albus*. Apparently, the muscle of *M. albus* has a relatively minor role in the increased synthesis of glutamine during aerial exposure (Tay *et al.*, 2003). The reason could be that the muscle needs to adopt the strategy of partial amino acid catabolism (see section V.B) to supply energy for locomotory activity in order to burrow into the mud.

Lim, Chew *et al.* (2004) injected NH$_4$HCO$_3$ (10 μmol g^{-1} fish) into the peritoneal cavity of *M. albus*, raising the level of ammonia in the body, in order to elucidate the strategies involved in defense against the toxicity of exogenous ammonia. During the subsequent 24 hours after NH$_4$HCO$_3$ injection, there is a significant increase in the ammonia excretion rate (Lim, Chew *et al.*, 2004), which indicates that the main strategy adopted by *M. albus* is to remove the majority of the exogenous ammonia through enhanced ammonia excretion. Six hours post-injection of NH_4HCO_3, ammonia contents in the tissues build up significantly, especially in the brain (Lim, Chew et al., 2004), which is in support of the conclusion that *M. albus* has high tolerance of ammonia toxicity at the cellular and sub-cellular levels. By hour 12 post-infusion, there are significant increases in the activities of glutamine synthetase in the muscle, liver and gut, accompanied by significant increases in glutamine contents in the muscle and the liver (Lim, Chew et al., 2004). There is also a significant increase in the glutamine content in the brain at hour 6 post-injection of NH_4HCO_3 (Lim, Chew et al., 2004). These results confirm the capability of *M. albus* to detoxify ammonia through glutamine synthesis.

3. GULF TOADFISH

The Gulf toadfish *O. beta*, belonging to the Family Batrachoididae, is a subtropical fish found in Florida (USA), Little Bahama Bank (Bahamas) and the entire Gulf of Mexico. It is commonly found in sea grass beds and rocky cuts in coastal bays and lagoons, and in shallows along open coast. The liver GS of *O. beta* is localized in both the mitochondria and in the cytosol (Anderson and Walsh, 1995). In confinement or crowding conditions, it shifts from ammonotely to ureotely. However, such a phenomenon results from a shutting down of ammonotely through the detoxification of endogenous ammonia to glutamine rather than an activation of ureogenesis (Walsh and Milligan, 1995). The activity of cytosolic GS increases more than 5-fold, but the mitochondrial GS and CPS activities remain relatively unchanged in specimens exposed to confinement or crowding conditions (Walsh et al., 1994; Walsh and Milligan, 1995; Julsrud et al., 1998). Under such conditions, GS mRNA also increases about 5-fold, as does the GS protein concentration (Kong et al., 2000). The increase in GS activity during confinement correlates with a surge in plasma cortisol, rendering cortisol a potential candidate for a role in the transcriptional regulation of GS expression in this fish (Kong et al., 2000). Wood, Hopkins et al. (1995) suggested that the induction of GS served to prevent ammonia excretion by converting it to glutamine, the amide group of which could be converted to urea. Urea can be stored momentarily and then released rapidly as a pulse, perhaps for reasons related to minimizing predation or nitrogen conservation when the fish is confined for periods of time to small spaces for shelter or during breeding (Walsh et al., 1994; Walsh, 1997). Recently, two GS genes have been isolated from *O. beta* (Walsh et al., 2003). RT PCR and RACE PCR revealed the presence of a second GS cDNA from gill tissue that shares only 73% nucleotide and amino acid sequence similarity with the cDNA previously cloned from liver. The original "liver" GS is expressed in all tissues, whereas the new "gill" GS shows expression primarily in the gill. The branchial GS activity apparently shows exclusive expression in the soluble compartment, while other tissues expressing the "liver" form possess both cytoplasmic and mitochondrial activities. *Opsanus beta* can remarkably shut down ammonia excretion at the gills, despite continued ventilation of the gill with water and perfusion with blood which has reasonably high ammonia concentration (Wang and Walsh, 2000). Gill GS may play an important role in trapping ammonia to minimize its leakage from the gills (Wood, Hopkins *et al.*, 1995).

Using $^{15}NH_4Cl$, Rodicia *et al.* (2003) investigated if *O. beta* would metabolize ammonia from their environment (3.8 mM NH_4Cl) into other, less toxic products. They (Rodicia *et al.*, 2003) observed that accumulation of ammonia into an amino acid pool was not a significant metabolic fate; protein synthesis was significantly enriched in all tissues. Therefore, they concluded that amino acid synthesis might be a pathway of ammonia detoxification en route to protein synthesis. However, if that was the case, the enrichment of the amino acid fraction had to be greater than the protein fraction, unless it was argued that protein synthesis in *O. beta* would favor, and selectively picked up, those ^{15}N-labeled amino acids. An alternative explanation would be the occurrence of a suppression of proteolysis (see section V.A) without a reduction in the rate of protein synthesis in the experimental animals; this would lead to the higher enrichment of ^{15}N in the protein fraction in the experimental fish compared with the control. The results obtained by Rodicia *et al.* (2003) actually suggest that ammonia was detoxified mainly to urea (see section V.D) and, to a lesser extent, glutamine when *O. beta* was exposed to HEA, and it was likely that both processes took place mainly in the liver.

4. MUDSKIPPERS

The amphibious mudskippers, *B. boddaerti* and *P. schlosseri*, do not accumulate glutamine during aerial exposure (Ip, Chew *et al.*, 2001b; Lim *et al.*, 2001). Instead, glutamine formation is only used as a means to detoxify exogenous and endogenous ammonia when these fishes are confronted with sub-lethal concentrations of environmental ammonia (Peng *et al.*, 1998). In an ammonia-loading situation, TFAA concentrations in the brain, liver, and muscle of the experimental fish increase, largely due to increases in glutamine levels (Peng *et al.*, 1998). For mudskippers exposed to sub-lethal concentrations of ammonia, the glutamine levels in the brain increase to 28 and 15 $\mu mol\, g^{-1}$ for *P. schlosseri* and *B. boddaerti*, respectively (Peng *et al.*, 1998). Although accumulation of glutamine also occurs in the liver, the levels attained are much lower than those in the brains (Peng *et al.*, 1998). Hence, *P. schlosseri* apparently varies its dependence on two different biochemical pathways, alanine formation (see section IV.B) and glutamine formation, to deal with activities on land and exposure to HEA, respectively. How these two biochemical pathways are regulated is not clear. For *P. schlosseri* exposed to 24 hours of terrestrial conditions, alanine accumulates (Ip *et al.*, 1993; Ip, Chew *et al.*, 2001b), but the aminating GDH activity in the muscle is unchanged (Ip *et al.*, 1993). The GDH activity (both aminating and deaminating) in the liver decreases, although there is an apparent increase in amination:deamination ratio (Ip *et al.*, 1993). This may represent a compromise between adaptations to aerial exposure (to decrease endogenous ammonia production) during an excursion on land at low tides and adaptations to ammonia loading (to detoxify endogenous and exogenous ammonia) during a stay in the burrow at high tide or breeding seasons.

D. Urea Synthesis through the Ornithine–Urea Cycle

Urea (and/or trimethylamine oxide, TMAO) forms an appreciable component of nitrogen output in fully aquatic fishes (Campbell and Anderson, 1991; Wood, 1993; Saha and Ratha, 1998). It can be produced via three pathways: (1) the ornithine–urea cycle (OUC) (Campbell and Anderson, 1991; Anderson, 2001); (2) routine turnover of arginine by argininolysis; and (3) the conversion of uric acid to urea by uricolysis. Only (1) is a synthetic (anabolic) pathway, (2) and (3) produce urea through catabolism instead. It is important to make a distinction between ureogenesis (Figure 8.1I) and ureotely (Figure 8.1J). Ureogenesis means the presence of OUC enzymes with activities that represent a functional urea cycle, sustaining at least a low rate of urea formation. On the other hand, ureotely means urea is the primary product of nitrogen excretion. Only a few teleosts are ureotelic, and the majority of tropical teleost fishes studied so far do not adopt ureogenesis as a major strategy to detoxify endogenous or exogenous ammonia. This is contrary to the belief that there is a tendency towards predominance of ureotely in amphibious species (Mommsen and Walsh, 1989, 1992; Wright, 1995; Walsh, 1997; Saha and Ratha, 1998; Wright and Land, 1998; Hopkins *et al.*, 1999), which are found mainly in the tropics.

Most of the earlier works depended on comparing the rates of urea and ammonia excretion in fishes that were first held in water for a control period, exposed to air for a set time, and then returned to water (Graham, 1997). The idea is that the post-emersion excretion pattern would indicate changes in nitrogen metabolism taking place during aerial exposure. Consequently, the focus was on whether the specimen switched to ureotely upon re-immersion. However, the major concern should be whether urea synthesis took a major role in detoxifying endogenous ammonia during aerial exposure, which actually requires a quantitative analysis of the nitrogen budget of the specimen. Even if the fish demonstrates ureotely during re-immersion, the role of urea synthesis in dealing with ammonia toxicity is unclear because other mechanisms (see sections V.A, B, C, and E) can be operating simultaneously and playing more significant roles. The same argument applies to fishes confronted with HEA. The emphasis should not be on whether ammonia exposure induces ureotely or ureogenesis, but whether urea synthesis takes the "major role" in detoxifying the bulk of the endogenous and exogenous ammonia accumulating in the body, and whether other strategies are also involved in defending against ammonia toxicity.

To date, it is certain that endogenously produced ammonia is not detoxified to urea through the OUC in many adult tropical teleost fishes during aerial exposure. These include: the mudskippers, *P. schlosseri, B. boddaerti* (Lim *et al.*, 2001), and *Periophthalmus chrysospilos* (Y. K. Ip and S. F. Chew, unpublished data); the marble goby, *O. marmoratus* (Jow *et al.*, 1999); the four-eyed sleeper, *B. sinensis* (Ip, Lim *et al.*, 2001); the weatherloach, *M. anguillicaudatus* (Chew *et al.*, 2001); the small snakehead, *C. asiatica* (Chew, Wong *et al.*, 2003); the African catfish, *Clarias gariepinus* (Ip and Chew, unpublished results); the swamp eel, *M. albus* (Tay *et al.*, 2003); and the mangrove killifish, *R. marmoratus* (Frick and Wright, 2002b) exposed to terrestrial conditions for various periods. Several of these (*P. schlosseri, O. marmoratus* and *B. sinensis*) possess all the OUC enzymes, but the activities of some of these enzymes, especially CPS, are just too low to render the cycle functional. Others do not have a full complement of OUC enzymes, but produce urea possibly through arginolysis and uricolysis (Wright and Land, 1998). The formation of urea in fishes is highly energy-dependent. A total of 5 mol of ATP (but 4 mol for some African lungfishes which possess carbamoyl phosphate synthetase I–CPS I) are hydrolyzed to ADP for each mole of urea synthesized, corresponding to 2.5 mol of ATP used for each mole of nitrogen assimilated. That may be the major reason why urea synthesis via OUC is rare in adult teleost fishes (Ip, Chew *et al.*, 2001a).

Although the alkaline Lake Magadi tilapia, *A. grahami*, detoxifies endogenous ammonia to urea in its natural habitat (Wilkie and Wood, 1996), detoxification of endogenous ammonia to urea (Figure 8.1I) as a result of the expression of OUC enzymes is apparently not a universal mechanism in teleosts adapting to an alkaline aqueous environment either. A study of four teleost fishes native to the alkaline (pH 9.5) Pyramid Lake in Nevada, America, showed that the processes for nitrogenous waste excretion were primarily ammonotelic (Wilkie *et al.*, 1993, 1994). Urea excretion is likely a consequence of uricolysis, because high levels of uricolytic enzymes and very low levels of OUC enzymes are present in the liver (McGeer *et al.*, 1994).

Similar results have been reported for the terek of Lake Van in eastern Turkey, where the pH is 9.8 (Danulat and Kempe, 1992). This terek has low OUC capacity, but high ammonia tolerance at the cellular and tissue level.

Although a complete OUC has been demonstrated in certain catfishes (*Heteropneustes fossilis* and *Clarias batrachus*) from India, which apparently possess CPS activities with characteristics of both type I and type III (Saha and Ratha, 1994; Saha *et al.*, 1997, 1999), it is debatable if urea synthesis takes a major role in conferring the high capacity of environmental ammonia tolerance (see air sac catfish and air-breathing catfishes below). Detoxifying the net influx of exogenous ammonia to urea and subsequently excreting it result in high expenditure of energy and the maintenance of an inwardly driven NH_3 gradient. It is probably because of this that ureogenesis in combination with ureotely is not commonly adopted by fish confronted with HEA, as seen in the mudskippers (Iwata, 1988; Iwata and Deguchi, 1995; Peng *et al.*, 1998; Randall *et al.*, 1999), the four-eyed sleeper (Anderson *et al.*, 2002), the marble goby (Ip and Chew, unpublished results), the weather-loach (Tsui *et al.*, 2002), the mangrove killifish (Frick and Wright, 2002a), the African catfish (Ip, Subaidah *et al.*, 2004), the small snakehead (Chew, Wong *et al.*, 2003), and the swamp eel (Ip, Tay *et al.*, 2004)

Be it for the detoxification of endogenous (in terrestrial conditions or alkaline pH_{env}) or exogenous plus endogenous (in HEA) ammonia, the dependence on ureogenesis implies the necessity to develop urea transporters to facilitate urea excretion (see Lake Magadi tilapia and Gulf toadfish below). This is in opposition to the dependence on ureogenesis for osmoregulation as seen in elasmobranchs, which have to depend on urea transporters to reduce the loss of urea instead (see section VI.).

1. LUNGFISHES

The fish-tetrapod transition represents one of the greatest events in vertebrate evolution. Air breathing evolved in fish (e.g., lungfishes), but prolonged terrestrial respiration is a tetrapod feature. Similarly, limbs with strong skeletal units appeared in Sarcopterygian fishes, but the loss of fin rays and appearance of digits are features of tetrapods (Forey, 1986). Lung-fishes depend entirely on aerial respiration and can live for an extended period out of water. There are few similarities between lungfishes and tetrapods, particularly amphibians, in aspects of gas exchange and excretory physiology, pulmonary circulation, and heart structure (Forey *et al.*, 1991; Schultze, 1994).

Unlike the Australian (*Neoceratodus forsteri*) and South American (*Lepidosiren paradoxa*) counterparts, the African lungfish, *P. aethiopicus* and *P. annectens,* can estivate in subterranean mud cocoons for long periods of time (Smith, 1935; Janssens and Cohen, 1968a, b), and have a greater OUC capacity (Janssens, 1964; Forster and Goldstein, 1966; Janssens and Cohen, 1966) than their non-estivating Australian counterpart (Goldstein *et al.*, 1967). On land, there is a lack of water to flush the branchial and cutaneous surfaces, impeding the excretion of ammonia, and consequently leading to the accumulation of ammonia in the body. Ammonia is toxic (Ip, Chew *et al.*, 2001a) and therefore African lungfishes have to avoid ammonia intoxication when out of water. Previous works on *P. aethiopicus* and *P. annectens* reveal that they are ureogenic (Janssens and Cohen, 1966; Mommsen and Walsh, 1989). Similar to tetrapods, they possess mitochondrial CPS I, which utilize NH_4^+ as a substrate, and an arginase which is present mainly in the cytosol, of the liver (Mommsen and Walsh, 1989). On the other hand, coelacanths, marine elasmobranchs, and some teleosts are known to have CPS III (Anderson, 1980; Mommsen and Walsh, 1989; Randall *et al.*, 1989), which utilizes glutamine as a substrate, and an arginase in the hepatic mitochondria. It has been suspected that the replacement of CPS III with CPS I, and mitochondrial arginase with cytosolic arginase, occurred before the evolution of the extant lungfishes (Mommsen and Walsh, 1989).

Since *P. dolloi* estivates within a dry layer of mucus on land (Brien, 1959; Poll, 1961) instead of in a cocoon inside the mud, like *P. aethiopicus* and *P. annectens*, it is likely that African lungfishes evolved through a sequence of events, i.e., air breathing, migrate to land, and then burrow into mud. Estivation can occur on land or in mud, but the latter must have certain advantages over the former, for instance, avoidance of predation. Therefore, Chew, Ong *et al.* (2003) speculated that burrowing into the mud could be a more advanced development during evolution. Indeed, they (Chew, Ong *et al.*, 2003) demonstrated that, like coelacanths, elasmobranchs, and some teleosts, *P. dolloi* possesses CPS III in the liver, and not CPS I as has been shown previously in other African lungfishes. However, similar to other African lungfishes and tetrapods, hepatic arginase is present mainly in the cytosol. To date, no GS activity has been detected in the liver of *P. aethiopicus* and *P. annectens* (Campbell and Anderson, 1991), probably because they possess CPS I and not CPS III. Since *P. dolloi* possessed CPS III, then it would be essential for it to have GS in the hepatic mitochondria to supply the glutamine needed for urea synthesis *de novo*. Indeed, GS activity is present in both the mitochondrial and cytosolic fractions of the liver of *P. dolloi*. Therefore, Chew, Ong *et al.* (2003) concluded that *P. dolloi* was a more primitive extant lungfish, and represents the missing link in the fish–tetrapod transition. However, it would be essential to re-examine the type of CPS present in *P. aethiopicus* and *P. annectens* to confirm that they indeed possess CPS I. If they actually possess CPS III, then the participation and the role of lungfishes in the evolution of CPS III to CPS I (Mommsen and Walsh, 1989) must be re-evaluated.

During 79–128 days of estivation out of water, *P. aethiopicus* accumulates urea in its body (Janssens and Cohen, 1968a). However, it was reported that urea accumulation did not involve an increased rate of urea synthesis (Janssens and Cohen, 1968a), even though the animals appear to be in continuous gluconeogenesis throughout estivation (Janssens and Cohen, 1968b). This apparent controversy arose because of two counteracting factors: (1) increase in the rate of urea production, and (2) decrease in the rate of ammonia production. During the initial phase of aerial exposure before the onset of a reduction in the rate of ammonia production, the rate of urea synthesis *de novo* theoretically has to be increased to detoxify ammonia which is produced at a normal (or slightly sub-normal) rate and cannot be excreted. After entering into estivation for a relative long period, ammonia production rate would have been suppressed (Smith, 1935; Janssens, 1964). This would subsequently result in a decrease in the rate of urea synthesis *de novo*, leading to those observations made in previous studies (Janssens and Cohen, 1968a, b). This analysis led Chew, Ong *et al.* (2003) to hypothesize that the rate of urea synthesis would increase in *P. dolloi* exposed to air without undergoing estivation.

During aerial exposure, the ammonia excretion rate in *P. dolloi* decreases significantly to 8–16% of the submerged control, because of a lack of water to flush the branchial and cutaneous surfaces. However, there are no significant increases in ammonia contents in the muscle, liver, brain, and plasma exposed to air for 6 days. In addition, the rate of ammonia excretion of the experimental animal remains low and does not return to the control level during the subsequent 24 hour period of re-immersion (Chew, Ong *et al.*, 2003). These results suggest that (1) endogenous ammonia production is drastically reduced (see section V.A) and (2) endogenous ammonia is detoxified effectively into urea. There are significant increases in the urea levels in the muscle (8-fold), liver (10.5-fold), and plasma (12.6-fold) of specimens exposed to terrestrial conditions for 6 days. Furthermore, there is a significant increase in the urea excretion rate in specimens exposed to terrestrial conditions for 3 days or more (Chew, Ong *et al.*, 2003). Taken together, it would mean *P. dolloi* increases the rate of urea synthesis during this 6 day period of aerial exposure. This is supported by the fact that aerial exposure leads to an increase in the hepatic OUC capacity, with significant increases in the activities of CPS III (3.8-fold), argininosuccinate synthetase + lyase (1.8-fold), and, more importantly, GS (2.2-fold), in *P. dolloi* (Chew, Ong *et al.*, 2003).

Upon re-immersion, the urea excretion rate in *P. dolloi* increases 22-fold as compared to the control specimen (Chew, Ong *et al.*, 2003), which is probably the greatest increase amongst fishes. These results suggest that, unlike marine elasmobranchs, *P. dolloi* probably possesses mechanisms that facilitate the excretion of urea in water, and that these mechanisms, in contrast to those of metamorphosed amphibian, do not function well on land.

Chew, Chan *et al.* (2004) also studied the strategies adopted by *P. dolloi* to ameliorate the toxicity of endogenous ammonia during short (6 day) or long (40 day) periods of estivation in a layer of dried mucus in air in the laboratory. Despite decreases in rates of ammonia and urea excretion, the ammonia contents in the muscle, liver, brain or gut of *P. dolloi* remain unchanged after 6 days of estivation. For specimens estivated for 40 days, the ammonia contents in the muscle, liver, and gut of specimens decrease significantly instead, which suggest the occurrence of a decrease in the rate of ammonia production. However, contrary to former reports on *P. aethiopicus* (Janssens and Cohen, 1968a, b), there is a significant increase in the rate of urea synthesis in *P. dolloi* during 40 days of estivation (Chew, Chan *et al.*, 2004). The excess urea formed is mainly stored in the body.

There is no doubt that the capability of detoxifying ammonia to urea contributes to the lungfishes' success in estivating on land. However, there is a dearth of information in the literature on the response of lungfishes in general to ammonia loading. In nature, *P. dolloi* encounters aerial exposure occasionally during drought. However, before the water totally dries up and leads to a reduction in ammonia excretion, the exogenous ammonia would be concentrated to high levels in the external medium, creating an ammonia-loading situation. Consequently, with a reversed ΔP_{NH3} gradient, exogenous ammonia may penetrate the skin and branchial surfaces into the body of the fish. In the laboratory, *P. dolloi* can tolerate HEA, up to $100 \, \text{mmol} \, l^{-1}$ NH_4Cl at pH 7 for at least 6 days (Chew, Ho *et al.*, 2004).

In an external medium containing 30 or $100 \, \text{mmol} \, l^{-1}$ NH_4Cl at pH 7, both ΔP_{NH3} and NH_4^+ concentration gradients are directed inward. Yet, the plasma ammonia concentrations in *P. dolloi* exposed to these two concentrations of NH_4Cl are very low, and the values (0.288 and $0.289 \, \text{mmol} \, l^{-1}$, respectively) are comparable. So, how does *P. dolloi* maintain such low levels of plasma ammonia despite the large inwardly directed NH_3 and NH_4^+ gradients? This can be achieved in part through the synthesis of urea *de novo* and its subsequent excretion. However, even then, the rate of ammonia removal must be fast enough to balance the rate of endogenous ammonia production and the influx of exogenous ammonia. An analysis of the nitrogen budget in a specimen exposed to $30 \, \text{mmol} \, l^{-1}$ NH_4Cl reveals that there is a reduction in ammonia production during the 6 day experimental period (Chew, Ho *et al.*, 2004). In addition, the rate of urea synthesis is up-regulated to detoxify both the endogenous and net influx of exogenous ammonia, which can be small due to the low NH_3 permeability of its skin (see below) and its being an air-breather (Chew, Ho et al., 2004).

It is because the plasma ammonia concentration is maintained at low levels that a continuous influx of NH_3 into the body of P. dolloi would occur. It is likely that P. dolloi could afford such a strategy because its body surfaces have low permeability to NH_3 (see section IV.C). Chew, Ho et al. (2004) estimated the flux of ammonia through the skin of P. dolloi down a 10-fold ammonia gradient at pH 7 as $0.003\,\mu\text{mol}\,\text{min}^{-1}\,\text{cm}^{-2}$, which is even lower than that of the giant mudskipper, P. schlosseri $(0.01\,\mu\text{mol}\,\text{min}^{-1}\,\text{cm}^{-2}$, Ip, Randall et al., 2004; see section IV.C). The branchial epithelial surface of aquatic teleost fishes has a higher permeability to NH_3 due to its major function in gaseous exchange. A reduction of the effective area of the branchial epithelium in P. dolloi also contributes to its ability to reduce the influx of exogenous NH_3. Both P. dolloi and P. schlosseri maintain low internal ammonia levels in HEA (30–100 mmol l^{-1} NH_4 Cl), although by different mechanisms, i.e., urea synthesis and active NH_4^+ transport, respectively. Thus, they share the common need to reduce the influx of NH_3 in order to render these mechanisms effective. At present, it is uncertain if P. dolloi or any other lungfishes can excrete acid to detoxify NH_3 externally as in P. schlosseri and B. boddaerti (see section IV.B).

Mommsen and Walsh (1991) postulated that since urea-N was much more costly to make than ammonia-N, marine elasmobranchs may excrete extra exogenous nitrogen, over and above the needs of osmoregulation, in the form of ammonia-N rather than urea-N. To date, the only information available to answer this hypothesis comes from a study in which dogfish shark were infused with ammonia at a rate of $1500\,\mu\text{mol}\,\text{kg}^{-1}\,\text{h}^{-1}$ for 6 h (Wood, Part et al., 1995). Both ammonia-N and urea-N excretion increased by similar extents during infusion, though the former more rapidly, and the entire ammonia-N load (actually 132%) was excreted within 18 hours. Hence, Mommsen and Walsh's (1991) hypothesis appears to be correct for marine elasmobranchs. However, since energy consumption is the major issue in whether ammonia would be detoxified to urea, the excretion of infused/injected ammonia as ammonia per se could be favored simply because of the lack of a simultaneous supply of extra energy resources. Moreover, unlike endogenous ammonia, ammonia infused/injected into the fish does not originate within cells, and has to penetrate through the plasmalemma and mitochondrial membranes in order to be accessible to the OUC enzymes. That means infused/injected ammonia is more likely to be excreted through the gills once it is absorbed into the blood.

Feeding, on the other hand, results in the catabolism of excess amino acids absorbed after food digestion. This would lead to the release of endogenous ammonia in sub-cellular compartments, specifically the hepatic mitochondria, via transdeamination (Campbell, 1991). Consequently, ammonia produced endogenously is more likely to activate the OUC present within liver cells. In addition, the problem of ureogenesis being energy-intensive would be circumvented by an ample supply of energy resources after feeding. Therefore, experiments on feeding might provide results different from those involving the infusion/injection of ammonia into the fish, and would offer new insights into the physiological role of OUC in ureogenic fishes.

Ip *et al.* (2005) injected NH_4Cl into the peritoneal cavity of *P. dollo* in the absence of a supply of food and confirmed that the majority of the injected ammonia was excreted as ammonia *per se* within 24 hours. This apparently contradicted the observations made by Chew, Ho *et al.* (2004) that *P. dolloi* was able to survive in high concentrations (30 or $100 \, mmol \, l^{-1}$ NH_4Cl) of environmental ammonia by up-regulating both the rates of urea synthesis and its excretion. The controversy arose because of one obvious reason – the excretion of endogenous ammonia by *P. dolloi* was impeded in the presence of high concentrations of environmental ammonia (Chew, Ho *et al.*, 2004), but was not affected at all when ammonia (as NH_4Cl) was injected into the fish peritoneally. In the latter conditions, the injected ammonia, although resulting in a momentary increase in the concentration of ammonia in the extracellular fluid, could be excreted easily because of the absence of a reversed ΔP_{NH3} (Ip *et al.*, 2005). These observations led Ip *et al.* (2005) to postulate that urea synthesis in *P. dolloi* could respond to intracellular (endogenous) ammonia concentration more readily than the extracellular ammonia concentration. In addition, they (Ip *et al.*, 2005) postulated that feeding might lead to an increase in urea synthesis in *P. dolloi*.

Indeed, there are significant increases in the rate of ammonia excretion in *P. dolloi* between hour 6 and hour 15 after feeding (Lim, Wong *et al.*, 2004). Simultaneously, there are significant increases in urea excretion rates between hour 3 and hour 18. As a result, there is a significant increase in the percentage of total nitrogen (N) excreted as urea-N, which exceeds 50%, between hour 12 and hour 21 post-feeding (Lim, Wong *et al.*, 2004). Therefore, it can be concluded that *P. dolloi* shifts from ammonotely to ureotely momentarily after feeding. At 12 hours post-feeding, the accumulation of urea-N is greater than the accumulation of ammonia-N in various tissues, which indirectly suggests that feeding leads to an increase in the rate of urea synthesis in *P. dolloi*. This is different from the results obtained by the injection of NH_4Cl into the peritoneal cavity of this fish; 80% of the injected ammonia is excreted within the subsequent 24 hours, a large portion of which was ammonia (Ip *et al.*, 2005). Feeding is more likely to induce urea synthesis because it provides an ample supply of energy resources and leads to the production of endogenous ammonia intracellularly in the liver. The urea synthetic capacity in *P. dolloi* is apparently adequate to prevent a big surge in plasma ammonia level (Lim, Wong *et al.*, 2004), as has been observed in other fishes (Wicks and Randall, 2002), although the brain of *P. dolloi* is likely to be confronted with ammonia toxicity as indicated by a significant increase in the glutamine content at hour 24 (Lim, Wong *et al.*, 2004).

2. LAKE MAGADI TILAPIA

The tilapia *A. grahami* is found in Lakes Magadi and Nakuru (introduced) in Kenya. It thrives at temperatures ranging from 16 to 40 °C, and in a highly alkaline environment (pH 10). This tilapia possesses a high capacity to detoxify endogenous ammonia to urea via the OUC (Walsh *et al.*, 1993). In fact, this is the first known example of complete ureotely in an entirely aquatic teleost fish (Randall *et al.*, 1989; Wood *et al.*, 1989, 1994). The liver contains significant levels of OUC enzymes, in conjunction with considerable GS activity. The latter delivers one of the substrates, glutamine, for CPS III. In addition, CPS III and all other OUC enzyme activities are present in the muscle of this species at levels more than sufficient to account for the rate of urea excretion (Lindley *et al.*, 1999). In addition, the muscle CPS can use NH_4^+ as a substrate. Hence, there is no need for GS and OUC to be tightly coupled, and GS is not well expressed in muscle (Lindley *et al.*, 1999). Wood *et al.* (1994) demonstrated that 80% of urea excretion occurred at the anterior end (gills) of *A. grahami*. The nature and concentration of urea transporters in the gills, as well as the lipid composition of branchial membranes of this fish, are unknown (see section IV.C).

When exposed to $500\,\mu mol\,l^{-1}$ NH_3 dissolved in Lake Magadi water, urea excretion increases 3-fold, which is a substantial amount due to the high baseline level of urea excretion rate (Walsh *et al.*, 1993). This is a rare observation, as urea synthesis is energy-intensive; detoxification of exogenous ammonia to urea for excretion would enhance a further net influx of ammonia. Similar to the case of *P. dolloi*, this strategy would function efficaciously if the skin has a low NH_3 permeability. However, at present, no such information is available, and it is uncertain if *A. grahami* would alter (decrease) the permeability of its skin to ammonia, as in the case of the giant mudskipper, *P. schlosseri* (see section IV.C), during long-term ammonia exposure.

In contrast, another tilapia, *Oreochromis nilotica*, which lives in Sagana Lake (pH 7.0), shows no increase in urea excretion (Wood *et al.*, 1989) when exposed to ammonia. No CPS III activity can be detected from this tilapia (Wright, 1993). Unlike the Lake Magadi tilapia, there is no change in activities of OUC enzymes, but an increase in the activity of allantoicase, an uricolytic enzyme is observed in *O. nilotica* exposed to HEA. Thus, unlike *A. grahami*, urea production in *O. nilotica* appears to occur via uricolysis and arginolysis.

3. GULF TOADFISH

Opsanus beta has a full complement of OUC enzymes in the liver at levels comparable to those in marine elasmobranchs (Mommsen and Walsh, 1989; Anderson and Walsh, 1995). However, it is primarily ammonotelic under conditions of minimal stress (Walsh *et al.*, 1990; Walsh and Milligan, 1995). High rates of urea excretion can be induced by feeding (Walsh and Milligan, 1995), long-term aerial exposure, or exposure to HEA (Walsh *et al.*, 1990, 1994). The capacity of *O. beta* to increase the rate of urea excretion does not seem to be related to aerial exposure during low tides in an intertidal environment (Hopkins *et al.*, 1999). Rather, the high levels of ammonia encountered in the seagrass beds appear to be an important trigger for increased urea excretion in its natural habitat. *Opsanus beta* exhibits a high 96 hour LC_{50} value of 9.75 mmol l^{-1} TAmm at pH 8.2 (or 519 μmol l^{-1} NH_3) (Wang and Walsh, 2000). However, the ability to synthesize and excrete urea does not appear to be the sole factor determining environmental ammonia tolerance in members of the Family Batrachoididae. The moderately ureotelic oyster toadfish *Opanus tau* has an even greater 96 hour LC_{50} value (19.72 mmol l^{-1} TAmm at pH 8.2 or 691.2 μmol l^{-1} NH_3) than the fully ureotelic *O. beta*. In addition, the ammonotelic plainfin midshipman, *Porichths notatus*, has a respectable 96 hour LC_{50} value of 6.0 mM TAmm (or 101.3 μmol l^{-1} NH_3) (Wang and Walsh, 2000). Therefore, it is possible that other strategies are involved in conferring this group of fish high tolerance to environmental ammonia.

Stressed *O. beta* also becomes ureotelic, especially in crowded conditions or when confined individually in small chambers (Walsh *et al.*, 1994; Walsh and Milligan, 1995). However, the apparent shift towards ureotely during confinement or crowding results from a shutting down of ammonotely rather than an activation of ureogenesis. It has been proposed that branchial GS may play a role in trapping ammonia to minimize its leakage through the gill (Wood, Hopkins *et al.*, 1995). The discovery of a second GS gene, other than the "liver" GS gene, with expression in the cytosol of branchial cells in *O. beta* (Walsh *et al.*, 2003), is in favor of such a proposal. Maximal *in vitro* CPS activities do not change during the transition from ammonotely to ureotely (Walsh and Milligan, 1995), suggesting a consistent level of CPS III protein. However, the level of its potent allosteric activator, N-acetyl glutamate, doubles during confinement (Julsrud *et al.*, 1998), indicating that CPS III is up-regulated by allosteric control. CPS III RNA also increased during the transition to ureotely (Kong *et al.*, 2000), which suggests that there may be an increase in turnover of protein in general via cortisol-induced proteolysis (Milligan, 1997; Mommsen et al., 1999), and that it is essential to maintain a certain concentration of enzyme to detoxify endogenous ammonia which is being produced at a slower rate.

The urea excretion event in *O. beta* is highly pulsatile, with virtually all the daily urea load being excreted in a single pulse through the gills lasting from 0.5 to 3 hours (see review by Wood, Hopkins et al., 1995; Walsh, 1997; Wood et al., 1997, 1998; Gilmour et al., 1998; Walsh and Smith, 2001). Urea excretion is facilitated by diffusion through an urea transporter (UT)-like protein, with high homology to the mammalian kidney UT-A2 and the dogfish shark kidney ShUT, in the toadfish gills. The permeability of the gills of *O. beta* to urea increases by 35-fold during the pulsatile period. It is possible that transporters are recruited to the pavement cell membrane by fusion with the plasma membrane of intracellular vesicles rich in tUT transporter (Walsh and Smith, 2001). The activation of the transporter appears to be preceded by an overall minimum in plasma cortisol levels, indicating hormonal involvement (Wood et al., 1998).

4. AIR SAC CATFISH

The stinging (or Singhi) catfish, *H. fossilis*, belonging to the Family Heteropneustidae (air sac catfishes), can be found in Pakistan, India, Bangladesh, Myanmar, and Thailand. It lives mainly in ponds, ditches, swamps, and marshes, but sometimes occurs in muddy rivers. It can survive in air for a long period and exhibits high tolerance to HEA. While *H. fossilis* is ureogenic and contains a functional OUC (Ratha et al., 1995; Saha and Ratha, 1987, 1989, 1998), which can be up-regulated during ammonia-loading, urea synthesis and ureotely do not appear to be the "major" contributor to its high tolerance to air or ammonia exposure.

Saha et al. (2001) concluded that *H. fossilis* was able to survive inside moist peat for months in a water-restricted condition because of ureogenesis, and possibly other physiological adaptations. However, there are discrepancies in their results, which shed doubt on ureogenesis being the major mechanism in dealing with ammonia toxicity in *H. fossilis* under conditions of water shortage. From their results (Saha et al., 2001, Figure 1), the total amounts of ammonia-N and urea-N excreted by a specimen during a 48 hour period after being kept in moist soil for 1 month were $34.5\,\mu\text{mol g}^{-1}$ and $25\,\mu\text{mol g}^{-1}$, respectively. Therefore, for a 50 g fish, the total-N excreted during this recovery period was $(34.5 + 25)\,\mu\text{mol g}^{-1} \times 50\,\text{g} = 2975\,\mu\text{mol}$. However, their results (Saha et al., 2001: Table 1) revealed that the amounts of ammonia-N and urea-N accumulated in a 50 g specimen kept in moist soil for 1 month were 117.03 and $212.98\,\mu\text{mol}$, respectively, based on the estimation that a 50 g fish would contain 1.5 g liver, 0.5 g kidney, 20 g muscle,

Table 8.1
A Summary of Strategies Adopted by Various Species of Fish to
Defend Against Ammonia (Both Endogenous and Exogenous) Toxicity

Fish species	Defense at the epithelial level				Defense at the cellular level				
	Active NH_4^+ pumping	Environmental acidification	NH_3 volatilization	Low membrane permeability to NH_3	Reduced proteolysis and/or amino acid catabolism	Alanine accumulation	Glutamine accumulation	Urea formation *de novo*	Ammonia tolerance in cells and tissues
Abehaze									
M. abei	—	—	—	—	—	—	—	Yes	Likely
Catfish									
C. batrachus	No	No	—	—	No	—	—	Yes?	Likely
C. gariepinus	Yes	No	No	Likely	Yes	No	No	No	Yes
H. fossilis	No	No	—	—	No	—	—	Yes?	Likely
Freshwater stingray									
H. signifer	—	No	—	—	Yes	No	No	Yes	Minor
P. motoro	—	No	—	—	Likely	No	No	No	Yes
Gulf toadfish									
O. beta	—	—	—	—	—	—	No	Yes	Minor
Lungfish									
P. dolloi	No	Yes	No	Likely	Yes	No	No	Yes	Yes

Species									
Magadi Lake tilapia									
A. grahami	No	Yes	No	No	—	Likely	—	—	—
Mangrove killifish									
R. marmoratus	No	Minor	Minor	Yes	—	Yes	—	—	—
Yes									
Mudskipper									
P. schlosseri	Yes	Minor (feeding)	Yes	Yes	Yes	Yes	No	Yes	Yes
B. boddaerti	Yes	No	Yes	Yes	Yes	—	No	Yes	Minor
Sleeper									
B. sinensis	Yes	No	Yes	No	Yes	—	No	No	No
O. marmoratus	No	No	Yes	No	Yes	—	No	No	No
Snakehead									
C. asiatica	Minor	No	No	Yes	No	—	No	No	No
C. gaucha	—	Yes?	—	—	—	—	—	—	—
Swamp eel									
M. albus	Yes	No	Yes	No	Yes	—	No	Yes	No
Weatherloach									
M. anguillicaudatus	No	Minor	Minor	Yes	—	Yes	No	Yes	No
Yes									

Yes = adopted as one of the major strategies; Yes? = needs to be confirmed as a major strategy adopted; Minor = adopted as a minor strategy; Likely = suspected to be one of the strategies adopted but needs verification; — = no information is available at present.

0.2 g brain, and 1.0 ml plasma. These add up to only $330 \mu mol$ (contrary to $18 \mu mol$ ammonia-N g^{-1} and $24 \mu mol$ urea-N g^{-1} as stated in the discussion of Saha *et al.*, 2001: 143) which could not account for the amount of $2975 \mu mol$ excreted during the subsequent 48 hour recovery period in water. Rather, their results (Saha *et al.*, 2001) revealed suppression of ammonia production (see section V.A) and tolerance of ammonia at the cellular and sub-cellular levels as primary strategies (see section V.E) adopted by this catfish to handle ammonia toxicity during water shortage. The content of ammonia in the muscle built up to $8.34 \mu mol\, g^{-1}$, and that of urea to only $4.96 \mu mol\, g^{-1}$, after being kept away from water for 1 month.

Saha and Ratha (1990) exposed *H. fossilis* to 25, 50, or $75\, mmol\, l^{-1}$ NH_4Cl for 28 days, and observed that the fish absorbed ammonia from the medium throughout the period in most cases. They reported that urea excretion increased by 1.5- to 2-fold between day 10 and day 12, and was maintained at that level thereafter. Therefore, they concluded: "prolonged hyper-ammonia stress induced the shift from ammonotelism to ureotelism in *H. fossilis.*" However, their results did not substantiate such a conclusion. They (Saha and Ratha, 1990: Table 1) reported the rate of ammonia excretion in control specimens fasted for 14 days as $7.82 \mu mol\, 48\, h^{-1}\, g^{-1}$. For a specimen being exposed to $75\, mmol\, l^{-1}$ NH_4Cl for 14 days, $56.37 \mu mol$ $48\, h^{-1}\, g^{-1}$ ammonia was absorbed. Assuming that ammonia excretion was totally impeded in a medium containing $75\, mmol\, l^{-1}$ $NH_4\, Cl$, the total ammonia credited was $7.82 + 56.37 = 64 \mu mol\, 48\, h^{-1}\, g^{-1}$. However, the increase in the rate of urea excretion was only $2.86-0.96 = 1.90 \mu mol$ $48\, h^{-1}\, g^{-1}$ (Saha and Ratha, 1990: Table 2). Hence, the increase in urea excretion represented only 2.96% of the $64 \mu mol\, 48\, h^{-1}\, g^{-1}$ of accumulated ammonia. Another enigma is that the rate of ammonia absorption from the medium increased with increasing concentrations of NH_4Cl in the external medium (Saha and Ratha, 1990: Table 1), but the rate of urea excretion remained relatively constant for all concentrations of NH_4Cl tested (Saha and Ratha, 1990: Table 2). If indeed ureogenesis and ureotely were the "major" strategies adopted by *H. fossilis* to survive ammonia exposure, one would expect an increase in urea excretion rate in proportion to the level of environmental ammonia that it was exposed to. As pointed out by Graham (1997), "paradoxically, some of the species reported to have the OUC enzymes are not ureotelic; urea, in fact, amounts to a quite small percentage of the total nitrogen excreted by *Heteropneustes* (even in $75\, mmol\, l^{-1}$ NH_4Cl)." Nonetheless, between tissue ammonia tolerance and detoxification of ammonia to urea, the former appears to be much more important than the latter in *H. fossilis* confronted with HEA. By day 14 of ammonia exposure ($75\, mmol\, l^{-1}$ NH_4Cl), the ammonia level in the muscle reaches $17.17 \, \mu\mathrm{mol} \, g^{-1}$, but the level of urea-N in the same tissue is only $4.32 \, \mu\mathrm{mol} \, g^{-1}$.

5. Air-Breathing Catfishes

The walking catfish, *Clarias batrachus*, belongs to the Family Clariidae. It can be found in Mekong and Chao Phraya basins, Malay Peninsula, Sumatra, and Java in Asia, inhabiting swamps, ponds, ditches, rice paddies, and pools left in low spots after rivers have been in flood. It has been reported that *C. batrachus* can undertake lateral migrations from the Mekong main stream, or other permanent water bodies, to flooded areas during the raining season, and returns to the permanent water bodies at the onset of the dry season. On land, it respires with auxiliary breathing organs.

Clarias batrachus is ammonotelic in water and can tolerate long-term aerial or HEA exposure. Saha and Ratha (1998) reported that *C. batrachus*, like *H. fossilis*, had a functional OUC with CPS with unique properties in the liver. Therefore, it was suggested that the capability of this catfish to tolerate HEA was due to the detoxification of ammonia to urea and the formation of FAAs (Saha and Ratha, 1998). While these are effective strategies to detoxify endogenous ammonia when confronted with alkaline pH_{env} (Saha, Kharbuli et al., 2002), they would be ineffective when the fish is confronted with HEA (see above). Indeed, there are apparent controversies in results reported by Saha and colleagues for *C. batrachus* in this regard.

By perfusing the liver with saline containing NH_4Cl, Saha and Das (1999) discovered that liver cells of *C. batrachus* could take up ammonia and release urea into the effluent. Both influx of ammonia and efflux of urea were saturable with respect to the rate of NH_4Cl infusion. In addition, they (Saha and Das, 1999) reported that hepatic OUC enzyme activities, with the exception of ornithine transcarbamylase, were up-regulated after the infusion of NH_4Cl. They concluded that 40–50% of the exogenous ammonia taken up by the liver was converted to urea-N. However, there are two apparent problems. First, ammonia accumulated to a very high level ($>28 \, \mu\mathrm{mol} \, g^{-1}$) in the liver, which actually confirms tissue ammonia tolerance (see section V.E), and not the formation of urea through OUC, as the major strategy to handle ammonia toxicity in the liver cells. Secondly, it was reported that the urea-N efflux from the liver was $0.44 \, \mu\mathrm{mol} \, min^{-1} \, g^{-1}$ liver (Saha and Das, 1999; Figure 8.1). This amounted to $0.22 \, \mu\mathrm{mol}$ urea $min^{-1} \, g^{-1}$, or $13.2 \, \mu\mathrm{mol}$ urea $h^{-1} \, g^{-1}$. Yet, the highest CPS activity ($8.68 \, \mu\mathrm{mol}$ urea $h^{-1} \, g^{-1}$; Saha and Das, 1999: Table 2) presented in the same report could not account for such a high rate of urea efflux. Therefore, it is doubtful if urea synthesis plays a "major" role in the defense against ammonia toxicity in the liver of *C. batrachus*.

In a separate report, Saha *et al.* (2000) concluded that the efflux of urea-N ($0.55\,\mu$mol min^{-1} g^{-1}) was much lower than the efflux of non-essential FAAs (4–$4.75\,\mu$mol min^{-1} g^{-1}) from the liver of *C. batrachus* perfused with high concentrations (5 or 10 mmol l^{-1}) of NH$_4$Cl. Therefore, they suggested that the synthesis of various non-essential FAAs (presumably from infiltrated exogenous ammonia) was the major strategy for ammonia detoxification in *C. batrachus*. However, their results did not support such a conclusion. Since ammonia uptake was a saturable process (Saha and Das, 1999), it can be deduced that the rate of ammonia uptake during the infusion of 10 mmol l^{-1} NH$_4$Cl was approximately $0.9\,\mu$mol min^{-1} g^{-1} (Saha and Das, 1999). Other than glutamine and asparagine, which contain 2 N each, 1 mol N can be found in every mole of FAA. Therefore, the rate of total-N excretion was $0.55 + 4.75$ or $5.3\,\mu$mol min^{-1} g^{-1}. This was much greater than the rate of ammonia uptake and indicates that the FAAs released could not be derived from the detoxification of exogenous ammonia. Moreover, there were significant increases in essential FAAs in the experimental liver, which actually suggest that FAAs were released through proteolysis. One possible explanation is that under such a non-physiological condition of ammonia (5 or 10 mmol l^{-1} NH$_4$Cl) infusion, amino acid catabolism was reduced in general to reduce the production of endogenous ammonia; but cells were in an adverse state and proteins began to break down through proteolysis at an abnormal rate leading to the release of FAAs. It is important to note that Saha *et al.* (2000) infused very high concentrations (5 or 10 mmol l^{-1} NH$_4$Cl) of ammonia into the liver of *C. batrachus*. Such experimental conditions were un-physiological because the brain would have been adversely affected (see section IV.A and D, and Figure 8.1), and the fish would have been killed long before the concentration of ammonia in the blood would reach such a high level.

Subsequently, working on live specimens, Saha, Dutta *et al.* (2002) concluded that *C. batrachus* depended mainly on urea production and not the synthesis of non-essential amino acids to survive in HEA. This is contrary to the conclusion made earlier by Saha *et al.* (2000). More importantly, there are apparent discrepancies in results reported by Saha, Dutta *et al.* (2002, between Figure 2 and Table 7). Taking the deficit in ammonia excretion plus ammonia influx to be $400\,\mu$mol h^{-1} kg^{-1} (Saha, Dutta *et al.*, 2002, Figure 2), that would give a total of ($400\,\mu$mol h^{-1} kg^{-1} × 24 h × 7 d × 100 g)/1000 g or $6720\,\mu$mol 100 g^{-1} fish in 7 d. Yet, the amount reported in Table 7 was 47.2 mmol or $47\,200\,\mu$mol 100 g^{-1} fish, which is 7-fold greater instead. The same applies to urea excretion. From Figure 2 (Saha, Dutta *et al.*, 2002), the excess urea-N excreted by a specimen exposed to NH$_4$Cl in a 7 day period can be calculated to be ($120 + 200 + 320 + 320 + 320 + 320 + 320 + 320$) μmol h^{-1} kg^{-1} × 24 h × 100 g/1000 g or $4608\,\mu$mol 100 g^{-1} fish.

However, the amount given in Table 7 was 35.3 mmol or 35 300 μmol 100 g^{-1} fish, which is more than 9-fold the amount reported in Figure 2. Based on the results reported in Table 7, Saha, Dutta *et al.* (2002) concluded that out of the 47.2 mmol of ammonia, which theoretically should have accumulated in the 100 g specimen, 35.5 mmol was excreted as urea-N, and therefore enhanced urea synthesis played a major role in ammonia detoxification in *C. batrachus*. For a 100 g specimen, an urea excretion rate of 35.5 mmol urea-N 7 d^{-1} is equivalent to a rate of 35 500 μmol/(2 N $\times$ 7 d $\times$ 24 h) or 105.7 μmol urea h^{-1}. Such a rate of urea synthesis could never be sustained by the hepatic CPS III activity of 17.36 μmol h^{-1} present in a 2 g liver of a 100 g fish (8.68 μmol h^{-1} g^{-1}; Saha and Das, 1999, Table 2). Furthermore, there are other problems associated with this report. First, they (Saha, Dutta *et al.*, 2002) reported that the plasma ammonia concentration was only 2.47 μmol ml^{-1} after 7 days of exposure to 25 mmol l^{-1} NH$_4$ Cl, confirming that the concentrations of NH$_4$ Cl used in an earlier study (Saha *et al.*, 2000) was unphysiological. How the fish maintain such a low plasma ammonia concentration when confronted with an inwardly driven ΔP_{NH3}, which is essential to understanding how *C. batrachus* deals with exogenous ammonia, was not discussed. The possibility of *C. batrachus* being able to actively excrete NH$_4^+$ was not considered (see below discussion for *C. gariepinus*). Secondly, the total non-essential amino acid drastically increased in the first day from 40.28 to 53.92 μmol g^{-1} and leveled off after reaching 58.07 and 58.25 μmol g^{-1} by day 3 and day 7, respectively (Saha, Dutta *et al.*, 2002, Table 3). Yet, ammonia continued to increase in the muscle from day 3 to day 7. Being the bulk of the body mass, any change in the content of TFAA in the muscle would reflect on the role of FAAs in ammonia detoxification. Hence, the conclusion made by Saha, Dutta *et al.* (2002) on ammonia being detoxified to FAAs in *C. batrachus* during 7 days of ammonia loading might not be valid.

Another member of the same Family Clarridae, the North African catfish, *C. gariepinus*, is found originally in the Niger and Nile Rivers, and in the Limpopo, Orange–Vaal, and Cunene River systems in Africa. It has been introduced into Europe and Asia, and is widely distributed and farmed in many parts of Asia at present. It can tolerate wide fluctuations in water availability and temperature (Donnely, 1973), but does not possess CPS in the liver (Terjesen *et al.*, 2001; Ip, Subaidah *et al.*, 2004). *Clarias gariepinus* does not detoxify ammonia to urea or amino acid during long-term aerial exposure or ammonia loading (100 mmol l^{-1} NH$_4$Cl, pH 7 for 5 days) (Ip, Subaidah *et al.*, 2004). Instead, it has the capability to excrete ammonia against a concentration gradient (Ip, Subaidah *et al.*, 2004), although the mechanisms involved appear to be different from those of *P. schlosseri*. When left in a small volume of freshwater for 4 days, it can excrete ammonia continuously and sequester it in the external medium, reaching $8 \, \mathrm{mmol \, l^{-1}}$ of TAmm (Ip, Subaidah *et al.*, 2004). It is uncertain why *C. gariepinus* and *C. batrachus* (as reported by Saha and colleagues) have totally different strategies to deal with ammonia toxicity although they both belong to the same genus and can crossbreed successfully. More importantly, Ip, Subaidah *et al.* (2004) were unable to detect CPS (I or III) activity in the liver of *C. batrachus* obtained from Singapore and Indonesia. Therefore, the role of urea in ammonia detoxification in air-breathing catfishes has to be re-evaluated.

6. ABEHAZE

Members of the genus *Mugilogobius* belonging to the Family Gobiidae are demersal, brackish or marine, and are widespread in the tropics. They can be found in Southeast Asia and Indo-West Pacific: Indonesia, Malaysia, Philippines, and Thailand. The abehaze, *Mugilogobius abei*, is found in Japan (Mukai *et al.*, 2000). Iwata *et al.* (2000) reported that *M. abei* possessed a functional OUC in the muscle, skin, and gills, and Kajimura *et al.* (2002) later discovered that this goby exhibited a diurnal nitrogen excretion rhythm. *Mugilogobius abei* can survive in $2 \, \mathrm{mmol \, l^{-1}}$ NH_4Cl at pH_{env} 7.6 for up to 8 days (Iwata *et al.*, 2000). It shifts from ammonotely to ureotely during ammonia loading and is able to synthesize urea using the OUC operating in multiple tissues. However, *M. abei* exhibits relative low tolerance ($2 \, \mathrm{mmol \, l^{-1}}$ at pH_{env} 7.6) to environmental ammonia in comparison to the Lake Magadi tilapia ($>75 \, \mathrm{mmol \, l^{-1}}$ at pH_{env} 10), the giant mudskipper ($100 \, \mathrm{mmol \, l^{-1}}$ at pH_{env} 7), the Gulf toadfish ($10 \, \mathrm{mmol \, l^{-1}}$ at pH_{env} 8.2) and catfishes ($30–75 \, \mathrm{mmol \, l^{-1}}$ at pH_{env} 7). Actually, it is doubtful if *M abei* detoxified the net influx of "exogenous" ammonia to urea as suggested by Iwata *et al.* (2000). The deficit of ammonia excretion during the first 4 days of exposure to $2 \, \mathrm{mmol \, l^{-1}}$ NH_4Cl amounts to $36.5 \, \mu\mathrm{mol \, N \, g^{-1}}$. However, the increase in excretion of urea-N during this period is only $20.5 \, \mu\mathrm{mol \, N \, g^{-1}}$, and the total urea-N accumulated in the body is $7.78 \, \mu\mathrm{mol \, N \, g^{-1}}$. Hence, the total of $28.28 \, \mu\mathrm{mol \, urea\text{-}N \, g^{-1}}$ accounts for only 78% of the endogenous ammonia, which would have been produced during the 4 day period, without considering the exogenous ammonia penetrated into the body.

Results obtained by Iwata *et al.* (2000) actually confirmed the tolerance of ammonia at the cellular and sub-cellular levels (see section V.E) as the major strategy of defence against environmental ammonia toxicity in *M. abei*. The ammonia content in the muscle increases from 4.98 to $13.95 \, \mu\mathrm{mol \, g^{-1}}$, and that in the whole body increases from 3.92 to $11.43 \, \mu\mathrm{mol \, g^{-1}}$ after 4 days of exposure to $2 \, \mathrm{mmol \, l^{-1}}$ NH_4Cl (pH_{env} 7.6). The blood pH and plasma ammonia concentration were not determined, probably due to the small size of the fish. However, it is logical to deduce that the ammonia level in the plasma must have increased simultaneously. With a normal blood pH of 7.3–7.6, an increase in the plasma TAmm to 1.5–2.0 mmol l^{-1} would have reduced or eliminated the ΔP_{NH3}, which is directed inward at the beginning of the experiment. If that was indeed the case, urea (and glutamine) is synthesized to detoxify mainly endogenous ammonia to maintain the newly established higher steady-state level of ammonia in the body. The advantage is obvious – the subsequent excretion of urea is completely independent of the presence of ammonia in the external medium. In this sense, the abehaze is adopting a strategy comparable to that of the Lake Magadi tilapia. However, no information is available on mechanisms involved in facilitating urea excretion in *M. albei* at present.

7. SNAKEHEAD

Earlier, Ramaswamy and Reddy (1983) concluded that the snakehead *Channa gachua* shifted toward ureotely during aerial exposure. However, there were problems in their experimental design and discrepancies in their results, which shed doubt on such a conclusion. First, they reported a 3-fold excess of urea-N excreted during re-immersion over the deficit in ammonia-N, which theoretically would have been retained during aerial exposure (Chew, Wong et al., 2003). Secondly, they did not examine the urea and ammonia contents in the muscle, which constitutes the bulk of the fish, nor assay for OUC enzymes.

Recent work (Chew, Wong et al., 2003) on the small snakehead *C. asiatica* reveals that it does not possess a functioning OUC and is incapable of detoxifying ammonia to urea through this pathway. In support of this proposition, there is no change in the rate of urea excretion in specimens exposed to terrestrial conditions. In addition, there is no major change in the urea contents in the muscle, liver, and plasma after 48 hours of aerial exposure. The ability to synthesize low levels of urea is not necessarily linked to a functional OUC. The activity of arginase in the liver of *C. asiatica* is sufficient to account for the amount of urea produced during the 48 hour experimental period.

8. MUDSKIPPERS

Gordon *et al.* (1969, 1978) reported that when the mudskipper, *Periophthalmus sorbinus*, was exposed to 12 hours of terrestrial conditions, urea production increased more than 3-fold. However, Gregory (1977) could not detect activities of some OUC enzymes, including CPS, argininosuccinate synthetase, and argininosuccinate lyase from the liver of *Periophthalmus expeditionium*, *Periophthalmus gracilis*, and *Scartelaos histophorous*. It was concluded that urea production in livers of *P. expeditionium* and *P. gracilis* occurred via uricolysis, involving urate oxidase, allantoinase, and allantoicase. The activities of arginase and urate oxidase are high enough to account for the amount of urea excreted in the liver of these two fishes (Gregory, 1977). Working on the mudskippers *Periophthalmus modestus* (previously as *P. cantonensis*) and *Boleophthalmus pectinitrostris*, Morii (1979) and Morii *et al.* (1978, 1979) reported that ammonia accumulated during aerial exposure was not detoxified to urea. Iwata *et al.* (1981) and Iwata (1988) reported that urea production in *P. modestus* remained unchanged after exposure to ammonia or air; and, when *P. modestus* was exposed to ^{15}N-labeled ammonia, urea-N was only slightly labeled (Iwata and Deguchi, 1995). Recently, Lim *et al.* (2001) confirmed that no N-acetylglutamate activated CPS activity could be detected (detection limit $= 0.001\,\mu\mathrm{mol}\,\mathrm{min}^{-1}\,\mathrm{g}^{-1}$) from the liver mitochondria of *Boleophthalmus boddaerti*. Taking all these results together, it can be concluded that urea synthesis *de novo* may not occur in *Periophthalmus* spp., *Scartelaos* spp. or *Boleophthalmus* spp.

To date, the only mudskipper that possesses a full complement of hepatic OUC enzymes, in spite of uncertainty on the type of mitochondrial CPS present, is the giant mudskipper *P. schlosseri* (Lim *et al.*, 2001). However, similar to other mudskipper species, detoxification of ammonia to urea does not occur in *P. schlosseri* confronted with adverse environmental conditions (aerial exposure, Ip *et al.*, 1993; Lim *et al.*, 2001; alkaline environmental pH, Chew, Hong *et al.*, 2003; environmental ammonia, Peng *et al.*, 1998; Randall *et al.*, 1999). Instead, *P. schlosseri* adopts other strategies to defend against ammonia toxicity (see section IV.A, B, C and V.A, C). However, with the successful adoption of these strategies, it remains an enigma as to why there is still the need to express the OUC in the liver of adult *P. schlosseri*. Being the only mudskipper that is carnivorous (other species are either herbivorous or omnivorous), Ip, Lim *et al.* (2004) speculated that the presence of the OUC in *P. schlosseri* might be related to its high protein diet (mangrove crabs and small fishes), and the OUC is involved in the defense against postprandial ammonia toxicity.

The ammonia and urea excretion rates of *P. schlosseri* increase 1.70- and 1.92-fold, respectively, within the first 3 hours post-feeding (Ip, Lim *et al.*, 2004). There are significant decreases in ammonia levels in the plasma and the brain, and in urea contents in the muscle and liver, of *P. schlosseri* at hour 3 post-feeding (Ip, Lim *et al.*, 2004). Taken together, these results indicate that, after feeding, *P. schlosseri* is capable of unloading ammonia originally present in some of its tissues in anticipation of the ammonia released from the catabolism of excess amino acids absorbed. In addition, there are significant increases in urea contents in the muscle, liver, and plasma (1.39-, 2.17- and 1.62-fold, respectively) at hour 6 post-feeding, and the rate of urea synthesis apparently increases 5.8-fold between hour 3 and hour 6 (Ip, Lim *et al.*, 2004). With the excess urea accumulated in the body at hour 6 being completely excreted between 6 and 12 hours, the percentage of N excreted as urea-N increases significantly to 26% during this period, but it never exceeds 50%, the criterion for ureotely (Ip, Lim *et al.*, 2004). Increased urea synthesis is likely to have occurred in the liver of *P. schlosseri* because the greatest increase in urea content is observed in the liver. These results suggest that an ample supply of energy resources, e.g. after feeding, is a prerequisite for the induction of urea synthesis. Together, increases in nitrogenous excretion and urea synthesis after feeding effectively prevent a postprandial surge of ammonia in *P. schlosseri*. Consequently, unlike other fish species (Wicks and Randall, 2002), there are significant decreases in the ammonia content in the brain of *P. schlosseri* throughout the 24-h period post-feeding, accompanied with a significant decrease in the brain glutamine content between hour 12 and hour 24 (Ip, Lim *et al.*, 2004).

9. FISH EMBRYOS

Griffith (1991) proposed that urea synthesis evolved in gnathostome fishes to avoid ammonia toxicity during protracted embryogenesis. Using radio-labeled bicarbonate, a substrate for OUC, Depeche *et al.* (1979) demonstrated urea synthesis in the rainbow trout and guppy during early embryonic development. Urea levels decrease towards the end of embryogenesis, and the incorporation of $^{14}C\text{-}HCO_3^-$ into urea is not found in isolated tissues of the adult trout (Depeche *et al.*, 1979). One reason why urea synthesis in rainbow trout is initiated soon after the eggs are fertilized is that embryos must rid themselves of nitrogenous wastes without convective mechanisms (Wright *et al.*, 1995). Unlike adult fish, embryos do not actively pump water over the gills, which have not yet developed. Urea synthesis therefore serves as a means to remove and maintain ammonia at levels below the threshold levels of toxicity (Wright and Land, 1998; Wright and Fyhn, 2001). Very low levels of CPS III have been shown to coincide with the expression of ornithine transcarbamylase (Wright *et al.*, 1995; Chadwick and Wright, 1999; Terjesen *et al.*, 2000). Since it would be essential for the developing embryos to retain the majority of the nitrogen released from amino acid catabolism for anabolic purposes (e.g. synthesis of purines and pyrimidines) during growth and development, the rate of urea synthesis is likely to be low compared with the rate of amino acid turnover. It is possible that partial amino acid catabolism (see section V.B) and/or glutamine formation (but without leading to urea production; see section V.C) take a more important role in avoiding ammonia toxicity than urea formation during fish embryo development. Thus, it would be essential to examine the type and compartmentation of GS isozymes in fish embryos, because it would reveal the intricate relationships between the cytosolic CPS II and the mitochondrial CPS III (both utilizing glutamine as a substrate), and between anabolic and catabolic processes, during fish embryo development. To date, no such information is available in the literature.

The African catfish, *C. gariepinus*, spawns shortly after heavy rainfall, and the embryos are deposited in a few centimeters of water on vegetation in temporarily flooded areas (Greenwood, 1955; Bruton, 1979). During early development of *C. gariepinus*, urea constitutes 62% of the total nitrogen excreted in embryos, 20% during yolk-sac and starved larval stages, and 44% after metamorphosis in fed larvae (Terjesen *et al.*, 1997). In contrast, immersed adult *C. gariepinus* (Eddy *et al.*, 1980) and adult *C. batrachus* (Saha and Ratha, 1989) excrete less urea, accounting for 13% and 15% of total nitrogen excreted, respectively. Based on arginine depletion rates in *C. gariepinus* yolk-sac and starved larvae *in vivo*, Terjesen *et al.* (1997) estimated that argininolysis could account for approximately one-third of the urea excreted, suggesting that other pathways for urea synthesis were also functional.

E. Extreme Ammonia Tolerance in Cells and Tissues

It would appear that high environmental ammonia tolerance is usually associated with high tolerance to ammonia at the cellular and sub-cellular levels in tropical fishes (Table 8.1). High ammonia tolerance in cells and tissues of these fishes results in their capacities to tolerate the accumulation of high levels of ammonia in their bodies. This is advantageous to fishes that are often exposed to HEA, because the high concentrations of ammonia in the blood would reduce the inwardly directed ΔP_{NH3} and therefore reduce the influx of NH_3 during ammonia loading. This phenomenon has been observed in various catfishes (see section V.D, Saha and Ratha, 1998; Saha, Dutta *et al.*, 2002), the abehaze (Iwata *et al.*, 2000), the swamp eel (Ip, Tay *et al.*, 2004), the weatherloach (Chew *et al.*, 2001; Tsui *et al.*, 2002), and the Asian freshwater stingray *Himantura signifier* (Ip *et al.*, 2003). Very often, the accumulated ammonia is not evenly distributed within the fish; some exhibit much higher level of ammonia in the muscle, whereas others, like the marble goby *O. marmoratus*, can tolerate a very high level of ammonia in the brain (Y. K. Ip and S. F. Chew, unpublished data). In the case of the weatherloach, it has the unique capability of maintaining the ammonia level in the brain lower than that in the blood after 48 hours of aerial exposure. How the cells and tissues, especially those in the brain, of these fishes tolerate high levels of internal ammonia (see deleterious effects in section IIID) is not clear at present. It is likely that they might have developed K^+ specific K^+-channels, K^+ specific Na^+, K^+-ATPase, and/or special NMDA receptors (Figure 8.1M). It is also possible that these fishes could tolerate the build-up of glutamine in astrocytes in the brain better than other fishes and mammals (Brusilow, 2002). It is well known that mammalian brains cannot tolerate ammonia levels >1–2 μmol g^{-1}, beyond which encephalopathy would develop (Cooper and Plum, 1987). Therefore, those tropical fishes capable of tolerating high levels of ammonia in the brain are ideal specimens to study mechanisms involved in defense against ammonia toxicity in the central nervous system, which might have been lost during the evolution of higher vertebrates.

1. WEATHERLOACH

   Although the plasma ammonia in submerged *M. anguillicaudatus* falls within the range for other facultative air-breathers, it is exceptionally high in aerially exposed specimens. The plasma ammonia concentrations in various air-tolerant teleosts are all less than 1.6 μmol ml^{-1} after aerial exposure (Ramaswamy and Reddy, 1983; Saha and Ratha, 1989). In contrast, the ammonia level in the plasma of *M. anguillicaudatus* rises (from 0.81 μmol ml^{-1}) to 2.46 μmol ml^{-1} after being exposed to terrestrial conditions for 6 hours (Chew *et al.*, 2001). The highest level (5.09 μmol ml^{-1}) is reached by hour 48. To our knowledge, no other fish accumulates such a high level of ammonia in the plasma during aerial exposure. Such a capability may be important not only for aerial exposure, but also for survival in HEA when fertilizers are added to the rice field. Indeed, the ammonia level in the plasma of *M. anguillicaudatus* rises (from 0.92 μmol ml^{-1}) to 4.2 μmol ml^{-1} after 48 hours of ammonia exposure (Tsui *et al.*, 2002). At hour 48, the NH$_3$ concentration in the external medium decreases slightly to 0.106 mmol l^{-1} due to a slight drop in pH$_{env}$ (from 7 to 6.8). Taking the blood pH and plasma ammonia concentration of specimens exposed to this medium as 7.349 and 4.2 mmol l^{-1}, respectively, it can be estimated that the NH$_3$ level in the plasma increases to 0.052 mmol l^{-1}. This would drastically reduce the NH$_3$ gradient from 17-fold to 2-fold. The elevated plasma ammonia level would reduce the influx of exogenous ammonia by restoring a relatively more favorable blood-to-water ΔP_{NH3} gradient (Figure 8.1L).

   There are significant increases in the ammonia contents in the muscle and liver of *M. anguillicaudatus* exposed to HEA. The ammonia levels in the muscle (18.9 μmol g^{-1}) and liver (17.5 μmol g^{-1}) are very high (Tsui *et al.*, 2002). Despite glutamine being accumulated in the muscle after 48 hours of aerial exposure, accumulation of glutamine does not occur when *M anguillicaudatus* is exposed to HEA. Hence, in spite of the presence of mechanisms to detoxify ammonia, they are not switched on when a net influx of exogenous ammonia occurs. Rather, *M. anguillicaudatus* adopts the strategy of allowing the internal ammonia concentration to build up in order to reduce the ammonia influx.

Ammonia toxicity in the weatherloach can be reduced by MK-801, which is a selective antagonist of the NMDA type of glutamate receptors (K. N. T. Tsui and D. J. Randall, unpublished results), as observed in mice (Marcaida *et al.*, 1992). Hence, NMDA receptors must be present in *M. anguillicaudatus*, and its activation by ammonia contributes to the toxicity of ammonia in this fish. However, the NMDA receptor of the weatherloach may have special properties that could lead to high ammonia tolerance in this fish. In addition, this loach, and others that adopt a similar strategy to handle ammonia toxicity, may have Na^+, K^+-ATPase, and K^+ channels with high substrate specificity for K^+ (Figure 8.1M), which facilitates the maintenance of the intracellular K^+ concentration and the resting membrane potential in the presence of high concentrations of extracellular ammonia. This is especially important to the brain and cardiac muscle cells, although skeletal muscles and some non-excitable cells would also be affected.

VI. ACCUMULATION OF NITROGENOUS END-PRODUCTS FOR OSMOREGULATION

Differences in water and ionic regulation in seawater as opposed to freshwater may have rendered the excretion of ammonia as a major end-product of nitrogen catabolism in the marine environment disadvantageous (Campbell, 1973). However, the accumulation of urea as an osmotic component was certainly advantageous, and the function of the OUC in this capacity is seen today in elasmobranchs, holocephans and coelacanths. The utilization of an end-product of nitrogen catabolism rather than amino acids themselves for osmotic purposes has an energetic advantage – the energy derived from carbon catabolism of the amino acids is not lost (Campbell, 1973). Moreover, unlike ions, urea can equilibrate across membranes speedily throughout various compartments of the fish without the needs of specific transporters (Ballantyne and Chamberlin, 1988). Indeed, specific urea transporters are lacking in elasmobranch red blood cells or parenchymal hepatocytes (Ballantyne, 2001). In addition, urea (and TMAO) contributes substantially to positive buoyancy in elasmobranchs (Withers *et al.*, 1994a, b).

Adoption of the ureosmotic strategy of osmoregulation requires some major changes in the biochemical architecture of the fish (Ballantyne, 1997). The perturbing effects of urea on macromolecules structure and function have to be counteracted by the simultaneous accumulation of stabilizing solutes (e.g. TMAO) (Hochachka and Somero, 1984; Yancey, 2001). In addition, urea might have affected the binding of non-esterified fatty acids to serum albumins, leading to the absence of albumins in elasmobranchs. This limits the role of lipids as a transportable catabolic fuel and results in their replacement by ketone bodies (Ballantyne *et al.*, 1987; Ballantyne, 1997, 2001).

Kirschner (1993) performed a detailed energetic comparison of ureosomotic (to avoid drinking seawater) and hyposmotic (to drink seawater and excrete the excess salt) strategies and reported the costs of osmoregulation via these two strategies to be comparable, comprising approximately 10–15% of standard metabolism. However, Walsh and Mommsen (2001) argued that uncertainties concerning urea transport and TMAO biochemistry would tip the balance toward the hyposmotic strategy being more economical. In addition, the ureosomotic strategy may offer other disadvantages: (1) it requires a commitment to carnivory for large amounts of nitrogen for urea synthesis, (2) it alters lipid and carbohydrate metabolism which become less compatible with herbivory and other dietary modes, and (3) it restricts the reproductive/development options of fish because of the potential solute loss during the high surface area-to-volume ratio period in young.

In order to be able to retain urea for osmoregulation in certain fishes, effective urea permeabilities would have had to decrease. In modern-day elasmobranchs, this effective decrease in urea permeability appears to be the result of the presence of specific secondarily active (Na^+ coupled) urea transporters in gills (Figure 8.1 N) and kidney, and modification of lipid composition of gills to achieve higher cholesterol-to-phospholipid ratios (Figure 8.1P) (Fines *et al.*, 2001; Walsh and Smith, 2001). In terms of strategy of adaptation, it shares the same principle with the reduction in the effective permeability to NH_3 (see section IV.C), which acts in preventing the entry of NH_3 instead.

1. MARINE ELASMOBRANCHS

Marine elasmobranchs (sharks, skates, and rays) are common in tropical waters. They are ureotelic, excreting the majority of their nitrogenous wastes as urea via the gills (Perlman and Goldstein, 1999; Shuttleworth, 1988; Wood, 1993; Wood, Hopkins *et al.*, 1995; Wood, Part *et al.*, 1995). They exhibit osmoconforming hypoionic regulation (Yancey, 2001) with body fluid osmolalities equal to or slightly higher than the environment. Their extracellular fluids are actively regulated to have considerably lower salt concentrations than the environment, with the osmotic difference balanced by extracellular (as well as intracellular) nitrogenous organic osmolytes. Unlike most teleost fishes, marine elasmobranchs are ureosmotic and have an active OUC (Anderson, 1980). Low permeability of ammonia at the gills may be important for maintaining urea levels, because ammonia is the major vehicle of nitrogen in the blood, and loss at the gills would preclude its use for urea synthesis in the liver. Wood, Part *et al.* (1995) estimated that brachial urea-N and branchial ammonia-N permeabilities in the dogfish shark were only about 7% and 4%, respectively, of that in a typical teleost. Marine elasmobranchs synthesize urea through the OUC with CPS III (Campbell and Anderson, 1991; Anderson, 1980, 1991, 1995, 2001), primarily for osmoregulation (Ballantyne, 1997; Perlman and Goldstein, 1999; Anderson, 2001). Urea is retained at high concentrations (300–600 mmol l^{-1}) in the tissues. This is accomplished by low permeability of the gills to urea and by urea reabsorption mechanisms in the gills (Smith and Wright, 1999) and kidney (see review by Walsh and Smith, 2001). In the little skate, *Raja erinacea*, the down-regulation of kidney urea transporter may play a key role in lowering tissue urea levels in response to external osmolality changes (Morgan *et al.*, 2003a). Morgan *et al.* (2003b) further demonstrated that urea uptake by brush border membrane vesicles is by a phloretin-sensitive, non-saturable uniporter in the dorsal section and a phloretin-sensitive, Na$^+$-linked urea transporter in the ventral section, which are critical for renal urea reabsorption in the little skate.

2. Freshwater Elasmobranchs

In tropical waters in Southeast Asia (Thailand, Indonesia, and Papua New Guinea) and South America (Amazon River basin), a number of elasmobranch species migrate into low salinity waters where they reduce plasma salt, urea, and TMAO levels. In muscle, urea and TMAO are reduced more than ions, usually maintaining a similar ratio between them (Yancey, 2001). For the Amazonian freshwater stingrays, urea reduction occurs as the result of reduced synthesis (Forster and Goldstein, 1976) and/or a higher renal clearance rate (Goldstein and Forster, 1971). *Potamotrygon* spp. are stenohaline Amazonian stingrays permanently adapted to freshwater. Although it has low levels of some of the enzymes related to urea synthesis (Anderson, 1980), it retains virtually no urea or TMAO *in situ* and cannot accumulate urea in laboratory salinity stress (Thorson *et al.*, 1967; Gerst and Thorson, 1977).

The river Batang Hari originates from the Barisan Range, flows eastwards through the whole of Jambi, Indonesia, and drains into the South China Sea. The white-edge freshwater whip ray, *Himantura signifier* (Family Dasyatidae), is a stingray found in the Batang Hari basin in Jambi, Sumatra. It is believed to occur only in freshwater. In the laboratory, *H. signifier* can survive in freshwater (0.7‰) indefinitely or in brackish water (20‰) for at least 2 weeks.

In freshwater, the blood plasma osmolality (416 mosmolal) of *H. signifier* is maintained hyperosmotic to that of the external medium (38 mosmolal)

(Tam *et al.*, 2003). There is approximately 44 mmol l^{-1} of urea in the plasma, with the rest of the osmolality made up mainly by Na$^+$ (167 mmol l^{-1}) and Cl$^-$ (164 mmol l^{-1}). In freshwater, it is not completely ureotelic, excreting at most 45% of its nitrogenous waste as urea. It has a functional OUC in the liver. The hepatic CPS III and GS activities are similar to those of other marine elasmobranchs.

When *H. signifier* is exposed to a progressive increase in salinity from 0.7‰ to 20‰ through an 8 day period, there is a continuous decrease in the rate of ammonia excretion (Tam *et al.*, 2003). After exposure to 20‰ for 4 days, the ammonia excretion rate is only one-fifth that of the freshwater control. In 20‰ water, there is no change in the ammonia content in the muscle and plasma, but a decrease is observed in the liver. Presumably, ammonia is used as a substrate for urea synthesis and storage for osmoregulation at higher salinities. Indeed, in 20‰ water, urea levels in the muscle, brain, and plasma increase significantly (Tam *et al.*, 2003). In addition, certain free amino acids are used as intracellular osmolytes in the muscle (ß-alanine, glycine, and sarcosine) and the brain (ß-alanine, glycine, glutamate, and glutamine).

In the plasma, osmolality increases to 571 mosmolal, in which, urea, Na$^+$ and Cl$^-$ contribute 83, 231, and 220 mmol l^{-1}, respectively (Tam *et al.*, 2003). This is almost isoosmotic to the external medium (540 mosmolal). The total amount of urea accumulated in the tissues of the specimen being exposed to 20‰ water is equivalent to the deficit in ammonia excretion through the 8 day period, indirectly indicating the occurrence of an increase in the rate of urea synthesis at higher salinities. There is also a significance decrease in the rate of urea excretion during passage through 5, 10, and 15‰ water (Tam *et al.*, 2003). However, the rate of urea excretion increases back to the control value (3.5 µmol day^{-1} g^{-1}) when the stingray reaches 20‰ water on day 5, presumably resulting from the steeper urea gradient built up between the plasma (83 mmol l^{-1}) and the external medium (0 mmol l^{-1}).

In comparison, the local marine stingray, *Taeniura lymma*, maintains a urea excretion rate of 4.7 µmol day^{-1} g^{-1} in full-strength seawater (30‰), with a plasma urea concentration of 380 mmol l^{-1} (Tam *et al.*, 2003). Therefore, *H. signifier* appears to have reduced its capacity to retain urea in order to survive in the freshwater environment. Consequently, it cannot survive well in full-strength seawater, although it is more euryhaline than the ammonotelic ocellate river stingray, *Potamotrygon motoro* (Family Potamotrygonidae) found in South America (Wood *et al.*, 2002). Unlike *P. motoro*, *H. signifier* retains the capacity to produce urea, as demonstrated by the capability of *H. signifier*, but not *P. motoro*, to detoxify ammonia to urea during ammonia loading (Ip *et al.*, 2003). *Potamotrygon motoro* does not have a functional OUC (Goldstein and Forster, 1971) because it inhabits in freshwater rivers in the Amazonas basins, including the ion-poor acidic blackwaters of Rio Negro, which have been secluded from the sea for millions of years in South America (Lovejoy, 1997). In freshwater containing $10 \text{ mmol} \, l^{-1}$ NH_4Cl at pH_{env} 7, ammonia accumulates in the muscle, brain, and plasma of *H. signifer*. The primary strategy adopted is to allow ammonia to build up internally, especially in the plasma, to slow down the net influx of exogenous ammonia (Ip *et al.*, 2003). This is reflected by the unaltered urea excretion rate ($3 \, \mu\text{mol day}^{-1} \text{g}^{-1}$) in specimens exposed to ammonia for the first day, during which ammonia excretion ($7.3 \, \mu\text{mol day}^{-1} \text{g}^{-1}$) is presumably impeded totally. Subsequently, the urea excretion rate increases continuously to $7.4 \, \mu\text{mol day}^{-1} \text{g}^{-1}$ by day 4 of ammonia exposure, with no change in the muscle urea content, indicating that it is able to release the excess urea without creating a problem for osmoregulation. Therefore, this is a species of fish ideal for studies on up- and down-regulation of urea transporters.

3. COELACANTHS

The coelacanth, *Latimeria chalumnae* (Class Sarcopterygii), is found off the islands of Grand Comoro and Anjouan in the Comoros or near South Africa. Another species, *Latimeria menadoensis* (Holder *et al.*, 1999), is found in the Celebes Sea, north of Sulawesi, Indonesia. Coelacanths inhabit steep rocky shores, sheltering in caves during the day and foraging for food only at night. They use urea in much the same way as elasmobranchs for ureosmotic adaptation (Griffith *et al.*, 1974). However, they do not have the same renal urea recovery abilities as elasmobranchs (Griffith *et al.*, 1976). The exact mechanisms involved in urea retention in coelacanths are uncertain at this moment. Interestingly, the total osmotic pressure reported is slightly less than seawater (Griffith and Pang, 1979), suggesting a tendency to lose water. Coelacanths appear to be the only bony fish to use urea extensively as an osmolyte and represent the most "primitive" bony fishes to have an extensive OUC. At present, there is conflicting information on the type of CPS in coelacanths in the literature. Goldstein *et al.* (1973) detected CPS, which utilize NH_4^+ as a substrate (hence, likely to be CPS I), in the liver of *L. chalumnae*, the level of which was equivalent to that of the CPS III in shark liver. Later, Mommsen and Walsh (1989) reported that *L. chalumnae* possesses CPS III activity in its liver instead (also see Walsh and Mommsen, 2001). Coelacanths are closely allied with lungfish (Benton, 1990; Yokobori *et al.*, 1994). In view of their important positions in the evolution of fish–tetrapod transition in the animal kingdom, it is important to re-evaluate the type of CPS present in coelacanths when future opportunity arises.

VII. SUMMARY

Most tropical teleost fishes are ammonotelic, producing ammonia and excreting it by diffusion of NH_3 across the gills. Accumulation of ammonia in the body can be due to either the inability to excrete or convert nitrogenous wastes or to a net influx of NH_3 from the environment. Although all three conditions could lead to increases in ammonia levels in the tissues of the fish, it is important to differentiate the source of the ammonia being detoxified, which is often neglected in the literature.

When confronted with alkaline pH_{env}, terrestrial conditions or low levels of environmental (exogenous) ammonia, fishes have difficulties in excreting ammonia that is endogenously produced (Figure 8.4, Table 8.1). Fishes, with few exceptions, are very susceptible to elevated tissue ammonia levels under adverse conditions. Some could, however, avoid endogenous ammonia toxicity by utilizing several physiological mechanisms, and consequently manifest high tolerance to aerial exposure. Suppression of proteolysis and/or amino acid catabolism may be a general mechanism adopted by some fishes during aerial exposure. Others, like the giant mudskipper, *P. schlosseri*, which uses amino acid as an energy source while active on land, reduces ammonia production by utilizing partial amino acid catabolism, leading to the accumulation of alanine. Some fishes convert excess endogenous ammonia to less toxic compounds, including glutamine and other amino acids for storage. A few species have active OUC and convert

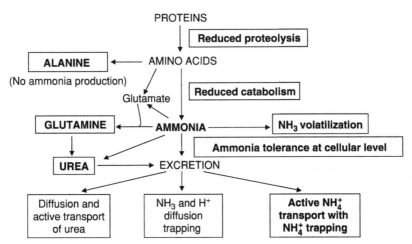

Fig. 8.4 A summary of strategies adopted by fishes in general to defend against toxicity of endogenous ammonia.

endogenous ammonia to urea for both storage and excretion. Under conditions of slightly elevated ambient ammonia, *P. schlosseri* can continue to excrete endogenous ammonia by active transport of ammonium ions. There are indications that some fishes can manipulate the pH of the body surface to facilitate NH_3 volatilization during aerial exposure or ammonia loading.

In contrast, fishes have to detoxify not only endogenous ammonia, but also exogenous ammonia that has penetrated into the body when they are confronted with HEA which results in a reversed P_{NH3} gradient (Figure 8.5, Table 8.1). To deal with exogenous ammonia, the most effective way is to manipulate the pH_{env} through increased CO_2 and acid (H^+) excretion to lower the concentration of NH_3 in the external medium. This means NH_3 is, in effect, detoxified to NH_4^+ externally, constituting a strategy of "environmental detoxification." Another way is to accumulate high levels of ammonia in the body, especially in the blood, to rebuild a more favorable P_{NH3} to reduce the influx of exogenous ammonia, or even to regain ammonia excretion. When ammonia builds up internally, as long as it is below a critical level, the brain is protected by the detoxification of ammonia to glutamine. It has been suggested that some fishes would "fix" ammonia to free amino acids during ammonia loading. However, the simultaneous build-up of essential amino acids, although to a different extent, in all cases suggests a reduction in amino acid catabolism to reduce endogenous ammonia production instead. Furthermore, this strategy is usually auxiliary to

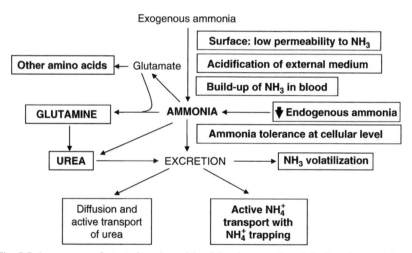

Fig. 8.5 A summary of strategies adopted by fishes in general to defend against toxicity of exogenous (environmental) ammonia.

the accumulation of ammonia in tissues and the blood. As an individual event, it does not seem appropriate to "fix" the penetrating exogenous ammonia because it would simply draw in exogenous ammonia continuously. The same argument would apply to urea formation, which is highly energy-dependent, in fish during severe ammonia loading.

Some fishes (elasmobranchs, holocephans, and coelacanths) evolved to synthesize and accumulate urea as an osmotic component (ureosmotic). The utilization of urea, an end-product of nitrogen catabolism, for osmotic purposes has an energetic advantage – the energy derived from carbon catabolism of the amino acids is not lost. In order to be able to retain urea for osmoregulation, effective urea permeabilities would have to decrease, as seen in extant marine elasmobranchs through modifications of the lipid composition of gills, and re-absorption of urea by specific secondarily active (Na^+-coupled) urea transporters in gills and kidney. However, for those elasmobranchs adapted secondarily back to a fresh- or brackish-water environment, there must be a reduction in the capacity of urea synthesis and/or the capacity of urea retention.

REFERENCES

Alderson, R. (1979). The effect of ammonia on the growth of juvenile Dover sole, *Solea solea* (L.) and turbot, *Seophthalmus maximus* (L.). *Aquaculture* **17**, 291–309.

Anderson, P. M. (1980). Glutamine- and N-acetylglutamate-dependent carbamoyl phosphate synthetase in elasmobranchs. *Science* **208**, 291–293.

Anderson, P. M. (1991). Glutamine-dependent urea synthesis in elasmobranch fishes. *Biochem. Cell Biol.* **69**, 317–319.

Anderson, P. M. (1995). Urea cycle in fish: Molecular and mitochondrial studies. *In* "Fish Physiology," Vol. 14. "Ionoregulation: Cellular and Molecular Approaches to Fish Ionic Regulation" (Wood, C. M., and Shuttleworth, T. J., Eds.), ch. 3, pp. 57–83. Academic Press, New York.

Anderson, P. M. (2001). Urea and glutamine synthesis : Environmental influences on nitrogen excretion. *In* "Fish Physiology," Vol. 20 (Wright, P. A., and Anderson, P. M., Eds.), ch. 7, pp. 239–277. Academic Press, New York.

Anderson, P. M., and Walsh, P. J. (1995). Subcellular localization and biochemical properties of the enzymes of carbamoyl phosphate and urea synthesis in the Batrachoidid fishes *Opsanus beta, Opsanus tau,* and *Porichthys notatus. J. Exp. Biol.* **198**, 755–766.

Anderson, P. M., Broderius, M. A., Fong, K. C., Tsui, K. N. T., Chew, S. F., and Ip, Y. K. (2002). Glutamine synthetase expression in liver, muscle, stomach and intestine of *Bostrichthys sinensis* in response to exposure to a high exogenous ammonia concentration. *J. Exp. Biol.* **205**, 2053–2065.

Arillo, A., Margiocco, C., Melodia, F., Mensi, P., and Schenone, G. (1981). Ammonia toxicity mechanisms in fish: Studies on rainbow trout (*Salmo gairdneri* Rich). *Ecotoxicol. Environ. Saf.* **5**, 316–325.

Avella, M., and Bornancin, M. (1989). A new analysis of ammonia and sodium transport through the gills of the freshwater rainbow trout (*Salmo gairdneri*). *J. Exp. Biol.* **142**, 155–175.

Ballantyne, J. S. (1997). Jaws: The inside story. The metabolism of elasmobranch fishes. *Comp. Biochem. Physiol.* **118B,** 703–742.

Ballantyne, J. S. (2001). Amino acid metabolism. *In* "Fish Physiology" Vol. 20. "Nitrogen Excretion" (Wright, P. A., and Anderson, P. M., Eds.), ch. 3, pp. 77–107. Academic Press, New York.

Ballantyne, J. S., and Chamberlin, M. E. (1988). Adaptation and evolution of mitochondria: Osmotic and ionic considerations. *Can. J. Zool.* **66,** 1028–1035.

Ballantyne, J. S., Moyes, C. D., and Moon, T. W. (1987). Compatible and counteracting solutes and the evolution of ion and osmoregulation in fishes. *Can. J. Zool.* **65,** 1883–1888.

Barzaghi, C. (2002).The mudskipper *Periophthalmodon schlosseri* increases acid (H^+) excretion and alters the skin lipid composition during short term and long term exposure to ammonia, respectively. MSc thesis, National University of Singapore.

Beaumont, M. W., Taylor, E. W., and Butler, P. J. (2000). The resting membrane potential of white muscle from brown trout (*Salmo trutta*) exposed to copper in soft, acidic water. *J. Exp. Biol.* **203,** 229–236.

Begum, S. J. (1987). Biochemical adaptive responses in glucose metabolism of fish (*Tilapia mossambica*) during ammonia toxicity. *Current Science* **56,** 705–708.

Benton, M. J. (1990). Phylogeny of the major tetrapod groups – morphological data and divergence dates. *J. Mol. Evol.* **30,** 409–424.

Binstock, L., and Lecar, H. (1969). Ammonium ion currents in the squid giant axon. *J. Gen. Physiol.* **53,** 342–361.

Boron, W. F., Waisbren, S. J., Modlin, I. M., and Geibel, J. P. (1994). Unique permeability barrier of the apical surface of parietal and chief cells in isolated perfused gastric glands. *J. Exp. Biol.* **196,** 347–360.

Bosman, D. K., Deutz, N. E. P., Maas, M. A. W., van Eijik, H. M. H., Smit, J. J. H., de Haan, J. G., and Chamuleau, R. A. F. M. (1992). Amino acid release from cerebral cortex in experimental acute liver failure, studied by *in vivo* cerebral cortex microdialysis. *J. Neurochem.* **59,** 591–599.

Boutilier, R. G., Heming, T. A., and Iwama, G. K. (1984). Physicochemical parameters for use in fish respiratory physiology. *In* "Fish Physiology," Vol. XA. "Gills: Anatomy, Gas Transfer, and Acid-Base Regulation" (Hoar, W. S., and Randall, D. J., Eds.), pp. 403–429. Academic Press, San Diego.

Brien, P. (1959). Ethologie du *Protopterus dolloi* (Boulenger) et de ses larves. Signification des sacs pulmonaires des Dipneustes. *Ann. Soc. R. Zool. Belg.* **89,** 9–48.

Brusilow, S. W. (2002). Reviews in Molecular Medicine – Hyperammonemic Encephalopathy. *Medicine* **81,** 240–249.

Bruton, M. N. (1979). The breeding biology and early development of *Clarias gariepinus* (Pisces: Clariidae) in Lake Sibaya, South Africa, with a review of breeding in species of the subgenus *Clarias* (*Clarias*). *Trans. Zool. Soc. Lond.* **35,** 1–45.

Buckley, J. A., Whitmore, C. M., and Liming, B. D. (1979). Effects of prolonged exposure to ammonia on the blood and liver glycogen of coho salmon (*Oncorhynchus kisutch*). *Comp. Biochem. Physiol.* **63C,** 297–303.

Burrows, R. E. (1964). Effects of accumulated excretory products on hatchery-reared salmonids. Research Report 66, P. 12. Fish and Wildlife Service, US Department of the Interior, Washington, DC.

Cameron, J. N., and Heisler, N. (1983). Studies of ammonia in the rainbow trout: Physicochemical parameters, acid-base behavior and respiratory clearance. *J. Exp. Biol.* **105,** 107–126.

Campbell, J. W. (1973). Nitrogen excretion. *In* "Comparative Animal Physiology" (Prosser, C. L., Ed.), 3rd edn., pp. 279–316. Saunders College Publishing, Philadelphia.

Campbell, J. W. (1991). Excretory nitrogen metabolism. *In* "Environmental and Metabolic Animal Physiology. Comparative Animal Physiology" (Prosser, C. L., Ed.), 4th edn., pp. 277–324. Wiley-Interscience, New York.

Campbell, J. W., and Anderson, P. M. (1991). Evolution of mitochondrial enzyme systems in fish: The mitochondrial synthesis of glutamine and citrulline. *In* "Biochemistry and Molecular Biology of Fishes. I. Phylogenetic and Biochemical perspectives" (Hochachka, P. W., and Mommsen, T. P., Eds.), pp. 43–75. Elsevier, Amsterdam.

Chadwick, T. D., and Wright, P. A. (1999). Nitrogen excretion and expression of urea cycle enzymes in the Atlanta cod (*Gadus morhus* L.): A comparison of early life stages with adults. *J. Exp. Biol.* **202,** 2653–2662.

Chang, A., Hammond, T. G., Sun, T. T., and Zeidel, M. L. (1994). Permeability properties of the mammalian bladder apical membrane. *Am. J. Physiol.* **267,** 1483–1492.

Chew, S. F., Chan, N. K. Y., Tam, W. L., Loong, A. M., Hiong, K. C., and Ip, Y. K. (2004). Nitrogen metabolism in the African lungfish (*Protopterus dolloi*) aestivating in a mucus cocoon on land. *J. Exp. Biol.* **207,** 777–786.

Chew, S. F., Gan, J., and Ip, Y. K. (2005). Nitrogen metabolism and excretion in the swamp eel, *Monopterus albus,* during 6 or 40 days of aestivation in mud. *Physiol. Biochem. Zool.* In press.

Chew, S. F., Ho, L., Ong, T. F., Wong, W. P., and Ip, Y. K. (2004). The African lungfish, *Protopterus dolloi,* detoxifies ammonia to urea during environmental ammonia exposure. *Physiol. Biochem. Zool.* **78,** 31–39.

Chew, S. F., Hong, L. N., Wilson, J. M., Randall, D. J., and Ip, Y. K. (2003). Alkaline environmental pH has no effect on the excretion of ammonia in the mudskipper *Periophthalmodon schlosseri* but inhibits ammonia excretion in the related species *Boleophthalmus boddaerti. Physiol. Biochem. Zool.* **76,** 204–214.

Chew, S. F., Jin, Y., and Ip, Y. K. (2001). The loach *Misgurnus anguillicaudatus* reduces amino acid catabolism and accumulates alanine and glutamine during aerial exposure. *Physiol. Biochem. Zool.* **74,** 226–237.

Chew, S. F., Ong, T. F., Ho, L., Tam, W. L., Loong, A. M., Hiong, K. C., Wong, W. P., and Ip, Y. K. (2003). Urea synthesis in the African lungfish *Protopterus dolloi* – hepatic carbamoyl phosphate synthetase III and glutamine synthetase are up-regulated by 6 days of aerial exposure. *J. Exp. Biol.* **206,** 3615–3624.

Chew, S. F., Wong, M. Y., Tam, W. L., and Ip, Y. K. (2003). The snakehead *Channa asiatica* accumulates alanine during aerial exposure, but is incapable of sustaining locomotory activities on land through partial amino acid catabolism. *J. Exp. Biol.* **206,** 693–704.

Claiborne, J. B., and Evans, D. H. (1988). Ammonia and acid-base balance during high ammonia exposure in a marine teleost (*Myoxocephalus octodecimspinosus*). *J. Exp. Biol.* **140,** 89–105.

Claiborne, J. B., Edwards, S. L., and Morrison-Shetlar, A. I. (2002). Acid-base regulation in fishes: Cellular and molecular mechanisms. *J. Exp. Zool.* **293,** 302–319.

Cooper, J. L., and Plum, F. (1987). Biochemistry and physiology of brain ammonia. *Physiol. Rev.* **67,** 440–519.

Crockett, E. L. (1999). Lipid restructuring does not contribute to elevated activities of Na^+/K^+-ATPase in basolateral membranes from the gill of seawater acclimated eel (*Anguilla rostrata*). *J. Exp. Biol.* **202,** 2385–2392.

Cutler, C. P., and Cramb, G. (2002). Branchial expression of an aquaporin 3 (AQP-3) homologue is down regulated in European eel *Anguilla anguilla* following seawater acclimation. *J. Exp. Biol.* **205,** 2643–2651.

Dabrowska, H., and Wlasow, T. (1986). Sublethal effect of ammonia on certain biochemical and haematological indicators in common carp (*Cyprinus carpio L.*). *Comp. Biochem. Physiol.* **83C**, 179–184.

Danulat, E., and Kempe, S. (1992). Nitrogenous waste and accumulation of urea and ammonia in *Chalcalburnus tarichi* (Cyprinidae), endemic to the extremely alkaline Lake Van (Eastern Turkey). *Fish Physiol. Biochem.* **9**, 377–386.

Daoust, P.-Y., and Ferguson, H. W. (1984). The pathology of chronic ammonia toxicity in rainbow trout, *Salmo gairdneri* Richardson. *J. Fish Diseases* **7**, 199–205.

Davenport, J., and Sayer, M. D. J. (1986). Ammonia and urea excretion in the amphibious teleost *Blennius pholis* (L.) in sea-water and in air. *Comp. Biochem. Physiol.* **84A**, 189–194.

de Vries, M. C., and Wolcott, D. L. (1993). Gaseous ammonia evolution is coupled to reprocessing of urine at the gills of ghost crabs. *J. Exp. Zool.* **267**, 97–103.

DeLong, D. C., Halver, J. E., and Mertz, E. T. (1958). Nutrition of salmonid fishes. VI. Protein requirements of chinook salmon at two water temperatures. *J. Nutr.* **65**, 589–599.

Denker, B. M., and Nigam, S. K. (1958). Molecular structure and assembly of the tight junction. *Am. J. Physiol.* **274**, F1–F9.

Depeche, J., Gilles, R., Daufresne, S., and Chapello, H. (1979). Urea content and urea production via the ornithine-urea cycle pathway during the ontogenic development of two teleost fishes. *Comp. Biochem. Physiol.* **63A**, 51–56.

Donnely, B. G. (1973). Aspects of behavior in the catfish, *Clarias gariepinus* (Pisces: Clariidae), during periods of habitat desiccation. *Arnoldia Rhod.* **6**, 1–8.

Driedzic, W. R., and Hochachka, P. W. (1976). Control of energy metabolism in fish white muscle. *Am. J. Physiol.* **230**, 579–582.

Eddy, F. B., Bamford, O. S., and Maloiy, G. M. O. (1980). Sodium and chloride balance in the African catfish *Clarias mossambicus. Comp. Biochem. Physiol.* **66A**, 637–641.

El-Shafey, A. A. M. (1998). Effect of ammonia on respiratory functions of blood of *Tilapia zilli. Comp. Biochem. Physiol.* **121A**, 305–313.

Evans, D. H., Piermarini, P. M., and Potts, W. T. W. (1999). Ionic transport in the fish gill epithelium. *J. Exp. Zool.* **283**, 641–652.

Felipo, V., Kosenko, E., Minana, M.-D., Marcaida, G., and Grisolia, S. (1994). Molecular mechanism of acute ammonia toxicity and of its prevention by L-carnitine. *In* "Hepatic Encephalopathy, Hyperammonemia and Ammonia Toxicity" (Felipo, V., and Grisola, S., Eds.), pp. 65–77. Plenum Press, New York.

Ferguson, H. W., Morrison, D., Ostland, V. E., Lumsden, J., and Byrne, P. (1992). Responses of mucus-producing cells in gill disease of rainbow trout (*Oncorhynchus mykiss*). *J. Comp. Path.* **106**, 255–265.

Fines, G. A., Ballantyne, J. S., and Wright, P. A. (2001). Active urea transport and an unusual basolateral membrane composition in the gills of a marine elasmobranch. *Am. J. Physiol.* **280**, R16–R24.

Flemström, G., and Isenberg, J. I. (2001). Gastroduodenal mucosal alkaline secretion and mucosal protection. *News Physiol. Sci.* **16**, 23–28.

Forey, P. L. (1986). Relationship of lungfishes. *In* "The Biology and Evolution of Lungfishes" (Bemis, W. E., Burggren, W. W., and Kemp, N. E., Eds.), pp. 75–92. Alan R. Liss, Inc., New York.

Forey, P. L., Gardiner, B. G., and Patterson, C. (1991). The lungfish, the coelacanth and the cow revisited. *In* "Origins of the Higher Groups of Tetrapods: Controversy and Consensus" (Schultze, H. P., and Trueb, L., Eds.), pp. 145–172. Cornell University Press, New York.

Forster, R. P., and Goldstein, L. (1966). Urea synthesis in the lungfish: Relative importance of purine and ornithine cycle pathways. *Science* **153**, 1650.

Forster, R. P., and Goldstein, L. (1976). Intracellular osmoregulatory role of amino acids and urea in marine elasmobranches. *Am. J. Physiol.* **230**, 925–931.

Frick, N. T., and Wright, P. A. (2002a). Nitrogen metabolism and excretion in the mangrove killifish *Rivulus marmoratus* I. The influence of environmental salinity and external ammonia. *J. Exp. Biol.* **205**, 79–89.

Frick, N. T., and Wright, P. A. (2002b). Nitrogen metabolism and excretion in the mangrove killifish *Rivulus marmoratus* II. Significant ammonia volatilization in a teleost during air-exposure. *J. Exp. Biol.* **205**, 91–100.

Gerst, J. W., and Thorson, T. B. (1977). Effects of saline acclimation on plasma electrolytes, urea excretion, and hepatic urea biosynthesis in a freshwater stingray, *Potamotrygon sp.* Garman, 1877. *Comp. Biochem. Physiol.* **56A**, 87–93.

Ghioni, C., Bell, J. G., Bell, M. V., and Sargent, J. R. (1997). Fatty acid composition, eicosanoid production and permeability in skin tissues of rainbow trout (*Oncorhynchus mykiss*) fed control or an essential fatty acid deficient diet. *Prostoglandins, Leukotrienes and Essential Fatty Acids* **56**, 479–489.

Gilmour, K. M., Perry, S. F., Wood, C. M., Henry, R. P., Laurent, P., Part, P., and Walsh, P. J. (1998). Nitrogen excretion and the cardiorespiratory physiology of the gulf toadfish, *Opsanus beta*. *Physiol. Zool.* **71**, 492–505.

Gluck, S. L., Nelson, R. D., Lee, B. S., Wang, Z. Q., Guo, X. L., Fu, J. Y., and Zhang, K. (1992). Biochemistry of the renal V-ATPase. *J. Exp. Biol.* **172**, 219–229.

Goldstein, L., and Forster, R. P. (1971). Osmoregulation and urea metabolism in the little skate *Raja erinacea*. *Am. J. Physiol.* **220**, 742–746.

Goldstein, L., Harley-Dewitt, S., and Forster, R. P. (1973). Activities of ornithine-urea cycle enzymes and of trimethylamine oxidase in the coelacanth, *Latimeria chalumnae*. *Comp. Biochem. Physiol.* **44B**, 357–362.

Goldstein, L., Janssens, P. A., and Forster, R. P. (1967). Lungfish *Neoceratodus forsteri*: Activities of ornithine-urea cycle enzymes. *Science* **157**, 316–317.

Gonzalez, R. J., and McDonald, D. G. (1994). The relationship between oxygen uptake and ion loss in fish from diverse habitats. *J. Exp. Biol.* **190**, 95–108.

Good, D. W. (1994). Ammonium transport by the thick ascending limb of Henle's loop. *Ann. Rev. Physiol.* **56**, 623–647.

Gordon, M. S., Boetius, I., Evans, D. H., McCarthy, R., and Oglesby, L. C. (1969). Aspects of the physiology of terrestrial life in amphibious fishes. I. The mudskipper, *Periophthalmus sobrinus*. *J. Exp. Biol.* **50**, 141–149.

Gordon, M. S., Ng, W. W. M., and Yip, A. Y. W. (1978). Aspects of the physiology of terrestrial life in amphibious fishes. III. The Chinese mudskipper *Periophthalmus cantonensis*. *J. Exp. Biol.* **72**, 57–75.

Goss, G. G., and Wood, C. M. (1990). Na^+ and Cl^- uptake kinetics, diffusive effluxes and acidic equivalent fluxes across the gills of rainbow trout I. Response to environmental hyperoxia. *J. Exp. Biol.* **152**, 521–547.

Graham, J. B. (1997)."Air-Breathing Fishes." Academic Press, San Diego.

Greenaway, P. (1991). Nitrogenous excretion in aquatic and terrestrial crustaceans. *Memoirs of the Queensland Museum* **31**, 215–227.

Greenaway, P., and Nakamura, T. (1991). Nitrogenous excretion in two terrestrial crabs (*Gecarcoidea natalis* and *Geograpsus grayi*). *Physiol Zool.* **64**, 767–786.

Greenwood, P. H. (1955). Reproduction in the cat-fish, *Clarias mossambicus* Peters. *Nature* **176**, 516–518.

Gregory, R. B. (1977). Synthesis and total excretions of waste nitrogen by fish of the *Periophthalmus* (mudskipper) and *Scartelaos* families. *Comp. Biochem. Physiol.* **57A,** 33–36.

Griffith, R. W. (1991). Guppies, toadfish, lungfish, coelacanths and frogs – a scenario for the evolution of urea retention in fishes. *Environ. Biol. Fish.* **32,** 199–218.

Griffith, R. W., and Pang, P. K. T. (1979). Mechanisms of osmoregulation of the coelacanth: evolution implications. *Occas. Papers Calif. Acad. Sci.* **134,** 79–92.

Griffith, R. W., Umminger, B. L., Grant, B. F., Pang, P. K. T., and Pickford, G. E. (1974). Serum composition of the coelacanth, *Latimeria chalumnae* Smith. *J. Exp. Zool.* **187,** 87–102.

Griffith, R. W., Umminger, B. L., Grant, B. F., Pang, P. K. T., Goldstein, L., and Pickford, G. E. (1976). Composition of the bladder urine of the coelacanth, *Latimeria chalmumnae*. *J. Exp. Zool.* **196,** 371–380.

Grizzle, J. M., and Thiyagarajah, A. (1987). Skin histology of *Rivulus ocellatus*: Apparent adaptation for aerial respiration. *Copeia* **1987,** 237–240.

Guillén, J. L., Endo, M., Turnbull, J. F., Kawatsu, H., Richards, R. H., and Aoki, T. (1994). Skin responses and mortalities in the larvae of Japanese croaker exposed to ammonia. *Fisheries Science* **60,** 547–550.

Hampson, B. L. (1976). Ammonia concentration in relation to ammonia toxicity during a rainbow trout rearing experiment in a closed freshwater–seawater system. *Aquaculture* **9,** 61–70.

Heitman, J., and Agre, P. (2000). A new face of the Rhesus antigen. *Nature Genetics* **26,** 258–259.

Henry, R. P., and Heming, T. A. (1998). Carbonic anhydrase and respiratory gas exchange. *In* "Fish Physiology" (Perry, S. F., and Tufts, B. L., Eds.), Vol. 17, pp. 75–111. Academic Press, San Diego.

Hermenegildo, C., Marcaida, G., Montoliu, C., Grisolia, S., Minana, M., and Felipo, V. (1996). NMDA receptor antagonists prevent acute ammonia toxicity in mice. *Neurochem. Res.* **21,** 1237–1244.

Hermenegildo, C., Monfor, C. P., and Felipo, V. (2000). Activation of *N*-methyl-D-aspartate receptors in rat brain *in vivo* following acute ammonia intoxication: Characterization by *in vivo* brain microdialysis. *Hepatol.* **31,** 709–715.

Hilgier, W., Haugvicova, R., and Albrecht, J. (1991). Decreased potassium-stimulated release of 3HD-aspartate from hippocampal slices distinguishes encephalopathy related to acute liver failure from that induced by simple hyperammonemia. *Brain Res.* **567,** 165–168.

Hill, W. G., Rivers, R. L., and Zeidel, M. L. (1999). Role of leaflet asymmetry in the permeability of model biological membranes to protons, solutes, and gases. *J. Gen. Physiol.* **114,** 405–414.

Hillaby, B. A., and Randall, D. J. (1979). Acute ammonia toxicity and ammonia excretion in rainbow trout (*Salmo gairdneri*). *J. Fish. Res. Board Can.* **36,** 621–629.

Hochachka, P. W., and Somero, G. N. (1984)."Biochemical Adaptation." Princeton University Press, Princeton, NJ.

Hogan, D. L., Crombie, D. L., Isenberg, J. I., Svendsen, P., Schaffalitzky de Muckadell, O. B., and Answorth, M. A. (1997). CFTR mediates cAMP- and Ca^{2+}-activated duodenal epithelial HCO_3^- secretion. *Am. J. Physiol.* **272,** G872–G878.

Holder, M. T., Erdmann, M. V., Wilcox, T. P., Caldwell, R. L., and Hillis, D. M. (1999). Two living species of coelacanths? *Proc. Natl. Acad. Sci. USA* **96,** 12616–12620.

Hopkins, T. E., Wood, C. M., and Walsh, P. J. (1999). Nitrogen metabolism and excretion in an intertidal population of the gulf toadfish (*Opsanus beta*). *Mar. Freshw. Behav. Physiol.* **33,** 21–34.

Howitt, S. M., and Udvardi, M. K. (2000). Structure function and regulation of ammonium transporters in plants. *Biochim. Biophys. Acta* **1465**, 152–170.

Ip, Y. K., Chew, S. F., and Randall, D. J. (2001a). Ammonia toxicity, tolerance, and excretion. *In* "Fish Physiology," Vol. 20. "Nitrogen Excretion" (Wright, P. A., and Anderson, P. M., Eds.), pp. 109–148. Academic Press, San Diego.

Ip, Y. K., Chew, S. F., Leong, I. W. A., Jin, Y., and Wu, R. S. S. (2001b). The sleeper *Bostrichthys sinensis* (Teleost) stores glutamine and reduces ammonia production during aerial exposure. *J. Comp. Physiol. B.* **171**, 357–367.

Ip, Y. K., Chew, S. F., Lim, A. L. L., and Low, W. P. (1990). The mudskipper. *In* "Essays in Zoology, Papers Commemorating the 40th Anniversary of Department of Zoology" (Chou, L. M., and Ng, P. K. L., Eds.), pp. 83–95. National University of Singapore Press.

Ip, Y. K., Lee, C. Y., Chew, S. F., Low, W. P., and Peng, K. W. (1993). Differences in the responses of two mudskippers to terrestrial exposure. *Zool. Sci.* **10**, 511–519.

Ip, Y. K., Lim, C. B., Chew, S. F., Wilson, J. M., and Randall, D. J. (2001). Partial amino acid catabolism leading to the formation of alanine in *Periophthalmodon schlosseri* (mudskipper): A strategy that facilitates the use of amino acids as an energy source during locomotory activity on land. *J. Exp. Biol.* **204**, 1615–1624.

Ip, Y. K., Lim, C. K., Wong, W. P., and Chew, S. F. (2004). Postprandial increases in nitrogenous excretion and urea synthesis in the giant mudskipper *Periophthalmodon schlosseri*. *J. Exp. Biol.* **207**, 3015–3023.

Ip, Y. K., Peh, B. K., Tam, W. L., Wong, W. P., and Chew, S. F. (2005). Effects of peritoneal injection with NH_4Cl, urea, or NH_4Cl^+ urea on the excretory nitrogen metabolism of the African lungfish *Protopterus dolloi*. *J. Exp. Zool.* **303A**, 272–282.

Ip, Y. K., Randall, D. J., Kok, T. K. T., Bazarghi, C., Wright, P. A., Ballantyne, J. S., Wilson, J. M., and Chew, S. F. (2004). The mudskipper *Periophthalmodon schlosseri* facilitates active NH_4^+ excretion by increasing acid excretion and decreasing NH_3 permeability in the skin. *J. Exp. Biol.* **207**, 787–801.

Ip, Y. K., Subaidah, R. M., Liew, P. C., Loong, A. M., Hiong, K. C., Wong, W. P., and Chew, S. F. (2004). The African catfish *Clarias gariepinus* does not detoxify ammonia to urea or amino acids during ammonia loading but is capable of excreting ammonia against an inwardly driven ammonia concentration gradient. *Physiol. Biochem. Zool.* **77**, 255–266.

Ip, Y. K., Tam, W. L., Wong, W. P., Loong, A. I., Hiong, K. C., Ballantyne, J. S., and Chew, S. F. (2003). A comparison of the effects of environmental ammonia exposure on the Asian freshwater stingray *Himantura signifier* and the Amazonian freshwater stingray *Potamotrygon motoro*. *J. Exp. Biol.* **206**, 3625–3633.

Ip, Y. K., Tay, A. S. L., Lee, K. H., and Chew, S. F. (2004). Strategies adopted by the swamp eel *Monopterus albus* to survive in high concentrations of environmental ammonia. *Physiol. Biochem. Zool.* **77**, 390–405.

Iwata, K. (1988). Nitrogen metabolism in the mudskipper, *Periophthalmus cantonensis*: Changes in free amino acids and related compounds in carious tissues under conditions of ammonia loading with reference to its high ammonia tolerance. *Comp. Biochem. Physiol.* **91A**, 499–508.

Iwata, K., and Deguichi, M. (1995). Metabolic fate and distribution of ^{15}N-Ammonia in an ammonotelic amphibious fish, *Periophthalmus modestus*, following immersion in ^{15}N-ammonium sulphate: A long term experiment. *Zool. Sci.* **12**, 175–184.

Iwata, K., Kajimura, M., and Sakamoto, T. (2000). Functional ureogenesis in the gobiid fish *Mugilogobius abei*. *J. Exp. Biol.* **203**, 3703–3715.

Iwata, K., Kakuta, M., Ikeda, G., Kimoto, S., and Wada, N. (1981). Nitrogen metabolism in the mudskipper, *Periophthalmus cantonensis*: A role of free amino acids in detoxification of ammonia produced during its terrestrial life. *Comp. Biochem. Physiol.* **68A**, 589–596.

Jacquez, J. A., Poppell, J. W., and Jeltsch, R. (1959). Partial pressure of ammonia in alveolar air. *Science* **129,** 269–270.

Jakubowski, M. (1958). The structure and vascularization of the skin of the pond-loach (*Misgurnus fossilis* L.). *Acta Biol. Cracoviensia* **1,** 113–127.

Janssens, P. A. (1964). The metabolism of the aestivating African lungfish. *Comp. Biochem. Physiol.* **11,** 105–117.

Janssens, P. A., and Cohen, P. P. (1966). Ornithine-urea cycle enzymes in the African lungfish, *Protopterus aethiopicas. Science* **152,** 358.

Janssens, P. A., and Cohen, P. P. (1968a). Biosynthesis of urea in the estivating African lungfish and in *Xenopus laevis* under conditions of water shortage. *Comp. Biochem. Physiol.* **24,** 887–898.

Janssens, P. A., and Cohen, P. P. (1968b). Nitrogen metabolism in the African lungfish. *Comp. Biochem. Physiol.* **24,** 879–886.

Jow, L. Y., Chew, S. F., Lim, C. B., Anderson, P. M., and Ip, Y. K. (1999). The marble goby *Oxyeleotris marmoratus* activates hepatic glutamine synthetase and detoxifies ammonia to glutamine during air exposure. *J. Exp. Biol.* **202,** 237–245.

Julsrud, E. A., Walsh, P. J., and Anderson, P. M. (1998). N-Acetyl-L-glutamate and the urea cycle in gulf toadfish (*Opsanus beta*) and other fish. *Arch. Biochem. Biophys.* **350,** 55–60.

Kajimura, M., Iwata, K., and Numata, H. (2002). Diurnal nitrogen excretion rhythm of the functionally ureogenic gobiid fish *Mugilogobius abei. Comp. Biochem. Physiol.* **131B,** 227–239.

Kikeri, D., Sun, A., Zeidel, M. L., and Hebert, S. C. (1989). Cell membranes impermeable to NH_3. *Nature* **339,** 478–480.

Kinsella, J. L., and Aronson, P. S. (1981). Interaction of NH_4^+ and Li^+ with the renal microvillus membrane Na^+ -H^+ exchanger. *Am. J. Physiol.* **241,** C220–C226.

Kirschner, L. B. (1993). The energetics of osmotic regulation in ureotelic and hyposmotic fishes. *J. Exp. Zool.* **267,** 19–26.

Knepper, M. A., Packer, R., and Good, D. W. (1989). Ammonium transport in the kidney. *Physiol. Rev.* **69,** 179–249.

Knoph, M. B., and Thorud, K. (1996). Toxicity of ammonia to Atlantic salmon (*Salmo salar* L.) in seawater – effects on plasma osmolality, ion, ammonia, urea and glucose levels and hematologic parameters. *Comp. Biochem. Physiol.* **113A,** 375–381.

Koh, K. T. (2000). Increases in carbon dioxide production and excretion in the mudskipper, *Boleophthalmus boddaerti,* in response to alkaline pH. MSc thesis, National University of Singapore.

Kok, W. K. (2000). Can the mudskipper *Periophthalmodon schlosseri* excrete NH_4^+ against a concentration gradient? MSc thesis, National University of Singapore.

Kok, W. K., Lim, C. B., Lam, T. J., and Ip, Y. K. (1998). The mudskipper *Periophthalmodon schlosseri* respires more efficiently on land than in water and vice versa for *Boleophthalmus boddaerti. J. Exp. Zool.* **280,** 86–90.

Kong, H., Kahatapitiya, N., Kingsley, K., Salo, W. L., Anderson, P. M., Wang, Y. S., and Walsh, P. J. (2000). Induction of carbamoyl phosphate synthetase III and glutamine synthetase mRNA during confinement stress in gulf toadfish (*Opsanus beta*). *J. Exp. Biol.* **203,** 311–320.

Korsgaard, B., Mommsen, T. P., and Wright, P. A. (1995). Urea excretion in teleostean fishes: Adaptive relationships to environment, ontogenesis and viviparity. *In* "Nitrogen Metabolism and Excretion" (Walsh, P. J., and Wright, P. A., Eds.), pp. 259–287. CRC Press, Boca Raton, FL.

Lande, M. B., Donovan, J. M., and Zeidel, M. L. (1995). The relationship between membrane fluidity and permeabilities to water, solute, ammonia, and protons. *J. Gen. Physiol.* **106**, 67–84.

Lande, M. B., Priver, N. A., and Zeidel, M. L. (1994). Determinants of apical membrane permeabilities of barrier epithelia. *Am. J. Physiol.* **267**, C367–C374.

Lang, T., Peters, G., Hoffmann, R., and Meyer, E. (1987). Experimental investigations on the toxicity of ammonia: Effects on ventilation frequency, growth, epidermal mucous cells, and gill structure of rainbow trout *Salmo gairdneri*. *Disease of Aquatic Organisms* **3**, 159–165.

Lee, C. G. L., Lam, T. J., Munro, A. D., and Ip, Y. K. (1991). Osmoregulation in the mudskipper, Boleophthalmus boddaerti II. Transepithelial potential and hormonal control. *Fish Physiol. Biochem.* **9**, 69–75.

Leong, A.-W. I. (1999). Effects of terrestrial exposure on nitrogen metabolism and excretion in *Bostrichthys sinensis*. Honours thesis, National University of Singapore.

Leung, K. M. Y., Chu, J. C. W., and Wu, R. S. S. (1999a). Effects of body weight, water temperature and ration size on ammonia excretion by the areolated grouper (*Epinephelus areolatus*) and mangrove snapper (*Lutjanus argentimaculatus*). *Aquaculture* **170**, 215–227.

Leung, K. M. Y., Chu, J. C. W., and Wu, R. S. S. (1999b). Nitrogen budgets for the areolated grouper *Epinephelus areolatus* cultured under laboratory conditions and in open-sea cages. *Mar. Ecol. Prog. Ser.* **186**, 271–281.

Levi, G., Morisi, G., Coletti, A., and Catanzaro, R. (1974). Free amino acids in fish brain: Normal levels and changes upon exposure to high ammonia concentrations *in vivo* and upon incubation of brain slices. *Comp. Biochem. Physiol.* **49A**, 623–636.

Lignot, J. H., Cutler, C. P., Hazon, N., and Cramb, G. (2002). Immunolocalization of aquaporin 3 in the gill and gastrointestinal tract of the European eel (*Anguilla anguilla* L.). *J. Exp. Biol.* **205**, 2653–2663.

Lim, C. K., Chew, S. F., Anderson, P. M., and Ip, Y. K. (2001). Mudskippers reduce the rate of protein and amino acid catabolism in response to terrestrial exposure. *J. Exp. Biol.* **204**, 1605–1614.

Lim, C. K., Chew, S. F., Tay, A. S. L., and Ip, Y. K. (2004). Effects of peritoneal injection of NH_4HCO_3 on nitrogen excretion and metabolism in the swamp eel *Monopterus albus* – increased ammonia excretion with an induction of glutamine synthetase activity. *J. Exp. Zool.* **301A**, 324–333.

Lim, C. K., Wong, W. P., Chew, S. F., and Ip, Y. K. (2004). The ureogenic African lungfish, *Protopterus dolloi*, switches from ammonotely to ureotely momentarily after feeding. *J. Comp. Physiol. B.* In press.

Lin, H., and Randall, D. J. (1995). Proton Pumps in Fish Gills. *In* "Fish Physiology" (Wood, C. M., and Shuttleworth, T. J., Eds.), Vol. 14, pp. 229–255. Academic Press, San Diego.

Lindley, T. E., Scheiderer, C. L., Walsh, P. J., Wood, C. M., Bergman, H. L., Bergman, A. L., Laurent, P., Wilson, P., and Anderson, P. M. (1999). Muscle as the primary site of urea cycle enzyme activity in an alkaline lake-adapted tilapia, *Oreochromis alcalicus grahami*. *J. Biol. Chem.* **274**, 29858–29861.

Liu, Z., Peng, J., Mo, R., Hui, C.-C., and Huang, C.-H. (2001). Rh type B glycoprotein is a new member of the Rh superfamily and a putative ammonia transporter in mammals. *J. Biol. Chem.* **276**, 1424–1433.

Lloyd, R., and Orr, L. (1969). The diuretic response by rainbow trout to sub-lethal concentrations of ammonia. *Wat. Res.* **3**, 335–344.

Lovejoy, N. R. (1997). Stingrays, parasites, and neotropical biogeography: A closer look at Brooks *et al.*'s hypotheses concerning the origins of neotropical freshwater rays (Potamotrygonidae). *Syst. Biol.* **46**, 218–230.

Low, W. P., Ip, Y. K., and Lane, D. J. W. (1990). A comparative study of the gill morphometry in three mudskippers – *Periophthalmus chrysospilos, Boleophthalmus boddaerti* and *Periophthalmodon schlosseri. Zool. Sci.* **7,** 29–38.

Low, W. P., Lane, D. J. W., and Ip, Y. K. (1988). A comparative study of terrestrial adaptations in three mudskippers – *Periophthalmus chrysospilos, Boleophthalmus boddaerti* and *Periophthalmodon schlosseri. Biol. Bull.* **175,** 434–438.

Madara, J. L. (1998). Regulation of the movement of solutes across tight junctions. *Ann. Rev. Physiol.* **60,** 143–159.

Maetz, J. (1973). Na^+/NH_4^+, Na^+/H^+ exchanges and NH_3 movement across the gill of *Carassius auratus. J. Exp. Biol.* **58,** 255–273.

Maetz, J., and Garcia-Romeu, F. (1964). The mechanisms of sodium and chloride uptake by the gills of a freshwater fish, *Carassius auratus*. II. Evidence for NH_4^+/Na^+ and HCO_3^-/Cl^- exchanges. *J. Gen. Physiol.* **47,** 1209–1227.

Mallery, C. H. (1983). A carrier enzyme basis for ammonium excretion in teleost gill-NH_4^+-stimulated Na^+-dependent ATPase activity in *Opsanus beta. Comp. Biochem. Physiol.* **74,** 889–897.

Marcaggi, P., and Coles, J. A. (2001). Ammonium in nervous tissue: Transport across cell membranes, fluxes from neurons to glial cells, and role in signalling. *Prog. Neurobiol.* **64,** 157–183.

Marcaida, G., Felipo, V., Hermenegildo, C., Minana, M. D., and Grisolia, S. (1992). Acute ammonia toxicity is mediated by NMDA type of glutamate receptors. *FEBS Lett.* **296,** 67–68.

Marini, A.-M., Matassi, G., Raynal, V., André, B., Cartron, J.-P., and Chérif-Zahar, B. (2000). The human Rhesus-associated RhAG protein and a kidney homologue promote ammonium transport in yeast. *Nature Genetics* **26,** 341–344.

Marshall, W. S. (2002). Na^+, Cl^-, Ca^{2+} and Zn^{2+} transport by fish gill: Retrospective review and prospective synthesis. *J. Exp. Zool.* **293,** 264–283.

Matei, V. E. (1983). The morpho functional adaptation to ammonium of the cell epithelium chloride cells in the freshwater stickleback *Pungitius pungitius. Tsitologiia* **25,** 661–666.

Matei, V. E. (1984). Changes in the ultrastructure of the gill epithelium in the nine-spined stickleback *Pungitius pungitius* under the action of ammonia in hypertonic medium. *Tsitologiia* **26,** 371–375.

McGeer, J. C., Wright, P. A., Wood, C. M., Wilkie, M. P., Mazur, C. F., and Iwama, G. K. (1994). Nitrogen excretion in four species of fish from an alkaline/saline lake. *Trans. Am. Fish. Soc.* **123,** 824–829.

McKenzie, D. J., Randall, D. J., Lin, H., and Aota, S. (1993). Effects of changes in plasma pH, CO_2 and ammonia on ventilation in trout. *Fish Physiol. Biochem.* **10,** 507–515.

McMahon, B. R., and Burggren, W. W. (1987). Respiratory physiology of intestinal air breathing in the teleost fish *Misgurnus anguillicaudatus. J. Exp. Biol.* **133,** 371–393.

Milligan, C. L. (1997). The role of cortisol in amino acid mobilization and metabolism following exhaustive exercise in rainbow trout (*Oncorhynchus mykiss* Walbaum). *Fish Physiol. Biochem.* **16,** 119–128.

Mommsen, T. P., and Walsh, P. J. (1989). Evolution of urea synthesis in vertebrates: The piscine connection. *Science* **243,** 72–75.

Mommsen, T. P., and Walsh, P. J. (1991). Urea synthesis in fishes: Evolutionary and Biochemical Perspectives. *In* "Biochemistry and Molecular Biology of Fishes, 1. Phylogenetic and Biochemical Perspectives" (Hochachka, P. W., and Mommsen, T. P., Eds.), pp. 137–163. Elsevier, Amsterdam.

Mommsen, T. P., and Walsh, P. J. (1992). Biochemical and environmental perspectives on nitrogen metabolism in fishes. *Experentia* **48**, 583–593.

Mommsen, T. P., Vijayan, M. M., and Moon, T. W. (1999). Cortisol in teleosts: Dynamics, mechanisms of action, and metabolic regulation. *Rev. Fish Biol. Fisheries* **9**, 211–268.

Morgan, R. L., Ballantyne, J. S., and Wright, P. A. (2003a). Regulation of a renal urea transporter with reduced salinity in a marine elasmobranch, *Raja erinacea. J. Exp. Biol.* **206**, 3285–3292.

Morgan, R. L., Wright, P. A., and Ballantyne, J. S. (2003b). Urea transport in kidney brush-border membrane vesicles from an elasmobranch, *Raja erinacea. J. Exp. Biol.* **206**, 3293–3302.

Morii, H. (1979). Changes with time ammonia and urea concentrations in the blood and tissue of mudskipper fish, *Periophthalmus cantonensis* and *Boelophthalmus pectinirostris* kept in water and on land. *Comp. Biochem. Physiol.* **64A**, 235–243.

Morii, H., Nishikata, K., and Tamura, O. (1978). Nitrogen excretion of mudskipper fish *Periophthalmus cantonensis* and *Boleophthalmus pectinirostris* in water and on land. *Comp. Biochem. Physiol.* **60A**, 189–193.

Morii, H., Nishikata, K., and Tamura, O. (1979). Ammonia and urea excretion from mudskipper fishes, *Periophthalmus cantonensis* and *Boleophthalmus pectinirostris* transferred from land to water. *Comp. Biochem. Physiol.* **63A**, 23–28.

Mukai, T., Kajimura, M., and Iwata, K. (2000). Evolution of a ureogenic ability of Japanese *Mugilogobius* species (Pisces: Gobidae). *Zool. Sci.* **17**, 549–557.

Murdy, E. O. (1989). A taxonomic revision and cladistic analysis of the oxudercine gobies (Gobiidae: Oxudercinae). *Rec. Australian Mus. Supp.* **11**, 1–93.

Murray, B. W., Busby, E. R., Mommsen, T. P., and Wright, P. A. (2003). Evolution of glutamine synthetase in vertebrates: Multiply glutamine synthetase genes expressed in rainbow trout (*Oncorhynchus mykiss*). *J. Exp. Biol.* **206**, 1511–1521.

Nakhoul, N. L., Hering-Smith, K. S., Abdulnour-Nakhoul, S. M., and Hamm, L. L. (2001a). Ammonium interaction with the epithelial sodium channel. *Am. J. Physiol.* **281**, F493–F502.

Nakhoul, N. L., Hering-Smith, K. S., Abdulnour-Nakhoul, S. M., and Hamm, L. L. (2001b). Transport of NH_3/NH_4^+ in oocytes expressing aquaporin-1. *Am. J. Physiol.* **281**, F255–F263.

Negrete, H. O., Lavelle, J. P., Berg, J., Lewis, S. A., and Zeidel, M. L. (1996). Permeability properties of the intact mammalian bladder epithelium. *Am. J. Physiol.* **271**, F886–F894.

Olson, K. R., and Fromm, P. O. (1971). Excretion of urea by two teleosts exposed to different concentrations of ambient ammonia. *Comp. Biochem. Physiol.* **40A**, 999–1007.

Paley, R. K., Twitchen, I. D., and Eddy, F. B. (1993). Ammonia, Na^+, K^+ and Cl^- levels in rainbow trout yolk-sac fry in response to external ammonia. *J. Exp. Biol.* **180**, 273–284.

Peng, K. W., Chew, S. F., Lim, C. B., Kuah, S. S. L., Kok, W. K., and Ip, Y. K. (1998). The mudskippers *Periophthalmodon schlosseri* and *Boleophthalmus boddaerti* can tolerate environmental NH_3 concentrations of 446 and 36 μM, respectively. *Fish Physiol. Biochem.* **19**, 59–69.

Pequin, L., and Serfaty, A. (1963). L'excretion ammoniacale chez un Teleosteen dulcicole *Cyprinius carpio. L. Comp. Biochem. Physiol.* **10**, 315–324.

Perlman, D. F., and Goldstein, L. (1999). Organic osmolyte channels in cell volume regulation in vertebrates. *J. Exp. Zool.* **283**, 725–733.

Person Le Ruyet, J., Galland, R., Le Roux, A., and Chartois, H. (1997). Chronic ammonia toxicity in juvenile turbot (*Scophthalmus maximus*). *Aquaculture* **154**, 155–171.

Poll, M. (1961). Révision systématique et raciation géographique des Protopteridae de l'Afrique centrale. *Ann. Mus. R. Afr. Centr. série in-8° Sci. Zool.* **103**, 3–50.

Praetorius, J., Friss, U. G., Ainsworth, M. A., Schaffalitzky de Muckadell, O. B., and Johansen, T. (2002). The cystic fibrosis transmembrane conductance regulator is not a base transporter in isolated duodenal epithelial cells. *Acta. Physiol. Scand.* **174**, 327–336.

Ramaswamy, M., and Reddy, T. G. (1983). Ammonia and urea excretion in three species of air-breathing fish subjected to aerial exposure. *Proc. Ind. Acad. Sci. (Anim. Sci.)* **92**, 293–297.

Randall, D. J., and Tsui, T. K. N. (2002). Ammonia toxicity in fish. *Marine Pollution Bull.* **45**, 17–23.

Randall, D. J., and Wicks, B. J. (2000). Fish: Ammonia production, excretion and toxicity. Paper presented in the Fifth International Symposium on Fish Physiology, Toxicology and Water Quality, 9–12 November, 1998, City University of Hong Kong.

Randall, D. J., Wilson, J. M., Peng, K. W., Kok, T. W. K., Kuah, S. S. L., Chew, S. F., Lam, T. J., and Ip, Y. K. (1999). The mudskipper, *Periophthalmodon schlosseri*, actively transports NH_4^+ against a concentration gradient. *Am. J. Physiol.* **277**, R1562–R1567.

Randall, D. J., Wood, C. M., Perry, S. F., Bergman, H., Maloiy, G. M., Mommsen, T. P., and Wright, P. A. (1989). Urea excretion as a strategy for survival in a fish living in a very alkaline environment. *Nature* **337**, 165–166.

Rao, V. L. R., Murthy, C. R. K., and Butterworth, R. F. (1992). Glutamatergic synaptic dysfunction in hyperammonemic syndromes. *Metab. Brain Dis.* **7**, 1–20.

Ratha, B. K., Sha, N., Rana, R. K., and Chaudhury, B. (1995). Evolutionary significance of metabolic detoxification of ammonia to urea in an ammoniotelic freshwater teleost, *Heteropneustes fossilis*, during temporary water deprivation. *Evol. Biol.* **8,9**, 107–117.

Reichenbach-Klinke, H. H. (1967). Untersuchungen uber die Einwirkung des Ammoniakgehalts auf den Fischorganismus. *Arch. Fischereiwissenschaft* **17**, 122–132.

Ritter, N. M., Smith, D. D. Jr., and Campbell, J. W. (1987). Glutamine synthetase in liver and brain tissues of the holocephalan *Hydrolagus colliei*. *J. Exp. Zool.* **243**, 181–188.

Robin, E. D., Travis, D. M., Bromberg, P. A., Forkner Jr., C. E., and Tyler, J. M. (1959). Ammonia excretion by mammalian lung. *Science* **129**, 270–271.

Rodicia, L. P., Sternberg, L., and Walsh, P. J. (2003). Metabolic fate of exogenous $^{15}NH_4Cl$ in the gulf toadfish (*Opsanus beta*). *Comp. Biochem. Physiol.* **136C**, 157–164.

Rozemeijer, M. J. C., and Plaut, I. (1993). Regulation of nitrogen excretion of the amphibious blenniidae *Alticus kirki* (Guenther, 1868) during emersion and immersion. *Comp. Biochem. Physiol.* **104A**, 57–62.

Saha, N., and Das, L. (1999). Stimulation of ureogenesis in the perfused liver of an Indian air-breathing catfish, *Clarias batrachus*, infused with different concentrations of ammonium chloride. *Fish Physiol. Biochem.* **21**, 303–311.

Saha, N., and Ratha, B. K. (1987). Active ureogenesis in a freshwater air-breathing teleost, *Heteropneustes fossilis*. *J. Exp. Zool.* **241**, 137–141.

Saha, N., and Ratha, B. K. (1989). Comparative study of ureogenesis in freshwater air-breathing teleosts. *J. Exp. Zool.* **252**, 1–8.

Saha, N., and Ratha, B. K. (1990). Alterations in the excretion pattern of ammonia and urea in a freshwater air-breathing teleost, *Heteropneustes fossilis* (Bloch) during hyper-ammonia stress. *Ind. J. Exp. Biol.* **28**, 597–599.

Saha, N., and Ratha, B. K. (1994). Induction of ornithine-urea cycle in a freshwater teleost, *Heteropneustes fossilis*, exposed to high concentrations of ammonium chloride. *Comp. Biochem. Physiol.* **108B**, 315–325.

Saha, N., and Ratha, B. K. (1998). Ureogenesis in Indian air-breathing teleosts: Adaptation to environmental constraints. *Comp. Biochem. Phsyiol.* **120A**, 195–208.

Saha, N., Das, L., and Dutta, S. (1999). Types of carbamyl phosphate synthetases and subcellular localization of urea cycle and related enzymes in air-breathing walking catfish, *Clarias batrachus* infused with ammonium chloride: A strategy to adapt under hyperammonia stress. *J. Exp. Zool.* **283,** 121–130.

Saha, N., Das, L., Dutta, S., and Goswami, U. C. (2001). Role of ureogenesis in the mud-dwelled Singhi catfish (*Heteropneustes fossilis*) under condition of water shortage. *Comp. Biochem. Physiol.* **128A,** 137–146.

Saha, N., Dkhar, J., Anderson, P. M., and Ratha, B. K. (1997). Carbamyl phosphate synthetase in an air-breathing teleost, *Heteropneustes fossilis. Comp. Biochem. Physiol.* **116B,** 57–63.

Saha, N., Dutta, S., and Bhattacharjee, A. (2002). Role of amino acid metabolism in an air-breathing catfish, *Clarias batrachus* in response to exposure to a high concentration of exogenous ammonia. *Comp. Biochem Physiol.* **133B,** 235–250.

Saha, N., Dutta, S., and Haussinger, D. (2000). Changes in free amino acid synthesis in the perfused liver of an air-breathing walking catfish, *Clarias batrachus* infused with ammonium chloride: A strategy to adapt under hyperammonia stress. *J. Exp. Zool.* **286,** 13–23.

Saha, N., Kharbuli, S. Y., Bhattacharjee, A., Goswami, C., and Haussinger, D. (2002). Effect of alkalinity (pH 10) on ureogenesis in the air-breathing walking catfish, *Clarias batrachus. Comp. Biochem. Physiol.* **132A,** 353–364.

Salama, A., Morgan, I. J., and Wood, C. M. (1999). The linkage between Na^+ uptake and ammonia excretion in rainbow trout: Kinetic analysis, the effects of $(NH_4)_2SO_4$ and NH_4HCO_3 infusion and the influence of gill boundary layer pH. *J. Exp. Biol.* **202,** 697–709.

Sardet, C., Pisam, M., and Maetz, J. (1979). The surface epithelium of teleostean fish gills. Cellular and junctional adaptations of the chloride cell in relation to salt adaptation. *J. Cell Biol.* **80,** 96–117.

Schmidt, W., Wolf, G., Grungreiff, K., and Linke, K. (1993). Adenosine influences the high-affinity-uptake of transmitter glutamate and aspartate under conditions of hepatic encephalopathy. *Metabol. Brain Dis.* **8,** 73–80.

Schultze, H. P. (1994). Comparison of hypotheses on the relationships of sarcopterygians. *Syst. Biol.* **43,** 155–173.

Sears, E. S., McCandless, D. W., and Chandler, M. D. (1985). Disruption of the blood–brain barrier in hyperammonemic coma and the pharmacologic effects of dexaethasone and difluoromethyl ornithine. *J. Neurosci. Res.* **14,** 255–261.

Shuttleworth, T. J. (1988). Salt and water balance – extrarenal mechanisms. *In* "Physiology of Elasmobranch Fishes" (Shuttleworth, T. J., Ed.), pp. 171–200. Springer-Verlag, Berlin.

Singh, S. K., Binder, H. J., Geibel, J. P., and Boron, W. F. (1995). An apical permeability barrier to NH_3/NH_4^+ in isolated, perfused colonic crypts. *Proc. Natl. Acad. Sci.* **92,** 11573–11577.

Smart, G. (1976). The effect of ammonia exposure on gill structure of the rainbow trout (*Salmo gairdneri*). *J. Fish Biol.* **8,** 471–475.

Smart, G. R. (1978). Investigations of the toxic mechanisms of ammonia to fish-gas exchange in rainbow trout (*Salmo gairdneri*) exposed to acutely lethal concentrations. *J. Fish Biol.* **12,** 93–104.

Smith, C. P., and Wright, P. A. (1999). Molecular characterization of an elasmobranch urea transporter. *Am. J. Physiol.* **276,** R622–R626.

Smith, D. D., Jr., Ritter, N. M., and Campbell, J. W. (1983). Glutamine synthetase isozymes in elasmobranch brain and liver tissues. *J. Biol. Chem.* **262,** 198–202.

Smith, H. W. (1930). Metabolism of the lungfish (*Protopterus aethiopicus*). *J. Biol. Chem.* **88,** 97–130.

Smith, H. W. (1935). Lung-fish. *Aquarium* **1**, 241–243.

Stewart, P. A. (1983). Modern quantitative acid-base chemistry. *Can. J. Physiol. Pharmacol.* **61**, 1444–1461.

Tam, W. L., Wong, W. P., Chew, S. F., Ballantyne, J. S., and Ip, Y. K. (2003). The osmotic response of the Asian freshwater stingray (*Himantura signifier*) to increased salinity: A comparison to a marine (*Taenima lymma*) and Amazonian freshwater (*Potamotrygon motoro*) stingrays. *J. Exp. Biol.* **206**, 2931–2940.

Tay, S. L. A., Chew, S. F., and Ip, Y. K. (2003). The swamp eel *Monopterus albus* reduces endogenous ammonia production and detoxifies ammonia to glutamine during aerial exposure. *J. Exp. Biol.* **206**, 2473–2486.

Taylor, E. (2000). *In* "Effects of exposure to sublethal levels of copper on brown trout: Mechanisms of ammonia toxicity." Paper presented in the Fifth International Symposium on Fish Physiology, Toxicology and Water Quality, 9–12 November, 1998, City University of Hong Kong, pp. 51–68.

Terjesen, B. F., Chadwick, T. D., and Verreth, J. A. J. (2001). Pathways for urea production during early life of an air-breathing teleost, the African catfish *Clarias gariepinus* Burchell. *J. Exp. Biol.* **204**, 2155–2165.

Terjesen, B. F., Ronnestad, I., Norberg, B., and Anderson, P. M. (2000). Detection and basic properties of carbamoyl phosphate synthetase III during teleost ontogeny: A case study in the Atlantic halibut (*Hippoglossus hippoglossus* L.). *Comp. Biochem. Physiol.* **126B**, 521–535.

Terjesen, B. F., Verreth, J., and Fyhn, H. J. (1997). Urea and ammonia excretion by embryos and larvae of the African Catfish *Clarias gariepinus* (Burchell 1822). *Fish Physiol. Biochem.* **16**, 311–321.

Thorson, T. B., Cowan, C. M., and Watson, D. E. (1967). *Potamotrygon* spp.: elasmobranchs with low urea content. *Science* **158**, 375–377.

Thurston, R. U., Russo, R. C., and Smith, C. E. (1978). Acute toxicity of ammonia and nitrite to cutthroat trout fry. *Trans. Am. Fish. Soc.* **107**, 361–368.

Thurston, R. V., Russo, R. C., and Vinogradov, G. A. (1981). Ammonia toxicity to fishes. The effect of pH on the toxicity of the un-ionized ammonia species. *Environ. Sci. Tech.* **15**, 837–840.

Tomasso, J. R., Goudie, C. A., Simco, B. A., and Davis, K. B. (1980). Effects of environmental pH and calcium on ammonia toxicity in channel catfish. *Trans. Am. Fish. Soc.* **109**, 229–234.

Trischitta, F., Denaro, M. G., Faggio, C., Mandolfino, M., and Schettino, T. (1998). H^+ and Cl^- secretion in the stomach of the teleost fish, *Anguilla anguilla*: Stimulation by histamine and carbachol. *J. Comp. Physiol. B.* **168**, 1–8.

Tsui, T. K. N., Randall, D. J., Chew, S. F., Jin, Y., Wilson, J. M., and Ip, Y. K. (2002). Accumulation of ammonia in the body and NH_3 volatilization from alkaline regions of the body surface during ammonia loading and exposure to air in the weather loach *Misgurnus anguillicaudatus*. *J. Exp. Biol.* **205**, 651–659.

Twitchen, I. D., and Eddy, F. B. (1994). Effects of ammonia on sodium balance in juvenile rainbow trout *Oncorhynchus mykiss* Walbaum. *Aqua. Toxicol.* **30**, 27–45.

USA Environmental Protection Agency (1998). Addendum to "Ambient Water Quality Criteria for Ammonia – 1984". National Technical Information Service, Springfield, VA.

Varley, D. G., and Greenaway, P. (1994). Nitrogenous excretion in the terrestrial carnivorous crab *Geograpsus greyi*: Site and mechanism of excretion. *J. Exp. Biol.* **190**, 179–193.

Vedel, N. E., Korsgaard, B., and Jenssen, F. B. (1998). Isolated and combined exposure to ammonia and nitrite in rainbow trout (*Oncorhynchus mykiss*): Effects on electrolyte status, blood respiratory properties and brain glutamine/glutamate concentrations. *Aqua. Toxicol.* **41**, 325–342.

Wall, S. M. (1996). Ammonium transport and the role of the Na$^+$, K$^+$-ATPase. *Min. Electrolyte Metabol.* **22**, 311–317.

Walsh, P. J. (1997). Evolution and regulation of ureogenesis and ureotely in (batrachoidid) fishes. *Ann. Rev. Physiol.* **59**, 299–323.

Walsh, P. J., and Milligan, C. J. (1995). Effects of feeding on nitrogen metabolism and excretion in the gulf toadfish (*Opsanus beta*). *J. Exp. Biol.* **198**, 1559–1566.

Walsh, P. J., and Mommsen, T. P. (2001). Evolutionary considerations of nitrogen metabolism and excretion. *In* "Fish Physiology" (Wright, P. A., and Anderson, P. M., Eds.), Vol. 20, ch.8, pp. 1–30. Academic Press, New York.

Walsh, P. J., and Smith, C. P. (2001). Urea transport. *In* "Fish Physiology," Vol. 20. "Nitrogen Excretion" (Wright, P. A., and Anderson, P. M., Eds.), ch. 8, pp. 279–307. Academic Press, New York.

Walsh, P. J., Bergman, H. L., Narahara, A., Wood, C. M., Wright, P. A., Randall, D. J., Maina, J. N., and Laurent, P. (1993). Effects of ammonia on survival, swimming and activities of enzymes of nitrogen metabolism in the Lake Magadi tilapia *Oreochromis alcalicus grahami*. *J. Exp. Biol.* **180**, 323–387.

Walsh, P. J., Danulat, E., and Mommsen, T. P. (1990). Variation in urea excretion in the gulf toadfish (*Opsanus beta*). *Mar. Biol.* **106**, 323–328.

Walsh, P. J., Mayer, G. D., Medina, M., Bernstein, M. L., Barimo, J. F., and Mommsen, T. P. (2003). A second glutamine synthetase gene with expression in the gills of the Gulf toadfish (*Opsanus beta*). *J. Exp. Biol.* **206**, 1523–1533.

Walsh, P. J., Tucker, B. C., and Hopkins, T. E. (1994). Effects of confinement/crowding on ureogenesis in the Gulf toadfish *Opsanus beta*. *J. Exp. Biol.* **191**, 195–206.

Walton, M. J., and Cowey, C. B. (1977). Aspects of ammoniagenesis in rainbow trout, *Salmo gairdneri*. *Comp. Biochem. Physiol.* **57**, 143–149.

Wang, X., and Walsh, P. J. (2000). High ammonia tolerance in fishes of the family batrachoididae (toadfish and midshipmen). *Aquat. Toxicol.* **50**, 205–221.

Wang, Z., Petrovic, S., Mann, E., and Soleimani, M. (2002). Identification of an apical Cl$^-$/HCO$_3^-$ exchanger in the small intestine. *Am. J. Physiol.* **282**, G573–G579.

Westhoff, C. M., Ferreri-Jacobia, M., Mak, D.-O. D., and Foskett, J. K. (2002). Identification of the erythrocyte Rh blood group glycoprotein as a mammalian ammonium transporter. *J. Biol. Chem.* **277**, 12499–12502.

Wicks, B. J., and Randall, D. J. (2002). The effect of feeding and fasting on ammonia toxicity in juvenile rainbow trout, *Oncorhynchus mykiss*. *Aquatic. Toxicol.* **59**, 71–82.

Wilkie, M. P. (2002). Ammonia excretion and urea handling by fish gills: Present understanding and future research challenges. *J. Exp. Zool.* **293**, 284–301.

Wilkie, M. P., and Wood, C. M. (1996). The adaptations of fish to extremely alkaline environments. *Comp. Biochem. Physiol.* **113**, 665–673.

Wilkie, M. P., Wright, P. A., Iwama, G. K., and Wood, C. M (1993). The physiological responses of the Lahontan cutthroat trout (*Oncorhynchus clarki henshawi*), a resident of highly alkaline Pyramid Lake (pH 9.4), to challenge at pH 10. *J. Exp. Biol.* **175**, 173–194.

Wilkie, M. P., Wright, P. A., Iwama, G. K., and Wood, C. M. (1994). The physiological adaptations of the Lahontan cutthroat trout (*Oncorhynchus clarki henshawi*) following transfer from well water to the highly alkaline waters of Pyramid Lake, Nevada (pH 9.4). *Physiol. Zool.* **67**, 355–380.

Wilson, J. M., Laurent, P., Tufts, B. L., Benos, D. J., Donowitz, M., Vogl, A. W., and Randall, D. J. (2000a). NaCl uptake by the branchial epithelium in freshwater teleost fish: An immunological approach to ion-transport protein localization. *J. Exp. Biol.* **203**, 2279–2296.

Wilson, J. M., Randall, D. J., Donowitz, M., Vogl, A. W., and Ip, Y. K. (2000b). Immunolocalization of ion-transport proteins to branchial epithelium mitochondria-rich cells in the mudskipper (*Periophthalmodon schlosseri*). *J. Exp. Biol.* **203,** 2297–2310.

Wilson, J. M., Randall, D. J., Kok, T. W. K., Vogl, W. A., and Ip, Y. K. (1999). Fine structure of the gill epithelium of the terrestrial mudskipper, *Periophthalmodon schlosseri. Cell Tissue Res.* **298,** 345–356.

Wilson, R. W., and Taylor, E. W. (1992). Transbranchial ammonia gradients and acid-base responses to high external ammonia concentration in rainbow trout (*Oncorhynchus mykiss*) acclimated to different salinities. *J. Exp. Biol.* **166,** 95–112.

Wilson, R. W., Wilson, J. M., and Grosell, M. (2002). Intestinal bicarbonate secretion by marine teleost fish - why and how? *Biochim. Biophys. Acta* **1566,** 182–193.

Wilson, R. W., Wright, P. A., Munger, S., and Wood, C. M. (1994). Ammonia excretion in freshwater rainbow trout (*Oncorhynchus mykiss*) and the importance of gill boundary layer acidification: Lack of evidence for $Na^+-NH_4^+$ exchange. *J. Exp. Biol.* **191,** 37–58.

Withers, P. C., Morrison, G., and Guppy, M. (1994a). Buoyancy role of urea and TMAO in an elasmobranch fish, the Port Jackson shark, *Heterodontus portusjacksoni. Physiol. Zool.* **67,** 693–705.

Withers, P. C., Morrison, G., Hefter, G. T., and Pang, T.-S. (1994b). Role of urea and methylamines in buoyancy of elasmobranches. *J. Exp. Biol.* **188,** 175–189.

Wood, C. M. (1993). Ammonia and urea metabolism and excretion. *In* "The physiology of fishes" (Evans, D. H., Ed.), pp. 379–423. CRC Press, Boca Raton.

Wood, C. M. (2001). Influence of feeding, exercise, and temperature on nitrogen metabolism and excretion. *In* "Fish Physiology," Vol. 20. "Nitrogen Excretion" (Wright, P. A., and Anderson, P. M., Eds.), ch. 6, pp. 201–238. Academic Press, New York.

Wood, C. M., Bergman, H. L., Laurent, P., Maina, J. N., Narahara, A., and Walsh, P. J. (1994). Urea production, acid-base regulation and their interactions in the Lake Magadi tilapia, a unique teleost adapted to a highly alkaline environment. *J. Exp. Biol.* **189,** 13–36.

Wood, C. M., Gilmour, K. M., Perry, S. F., Part, P., Laurent, P., and Walsh, P. J. (1998). Pusatile urea excretion in gulf toadfish (*Opsanus beta*): Evidence for activation of a specific facilitated diffusion transport system. *J. Exp. Biol.* **201,** 805–817.

Wood, C. M., Hopkins, T. E., Hogstrand, C., and Walsh, P. J. (1995). Pulsatile urea excretion in the ureagenic toadfish *Opsanus beta*: An analysis of rates and routes. *J. Exp. Biol.* **198,** 1729–1741.

Wood, C. M., Hopkins, T. E., and Walsh, P. J. (1997). Pulsatile urea excretion in the toadfish (*Opsanus beta*) is due to a pulsatile excretion mechanism, not a pulsatile production mechanism. *J. Exp. Biol.* **200,** 1039–1046.

Wood, C. M., Matsuo, A. Y. O., Gonzalez, R. J., Wilson, R. W., Patrick, M. L., and Val., A. L. (2002). Mechanisms of ion transport in *Potamotrygon*, a stenohaline freshwater elasmobranch native to the ion-poor blackwaters of the Rio Negro. *J. Exp. Biol.* **205,** 3039–3054.

Wood, C. M., Part, P., and Wright, P. A. (1995). Ammonia and urea metabolism in relation to fill function and acid-base balance in a marine elasmobranch, the spiny dogfish (*Squalus acanthias*). *J. Exp. Biol.* **198,** 1545–1558.

Wood, C. M., Perry, S. F., Wright, P. A., Bergman, H. L., and Randall, D. J. (1989). Ammonia and urea dynamics in the Lake Magadi tilapia, a ureotelic teleost fish adapted to an extremely alkaline environment. *Respir. Physiol.* **77,** 1–20.

Wright, P. A. (1993). Nitrogen excretion and enzyme pathways for ureagenesis in freshwater tilapia (*Oreochromis niloticus*). *Physiol. Zool.* **66,** 881–901.

Wright, P. A. (1995). Nitrogen excretion: Three end products, many physiological roles. *J. Exp. Biol.* **198,** 273–281.

Wright, P. A., and Fyhn, H. J. (2001). Ontogeny of nitrogen metabolism and excretion. *In* "Fish Physiology" Vol. 20. "Nitrogen Excretion" (Wright, P. A., and Anderson, P. M., Eds.), ch. 55, pp. 149–200. Academic Press, New York.

Wright, P. A., and Land, M. D. (1998). Urea production and transport in teleost fishes. *Comp. Biochem. Physiol.* **119A,** 47–54.

Wright, P. A., Iwama, G. K., and Wood, C. M. (1993). Ammonia and urea excretion in Lahontan cutthroat trout (*Oncorhynchus clarki henshawi*) adapted to the highly alkaline Pyramid Lake (pH 9.4). *J. Exp. Biol.* **175,** 153–172.

Wright, P. A., Randall, D. J., and Perry, S. F. (1989). Fish gill boundary layer: A site of linkage between carbon dioxide and ammonia excretion. *J. Comp. Physiol.* **158,** 627–635.

Wright, P. A., Felskie, A. K., and Anderson, P. M. (1995). Induction of ornithine-urea cycle enzymes and nitrogen metabolism and excretion in rainbow trout (*Oncorhynchus mykiss*) during early life stage. *J. Exp. Biol.* **198,** 127–135.

Yancey, P. H. (2001). Nitrogen compounds as osmolytes. *In* "Fish Physiology" Vol. 20. "Nitrogen Excretion" (Wright, P. A., and Anderson, P. M., Eds.), ch. 9, pp. 309–341. Academic Press, New York.

Yap, A. S., Mullin, J. M., and Stevenson, B. R. (1998). Molecular analysis of tight junction physiology: Insights and paradoxes. *J. Membrane Biol.* **163,** 159–167.

Yip, K. P., and Kurtz, I. (1995). NH_3 permeability of principal cells and intercalated cells measured by confocal fluorescence imaging. *Am. J. Physiol.* **369,** F545–550.

Yokobori, S., Hasegawa, M., and Ueda, T. (1994). Relationship among coelacanths, lungfishes, and tetrapods – A phylogenetic analysis based on mitochondrial cytochrome-oxidase-I gene-sequence. *J. Mol. Evol.* **38,** 602–609.

Zeidel, M. L. (1996). Low permeabilities of apical membranes of barrier epithelia: What makes watertight membranes watertight? *Am. J. Physiol.* **271,** F243–F245.

9

IONOREGULATION IN TROPICAL FISHES
FROM ION-POOR, ACIDIC BLACKWATERS

RICHARD J. GONZALEZ
ROD W. WILSON
CHRISTOPHER M. WOOD

I. INTRODUCTION

Most of what we currently understand about the mechanisms of ion regulation in freshwater fishes comes from a wealth of studies on a relatively small number of model species from temperate climates, primarily from the Family Salmonidae and to a lesser extent Cyprinidae. The overall picture that has emerged is that to maintain internal Na^+ and Cl^- levels higher than the surrounding fresh waters, fish must balance diffusive losses with active uptake and that the gills are the main site for both of these processes, although the gut and kidney play roles too. The gut can be a site of ion uptake from food (Smith *et al.*, 1989; D'Cruz and Wood, 1998) and the kidney, which produces dilute urine to conserve ions, is responsible for less than 10% of ion loss (McDonald and Wood, 1981; McDonald, 1983b).

The model species approach has been very successful in providing a clear picture of the basic mechanisms of ion regulation, and a deeper understanding of how systems are integrated, but it is not without limitations. In

The Physiology of Tropical Fishes: Volume 21
FISH PHYSIOLOGY

particular, it is impossible to fully appreciate the range of physiological adaptations that freshwater fishes possess from the study of just a few species with fairly narrow habitat characteristics. For example, recent studies on the cyprinodont killifish in freshwater indicate that the ionic transport systems are fundamentally different from those of the model species (Patrick and Wood, 1999; Katoh *et al.*, 2003). Consequently, there is much to be gained by expanding our examination to other species, particularly tropical fishes, which face virtually all the environmental challenges temperate fish species do. Indeed, in many instances the challenges in tropical systems can be much more extreme, so studies on tropical fishes will expand our understanding of how physiology is affected by the environment and the range of physiological adaptations that are employed. Further, there is an enormous diversity of tropical species, and examination of their physiology within a phylogenetic context holds great promise for the expansion of our understanding of the evolution of physiological adaptation.

   One such system involves the fishes that inhabit the extremely ion-poor, acidic Rio Negro, of the Amazon River system. The Rio Negro, the largest tributary of the Amazon, is approximately 1700 km long and drains an area of about 700 000 km^2 in the northwestern region of the Amazon basin. The soils of this region are the remains of an ancient alluvial flood plain and are made up largely of silicate sand (mainly podzols), which bind nutrients very loosely (Val and Almeida-Val, 1995). Consequently, the soils have long been stripped of nutrients and the waters draining this region are extremely ion-poor. As seen in Table 9.1, typical Na$^+$ and Cl$^-$ levels in the main channel of the Rio Negro are one-sixteenth and one-fifth of global river averages, respectively; water Ca^{2+} levels are even lower, about one-seventieth global river average. In smaller forest streams that feed into the Rio Negro ion

Table 9.1

Ion Concentrations, Dissolved Organic Material (DOM; as Carbon Units), and pH of Rio Negro, Forest Streams in Rio Negro Drainage, and World River Average Values

	Rio Negro	Forest stream	World rivers
Na$^+$(μmol l^{-1})	16.5 ± 5.3	9.4 ± 2.5	270
Cl$^-$(μmol l^{-1})	47.9 ± 19.5	59.4 ± 10.8	220
Ca^{2+}(μmol l^{-1})	5.3 ± 1.6	1.0 ± 0.9	370
Mg^{2+}(μmol l^{-1})	4.7 ± 1.4	1.5 ± 0.6	170
DOM (mg C l^{-1})	10	—	0
pH	5.1 ± 0.6	4.5 ± 0.2	7.8–8.0

   Values for Rio Negro and forest stream from Furch, 1984. Values for world rivers from Wetzel, 1983.

levels can be still more dilute. As a consequence of the low nutrient levels in the water, bacterial diversity is reduced and plant materials are broken down very slowly. The large quantities of organic acids, primarily humic and fulvic acids (Leenheer, 1980; Ertel *et al.*, 1986) from partially decayed plant material, produce the characteristic tea color of the water that give the river its name. These substances, in combination with the very low buffering capacity resulting from the generally low ion levels, produce pH levels that are often very acidic. The pH of the main river can vary between 4.5 and 6.0, but forest streams and flooded forest waters can be 2 or more units lower (Walker and Henderson, 1996).

Extremely ion-poor, acidic waters, such as those found in the Rio Negro, pose a variety of challenges for ion regulation in freshwater fish. In fact, a typical North American fish, such as a salmonid or cyprinid, would find exposure to Rio Negro water toxic, and die within a few hours. Yet, the icthyofauna of the Rio Negro is exceptionally rich and diverse. Recent estimates indicate that over 1000 species of fish from over 40 different families inhabit the Rio Negro. Species present in the Rio Negro include a lungfish *Lepidosiren*, two Osteoglossiformes *Arapaima* and *Osteoglossum*, several dozen gymnotids, and many cichlids. The order Characiformes is particularly diverse, represented by 12 families, including the Characidae with several hundred species (Val and Almeida-Val, 1995).

Thus the fish of the Rio Negro pose a superb opportunity to explore the range of physiological adaptations for ion regulation in ion-poor, acidic waters as well as the evolution of these specializations within a phylogenetic framework. In order to illustrate this potential, we will first describe our current understanding regarding mechanisms of ion regulation in freshwater fishes and what we have learned about the effects of ion-poor, acidic waters on ion regulation. This understanding is based largely upon studies of model salmonids and a few acidophilic species from North America. We will then show how recent studies of ion regulation in fishes native to the ion poor, acidic Rio Negro have changed and extended our understanding.

II. GENERAL MECHANISMS OF ION REGULATION IN FRESHWATER FISHES

Extracellular fluids of freshwater fishes typically have internal Na^+ concentrations of around 130–160 mmol l^{-1} and Cl^- levels of around 120–140 mmol l^{-1}. These concentrations are much greater than levels in the surrounding freshwater (usually ≤ 1 mmol l^{-1}), which causes a constant, diffusive loss of Na^+ and Cl^- across the gills, largely through paracellular tight junctions. Branchial epithelia of freshwater fishes are characterized as "tight" epithelia (Isaia, 1984; Madara, 1988); indeed the electrical resistance of the branchial epithelium is among the highest measured for any epithelium (Wood *et al.*, 2002a). A number of studies that correlate the magnitude of ion efflux with water Ca^{2+} concentration indicate that the low permeability of gill epithelia is governed, at least in part, by the binding of Ca^{2+} to tight junction proteins (Hunn, 1985; Freda and McDonald, 1988; Gonzalez and Dunson, 1989a).

To maintain internal Na^+ and Cl^- levels in the face of diffusive losses, both ions are actively transported from the water into the blood across the gills in two-step processes (Figure 9.1). The mechanisms involved have been addressed in a number of recent reviews (Potts, 1994; Goss *et al.*, 1995; Perry, 1997; Kirschner, 1997; Evans *et al.*, 1999; Marshall, 1995, 2002), which have highlighted both our current knowledge and the weaknesses of existing theories. First, Na^+ and Cl^- are transported across the apical membrane of the branchial epithelium in exchange (direct or indirect exchange, as discussed subsequently) for acid and base equivalents, respectively. The

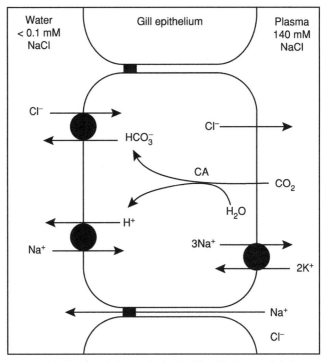

Fig. 9.1 General model of Na^+ and Cl^- regulation in gills of freshwater fish. See text for details.

catalytic action of intracellular carbonic anhydrase is critical in providing these acid (H^+) and basic (HCO_3^-, OH^-) equivalents. Once inside the cell, Na^+ is actively extruded across the basolateral membrane by the action of Na^+/K^+-ATPase. The mechanism of Cl^- transport remains uncertain. Commonly, it is assumed to be linked in some way to energy provided by the Na^+/K^+-ATPase pump, but this has never been satisfactorily demonstrated. Likely, once Cl^- enters across the apical membrane, it can exit passively through basolateral channels, as Cl^- will be above electrochemical equilibrium in the intracellular compartment of the ionocytes relative to the extracellular fluids.

Many more questions remain regarding transfer of Na^+ and Cl^- across the apical membrane. For many years the prevailing model for Na^+ uptake (option A in Figure 9.2) called for an electro-neutral exchange of Na^+ for H^+, or possibly NH_4^+, across the apical membrane by an antiporter (Maetz and Garcia-Romeu, 1964). According to this model, if water pH was equal to or less than intracellular pH, the driving force for this exchange must be a Na^+ concentration gradient produced by the low intracellular Na^+ concentration brought about by the action of Na^+/K^+-ATPase on the basolateral membrane. However, this aspect of the model poses a significant problem.

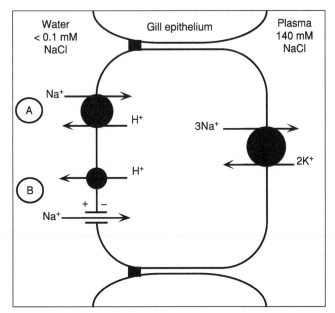

Fig. 9.2 Two models of Na^+ transport across the apical membrane of the branchial epithelium of freshwater fish. See text for details.

Measurements of intracellular Na^+ (and Cl^-) levels indicate concentrations $>30\,mmol\,l^{-1}$ (Wood and LeMoigne, 1991; Morgan *et al.*, 1994). This means that in waters with a Na^+ concentration $\leq 1\,mmol\,l^{-1}$, there is no gradient to drive exchange. This fundamental flaw in the original model led to the proposal of an alternate mechanism for Na^+ transport across the apical membrane (Avella and Bornancin, 1989; Lin and Randall, 1991, 1993, 1995). This newer model involves the active extrusion of H^+ across the apical membrane by H^+-ATPase creating a local, negative, potential across the apical membrane that draws in Na^+ through an ion-specific channel (option B in Figure 9.2), as was first described in the salt-absorbing cells of amphibian epithelia (Ehrenfeld *et al.*, 1985). The obvious benefit of this model is the lack of a requirement for a sufficient H^+ or Na^+ chemical gradient to drive transport, instead it relies on a highly localized electrochemical gradient for Na^+ entry generated by primary active, electrogenic H^+ extrusion at the apical membrane.

Evidence for this new model has been accumulating from two lines of research. Immunolocalization and molecular techniques have confirmed the presence of H^+-ATPase and Na^+ channel proteins in gill epithelia (Lin *et al.*, 1994; Sullivan *et al.*, 1995; Wilson *et al.*, 2000), and work with a pharmacological inhibitor of H^+-ATPase, bafilomycin, has shown strong inhibition of Na^+ uptake in larval trout (Bury and Wood, 1999), carp and tilapia (Fenwick *et al.*, 1999). Yet, while the H^+-ATPase/Na^+ channel model is clearly gaining support, the issue is not completely settled. Indeed, some immunolocalization and pharmacological blocker studies indicate the presence of Na^+/H^+ exchangers in branchial epithelia of certain species (Wilson *et al.*, 2000; Wood *et al.*, 2002b). Further, it should be noted that the link between H^+ extrusion and Na^+ uptake has not yet been confirmed.

As for Cl^- uptake across the apical membrane there is considerable evidence that it occurs via a Cl^-/HCO_3^- exchanger (which obviates the electrical gradient). However, intracellular Cl^- and HCO_3^- levels are $>30\,mmol\,l^{-1}$ and about $2\,mmol\,l^{-1}$, respectively (Wood and LeMoigne, 1991). So, as with Na^+, the global chemical gradients are clearly unfavorable and the mechanism(s) energizing this process remains unknown.

The transports of Na^+ and Cl^- across the apical membrane are not directly linked, and indeed Na^+/H^+ exchange and Cl^-/HCO_3^- exchange may change in opposite directions in compensation for systemic acid-base disturbances. Nevertheless, there does appear to be the possibility of some degree of connection. If HCO_3^- and H^+, the counter-ions required for transport of Cl^- and Na^+, respectively, are produced in the branchial epithelial cells by the hydration of CO_2, a reaction catalyzed by carbonic anhydrase, then it is conceivable that the rate of Cl^- uptake in exchange for HCO_3^- could influence H^+ availability for Na^+ uptake. This indirect linkage, however, has not been clearly demonstrated, and suggestions that Na^+ and Cl^- transport occur in different cell types (Goss *et al.*, 1995; Marshall, 2002) may obviate such an association.

III. EFFECTS OF ION-POOR, ACIDIC WATERS ON MODEL TELEOSTS

Until relatively recently, most of what we know about the physiological responses of freshwater fishes to soft water with low environmental pH has resulted from studies, stimulated by the acid rain problem, largely on salmonid species, and to a lesser extent cyprinids and cichlids. These species do not naturally occur in acidic waters and are not particularly tolerant of low pH. The effects of acid waters on salmonids have been thoroughly reviewed by Wood (1989) and Reid (1995) and will be briefly summarized here. In water with an extremely low pH, death occurs rapidly due to problems associated with extensive breakdown of gill structure and suffocation due to massive mucus production (Packer and Dunson, 1972). At more moderate pH levels, however, a number of studies have shown that the primary toxic action is a disturbance of ion regulation. High H^+ concentrations in the water inhibit active uptake of Na^+ and Cl^- (Packer and Dunson, 1970; Maetz, 1973; McWilliams and Potts, 1978) and at the same time massively stimulate diffusive ion loss (McWilliams, 1982; McDonald, 1983a). The ionic imbalance leads to a drop in plasma Na^+ and Cl^- levels and a reversal in the transepithelial potential across the gills (McWilliams and Potts, 1978). The fall in plasma ion levels causes a substantial shift of fluid from the extracellular to intracellular compartment for osmotic re-equilibration, resulting in a reduced blood volume and elevated blood viscosity and blood pressure. Once the plasma ion losses surpass about 30%, associated cardiovascular failure kills the fish (Milligan and Wood, 1982).

Exactly how low pH is thought to inhibit Na^+ and Cl^- transport depends upon the mechanism involved. When Na^+/H^+ exchange was thought to be the mechanism responsible for Na^+ uptake it was generally believed that high levels of H^+ in the water competitively inhibited access of Na^+ to the binding site of the antiporter or perhaps, at very low pH, damaged the protein (Wood, 1989). With the recently proposed H^+-ATPase/Na^+ channel model, the mechanism of H^+ inhibition had to be re-evaluated. As external pH falls, it is thought that an increased rate of H^+ flux into the cell would make it difficult for the H^+-ATPase to generate a sufficient potential to drive Na^+ uptake (Lin and Randall, 1991). Regardless of the mechanism, the theoretical low pH limit for active uptake is believed to be about pH 4.5–5.5 (McDonald, 1983b; Lin and Randall, 1991, 1993). Inhibition of

Cl^- uptake is even sketchier. Unless H^+ ions somehow damage Cl^-/HCO_3^- exchangers, an unlikely event at moderately low pH levels, the inhibitory effect is most likely indirect through the reduction of Na^+ transport, as suggested by Wood (1989). Reduced Na^+ uptake would presumably lead to a build-up of H^+ in the cell, which would inhibit the further production of HCO_3^-. In turn, lowered intracellular HCO_3^- levels would mean less is available for exchange, thus slowing Cl^- transport.

At the same time that low pH inhibits ion uptake it also stimulates diffusive losses. There are substantial data suggesting that the primary mechanism for stimulation of ion efflux at low pH is leaching of Ca^{2+} from paracellular tight junctions, which renders the branchial epithelium much more permeable. This assertion is supported by studies showing that elevated water Ca^{2+} levels limit the degree of stimulation at low pH (McDonald *et al.*, 1980; McWilliams, 1982). In fact, in water with very high Ca^{2+} levels the ion disturbance is virtually eliminated and acid-base disturbances take precedence (McDonald *et al.*, 1980). Quantitatively, stimulated efflux is by far the more serious consequence of low pH exposure. While complete inhibition of uptake yields an imbalance of about 400 nmol g^{-1} h^{-1}, efflux can be stimulated to a point at least five times higher.

If the stimulation of diffusive efflux is not too severe then fish may be able to initiate adjustments to restore some semblance of ion balance by lowering diffusive efflux but not by restoring uptake (Audet and Wood, 1988). Diffusive losses drop partially due to the diminished gradient as plasma ion levels fall, but mainly it seems, due to a reduction of branchial permeability. It has been suggested that this may be achieved by adjustments to cell volume and/or tight junctions, and the latter may involve hormones such as prolactin or cortisol, but there is no firm evidence on the exact mechanism. Given enough time, the degree of reduction of efflux coupled with some minor restoration of influx can restore ion balance. However, ion balance cannot be regained until uptake is restored completely upon return to neutral pH.

IV. EFFECTS OF ION-POOR, ACIDIC WATERS ON NORTH AMERICAN ACIDOPHILIC TELEOSTS

Some work has been done on a few North American species that inhabit naturally soft, acidic waters in the eastern United States, and the bulk of it is qualitatively similar to findings for sensitive species. Two species in particular, banded sunfish (*Enneacanthus obesus*) and yellow perch (*Perca flavescens*), show superior low pH tolerance compared to salmonids or cyprinids. Banded sunfish survive direct transfer to pH 3.5 and reproducing populations have been found in waters with a pH of 3.7 (Graham and Hastings,

1984). While not as tolerant as banded sunfish, yellow perch populations are apparently unaffected in pH 4.3 waters (Harvey, 1979), which makes them significantly more tolerant than salmonids. Both sunfish and perch appear to utilize a strategy for maintenance of ion balance in low pH waters centered on low branchial ion permeability and resistance to stimulation of efflux by low pH. Isotopic measurements of unidirectional Na^+ efflux of sunfish in water with circum-neutral pH show that they lose Na^+ at a rate less than 50 nmol g^{-1} h^{-1}, a value that is only about one-tenth the rate of loss for the acid-sensitive rainbow trout and common shiners. Perch do not have efflux rates as low as sunfish, but they are still well below rates of the acid-sensitive species. This suggests that gills of perch and especially sunfish have much lower intrinsic permeability than gills of shiners or trout.

The ion efflux rates of sunfish and perch continue to remain low during exposure to low pH (Figure 9.3A). In sunfish, for example, Na^+ efflux is only mildly stimulated upon transfer to pH 4.0 and returns to low control levels within 24 hours (Gonzalez and Dunson, 1987, 1989a, b). Efflux of sunfish is not stimulated to a sufficient level to cause mortality until water pH drops to about 3.25 or lower. Resistance to stimulation of efflux in sunfish, and probably perch, is correlated with a very high branchial affinity for Ca^{2+}. In an effort to evaluate the affinity of sunfish gills for Ca^{2+}, Gonzalez and Dunson (1989a) measured Na^+ efflux in pH 3.25 water at a range of different Ca^{2+} concentrations. They estimated that a Ca^{2+} concentration of only 19 μmol l^{-1} was sufficient to reduce efflux by 50% relative to measurements in Ca^{2+}-free water. In contrast, estimates for rainbow trout at higher pH levels are 20–35 times higher (Table 9.2). A high branchial affinity of tight junctions for Ca^{2+} would act to maintain junction integrity in the face of elevated H^+ levels in the surrounding water.

During chronic low pH exposure, sunfish, like trout, are able to reduce diffusive ion losses significantly. But unlike trout they appear to be able to do it to a much greater degree. Indeed, sunfish can lower efflux to virtually zero during continued exposure. During 5 weeks at pH 4.0, where uptake is fully inhibited (see below), sunfish show no decline in body Na^+ concentration, a direct indicator of overall ion balance. Further, adjustments made to reduce efflux at pH 4.0 can reduce rates of ion loss during exposure to even lower pH levels (Figure 9.4).

Low intrinsic branchial permeability and control of efflux are crucial at low pH because Na^+ uptake shows little specialization for operation in ion-poor, acidic waters. Both species have low maximum uptake capacities and sunfish have a very low affinity (i.e. high K_m), compared to trout (Table 9.3). This means that at typical environmental concentrations of less than 50 μmol l^{-1}, sunfish would take up only around 35–40 nmol g^{-1} h^{-1}. Perch, because of their higher affinity, would have an uptake rate about four times

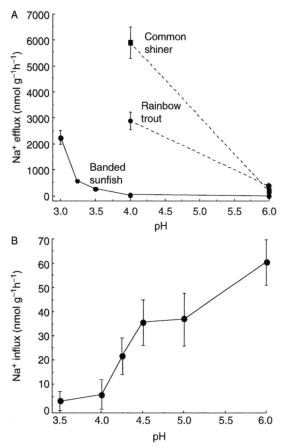

Fig. 9.3 The influence of water pH on (A) Na^+ efflux and (B) Na^+ influx of banded sunfish during the first hour of exposure. Efflux of rainbow trout and common shiners at pH 6.0 and 4.0 are shown for comparison. Values are means $\pm$ SE. (Trout and shiner data adapted from Freda and McDonald, 1988; sunfish data are from Gonzalez and Dunson, 1987.)

Table 9.2
Estimated Water Ca^{2+} Concentration that Reduces Na^+ Efflux at
Low pH by 50% from Efflux in Ca^{2+}-free Water (Means $\pm$ SE)

Species	Ca^{2+} concentration ($\mu mol\, l^{-1}$)
Rainbow trout – pH 4.0	375 ± 110
Rainbow trout – pH 3.5	675 ± 95
Banded sunfish – pH 3.25	19 ± 22

Sunfish data from Gonzalez and Dunson, 1989a.
Trout data from Freda and McDonald, 1988.

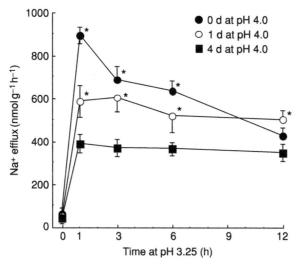

Fig. 9.4 The effect of pre-exposure to pH 4.0 for 0, 1 or 4 day on rate of Na$^+$ efflux upon transfer to pH 3.25 in banded sunfish. Values are means ± SE. (Data from Gonzalez and Dunson, 1989a.)

Table 9.3

Kinetic Parameters for Na$^+$ Transport in Two Species of Acidophilic Fish, Sunfish and Perch, in Comparison to Acid-sensitive Rainbow Trout and Common Shiners (Means ± SE)

Species	$K_m(\mu moll^{-1})$	$J_{max}(nmolg^{-1}h^{-1})$
Common shiner	158 ± 54	460 ± 74
Rainbow trout	48 ± 13	379 ± 35
Yellow perch	21 ± 8	249 ± 24
Banded sunfish	125 ± 60	128 ± 26

Sunfish data from Gonzalez and Dunson, 1989a.

Shiner, Trout and Perch data from Freda and McDonald, 1988.

higher. However, since efflux is about the same (and still low relative to trout) under these conditions, both species are in ionic equilibrium under these conditions.

Ion uptake also showed little resistance to low pH. Na$^+$ uptake of sunfish, which is low to begin with, fell progressively as pH dropped (Figure 9.3B), and was completely inhibited at pH 4.0 and below. Further, during prolonged exposure to low pH, uptake showed no ability to recover at all, even after weeks of exposure (Gonzalez and Dunson, 1989a). This further emphasizes that survival at this pH and below is based upon ability to keep diffusive ion losses very low. Ion uptake in perch is also inhibited by low pH, but unlike sunfish, perch exhibit some ability to restore uptake during chronic exposure. During 48 hours at pH 4.0, for instance, they completely restored uptake to neutral pH levels. It is not known if perch are able to maintain this uptake at lower pH levels.

The ionoregulatory pattern displayed by sunfish, and to a lesser degree perch, of low intrinsic permeability and low rates of uptake, reflects the overriding necessity to prevent the stimulation of efflux at low pH, which produces much larger losses of Na^+ than does inhibition of uptake. Inhibition of uptake with no change in efflux would only produce a net loss of around $50 \, nmol \, g^{-1} \, h^{-1}$ in these species while efflux can rise as high as $600 \, nmol \, g^{-1} \, h^{-1}$ at pH 3.25 (Gonzalez and Dunson, 1989a). Even if influx were maintained at such a low pH, it would make little quantitative difference. Taken together these characteristics suggest a strategy of waiting out low pH episodes with no uptake by preventing ion losses.

Examination of ion regulation in these two acid-tolerant species raises several basic questions regarding requirements for maintenance of ion balance at low pH. For instance, is the low permeability/low uptake pattern of ion regulation observed in these species the only way to be acid tolerant? Is low intrinsic permeability a vital necessity for controlling rates of Na^+ loss at low pH, and is specialization of uptake impossible? Or instead can fish somehow circumvent the theoretical pH limitations for uptake and continue to transport ions?

V. ADAPTATION TO ION-POOR, ACIDIC WATERS IN RIO NEGRO TELEOSTS

An early study by Dunson *et al.* (1977) first showed that Rio Negro species were exceptionally tolerant to low pH water, but did not examine the mechanisms of ion regulation that these fishes employed. Twenty years later new studies of the physiology of Amazonian species in low pH water confirmed that acid tolerance was indeed correlated with occurrence in the Rio Negro and provided insight into the mechanisms involved. Wilson *et al.* (1999) compared responses to low pH in ion and acid-base exchange with the water for three species that differ in their degree of natural occurrence in the waters of the Rio Negro. The armored catfish tamoata, *Hoplosternum littorale*, does not naturally occur in the Rio Negro, while matrincha (*Brycon erythropterum*) spend most of their life in the whitewater Rio Branco but migrate annually through the Rio Negro and tambaqui (*Colossoma macropomum*) spend extended periods of time in the Rio Negro, even entering dilute acidic waters of flooded jungles to feed during the rainy season. During serial exposure to pH levels down to 3.5, the non-resident tamoata proved to be the most sensitive, dying within 24 hours during exposure to pH 3.5. Matrincha were more tolerant, surviving 24 hours' exposure to pH 3.5, but most dying during recovery at pH 6.0. Tambaqui were the most tolerant, surviving 24 hours at pH 3.5 with no mortality at all during exposure or subsequent recovery in pH 6 water. These findings suggest that exceptional low pH tolerance is a characteristic of Rio Negro species, not all Amazonian species in general.

Examination of the results confirms that the basis of low pH tolerance in Rio Negro species is the ability to avoid ionic disturbances. For example, during the first hour at pH 3.5 tamoata had rates of net Na^+ and Cl^- loss at least twice as high as the least sensitive tambaqui and they only survived the first few hours (Figure 9.5). Measurements of acid-base exchange with the water showed that during exposure to low pH in extremely dilute waters like the Rio Negro, no acid-base disturbance results. Even the acid-sensitive tamoata do not suffer any acid-base disturbance at pH 3.5, a pH in which they die quickly. It appears that the results for these fish were qualitatively similar to previous findings for less tolerant salmonids (e.g. Wood, 1989). The difference is mainly quantitative, with the acid-tolerant tambaqui having a threshold for acute toxicity at least 1 pH unit lower than salmonids, representing a 10-fold lower threshold to H^+ ions.

A companion study (Wood *et al.*, 1998), using vascular catheterization for repetitive blood sampling, examined the internal physiology of the acid-tolerant tambaqui, supplied detailed support for the flux results and additional insights as to the mechanisms involved. During serial exposure to pH levels down to 3.0, tambaqui never developed a substantial blood acid-base disturbance (Figure 9.6), in accord with the measurements of acid-base exchange by Wilson *et al.* (1999; Figure 9.5). Blood pH remained virtually constant at about 7.8 (Figure 9.6C). In addition, there was no disturbance in blood gases. Throughout the duration of the exposure, blood PO_2 remained at about 35 Torr and PCO_2 at around 5 Torr (Figure 9.6A, B). Even at pH 3.0, when tambaqui were exhibiting a strong stress response, as evidenced by significantly elevated plasma cortisol, glucose, and ammonia (Figure 9.7A, B, C) there was no acid-base disturbance or problems with O_2 delivery since blood lactate levels remained low.

The primary cause of stress at pH 3.0 appeared to be a sizeable ionic disturbance as indicated by a significant drop in plasma Na^+ and Cl^- concentrations (Figure 9.7D). Both Na^+ and Cl^- fell by about 25% from levels at pH 6.5. At the same time plasma protein concentration increased markedly, indicating that an osmotic shift of water was occurring, which

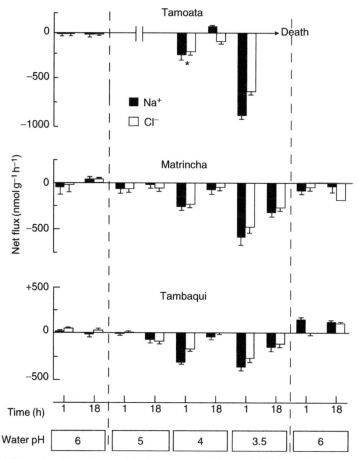

Fig. 9.5 Effect of graduated water pH changes on net Na^+ and Cl^- fluxes in three species of Amazonian fish. Fish were exposed to each new pH for 24 h, and fluxes were measured after 1 and 18 h of exposure to each new pH. Values are means ± SE. No tamoata survived beyond the first hour of exposure to pH 3.5. Asterisk indicates significant difference from control fluxes at pH 6. (Data from Wilson *et al.*, 1999.)

caused shrinkage of the plasma volume. These disturbances, which are very similar to those observed in salmonids, clearly confirm that the primary toxic effect of low pH on this Rio Negro fish is interference with ion regulation. The only difference is that a much lower pH (3.0 *vs* 4.5) is required to elicit the disturbance (i.e. >30-fold greater H^+ concentration).

During recovery at pH 6.5 after exposure to pH 3.0 there was no recovery of plasma ions or protein concentration (Figure 9.7D, E), suggesting that at this low pH level more severe damage occurred to the branchial

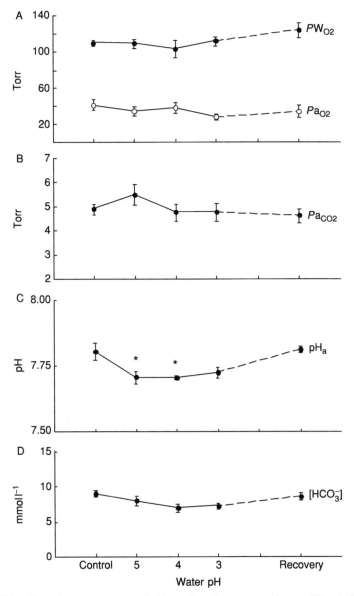

Fig. 9.6 The effect of exposure to a graded low pH and recovery regime on (A) arterial Pa_{O_2} relative to water Pw_{O_2}; (B) arterial Pa_{CO_2}; (C) arterial (extracellular) pHa; and (D) arterial plasma $[HCO_3^-]$ of tambaqui. Values are means $\pm$ SE. Asterisk indicates significant differences from control fluxes at pH 6.5. (Data from Wood *et al.*, 1998.)

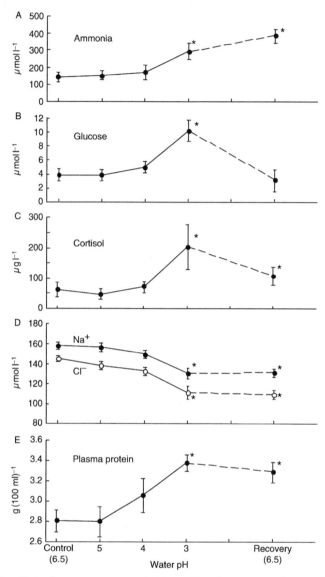

Fig. 9.7 The effect of exposure to a graded low pH and recovery regime on (A) plasma ammonia; (B) plasma glucose and whole blood lactate; (C) plasma cortisol; (D) plasma $[Na^+]$ and $[Cl^-]$; and (E) plasma protein concentration in tambaqui. Values are means ± SE. Asterisk indicates significant difference from control values at pH 6.5. (Data from Wood *et al.*, 1998.)

epithelium of tambaqui. Interestingly, in the flux study with tambaqui (Wilson *et al.*, 1999), after exposure to pH 3.5 rather than 3.0, upon return to neutral pH, recovery occurred quickly (Figure 9.5). Taken together, these results indicate that there is a threshold between pH 3.5 and 3.0 for branchial damage in tambaqui. In comparison, results from the North American banded sunfish indicate that the threshold for branchial damage occurred between pH 4.0 and 3.5 (Gonzalez and Dunson, 1989a) and that pathological ion losses do not occur until such damage is initiated. Thus, it appears that tambaqui are even more tolerant of low pH than banded sunfish.

While unidirectional ion fluxes could not be measured in these early studies, due to restrictions on isotope use, some of the results provided clues as to the underlying mechanism responsible for such a high degree of low pH tolerance. Measurements of the PO_2 and PCO_2 gradient across the gills of tambaqui indicated unusually large gradients. For example, in water with a PO_2 of about 110 Torr, arterial blood has a PO_2 of only about 35 Torr (Figure 9.6A); such a large gradient (75 Torr) suggests that tambaqui gills have a very low diffusing capacity. Despite having unusually high total gill surface area (Saint-Paul, 1984), low diffusion capacity would minimize ionic losses in dilute, acidic environments. However, it has not been demonstrated whether low permeability to gases necessarily is equivalent to a low permeability to ions. Alternatively, the relatively large transbranchial PO_2 and PCO_2 gradients in tambaqui may simply result from reduced gill ventilatory water flow rates. This in itself would represent a strategy to minimize ionic losses that does not necessitate lower absolute permeabilities to gases or ions. This strategy would require tambaqui hemoglobin to have a relatively high affinity for O_2 to maintain effective arterial blood oxygen content in the face of low arterial PO_2 levels. This indeed appears to be the case, based upon *in vivo* measurements of arterial and venous PO_2 and oxygen content values (Figure 9.8, see Chapters 6 and 7). Nonetheless, either strategy (low permeability to ions/gases or reduced ventilatory flow rates) would point, once again, to the primacy of control of ion loss in conveying tolerance to low pH in freshwater fish.

Measurements of trans-epithelial potential (TEP) of tambaqui confirmed the effects of pH on branchial permeability. Qualitatively, the TEP results on tambaqui, for the effects of both low pH and elevated Ca^{2+}, were very similar to those reported earlier for a salmonid by McWilliams and Potts (1978), suggesting that the basic mechanisms are similar in the two species. Water pH had a significant effect on TEP across the gills (Figure 9.9A). At pH 6.5 and $20 \mu mol \ l^{-1} \ Ca^{2+}$, TEP in the tambaqui was about $-25 \, mV$ (inside referenced to outside as $0 \, mV$) and it became gradually more positive as pH declined; at pH 3.0, TEP was $+35 \, mV$. While there is some question regarding the origin of branchial TEP (McWilliams and Potts, 1978; Potts,

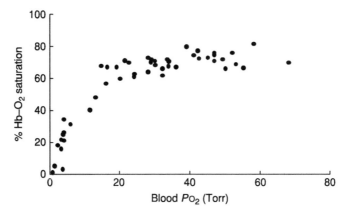

Fig. 9.8 The calculated % saturation of hemoglobin with oxygen for *in vivo* samples of tambaqui whole blood compared with their respective Po_2 values. Blood samples were taken from tambaqui with catheters implanted in either the caudal artery or caudal vein (see Wood *et al.*, 1998), giving a range of *in vivo* arterial and venous oxygen tensions under resting conditions. The % saturation of hemoglobin with oxygen was calculated assuming a molecular weight of 68 K for hemoglobin, and an oxygen solubility for plasma of 0.00148 μmol l^{-1} mmHg^{-1} (using the formula of Boutilier *et al.*, 1984 for human plasma at 28 °C). Note that the blood seems to plateau at just above 80% of the available O_2 binding sites on hemoglobin. This is probably due to the presence of low levels of nitrite in the ambient water during experiments, resulting in a fraction of the hemoglobin being converted to the non-functional methemoglobin. Based on these *in vivo* blood samples the affinity of tambaqui hemoglobin for oxygen would appear to be high (i.e. $P_{50} < 8$ mmHg).

1984; Lin and Randall, 1993; Kirschner, 1994), the simplest interpretation of these findings is that as pH falls, overall branchial permeability rises, but the branchial epithelium becomes more permeable to Cl$^-$ relative to Na$^+$, thus raising the TEP.

Measurements of TEP also shed light on the role of Ca^{2+} in determining branchial Na$^+$ and Cl$^-$ permeability in tambaqui. Increased water Ca^{2+} concentration had two effects on TEP. At pH 6.5, increasing Ca^{2+} levels on a logarithmic scale caused the TEP to become more positive (Figure 9.9B). In addition, higher water Ca^{2+} concentrations attenuated the effects of low pH on TEP (Figure 9.9C). At the highest Ca^{2+} concentration tested, 10 mmol l^{-1}, falling pH had virtually no effect on TEP. Thus it appears that high Ca^{2+} levels have a similar effect as low pH (high H$^+$ concentration). Wood *et al.* (1998) proposed that H$^+$ and Ca^{2+} titrate the negative charge of paracellular channels and therefore modify their relative permeability to Cl$^-$ and Na$^+$ in a similar manner, although they have very different effects on absolute permeability; low pH acts to raise permeability, elevated Ca^{2+} levels act to lower it.

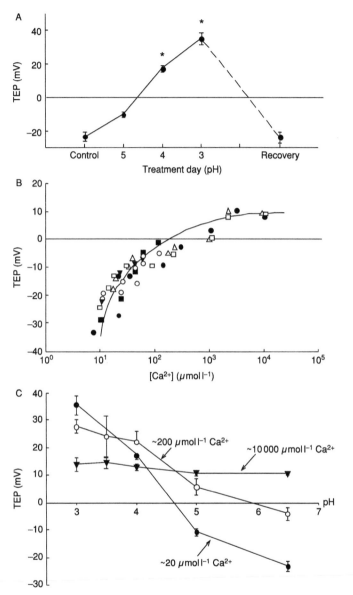

Fig. 9.9 (A) The effect of exposure to a graded low pH and recovery regime on transepithelial potential (TEP, inside relative to outside as 0 mV) in tambaqui. Values are means ± SE. Asterisk indicates significant difference from control values at pH 6. (B) The effect of acute changes in water Ca^{2+} concentration at circumneutral pH on TEP in tambaqui. Each symbol represents a different fish. (C) The effect of acute changes in water pH on TEP at three different water Ca^{2+} concentrations in tambaqui. Values are means ± SE. (Data from Wood et al., 1998.)

Considering the effects of water Ca^{2+} on TEP, in tambaqui (Figure 9.9B, C) it is not surprising that elevated water Ca^{2+} levels acted to reduce rates of Na^+ and Cl^- loss at low pH. At pH 3.5, net ion losses were reduced by about 70% as water Ca^{2+} levels rose from 20 to 700 μmol l^{-1}. Interestingly, different results were seen in three species collected directly from the Rio Negro and studied in Rio Negro water (Gonzalez *et al.*, 1998). Based upon rates of net Na^+, K^+, and Cl^- loss at low pH, all three species proved to be much more tolerant of low pH than tambaqui, but raising the concentration of Ca^{2+} in the water from 10 to 100 μmol l^{-1} had no effect on loss rates at low pH (Figure 9.10). These surprising findings could indicate an extremely high branchial affinity for Ca^{2+}, or alternatively they could mean that water Ca^{2+} plays no role in determining branchial permeability in these Rio Negro natives. Since the measurements were made in Rio Negro water, which contains high levels of dissolved organic matter (DOM), it also raised the possibility that the dissolved organics may somehow be a factor in determining branchial permeability, a point discussed below.

Thus, the picture that emerges from these early studies is that Rio Negro species are significantly more tolerant than salmonids (and even North American acidophilic species), but that the basic responses to low pH that Rio Negro fishes display are the same as those displayed by salmonids. Low pH tolerance appears to be a function of ability to control rates of ion loss at low pH, which is facilitated by low intrinsic permeability of the gills and possibly high affinity of tight junctions for water Ca^{2+}, at least for some species. However, due to restrictions on isotope use, we had no detailed information regarding effects of low pH on ion transport or details of the actual ion permeability of the branchial epithelium of Rio Negro fish. The next series of studies, performed on Rio Negro fishes acquired from aquarium suppliers in North America, were able to utilize radioisotopes to evaluate unidirectional ion fluxes during exposure to low pH. These studies shed new light on the specializations required for maintenance of ion balance in ion-poor, acidic waters and showed that the picture was not as simple as initially thought.

Examination of diffusive efflux in three characid species and a cichlid from the Rio Negro revealed, contrary to previous findings, that tolerance of low pH does not necessarily require a low intrinsic branchial permeability. For example, neon tetras (*Paracheirodon innesi*) had rates of Na^+ efflux at pH 6.5 about 7-fold greater than the banded sunfish (Figure 9.11) and similar in magnitude to acid-sensitive salmonids. Yet during exposure to low pH, efflux does not greatly increase and they can survive indefinitely at pH 3.5 with no significant disruption of ion balance (Gonzalez and Preest, 1999). Similar high rates of efflux at pH 6.5 were seen in blackskirt tetras (*Gymnocorymbus ternetzi*) and cardinal tetras (*Parcheirodon axelrodi*)

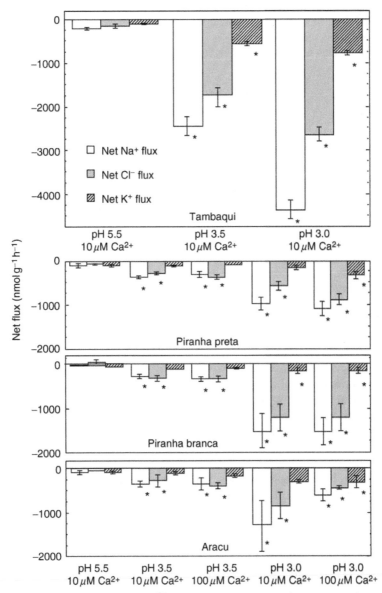

Fig. 9.10 Effect of water pH and Ca^{2+} concentration on net Na^+ Cl^-, and K^+ fluxes of tambaqui (not Ca^{2+}) from aquaculture and piranha preta (*Serrasalmus rhombeus*), piranha branca (*Serrasalmus rhombeus*), and aracu (*Leporinus fasciatus*) collected from the Rio Negro. Values are means ± SE. Asterisk indicates significant difference from pH 5.5, $10 \, \mu mol \, l^{-1}$ Ca^{2+} treatment. (Data from Gonzalez *et al.*, 1998.)

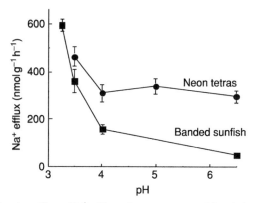

Fig. 9.11 Effect of water pH on Na^+ efflux of neon tetras and banded sunfish. Values are means $\pm$ SE. (Data for banded sunfish adapted from Gonzalez and Dunson, 1987; data for neon tetras adapted from Gonzalez and Preest, 1999.)

(Gonzalez *et al.*, 1997; Gonzalez and Wilson, 2001). In contrast, the lone cichlid examined, angelfish (*Pterophyllum scalare*), did have low rates of Na^+ efflux, about 50 nmol g^{-1} h^{-1}. In this regard, angelfish were very similar to North American banded sunfish. That the three species of characids do not possess low intrinsic branchial permeability, yet are very tolerant of low pH, is the first indication that this characteristic is not an absolute prerequisite for acid tolerance. Rather the key is the ability to prevent stimulation of the efflux component, and this can be achieved regardless of the intrinsic level that permeability starts from.

In another departure, the role of water Ca^{2+} concentration in the control of diffusive efflux was not clearly supported in these studies. In blackskirt tetras and angelfish, raising water Ca^{2+} levels up to 500 μmol l^{-1} moderately reduced the stimulation of Na^+ efflux at low pH (Figure 9.12). However, with neon tetras, raising Ca^{2+} levels to the same degree had no discernible affect on Na^+ efflux at pH 3.5 (Figure 9.12) and the same was true for cardinal tetras. Interestingly, elevated water Ca^{2+} levels did reduce Na^+ efflux at neutral pH in neon tetras and exposure to an equimolar level of $LaCl_3$, a strong Ca^{2+} competitor, stimulated Na^+ losses significantly. So it seems in these species that water Ca^{2+} is playing some role in governing branchial permeability, but the role may become less important or disappear at low pH, and other, Ca^{2+}-independent mechanisms of some sort are employed. It is important to remember that the waters these species natural-ly inhabit can have Ca^{2+} levels as low as 1 μmol l^{-1}, and that as discussed subsequently, the presence of dissolved organic matter may reduce the bioavailability of this Ca^{2+}. This could represent a strong selective pressure for development of alternate mechanisms of control of ion efflux at low pH.

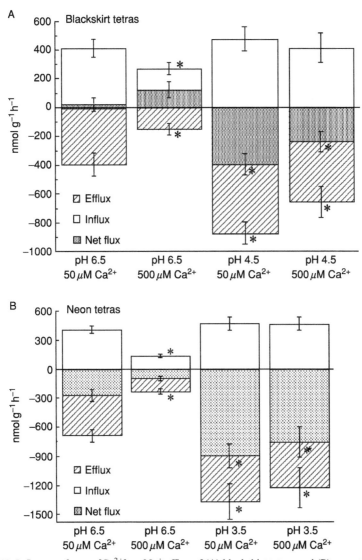

Fig. 9.12 Influence of water $[Ca^{2+}]$ on Na^+ efflux of (A) blackskirt tetras and (B) neon tetras at pH 6.5 and pH 4.0 (blackskirts) or 3.5 (neons). Values are means ± SE. Asterisks indicate significant differences from pH 6.5, $50\,\mu mol\,l^{-1}$ Ca^{2+} treatment. (Data for blackskirt and neon tetras from Gonzalez *et al.*, 1997 and Gonzalez and Preest, 1999, respectively.)

As with ion efflux, patterns of Na^+ transport by the four species examined fall into two categories. Angelfish (*P. scalare*), which have very low rates of Na^+ efflux, have an uptake mechanism with little specialization for ion-poor waters. Kinetic analysis of Na^+ uptake reveals a very low affinity (high K_m), which means that in the ion-poor waters that they inhabit, their rate of uptake is very low (Table 9.4). Despite these low rates of uptake they maintain ion balance because of their equally low efflux. In addition, ion uptake in angelfish shows little specialization for low pH. Na^+ uptake is progressively inhibited by declining pH and completely shut down at pH 3.5. Again angelfish appear to be very similar to banded sunfish, possessing low intrinsic permeability, no specializations for ion transport, and utilizing a strategy of waiting out episodes of low pH by reducing rates of loss during the exposure.

In contrast to angelfish, all species of characids possess ion uptake mechanisms that are highly specialized for operation in ion-poor and

Table 9.4
Kinetic Parameters of Na^+ Transport for Fish from the Rio Negro

Species	$K_m (\mu mol\ l^{-1})$	$J_{max} (nmol\ g^{-1} h^{-1})$	Na^+ uptake*\ $(nmol\ g^{-1} h^{-1})$
Characidae			
P. innesi pH 6.5[1]	12.9 ± 5.8	448.2 ± 43.5	272
P. innesi pH 3.5[1]	17.6 ± 10.3	513.3 ± 60.3	273
P. axelrodi[2]	53.7 ± 7.8	773.0 ± 38.2	210
G. ternetzi[3]	27.7 ± 2.7	691.3 ± 19.9	290
Hemigrammus[4]	30.9 ± 5.5	1440.0 ± 75.9	566
Gasteropelecidae			
C. strigata[4]	32.5 ± 6.4	1225.0 ± 74.5	467
Siluridae			
Pimelodes[4]	29.7 ± 7.3	1263.9 ± 82.2	509
Clariidae			
C. julii[4]	147.8 ± 38.3	3604.6 ± 449.9	430
Cichlidae			
Apistogramma A[4]	258.5 ± 46.1	1752.5 ± 181.2	126
Geophagus[4]	111.8 ± 30.9	1154.5 ± 135.5	175
S. jurupari[4]	276.7 ± 149.3	457.1 ± 102.9	33
P. scalare[2]	136.1 ± 66.4	428.0 ± 96.7	55
Potamotrygonidae			
Potamotrygon sp.[5]	468.0 ± 100.0	443.0 ± 67.0	19

*Rate of Na^+ uptake in water with a Na^+ concentration of $20\ \mu mol\ l^{-1}$ calculated using the kinetic parameters.

Sources: [1]Gonzalez and Preest, 1999; [2]Gonzalez and Wilson, 2001; [3]Gonzalez *et al.*, 1997; [4] Gonzalez *et al.*, 2002; [5]Wood *et al.*, 2002b.

low-pH waters. Along with high maximum transport capacities, all four species have very high affinities. In fact, neon tetras (*P. innesi*) have the lowest K_m value for Na^+ uptake recorded for a freshwater fish. Such high affinity and transport capacity ensure that the rate of uptake will be high even in extremely ion-poor water. In this regard, the tetras (*P. innesi, P. axelrodi, G. ternetzi, Hemigrammus*) are somewhat similar to North American yellow perch (Table 9.2), though maximum transport capacities are much higher in the tetras.

The ion transport mechanisms of characids possess several additional specializations. Perhaps the most impressive trait is seen in the con-generic neon tetras and cardinal tetras. Both species display Na^+ and Cl^- uptake systems that are completely insensitive to low pH (Figure 9.13). Kinetic parameters of Na^+ and Cl^- uptake measured after 24 hours at pH 3.5 in neon tetras are not different from parameters measured at pH 6.5 (Table 9.4). Even upon direct transfer of neon tetras from pH 6.5 to pH 3.25, which represents an instant, almost 2000-fold rise in ambient H^+ concentration, Na^+ uptake is completely unaffected. In addition, during low pH exposure when diffusive efflux is stimulated, ion uptake actually can up-regulate rapidly. Both neon and blackskirt tetras demonstrate this ability to quickly up-regulate Na^+ transport. Neon tetras are particularly impressive, being able to up-regulate within 6 hours of initiation of higher loss rates after direct transfer to pH 3.5. So not only is Na^+ transport completely resistant to low pH, but neon tetras also are able to elevate uptake quickly in response to increased ion losses. Such a high degree of pH insensitivity and rapid stimulation of Na^+ transport at low pH, neither of which have been observed before in any species of freshwater fish, are clearly adaptive for fishes that inhabit extremely acidic waters. Together, these specializations would act to prevent significant net losses of Na^+ in low pH water.

While the adaptive significance of pH insensitive ion transport in these fishes is clear, the underlying mechanism responsible for transport is more difficult to explain. Both of the current models for Na^+ transport, reviewed earlier (Figure 9.2), require the extrusion of H^+, either by H^+-ATPase or Na^+/H^+ exchange. One can perhaps envision how a reduced sensitivity could be brought about with these mechanisms. In the case of Na^+/H^+ exchange, a high Na^+ affinity would reduce competitive inhibition by H^+. The problem for the H^+-ATPase/Na^+ channel arrangement is more complicated. In order to take in Na^+, the H^+ pump needs to establish an apical polarization of about 58 mV for each log unit of pH gradient. At water pH 3.5, there are 3–4 log units of H^+ concentration from ionocyte cytoplasm pH to water pH. This means an exceptionally large apical potential of -174 to -232 mV is required for continued uptake.

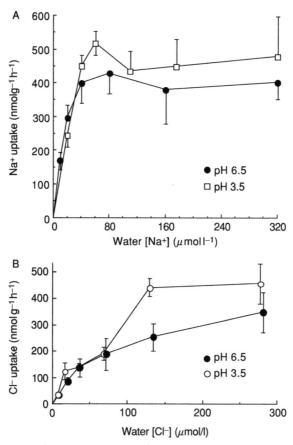

Fig. 9.13 Relationship between (A) water Na^+ concentration and rate of Na^+ uptake and (B) water Cl^- and rate of Cl^- uptake in neon tetras at pH 6.5 and 3.5. Values are means $\pm$ SE. (Data for Na^+ uptake from Gonzalez and Preest, 1999; data for Cl^- uptake from Preest, Gonzalez and Wilson, unpublished observations.)

In an effort to examine the nature of the Na^+ and Cl^- transport mechanisms of neon tetras, the most specialized species examined so far, we measured ion uptake in the presence of a range of pharmacological inhibitors that block different components of transport (M. Preest *et al.*, in press). Interestingly, neon tetras prove virtually completely insensitive to every blocker used. Even amiloride, a relatively non-specific blocker of both Na^+/H^+ exchangers and Na^+-channels, had very little effect on Na^+ transport. Exposure to $100\,\mu mol\,l^{-1}$, a concentration that inhibits Na^+ uptake in rainbow trout by more than 90% (Wilson *et al.*, 1994), inhibited uptake in neons by only 12%. Likewise, exposure to amiloride analogs, which are highly specific inhibitors of Na^+/H^+ exchangers (DMA, MIA, HMA, EIPA) and Na^+-channels (Benzamil, Phenamil), had virtually no effect on Na^+ transport. These findings suggest that tetras may possess a very different transport mechanism, one that does not involve extrusion of H^+ in some form. Alternatively, the Na^+ uptake mechanism that tetras possess may have evolved to the point that it is no longer sensitive to the blockers.

From these studies it appears that there are two distinct patterns of ion regulation in Rio Negro fishes. The first pattern, displayed by the three species of characids, is one of high rates of ion uptake and efflux, high affinity for Na^+ uptake, and strong resistance of both to effects of low pH. The alternate pattern, seen in the cichlid angelfish, is similar to banded sunfish, with very low intrinsic permeability that is resistant to stimulation by low pH but an unspecialized (acid-sensitive) ion uptake system. While the tetra pattern allows for "business as usual" even at low pH, the angelfish

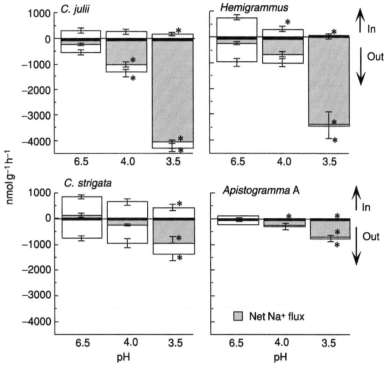

Fig. 9.14 Effect of water pH on Na^+ fluxes of four species of fish from the Rio Negro. Values are means ± SE. Asterisk indicates significant difference from pH 6.5. (Data from Gonzalez et al., 2002.)

pattern is designed for "waiting out" low pH exposure by greatly limiting loss rates.

The two patterns of ion regulation were confirmed by a survey of Na^+ fluxes of eight species of fish collected directly from the Rio Negro (Gonzalez *et al.*, 2002). Kinetic analysis of Na^+ transport revealed that while all eight species had similarly high transport capacities, the four cichlid species included had very low transporter affinities (high K_m) for Na^+, and a tetra (*Hemigrammus*), catfish (*Pimelodes*), and hatchet fish (*C. strigata*) had much higher affinities (Table 9.4). Thus it appears that low transporter affinity for Na^+ is a general characteristic of Rio Negro cichlids and suggests that it may have evolved in the group before colonization of the ion-poor waters of the river. At the same time, high affinity for Na^+ is a characteristic of characids and several other families.

Four of the eight species collected from the Rio Negro were also exposed to low pH (4 and 3.5) and they proved to vary in their tolerance (Figure 9.14). Based on degree of stimulation of Na^+ efflux, the most tolerant were hatchet fish (*C. strigata*) and a cichlid, *Apistogramma*. Despite the high level of acid tolerance, *Apistogramma*, like angelfish (*P. scalare*), was unable to take up Na^+ at low pH. In contrast, both hatchet fish and armored catfishes, *Corydoras*, had Na^+ transport systems that were very resistant to inhibition by low pH. Even at pH 3.5, Na^+ uptake was only mildly inhibited. Still, neither species exhibited pH insensitivity like the neon and cardinal tetras. Thus, so far we only have evidence that the pH insensitive Na^+ transport mechanism has evolved in the genus *Paracheirodon*.

VI. ADAPTATION TO ION-POOR, ACIDIC WATERS IN
RIO NEGRO ELASMOBRANCHS

An additional species collected directly from the Rio Negro was examined – the freshwater Amazonian stingray (*Potamotrygon sp.*). The potamotrygonid rays entered the Amazon at least 15 million years ago (Lovejoy, 1996), they are the only family of elasmobranch to speciate in freshwater (Lovejoy *et al.*, 1998), and several species inhabit even the extremely ion-poor, acidic waters of the Rio Negro. Unlike their marine and euryhaline relatives, *Potamotrygon* are ammonotelic not ureotelic (Gerst and Thorson, 1977), and regulate blood ions and urea at levels similar to freshwater teleosts (Griffith *et al.*, 1973; Bittner and Lang, 1980). Characterization of Na^+ uptake (Figure 9.15) revealed a transport system with very low affinity and only modest capacity (Table 9.4) that was strongly inhibited at pH 4.0 to <20% of control values (Wood *et al.*, 2002b). At circumneutral pH, Na^+ efflux was low, almost balancing the very low rate of influx (about,

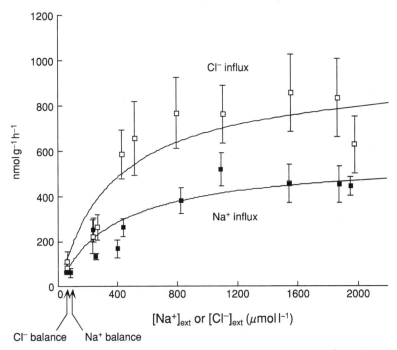

Fig. 9.15 Relationships between water ion concentrations and the rates of Na^+ and Cl^- uptake in potomotrygonid rays from the Rio Negro. Values are means ± SE. Note the higher exchange rates for Cl^-, reflecting a significant exchange diffusion component. Arrows indicate the balance points where influx and efflux rates would be equal. (Data from Wood *et al.*, 2002b.)

20–50 nmol g^{-1} h^{-1}), and did not increase at all at pH 4.0 in Rio Negro water. Therefore rays display an ionoregulatory pattern similar to cichlids but with even more pronounced characteristics (low uptake, low permeability resistant to low pH). Indeed, the kinetic analyses of influx and efflux suggested that these freshwater elasmobranchs are unable to achieve perfect ionic balance by gill mechanisms alone in these very dilute waters, such that uptake through the diet may be critical to survival.

Chloride exchange (Figure 9.15) was also measured in *Potamotrygon*, the only fish taken directly from the Rio Negro where this has been done (Wood *et al.*, 2002b). As discussed subsequently, the mechanism of Cl^- uptake, as well as the mechanism of Na^+ uptake, may well be different from those in teleosts. Nevertheless, it is reassuring to see that Cl^- fluxes exhibited the same general patterns as Na^+ fluxes, with very low affinity uptake and low efflux rates, with the latter being entirely resistant to pH 4.0. Interestingly though, both the influx and efflux rates of Cl^- were greater than those of

Na$^+$, reflecting a substantial exchange diffusion component in the former, such that a significant portion of Cl$^-$ efflux simply paralleled Cl$^-$ influx. This is probably an evolutionary remnant from the marine ancestry of these elasmobranchs (Bentley *et al.*, 1976).

Pharmacological analysis of ion transport in *Potamotrygon* (Wood *et al.*, 2002b) revealed that Na$^+$ influx was strongly inhibited by amiloride (10^{-4} M) and its analogs, with HMA (4×10^{-5} M), a preferential Na$^+$/H$^+$ exchange blocker, being slightly more potent than phenamil (4×10^{-5} M), a preferential Na$^+$ channel blocker. Chloride uptake was insensitive to stilbenes such as DIDS (10^{-4} M) and SITS (10^{-4} M), but both influx and efflux rates of Cl$^-$ were strongly inhibited by DPC (10^{-4} M) and thiocyanate (10^{-4} M). The latter data suggested that apical Cl$^-$ exchange occurred via a Cl$^-$ channel or an unusual anion exchange mechanism.

Overall, these results were in accord with a model recently proposed for the dasyatid rays, euryhaline cousins of the potamotrygonids, where immunocytochemistry indicated two types of gill ionocytes, one with abundant basolateral Na$^+$/K$^+$-ATPase, and the other with abundant basolateral H$^+$-ATPase (Piermarini and Evans, 2000, 2001). In the latter, an unusual anion exchanger (pendrin) has been immunolocalized to the apical membrane (Piermarini *et al.*, 2002). Piermarini and Evans (2001) and Piermarini *et al.* (2002) proposed that basolateral Na$^+$/K$^+$-ATPase energizes an apical Na$^+$/H$^+$ exchanger in one cell type, while basolateral H$^+$-ATPase energizes apical Cl$^-$ and HCO$_3^-$ movements via pendrin in the other. This scheme is entirely different from that discussed earlier (Figure 9.2) for "model" freshwater teleosts where H$^+$-ATPase is apical and energizes a Na$^+$ channel. However, recently, Katoh *et al.* (2003) reported that H$^+$-ATPase occurs on the basolateral side of the gill ionocytes in the killifish (*Fundulus heteroclitus*) adapted to ion-poor freshwater, a species which exhibits vigorous Na$^+$ uptake but negligible Cl$^-$ uptake in freshwater (Patrick *et al.*, 1997). Fenwick *et al.* (1999) reported that blockade of H$^+$-ATPase with bafilomycin inhibits Cl$^-$ uptake as well as Na$^+$ uptake in larval tilapia and carp. Clearly these reports indicate that a variety of different schemes may exist for ionoregulation by teleosts and elasmobranchs in dilute waters.

VII. THE ROLE OF ORGANIC MATTER IN ADAPTATION OF FISH TO ION-POOR, ACIDIC WATERS

The Rio Negro gets its name from its dark tea color, caused by a high content of dissolved organic matter (DOM) comprising humic, fulvic, and other organic acids plus assorted other compounds derived from the breakdown of jungle vegetation (Leenheer, 1980; Thurman, 1985; Ertel *et al.*, 1986; Kuchler *et al.*, 1994). Total concentrations may be as high as 30 mg C l^{-1}, and average about half of this value in the whole watershed. Blackwaters are widespread throughout the world in tropical areas and are notable for their low biological productivity. Janzen (1973) has speculated, largely on the basis of anecdotal evidence, that the plant "secondary compounds", many of which are phenolics with known bacteriostatic or insecticidal properties, may account for the generally low abundance and diversity of animal life in tropical blackwaters. However, in view of the very low nutrient, ion, and pH levels in these waters, it would be difficult to attribute these phenomena only to the toxic characteristics of the DOM. Further, the exceptionally high fish diversity of the Rio Negro argues against this interpretation. Certainly, it is possible that DOM components may be toxic to foreign species, but it seems likely that native species would have evolved mechanisms to co-exist with them. Indeed, Gonzalez *et al.* (1998, 2002) speculated that in native fish, the high DOM content of blackwaters might actually be protective against dilute and acidic conditions, especially in limiting gill permeability. Kullberg *et al.* (1993) had earlier predicted that the binding of DOM to biological surfaces would increase at low pH, because water acidity would reduce the negative charge on DOM. Campbell *et al.* (1997) provided direct evidence of increased DOM binding to both algae and isolated fish gill cells at low pH and speculated that it occurred by either hydrogen bonding or hydrophobic bonding. Both Kullberg *et al.* (1993) and Campbell *et al.* (1997) hypothesized that the resulting effects on membrane permeability could be either stabilizing or destabilizing, depending on the exact mechanism of the interaction.

In support of the stabilizing idea, the effects of low pH exposure in two teleost species were less severe in natural Rio Negro water than in an artificial medium made from distilled water with the same concentrations of Na^+, Cl^-, and Ca^{2+} but no DOM (Gonzalez *et al.*, 2002). Specifically, in the cichlid *Geophagus*, Na^+ influx was less inhibited by pH 3.75 and Na^+ efflux was completely resistant to pH 3.75 in Rio Negro water, but efflux increased greatly in synthetic water at the same pH with no DOM. In the silurid catfish, *Pimelodes*, similarly exposed to pH 3.75, this protective effect of blackwater was delayed, but seen upon recovery at pH 6.5.

Wood *et al.* (2003) further investigated these protective actions of DOM in the elasmobranch *Potamotrygon sp.* during exposure to pH 4.0 in either Rio Negro water or a natural water ("reference water") with similarly low ion levels but negligible DOM. Interestingly, blackwater appeared to have only slight influence at circumneutral pH, causing modest shifts in the influx and efflux kinetic relationships so that balance could occur at lower environmental Na^+ and Cl^- concentrations. However, at pH 4.0, the effects were pronounced. Blackwater exerted a dual protective action, providing modest support to Na^+ and Cl^- influxes so that they were significantly less inhibited by low pH, and more importantly, preventing the doubling of both Na^+ and Cl^- effluxes seen in reference water at low pH, as well as allowing faster recovery upon return to circumneutral pH (Figure 9.16). Furthermore, greatly increased ammonia excretion occurred at low pH in the presence of blackwater, but not in the presence of reference water. This may be of adaptive value during low pH exposure in alkalinizing the boundary layer next to the gill surface, as documented in trout by Playle and Wood (1989). Clearly, there is something about natural blackwater that is beneficial to stingrays as well as teleosts in tolerating periods of low pH, which are common in this dilute environment.

To a certain extent, the actions of blackwater appear to be similar to those of elevated Ca^{2+}. In reference water, a 10-fold elevation in Ca^{2+} (to $100 \, \mu mol \, l^{-1}$) completely protected the efflux components and allowed ammonia excretion to increase at pH 4.0, yet when the same amount of Ca^{2+} was added to Rio Negro water, there was no additional protective action above that already exerted by the blackwater (Wood *et al.*, 2003). Similarly, large increases in water Ca^{2+} concentration, which are normally very protective against the stimulatory effect of low pH on ion leakage in "model" teleosts (McDonald *et al.*, 1983; Wood, 1989) and in at least one Rio Negro native in reference water (tambaqui–Wood *et al.*, 1998), had no protective effect in three acid-tolerant teleost species exposed to low pH in their native Rio Negro water (Gonzalez *et al.*, 1998). Wood *et al.* (2003) performed simple biogeochemical modeling on these data and concluded that blackwater DOM does not act by preventing H^+ binding to the gills, and that DOM reduces but does not completely prevent Ca^{2+} binding to the gills at low pH. In sum, these results suggest that DOM may act by a similar mechanism to Ca^{2+} in stabilizing the gill epithelium at low pH. Clearly, it would be informative to measure the effects of blackwater and its constituents on transepithelial potential to see how similar the effects are to those of Ca^{2+} (cf. Figure 9.9).

Other "beneficial" features of DOM at low pH are its important contribution to water buffer capacity and its well-known ability to bind up potentially toxic metals, rendering them non-bioavailable (Leenheer, 1980; Thurman, 1985; Kullberg *et al.*, 1993; Kuchler *et al.*, 1994). Aluminum and iron levels are particularly high in the Rio Negro (Val and Almeida-Val, 1995), and these metals, especially aluminum, are highly toxic to fish at low pH (Wood, 1989). Based on geochemical modeling, Wood *et al.* (2003) concluded that measured levels of DOM in the Rio Negro were more than sufficient to prevent any aluminum or iron from binding to fish gill surfaces.

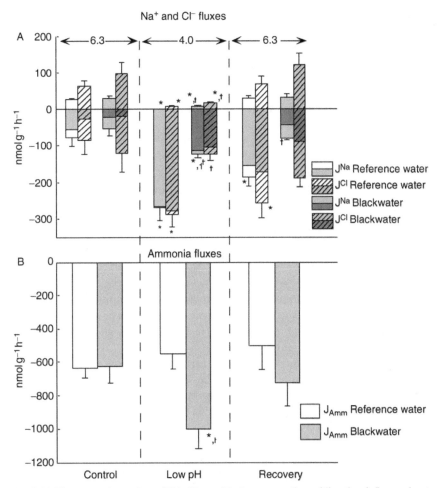

Fig. 9.16 The protective action of Rio Negro blackwater on the unidirection influx and net ammonia responses of potomotrygonid rays to exposure to pH 4.0 for 2 h. Upward bars are influxes, downward bars are effluxes, and solid bars are net fluxes. Reference water was a natural water with comparable low ion levels but negligible DOM. Values are means ± SE. Asterisk indicates significant difference from comparable values in the same water in the control period at circumneutral pH; dagger indicates significant difference for fish in reference water relative to comparable values for fish in blackwater during the same treatment period. Note the modest protective action of blackwater against ion influx inhibition by low pH, and marked protective action against ion efflux stimulation at low pH, whereas blackwater facilitated increased ammonia efflux at low pH. (Data from Wood *et al.*, 2003.)

Humic acid is often considered to represent DOM, but when a commercially prepared humic acid (derived from peat) was added to reference water at the same concentration as measured total DOM in blackwater, it was not protective, but rather greatly exacerbated ion losses at low pH in *Potamotrygon* (Wood *et al.*, 2003). However, Thurman (1985) has pointed out that humic acid derived from peat may have very different properties than humic acids of natural waters. Furthermore, humic acids make up only about 10% of the DOM in typical blackwater, while the smaller fulvic acids are much more numerous (40%), and fully half of the DOM is composed of hydrophilic acids, carboxylic acids, amino acids, carbohydrates, and other hydrocarbons, many of which remain uncharacterized. Conceivably, any or several of these fractions in combination could be the important protective agent.

VIII. AMMONIA EXCRETION IN ION-POOR, ACIDIC WATERS IN RIO NEGRO TELEOSTS

Ammonia exists in aqueous solutions as either ammonia gas (NH_3) or ionized ammonium ions (NH_4^+). The pK of the reaction linking these two forms in water ($NH_4^+ + H_2O \leftrightarrow NH_3 + H_3O^+$) is approximately 9.5 (Boutilier *et al.*, 1984), high enough above the pH of fish body fluids (plasma ~7.8) to maintain the gaseous form as the minor fraction ($<5\%$) within the circulation. In addition, the circulating levels of total ammonia ($T_{Amm} = [NH_3] + [NH_4^+]$) in fish are usually very low (0.05–0.5 mM) due to the rapid excretion across the gills into aquatic environment. Despite the low total ammonia and NH_3 partial pressures in particular, it has frequently been assumed that the permeability of gill cell membranes to NH_3 is sufficient to allow the majority of ammonia excretion across the gills of freshwater fish by simple gaseous diffusion (see Chapter 8). On the other hand, more than 95% of the total ammonia in the circulation exists as NH_4^+, and a direct link between Na^+ uptake and NH_4^+ excretion via an apical Na^+/NH_4^+ exchange mechanism has been debated ever since it was first postulated by Krogh (1939). The idea of this exchange is intuitively attractive as it utilizes the excretion of an endogenous toxic nitrogenous waste product as a counter ion in the absorption of essential Na^+ ions from the environment. Numerous studies since Krogh have attempted to confirm or rule out a direct link between sodium uptake and ammonia excretion. However, despite efforts over more than six decades, it still remains controversial whether the majority of ammonia excretion across the gills of freshwater fish is driven solely by diffusion of NH_3 gas down a partial pressure gradient or via a flexible combination involving apical and/or basolateral exchange of endogenous

NH_4^+ ions for Na^+ (Cameron and Heisler, 1983; Wright and Wood, 1985; Balm et al., 1988; Wood, 1993; Wilson et al., 1994; Wilkie, 1997; Salama et al., 1999; Kelly and Wood, 2002).

Fish living in permanently acidic and ion-poor waters offer an interesting tool for comparative physiologists to study this controversy. On the one hand, the average acidity in the Rio Negro, being more than 4 pH units below the ammonia/ammonium pK, will result in almost negligible NH_3 partial pressures in the ambient water, thus serving as a perfect sink for excretion via NH_3 diffusion from blood to water. If this provides the primary efflux pathway, one would predict that ammonia excretion in acidophilic fishes is a simple story with no particular limitations and the rate being governed solely by the partial pressure of NH_3 in plasma. On the other hand, the requirement for Na^+ uptake in the face of very high ambient H^+ concentrations would make utilizing Na^+/NH_4^+ exchange an intriguing proposition for by-passing the limitations of typically acid-sensitive Na^+ uptake mechanisms (Na^+/H^+ exchange or H^+-ATPase linked Na^+ channels). Acid-tolerant neon tetras represent a particularly useful model species for studying these mechanisms. Experimenters have frequently studied modes of ammonia excretion by manipulating the transbranchial PNH_3 gradients through changes in the ambient pH (and hence the relative quantities of NH_3 and NH_4^+ in the external water). However, this can be complicated by the pH-sensitivity of Na^+ uptake in most freshwater fishes. Thus, interpreting changes in ammonia excretion at low pH are complicated by the potential for combined effects of altered NH_3 diffusion gradients and/or inhibition of Na^+ uptake (and the potential for subsequent NH_4^+ exchange). The acid-insensitivity of the Na^+ uptake mechanism in neon tetras negates this complication and should allow a clearer interpretation of such experiments.

At face value, the acute responses of ammonia excretion in neon tetras to decreased and increased pH are straightforward and as predicted (Figure 9.17). Acidification of the external water from pH 6.0 to pH 5.0 resulted in an almost 2-fold stimulation of ammonia excretion (Wilson, 1996), whereas raising the ambient pH from 6.8 to 7.9 resulted in a small (12%) inhibition (R. W. Wilson, D. L. Snellgrove and D. M. Scott, unpublished data). Qualitatively, this follows the expected trend based upon alterations of the NH_3 fraction in the external water; i.e. increased or decreased transbranchial PNH_3 gradients at low and high water pH, respectively. However, on closer inspection the magnitude of these results is actually far from expected. The normal PNH_3 in the blood plasma of teleost fish is between 50 and 100 μTorr (Cameron and Heisler, 1983; Wright and Wood, 1985; Wilson et al., 1994). With an average total ammonia concentration of $\sim 11 \mu M$ in the experimental water, the PNH_3 concentration in the bulk

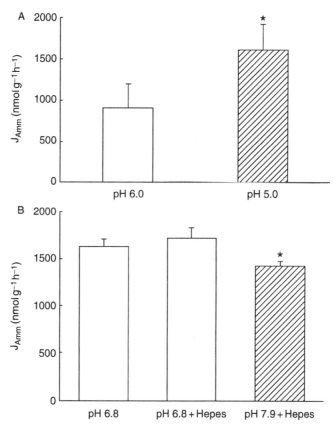

Fig. 9.17 (A) The effect of acute (1 h) acid exposure (pH 5.0) on ammonia excretion in neon tetras held in artificial ion-poor water (nominal NaCl = 50 μM, CaCl$_2$ = 50 μM; data from Wilson, 1996). (B) The effect of acute (1 h) exposures to 5 mM Hepes-buffered water at either pH 6.8 (same pH as unbuffered control water) or pH 7.9 on ammonia excretion in neon tetras held in artificial ion poor water (nominal NaCl = 50 μM, CaCl$_2$ = 50 μM); Wilson, Snellgrove and Scott, unpublished data). Asterisk indicates significant difference from control flux at pH 6.0 or 6.8 + Hepes.

water under control conditions (pH 6.0) should already be extremely low (<0.2 μTorr), and essentially not limiting to NH$_3$ diffusion across the gills. A further 10-fold reduction in bulk water PNH$_3$ (to <0.02 μTorr) would therefore not be expected to have a measurable effect on ammonia excretion and yet a very large increase is observed. A similar stimulatory effect on ammonia excretion was observed in another Rio Negro inhabitant, the tambaqui (*Colossoma macropomum*) when acutely transferred from pH 5.0 to 4.0 (Wilson *et al.*, 1999). This is even more surprising given the minute background levels and changes in bulk water PNH$_3$ when pH is changed from 5 to 4. Conversely, increasing pH from 6.8 to 7.9 (Figure 9.17B) would cause the bulk water PNH_3 to increase from ~ 1 to $15\,\mu$Torr, and substantially reduce the transbranchial PNH_3 gradient, yet ammonia excretion was only inhibited by 12%. These results would only be compatible with NH_3 diffusion as the primary mechanism of excretion if neon tetras have dramatically lower plasma PNH_3 levels and gill NH_3 permeabilities than other teleosts, and/or much higher concentrations of total ammonia in the gill boundary layer compared to the bulk water.

The data in Figure 9.17B reveal a further departure from the normal teleost in neon tetras. It is well established that when inspired water is >pH 6, excretion of CO_2 and/or H^+ across the gills creates a relatively acidic boundary layer adjacent to the gill surface (Playle and Wood, 1989). Under normal conditions this localized acidification facilitates ammonia excretion by lowering the PNH_3 in the boundary layer thus enhancing the transbranchial PNH_3 gradient (Wilson et al., 1994). The fact that ammonia excretion in neon tetras was unaffected by buffering the boundary layer to pH 6.8 (Figure 9.17B) suggests that boundary layer acidification either does not occur, or plays little role in facilitating ammonia excretion in these fish.

Yet another deviation from the standard freshwater teleost model arises from examination of the sensitivity of ammonia excretion to external Na^+ concentrations and conversely, the sensitivity of Na^+ influx to external NH_4^+ concentrations, in neon tetras (Figure 9.18). Ammonia excretion was unaffected by a reduction of external Na^+ concentration from $64\,\mu$M to close to the K_m for Na^+ uptake ($12\,\mu$M) (Figure 9.18A; Wilson, 1996). In contrast, Na^+ uptake was inhibited by 27% when external NH_4^+ concentration was raised from 11 to $500\,\mu$M (Figure 9.18B). Water pH was kept constant at 6.5 during the latter experiment, so ambient PNH_3 would also be elevated but only to $\sim 25\,\mu$Torr, in theory insufficient to reverse NH_3 diffusion across the gills if plasma PNH_3 in neons is similar to other freshwater teleosts (see above). It is therefore tempting to speculate that this inhibition of Na^+ uptake was due to a direct effect of elevated external NH_4^+ on apical Na^+/NH_4^+ exchange rather than the indirect effect of inward NH_3 diffusion (and subsequent alkalinization of either gill ionocytes or blood plasma) upon Na^+ uptake linked to H^+ excretion.

Given the remarkable insensitivity of Na^+ uptake in neon tetras to ambient acidity and the standard pharmacological blockers, the possibility of alternatives to the standard model of Na^+ uptake in exchange for H^+ excretion requires further investigation. A direct link between Na^+ uptake and ammonia excretion via either apical or basolateral Na^+/NH_4^+ exchange cannot be ruled out although the dependency of such an exchange system may not work in both directions (i.e. ammonia excretion may help drive some Na^+ uptake, but not vice versa). With the recent reports of Rhesus

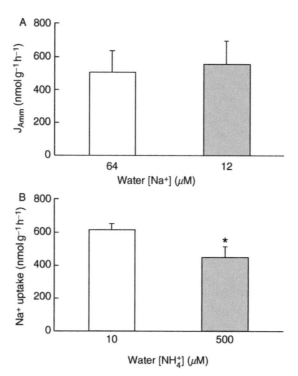

Fig. 9.18 (A) The effect of an acute (1 h) reduction in the ambient sodium concentration (from 64 μM to 12 μM) on ammonia excretion in neon tetras held in artificial ion poor water (data from Wilson, 1996). Note that water [Na⁺] was reduced to a value that is close to the K_m for Na⁺ uptake in neon tetras. (B) The effect of an acute (1 h) increase in the ambient NH₄⁺ concentration (from 10 μM to 500 μM) on unidirectional sodium uptake rate in neon tetras held in artificial ion-poor water (see Figure 9.17 for nominal ion concentrations; Wilson, Snellgrove and Scott, unpublished data). Water pH was maintained constant throughout at pH 6.5 at which the ambient NH₃ fraction will never exceed 25 μTorr. Asterisk represents significant difference from the control flux.

proteins acting as high affinity ammonium transporters in micro-organisms, plants, and animals (Marini *et al.*, 2000) and of novel active mechanisms for active ammonium ion excretion in various fish and crustaceans (Randall *et al.*, 1999; Weihrauch *et al.*, 2002), it is clear that the traditional views of ammonia excretion (and active ion uptake) need to be questioned. With the atypical properties of these systems in fishes like the neon tetra, further study of these intriguing animals should lead to the emergence of new ideas that help explain the ionoregulatory mechanisms underlying the adaptation to acidic, ion-poor waters as well as in freshwater fishes in general.

IX. FUTURE DIRECTIONS

We have only just begun to scratch the surface in understanding the ionoregulatory adaptations of tropical fishes to ion-poor waters, which are often acidic, and often rich in DOM. The following should prove to be fruitful areas for future research:

1. The acquisition of Ca^{2+} for normal growth would appear to be a critical problem for these fishes, in view of the very low Ca^{2+} concentrations in the water (Table 9.1) and the ability of DOM to reduce its bioavailability. To date, water-borne Ca^{2+} uptake has been measured only in *Potamotrygon*, where it was completely inhibited by exposure to pH 4.0 (Wood *et al.*, 2002b). A thorough investigation of Ca^{2+} metabolism in relevant teleosts is needed.

2. Chloride is the other major electrolyte in blood plasma besides Na^+, and it is almost as low in the water (Table 9.1). However, the only species in which Cl^- uptake has been measured are neon tetras, and *Potamotrygon*. The mechanisms of Cl^- uptake and loss should be assessed in comparable detail to those of Na^+ transport in relevant teleosts, especially since we now have several documented examples of temperate teleosts essentially lacking the capability for Cl^- uptake at the gills (*Anguilla* – Hyde and Perry, 1987; *Fundulus* – Patrick *et al.*, 1997).

3. In light of the above, and reports that some Rio Negro teleosts and an elasmobranch cannot achieve ionic balance by transport of water-borne ions alone at the very low Na^+ and Cl^- concentrations normally present in the water (Gonzalez *et al.*, 2002; Wood *et al.*, 2002b), the importance of food in ion balance should be investigated. Indeed food as a possible source of electrolytes for blackwater fishes has been discussed by several authors (e.g. Janzen, 1973; Val and Almeida-Val, 1995), but never critically assessed. The importance of food as a source of electrolytes for trout under sub-lethal acid stress in dilute environments has been highlighted by the work of D'Cruz and Wood (1998).

4. Virtually nothing is known about the other side of osmoregulation, water balance, in fish of this region, or about kidney function. Early reports indicated that reabsorption of electrolytes from the urine might be exceptionally efficient (Mangum *et al.*, 1978; Cameron and Wood, 1978), but there have been no systematic studies. Studies addressing water permeability and renal function, and particularly the effects of water Ca^{2+}, DOM, and pH thereon, are needed.

5. A challenge in future research will be to work out the exact mechanism(s) by which DOM exerts its protective effects on ionoregulation in native species, which particular fractions of the DOM are specifically responsible, and whether the same effects are seen in foreign species which do not normally encounter blackwater DOM. Particular attention should be devoted to its apparent Ca^{2+}-like effect in stabilizing membrane permeability.

6. All studies to date have examined ionoregulatory physiology in these fishes under well-aerated laboratory conditions. However, in nature, these fishes encounter frequent bouts of environmental hypoxia, often complicated by environmental hypercapnia to the point that many of these species have evolved supplementary air breathing or surface-skimming adaptations (Val and Almeida-Val, 1995; see Chapters 6 and 7). Ion transport is costly, and one might predict that the greatest impact of oxygen limitation would occur in the gill tissue, which is directly exposed to the external water. Investigations of the interactive effects of hypoxia and hypercapnia on ionoregulation in these species would be of great interest.

7. The explosion of molecular techniques in the last decade has greatly illuminated, and simultaneously complicated, our understanding of ionoregulatory physiology in non-tropical species (e.g. Lin *et al.*, 1994; Sullivan *et al.*, 1995; Wilson *et al.*, 2000; Piermarini and Evans, 2000, 2001; Piermarini *et al.*, 2002; Katoh *et al.*, 2003). The time is now ripe to apply techniques such as immunocytochemical localization for identification and western blotting for quantification of specific transporters, and in situ hybridization and semiquantitative northern blotting or real time PCR for quantification of transporter mRNA levels, to the ionoregulatory organs of fishes living in these unusual tropical environments. In particular we should focus on the remarkable mechanism whereby Na^+ and Cl^- uptake can be completely insensitive to water pHs as low as 3.25 in at least two *Paracheirodon* species.

8. Many of the above questions can be addressed within a phylogenetic context, which will yield valuable information regarding how physiological specializations for extremely dilute, acidic waters evolved in these fishes. For example, ion transport in close, and not so close, relatives of neon and cardinal tetras can be examined to determine whether pH insensitive transport arose once in an ancestor of the two species, in an even earlier ancestor within the genus, or if it is much more widely found, possibly arising multiple times in the evolutionary history of this group. Such information will enrich our understanding of how physiological traits evolve.

ACKNOWLEDGMENTS

Some of the work reviewed here was supported by NSERC (Canada) grants to CMW, and by Royal Society Research Grants to RWW. CMW is supported by the Canada Research Chair Program. Our warmest thanks to Dal and Vera for attracting us to these interesting problems and graciously hosting us over the years. Also thanks to the numerous students in the lab of Dal and Vera for their technical and not so technical support.

REFERENCES

Audet, C., and Wood, C. M. (1988). Do rainbow trout (*Salmo gairdneri*) acclimate to low pH? *Can. J. Fish. Aquat. Sci.* **45,** 1399–1405.

Avella, M., and Bornancin, M. (1989). A new analysis of ammonia and sodium transport through the gills of the freshwater rainbow trout (*Salmo gairdneri*). *J. Exp. Biol.* **142,** 155–175.

Balm, P., Goosen, N., van der Rijke, S., and Wendelaar Bonga, S. (1988). Characterization of transport Na^+-ATPases in gills of freshwater tilapia. Evidence for branchial $Na^+/H^+(NH_4^+)$, ATPase activity in fish gills. *Fish Physiol. Biochem.* **5,** 31–38.

Bentley, P. J., Maetz, J., and Payan, P. (1976). A study of the unidirectional fluxes of Na^+ and Cl^- across the gills of the dogfish *Scyliorhinus canicula* (Chondrichthyes). *J. Exp. Biol.* **64,** 629–637.

Bittner, A., and Lang, S. (1980). Some aspects of osmoregulation of Amazonian freshwater stingrays (*Potamotrygon hystrix*). I. Serum osmolality, sodium and chloride content, water content, hematocrit, and urea level. *Comp. Biochem. Physiol.* **67A,** 9–13.

Boutilier, R. G., Heming, T. A., and Iwama, G. K. (1984). Physicochemical parameters for use in fish respiratory physiology. *In* "Fish Physiology" (Hoar, W. S., and Randall, D. J., Eds.), Vol. 10A, pp. 403–430. Academic Press, New York.

Bury, N. R., and Wood, C. M. (1999). Mechanism of branchial apical silver uptake by rainbow trout is via the proton-coupled Na^+ channel. *Am. J. Physiol.* **277,** R1385–R1391.

Cameron, J. N., and Heisler, N. (1983). Studies of ammonia in the trout: Physico-chemical parameters, acid-base behaviour, and respiratory clearance. *J. Exp. Biol.* **105,** 107–125.

Cameron, J. N., and Wood, C. M. (1978). Renal function and acid-base regulation in two Amazonian Erythrinid fishes: *Hoplias malabaricus*, a water-breather, and *Hoplerythrinus unitaeniatus*, a facultative air-breather. *Can. J. Zool.* **56,** 917–930.

Campbell, P. G. C., Twiss, M. R., and Wilkinson, K. J. (1997). Accumulation of natural organic matter on the surfaces of living cells: Implications for the interaction of toxic solutes with aquatic biota. *Can. J. Fish. Aquat. Sci.* **54,** 2543–2554.

D'Cruz, L. M., and Wood, C. M. (1998). The influence of dietary salt and energy on the response to low pH in juvenile rainbow trout. *Physiol. Zool.* **71,** 642–657.

Dunson, W. A., Swarts, F., and Silvestri, M. (1977). Exceptional tolerance to low pH of some tropical blackwater fish. *J. Exp. Zool.* **201,** 157–162.

Ehrenfeld, J., Garcia-Romeau, F., and Harvey, B. J. (1985). Electrogenic active proton pump in *Rana esculenta* skin and its role in sodium ion transport. *J. Physiol. Lond.* **359,** 331–355.

Ertel, J. R., Hedges, J. I., Devol, A. H., Richey, J. E., and de Nazare Goes Ribeiro, M. (1986). Dissolved humic substances of the Amazon River system. *Limn. Oceanogr.* **31,** 739–754.

Evans, D. H., Piermarini, P. M., and Potts, W. T. W. (1999). Ionic transport in the fish gill epithelium. *J. Exp. Zool.* **283,** 641–652.

Fenwick, J. C., Wendelar-Bonga, S. E., and Flik, G. (1999). *In vivo* bafilomycin-sensitive Na^+ uptake in young freshwater fish. *J. Exp. Biol.* **202,** 3659–3666.

Freda, J., and McDonald, D. G. (1988). Physiological correlates of interspecific variation in acid tolerance in fish. *J. Exp. Biol.* **136**, 243–258.

Furch, K. (1984). Water chemistry of the Amazon basin: The distribution of chemical elements among freshwaters. *In* "The Amazon. Limnology and Landscape Ecology of a Mighty Tropical River and Its Basin" (Sioli, H., Ed.), pp. 167–199. W. Junk, Dordrecht.

Gerst, J. W., and Thorson, T. B. (1977). Effects of saline acclimation on plasma electrolytes, urea excretion, and hepatic urea biosynthesis in a freshwater stingray, *Potamotrygon sp.* Garman 1877. *Comp. Biochem. Physiol.* **56A**, 87–93.

Gonzalez, R. J., and Dunson, W. A. (1987). Adaptations of sodium balance to low pH in a sunfish (*Enneacanthus obesus*) from naturally acidic waters. *J. Comp. Phys.* **157**, 555–566.

Gonzalez, R. J., and Dunson, W. A. (1989a). Acclimation of sodium regulation to low pH and the role of calcium in the acid-tolerant sunfish *Enneacanthus obesus. Physiol. Zool.* **62**, 977–992.

Gonzalez, R. J., and Dunson, W. A. (1989b). Differences in low pH tolerance among closely related sunfish of the genus *Enneacanthus. Env. Biol. Fishes.* **26**, 303–310.

Gonzalez, R. J., and Preest, M. (1999). Mechanisms for exceptional tolerance of ion-poor, acidic waters in the neon tetra (*Paracheirodon innesi*). *Physiol. Biochem. Zool.* **72**, 156–163.

Gonzalez, R. J., and Wilson, R. W. (2001). Patterns of ion regulation in acidophilic fish native to the ion-poor, acidic Rio Negro. *J. Fish Biol.* **58**, 1680–1690.

Gonzalez, R. J., Wood, C. M., Wilson, R. W., Patrick, M., Bergman, H., Narahara, A., and Val, A. L. (1998). Effects of water pH and Ca^{2+} concentration on ion balance in fish of the Rio Negro, Amazon. *Physiol. Zool.* **71**, 15–22.

Gonzalez, R. J., Dalton, V. M., and Patrick, M. L. (1997). Ion regulation in ion-poor, acidic water by the blackskirt tetra (*Gymnocorymbus ternetzi*) a fish native to the Amazon River. *Physiol. Zool.* **70**, 428–435.

Gonzalez, R. J., Wood, C. M., Wilson, R.W, Patrick, M. L., and Val, A. L. (2002). Diverse strategies of ion regulation in fish collected from the Rio Negro. *Physiol. Biochem. Zool.* **75**, 37–47.

Goss, G. G., Perry, S. F., and Laurent, P. (1995). Ultrastructural and morphometric studies on ion and acid-base transport processes in freshwater fish. *In* "Cellular and Molecular Approaches to Fish Ionic Regulation, Fish Physiology" (Wood, C. M., and Shuttleworth, T. J., Eds.), Vol. 14, pp. 257–289. Academic Press, London.

Graham, J. H., and Hastings, R. W. (1984). Distributional patterns of the sunfish on the New Jersey coastal plain. *Env. Biol. Fishes.* **10**, 137–148.

Griffith, R. W., Pang, P. K. T., Srivastava, A. K., and Pickford, G. E. (1973). Serum composition of freshwater stingrays (Potamotrygonidae) adapted to fresh and dilute sea water. *Biol. Bull.* **144**, 304–320.

Hunn, J. B. (1985). Role of calcium in gill function in freshwater fishes. *Comp. Biochem. Physiol.* **82A**, 543–547.

Hyde, D. A., and Perry, S. F. (1987). Acid-base and ionic regulation in the american eel (*Anguilla rostrata*) during and after prolonged aerial exposure: Branchial and renal adjustments. *J. Exp. Biol.* **133**, 429–447.

Isaia, J. (1984). Water and non-electrolyte permeation. *In* "Fish Physiology" (Hoar, W. S., and Randall, D. J., Eds.), Vol. 10B, pp. 1–38. Academic Press, New York.

Janzen, D. H. (1973). Tropical blackwater rivers, animals, and mast fruiting by the Dipoterocarpaceae. *Biotropica* **6**, 69–103.

Katoh, F., Hyodo, S., and Kaneko, T. (2003). Vacuolar-type proton pump in the basolateral plasma membrane energizes ion uptake in branchial mitochondria-rich cells of killifish *Fundulus heteroclitus* adapted to a low ion environment. *J. Exp. Biol.* **206**, 793–803.

Kelly, S. P., and Wood, C. M. (2002). The cultured branchial epithelium of the rainbow trout as a model for diffusive fluxes of ammonia across the fish gill. *J. Exp. Biol.* **204**, 4115–4124.

Kirschner, L. B. (1994). Electrogenic action of calcium on crayfish gill. *J. Comp. Physiol. B.* **164**, 215–221.

Kirschner, L. B. (1997). Extrarenal mechanisms in hydromineral and acid-base regulation in aquatic vertebrates. *In* "The Handbook of Physiology" (Dantzler, W. H., Ed.), pp. 577–622. American Physiological Society, Bethesda, MD.

Krogh, A. (1939). "Osmotic Regulation in Aquatic Animals." Cambridge University Press, London.

Kuchler, I. L., Miekely, N., and Forsberg, B. (1994). Molecular mass distributions of dissolved organic carbon and associated metals in waters from Rio Negro and Rio Solimoes. *Sci. Total Environ.* **156**, 207–216.

Kullberg, A., Bishop, K. H., Hargeby, A., Janssen, M., and Petersen, R. C. (1993). The ecological significance of dissolved organic carbon in acidified waters. *Ambio.* **22**, 331–337.

Leenheer, J. A. (1980). Origin and nature of humic substances in the waters of the Amazon River Basin. *Acta Amazonica* **10**, 513–526.

Lin, H., and Randall, D. J. (1991). Evidence for the presence of an electrogenic proton pump on the trout gill epithelium. *J. Exp. Biol.* **161**, 119–134.

Lin, H., and Randall, D. J. (1993). H^+-ATPase activity in crude homogenates of fish gill tissue: Inhibitor sensitivity and environmental and hormonal regulation. *J. Exp. Biol.* **180**, 163–174.

Lin, H., and Randall, D. J. (1995). Proton pumps in fish gills. *In* "Cellular and Molecular Approaches to Fish Ionic Regulation, Fish Physiology" (Wood, C. M., and Shuttleworth, T. J., Eds.), Vol. 14, pp. 229–255. Academic Press, London.

Lin, H., Pfeiffer, D. C., Vogl, A. W., Pan, J., and Randall, D. J. (1994). Immunolocalization of H^+-ATPase in the gill epithelia of rainbow trout. *J. Exp. Biol.* **195**, 169–183.

Lovejoy, N. R. (1996). Systematics of myliobatoid elasmobranches: With emphasis on the phylogeny and historical biogeography of the neotropical freshwater stingrays (Potamotrygonidae: Rajiformes). *Zool. J. Linn. Soc.* **117**, 207–257.

Lovejoy, N. R., Bermingham, E., and Martin, A. P. (1998). Marine incursions into South America. *Nature* **396**, 421–422.

Madara, J. L. (1988). Tight junction dynamics: Is paracellular transport regulated? *Cell* **53**, 497–498.

Maetz, J. (1973). Na^+/NH_4^+, Na^+/H^+ exchanges and NH_3 movement across the gill of *Carassius auratus*. *J. Exp. Biol.* **58**, 255–275.

Maetz, J., and Garcia-Romeau, F. (1964). The mechanism of sodium and chloride uptake by the gills of a fresh-water fish, *Carassius auratus*. *J. Exp. Biol.* **58**, 255–275.

Mangum, C. P., Haswell, M. S., Johansen, J., and Towle, D. W. (1978). Inorganic ions and pH in the body fluids of Amazon animals. *Can. J. Zool.* **56**, 907–916.

Marini, A. M., Matassi, G., Raynal, V., Andre, B., Cartron, J. P., and Cherif-Zahar, B. (2000). The human Rhesus-associated RhAG protein and a kidney homologue promote ammonium transport in yeast. *Nature Genet.* **26**, 341–344.

Marshall, W. S. (1995). Transport processes in isolated teleost epithelia: Opercular epithelium and urinary bladder. *In* "Cellular and Molecular Approaches to Fish Ionic Regulation, Fish Physiology" (Wood, C. M., and Shuttleworth, T. J., Eds.), Vol. 14, pp. 1–23. Academic Press, London.

Marshall, W. S. (2002). Na^+, Cl^-, Ca^{2+}, and Zn^{2+} transport by fish gills: Retrospective review and prospective synthesis. *J. Exp. Zool.* **293**, 264–283.

McDonald, D. G. (1983a). The interaction of environmental calcium and low pH on the physiology of the rainbow trout, *Salmo gairdneria*. I. Branchial and renal net ion and H$^+$ fluxes. *J. Exp. Biol.* **102**, 123–140.

McDonald, D. G. (1983b). The effects of H$^+$ upon the gills of freshwater fish. *Can. J. Zool.* **61**, 691–703.

McDonald, D. G., and Wood, C. M. (1981). Branchial and renal acid and ion fluxes in the rainbow trout, *Salmo gairdneri*, at low environmental pH. *J. Exp. Biol.* **93**, 101–118.

McDonald, D. G., Hobe, H., and Wood, C. M. (1980). The influence of environmental calcium on the physiological responses of the rainbow trout, *Salmo gairdneri*, to low environmental pH. *J. Exp. Biol.* **88**, 109–131.

McDonald, D. G., Walker, R. L., and Wilkes, R. L. K. (1983). The interaction of environmental calcium and low pH on the physiology of the rainbow trout, *Salmo gairdneri*. II. Branchial inoregulatory mechanisms. *J. Exp. Biol.* **102**, 141–155.

McWilliams, P. G. (1982). The effects of calcium on sodium fluxes in the brown trout, *Salmo trutta*, in neutral and acid media. *J. Exp. Biol.* **96**, 439–442.

McWilliams, P. G., and Potts, W. T. W. (1978). The effects of pH and calcium concentration on gill potentials in the brown trout, *Salmo trutta*. *J. Comp. Physiol.* **126**, 277–286.

Milligan, C. L., and Wood, C. M. (1982). Disturbances in haematology, fluid volume distribution and circulatory function associated with low environmental pH in the rainbow trout, *Salmo gairdneri*. *J. Exp. Biol.* **99**, 397–415.

Morgan, I. J., Potts, W. T. W., and Oates, K. (1994). Intracellular ion concentration in branchial epithelial cells of brown trout (*Salmo trutta* L.) determined by X-ray microanalysis. *J. Exp. Biol.* **194**, 139–151.

Packer, R. K., and Dunson, W. A. (1970). Effects of low environmental pH on blood pH and sodium balance of brook trout. *J. Exp. Zool.* **174**, 65–72.

Packer, R. K., and Dunson, W. A. (1972). Anoxia and sodium loss associated with death of brook trout at low pH. *Comp. Biochem. Physiol.* **41A**, 17–26.

Patrick, M. L., and Wood, C. M. (1999). Ion and acid-base regulation in the freshwater mummichog (*Fundulus heteroclitus*): A departure from the standard model for freshwater teleosts. *Comp. Biochem. Physiol.* **122A**, 445–456.

Patrick, M. L., Pärt, P., Marshall, W. S., and Wood, C. M. (1997). The characterization of ion and acid-base transport in the freshwater-adapted mummichog (*Fundulus heteroclitus*). *J. Exp. Zool.* **279**, 208–219.

Perry, S. F. (1997). The chloride cell: Structure and function in the gills of freshwater fishes. *Ann. Rev. Physiol.* **59**, 325–347.

Piermarini, P. M., and Evans, D. H. (2000). Effects of environmental salinity on Na$^+$/K$^+$-ATPase in the gills and rectal gland of a euryhaline elasmobranch (*Dasyatis sabina*). *J. Exp. Biol.* **203**, 2957–2966.

Piermarini, P. M., and Evans, D. H. (2001). Immunochemical analysis of the vacuolar proton-ATPase B-subunit in the gills of a euryhaline stingray (*Dasyatis sabina*): Effects of salinity and relation to Na$^+$/K$^+$-ATPase. *J. Exp. Biol.* **204**, 3251–3259.

Piermarini, P. M., Verlander, J. W., Royaux, I. E., and Evans, D. H. (2002). Pendrin immunoreactivity in the gill epithelium of a euryhaline elasmobranch. *Am. J. Physiol.* **283**, R983–R992.

Playle, R. C., and Wood, C. M. (1989). Water chemistry changes in the gill microenvironment of rainbow trout: Experimental observations and theory. *J. Comp. Physiol. B.* **159**, 527–537.

Potts, W. T. W. (1984). Transepithelial potentials in fish gills. *In* "Fish Physiology" (Hoar, W. S., and Randall, D. J., Eds.), Vol. 10B, pp. 105–128. Academic Press, New York.

Potts, W. T. W. (1994). Kinetics of sodium uptake in freshwater animals: A comparison of ion exchange and proton pump hypotheses. *Am. J. Physiol.* **266**, R315–R320.

Preest, M., Gonzalez, R. J., and Wilson, R. W. (2005). A pharmacological examination of the Na^+ and Cl^- transport mechanisms in freshwater fish. *Physiol. Biochem. Zool.,* in press.

Randall, D. J., Wilson, J. M., Peng, K. W., Kok, T. W. K., Kuah, S. S. L., Chew, S. F., Lam, T. J., and Ip, Y. K. (1999). The mudskipper, *Periophthalmodon schlosseri*, actively transports NH_4^+ against a concentration gradient. *Am. J. Physiol. Physiol. – Reg. Int. Comp. Physiol.* **277**, R1562–R1567.

Reid, S. D. (1995). Adaptation to and effects of acid water on the fish gill. *In* "Biochemistry and Molecular Biology of Fishes" (Hochachka, P. W., and Mommsen, T. P., Eds.), Vol. 5, pp. 213–227. Elsevier, Amsterdam.

Saint-Paul, U. (1984). Physiological adaptations to hypoxia of a neotropical characoid fish *Colossoma macropomum*, Serrasalmidae. *Environ. Biol. Fishes.* **11**, 53–62.

Salama, A., Morgan, I. J., and Wood, C. M. (1999). The linkage between Na^+ uptake and ammonia excretion in rainbow trout: Kinetic analysis, the effects of $(NH_4)_2SO_4$ and NH_4HCO_3 infusion and the influence of gill boundary layer pH. *J. Exp. Biol.* **202**, 697–709.

Smith, N. F., Talbot, C., and Eddy, F. B. (1989). Dietary salt intake and its relevance to ionic regulation in freshwater salmonids. *J. Fish Biol.* **35**, 749–753.

Sullivan, G. V., Fryer, J. N., and Perry, S. F. (1995). Immunolocalization of proton pumps (H^+-ATPase) in pavement cells of rainbow trout gill. *J. Exp. Biol.* **198**, 2619–2629.

Thurman, E. M. (1985). Organic Geochemistry of Natural Waters. Martinus Nijhoff/W. Junk, Dordrecht.

Val, A. L., and de Almeida-Val, V. M. F. (1995). Fishes of the Amazon and Their Environment. Springer, Berlin.

Walker, I., and Henderson, P. A. (1996). Ecophysiological aspects of Amazonian blackwater litterbank fish communities. *In* "Physiology and Biochemistry of Fishes of the Amazon" (Val, A. L., Almeida-Val, V. M. F., and Randall, D. J., Eds.), pp. 7–22. Instituto Nacional de Pesquisas da Amazonia, Manaus.

Weihrauch, D., Ziegler, A., Siebers, D., and Towle, D. W. (2002). Active ammonia excretion across the gills or the green shore crab *Carcinus maenas:* Participation of Na^+/K^+-ATPase, V-type H^+-ATPase and functional microtubules. *J. Exp Biol.* **205**, 2765–2775.

Wetzel, R. G. (1983). Limnology. Saunders College Publishing, Philadelphia.

Wilkie, M. (1997). Mechanisms of ammonia excretion across fish gills. *Comp. Biochem. Physiol.* **118A**, 39–50.

Wilson, J. M., Laurent, P., Tufts, B. L., Benos, D. J., Donowitz, M., Vogl, A. W., and Randall, D. J. (2000). NaCl uptake by the branchial epithelium in freshwater teleost fish: An immunological approach to ion-transport protein localization. *J. Exp. Biol.* **203**, 2279–2296.

Wilson, R. W. (1996). Ammonia excretion in fish adapted to an ion-poor environment. *In* "Physiology and Biochemistry of Fishes of the Amazon" (Val, A. L., Almeida-Val, V. M. F., and Randall, D. J., Eds.), pp. 123–128. Instituto Nacional de Pesquisas da Amazonia, Manaus.

Wilson, R. W., Wood, C. M., Gonzalez, R. J., Patrick, M. L., Bergman, H., Narahara, A., and Val, A. L. (1999). Net ion fluxes during gradual acidification of extremely soft water in three species of Amazonian fish. *Physiol. Biochem. Zool.* **72**, 277–285.

Wilson, R. W., Wright, P. M., Munger, S., and Wood, C. M. (1994). Ammonia excretion in freshwater rainbow trout (*Oncorynchus mykiss*) and the importance of gill boundary layer acidification: Lack of evidence for Na^+/H^+ exchange. *J. Exp. Biol.* **191**, 37–58.

Wood, C. M. (1989). The physiological problems of fish in acid waters. *In* "Acid Toxicity and Aquatic Animals" (Morris, R., Brown, D. J. A., Taylor, E. W., and Brown, J. A., Eds.), pp. 125–152. Society for Experimental Biology Seminar Series, Cambridge University Press, Cambridge.

Wood, C. M. (1993). Ammonia and urea excretion and metabolism. *In* "The Physiology of Fishes," pp. 379–425. CRC Press, New York.

Wood, C. M., and LeMoigne, J. (1991). Intracellular acid-base responses to environmental hyperoxia and normoxic recovery in rainbow trout. *Resp. Physiol.* **86,** 91–113.

Wood, C. M., Kelly, S. P., Zhou, B., Fletcher, M., O'Donnell, M., Eletti, B., and Pärt, P. (2002a). Cultured gill epithelia as models for the freshwater fish gill. *Biochim. Biophys. Acta Biomembranes* **1566,** 72–83.

Wood, C. M., Matsuo, A. Y. O., Gonzalez, R. J., Wilson, R. W., Patrick, M. L., and Val, A. L. (2002b). Mechanisms of ion transport in *Potamotrygon*, a stenohaline freshwater elasmobranch native to the ion-poor blackwaters of the Rio Negro. *J. Exp. Biol.* **205,** 3039–3054.

Wood, C. M., Matsuo, A. Y. O., Wilson, R. W., Gonzalez, R. J., Patrick, M. L., Playle, R. C., and Val, A. L. (2003). Protection by natural blackwater against disturbances in ion fluxes caused by low pH exposure in freshwater stingrays endemic to the Rio Negro. *Physiol. Biochem. Zool.* **76,** 12–27.

Wood, C. M., Wilson, R. W., Gonzalez, R. J., Patrick, M. L., Bergman, H. L., Narahara, A., and Val, A. L. (1998). Responses of an Amazonian teleost, the tambaqui (*Colossoma macropomum*) to low pH in extremely soft water. *Physiol. Zool.* **71,** 658–670.

Wright, P. A., and Wood, C. M. (1985). An analysis of branchial ammonia excretion in the freshwater rainbow trout: Effects of environmental pH changes and sodium uptake blockade. *J. Exp. Biol.* **114,** 329–353.

10

METABOLIC AND PHYSIOLOGICAL ADJUSTMENTS TO LOW OXYGEN AND HIGH TEMPERATURE IN FISHES OF THE AMAZON

VERA MARIA F. DE ALMEIDA-VAL
ADRIANA REGINA CHIPPARI GOMES
NÍVIA PIRES LOPES

I. INTRODUCTION

Adaptations of organisms to long- and short-term environmental changes are one of the basic concepts of evolution (Futuyma, 1986; Pigliucci, 1996; Rose and Lauder, 1996; Hochachka and Somero, 2002). These adaptations involve genetic changes that will result in either metabolic/physiological

The Physiology of Tropical Fishes: Volume 21
FISH PHYSIOLOGY

adjustment to short-term changes (e.g. gene regulation of LDH isoforms), or in changes at population and species levels (e.g. overall down-regulation of energy metabolism). During the evolution, individuals must cope with short- and long-term variations of the same physical parameters, i.e., temperature, pressure, and oxygen. In most cases, functional responses involve adjustments in metabolic processes which depend on the genetic make-up and may, in addition, result in anatomical and morphological variation (Almeida-Val, Val *et al.*, 1999). Evolutionary changes rely on genetic mutation and selection (in the broad sense), but a quantitative assessment of genetic variation alone fails to consider the phenotype range of variation of any given genotype (Schichting and Pigliucci, 1993, 1995). Thus, these two adaptation processes are interdependent: metabolic adaptation and (long-term) genetic changes will alter different spectra – the spectrum of selection is altered by physiological changes, and the spectrum of physiological and metabolic patterns will be altered by genetic mutation over evolutionary time (Walker, 1983; Walker, 1997; Almeida-Val, Val *et al.*, 1999). Currently, the interplay between metabolic and genetic adaptation may be the reflection of gene regulation processes: regulatory loci directly respond to specific environmental stimuli by triggering a specific series of "changes" (Pigliucci, 1996) and, in consequence, induce metabolic adjustments during the transcriptional phase. Subsequently, other changes in metabolism, which may be post-transcriptional, may take place to allow the fine adjustments that allow for a perfect interaction between organisms and environment. In fact, the discovery of regulation of many genes reconciles the apparent paradox between "unity *versus* diversity" suggested by Hochachka (1988), in which the relative constancy of chemical structure opposes with genotypic and phenotypic diversity within and between species.

The adaptation of Amazonian fishes to warm and hypoxic waters is of special interest, primarily because of the number of commercially valuable fish that can die in the course of a day as a consequence of occasional cold fronts that break into the Amazon during the southern winter. These cold fronts cause a turnover between the more oxygen-rich, cooled down surface water and warmer, oxygen-depleted deep waters in *várzea* (floodplain) lakes and *igapós* (flooded forests) during the periods of high water levels (Junk *et al.*, 1983; Val and Almeida-Val, 1995; Almeida-Val, Val *et al.*, 1999). Historically, floodplains of the Amazon were deforested for agricultural production by Pre-Colombian natives and later by European immigrants. Thus, this phenomenon seems to be the result of habitat destruction by humans since under intact forest canopies, no (or only insignificant) cold-front fish kills occur, because temperature differences between surface and deep water are less pronounced. However, this phenomenon is not exclusive from Amazonian *várzea* lakes and may also occur in other kinds of water bodies, in which thermoclines occur.

Amazonian ecosystems are specially suited for studies of individual metabolic adaptations in relation to genetic speciation, because along with tremendous species richness (estimated *ca.* 3000 species of fish, see Chapter 1, this volume – after Roberts, 1972; Böhlke *et al.*, 1978; Rapp-Py-Daniel and Leão, 1981), there are vast areas of pristine habitats in an "equatorial hot climate", with open lakes partly covered by aquatic macrophytes, and inundation of high-canopy forests that are subject to drastic diurnal and seasonal changes of oxygen levels.

Earlier studies showed that long- and short-term changes in oxygen are both determinants of fish distribution in Amazonian water bodies and that hypoxia tolerance is particularly common in fishes of the Amazon. This fact, along with associated tropical temperature regimes, has driven fish species through a series of adjustments at different levels of biological organization, e.g., ethological, morphological, anatomical, physiological, metabolic, and molecular, which are combined to produce phenotypic plasticity that allows them to adapt to the pulsating nature of the basin (Junk *et al.*, 1983; Val and Almeida-Val, 1995; Almeida-Val and Farias, 1996).

II. ENVIRONMENTAL CHALLENGES

Past and present patterns of oxygen availability in Amazonian water bodies may be seen as putative escalators of evolutionary selective pressure. Hypoxic and anoxic conditions were prevalent in the aquatic environment during the Cambrian period, owing to the low atmospheric oxygen levels at that time. At present, the poorly oxygenated waters of the Amazon basin result from different phenomena. Thus, since the Cambrian geological period, oxygen depletion has been a limiting factor for aquatic life in general (Randall *et al.*, 1981; Almeida-Val and Farias, 1996; Almeida-Val, Val *et al.*, 1999). South America and Africa appeared after Gondwanaland broke up in the southern hemisphere during the Cretaceous. During the Tertiary period the Andean Mountains were folded up on the western part of South America, inducing remarkable changes in the Amazon basin. The Pacific drainage of the upper tributaries of the Amazon River was cut, and the whole Amazon basin became oriented towards the Atlantic. Consequently, a completely new set of habitats became available, and seasonal oscillations in river water levels became the main driving force of the Amazon basin (reviewed by Val and Almeida-Val, 1995). Flood pulses (Junk *et al.*, 1989), i.e., annual oscillations of water levels, produce an average crest of 10 m between November and June in central Amazonian (see Chapter 1, this volume).

These flood pulses inundate a large area and make several new habitats available to fish in the flooded forest (*igapó*) and in the floodplain areas (*várzea*). Such flood pulses also cause the appearance and disappearance of many other aquatic formations such as *paranás* (channels), *igarapés* (small streams), and beaches. During low water levels, the receding water leaves behind small discrete water bodies, i.e., temporary lakes, while during high water levels, these water bodies are all interconnected. Changes in chemical, physical, and biological parameters occur, and such predictable inundations affect virtually all living organisms of the Amazon.

Although attention must be drawn to the oscillations in dissolved oxygen in water, investigations of temperature effects on metabolism in Amazonian fish are unavoidable. The main reason for this is that these fish spend their life cycles at high temperatures and have reorganized their metabolism in different organs to acclimate to these conditions, as well as undergone adjustments at the level of the whole organism, e.g., differential metabolic rates and differential scaling patterns. Water temperatures in the Amazon basin typically range from ~25 to 32 °C, although under some conditions temperatures may be higher (Kramer *et al.*, 1978). This is in marked contrast to water in the north-temperate zone, which approaches 0 °C in winter and does not exceed 25 °C in summer. A life cycle spent at low temperatures is often associated with elevated activities of key enzymes of energy metabolism. In contrast, acclimation of fish to high temperatures always results in lower activity levels of enzymes associated with aerobic energy (Guderley and Gawlicka, 1992; Johnston *et al.*, 1985; Jones and Sidell, 1982; Way-Kleckner and Sidell, 1985). Furthermore, there is also a propensity for aerobic oxidation to be enhanced, especially the use of fatty acids, to meet energy demands at low temperatures (West *et al.*, 1999). In addition to temperature differences, fishes living in the Amazon drainage basin may cope with lower levels of oxygen. The bottom of the water column in the Amazon *várzea* lakes and *igapós* may be hypoxic or even anoxic, and dissolved oxygen can reach 0 ppm (mg/l) at night and oversaturated values during the following day in floodplain areas (Almeida-Val, Val *et al.*, 1999; Val, 1996). In general, Amazonian fishes deal with these conditions through metabolic depression, anaerobic metabolism, surface-skimming, air-breathing, or some combination of these adaptations (Almeida-Val *et al.*, 1993; Almeida-Val and Hochachka, 1995). Aerial respiration and a high anaerobic potential are not mutually exclusive; there are occasions when air breathing is inappropriate and the solution may be anaerobic energy production. Furthermore, the combination of anaerobic power and metabolic suppression, as evidenced by other good anaerobes (Hochachka, 1980) can also be found in Amazonian fishes (Almeida-Val *et al.*, 2000).

This chapter will focus on the main effects of environmental changes on energy metabolism of fishes of the Amazon. To accomplish this goal, we will review both the effects of natural environmental changes and discuss some of the metabolic responses of fish to different thermal regimes (temperature changes) and oxygen depletion (naturally occurring hypoxia and anoxia). This chapter will also consider the differential scaling properties of metabolism, and transcriptional and post-transcriptional changes in enzymes with key metabolic functions, such as LDH, in response to the main aquatic parameters, which are considered of great importance in regard to fish metabolism and aquatic ecosystems of the Amazon: temperature and oxygen. As will become evident throughout this chapter, it is not always possible to deal with each parameter separately, since aquatic ecosystems of the Amazon comprise a mosaic of situations and environmental conditions. This impairs considerations of just one phenomenon by itself and necessitates a more integrative treatment of many of the issues that will be addressed in the following sections. Chapter 9 deals with adaptations to acidic and ion-poor water conditions, these being other important environmental challenges in the Amazon basin.

III. EFFECTS OF TEMPERATURE ON FISH METABOLISM

Hochachka and Somero (2002) stated that "biogeographic patterning indicates that temperature is a major determinant of habitat suitability." In fact, distribution patterns of organisms reflect temperature gradients or discontinuities and can be observed in both aquatic and terrestrial habitats, spatially and temporally, influencing all types of organisms. Species replacement can be seen with changes in latitude as well along vertical gradients in temperature at given latitude; good examples are temperature changes in the transitions from subtidal to intertidal marine habitats and from low to high elevations in mountainous regions. Some aquatic animals typically show diurnal vertical migrations and select the appropriate time of the day for foraging. The effects of temperature on organisms are universal among different species, but vary according to the thermal regime to which the species is acclimated. Basically, temperature affects every aspect of the organism's physiology, and consequently imposes strict limits on where life can occur (Hochachka and Somero, 1984, 2002; Prosser, 1991). The dynamics of temperature changes and their consequences differ between temperate climates and tropical regions, particularly in aquatic ecosystems. As water temperature from temperate regions rises towards the upper tolerance limits, the animals must cope with decreased dissolved oxygen and increased oxygen demand due to elevated metabolic rates and, consequently, elevated maintenance costs. Thus, regulated physiological and metabolic parameters such as oxygen consumption rates, blood and tissue oxygenation, acid-base status and cellular energy levels may show substantial changes before harmful effects occur (Pörtner, 1993; Pörtner and Grieshaber, 1993; Sartoris *et al.*, 2003). In tropical regions, the dynamics of temperature oscillations at different times of the day and throughout the year are different. As mentioned above, temperatures may drop overnight and increase during the day in *várzea* lakes and *igapós*. However, oxygen levels do not change inversely, as occurs in temperate lakes. At night, when temperature drops, other phenomena take place, and the water column may become completely anoxic (Val and Almeida-Val, 1995; Almeida-Val, Val *et al.*, 1999; Chippari Gomes, 2002). Therefore, the imbalance that occurs between energy consumption and oxygen-dependent energy production in temperate fishes at extreme temperatures may be aggravated in tropical species due to the complexity of changes in physical-chemical water parameters.

Most fishes are organisms whose body temperatures conform to the temperature of the aquatic environment and are thus considered ectotherms (Hochachka and Somero, 1973, 1984, 2002). Ectotherms lack anatomical and physiological means for maintaining a thermal gradient between the external medium and the body. As stated by Hochachka and Somero (2002), a primary source of the difficulty faced by aquatic ectotherms in avoiding thermal equilibration with their medium is the requirement for gas exchange at respiratory surfaces. Metabolically produced heat is lost rapidly at respiratory surfaces during uptake of O_2 and elimination of CO_2 and other waste products, such as ammonia. This phenomenon may be less pronounced in tropical fishes due to higher medium temperatures. However, other environmental challenges may impose energy expenditures at the respiratory surfaces, e.g. low pH, ion-poor waters, and diurnal oxygen depletion (see Chapters 6, 7 and 8 for details on this issue). Thus, Amazonian fishes face other problems regarding temperature, since many water- and air-breathing fishes depend on the water surface layer to breathe, which imposes increased exposure to higher temperatures as well as higher radiation levels. For ectotherms, rates of respiration, feeding, growth, and locomotion are strongly influenced by changes in environmental temperature on both daily and on a seasonal basis. Such effects of temperature on rates of biological activity can be quantified by determining the specific temperature coefficient (Q_{10}) of a process, i.e., the effect that a change in $10\,°C$ will have on the rate being measured. As repeatedly described in the literature for many processes such as rates of respiration and enzymatic activity, Q_{10} values near 2.0 or slightly higher are observed when thermal effects are studied within the normal ranges of body temperatures. Outside this range, Q_{10} values may deviate sharply from 2.0 (reviewed by Hochachka and Somero, 2002). For fishes, the temperature coefficients of metabolic rates lie mostly within a range of 0.05 to 0.10, which corresponds to Q_{10} values of 1.65 to 2.70 (Jobling, 1994).

The effects of temperature on fish metabolism have been studied extensively, and a vast body of literature exists regarding metabolic thermal compensation, which quantifies the effects of temperature on rates of oxygen consumption by differently adapted and differently acclimated ectotherms (see Pauly, 1998; Hölker, 2003 for different reviews on this subject). According to Hölker (2003), the patterns generated by growth-related processes, such as mortality, reproduction and rate of food consumption, can be explained by acclimation temperatures, i.e., latitudinal fish distribution. Reviewing the FishBase 98 for several parameters, Hölker (2003) stated that tropical fishes, which, on average, live in a range of *ca.* 20–30 °C, occupy higher trophic levels and consume more food than their colder-water counterparts. This author suggested that, on average, tropical fishes tend to be smaller than temperate fishes, and that metabolic rates will be higher in warm-water fishes. However, studies have revealed a different picture for Amazonian fish species.

Among fishes of the Amazon, metabolic rates of exclusively water-breathing species vary along a spectrum from "sluggish"- to "athletic"-type behavior patterns. As expected, comparisons between Amazon and temperate fish species suggest that the more sluggish a fish is, the less oxygen is consumed per unit weight (reviewed in Val and Almeida-Val, 1995). Regardless of the high acclimation temperatures, Amazon fishes will experience a suppressed aerobic capacity (Almeida-Val and Hochachka, 1995; Driedzic and Almeida-Val, 1996; West *et al.*, 1999), and this is directly reflected in their metabolic rates (Figure 10.1). A simple comparison among three sets of data obtained with respiration rates (whole organism oxygen consumption) as a function of increased body mass was plotted according to the allometric relationship $VO_2 = aM^b$, where a = log mass coefficient, M = log-body mass, and b = mass exponent. Variations of b were adopted as described by Wootton (1990) for fishes in general, including Antarctic, temperate, and tropical fishes; by Hammer and Purps (1996) for tropical fishes, particularly facultative air-breathing species; and for the anoxia-tolerant Oscar (*Astronotus ocellatus*), as described in Almeida-Val *et al.* (2000). Figure 10.1 shows a clear trend towards metabolic suppression rates in fishes as they become naturally acclimated nearer the Equator. These data can be interpreted as being temperature-related or as being an adaptive characteristic to typically hypoxic waters of the tropics. In either case, there is a good reason to believe that a metabolic gradient of adaptation occurs and may be biogeographically linked to latitude.

According to Hochachka and Somero (2002), the temperature-linked biogeographic patterning found in nature is a clear manifestation not only

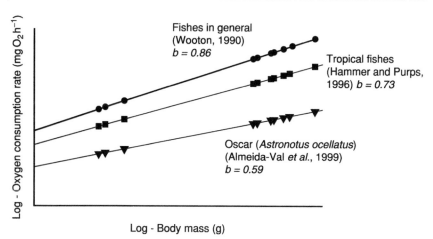

Fig. 10.1 Relationship between body mass (g) and mass-specific oxygen uptake (specific routine metabolic rate – $VO_2 = mg\,O_2\,g^{-1}\,h^{-1}$) from *Astronotus ocellatus* (Almeida-Val, Paula-Silva *et al.*, 1999), fishes in general (Wootton, 1990), and tropical fishes (Hammer and Purps, 1996). Values were determined by measuring oxygen consumption in a completely sealed chamber and recording the change in oxygen concentration within the chamber over a period of time (approximately 2 hours) as described by Almeida-Val *et al.* (1999). Values refer to each individual and are plotted against its respective body mass, according to the allometric relationship described as follows: $VO_2 = aM^b$, where: a = log mass coefficient, M = log-body mass, and b = mass exponent. Comparison of b exponents from *Astronotus* with other groups of fishes reveals a decrease in the reliance on aerobic metabolism as the animal grows when compared with b exponents obtained for fishes in general and tropical fishes. There are two possible (although not exclusive) explanations for this finding: (i) *A. ocellatus* retains a higher ability in suppressing its metabolic rate as an evolutionary adaptive response to the chronically hypoxic environment; and (ii) this is an adaptation to different thermal regimes over evolutionary time. See text for more details.

of the pervasive effects of temperature on all organisms, but also of the success of organisms in adapting to thermal variation. In fact, investigations of activity levels of enzymes of energy metabolism in heart and muscle of Amazonian teleosts in comparison with north-temperate teleosts revealed that the general differences between these two groups of fishes are in fact due to their thermal history (West *et al.*, 1999).

IV. ENZYME LEVELS REFLECT THE NATURAL HISTORY OF FISH

Genes for glycolytic enzymes are thought to be regulated by being linked to common inducing or repressing signals, during long-term evolutionary processes (Hochachka *et al.*, 1996). Thus, the capacity of glycolytic regulation of organisms will be indicated by enzyme levels of organs and tissues, and will reflect life style and respiration patterns of the organism. In an earlier review, we compared the metabolic profiles of Amazon fishes to those of temperate species and suggested that fish of the Amazon generally show down-regulated enzymes, whereas an up-regulation of relative glycolytic capacity can be observed in both anaerobic and aerobic tissues, regardless of respiration type (Almeida-Val and Hochachka, 1995). Comparing rates of fish heart enzyme activities of anaerobic and oxidative metabolism, i.e., the ratio of lactate dehydrogenase to citrate synthase (LDH/CS), north-temperate teleosts show lower anaerobic power than tropical fishes, regardless of their respiratory type (reviewed by Almeida-Val, Val *et al.*, 1999). Besides the fact that such different patterns occur in different species, one might hypothesize that phenotypic and gene regulatory plasticity may have preceded speciation during the evolution of these species, thus allowing for down-regulation of enzyme levels in tropical fish. To better assess this hypothesis, it is necessary to review some comparative studies performed on Amazonian and north-temperate teleosts using enzyme activity levels of energy metabolism of different tissues, in particular, the heart (Driedzic and Almeida-Val, 1996; West *et al.*, 1999).

Among the Amazonian fishes, the activity levels of enzymes required for anaerobic glycolytic metabolism do not correlate with the ability of ventricular strips to maintain force development in the face of an anoxic challenge (Bailey *et al.*, 1999; West *et al.*, 1999). Furthermore, the activity levels of these enzymes (i.e. phosphofructokinase – PFK, pyruvate kinase – PK, and LDH) are not higher in the water-breathing, anoxia-tolerant oscar (*Astronotus ocellatus*) and tambaqui (*Colossoma macropomum*) relative to the facultative air-breathers acari-bodó (*Liposarcus pardalis*) and tamoatá (*Hoplosternun littorale*) and the obligate air-breather pirarucu (*Arapaima gigas*). The activity levels of hexokinase (HK), which catalyzes the first step in glucose breakdown, was higher in the two above-mentioned armored catfish species, which showed a greater capacity to maintain ventricular-strip contraction under cyanide poisoning than did the oscar and tambaqui, which failed (Bailey *et al.*, 1999). Thus, this data set revealed that enzyme activity neither presented a consistent pattern that could be associated with anaerobic glycolysis and whole-animal hypoxia tolerance, nor with the ability of isolated hearts to maintain performance under conditions of impaired oxidative phosphorylation. Similar results were achieved in red muscle, which will deserve a special section further. Because levels of enzymes associated with anaerobic glycolysis showed no elevated values in hearts with high resistance to anoxia in those studies, the final conclusion was that the general differences in enzyme activities between Amazonian and north-temperate teleosts, i.e., higher levels in the latter group, are most likely due to their thermal histories. Here, long-term changes in environmental temperature have driven the evolutionary changes among groups of fishes. Enzyme activity levels are often higher in Antarctic species than in north-temperate teleosts (Crockett and Sidell, 1990; Driedzic, 1992) and within north-temperate teleosts when they are acclimated to low as opposed to high temperature (Jones and Sidell, 1982; Johnston *et al.*, 1985; Way-Kleckner and Sidell, 1985; Guderley and Gawlicka, 1992). The enzyme activity levels reported by West *et al.* (1999) were determined at a common assay temperature of about 25 °C, which is 10 °C higher than the acclimation temperature of north-temperate teleosts. Their enzyme activity levels are typically 1.5- to 2.2-fold higher in hearts and 3- to 4-fold higher in red muscle than those from Amazonian species. Q_{10} values for the activity of the enzymes are highly variable with respect to enzyme and species (Crockett and Sidell, 1990; Bailey *et al.*, 1991; Sephton and Driedzic, 1991). According to West *et al.* (1999), if the *in vitro* enzyme activity levels of north-temperate teleosts were decreased by *ca.* 2-fold to bring them into the physiological range they were acclimated, the resultant activity levels would be similar in the two groups of animals and they should therefore have the same metabolic capacity in their tissues at their respective acclimation temperatures. Thus, thermal regimes and fish life history are both determinants of enzyme activity levels: the lower the thermal regime of a species, the higher the expression of many of the enzymes associated with energy metabolism.

The above discussion refers to mean activity levels of enzymes involved in energy metabolism in the heart of fishes from different thermal zones, regardless of phylogeny and respiration types among species (Driedzic and Almeida-Val, 1996; West *et al.*, 1999). As mentioned above, a short literature review on metabolic characteristics of air- versus water-breathers revealed similar results that can be summarized as follows: (i) enzyme-level adjustments in air-breathing fishes are lower in terms of their absolute enzyme activity levels in pathways of both aerobic and anaerobic metabolism, and (ii) glycolytic rates are up-regulated in relation to oxidative tissue capacities of air-breathers, which varies on a tissue basis (Almeida-Val and Hochachka, 1995).

However, after reviewing those absolute enzyme values we suggested, based on a tissue-by-tissue comparison, the heart is a metabolic "hot spot" in many Amazonian air-breathers (Almeida-Val and Hochachka, 1995). Hearts of the species *Lepidosiren paradoxa* (Amazonian lungfish), which are very oxidative and have a high mitochondrial content, contrast with those of *Arapaima gigas* (pirarucu), whose hearts show lower oxidative power. However, while *A. gigas* hearts are less oxidative than lungfish hearts, they maintain an impressively high oxidative potential in their muscles, compared to muscles of most water-breathers (reviewed by Almeida-Val and Hochachka, 1995). Although lungfishes are phylogenetically specialized compared to teleosts, they show similar metabolic patterns when compared to the advanced teleost *Symbranchus marmoratus* (the South American swamp eel, or muçum), which has a similar life style and burrows into the mud during the dry season. Both species sustain high anaerobic potential in their hearts, with LDH activity levels up to 5-fold higher than those of most mammalian hearts and higher than most fish hearts. Heart LDH activities in *Arapaima gigas* and *Osteoglossum bicirrhosun* (aruanã) are comparable to tuna heart LDH levels, which are half of lungfish levels (reviewed by Almeida-Val and Hochachka, 1995). The reason for such high LDH levels in hearts of lungfish and muçum, different from other interpretations, where LDH is viewed just as an anaerobic power enzyme, is thought to be related to their recovery from estivation, since their heart type kinetics (LDH-B_4 orthologs – isoform predominant in vertebrate heart muscles), or high pyruvate inhibition rates, and higher trends towards lactate–pyruvate reversible conversion, will guarantee an efficient lactate back-conversion when oxygen becomes available upon arousal from estivation. In fact, during estivation, *Lepidosiren paradoxa* shows a suppression of metabolic characteristics, slowing down LDH levels in the heart from >1000 units per gram of wet tissue when awake (Hochachka and Hulbert, 1978) to <100 units per gram of wet tissue when estivating (Mesquita-Saad *et al.*, 2002). Thus, one may conclude that the substrate preference of the heart and the muscle is carbohydrate, not lipid-based (Hochachka, 1979; Hochachka and Hulbert, 1978).

Although air-breathing fishes have certain characteristics that resemble the biochemical strategies of diving aquatic mammals, mainly with regard to relative hypometabolism during diving (Burggren *et al.*, 1985; Dunn *et al.*, 1983) or estivation (Mesquita-Saad *et al.*, 2002), the preferential use of oxygen-efficient carbohydrate metabolism, which appears to be fundamental for some air-breathing fishes, may become less significant when fat fuels become more important under particular conditions. At this point, it is obvious that substrate (fuel) preferences deserve some attention.

V. FUEL PREFERENCES IN TROPICAL VERSUS TEMPERATE FISHES

In most fishes, cardiac ATP production under aerobic conditions is usually supported by a mixed catabolism of exogenous glucose and fatty acids. Studies on fuel availability, performance of isolated hearts, rates of oxygen consumption and $^{14}CO_2$ production by intact hearts and isolated mitochondria, along with *in vitro* enzyme activity levels support this idea (Driedzic, 1992; Driedzic and Gesser, 1994). In many species of north-temperate teleosts, seasonal decreases in temperature from about ~15–25 °C in summer to ~0–5 °C in winter result in an enhancement of aerobic-based fatty acid metabolism in heart (Way-Kleckner and Sidell, 1985; Sephton and Driedzic, 1991; Bailey and Driedzic, 1993). An extension of this characteristic can be seen in Antarctic fishes, which spend their life cycle at 0 °C. Crockett and Sidell (1990) and Sidell and Crockett (1995) described a strong reliance of cardiac energy metabolism on fatty acids associated with substantially higher *in vitro* activities of mitochondrial enzymes. A comparative analysis of activity levels of enzymes required for the use of glucose and glycogen in anaerobic metabolism (Table 10.1) suggests that within the Amazonian fishes, the enzymes PK and LDH do not differ between facultative air- and water-breathing species. However, the activity level of HK is higher, which may be correlated with the use of glucose as fuel for aerobic-based metabolism in the heart. According to Bailey *et al.* (1991), high activity levels of HK are consistent with the indirect arguments that extracellular glucose is used by fish heart as a metabolic fuel under oxygen limitation. Driedzic and Bailey (1999) suggested that hearts of fish have the ability to maintain high levels of performance in the absence of oxidative metabolism. According to these authors, under oxygen-limiting conditions, lactate is produced and, similar to other tissues, intracellular glycogen is mobilized (Dunn *et al.*, 1983; van Waarde *et al.*, 1983; Driedzic, 1988). However, extracellular glucose is also critical in extending heart viability. Although intracellular glucose levels are generally maintained at very low levels in fish hearts, during periods of oxygen limitation, an increase in glucose in the heart was described in lungfish (Dunn *et al.*, 1983), in goldfish (Shoubridge and Hochachka, 1983) and in the small Amazon cichlid (Almeida-Val and Farias, 1996), achieving levels similar to blood. HK may be an important rate-controlling site to achieve the right energy production under anaerobiosis. The results obtained by (West *et al.*, 1999) with regards to enzyme activity levels in the hearts of Amazonian fish species, combined with the results obtained by Bailey *et al.* (1999) on the cardiac performance of isolated hearts submitted to anoxia, using the same species as study models, have revealed that the ability of heart strips of the armored catfish *Hoplosternum littorale* (tamoatá) to recover from cyanide poisoning may be related to translocation of glucose transporters from intracellular sites to the plasma membrane. This in turn allows a greater uptake of glucose and support of the energy demands under anoxic conditions, e.g., cyanide poisoning (Driedzic and Bailey, 1999). In subsequent studies, no other armored catfish showed similar levels for HK (Lopes, 2003), suggesting that tamoatá deserves further attention regarding fuel preference studies and anoxia tolerance in their tissues and organs.

Another point that emerges from this discussion is that, regardless of temperature acclimation and the differential means of enzyme activity levels,

Table 10.1
Enzyme Activity Levels in Heart Muscle of Temperate and Tropical Fish Species

Species	HK	PK	LDH	HOAD	CS	References
Temperate fishes						
Perca flavescens	23.5 ± 2.4	158.7 ± 9.9	255 ± 11	7.6 ± 0.7	26.2 ± 3	West et al., 1999
Oncorhynchus mykiss	25.6 ± 1.5	71.2 ± 2.6	307 ± 18	17.8 ± 2.3	47.8 ± 3.1	West et al., 1999
Anguilla rostrata	14.7 ± 0.5	90.0 ± 0.7	726 ± 55	10.3 ± 0.5	38.9 ± 3.0	West et al., 1999
Myxine glutinosa	1.70	35.9	114.4	1.78	6.92	Driedzic, 1988
Squalus acanthias	3.81 ± 0.18	8.85 ± 0.2	—	—	21.12 ± 0.8	Sidell and Driedzic, 1985
Gadus morhua	4.92 ± 0.15	46.5 ± 0.9	—	—	9.6 ± 0.4	Sidell and Driedzic, 1985
Morone saxatilis	14.8 ± 1.2	37 ± 2.6	—	—	5.94 ± 0.4	Sidell and Driedzic, 1985
Macrozoarces americanus	2.45	36.34	127.8	1.79	12.78	Driedzic, 1988
Ictalurus punctatus	13.2 ± 14	58.8 ± 3.6	423 ± 11	5.6 ± 0.1	12.5 ± 0.3	West et al., 1999
Tropical fishes						
Colossoma macropomum	3.0 ± 0.3	76.5 ± 6.9	573 ± 11	8.7 ± 1.0	13.9 ± 3.0	West et al., 1999
Hoplosternum littorale	20.2 ± 2.1	60.0 ± 2.0	235 ± 14	10.7 ± 0.4	20.3 ± 1.5	West et al., 1999
Arapaima gigas	10.1	27.3 ± 2.8	256 ± 30	2.4 ± 0.4	23.0 ± 2.0	West et al., 1999
Satanoperca aff. jurupari	—	96.6 ± 5	52 ± 6	—	—	Chippari Gomes, 2002
Cicla monoculus	—	12.1 ± 1	13 ± 1	—	—	Chippari Gomes, 2002
Geophagus aff altifrons	—	113.5 ± 6	89.2 ± 14	7.6 ± 0.5	—	Chippari Gomes et al., 2000
Astronotus ocellatus	2.4 ± 0.3	46.2 ± 8.1	134 ± 3	5.8 ± 0.4	10.4 ± 0.7	West et al., 1999
Astronotus crassipinnis	—	55 ± 0.5	8.6 ± 0.5	—	19.8 ± 5	Chippari Gomes et al., 2000
Symphysodon aequifasciatus	—	41.9 ± 5	33.4 ± 2	—	20.3 ± 2	Chippari Gomes et al., 2000
Glyptoperichthys gibbceps	1.4 ± 0.1	27.9 ± 0.87	21.6 ± 0.5	6.8 ± 0.5	79.6 ± 3.7	Lopes, 2003
Platydoras costatus	2.2 ± 0.7	66.7 ± 9.6	492 ± 114	1.36 ± 0.2	—	Lopes, 2003
Calophysus macropterus	1.3 ± 0.17	57.2 ± 1.4	635 ± 15	18.8 ± 0.7	32.7 ± 1.9	Lopes, 2003
Prochilodus nigricans	1.46	29.1	135.3	6.36	19.98	Lopes, 1999
Comparison between the two groups						
Range (temperate)	2.45–25.6	8.4–158.7	114–726	1.78–17.8	6.92–47.8	
Range (tropical)	1.3–20.2	12.1–113.5	8.6–635	1.36–18.8	10.4–79.6	

Enzyme activity is expressed as μmol min^{-1} g^{-1} wet weight (mean ± SEM).

as described by West *et al.* (1999), the magnitude of ranges of enzyme activity levels is very similar between tropical and temperate teleosts (Table 10.1). This reveals once again that inspecting these data very closely may reveal more important characteristics than an attempt to establish general patterns will, particularly when dealing with the mosaic of changes in aquatic ecosystems that the animals face in tropical areas. In fact, most studies on whole animals that have been done in our laboratory with Amazonian fish species subjected to some level of oxygen depletion (acute hypoxia, graded hypoxia or anoxia) revealed that the animals showed alterations in plasma glucose and lactate levels resulting from the usage of glucose reserve mobilization and anaerobic-based lactate production (Table 10.2). Furthermore, after these experimental procedures with different Amazonian fish species, most of which were considered hypoxia-tolerant, it became clear that anaerobic glycolysis took place in most species and that this response is combined with metabolic depression in some species, mainly in those that were already known to be hypoxia-tolerant (Table 10.2).

Glucose mobilization may occur even in facultative air-breathers such as the armored catfish *Glyptoperychthys gibbceps*, which is not necessarily related to anaerobic metabolism activation, since lactate levels are significantly decreased (Lopes, 2003). Following these studies, a detailed study using *G. gibbceps* as a model was performed in our laboratory. This species was compared to a closely related armored catfish, *Lipossarcus pardalis*, by MacCormack *et al.* (2003). We observed that no changes in heart rate occurred under conditions of controlled hypoxia in aquaria, under natural hypoxia in a simulated pond, and in field cage sites. When denied aerial respiration under hypoxia in laboratory aquaria, *G. gibbceps* increased gill ventilation rates, but neither *G. gibbceps* nor *L. pardalis* exhibited alterations in heart rate, suggesting that bradycardia is not one of their strategies against hypoxia. On the other hand, *G. gibbceps* was hyperglycemic under normoxia (mean plasma glucose ranging from 16.88 to 31.24 μmol ml^{-1}) and extremely large increases were observed under hypoxia (29.84 to 51.11 μmol ml^{-1}). Unlike graded hypoxia responses, where lactate decreases (Lopes, 2003; Table 10.2) or natural hypoxia in a simulated pond, where lactate does not change (MacCormack *et al.*, 2003), the plasma lactate levels of this species increased from 1.55±0.81 to 65.91±7.48 μmol ml^{-1} when submitted to acute hypoxia in indoor aquaria and denied access to air (MacCormack *et al.*, 2003). Therefore, the reliance on extracellular glucose as a metabolic fuel under oxygen limitations may constitute one of the strategies to better deal with oxygen limitation and may occur in addition to other responses. It may be dependent upon respiratory pattern and species phylogeny.

After investigations of metabolic responses of the common carp to prolonged hypoxia, Zhou *et al.* (2000) concluded that metabolic depression

Table 10.2

Changes in the Plasma Amounts of Glucose and Lactate After
Different Levels of Oxygen Deprivation

Species	Glucose	Lactate	References
Colossoma macropomum	(N) 75.68	1.60	Chagas, 2001
	(GH) 135.13*	4.55*	
Platydoras costatus	(N) 36.83 ± 6.48	0.47 ± 0.01	Lopes, 2003
	(AH) 43.86 ± 1.44	0.89 ± 0.09*	
Calophysus macropterus	(N) 50.87 ± 12.04	0.37 ± 0.06	Lopes, 2003
	(AH) 196.64 ± 26.42*	0.35 ± 0.05	
Hoplosternum littorale	(N) 57.73 ± 4.94	3.60 ± 0.47	Lopes, 2003
	(GH) 70.61 ± 8.40*	4.43 ± 1.02*	
Glyptoperichthys gibbceps	(N) 37.26 ± 2.61	0.91 ± 0.03	Lopes, 2003
	(AH) 174.31 ± 30.53*	0.20 ± 0.11*	
Liposarcus pardalis	(N) 59.46 ± 3.74	1.62 ± 0.11	Lopes, 2003
	(GH) 95.54 ± 11.70*	14.11 ± 1.47*	
Astronotus ocellatus	(N) 41.08	1.50	Muusze et al., 1998
	(GH) 29.01	16.5	
Astronotus crassipinnis	(N) 67.73 ± 4.70	1.49 ± 0.33	Chippari Gomes, 2002
	(GH) 232.26 ± 16.80*	17.11 ± 0.73*	
Symphysodon aequifasciatus	(N) 23.64 ± 4.81	1.52 ± 0.44	Chippari Gomes, 2002
	(GH) 93.64 ± 23.91*	7.61 ± 0.69*	
Heros sp.	(N) 89.00± 24.73	1.93 ± 0.36	Chippari Gomes et al., 2000
	(AH) 85.87 ± 9.88	11.14 ± 1.08*	
Uaru amphiacanthoides	(N) 88.02 ± 17.99	2.63 ± 0.93	Chippari Gomes et al., 2000
	(AH) 70.56 ± 18.88	14.52 ± 0.64*	
Satanoperca jurupari	(N) 111.58 ± 37.29	0.91 ± 0.05	Chippari Gomes et al., 2000
	(AH) 156.04 ± 37.37	9.69 ± 0.09*	
Geophagus altifrons	(N) 147.05 ± 16.29	1.44 ± 0.57	Chippari Gomes et al., 2000
	(AH) 139.20 ± 21.93	10.17 ± 1.26*	

Glucose content is expressed as mg/plasma dl, lactate levels is expressed as μmol/l.
*$P < 0.05$; (N) normoxia; (AH) acute hypoxia; (GH) graded hypoxia.

allows this species to reduce accumulation of lactate and save on the use of energy reserves in the face of hypoxic stress. It is important to consider that for fish inhabiting water bodies that are frequently hypoxic, as in the Amazon, natural selection will favor the evolution of adaptive strategies such as varying enzyme activity levels, modifications in enzyme kinetics, and metabolic depression. These biochemical adaptations, together with a decrease in locomotor activity, occur along with a reduction of oxygen consumption rates, as observed in several experiments realized in our laboratory (Muusze *et al.*, 1998; Chippari Gomes, 2002; Lopes, 2003) as well as in others (Dunn and Hochacka, 1986; Zhou *et al.*, 2000). Chronic exposure to hypoxia could, therefore, induce a suppressed metabolic rate in tropical animals, as already mentioned above (Figure 10.1).

Fuel utilization under metabolic suppression and under oxygen deprivation in systems that use metabolic arrest involves the mobilization of anaerobic pathways. The use of carbohydrate reserves such as glycogen or amino acids, which are the main fuels available for fermentation in animals, is almost always necessary (Hochachka and Somero, 2002). Lipid reserves are used when oxygen is not necessarily limiting, such as in the burrowing lungfish (Dunn *et al.*, 1983), which uses lipid as the major fuel during the early phases of estivation. When this reserve becomes depleted, proteins are mobilized and amino acids serve as precursors for gluconeogenesis and for catabolic substrates. Dunn *et al.* (1983) suggested that glycogen reserves are conserved to save muscle energy during arousal and escape.

According to Hochachka and Somero (2002), most biologists realized early that fishes differ greatly in their patterns and capacities of locomotion, ranging from fast-start, burst-swim specialists, to species that can swim steadily but slowly for intermediate or long periods, and finally to species that can swim for long periods of time and long distances. Extensive studies have shown that the biochemical machinery of red and white muscles is adapted and coordinated in its physiological adjustments, affecting fuel and oxygen supply capacities, and can be so extensive that it can mix-up the distinction between white and red muscle. For some groups, such as scombroid fishes, mitochondrial enzyme concentrations per gram of white muscle can be higher than its homologous (orthologous) enzyme levels in red muscle of some sluggish Amazon fishes. In red muscle of some Amazon fishes, the concentrations of enzymes in anaerobic metabolism may be higher than in the white muscle of more hypoxia-sensitive fishes from more oxygen-rich, usually colder, waters. To better address this issue, we will review some data on red and white muscle biochemical machinery in the following section, presenting new data about neotropical fishes, particularly Amazon fishes.

VI. RELATIVE AMOUNT OF RED MUSCLE IN FISH AND ITS ADAPTIVE ROLE

Red and white muscle fibers are often spatially differentiated in fish. The slow oxidative fibers in red muscle contrast with white muscle, which is anaerobic and displays exceptional compositional homogeneity. The relative amount of these fibers changes according to several characteristics of the species, such that in some species, like the fast pike, the entire swimming musculature is a uniformly white (fast-twitch muscle) glycolytic system, with red muscle fibers (slow-twitch muscle) extremely reduced (Moyes *et al.*, 1992). Among fishes of the Amazon, some of the highest relative anaerobic rates in metabolism occur in *Osteoglossum bicirrhosum* (aruanã, or water monkey), i.e., LDH/CS (Citrate Synthase) ratios can reach 800. The aruanã has one of the lowest ratios of red to white muscle among studied Amazon species. A cursory inspection of red:white muscle ratios among fishes of the Amazon reveals an interesting picture regarding environmental adaptations (Table 10.3). It is clear that fishes with long time/long distance swimming habits have more slow-twitch red muscle and, therefore lower ratios of white to red muscle (WM/RM). On the other hand, fish species with fast-twitch, burst-swimming activity show such a low amount of red muscle that ratios between fast and slow muscle fibers can reach the extreme of nearly absent red muscle fibers in the total skeletal muscle. This is the case in an ornamental fish, the cichlid discus *Symphysodon aequifasciatus* (Table 10.3). In fact, high WM/RM ratios occur in the whole cichlid family, with the exception of *Satanoperca acuticepts*. Cichlids constitute a family that sustains highly specialized reproductive habits. Most species retain territorial and show aggressive parental care, and consequently require a high capacity for fast-twitch muscle fibers to allow for bouts of burst swimming. The cichlids are also considered a hypoxia-tolerant family as a whole; this characteristic may occur in different degrees in each group of species (Chippari Gomes, 2002). Fishes with a higher proportion of red muscles fit into the category of either active species, such as the Serrasalmidae (piranhas and pacus), or species with long and continuous swimming habits, such as fish occurring in large schools in the Amazon region, e.g., characids and prochilodontids. In an intermediary position, other fish families with moderately active lives or facultative air-breathing habits will have intermediate WM/RM ratios (Table 10.3). Red muscle fibers have a well-developed blood supply, high myoglobin and mitochondria contents, high concentrations of lipids and cytochromes, and high activities of respiratory and citric acid cycle enzymes. Therefore, red muscle fibers show active aerobic metabolism, using both carbohydrates and lipids as substrates (reviewed by Van Ginneken *et al.*, 1999). However, the bulk of muscle tissue in most fish species consists of white muscle, which depends mainly on anaerobic glycolysis for its energy supply.

Biochemical comparisons between both kinds of muscle fibers are well described with regard to amino acids, phosphorylated compounds, and enzyme activity levels and most authors refer to them as clearly different when responding to a stressor, particularly hypoxia (Van Ginneken *et al.*, 1999; Hochachka and Somero, 2002). Comparative analyses of pairs of species,

Table 10.3

Relative Amounts of Red Muscle and White Muscle,
Represented as the % of Total Weight, and the Ratio Between the Two

Species	Red muscle somatic index	White muscle somatic index	WM/RM ratio
Osteoglossidae			
Osteoglossum bicirrhosum ($n = 1$)	0.82	30.94	37.7
Characidae			
Triportheus flavus ($n = 1$)	2.20	26.8	12.2
Triportheus albus ($n = 1$)	2.92	28.99	9.9
Curimatidae			
Curimata inornata ($n = 4$)	3.24	33.89	10.5
Psectrogaster amazonica ($n = 3$)	1.77	31.14	17.6
Psectrogaster rutiloides ($n = 2$)	2.14	23.57	11.0
Serrasalmidae			
Pigocentrus nattereri ($n = 3$)	1.95	18.74	9.6
Mylossoma duriventre ($n = 4$)	4.04	22.41	5.6
Metynnis hypsauchen ($n = 4$)	3.91	19.17	4.9
Anostomidae			
Leporinus friderici ($n = 2$)	2.58	33.94	13.2
Rhytiodus microlepis ($n = 1$)	2.11	25.80	12.2
Callichthyidae			
Hoplosternum litoralle ($n = 12$)	0.71	10.49	14.77
Loricariidae			
Liposarcus pardalis ($n = 8$)	1.57	11.36	7.24
Doradidae			
Corydoras sp ($n = 7$)	2.19	9.15	4.18
Cichlidae			
Satanoperca acuticeps ($n = 2$)	1.26	14.36	11.4
Cichlassoma severum ($n = 1$)	0.84	22.71	27.0
Cichla monoculus ($n = 4$)	0.69	33.87	40.1
Geophagus altifrons ($n = 15$)	0.50	25.28	50.6
Uaru amphiacanthoides ($n = 3$)	0.64	27.67	43.1
Astronotus crassipinnis ($n = 23$)	0.39	22.73	58.3
Satanoperca jurupari ($n = 15$)	0.77	40.46	52.5
Symphysodon aequifasciatus ($n = 19$)	0.06	28.18	469.1

even those showing similar metabolic trends, suggest they vary on a species-specific basis, reflecting many characteristics, including the type of swimming performance and particular mode of life (reviewed by Johnston, 1977). As shown in Table 10.3, it is clear that a correlation exists between fish mode of life and the degree of red muscle development in the myotome, and such correlation may occur in nature due to the fact that the basic metabolic differentiation of red and white muscles in fish may be under environmental evolutionary pressure.

To better address this issue, we have compared two closely related fish species that were acclimated differently: *Prochilodus scrofa* (curimbatá) and *Prochilodus nigricans* (curimatã). These two congeneric species have different life styles, live under different thermal regimes (i.e., different latitudes; the former species lives close to the Tropic of Capricorn and the latter is closer to the Equator), and have different migratory habits. *Prochilodus scrofa* inhabits Paraná-Pardo basin located at the southeast region of Brazil and has short-distance, fast-swimming habits, "running" upstream large rivers during the spawning season, which requires a high anaerobic potential. On the other hand, *Prochilodus nigricans* inhabits the Amazonian basin and has long-distance, low-speed no-stopping swimming habits (year-round migration habits), requiring more of an endurance-type, low-twitch, oxidative muscle fiber, similar to red muscle. Evolution has driven these two species to develop different amounts of muscle fibers, which is illustrated in Figure 10.2. Once again, a long-term environmental pressure, combined with the development of different habits and differential adaptation to different kinds of habitats can result in long-term metabolic and morphological adjustments, both depending on the evolutionary genetic processes of adaptation, as discussed at the beginning of this chapter. Evolutionary history has thus played an important role in establishing biochemical characteristics in the muscle fibers of these species.

The analysis of muscle enzyme activity levels fits perfectly within their modes of life as well. Absolute activity levels of enzymes of glycolytic pathway (HK, pyruvate kinase – PK, and LDH), citric acid cycle (CS, citrate synthase), mixed functions (malate dehydrogenase – MDH), and lipid metabolism (beta-hydroxyacyl-CoA dehydrogenase – HOAD) in heart, red muscle, and white muscle of these two closely related species are summarized in Table 10.4. The activity levels of these enzymes were measured under saturated conditions of substrates as described elsewhere (Moyes *et al.*, 1989, 1992; Driedzic and Almeida-Val, 1996). Considerable species differences can be observed in the absolute activity levels of the enzymes, reflecting their different metabolic profiles, which can be associated with the life history of each species. HK activities are higher in heart compared to white muscle levels in both species. Except for the heart, the long-swimming species *P. nigricans* shows lower HK values than the higher-speed, burst-swimming species *P. scrofa*. HK, as already mentioned, is unique among glycolytic enzymes because its activity in vertebrate muscles is directly related to the preferential utilization of glucose from liver rather than muscle glycogen for energy production, i.e., a preferential usage of free glucose. For fish species swimming upstream, such as *P. scrofa,* anaerobic power is very important, since the animal must maintain high energy production in a low oxygen-loading environment, the white muscle fibers. Indeed, white muscle

Fig. 10.2 Comparison of two congeneric species acclimated to different thermal regimes (i.e., different latitudes), showing differences in relative amount of red muscle: *Prochilodus scrofa* (curimbatá) bottom and *Prochilodus nigricans* (curimatã) top. The former has short-distance, fast-swimming habits, "running" upstream in large rivers during the spawning season, which requires a high anaerobic potential (higher amount of white muscle); the latter has long-distance, low-speed no-stopping swimming habits (year-round migration habits), requiring more of an endurance-type, low-twitch, oxidative muscle fiber, similar to red muscle. See Table 10.4 and text for differences in metabolic profiles of red muscles of the two species.

fibers from this species retain higher anaerobic capacities (Table 10.4), which can be seen through the differences in glycolytic enzymes, mainly PK and LDH activity levels.

The picture that emerges from these data is that *P. nigricans* has the ability to endure continuous swimming with a higher amount of mito-chondria-rich red muscle, metabolically adapted to preferential usage of carbohydrate in an oxygen-richer environment, since red muscle fibers are up-loaded with blood, as mentioned earlier (Figure 10.2 and Table 10.4).

Table 10.4
Absolute Enzyme Activities (V_{max}) of Heart, Red Muscle, and
White Muscle of the Congeneric Species of *Prochilodus*

Enzyme levels	Prochilodus nigricans	Prochilodus scrofa	P values (t student)
Heart			
HK	1.47 ± 0.12	1.001 ± 0.04	0.0017
PK	29.68 ± 1.09	25.78 ± 0.5	0.0043
LDH (1 mM)	137.52 ± 5.27	123.98 ± 2.93	0.038
LDH (10 mM)	246.64 ± 10.03	248.18 ± 7.92	0.905
CS	18.85 ± 0.44	13.66 ± 0.22	<0.001
MDH	190.14 ± 2.87	185.99 ± 4.07	0.415
HOAD	6.58 ± 0.24	7.98 ± 0.18	0.0002
Red muscle			
HK	1.53 ± 0.09	5.05 ± 0.31	<0.001
PK	11.85 ± 0.54	25.63 ± 0.6	<0.001
LDH (1 mM)	143.96 ± 21.89	174.77 ± 6.31	0.003
LDH (10 mM)	127.91 ± 5.65	236.07 ± 11.76	<0.001
CS	23.87 ± 0.41	13.03 ± 0.36	<0.001
MDH	105.33 ± 1.19	67.76 ± 1.7	<0.001
HOAD	10.03 ± 0.34	5.8 ± 0.19	<0.001
White muscle			
HK	0.016 ± 0.002	0.936 ± 0.034	<0.001
PK	406.24 ± 3.93	475.85 ± 5.73	<0.001
LDH (1 mM)	561.96 ± 21.89	951.82 ± 28.97	<0.001
LDH (10 mM)	573.12 ± 21.32	519.43 ± 13.32	0.046
CS	3.51 ± 0.12	3.09 ± 0.12	0.023
MDH	47.71 ± 1.79	82.74 ± 2.45	<0.001
HOAD	9.71 ± 0.45	1.49 ± 0.14	<0.001

Enzyme activity is expressed as μmol min^{-1} g^{-1} wet weight (mean ± SEM).
Source: After Lopes, 1999.

Also, a higher level of lipid metabolic enzyme HOAD is present in the oxidative red fibers, as is a higher amount of CS, which is also higher in white muscle of *P. nigricans*. The result of a lower blood perfusion and consequently lower oxygen loading is evident in *P. scrofa,* which retains a higher anaerobic capacity in both red and white muscles (Table 10.4). In some way, these results change the notion that a life cycle spent at low temperatures is often associated with elevated activities of key enzymes in energy metabolism, i.e., the higher the latitude at which a fish occurs, the higher its enzyme activity levels, as discussed in previous sections. When the comparison is made between two closely related species living at different latitudes, the species adapted to higher latitudes and lower temperature regimes generally shows lower absolute enzyme levels than the one living closer to the Equator at higher temperature regimes. Furthermore, the absolute enzyme levels are tissue-specific and are adapted to the metabolic profile that fits in their evolutionary life histories, which suggests that in terms of general patterns, congeneric or confamilial species should be compared to generate a better picture of the role of evolution in the establishment of metabolic profiles and morphological adjustments, which depend upon long-term, structural genetic changes during evolution.

VII. OXYGEN DEPRIVATION AND ITS CONSEQUENCES IN AMAZON FISHES

Oxygen depletion in water (hypoxia) is a common phenomenon in nature, which may be caused by human activities or may have natural origins. Acute pollution episodes may cause mortality and/or permanent damage to aquatic organisms. On the other hand, constant pollution activities may induce a decrease in oxygen availability in water bodies over several years and can result in changes in species distribution and cause severe decreases in population sizes. Paradoxically, this may result in animals adapted to the new hypoxic conditions. This may be the case because some fishes have adapted to hypoxic conditions due to their evolutionary history, which will enable them to better cope with the new situation and have better chances of survival in a polluted site. Furthermore, natural episodes of hypoxia occur globally in different environments and may have different causes and effects in aquatic organisms that inhabit different ecosystems. Hypoxia occurs naturally below the frozen surface of lakes, especially when photosynthesis decreases due to snowfall and oxygen consumption decreases under winter ice due to nitrification (Van Ginneken, 1996); hypoxia also occurs in some large lakes such as Lake Tanganyika due to strict stratification (Coulter, 1991); and hypoxic conditions may also occur in aquaculture ponds due to overstocking (Boyd and Schmitton, 1999). In the waters of the Amazon, flood pulses occur annually (Junk *et al.*, 1989; Val *et al.*, 1998) and cause oscillations of several physical-chemical factors resulting in the seasonal variance in oxygen availability (see Chapter 1). In aquatic ecosystems of the Amazon, episodes of severe hypoxia can occur, and oxygen levels may drop down to values below $2.0\,\text{mg}\,O_2\,l^{-1}$, lasting up to several months at a time (Val *et al.*, 1986; Val *et al.*, 1998). To survive such conditions, fish of the Amazon have developed a series of coordinated metabolic adjustments, which, combined with morphological and anatomical changes, have resulted in a number of solutions to avoid stress caused by hypoxia (Val and Almeida-Val, 1995). In the case of the Amazon, the amount of oxygen is altered due to the interactions of many characteristics and processes, including photosynthesis, respiration of aquatic macrophytes and phytoplankton, light penetration, organic decomposition, molecular oxygen diffusion, wind, water body depth and shape, and temperature. Both long- and short-term changes in oxygen are determinants of fish distribution in Amazonian water bodies (Almeida-Val, Paula-Silva *et al.*, 1999; Almeida-Val, Val *et al.*, 1999).

The critical oxygen tension (Pc – critical pressure) is the tension of oxygen in water below which oxygen consumption by the fish begins to drop. The critical oxygen tension, or threshold, varies among fish species. In the past, two definitions were given for critical oxygen tension: "incipient limiting level" and "no-effect oxygen threshold," which reflected the level at which fish populations were not impaired in their growth – development – activities (Davis, 1975). Minimal values were described for populations of trout (55% air saturation, AS), carp (50% AS), and eel (35% AS). Some authors have proposed the concept of "incipient lethal level", or the level that the animals can resist for some time but eventually die, probably because of the activation of anaerobic metabolism, which would result in metabolic imbalance (Fry *et al.*, 1947; Davis, 1975). Other studies have described Pc as a parameter that is dependent on metabolic demands and the ability of the animal to supply oxygen to its tissues. Thus, all of these indices (Pc, incipient limiting level, and incipient lethal level) should be described under defined experimental conditions. Today, many more accurate approaches are used in studies of hypoxia tolerance, particularly in fish, and it is well known that conformity and regulation are two mutually exclusive metabolic conditions in fish when exposed to graded hypoxic conditions. Studies on the effects of hypoxia on fish have increased markedly in the last decades and many reviews are available (Nikinmaa, 2002; Wu, 2002). Most processes are now elucidated and may be summarized as follows: adaptive responses to hypoxia shown by fish are (i) escape reaction; (ii) adaptations of circulatory and ventilation systems; (iii) reduction of activity to standard metabolic rates (SMR); (iv) activation of anaerobic glycolysis; (v) phosphocreatine depletion; (vi) metabolic depression (reduction of activity bellow standard metabolic rates); and (vii) release of inhibitory neurotransmitters in brain (reviewed in Van Ginneken, 1996). Recent molecular studies have revealed that gene regulation and signal transduction are common in vertebrates exposed to hypoxia and also occurs in fish (Soitamo *et al.*, 2001; Powell and Hahn, 2002). Considering the current literature, we can affirm that evolution played a crucial role shaping many mechanisms to help fishes to better cope with low oxygen environments.

A. How Fish Became Tolerant to Hypoxia

The evolution of fishes occurred independently from other vertebrates for several hundred millions of years, and the periods of low oxygen pressure are coincident with the appearance of important groups of fishes. Among these are groups with adaptations to breathe air, such as the lungfishes (Dipnoan). Moreover, modern groups like the Osteoglossomorpha, Ostariophysi and Plecomorpha, which have representatives living in natural hypoxic waters, also developed some adaptations to breathe air and tolerate hypoxia. Among modern Ostariophysi, the current hypoxic environments of tropical waters have resulted, in recent history, in the development of facultative air breathing. Three families of neotropical armored catfishes have developed adaptations in their digestive system to breathe air – these adaptations are coincident with the encapsulation of the swim bladder (Rapp Py-Daniel, 2000). The two main groups of fishes have appeared in periods of limiting oxygen availability. Both Chrossopterygian (lobed-fin fishes) and Actinopterygian (ray-fined fishes) are thought to have developed air-breathing habits and hypoxia tolerance since early in their evolutionary history. Fish arose more than 500 MYA. As reviewed recently, the evolution of fish occurred independently from other vertebrates for several hundred millions of years (Nikinmaa, 2002) and hypoxic environments dominated the first geological eras after fish arose (Berner and Canfield, 1989; Graham, 1997). The current values of atmospheric oxygen were reached around 200 MYA. Thus, the origin of fishes and half of their evolutionary history occurred under conditions of oxygen depletion.

From that period through to the present, some environments on earth have remained hypoxic, if not constantly, at least periodically. Aquatic ecosystems such as shallow and warm waters of the tropics are commonly found to be hypoxic and even anoxic during some periods of the day (Junk, 1984; Val, 1995; Almeida-Val, Val *et al.*, 1999). Not surprisingly, these environments are among the most diverse in the world (see Chapters 1 and 2).

The amount of dissolved oxygen varies in response to the interactions of many characteristics and processes in the waters of the Amazon. Seasonal variations and daily and spatial oscillations may occur and induce complex patterns of oxygen distribution. During most of the year, anoxic conditions are observed at night (Figure 10.3). However, plant cover, depth and sunlight play important roles in oxygen availability during the day. Hypoxia tolerance therefore became particularly common among fishes of the Amazon, and we believe that such chronic hypoxic conditions drove fish species through a series of adjustments at different levels of biological organization (ethological, morphological, anatomical, physiological, metabolic, and molecular) that were combined to produce phenotypic plasticity, allowing them to survive the pulsating nature of the basin (reviewed in Almeida-Val, Val *et al.*, 1999).

The appearance, diversification, and evolution of fish fauna in the Amazon are all associated with hydrographic basin formation (Lundberg,

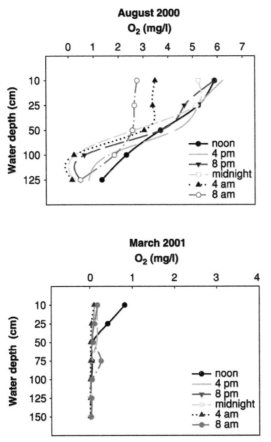

Fig. 10.3 Oxygen distribution in the water column of *várzea* lakes near Catalão Lake at two different times of the year. Data were collected in the right side of the Rio Negro, inside the lake just before where the Rio Negro meets the Rio Solimões. The two boxes show that the amount of dissolved oxygen varies with depth, time of the day, and time of the year. These seasonal variations and daily and spatial oscillations may induce complex patterns of oxygen distribution. Also, during different times of the year, anoxic conditions can always be observed at night. (Original data obtained by Chippari Gomes, 2002.)

1998). Current environmental heterogeneity, caused by flood pulses, different water types, and physico-chemical parameters, are the main causes of recent adaptive radiation in fish of the Amazon (Junk *et al.*, 1989; Val, 1993). The 3000 fish species described to date in the Amazon display a variety of adaptations to their environments that include behavioral, physiological, biochemical, and genetic changes. The time course of the appearance of such adaptive traits seems to be related to the intensity and periodicity of the constraints imposed upon each individual, population, species, and group of species. However, the description of several adaptive strategies at numerous taxonomic levels has revealed that the selective pressure during evolution may be caused by several chronic constraints such as short- and long-term changes in oxygen, water pH, ion-poor waters, acidity, and daily and spatial temperature oscillations, among others (Almeida-Val, Val *et al.*, 1999). Thus, adapting to such ever-changing environments is probably the main cause of fish diversity in the Amazon.

B. Levels of Responses to Hypoxia

Five levels of response to hypoxia are commonly described in aquatic ecosystems: ecological, behavioral, physiological, biochemical, and molecular. At the ecological level, results may be different according to environment characteristics. However, for most environments, an increase in hypoxia episodes may be devastating because it may cause mass mortality, defaunation of benthic populations, declines in fisheries production, permanent damage to part of the aquatic environment, changes in community composition, and, as an ultimate consequence, a decrease in animal diversity. On the other hand, chronic hypoxic situations, such as are common in the Amazon basin, have caused several adaptations at different levels of biological organization, thereby inducing increased species diversity. Also, seasonal changes in species composition may occur as a result of different oxygen availabilities. The classic study of Junk *et al.* (1983) showed that during low oxygen conditions, the species remaining in the lake were mostly those that could breathe air or those with some kind of hypoxia tolerance. Among water-breathers, cichlids always remain inside hypoxic lakes, and are a group considered to be tolerant of hypoxia and anoxia (Junk *et al.*, 1983). The occurrence of cichlid species in Catalão Lake was investigated during one year and was correlated with oxygen availability (Chippari Gomes, 2002). Variance in oxygen availability occurs throughout the year. During periods of low water, oxygen levels drop substantially, reaching concentrations below 1 ppm. Although the number of captured species varied during the year, we cannot imply that cichlids species richness was related to oxygen availability (Chippari Gomes, 2002). However, the abundance of cichlids was higher when oxygen reached the lowest level of the year (Figure 10.4). As hypoxia-tolerant fishes, cichlids remain in hypoxic and anoxic waters, while more sensitive species escape. Other species that may remain in hypoxic lakes have the option to breathe air or skim the water surface. The tambaqui has the ability to expand its inferior lips to help direct more oxygenated water through its gills. In the wild, the occurrence of lips is inversely related to oxygen availability (Figure 10.5).

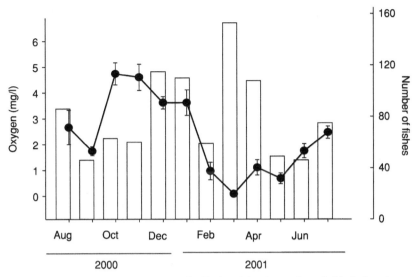

Fig. 10.4 Relationship between oxygen distribution (closed round symbols) during the year (values obtained at noon) and the abundance (number of fishes – open bars) of cichlids captured near Catalão Lake. The highest abundance of fishes is coincident with the lowest oxygen availability, which can be explained as their ability to occupy and stay in hypoxic environments, which is not observed in any other group of fish. (Original data obtained by Chippari Gomes, 2002.)

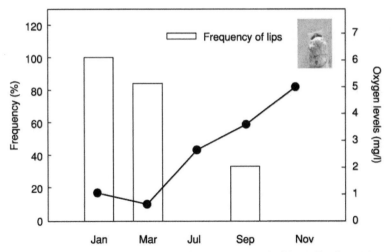

Fig. 10.5 Relationship between frequencies of animals captured with swollen lips and oxygen availability where they were captured. Tambaqui swells its inferior lips (inlet) to better capture the surface water layer, which is richest in oxygen. (Redrawn from Val and Almeida-Val, 1995.)

Field studies of fish distribution in aquatic ecosystems of the Amazon basin have always correlated oxygen availability with preferred species habitats, migration (particularly lateral migration), and adaptive characteristics, namely air breathing or aquatic surface respiration (ASR) (Junk *et al.*, 1983; Cox-Fernandes, 1989; Crampton, 1998). The pioneering work of Junk and collaborators (1983) at Camaleão Lake on Marchantaria Island described fish movements throughout the year that were related to changes in environmental oxygen levels. Another interesting study by Crampton (1998) investigated the distribution, migratory behavior, and respiratory adaptations of electric eels (Gymnotiformes) in Rio Tefé, located in the upper Amazon basin. This author suggested that all these parameters were directly related to oxygen availability, and that oxygen availability may have an important influence on the design of electric signals in the eels. Regardless of whether this is related to oxygen distribution in the water, electric eels of the Amazon have a variety of adaptive strategies to survive episodes of hypoxia. Some species were even described as anoxia-tolerant under field and experimental conditions. Among the studied species, Crampton (1998) described air-breathing organs in two species, air-gulping in five species, and ASR in 12 species. However, these data were based on chemically induced hypoxia (forced oxygen reduction by using sodium sulfite in the metabolic chamber water). Thus, these findings must be confirmed under different conditions, since sodium sulfite may cause drastic changes in ionic homeostasis, and can induce sulfohemoglobins formation, which results in other physiological effects such as physiological impairment of oxygen transfer to tissues. In spite of this, the fact that some water-breathing species could survive anoxia for 6 hours without ASR, which was inhibited by the experimental design, is good evidence that those species are tolerant of hypoxia. As with air breathing and ASR, hypoxia tolerance appears in fish as a homoplasic characteristic, which, different from other vertebrate lineages, has multiple independent evolutionary origins and is the consequence of multiple environmental pressures, causing adaptive radiation.

Reviewing the subject of hypoxia tolerance in fishes, Hochachka and Somero (2002) suggested that except for the Ostariophysi, within which are several groups known to be notably hypoxia-tolerant, the distribution of hypoxia tolerance within the phylogeny appears sporadic, which seems to be consistent with multiple independent origins of hypoxia-tolerance mechanisms in teleost fishes, different from other hypoxia-tolerant vertebrate groups such as turtles and pinnipeds (reviewed by Hochachka and Somero, 2002). When considering air breathing as a defense adaptation against hypoxia, many (and possibly all) air-breathing fishes – described as "the first diving vertebrates" – display impressive hypoxia defense mechanisms similar to those described for aquatic turtles (Almeida-Val and Hochachka, 1995).

The Ostariophysi comprise four fish Orders that are predominantly tropical. Three out of four Orders occur in tropical and subtropical hypoxic waters. The Gymnotiformes (electric eels), Siluriformes (catfishes), and Characiformes (suckers, tambaqui, pacu) retain hypoxic-tolerant species with many representatives in the Amazon basin. Such hypoxia tolerance results from different kinds of adaptational changes. The fourth group, Cypriniformes (goldfish, carp), does not occur in South American basins but is strongly hypoxia-tolerant due to metabolic adaptations (Van Waarde et al., 1993). Many authors have suggested that air breathing is an indicator of hypoxia defense adaptations and of hypoxia tolerance. As such, the examination of the phylogeny of the group also suggests that this trait has evolved numerous times within the fishes (Val and Almeida-Val, 1995). According to Hochachka and Somero (2002) and Val and Almeida-Val (1995), while in some lineages this may have involved common descendants, in many others it is clear that hypoxia tolerance has evolved independently. Because of the large number of fish species that are known to be hypoxia-tolerant, these authors suggested that it is clear and compelling that hypoxia tolerance has arisen independently many times within this group of organisms. What we now need is a detailed evolutionary analysis of hypoxia adaptations in fishes analogous to that presented for diving in pinnipeds (Mottishaw et al., 1999). However, the possibility of many traits originating by parallel convergent evolution in response to similar environmental challenges appears to be the key for beginning to understand how comparable complex physiologies independently evolved within a group as diverse as fishes. One of the best examples is the development of the ASR in response to the chronic hypoxic waters of Amazonian várzea lakes (Almeida-Val, Val et al., 1999).

C. Appearance of Aquatic Surface Respiration and the Development of Air-Breathing Organs

While these subjects have already been fully investigated by several authors through a description of the regulation of blood physiological parameters, enzyme levels and their tissue expression, ventilation adjustments, adjustments of hematological parameters, ion regulation, and behavior; the relationship between these adaptive strategies and their occurrence among related fish groups is poorly understood. Studies on the relationship between aquatic surface respiration (ASR), an innate behavior, and the physiological responses that follow such a strategy have been addressed to evaluate its efficiency (Almeida-Val et al., 1993; Val, 1995). This adaptation may be considered a homoplasic characteristic among fish since it has appeared in different phylogenetic groups in response to the same environmental pressures. Such behavior is useful for fish inhabiting hypoxic várzea lakes, and its evolution among several groups may be considered one of the main surviving hypoxia strategies among fishes of the Amazon. The cichlids (Perciformes) are among the most advanced teleosts that occur in the Amazon basin; they are considered a highly specialized group with a high degree of adaptive radiation and are considered to have a faster evolutionary rate compared to their African counterparts (Farias *et al.*, 1998). Several cichlid species show ASR, a behavior that is more noticeable in young of the species. During growth, some species, e.g., *Astronotus ocellatus*, reduce the number of incursions to the water surface and increase their anaerobic glycolytic power as a result of increased mass-specific LDH levels, improving their survivorship under hypoxic conditions (Almeida-Val, Paula-Silva *et al.*, 1999; Almeida-Val *et al.*, 2000). LDH isoforms also change accordingly in the brain of this species. It is important to state, however, that LDH scaling properties have also been described in non-hypoxia-tolerant fish groups, apparently due to an increase in burst swimming capacity (Burness and Leary, 1999).

The implications of ASR are numerous, as are the implications of air breathing, which may be considered another homoplasic characteristic among fishes. However, in several cases, the diversification of air-breathing fishes reflects the successful adaptive radiation of a particular air-breathing type, as in the groups Callichthyidae and Clariidae (Graham, 1997). Fossil records indicate that one of the highly derived living catfishes, *Corydoras*, belongs to the family Callichthyidae, indicating an early Cenozoic differentiation between callichthyids and loricarioids (Lundberg, 1998). The latter is another air-breathing type, which has not been considered to have spread as successfully as the callichthyids. In fact, air breathing occurs 28 times in the Callichthyidae, while only seven species in the Loricariidae have been described as air-breathers.

Similar to ASR, ABO structure cannot be categorized, because no one can distinguish its homology or convergence. The presence of ABO has been described in 49 fish families, all of them presenting different solutions for breathing air. It is most probable that these solutions appeared among ray-finned fishes in response to the same environmental pressure: hypoxia. Air-breathing habits were described early in the literature as a widespread adaptive trait. In 1910, Rauther described them as respiratory adaptations, and subsequent authors have done the same (reviewed by Graham, 1997). According to many authors, the development of air breathing among fish is the result of both habitat and behavioral factors: hypoxia and emergence. Both traits have influenced the origin of this characteristic. It is postulated that no other environmental pressure has been so widespread in the aquatic environment or has occurred throughout the vertebrate evolutionary history that could lead to so many episodes of air breathing as low oxygen availability (Johansen, 1970; Graham *et al.*, 1978). Some researchers have suggested that air breathing in fish arose accidentally (by chance) in fish that were skimming water surfaces (Gans, 1970), or that air breathing was precipitated by changes in water flow (Hora, 1935). However, these interpretations are rather rare.

Regardless of why or how ABO appeared and how ASR developed, fish have found a way to live in environments with low oxygen content. These adaptations allowed them to explore a wide range of ecological niches, and yet most air-breathing species must deal with other types of constraint. Almeida-Val and Hochachka (1995) pointed out that diving into the water bodies and holding their breath for long periods induces metabolic changes in fish, i.e., slowing down total metabolic rates, decreasing oxidative enzyme rates, and increasing anaerobic ability. While these characteristics were first noticed in air-breathing species, further investigations have shown that low metabolic profiles are common in fishes of the Amazon, independent of respiration patterns or life style (Driedzic and Almeida-Val, 1996; West et al., 1999; Table 10.1). Thus, the environmental pressure imposed by low oxygen availability may be considered the main driving force in the development of long-term metabolic adjustments.

Hypoxia affects air-breathing fishes in different ways. Obligate air-breathing species are not strongly influenced by water oxygen availability, since they have reduced gill surface areas. Other air-breathing fishes are affected by hypoxia in different ways; the threshold for oxygen content in the water varies by species. For example, among Amazonian fishes, the jeju (Hoplerythrinus unitaeniatus), a facultative air-breather, starts breathing air when oxygen drops to 81 mmHg (Stevens and Holeton, 1978a), while the armored catfishes (Hypostomus spp.) may seek air when oxygen drops down to 21, 35, or 60, dependent upon the experimental temperature (Gee, 1976; Graham and Baird, 1982; Fernandes and Perna, 1995). The swamp eel (Symbranchus marmoratus), an advanced teleost, may tolerate 33–69 mmHg before starting to breathe air; but in addition to the fact that this species estivates during dry seasons, these thresholds may vary according to body size and hypoxia acclimation (Bicudo and Johansen, 1979; Graham et al., 1987). When we consider respiratory partitioning of oxygen uptake, the so-called facultative air-breathing fishes (Graham, 1997) show a variety of patterns, which are affected by age, water oxygen partial pressure, body size, and temperature. Some Amazon fishes, such as Arapaima gigas, which is considered an obligate air-breather, may breathe 50 to 100% oxygen via air depending on body size and oxygen content in the water (Stevens and Holeton, 1978b). Following these changes in respiration patterns during growth, changes in physiological and biochemical parameters have been observed in Arapaima gigas (Salvo-Souza and Val, 1990). Recent investigations in our laboratory revealed that the relative heart mass of this species also changes, scaling negatively with total body mass; a strong shift is observed when the small juveniles begin to search for air (V. M. F. Almeida-Val and C. Moyes, unpublished data).

ASR is also considered to be affected by aquatic oxygen availability. All of the above-described factors that affect air-breathing behavior also affect ASR. Therefore, decreasing oxygen availability in the water induces ASR in most observed fish species, and the efficiency of this innate behavior in terms of oxygen blood loading was found to be enough to guarantee tambaqui survivorship during long episodes of oxygen depletion (Val, 1995). Juveniles of *Astronotus ocellatus*, a cichlid fish that, in adulthood, tolerates 6 hours under anoxia at 28 °C (Muusze *et al.*, 1998) are able to survive hypoxia indefinitely if allowed to practice ASR (S. C. Land, personal communication), but are not able to tolerate long-term hypoxia if denied access to the water surface (Almeida-Val, Paula-Silva *et al.*, 1999).

D. Cichlids: The Good "Strategists" against Environmental Hypoxia

The cichlids are advanced teleosts that belong to the Perciformes, superorder Acantopterygii (Nelson, 1994). This family is a diversified group (*ca.* 1300 species – (Kullander, 1998) distributed across Africa, Madagascar, Central and South America, Mexico, southern India and Sri Lanka (Kullander and Nijssen, 1989; Kullander, 1998). The vast majority of species is found in African lakes (Lowe-McConnell, 1987; Kullander, 1998). Following the African species, the greatest diversity of cichlids is found in South America (Nelson, 1994), the home of nearly 300 species, which corresponds to 6–10% of the freshwater fish fauna. Approximately 150 species are found in the Amazon basin (Lowe-McConnell, 1991), representing the third most abundant family of fish in the Amazon (Géry, 1984). The plasticity of this group is noticeable in color patterns, shape, feeding behavior, reproduction, and ability to adapt to the most diverse environments. The geographic distribution and genetic-evolutionary characteristics of this family are well documented in both Africa and in South America (Kornfield and Smith, 1982; Kornfield, 1984; Greenwood, 1991; Lowe-McConnell, 1991; Ribbink, 1991; Stiassny, 1991).

In recent studies, Farias and co-workers suggested that neotropical cichlids constitute a monophyletic clade and show rapid rates of evolution, with a significantly higher genetic variability compared to their African counterparts (Farias *et al.*, 1999). Many authors attribute this ability of adaptation to heterogeneous habitats, as well as their rapid adaptive radiation as the causes of the many events of speciation (Fryer and Illes, 1972; Kornfield, 1979, 1984; Stiassny, 1991). Therefore, this group of fishes has been considered extremely plastic (Stiassny, 1991; Almeida-Val, Paula-Silva *et al.*, 1999; Almeida-Val, Val *et al.*, 1999).

Isozymes have proven to be excellent tools to understand the relationship between an animal and its habitat (Kettler and Whitt, 1986; Whitt, 1987). The preferential distribution of isozymes in different organs and tissues of fishes reflects the metabolic adjustments that occur during certain periods, such as stress, growth, migration, or sexual maturation. The choice for a specific isozyme system is normally based on the parameters and metabolic adjustments to be analyzed. The isozyme system of LDH (LDH; E.C. 1.1.1.27) is one of the most studied in vertebrates and has been considered one of the best tools with which to study metabolic adjustments to environmental changes or intrinsic adjustments of animals (Almeida-Val et al., 1995; Almeida-Val, Paula-Silva et al., 1999; Almeida-Val, Val et al., 1999). In vertebrates, LDH is a tetramer composed of two subunits types, LDH-A and LDH-B, which are encoded by two genes. Subunit A is found primarily in skeletal muscle and is very efficient in the conversion of pyruvate to lactate under anaerobiosis. Subunit B is typical of cardiac muscle and is inhibited by high pyruvate concentrations, preventing accumulation of lactate in this organ. A random combination of these subunits results in five isozymes that retain different properties. These isozymes are expressed differentially in different tissues according to energetic needs, oxygen availability and physiological functions (Markert and Holmes, 1969; see also Almeida-Val and Val, 1993 for a review of this subject). An increase in isozyme LDH-A_4 (isoform predominant in vertebrate skeletal muscle) expression in tissues considered typically aerobic, such as the heart and brain, can be detected in animals exposed to hypoxia or anoxia (Hochachka and Storey, 1975). In studies of cichlids exposed to hypoxia, we found that a decrease in LDH-B expression in aerobic tissues may occur after hypoxia exposure, which is combined with an increase in LDH-A, resulting in increased anaerobic capacity (Almeida-Val et al., 1995).

The isozyme systems, especially LDH, have been the subject of many studies on Amazon fishes (Almeida-Val et al., 1992; Almeida-Val and Val, 1993). Two different models have been suggested for LDH metabolic distribution in fish species: (i) predominance of isozyme B_4 in the heart, indicating the maintenance of the aerobic metabolism at low rates during hypoxia episodes; and (ii) low expression of the LDH-B^* gene combined with a strong expression of the LDH-A^* gene in all tissues, suggesting activation of anaerobic metabolism during hypoxia (Almeida-Val et al., 1993). Reduction in LDH-B^* gene expression in the heart was first observed in wild flatfish (Markert and Holmes, 1969) and in sticklebacks (Rooney and Ferguson, 1985). Previous research on rainbow trout LDH (Moon and Hochachka, 1971) and other isozyme systems in fish (Hochachka, 1965; Baldwin and Hochachka, 1970; Schwantes and Schwantes, 1982a,b; De Luca et al., 1983; Almeida-Val et al., 1995; Farias et al., 1997) indicate that

the expression of different enzyme variants depends on environmental parameters. We have shown that LDH tissue distribution in Amazon cichlids is
related to the ability of these animals to tolerate hypoxic environments, and
express some degree of phenotypic plasticity in the heart, therefore revealing
their preferential habitats (Almeida-Val *et al.*, 1995). One good example is the
presence of two models of LDH distribution in the heart of *Cichasoma
amazonarum*. Depending on the availability of oxygen in their habitat, these
animals will show a predominance of isozyme B_4 or A_4 in the heart (Almeida-
Val *et al.*, 1995). When this species is exposed to severe hypoxia (about
30 mmHg) for a long period of time (51 days), significant changes in LDH
distribution may be observed. Isozyme A_4 expression increases in heart and
brain, while isozyme B_4 increases in the liver and "disappears" in the skeletal
muscle. However, the most significant change can be observed through enzyme
assays in the brain, which adopted muscle-type kinetics that caused by the
new LDH isozyme distribution (Almeida-Val *et al.*, 1995; Val *et al.*, 1998).

Other species of Amazon cichlids (*Astronotus ocellatus*, *Cichla monocu-
lus*, *Satanoperca* aff *jurupari*) show the same ability to change LDH distribution as described for *Cichlassoma amazonarum* exposed to hypoxia (Chippari
Gomes *et al.*, 2003). On the other hand, species such as *Geophagus sp*,
Pterophylum sp, *Acarichthys heckelli*, *Crenicichla sp*, *Hypselecara sp*, and
Symphysodon sp present the distribution considered common for all vertebrates: predominance of isozyme A_4 in skeletal muscle and predominance of
isozyme B_4 in the heart. Reduction in cardiac muscle LDH-B* expression is
evident in *Astronotus crassipinnis*, *Heros* sp, *Heros severum*, *Acaronia nassa*
and *Geophagus* cff. *harreri*.

VIII. THE LDH GENE FAMILY AS A STUDY MODEL: REGULATORY AND STRUCTURAL CHANGES AND THEIR EVOLUTIONARY ADAPTIVE ROLES

The radiation of fishes into the richest vertebrate group in terms of
number of species is thought to be an evolutionary success that began
approximately 500 MYA. Successive genome duplication events during the
first radiation episode in vertebrate evolution gave rise to many and multiple
new types of proteins and, therefore, new metabolic possibilities and adaptive opportunities. Gene duplication occurred and continues to occur in
restricted regions of the DNA, resulting in the appearance of new proteins
or new ways of regulating their transcription. During evolution, duplicated
genes may remain similar to the original gene, without any specialization, or
they may differentiate from the original gene, producing proteins for a
specialized new metabolic function. These duplicated and differentiated
genes are homologous, and the proteins produced are termed isoforms, or

isozymes if they are related to enzymes. A third potential fate of duplicated genes is that they may become silenced, i.e., they remain in the genome but are not translated. The occurrence of true novelties, i.e., the appearance of completely new genes, is an exceptional event since there is a strong trend for the preservation of critical sequences of both structural and regulatory genes. Thus, structural and functional properties of single enzymes are helpful in evaluating evolutionary aspects of a given animal group.

Isoforms, or isozymes, are different forms of a single enzyme, which show exactly the same specificity and catalyze the same reaction. No other isozyme system or gene family has been so much investigated as extensively as LDH, and studies have elucidated many metabolic adaptations to different environments as well as many mechanisms that have occurred during evolutionary history of fish groups (Almeida-Val and Val, 1993). The current distribution of LDH isozymes among teleost fishes reflects the presence of three duplicated genes that originated at different times. Recent studies show that an ancient gene (LDH-C), first thought to be the most recent one, gave rise to LDH-A gene, which is highly specialized in anaerobic metabolism, at the beginning of vertebrate evolution (*ca.* 500 MYA). After successive gene duplications, LDH-A gave rise to LDH-B, which is adapted to recovering periods of anaerobiosis, as it has the ability to convert lactate to pyruvate during recovery from periods of hypoxia or estivation (Hochachka, 1980; Whitt, 1984; Crawford *et al.*, 1989).

The distribution of LDH genes in fish is tissue-specific and varies with phylogeny, and the analysis of homologies between them may elucidate several adaptive processes that fish have undergone. Genes generated by evolutionary duplication are referred to as paralogs and are studied from the functional point of view in order to identify preferential metabolism in different organs of the same species. Genes generated by speciation events during evolution of taxonomic groups are referred to as orthologs (same gene in different species) and these are used in comparative metabolic studies (Powers *et al.*, 1983; Powers and Schulte, 1998; Hochachka and Somero, 2002). The following sections summarize most of the investigations on fishes of the Amazon regarding this enzyme family, which has shown to be an excellent tool for evolutionary studies as well as for analysis of metabolic adaptations to environmental changes. The regulation of its genes is described as one of the best processes with which to deal with short-term oscillations in oxygen availability in aquatic environments.

A. Increase in Anaerobic Power during Development

The plasticity in regulating the expression of genes LDH-A* and LDH-B* found in Amazon cichlids indicates the capacity of these animals to base their metabolism on anaerobic glycolysis and in the expression of LDH-A* when oxygen availability is low. Therefore, such plasticity offers these animals the ability to visit localities with low oxygen concentration for feeding or breeding. Cichlids are territorial fish with very aggressive behavior and strong parental care (Chellapa *et al.*, 1999). Therefore, burst swimming is common, and a strong anaerobic power is useful in their tissues. Although the hearts are relatively smaller in cichlids compared to other species (personal observations), it is possible that anaerobic glycolysis takes place during limited oxygen availability and that heart work is sustained for short periods at glucose expenses. The LDH-A isozyme will be more helpful in such situations than LDH-B isozyme. In fact, species in the genus *Astronotus* do tolerate anoxia as adults, and their increased hypoxia tolerance with fish growth may be explained as an increase in their anaerobic power (represented by absolute LDH levels in their tissues and organs) rather than a decrease in specific metabolic rates (Almeida-Val, Paula-Silva *et al.*, 1999; Almeida-Val *et al.*, 2000).

Studies conducted in our laboratory have demonstrated that both LDH enzyme levels and hypoxia survivorship are a function of body mass (Figure 10.6; Almeida-Val, Paula-Silva *et al.*, 1999; Almeida-Val *et al.*, 2000). As fish size increases, the ability to survive under severe hypoxia increases as well. The total amount of time required to reach disequilibrium (i.e., loss of orientation preceding death) increases as the animal increases in size, suggesting that *Astronotus ocellatus* shows an increase in hypoxia tolerance with age (Figure 10.6; Almeida-Val *et al.*, 2000).

Scaling effects on hypoxia tolerance in fishes have not been described for any other fish species, and it is interesting to explore the physiological

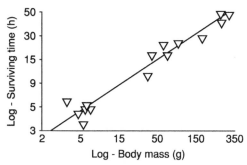

Fig. 10.6 Relationship between body mass (g) of *Astronotus ocellatus* and its ability to survive hypoxia (hours). The log-log regression shows a close relationship ($r = 0.98$). (Redrawn from Almeida-Val *et al.*, 1999, using the following regression coefficients: $a = 1.74$ and $b = 0.59$, derived from $Y = 1.74W^{0.59}$, where Y is the time (h) required for each fish to reach the disequilibrium that immediately precedes death. This value was interpreted as hypoxia survivorship. W = weight in grams.

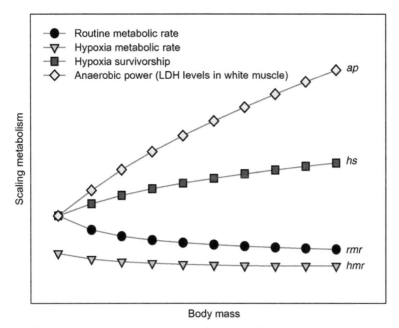

Fig. 10.7 Diagrammatic representation of the relationships between body mass (g) and mass-specific metabolic rate (*rmr*); hypoxia suppressed metabolic rate (*hmr*); hypoxia survivorship (*hs*) and mass-specific anaerobic power (*ap*) obtained from white muscle LDH levels (U.gwt^{-1}) of different-sized *Astronotus ocellatus*. Values for each variable (*Y*) were analyzed in relation to body mass (*X*) using exponential equations where *a* and *b* values were obtained using Sigma Plot software (first described in Almeida-Val, Paula-Silva *et al.*, 1999 and Almeida-Val *et al.*, 2000). As the contribution of anaerobic power increases, there is a decrease in the contribution of suppression of metabolic rate to facilitate survival of hypoxia. These results can be interpreted as an increase in the white muscle somatic index related to liver and heart somatic indices, which remain lower for this species with increased growth.

meaning of these results considering hypoxia defense mechanisms in these groups. Three key factors determine the hypoxia tolerance of any given fish species: absolute *rmr* (routine metabolic rate), hypoxia-mediated suppression rate *hmr* (hypoxia suppressed metabolic rate), and back-up anaerobic mechanisms *ap* (anaerobic power – mainly making up any remaining energy deficit by anaerobic glycolysis). If we assume that *hmr* scales in parallel with *rmr* and that LDH activity *vs.* log M (body mass) is an expression of glycolytic scaling or *ap*, then we can illustrate the relationships of all three above parameters with body mass in a diagram, as in Figure 10.7. These results were first presented in 2000, and since then, we have been investigating other parameters regarding gene expression in different-sized cichlids, mainly *Astronotus* species (Almeida-Val and co-workers, unpublished data).

B. Thermal Properties of Lactate Dehydrogenase

Lactate dehydrogenase (LDH) is a glycolytic enzyme that catalyzes the interconvertion of pyruvate and lactate, using the cofactors NADH and NAD^+, respectively. The temperature-adaptive interspecific variations in its kinetic properties reflect species thermal acclimation in the same way temperature affects enzyme levels. According to Hochachka and Somero (2002) there is a temperature-compensatory modification of enzymatic function that can be observed considering how the catalytic performance of orthologs of A_4-LDH varies among species that have evolved at widely different temperatures. Reviewing data from 18 different species of vertebrates (fish, amphibian, reptilian, mammals, and avian), these authors concluded that interspecific differences in kinetic properties are a reflection of evolution under different thermal conditions, rather than a consequence of the evolutionary lineage to which a species belongs. The relationship between adaptation temperature and catalytic rate constant (k_{cat}) is found in all branches of the evolutionary tree (Hochachka and Somero, 2002). These data are in accordance with the idea that there is a relationship between acclimation temperature and enzyme activity levels, which ultimately reflects metabolic profiles and whole-body metabolic rates of vertebrate species.

Comparisons of K_m (apparent Michaelis–Menten constant) values for orthologs of enzymes from differently adapted species also reveal a common pattern: increases in measurement temperature cause increases in K_m for all orthologs, but at the normal acclimation temperature of differently adapted species, K_m values are highly conserved. Apparent K_m can be used as an approximate index of the affinity between enzyme and substrate or cofactor. A low K_m indicates high affinity for the ligand in question; a high value denotes weak binding. The comparison among different fish species of apparent K_m values of pyruvate for orthologs of LDH-A_4 reveals that higher enzyme affinity occurs at lower temperatures and is distributed in accordance with their thermal ranges (reviewed in Hochachka and Somero, 2002). Studies of A_4-LDH orthologs of congeneric fishes are useful to demonstrate that small differences in maximal habitat temperature are sufficient to favor adaptive changes. Holland *et al.* (1997) studied LDH orthologs of barracudas (genus *Sphyraena*) and suggested that differences in apparent K_m (pyruvate) values are adaptive and related to the thermal regimes to which they are acclimated, i.e., temperate barracuda species show higher K_m values than subtropical and tropical congeners. The number of sequence changes that underlie the different kinetic properties of A_4-LDH orthologs of barracuda congeners is small, since only a single difference at position 8 distinguishes the orthologs of the south temperate species and the subtropical species (reviewed by Hochachka and Somero, 2002).

Another extremely interesting study has been performed by Powers and co-workers, and relates LDH-B* alleles to the temperature gradient in which the killifish *Fundulus heteroclitus* is distributed. Powers *et al.* (1993) showed that in this species only one or two amino acid substitutions are sufficient to explain the differences in kinetics and thermal stability of LDH-B* alleles in populations occurring in the north and south of the north American Atlantic coast. Further studies, based in molecular tools, indicated that a limited number of mutations in the regulatory sequence of LDH-B* gene from *Fundulus heteroclitus* resulted in changes in its expression. Schulte (2001) suggested that increased LDH activity and changes in LDH-B* regulatory sequences, respectively phenotypic and genotypic differences between populations, have been affected by natural selection, rather than genetic drift, in response to the impact of environmental thermal adaptation. Thus, studies of differently adapted confamilial or congeneric and conspecific organisms have shown that temperature differences of a few degrees have sufficient effects on proteins to favor adaptive change (Hochachka and Somero, 2002).

Thermal stability of proteins is linked to conformational flexibility, which in turn is linked to protein function. Because proteins cannot become too rigid, mainly in those regions of the molecule involved in recognizing ligands, thermal stability properties are an important issue from evolutionary point of view. Also, a relative maintenance of kinetic properties regardless temperature acclimation should be advantageous from ecological point of view, particularly in fishes living at high thermal regimes, since adaptations to higher temperatures may induce increases in structural stability of proteins (Hochachka and Somero, 1984). One of the consequences of increasing structural stability by increasing the number of weak bonds is the decrease of catalytic efficiency, increasing the energy of activation. Furthermore, there is a loss in catalytic efficiency as protein stability is increased. If body temperature is related to heat stability, experiments with fish exposed to different thermal regimes should provide some idea about this relationship. For fish of the Amazon, life at a higher temperature range may have selected more stable proteins. Optimal K_m values for fish of the Amazon were suggested to fit between 25 and 30 °C. In fact, the initial measurements by Hochachka and Somero (1968) showed that lungfish LDH has optimal K_m values between 30 and 35 °C. Apparent K_m for pyruvate from muscle LDH of fish species acclimated to extreme thermal regimes showed that optimal K_m values were found to occur in the range of environmental temperatures (reviewed by Val and Almeida-Val, 1995).

In summary, there are at least five different processes by which fish can compensate for temperature changes: (i) changes in substrate concentrations and products in a single pathway; (ii) changes in modulator concentrations affecting the enzymatic reactions; (iii) changes in the enzyme conformations that affects its substrate affinity (K_m) and its velocity (V_{max}); (iv) quantitative changes in the enzyme synthesis by gene regulation; and (v) qualitative changes in isoforms (isozymes). Thus, adaptation may be achieved in different metabolic steps, from transcriptional to post-transcriptional phases. In general, all these mechanisms strongly influence thermal acclimatization or acclimation. Long-term temperature changes are not observed in the current climate of the Amazon region and temperature effectiveness in inducing metabolic changes is probably reduced. Pronounced temperature drops may occur annually, but tend to last no longer than 3 days. Short-term oscillations are mainly compensated for by behavioral responses, e.g., avoiding daily oscillations by migrating to different microhabitats. Long-term climatic changes, however, provide enough time for phenotypic adjustments via transcriptional adjustments, i.e., gene regulation, which result in changes in quantities or types of enzymes (or isozymes). The enzyme properties currently observed in nature are thus the result of genetic modifications over the course of evolution, i.e., in the course of many generations through geological time. These changes include structural changes in enzymes and, consequently, changes in physical and chemical catalytic properties. Current thermal stability of many enzymes should then result from adjustments developed through their past experiences regarding environmental changes and, particularly, thermal regimes.

1. LDH THERMAL SENSIBILITY FOLLOWS THE PHYLOGENY OF MAJOR GROUPS OF AMAZON FISHES

Most enzymes can be characterized by their differential properties of thermal stability. Different isozymes show different thermal stabilities, which may be used to characterize their origins. For more than three decades, these thermal properties were used as powerful tools in identifying gene orthology, i.e., homology between two genes generated by speciation (Wilson *et al.*, 1964; Goldberg and Wuntch, 1967; Hauss, 1975; D'Ávila-Limeira, 1989). Thermal stability properties differ for isozymes encoded at MDH genes, LDH genes, PGI genes, for example. An interesting picture has emerged from the studies by D'Ávila-Limeira (1989) while searching for properties of homology among LDH orthologs from fish of the Amazon. The differences obtained for thermal stability of LDH-B_4 isozyme from 27 different fish species indicate a phylogenetic relationship between maximal thermal stability and species/orders phylogeny. LDH-B_4 thermal stability was found to be lower in non-specialized groups compared to highly specialized groups, such as the advanced teleosts. Further studies have completed this picture, and we have found that among these two extremes, some orders, such as the Curimatidae, that have originated prior to the Andes uplift and have

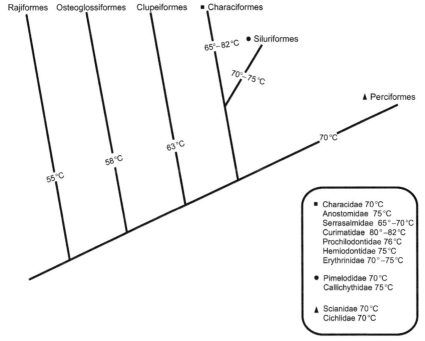

Rajiformes　Osteoglossiformes　Clupeiformes　■ Characiformes

● Siluriformes

65°–82°C

70°–75°C

▲ Perciformes

70°C

63°C

58°C

55°C

■ Characidae 70°C
　Anostomidae 75°C
　Serrasalmidae 65°–70°C
　Curimatidae 80°–82°C
　Prochilodontidae 76°C
　Hemiodontidae 75°C
　Erythrinidae 70°–75°C

● Pimelodidae 70°C
　Callichythidae 75°C

▲ Scianidae 70°C
　Cichlidae 70°C

Fig. 10.8 Simplified phylogenetic diagram of the main groups of fish, modified from Val and Almeida-Val (1995), showing the relationship between the thermal stability of LDH-B* orthologs and the phylogenetic positions of specific fish orders. The box details the amplitude of the range of values obtained for the thermal stability of LDH-B* orthologs for each studied family. These results can be related to thermal history of each group (see text for a detailed explanation).

probably experienced a variety of thermal regimes, show exceptionally high thermal stability temperatures (Figure 10.8; Val and Almeida-Val, 1995).

The main role of enzymes is to decrease the activation energy of a chemical reaction, enabling it to occur at moderate temperatures. The flexibility of the enzyme structure probably occurs during binding events; this modification is known as "induced fit" (Koshland, 1973) or "hand to glove" model (Hochachka and Somero, 1984). Changes in enzyme conformation will be accompanied by changes in the energy inputs or outputs that are associated with conformational changes and are probably responsible for changes in the activation energy. Wilson *et al.* (1964) investigated the thermal stability of LDH-B_4 isozymes in 55 species of vertebrates (from lower to higher taxa, including mammals) and suggested that the increases in heat stability of this isozyme are related to the phylogenetic status of the species. Thus, mammals will be found in one branch of the phylogenetic tree and reptiles and birds on the other branch, the latter having LDHs with 20 °C difference in resistance compared to fish group. Thermal history is thus the most probable pressure determining thermal stability in protein structure, mainly in enzymes that play important roles in metabolism. Endotherms such as mammals and birds, which maintain their body temperature at 37 °C and 39 °C, respectively, may then present proteins with higher heat stability. According to Hochachka and Somero (1984), adaptations to higher temperature regimes induce increases in structural stability of proteins. One of the consequences of such increase is a higher structural stability, which increases the number of weak bonds and reduces catalytic efficiency. There is, however, a loss of catalytic efficiency as protein stability increases. However, for fish of the Amazon acclimated to different thermal ranges, we have found that as acclimation temperature increases, LDH thermal resistance decreases, indicating a compensation for obtaining more efficient catalytic properties (Val and Almeida-Val, 1995). It is clear that the acclimation process affects thermal resistance and that LDH-B_4 orthologs lose their thermal stability at higher temperatures and tend to behave as a thermal independent isozyme.

As stated above, for fish of the Amazon, life at higher temperatures may have imposed compensatory adjustments in their proteins. The catalytic efficiency of many enzymes may be investigated more effectively through their functional properties rather than through their structural properties. Such considerations lead us back to kinetic properties such as K_m, which can be thermally modulated and reflect thermal regimes of species. For Amazonian fishes, optimal K_m for enzymes should lie within the range of 25–30 °C. The comparison of lungfish and tuna muscle LDH suggests the importance of acclimation temperature for metabolic functions. In fact, it is important to keep K_m values close to the substrate concentrations in the tissues; otherwise, the enzyme will not be able to work properly or will require higher activation energy to reach its optimum rates.

2. Why Does Tambaqui LDH not Respond to Temperature Oscillations?

Intracellular pH varies according to temperature changes (Reeves, 1977; Somero, 1981). The combination of pH and temperature has a significant effect on the post-transcriptional regulation of enzyme activity levels in all organisms. According to Hochachka and Lewis (1971), LDH temperature modulation in trout liver may be severely affected in alkaline pHs. Changes in intracellular pH can be harmful to the organism, and metabolic regulation decreases the effects of temperature on physiological pH changes and therefore decreases the effects of temperature on enzyme K_m.

The LDH of Amazonian fishes is similarly affected by temperature and pH, despite the small oscillations in environmental temperatures in the

Amazon basin (Val and Almeida-Val, 1995). Comparative studies of temperature and pH effects on LDH in tambaqui and pacu, *Mylossoma durriventris,* showed that the heart muscle LDH is differently affected by temperature (Almeida-Val *et al.*, 1991) and is minimized when pH is close to the physiological norm. The heart of tambaqui shows no variation in K_m values and substrate saturation values under different assay temperatures when pH is 7.5. Keeping K_m values constant at physiological pH is almost mandatory in this species, since the combination of high temperatures and low oxygen availability requires it to increase anaerobic glycolysis and go to the water surface, which is richer in oxygen but retains higher temperatures. Very low Q_{10} values are evident in heart LDH of tambaqui at different temperature ranges, varying from 0.9 to 1.5, compared to 1.0–2.7 in *Mylossoma durriventris* LDH (Val and Almeida-Val, 1995). Similar results were found after thermal acclimation by tambaqui (Farias, 1992). However, temperate fish species living in different thermal regimes keep their LDH K_m values for pyruvate practically constant when facing a large temperature range (eurythermic), while species facing a narrow temperature range (stenothermic) exhibit compensatory responses to thermal acclimation (Coppes and Somero, 1990). This thermal independence of LDH in tambaqui reflects a reduced change in metabolism and, consequently, a compensatory response to short-term daily oscillations in water surface temperatures. These preliminary results therefore warrant further investigation.

C. Control of LDH and Other Genes Under Hypoxia – Fine Adjustments

As indicated earlier, the ability of organisms to deal with environmental changes depends on the magnitude of the change, the time frame in which the change occurs, and the individual genetic constitution, which may be altered over generations by the selection of genetic variants that are better suited to cope with the new environmental situation. As a consequence, environmental stress has been considered to be among the most important triggers of change in biological organization and functioning during evolution (Almeida-Val, Val *et al.*, 1999). As far as morphology and anatomy are concerned, changes in the structure of DNA and proteins may be tolerated without phenotypic effects, i.e., they may be neutral. This invariance may be the result of chemical redundancy (degeneracy of genetic code, DNA repair, repeated genes, exchangeable amino acids within protein domains), or the result of homeostatic reactions (gene regulation via negative feedback at the level of transcription and translation, physiological homeostasis, pH-buffering). Wilson (1976) called attention to the importance of gene regulation events during evolution of plants and animals. This author stated that "although definitive conclusions are not possible at present, it seems likely that evolution at the organismal level depends predominantly on regulatory gene mutations. Structural gene mutations may have a secondary role in organism evolution." Thus, changes in form, color, morphology, physiology, and metabolism of many organisms may occur according to environmental changes and the investigations about the kind of genetic (or metabolic) control over phenotypes under different environmental conditions have revealed that some genes are turned on or off accordingly (Walker, 1979; Smith, 1990; De Jong, 1995; Land and Hochachka, 1995; Hochachka, 1996; Walker, 1997; Hochachka *et al.*, 1998). As we have seen in the previous sections, long-term adaptational responses to the low-oxygen environments involved oxidative metabolic suppression in fish of the Amazon, as first suggested by Hochachka and Randall (1978) and corroborated by Driedzic and Almeida-Val (1996) and West *et al.* (1999). However, the immediate hypoxia responses from fish of the Amazon have been poorly studied from the evolutionary point of view (reviewed by Almeida-Val, Val *et al.*, 1999).

Oxygen sensing and its physiological and biochemical consequences in cells are not fully understood yet, despite the fact that some mechanisms have been extensively studied in isolated cell models, e.g., rat cardiac myocytes (Webster *et al.*, 1994), rat liver hepatocytes (Keitzmann *et al.*, 1992, 1993), or aquatic turtle hepatocytes (Land and Hochachka, 1995). The best studied system is the type I cell of the carotid body of mammals. All these studies suggest that some DNA sites are suppressed and some are activated when cells are exposed to hypoxia. Hochachka (1996) summarized these data and suggested that up- or down-regulation of certain genes or group of genes is dependent on the intensity of hypoxia constraint and the ability of the model to tolerate this constraint. According to his review, a series of messengers (first and second) will be activated by an oxygen-sensing mechanism that will affect several hundred nuclear genes and 13 mitochondrial genes when the cells are exposed to moderate hypoxia. However, the exposure to severe hypoxia will down-regulate most DNA sites, inducing a decrease in mitochondrial volume densities, a decrease in Krebs cycle enzyme rates, and an increase in the ratios of anaerobic to aerobic pathways. Thus, up-regulation of glycolytic rates is considered to be certain in most hypoxia-responsive tissues. In the last decade, studies on mammalian cells have described a transcriptional factor that coordinates the increased expression of glycolytic enzymes and the decreased expression of aerobic metabolism pathways whose expression is induced by hypoxia: the hypoxia-inducible factor 1 (HIF 1) (Firth *et al.*, 1995; Wang *et al.*, 1995; Ebert *et al.*, 1996; Jiang *et al.*, 1996). These studies were summarized by Hochachka *et al.*

(1998) and showed that most glycolytic enzymes are induced, in a second round of gene expression, by HIF 1. The activation of PFK, PGK (phosphoglycerate kinase), and LDH-A is induced by HIF 1, which in turn is synthesized after a signal transduction pathway is activated by oxygen-sensing mechanisms. In a recent review, Nikinmaa (2002) summarized the most important developments in the field of hypoxia adaptation in fish; more than 120 genes are hypoxia-regulated (Gracey *et al.*, 2001), and up to 40 genes are known to be induced by hypoxia in mammals (Semenza, 1999). A comparison between studies conducted in fish and mammals shows that at least some genes are commonly up-regulated after hypoxia exposure, e.g. glycolytic enzymes such as LDH, enolase, and triosephosphate isomerase. Only recently, the putative O_2 sensor molecule was identified in mammals as a protein belonging to the family of the prolylhydroxilases, which catalyze the hydroxylation of proline 564 of HIF1α, promoting its stabilization in hypoxic cells (Bruick and McKnight, 2001; Yu *et al.*, 2001). Another protein, asparagynyl hydroxylase, may also be involved in HIF1α stabilization and oxygen sensing (Lando *et al.*, 2002).

Studies conducted with cDNA microarrays in the mudsucker *Gillichthys mirabili* by Gracey and co-workers (2001) showed that surviving hypoxia involves three molecular strategies: (i) down-regulating genes for protein synthesis and locomotion to reduce energy consumption; (ii) up-regulating genes for anaerobic ATP production and gluconeogenesis, and (iii) suppressing cell growth and channeling energy to essential metabolic processes. Observed gene expression was tissue-specific and reflected different metabolic roles when the animals experienced hypoxia. Reviewing the evolution of the coordinated regulation of glycolytic enzyme genes by hypoxia, Webster (2003) suggested that the regulation of these genes by hypoxia in insects, fish, reptiles, birds and mammals and possibly all mobile multicellular species is multifactorial, with clear origins in the prokaryotic and fungal regulatory systems. These genes are coordinately and individually regulated by a variety of hypoxia-responsive transcription factors including HIF−1α. HIF−1α is probably the main component and is largely responsible for coordinating the induction. HIF−1α biding sites have been reported in at least eight glycolytic enzyme genes, including PFK, aldolase, PK, PGK, enolase, LDH, HK, and glyceraldehyde phosphate dehydrogenase (GAPDH). Webster (2003) also mentioned that HIF−1 pathway has been described in fish, and it is possible that this pathway developed in the Silurian period about 500 MYA, when highly mobile sea and land species were evolving. In fact, this evolutionary period coincides with high DNA duplication rates due to polyploidy events, which induced the radiation of vertebrate groups and the appearance of new duplicated genes, which gave rise to most gene families currently known to exist.

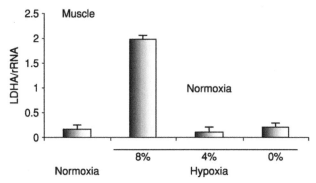

Fig. 10.9 Diagram showing the ability of juvenile *Astronotus* to regulate LDH synthesis in skeletal muscle to differentially respond to hypoxia and anoxia. Experiments were performed over 8 hours ($n = 5$). Results are obtained after RT-PCR, followed by northern-blotting detection. The ability to depress metabolism and, consequently, suppress LDH-A expression is related to the time fish is allowed to acclimate to the new condition. (Redrawn from Oliveira *et al.*, 2002.)

Recent experiments conducted in our laboratory have revealed that, at least for the gene LDH-A*, hypoxia may induce or suppress its expression, according to animal size, and time and intensity of exposure to oxygen depletion. Juveniles *Astronotus* show differential responses, which are related to their ability to tolerate hypoxia and anoxia. Furthermore, the ability to depress metabolism and, consequently, suppress LDH-A expression is related to the time a fish is allowed to acclimate to the new hypoxic condition (Figure 10.9).

In summary, the diversity of strategies adopted by fish during their evolution to cope with hypoxic environments has added to species diversity, making this subject one of the most interesting in the field of comparative physiology. In addition, our knowledge about the mechanisms involved in metabolic adaptations to tolerate hypoxia in fish is far from being complete; as more sophisticated tools become available, the more we realize the size of our knowledge gap regarding adaptations by fish to hypoxia.

IX. WHAT THE FUTURE HOLDS

Two major concerns are worth mentioning in this section. The first is the occurrence of global warming caused by the effects of increased CO_2 release in the atmosphere that is a result of fossil fuel burning and deforestation. The second is the speed with which hypoxia is spreading globally throughout water bodies as a result of pollution. Both problems are related to the main issues discussed in this chapter: temperature and oxygen availability. According to Hochachka and Somero (2002), global warming will certainly affect the strong role of temperature in governing the distribution patterns of organisms in the planet. Thus, understanding the effects of temperature changes in metabolism assumes a practical and economical significance. Some of the questions that may be important to investigate are: (i) what is the minimum amount of change in temperature that will perturb a system enough to induce adaptive responses in a population or species?; (ii) what are the tolerance limits of extant species?; and (iii) at what temperature does a system (whole body or specific metabolic pathway) show optimum performance? These questions remain unanswered for most vertebrate groups, including fishes. Amazonian fishes are already subject to changes in their environment, not only in terms of temperature *per se*, but also in terms of the effects that global warming has on the annual rhythms of flood pulses (e.g., effects of El Niño and of La Niña), which are strongly related to all biological rhythms and processes of organisms living in the Amazon.

On the other hand, hypoxia is a natural phenomenon in the Amazon basin and has imposed so many pressures throughout evolution that most fish groups have become dependent on surface waters, either during their entire life span or during the first part of it. However, healthy environments have been transformed into poorly oxygenated waters due to excessive anthropogenic input of nutrients and organic material, particularly in poorly circulated waters, and this will soon become a problem in the Amazon region. Hypoxia and anoxia currently affect thousands of square kilometers of marine and freshwater environments of North and South America, Africa, Europe, India, Southeast Asia, Australia, Japan, and China.

Although it is a typically hypoxic environment, the Amazon basin has its own rhythms, and any changes in oxygen content of the system will disturb the balance between fish and their environment, just as in any other ecosystem. In addition, some chemical pollutants associated with our modern way of life, including polymers, metals, petroleum, aquaculture nutrients, pesticides, and herbicides, will continuously present new metabolic challenges to fishes of the Amazon. In the past few decades, increases in nutrient levels have become evident in coastal and inland waters, and such increases are caused primarily by intensive farming, application of fertilizers, deforestation, and discharge of domestic wastewaters coincident with human population growth. Thus, hypoxia caused by eutrophication and organic pollution is now considered to be among the most pressing water pollution problems in the world (Wu, 2002).

With regard to promising progress in fish metabolic adaptation, future work will include molecular approaches to investigate gene regulation of proteins involved in adapting to environmental change. Vis-à-vis the above-mentioned review of glycolytic gene regulation (Webster, 2003), it is worth mentioning that HIF-1α pathway evolved concurrently with most of the genes it regulates. The same recognition sequence is essential (although not sufficient) for gene activation by HIF−1, and it is known that this same sequence is required for hypothermia, dehydration, and the UV response of *Arabidopsis* genes involved in carbohydrate metabolism. According to Webster (2003), it is possible that this recognition sequence and the protein that binds it are related in plants and animals, and this may provide the link between gene regulation in hypoxic root tips and the HIF−1 pathway of insects, fish, birds, and mammals. At this point, we can still recommend studies on LDH gene family, which continues to be one of the best studied enzyme systems and one of the most responsive enzymes to several environmental challenges. Changes in protein sequence and expression of LDH-B in *Fundulus heteroclitus*, which are correlated with populations adapted to different thermal regimes, show that there are many ways to approach the role of evolution and natural selection on fish metabolic adaptation: quantitative genetics, comparative methods, and molecular population genetics (Schulte, 2001).

In fact, the question is the same as that posed by Darwin and rephrased by Peter Hochachka in the 1960s: "How have living systems, which are based on a common set of biochemical structures and processes and subject to a common set of physical-chemical laws, been able to adapt to the enormously wide spectrum of environmental conditions found in the biosphere?" Some answers to this question are found in the directions for research on biochemical adaptation registered in the literature produced by Peter Hochachka and his co-workers. In particular, his earlier research insights about Amazon fish adaptations was already very clear in his mind when he wrote the book *Living without Oxygen* in 1980. In his preface to this book he states: "the seed for this book was first planted in my mind during the 1976 expedition to the Amazon, where oxygen is one of the most vicarious of wild environmental parameters faced by aquatic organisms." One of the main legacies Peter Hochachka leaves is his suggestion to follow and pursue the elucidation of mechanisms and their adaptive significance, and to understand the evolution of such mechanisms and adaptations. According to Mangum and Hochachka (1998), what we have missed in early attempts to analyze the evolution of physiological systems was hard evidence, including ways how to test, that the physiological adaptations arose from and were maintained by natural selection, thereby enhancing species fitness. Currently, the most penetrating studies in evolutionary physiology and biochemistry appear to be those that successfully integrate mechanistic/adaptational knowledge with larger evolutionary issues, often using molecular biology techniques to clarify process and phylogeny. Thus, it is important to keep in mind that while the tools do matter, what matters most is the problem to be addressed and the integrative approach to solve it, without neglecting the local issues regarding the evolutionary costs of metabolic and genetic adaptations when facing new environmental challenges.

In conclusion, we suggest that future approaches should take all the above advice into account, but should also consider both the social and economic consequences of human activities before planning the direction of new investigations regarding metabolic adaptations of fish.

ACKNOWLEDGMENTS

We thank Dr William Driedzic for his critical review and valuable suggestions for the manuscript; Mrs Vicky Kjoss for language review; and the National Brazilian Research Council (CNPq) for financial support. Somatic indexes were obtained by Hevea Monteiro Maciel (undergraduate fellowship supported by CNPq). ARCG and NPL were the recipients of PhD fellowships from CNPq. VMFAV is a Research Fellow from CNPq.

REFERENCES

Almeida-Val, V. M. F., and Farias, I. P. (1996). Respiration in fish of the Amazon: Metabolic adjustments to chronic hypoxia. *In* "Physiology and Biochemistry of the Fishes of the Amazon" (Val, A. L., Almeida-Val, V. M. F., and Randall, D. J., Eds.), pp. 257–271. INPA, Manaus.

Almeida-Val, V. M. F., and Hochachka, P. W. (1995). Air-breathing fishes: Metabolic biochemistry of the first diving vertebrates. *In* "Biochemistry and Molecular Biology of Fishes" (Hochachka, P. W., and Mommsen, T., Eds.), Vol. 5, pp. 45–55. Elsevier, Amsterdam.

Almeida-Val, V. M. F., and Val, A. L. (1993). Evolutionary trends of LDH isozymes in fishes. *Comp. Biochem. Physiol.* **105B**(1), 21–28.

Almeida-Val, V. M. F., Farias, I. P., Silva, M. N. P., Duncan, W. P., and Val, A. L. (1995). Biochemical adjustments to hypoxia by Amazon cichlids. *Braz. J. Med. Biol. Res.* **28,** 1257–1263.

Almeida-Val, V. M. F., Paula-Silva, M. N., Caraciolo, M. C. M., Mesquita, L. S. B., Farias, I. P., and Val, A. L. (1992). LDH isozymes in Amazon fish – III. Distribution patterns and functional properties in Serrasalmidae (Teleostei: Ostariophysi). *Comp. Biochem. Physiol.* **103B**(1), 119–125.

Almeida-Val, V. M. F., Paula-Silva, M. N., Duncan, W. P., Lopes, N. P., Val, A. L., and Land, S. C. (1999). Increase of anaerobic potential during growth of an Amazonian cichlid, *Astronotus ocellatus.* Survivorship and LDH regulation after hypoxia exposure. *In* "Biology of Tropical Fishes" (Val, A. L., and Almeida-Val, V. M. F., Eds.), pp. 437–448. INPA, Manaus.

Almeida-Val, V. M. F., Schwantes, M. L. B., and Val, A. L. (1991). LDH isozymes in Amazon fish – II. Temperature effects in LDH kinetic properties from Mylossoma duriventris and Colossoma macropomum (Serrasalmidae). *Comp. Biochem. Physiol.* **98B**, 79–86.

Almeida-Val, V. M. F., Val, A. L., Duncan, W. P., Souza, F. C. A., Paula-Silva, M. N., and Land, S. (2000). Scaling effects on hypoxia tolerance in the Amazon fish *Astronotus ocellatus* (Perciformes: Cichlidae): Contribution of tissue enzyme levels. *Comp. Biochem. Physiol.* **125B**, 219–226.

Almeida-Val, V. M. F., Val, A. L., and Hochachka, P. W. (1993). Hypoxia tolerance in Amazon fishes: Status of an under-explored biological "goldmine." *In* "Surviving Hypoxia: Mechanisms of Control and Adaptation" (Hochachka, P. W., Lutz, P. L., Sick, T., Rosenthal, M., and Van den Thillart, G., Eds.), pp. 435–445. CRC Press, Boca Raton, FL.

Almeida-Val, V. M. F., Val, A. L., and Walker, I. (1999). Long- and short-term adaptation of Amazon fishes to varying O_2-levels: Intra-specific phenotypic plasticity and inter-specific variation. *In* "Biology of Tropical Fishes" (Val, A. L., and Almeida-Val, V. M. F., Eds.), pp. 185–206. INPA, Manaus.

Bailey, G. S., and Driedzic, W. R. (1993). Influence of low temperature acclimation on fate of metabolic fuels in rainbow trout (*Onchorhynchus mykiss*) hearts. *Can J. Zool.* **71**, 2167–2173.

Bailey, J. R., Sephton, D. H., and Driedzic, W. R. (1991). Impact of an acute temperature change on performance and metabolism of pickerel (*Esox niger*) and eel (*Anguilla rostrata*) hearts. *Physiol. Zool.* **64**, 697–716.

Bailey, J. R., Val, A. L., Almeida-Val, V. M. F., and Driedzic, W. R. (1999). Anoxic cardiac performance in Amazonian and north-temperate-zone teleosts. *Can. J. Zool.* **77**(5), 683–689.

Baldwin, J., and Hochachka, P. W. (1970). Functional Significance of Isoenzymes in Thermal Acclimatization. *Biochem. J.* **116**, 883–887.

Berner, R. A., and Canfield, D. E. (1989). A new model for atmospheric oxygen over Phanerozoic time. *Am. J. Sci.* **289**, 333–361.

Bicudo, J. E. P. W., and Johansen, K. (1979). Respiratory gas exchange in the air-breathing fish, *Synbranchus marmoratus*. *Environ. Biol. Fishes* **4**(1), 55–64.

Böhlke, J. E., Weitzman, S. H., and Menezes, N. A. (1978). Estado atual da sistemática dos peixes de água doce da América do Sul. *Acta Amazonica* **8**(4), 657–677.

Boyd, C. E., and Schmitton, H. R. (1999). Achievement of sustainable aquaculture through environmental management. *Aqua. Econ. Manag.* **3**(1), 59–70.

Bruick, R. K., and McKnight, S. L. (2001). A conserved family of prolyl-4-hydroxilases that modify HIF. *Science* **294**, 1337–1340.

Burggren, W., Johansen, K., and McMahon, B. (1985). Respiration in phyletically ancient fishes. *In* "Evolutionary Biology of Primitive Fishes" (Foreman, R. E., Gorbman, A., Dodd, J. M., and Olsson, R., Eds.), pp. 217–252. Plenum Press, New York.

Burness, G. P., and Leary, S. C. (1999). Allometric scaling of RNA, DNA, and enzyme levels: An intraspecific study. *Am. J. Physiol. Regulatory Integrative Comp. Physiol.* **46**, R1164–R1170.

Chagas, E. C. (2001). Influência da suplementação de ácido ascórbico (vitamina C) sobre o crescimento e a resistência ao estresse em tambaqui (*Colossoma macropomum*, Cuvier, 1818). MSc thesis INPA-UFAM Pos Graduate Program, Manaus, 80 p.

Chellapa, S., Yamamoto, M. E., and Cacho, M. S. R. F. (1999). Reproductive behavior and ecology of two species of Cichlid fishes. *In* "Biology of Tropical Fishes" (Val, A. L., and Almeida-Val, V. M. F., Eds.), pp. 113–126. INPA, Manaus.

Chippari Gomes, A. R., Paula-Silva, M. N., Val, A. L., Bicudo, J. E. P., and Almeida-Val, V. M. (2000). Hypoxia tolerance in amazon cichlids. *In* "Proceedings of the IV International Congress on the Biology of Fish," pp. 43–54. Aberdeen.

Chippari Gomes, A. R. (2002). *Adaptações metabólicas dos ciclídeos aos ambientes hipóxicos da Amazônia*. Tese de Doutorado, Programa Integrado de Pós-Graduação em Biologia Tropical e Recursos Naturais, Programa de Biologia de Água Doce e Pesca Interior, INPA-UFAM, Manaus.

Chippari Gomes, A. R., Lopes, N. P., Paula-Silva, M. N., Oliveira, A. R., and Almeida-Val, V. M. F. (2003). Hypoxia tolerance and adaptations in fishes: The case of Amazon cichlids. *In* "Fish Adaptations" (Val, A. L., and Kapoor, B. G., Eds.), pp. 37–54. Science Publishers, Inc. Enfield, NH.

Coppes, Z. L., and Somero, G. N. (1990). Temperature-adaptive differences between the M_4 lactate dehydrogenases of stenothernal and eurythermal sciaenid fishes. *J. Exp. Zool.* **254**, 127–131.

Coulter, G. H. (1991). "Lake Tanganyika and Its Life." Oxford University Press, New York.

Cox-Fernandes, C. (1989). *Estudo das migrações laterais de peixes do sistema lago do Rei (Ilha do Careiro).* Dissertação de Mestrado, Programa Integrado de Pós-Graduação em Biologia Tropical e Recursos Naturais, Programa de Biologia de Água Doce e Pesca Interior INPA-UFAM, Manaus.

Crampton, W. G. R. (1998). Effects of anoxia on the distribution, respiratory strategies and electric signal diversity of gymnotiform fishes. *J. Fish Biol.* **53**, 307–330.

Crawford, D. L., Constantino, H. R., and Powers, D. A. (1989). Lactate dehydrogenase B c-DNA from the teleost *Fundulus heteroclitus*: Evolutionary implications. *Mol. Biol. Evol.* **6**, 369–383.

Crockett, E. L., and Sidell, B. D. (1990). Some pathways of energy metabolism are cold adapted in Antarctic fishes. *Physiol. Zool.* **63**, 472–488.

D'Ávila-Limeira, N. (1989). *Estudos sobre a lactato desidrogenase (LDH) em 27 espécies de peixes da bacia amazônica: Aspectos adaptativos e evolutivos.* Dissertação de Mestrado, Programa Integrado de Pós-Graduação em Biologia Tropical e Recursos Naturais, Programa de Biologia de Água Doce e Pesca Interior, INPA-UFAM, Manaus.

Davis, J. C. (1975). Minimal dissolved oxygen requirements of aquatic life with emphasis on Canadian species: A review. *J. Fish. Res. Board Can.* **32**, 2295–2332.

De Jong, G. (1995). Phenotypic plasticity as a product of selection in a variable environment. *Am. Nat.* **145**, 493–512.

De Luca, P. H., Schwantes, M. L. B., and Schwantes, A. R. (1983). Adaptive features of ectothermic enzymes – IV. Studies on malate dehydrogenase of *Astyanax fasciatus* (Characidae) from Lobo Reservoir (São Carlos, São Paulo, Brasil). *Comp. Biochem. Physiol.* **74B**, 315–324.

Driedzic, W. R. (1988). Matching of cardiac oxygen delivery and fuel supply to energy demand in teleosts and cephalopods. *Can J. Zool.* **66**, 1078–1083.

Driedzic, W. R. (1992). Cardiac Energy Metabolism. *In* "Fish Physiology" (Hoar, W. S., Randall, D. J., and Farrell, A. P., Eds.), Vol. XIIA, pp. 219–266. Academic Press, New York.

Driedzic, W. R., and Almeida-Val, V. M. F. (1996). Enzymes of cardiac energy metabolism in Amazonian teleosts and the fresh-water stingray (*Potamotrygon hystrix*). *J. Exp. Zool.* **274**, 327–333.

Driedzic, W. R., and Bailey, G. S. (1999). Anoxia cardiac tolerance in Amazonian and North temperate teleosts is related to the potential to utilize extracellular glucose. *In* "Biology of Tropical Fishes" (Val, A. L., and Almeida-Val, V. M. F., Eds.), pp. 217–227. INPA, Manaus.

Driedzic, W. R., and Gesser, H. (1994). Energy metabolism and contractility in ectothermic vertebrate hearts: Hypoxia, acidosis, and low temperature. *Physiol. Rev.* **74**, 221–258.

Dunn, J. F., Hochachka, P. W., Davison, W., and Guppy, M. (1983). Metabolic adjustments to diving and recovery in the African lungfish. *Am. J. Physiol.* **245**, R651–R657.

Dunn, J. F., and Hochacka, P. W. (1986). Metabolic responses of trout (*Salmo Gairdneri*) to acute Environmental Hypoxia. *J. Exp. Biol.* **123**, 229–242.

Ebert, B. L., Gleadle, J. M., O'Rourke, J. F., Bartlett, S. M., Poulton, J., and Ratcliffe, P. J. (1996). Isoenzyme-specific regulation of genes involved in energy metabolism by hypoxia: Similarities with the regulation of erythropoietin. *Biochem. J.* **313**, 809–814.

Farias, I. P. (1992). *Efeito da aclimatação térmica sobre a lactato desidrogenase de Colossoma macropomum e Hoplosternun littorale (Amazonas, Brasil).* Dissertação de Mestrado, Programa Integrado de Pós-Graduação em Biologia Tropical e Recursos Naturais, Programa de Biologia de Água Doce e Pesca Interior INPA-UFAM, Manaus.

Farias, I. P., Ortí, G., Sampaio, I., Schneider, H., and Meyer, A. (1999). Mitochondrial DNA phylogeny of the family Cichlidae: Monophyly and fast molecular evolution of the neotropical assemblage. *J. Mol. Evol.* **48**, 701–711.

Farias, I. P., Paula-Silva, M. N., and Almeida-Val, V. M. F. (1997). No co-expression of LDH-C in Amazon cichlids. *Comp. Biochem. Physiol.* **117B**, 315–319.

Farias, I. P., Schneider, H., and Sampaio, I. (1998). Molecular phylogeny of neotropical cichlids: The relationships of cichlasomines and heroines. *In* "Phylogeny and Classification of Neotropical Fishes. Part 5 – Perciformes" (Malabarba, L. R., Reis, R. E., Vari, R. P., Lucena, Z. M., and Lucena, C. A. S., Eds.), pp. 499–508. Edipucrs, Porto Alegre.

Fernandes, M. N., and Perna, S. A. (1995). Internal morphology of the gill of a loricariid fish, *Hypostomus plecostomus*: Arterio-arterial vasculature and muscle organization. *Can. J. Zool.* **73**, 2259–2265.

Firth, J. D., Ebert, B. L., and Ratclife, P. J. (1995). Hypoxic regulation of LDH A: Interaction between hypoxia inducible factor I and camp response elements. *J. Biol. Chem.* **270**, 21021–21027.

Fry, F. E. J., Black, V. S., and Black, E. C. (1947). Influence of temperature on the asphyxiation of young goldfish (*Carassius auratus* L.) under various tensions of oxygen and carbon dioxide. *Biol. Bull.* **92**, 217–224.

Fryer, G., and Illes, T. D. (1972). "The Cichlid Fishes of the Great Lakes of Africa: Their Biology and Evolution." Oliver and Boyd, London.

Futuyma, D. (1986). "Evolutionary Biology" 2nd edn. Sinauer, Sunderland.

Gans, C. (1970). Strategy and sequence in the evolution of the external gas exchangers of ectothermal vertebrates. *Forma et Function* **3**, 61–104.

Gee, J. H. (1976). Buoyancy and aerial respiration: Factors influencing the evolution of reduced swim-bladder volume of some Central American catfishes (Trichomycteridae, Callichthyidae, Loricariidae, Astroblepidae). *Can. J. Zool.* **54**, 1030–1037.

Géry, J. (1984). The fishes of the Amazonia. *In* "The Amazon. Limnology and Landscape Ecology of a Mighty Tropical River and its Basin" (Sioli, H., Ed.), pp. 15–46. W. Junk, Dordrecht.

Goldberg, E., and Wuntch, T. (1967). Electrophoretic and kinetic properties of *Rana Pipiens* LDH isozymes. *J. Exp. Zool.* **165**, 101–110.

Gracey, A. Y., Troll, J. V., and Somero, G. N. (2001). Hypoxia-induced gene expression profiling in the euryoxic fish *Gillichthys mirabilis*. *Proc. Natl Acad. Sci. USA* **98**, 1993–1998.

Graham, J. B. (1997). "Air-Breathing Fishes: Evolution, Diversity and Adaptation." Academic Press, San Diego.

Graham, J. B., and Baird, T. A. (1982). The transition to air-breathing in fishes: I. Environmental effects on the facultative air-breathing of *Ancistrus chagresi* and *Hypostomus plecostomus* (Loricariidae). *J. Exp. Biol.* **96**, 53–67.

Graham, J. B., Baird, T. A., and Stockmann, W. (1987). The transition to air-breathing in fishes. *J. Exp. Biol.* **129**, 81–106.

Graham, J. B., Rosenblatt, R. H., and Gans, C. (1978). Vertebrate air-breathing arose in fresh waters and not in the oceans. *Evolution* **32**, 459–463.

Greenwood, P. H. (1991). Speciation. *In* "Cichlid Fishes – Behaviour, Ecology, and Evolution" (Keenleyside, M. H. A., Ed.), pp. 86–102. Chapman & Hall, London.

Guderley, H. E., and Gawlicka, A. (1992). Qualitative modification of muscle metabolic organization with thermal acclimation of rainbow trout, *Oncorrhynchus mykiss*. *Fish Physiol. Biochem.* **10**, 123–132.

Hammer, C., and Purps, M. (1996). The metabolic exponent of *Hoplosternun littorale* in comparison with Indian air-breathing catfish, with methodological investigation on the nature of metabolic exponent. *In* "Physiology and Biochemistry of the Fishes of the

Amazon" (Val, A. L., Almeida-Val, V. M. F., and Randall, D. J., Eds.), pp. 283–297. INPA, Manaus.

Hauss, R. (1975). Wirkungen der Temperaturen auf Proteine. Enzyme und Isozenzyme aus organen des fisches Rhodeus amarus. Bloch. I: Einfluβ der Adaptationstemperatur auf die weiβe Ruckenmuskulatur. *Zool. Anz.* **194,** 243–261.

Hochachka, P. W. (1965). Isoenzymes in Metabolic Adaptation of a Poikilotherm: Subunit Relationships in Lactic Dehydrogenase of Goldfish. *Arch. Biochem. Biophys.* **111,** 96–103.

Hochachka, P. W. (1979). Cell metabolism, air-breathing, and the origins of endothermy. *In* "Evolution of Respiratory Process" (Wood, S. C., and Lenfant, C., Eds.), pp. 253–288. Marcel Dekker Inc., New York.

Hochachka, P. W. (1980). "Living Without Oxygen." Harvard University Press, Cambridge, MA.

Hochachka, P. W. (1988). The nature of evolution and adaptation: Resolving the unity-diversity paradox. *Can. J. Zool.* **66,** 1146–1152.

Hochachka, P. W. (1996). Oxygen sensing and metabolic regulation: Short, intermediate, and long term roles. *In* "Physiology and Biochemistry of the Fishes of the Amazon" (Val, A. L., Almeida-Val, V. M. F., and Randall, D. J., Eds.), pp. 233–256. INPA, Manaus.

Hochachka, P. W., and Hulbert, W. C. (1978). Glycogen "seas" glycogen bodies, and glycogen granules in heart and skeletal muscle of two air-breathing, burrowing fishes *Can. J. Zool.* **56,** 774–786.

Hochachka, P. W., and Lewis, J. K. (1971). Interacting effects of pH and temperature on the Km values for fish tissue lactate dehydrogenases. *Comp. Biochem. Physiol.* **39B,** 925–933.

Hochachka, P. W., and Randall, D. J. (1978). Alpha-Helix Amazon expedition, September–October 1976. *Can. J. Zool.* **56,** 713–716.

Hochachka, P. W., and Somero, G. N. (1968). The adaptation of enzymes to temperature. *Comp. Biochem. Physiol.* **27,** 659–663.

Hochachka, P. W., and Somero, G. N. (1973). "Strategies of Biochemical Adaptation." W. B. Saunders, Philadelphia.

Hochachka, P. W., and Somero, G. N. (1984). "Biochemical Adaptation." Princeton University Press, Princeton, NJ.

Hochachka, P. W., and Somero, G. N. (2002). "Biochemical Adaptation. Mechanism and Process in Physiological Evolution." Oxford University Press, New York.

Hochachka, P. W., and Storey, K. B. (1975). Metabolic consequences of diving in animals and man. *Science* **187,** 613–621.

Hochachka, P. W., Buck, L. T., Doll, C. J., and Land, S. C. (1996). Unifying theory of hypoxia tolerance: Molecular/metabolic defense and rescue mechanisms for surviving oxygen lack. *Proc. Natl Acad. Sci. USA* **93,** 9493–9498.

Hochachka, P. W., McClelland, G. B., Burness, C. P., Staples, J. F., and Suarez, R. K. (1998). Integrating metabolic pathway fluxes with gene-to-enzyme expression rates. *Comp. Biochem. Physiol.* **120B,** 17–26.

Hölker, F. (2003). The metabolic rate of roach in relation to body size and temperature. *J. Fish Biol.* **62,** 565–579.

Holland, L. Z., McFall-Ngai, M., and Somero, G. N. (1997). Evolution of lactate dehydrogenase-A homologs of barracuda fishes (genus *Sphyraena*) from different thermal environments: Differences in kinetic properties and thermal stability are due to amino acid substitutions outside the active site. *Biochemistry* **36,** 3207–3215.

Hora, S. L. (1935). Physiology, bionomics, and evolution of air-breathing fishes of India. *Trans. Nat. Inst. Sci. India* **I**(1), 1–16.

Jiang, B.-H., Rue, E. A., Wang, G. L., Roe, R., and Semenza, G. L. (1996). Dimerization, DNA biding, and transactivation properties of hypoxia-inducible factor 1. *J. Biol. Chem.* **271,** 17771–17778.

Jobling, M. (1994). Fish Bioenergetics. Chapman & Hall, London.

Johansen, K. (1970). Cardiorespiratory adaptations in the transition from water breathing to air breathing. Introduction. *Fed. Proc.* **29**(3), 1118–1119.

Johnston, I. A. (1977). A comparative study of glycolysis in red and white muscles of the trout (*Salmo gairdneri*) and mirror carp (*Cyprinus carpio*). *J. Fish Biol.* **11**, 575–588.

Johnston, I. A., Sidell, B. D., and Driedzic, W. R. (1985). Force-velocity characteristics and metabolism of carp muscle fibres following temperature acclimation. *J. Exp. Biol.* **119**, 239–249.

Jones, P., and Sidell, B. (1982). Metabolic responses of striped bass (*Morone saxatilis*) to temperature acclimation. II. Alterations in metabolic carbon sources and distribution of fiber types in locomotory muscle. *J. Exp. Zool.* **219**, 163–171.

Junk, W. J. (1984). Ecology of the várzea, floodplain of Amazonian whitewater rivers. *In* "The Amazon. Limnology and Landscape Ecology of a Mighty Tropical River and Its Basin" (Sioli, H., Ed.), pp. 215–244. W. Junk, Dordrecht.

Junk, W. J., Bayley, P. B., and Sparks, R. E. (1989). The flood pulse concept in river-floodplain systems. *In* "Proceedings of the International Large River Symposium" (Dodge, D. P., Ed.), Vol. 106, pp. 110–127. Can. Spec. Publ. Fish. Aquat. Sci., Canada.

Junk, W. J., Soares, M. G., and Carvalho, F. M. (1983). Distribution of fish species in a lake of the Amazon river floodplain near Manaus (lago Camaleao), with special reference to extreme oxygen conditions. *Amazoniana* **7**, 397–431.

Keitzmann, T., Schimidt, H., Probst, I., and Jungermann, K. (1992). Modulation of the glucagons-dependent activation of the PEPCK gene by oxygen in rat hepatocyte cultures. *FEBS Lett.* **311**, 251–255.

Keitzmann, T., Schmidt, H., Unthan-Feschner, K., Probst, I., and Jungermann, K. (1993). A ferro-heme protein senses oxygen levels which modulate the glucagons dependent activation of the PEPCK gene in rat hepatocyte cultures. *Biochem. Biophys. Res. Comm.* **195**, 792–798.

Kettler, M. K., and Whitt, G. S. (1986). An apparent progressive and current evolutionary restriction in tissue expression of a gene, the lactate dehydrogenase-C gene, within a family of bony fish (Salmoniformes: Umbridae). *J. Mol. Evol.* **23**, 95–107.

Kornfield, I. L. (1979). Evidence for rapid speciation in African cichlid fishes. *Experientia* **34**, 335–336.

Kornfield, I. (1984). Descriptive genetics of cichlid fishes. *In* "Evolutionary Genetics of Fishes" (Turner, U. B., Ed.), pp. 591–617. Plenum Press, New York.

Kornfield, I., and Smith, D. C. (1982). The Cichlid fish of cuatro ciénegas, México: Direct Evidence of Conspecificity Among Distinct Trophic Morphes. *Evolution* **36**, 658–664.

Koshland, H. A. (1973). Protein shape and biological control. *Sci. Am.* **229**, 52–64.

Kramer, D. L., Lindsey, C. C., Moodie, G. E. E., and Stevens, E. D. (1978). The fishes and the aquatic environment of the central Amazon basin, with particular reference to respiratory patterns. *Can. J. Zool.* **56**, 717–729.

Kullander, S. O. (1998). A phylogeny and classification of the South American Cichlidae (Teleostei: Perciformes). *In* "International Symposium on Phylogeny and Classification of Neotropical Freshwater Fishes," pp. 1–52. Porto Alegre, Brazil.

Kullander, S. O., and Nijssen, H. (1989). "The Cichlid of Surinam (*Teleostei: Labrodei*)." E. J. Brill, Leiden.

Land, S. C., and Hochachka, P. W. (1995). A heme-protein-based oxygen-sensing mechanism controls the expression and suppression of multiple proteins in anoxia – tolerant turtle hepatocytes. *Proc. Natl Acad. Sci. USA* **92**, 7505–7509.

Lando, D., Peet, D. J., Whelan, D. A., Gorman, J. J., and Whitelaw, M. L. (2002). Asparagine hydroxylation of the HIF transactivation domain: A hypoxic switch. *Science* **295**, 858–861.

Lopes, N. P. (1999). Perfil metabólico de duas espécies de peixes migradores do gênero *Prochilodus* MSc thesis INPA-UFAM Pos Graduate Program, Manaus, 66 p.

Lopes, N. P. (2003). Ajustes metabólicos em sete espécies de Siluriformes sob condições hipóxicas: Aspectos adaptativos. Dissertação de Mestrado, Programa Integrado de Pós -Graduação em Biologia Tropical e Recursos Naturais, Programa de Biologia de Água Doce e Pesca Interior INPA-UFAM, Manaus.

Lowe-McConnell, R. H. (1987). "Ecological Studies in Tropical Fish Communities." Cambridge University Press, Cambridge.

Lowe-McConnell, R. H. (1991). Ecology of cichlids in South American and African waters excluding the African Great Lakes. *In* "Cichlid Fishes: Behavior, Ecology and Evolution" (Keenleyside, M. H. A., Ed.), Vol. 3, pp. 61–85. Croom Helm, London.

Lundberg, J. G. (1998). The temporal context for the diversification of neotropical fishes. *In* "Phylogeny and Classification of Neotropical Fishes. Part 1 – Fossils and Geological Evidence" (Malabarba, L. R., Reis, R. E., Vari, R. P., Lucena, Z. M., and Lucena, C. A. S., Eds.), pp. 49–68. Edipucrs, Porto Alegre.

MacCormack, T. J., Lindsey, R. S., Roubach, R., Almeida-Val, V. M. F., Val, A. L., and Driedzic, W. R. (2003). Changes in ventilation, metabolism, and behaviour, but not bradycardia, contribute to hypoxia survival in two species of Amazonian armoured catfish. *Can. J. Zool.* **81,** 272–280.

Mangum, C. P., and Hochachka, P. W. (1998). New directions in comparative physiology and biochemistry: Mechanisms, adaptations, and evolution. *Physiol. Zool.* **71,** 471–484.

Markert, C. L., and Holmes, R. S. (1969). Lactate dehydrogenase isozymes of the flatfish, pleuronectiformes: Kinetic, molecular, and immunochemical analysis. *J. Exp. Zool.* **171,** 85–104.

Mesquita-Saad, L. S. B., Leitão, M. A. B., Paula-Silva, M. N., Chippari Gomes, A. R., and Almeida-Val, V. M. F. (2002). Specialized metabolism and biochemical suppression during aestivation of the extant South American lungfish – *Lepidosiren paradoxa. Braz. J. Biol.* **62,** 495–501.

Moon, T. W., and Hochachka, P. W. (1971). Effect of thermal acclimation on multiple forms of the liver-soluble NADP-linked isocitrate dehydrogenase in the Family Salmonidae. *Comp. Biochem. Physiol.* **40B,** 207–213.

Mottishaw, P. D., Thornton, S. J., and Hochachka, P. W. (1999). The diving response mechanism and its surprising evolutionary path in seals and sea lions. *Am. Zool.* **39,** 434–450.

Moyes, C. D., Buck, L. T., Hochachka, P. W., and Suarez, R. K. (1989). Oxidative properties of Carp Red and White Muscle. *J. Exp. Biol.* **143,** 321–331.

Moyes, C. D., Mathieu-Costello, O., Brill, R. W., and Hochachka, P. W. (1992). Mitochondrial metabolism of cardiac and skeletal muscles from a fast and a slow fish. *Can. J. Zool.* **70,** 1246–1253.

Muusze, B., Marcon, J., Van den Thillart, G., and Almeida-Val, V. (1998). Hypoxia tolerance of Amazon fish: Respirometry and energy metabolism of the cichlid *Astronotus ocellatus. Comp. Biochem. Physiol.* **120A,** 151–156.

Nelson, J. S. (1994). "Fishes of the World" 3rd edn. Wiley, New York.

Nikinmaa, M. (2002). Oxygen-dependent cellular functions – why fishes and their aquatic environment are a prime choice of study. *Comp. Biochem. Physiol.* **133A,** 1–16.

Oliveira, A. R., Val, A. L., Paula-Silva, M. N., *et al.* (2002). LDH gene responses to hypoxia and anoxia in *Astronotus crassipinus. In* "Responses of Fish to Aquatic Hypoxia" (Randall, D., and MacKinlay, D., Eds.). *Proc. Int. Symp. Congress on the Biology of Fish,* July 2002. Vancouver, Canada, pp. 65–69.

Pauly, D. (1998). Tropical fishes: Patterns and propensities. *J. Fish Biol.* **53**(Supplement A), 1–17.

Pigliucci, M. (1996). How organisms respond to environmental changes: From phenotypes to molecules (and vice versa)? *Tree* **11,** 168–173.

Pörtner, H. (1993). Multicompartmental analyses of acid-base and metabolic homeostasis during anaerobiosis: Invertebrate and lower vertebrate examples. *In* "Surviving Hypoxia:

Mechanisms of Control and Adaptation" (Hochachka, P., Lutz, P., Sick, T. J., Rosenthal, M., and Van DenThillart, G. E. E. J. M., Eds.), pp. 330–357. CRC Press, Boca Raton, FL.

Pörtner, H., and Grieshaber, M. (1993). Characteristics of the critical PO_2 (s): Gas exchange, metabolic rate and the mode of energy production. *In* "The Vertebrate Gas Transport Cascade: Adaptations to Environment and Mode of Life" (Bicudo, J. E. P. W., Ed.), pp. 330–357. CRC Press, Boca Raton, FL.

Powell, W. H., and Hahn, M. E. (2002). Identification and functional characterization of hypoxia-inducible Factor 2x from the estuarine teleost, *Fundulus heteroclitus*: Interaction of HIF−2x with two ARNT2 splice variants. *J. Exp. Zool.* **294**, 17–29.

Powers, D. A., DiMichele, L., and Place, A. R. (1983). The use of enzyme kinetics to predict differences in cellular metabolism, developmental rate, and swimming performance between LDH-B genotypes of the fish, *Fundulus heteroclitus*. *In* "Isozymes: Current Topics in Biological and Medical Research," Vol. 10, "Genetics and Evolution," pp. 147–170. Alan R. Liss, New York.

Powers, D. A., and Schulte, P. M. (1998). Evolutionary adaptations of gene structure and expression in natural populations in relation to changing environment: A multidisciplinary approach to address the million-year saga of a small fish. *J. Exp. Zool.* **282**, 71–94.

Powers, D. A., Smith, I., Gonzalez-Villasenor, L., DiMichelle, L., and Crawford, D. L. (1993). A multidisciplinary approach to the selectionist/neutralist controversy using the model teleost *Fundulus heteroclitus*. *In* "Oxford Surveys in Evolutionary Biology" (Futuyma, D., and Antonovics, J., Eds.), Vol. 9, pp. 43–107. Oxford University Press, Oxford.

Prosser, C. L. (1991). Temperature. *In* "Environmental and Metabolic Animal Physiology" (Prosser, C. L., Ed.), pp. 109–165. Wiley, New York.

Randall, D. J., Burggren, W. W., Farrell, A. P., and Haswell, M. S. (1981). "The Evolution of Air-breathing Vertebrates." Cambridge University Press, Cambridge.

Rapp Py-Daniel, L. (2000). Tracking evolutionary changes in Siluriformes (Teleostei: Ostariophysi). *In* "International Congress on the Biology of Fish," pp. 57–72. Physiology Section, American Fisheries Society, Aberdeen.

Rapp-Py-Daniel, L. H., and Leão, E. L. M. (1981). A coleção de peixes do INPA: Base do conhecimento científico sobre a ictiofauna amazônica gerada pelo Instituto Nacional de Pesquisas da Amazônia. *In* "Bases científicas para estratégias de preservação e desenvolvimento da Amazônia: Fatos e perspectivas" (Val, A. L., Feldberg, E., and Figliuolo, R., Eds.), pp. 299–312. INPA, Manaus.

Reeves, R. B. (1977). The interaction of body temperature and acid-base balance in ectothermic vertebrates. *Ann. Rev. Physiol.* **39**, 559–586.

Ribbink, A. J. (1991). Distribution and ecology of the cichlids of the African Great Lakes. *In* "Cichlid Fishes – Behaviour, Ecology and Evolution" (Keenleyside, M. H. A., Ed.), pp. 36–59. Chapman & Hall, London.

Roberts, T. R. (1972). Ecology of the fishes in the Amazon and Congo basins. *Bull. Mus. Comp. Zool.* **143**, 117–147.

Rooney, C. H. T., and Ferguson, A. (1985). Lactate dehydrogenase isozymes and allozymes of nine-spined stickleback *Pungitius pungitius* (L.) (Osteichthyes: Gasteroteidae). *Comp. Biochem. Physiol.* **81B**, 711–715.

Rose, M., and Lauder, G. (1996). "Adaptation." Academic Press, San Diego.

Salvo-Souza, R. H., and Val, A. L. (1990). O gigante das águas amazônicas. *Ciência Hoje* **11**(64), 9–12.

Sartoris, F. J., Bock, C., Serendero, G., Lannig, G., and Pörtner, H. (2003). Temperature-dependent changes in energy metabolism, intracellular pH and blood oxygen tension in the Atlantic cod. *J. Fish Biol.* **62**, 1239–1253.

Schichting, C., and Pigliucci, M. (1993). Evolution of phenotypic plasticity via regulatory genes. *Am. Nat.* **142**, 366–370.

Schichting, C., and Pigliucci, M. (1995). Gene regulation, quantitative genetics and the evolution of reaction norms. *Evol. Ecol.* **9**, 154–168.

Schulte, P. M. (2001). Environmental adaptations as windows on molecular evolution. *Comp. Biochem. Physiol.* **128B**, 597–611.

Schwantes, M. L. B., and Schwantes, A. R. (1982). Adaptive features of ecothermic enzymes: II The effects of acclimation temperature on the malate dehydrogenase of the spot, *Leiostomus xanthurus. Comp. Biochem. Physiol.* **72B**, 59–64.

Schwantes, M. L. B., and Schwantes, A. R. (1982). Adaptive features of ectotermic enzymes – I. Temperature effects on the malate dehydrogenase from a temperate fish *Leiostomus xanthurus. Comp. Biochem. Physiol.* **72B**, 49–58.

Semenza, G. L. (1999). Regulation of mammalian O_2 homeostasis by hypoxia-inducible factor 1. *Annu. Rev. Cell Develop. Biol.* **15**, 551–578.

Sephton, D. H., and Driedzic, W. R. (1991). Effect of acute and chronic temperature transition on enzymes of cardiac metabolism in white perch (*Morone americana*), yellow perch (*Perca flavescens*), and smallmouth bass (*Micropterus dolumieui*). *Can. J. Zool.* **69**, 258–262.

Shoubridge, E. A., and Hochachka, P. W. (1983). The integration and control of metabolism in the anoxic goldfish. *Mol. Physiol.* **4**, 165–195.

Sidell, B. D., Crockett, E. L., *et al.* (1995). Antarctic fish tissues preferentially catabolize monoenoic fatty acids. *J. Exp. Zool.* **271**, 73–81.

Sidell, B. D., and Driedzic, W. R. (1985). Relationship between cardiac energy metabolism and work demand in fishes. *In* "Respiration, Metabolism, and Circulation" (Gilles, R., Ed.), pp. 386–401. Springer-Verlag.

Smith, H. (1990). Signal perception, differential expression within multigene families and the molecular basis of phenotypic plasticity. *Plant Cell Environ.* **13**, 585–594.

Soitamo, A. J., Rabergh, C. M. I., Gassmann, M., Sistonen, L., and Niknmaa, M. (2001). Accumulation of protein occurs at normal venous oxygen tension. *J. Biol. Chem.* **276**, 20.

Somero, G. N. (1981). pH-temperature interactions on proteins: Principles of optimal pH and buffer system design. *Mar. Biol. Lett.* **2**, 163–178.

Stevens, E. D., and Holeton, G. F. (1978). The partitioning of oxygen uptake from air and from water by erythrinids. *Can. J. Zool.* **56**, 965–969.

Stevens, E. D., and Holeton, G. F. (1978). The partitioning of oxygen uptake from air and from water by the large obligate air-breathing teleost pirarucu (*Arapaima gigas*). *Can. J. Zool.* **56**, 974–976.

Stiassny, M. L. J. (1991). Phylogenetic interrelationships of the family Cichlidae: An overview. *In* "Cichlid Fishes – Behaviour, Ecology and Evolution" (Keenleyside, M. H. A., Ed.), pp. 1–31. Chapman & Hall, London.

Val, A. L. (1993). Adaptations of fishes to extreme conditions in fresh waters. *In* "The Vertebrate Gas Transport Cascade. Adaptations to Environment and Mode of Life" (Bicudo, J. E. P. W., Ed.), pp. 43–53. CRC Press, Boca Raton, FL.

Val, A. L. (1995). Oxygen transfer in fish: Morphological and molecular adjustments. *Brazil. J. Med. Biol. Res.* **28**, 1119–1127.

Val, A. L. (1996). Surviving low oxygen levels: Lessons from fishes of the Amazon. *In* "Physiology and Biochemistry of the Fishes of the Amazon" (Val, A. L., Almeida-Val, V. M. F., and Randall, D. J., Eds.), pp. 59–73. INPA, Manaus.

Val, A. L., and Almeida-Val, V. M. F. (1995). "Fishes of the Amazon and their Environment. Physiological and Biochemical Aspects." Springer-Verlag, Heidelberg.

Val, A. L., Schwantes, A. R., and Almeida-Val, V. M. F. (1986). Biological aspects of Amazonian fishes. VI. Hemoglobins and whole blood properties of *Semaprochilodus* species (Prochilodontidae) at two phases of migration. *Comp. Biochem. Physiol.* **83B**, 659–667.

Val, A. L., Silva, M. N. P., and Almeida-Val, V. M. F. (1998). Hypoxia adaptation in fish of the Amazon: A never-ending task. *S. Afr. J. Zool.* **33**, 107–114.

Van Ginneken, V. J. T. (1996). Influence of hypoxia and acidification on the energy metabolism of fish. An *In vivo* 31P-MNR and calorimetric study. PhD thesis, State University of Leiden, Leiden.

Van Ginneken, V. J. T., Van Den Thillart, G. E. E. J. M., Muller, J., Van Deursen, S., Onderwater, M., Visée, J., Hopmans, V., Van Vliet, G., and Nicolay, K. (1999). Phosphorylation state of red and white muscle in tilapia during graded hypoxia: An *in vivo* 31P-NMR study. *Am. J. Physiol.* **277**, R1501–R1512.

Van Waarde, A., Van den Thillart, G., and Verhagen, M. (1993). Ethanol formation and pH-regulation in fish. *In* "Surviving Hypoxia. Mechanisms of Control and Adaptation" (Hochachka, P. W., Lutz, P. L., Sick, T., Rosenthal, M., and denThillart, G. van, Eds.), pp. 157–170. CRC Press, Boca Raton, FL.

Van Waarde, A., Van den Thillart, G. V. D., and Kesbeke, F. (1983). Anaerobic energy metabolism of the european eel, *Anguilla anguilla* L. *J. Comp. Physiol.* **149**, 469–475.

Walker, I. (1983). Complex-irreversibility and evolution. *Experientia* **39**, 806–813.

Walker, I. (1979). The mechanical properties of proteins determine the laws of evolutionary change. *Acta Biotheoret.* **28**, 239–282.

Walker, I. (1997). Prediction or evolution? Somatic plasticity as a basic, physiological condition for the viability of genetic mutations. *Acta Biotheoret.* **44**, 165–168.

Wang, G. L., Jiang, B.-H., Rue, E. A., and Semenza, G. L. (1995). Hypoxia inducible factor 1 is a basic-helix-loop-PAS heterodimer regulated by cellular oxygen tension. *Proc. Natl Acad. Sci. USA* **92**, 5510–5514.

Way-Kleckner, N., and Sidell, B. D. (1985). Comparison of maximal activities of enzymes from tissues of thermally acclimated and naturally acclimatized chain pickerel (*Esox niger*). *Physiol. Zool.* **58**, 18–28.

Webster, K. A. (2003). Evolution of the coordinate regulation of glycolytic enzyme genes by hypoxia. *J. Exp. Biol.* **206**, 2911–2922.

Webster, K. A., Discher, D. J., and Bishopric, N. H. (1994). Regulation of *fos* and *jun* immediate-early genes by redox or metabolic stress in cardiac myocytes. *Circ. Res.* **74**, 679–686.

West, J. L., Bailey, J. R., Almeida-Val, V. M. F., Val, A. L., Sidell, B. D., and Driedzic, W. R. (1999). Activity levels of enzymes of energy metabolism in heart and red muscle are higher in north-temperate-zone than in Amazonian teleosts. *Can. J. Zool.* **77**, 690–696.

Whitt, G. S. (1984). Genetic, developmental and evolutionary aspects of the lactate dehydrogenase isozyme system. *Cell Biochem. Funct.* **2**, 134–137.

Whitt, G. S. (1987). Species differences in isozyme tissue patterns: Their utility for systematic and evolutionary analyses. *In* "Izosymes: Current Topics in Biological and Medical Research" (Ratazzi, M. C., Scandalios, J. G., and Whitt, G. S., Eds.), pp. 2–23. Alan R. Liss, New York.

Wilson, A. C. (1976). Gene Regulation in Evolution. *In* "Molecular Evolution" (Ayala, F. J., Ed.), pp. 225–235. Sinauer Associates Inc., Sunderland.

Wilson, A. C., Kaplan, N. O., Levine, L., Pesce, A., Reichlin, M., and Allinson, W. (1964). Evolution of lactate dehydrogenase. *FASEB Fed. Proc.* 1258–1266.

Wootton, R. J. (1990). "Ecology of Teleost Fishes." Chapman & Hall, New York.

Wu, R. S. S. (2002). Hypoxia: From molecular responses to ecosystem responses. *Mar. Poll. Bull.* **45**, 35–45.

Yu, F., White, S. B., Zhao, Q., and Lee, F. S. (2001). HIF-1α binding to VHL is regulated by stimulus-sensitive proline hydroxylation. *Proc. Natl Acad. Sci. USA* **98**, 9630–9635.

Zhou, B. S., Wu, R. S. S., Randall, D. J., Lam, P. K. S., Ip, Y. K., and Chew, S. F. (2000). Metabolic adjustments in the common carp during prolonged hypoxia. *J. Fish Biol.* **57**, 1160–1171.

11

PHYSIOLOGICAL ADAPTATIONS OF FISHES TO TROPICAL INTERTIDAL ENVIRONMENTS

KATHERINE LAM
TOMMY TSUI
KAZUMI NAKANO
DAVID J. RANDALL

The Physiology of Tropical Fishes: Volume 21
FISH PHYSIOLOGY

I. INTRODUCTION

The intertidal zone is a unique environment between the land and the sea. It is characterized by changes in physical and chemical factors brought about by the day/night and tidal cycles. At high tide, conditions in the intertidal zone are similar to the sub-tidal. During low tide, the shore is exposed to the air as the environment gradually changes from an aquatic to a virtually terrestrial one. Mangroves, water-filled burrows on mudflats, tide-pools on rocky shores, shallow seagrass beds and coral reef lagoons are typical habitats for intertidal fishes. Physical factors in the intertidal environment, such as temperature, salinity, pH, oxygen and carbon dioxide content of the water change according to day/night, the tidal cycle, and climatic change. Fluctuations in these physical factors in tropical intertidal habitats are most dramatic. During sunny daylight, animals living there have to face problems of heat, desiccation, and increases in salinity. During rainy times, the salinity and pH of the habitat is decreased. At night, pH and oxygen content may be drastically decreased by respiring algae (Truchot and Duhamel-Jouve, 1980; Morris and Taylor, 1983). Seasonal climatic patterns may bring about wind stress and evaporation. Fishes that have evolved in this extreme habitat have special physiological, behavioral, and morphological adaptations to it. The intertidal zone is occupied by fishes with differing degrees of affinity to the habitat. There are two types: some are residents spending most or all of their life in the intertidal zone. Others are visitors or transients are present for part of the day, year or tidal cycle.

Resident fishes remain in the intertidal zone during low tide. They mostly make use of shelters to avoid exposure to predation and desiccation. Such shelters include water-filled tidepools or rockpools on rocky shores and coral reef lagoons. On mangroves, mudflats or seagrass beds, shelters may be in the form of mud burrows filled with seawater. These resident fishes are more specialized for intertidal life and have pronounced physiological adaptations to intertidal life. Most other fishes, especially juveniles, migrate in and out with the tide in search of food and to avoid deep-water predators. Such juveniles are found in the intertidal zone in the first few months of their life but gradually migrate to deeper waters as they grow. The intertidal and shallow sub-tidal zone is thus an important nursery ground. As a result, many species are only present in intertidal regions at certain seasons, including the killifishes, *Fundulus* spp., in salt marshes (Kneib, 1987), the lumpfish, *Cyclopterus lumpus* in tide pools (Moring, 1990) and flatfishes, e.g., *Pleuronectes platessa*, *Platichthys flesus* and *Limanda limanda*, on sandy shores (Ansell and Gibson, 1990).

Bridges (1993) and Gibson (1993) reviewed the ecophysiology, behavioral, and morphological adaptations of intertidal fishes. Blaber (2000) reviewed the ecology and adaptation of tropical and subtropical estuarine fishes. This chapter focuses on the tropical intertidal zone and begins with habitat types, the nature of environmental changes which fishes are likely to experience therein, and the major groups of resident fish. Five commonest intertidal habitats in the tropics, i.e., mangroves, mudflats, seagrass beds, tidepools on rocky shore and shallow coral reefs, will be examined in detail.

A. Environmental Conditions

Figure 11.1 shows the global distribution of mangroves, estuaries, seagrass beds, and coral reefs. The geographical pattern of species richness seen in mangroves, seagrasses, and corals are parallel, that is, they are all highest in the tropics framed by the $>20\,°C$ seawater winter isotherm. This suggests that these communities require similar environmental factors and processes for maintaining such species diversity. Mangroves occupy about $181\,000\,km^2$ of the tropical and subtropical coastline (Alongi, 2002). An estimate of global seagrass bed area is $0.6 \times 10^6\,km^2$ (Charpy-Roubaud and Sournia, 1990), which is equivalent to 10% of the global coastline or 0.15% of the global sea area. The area covered by seagrasses is comparable to that of coral reefs, macroalgae, and mangroves (Duarte, 2000; Hemminga and

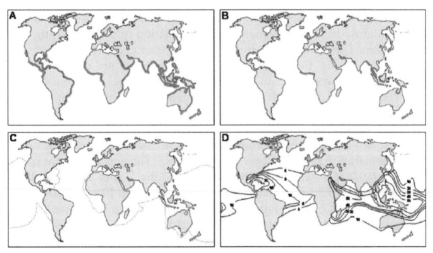

Fig. 11.1 Global distribution of (A) mangroves, denoted by thick grey lines; (B) major estuaries, denoted by thick grey lines; (C) seagrass beds, the limit of which is denoted by dotted lines; and (D) coral reefs, with the figures being the number of reef coral genera on the contours.

Duarte, 2000). At present, coral reefs cover some $2 \times 10^6 \, \text{km}^2$ of tropical oceans, the longest of them, the Great Barrier Reef, extending 2000 km along the eastern coast of Australia (Achituv and Dubinsky, 1990).

1. MANGROVES

Mangroves are intertidal, sediment-based and mangal plant-dominated habitats. They are common at the entrances to many tropical estuaries. These lands together with mudflats on the seaward side, when covered by high tide, are integral part of tropical estuaries, i.e., a partially enclosed coastal body of water which is either permanently or periodically open to the sea and within which there is a measurable variation of salinity due to the mixture of seawater with freshwater derived from land drainage (Blaber, 2000). Mangrove trees usually grow intensively and cover an extensive area. It is considered that this vegetation provides shading for organisms, especially larval and juvenile forms, from direct sunlight (MacNae and Kalk, 1962). The complex and entangled roots and stems provide fish with refugia from predators (Thayer *et al.*, 1987). Decaying leaf matter may be used as a direct food source for fish or other detritus feeders such as harpacticoid copepods, amphipods, and polychaetes, which in turn are preyed upon by fish (Heald and Odum, 1970; Odum and Heald, 1972). Detailed information on mangrove taxonomy and ecology is given by MacNae (1968), Hutchings and Saenger (1987), and Robertson and Alongi (1992).

There is considerable variation in the physical parameters of the aquatic component of mangrove stands. These can change diurnally, seasonally, and according to the weather, making the habitat variable. Freshwater influx causes salinity declines, and high incident radiation results in heating and evaporation. These, in turn, increase temperature and salinity (Gundermann and Popper, 1984). For example, the annual temperature in an Arabian mangrove is between 12 and 35 °C with high salinities between 40 and 50‰ (Sheppard *et al.*, 1992). Extreme fluctuations in salinity occur in some mangrove systems which are in the vicinity of large river estuaries and this is regulated mainly by rainfall and river flow. Blaber and Milton (1990) recorded a 0–35‰ range in annual salinity in the Solomon Islands and where temperature ranged between 27 and 32 °C. The normal current speed of 0.008–$0.078 \, \text{ms}^{-1}$ can be increased to $0.281 \, \text{ms}^{-1}$ after heavy rainfall. A salinity range of 5.5–20‰ has been recorded for South Florida mangroves (Thayer *et al.*, 1987). The south and north coast of Puerto Rico have salinity ranges of 0–46 and 0–30‰, respectively (Austin, 1971). Chong *et al.* (1990) recorded salinities of 20–32‰ and temperatures of 24–28 °C in a Malaysian mangrove. In Charlotte Harbor, Florida, temperature ranges between 16.5 and 39 °C, salinity from 0 to 30‰, pH between 4.4 and 8.5 and dissolved oxygen levels of 1.0–$7.0 \, \text{mg} \, \text{l}^{-1}$ were recorded. Mean dissolved oxygen in pore water of mangroves can be very low, such as $2.6\,\mathrm{mg\,l^{-1}}$ in winter and $4.2\,\mathrm{mg\,l^{-1}}$ in summer in a Hong Kong mangrove (Zhou, 2001). Mangrove water is usually slightly acidic, e.g., 6.8 (Zhou, 2001), because of decaying processes producing plant detritus. In small tidal pools in a Florida mangrove where seven species of fish occur, H_2S values were $<50\,\mathrm{ppb}$ when oxygen levels are low at 1–3 ppb (Abel et al., 1987). Table 11.1 provides temperature, salinity, and dissolved oxygen values for some mangroves where fish assemblages have been studied.

2. MUDFLATS

Mudflats are areas in tidal estuaries composed of sediments ranging from fine silt to coarse sand. Most mudflats are at the seaward side of mangroves. Similar to mangroves, the physical and biochemical parameters of this habitat fluctuate diurnally and seasonally and are subjected to climatic changes. Without shading by the vegetation canopy, mudflats are subjected to heat and evaporation, resulting in increases in temperature and salinity. Temperature on a mudflat at Deep Bay, Hong Kong, increased to 34 °C on a sunny day, salinity ranged from 2 to 30‰, and dissolved oxygen from <0.1 to $4.5\,\mathrm{mg\,l^{-1}}$ (Morton and Lee, 2003). On another intertidal muddy sandflat at Ponggol Estuary in Singapore, mean temperature ranged between 27.9 and 32 °C, salinity between 26.9 and 29.4‰, pH between 7.2 and 8.4, ammonia content between 0.02 and 0.8 ppm and dissolved oxygen between 1.77 and $5.26\,\mathrm{ml\,l^{-1}}$ (Chua, 1973). Table 11.2 provides temperature and salinity values for some mudflats where fish assemblages have been studied.

3. SEAGRASS BEDS

Seagrass beds are dense beds of short, marine angiosperms which belong to the Potamogetonaceae and Hydrocharitaceae in low salinity zones (Figure 11.2). These habitats are vital components of coastal ecosystems because of their trophic and nursery importance for invertebrates and fishes. The grasses can be consumed as either fresh leaves or detritus. Seagrasses also provide shelter for small animals from larger predators, thereby increasing the diversity of this habitat. Detailed information on seagrass taxonomy and ecology is given by Larkum and den Hartog (1989) and Hemminga and Duarte (2000). In contrast to mangroves, most seagrass beds are sub-tidal (Blaber, 2000). In this study, only tropical intertidal ones will be discussed. Since seagrass beds usually occur near inner bays with a freshwater influx and the water column is not shaded, the salinity and temperature of these habitats have a wide seasonal range. Table 11.3 provides temperature and salinity values from some seagrass beds, where fish assemblages have been studied. Wide daily fluctuations in oxygen, carbon dioxide and thus pH are

Table 11.1

Fish Families Recorded from Tropical Mangroves. Both True and Associated Mangrove Residents are Included (Figures Indicate Number of Species)

Geographic location	Hong Kong	Solomon Islands	Trinity, Australia	Dampier, Western Australia	Embley estuary, Gulf of Carpentaria, Northern Australia	Alligator Creek, Northern Queensland, Australia	Vellar-Coloroon (Pichavaram), India	Selangor, Malaysia	Florida Bay	South Florida	Indian River Lagoon, Central Florida	Bahia de la Paz, Mexico	Caeté Estuary, Brazil	Gulf of Nicoya, Costa Rica	East Africa
Monthly temperature range (°C)	26.2–27	27–32	26–29	17–31.3	~30	21–31, may increase by 8°C for ~6 h in hot days	25–32	24–28	21–~31		>20		Mean = 25.7	27–31	
Monthly range of salinity (‰)	6–10	0–35	<20	35.7–39.6	0–35	30–38	15–35	20–32	0.05–59.3	5.5–20 and 27–42, depending on sites	May <20 and extreme up to 200		6 PSU– >35 PSU	15—29	
Dissolved oxygen/mg l^{-1}														5.2–5.7	
Depth (m)			Exposed at low tide	<0.2 m	<2	0.5–5		3 m at high tide to 5 m at low tide		~1			4–5		
Reference	Vance, 1999	Blaber and Milton, 1990	Blaber, 1980	Blaber et al., 1985	Blaber et al., 1989	Robertson and Duke 1987, 1990	Blaber, 2000	Chong et al., 1990; Sasekumar et al., 1992	Ley et al., 1999	Thayer et al., 1987	Gilmore, 1995	Macda-Martinez et al., 1982	Barletta et al., 2000	Rojas et al., 1994	Little et al., 1988
Remarks							Only top 17 families				Top 25 species only	Only major families			

Family	Common name	Hong Kong	Solomon Islands	Trinity, Australia	Dampier, Western Australia	Embley estuary	Alligator Creek	Vellar-Coloroon	Selangor	Florida Bay	South Florida	Indian River Lagoon	Bahia de la Paz	Caeté	Gulf of Nicoya	East Africa
Acanthuridae	Sailfin tang		1													
Achiridae	American sole									1	1	1				

Family	Common name										Total
Ambassidae	Glassfish			1							
Anguillidae	Freshwater eel		1			1	1	1			
Apogonidae	Cardinalfish	9	1	1		1	1	1			12
Ariidae	Sea catfish				3	8	4	4	1	2	1
Atherinidae	Silversides	4		1	1	4	4				
Bagridae	Bagrid catfish				3	3	1				
Balistidae	Leatherjack					1					
Batrachoididae	Toadfish					1	2	1			
Belonidae	Needlefish	1	2	1		4	2	3		1	1
Blenniidae	Blenny	3		1			1	1			
Bothidae	Lefteye flounder		1								1
Bythitidae	Viviparous brotula	4				2					
Callionymidae	Dragonet	1					1				
Carangidae	Jack	7	4	3	11	5	5	2			5
Carcharhinidae	Requiem shark	4	1	1			1				1
Centrarchidae	Sunfish						1	2			
Centriscidae	Shrimpfish										
Centropomidae	Snook		2		3	1	1	1		1	6
Chaedodontidae	Butterflyfish	1									
Chandidae	Milkfish	2	3	2	3						
Chirocentridae	Tangs	1	1			1					
Cichlidae	Tilapia				3	1	2				2
Clinidae	Clinid						1				
Clupeidae	Sardine	4	2	2	6	7	3	3	4		2
Cynoglossidae	Tonguefish	1	2				1	1			
Cyprinodontidae	Killifish						9	7		1	
Dasyatidae	Stingray	1	1	2	3	3	2	1			
Diodontidae	Spiny puffer						1				
Drepaneidae	Sicklefish	1			2						

Table 11.1 (*cont.*)

Geographic location	Hong Kong	Solomon Islands	Trinity, Australia	Dampier, Western Australia	Embley estuary, Gulf of Carpentaria, Northern Australia	Alligator Creek, Northern Queensland, Australia	Vellar-Coloroon (Pichavaram), India	Selangor, Malaysia	Florida Bay	South Florida	Indian River Lagoon, Central Florida	Bahia de la Paz, Mexico	Caeté Estuary, Brazil	Gulf of Nicoya, Costa Rica	East Africa
Echeneididae Remora			1						1						
Eleotridae Sleeper		6			1			3			1		2	2	
Elopidae Tarpon			1					1	1		1			1	
Engraulidae Anchovy		3	4	1		3	6	8		2		✓		4	
Ephippidae Batfish				2					2					1	
Exocoetidae Flyingfish			3	2					1	1					
Fistulariidae Flutemouth											1			1	
Funduclidae Killifish											1				
Gerreidae Mojarra		4	1	4	2	1	4	2	5	2	2	✓		4	
Gobiesocidae Clingfish									1	1					
Gobiidae Goby	4	25			11	4	>20	8	4	5	3	✓	6		
Grammistidae Basslet											1			1	
Haemulidae Sweetlip/grunt					1			2	3	2				6	
Hemiramphidae Halfbeak		3		4	1	2		3			1	✓			
Kyphosidae Ridderfish		1													
Labridae Wrasse															
Lactaridae False trevally		1													
Leiognathidae Ponyfish		7	2	3	4	5	6	7							
Lepisostidae Gar									1						
Lethrinidae Emperorfish															
Lobotidae Tripletail									1						
Lophiidae Crestfish														1	
Loricariidae Armored catfish													>1		
Lutjanidae Snapper		7	2	2	2		4	2	3	3	1			4	
Megalopidae Tarpon											1				

Family	Common name												
Melanotaeniidae	Rainbowfish												
Microdesmidae	Wormfish	1											
Monacanthidae	Japanfilefish												
Monodactylidae	Moonfish	1					1						
Mugilidae	Mullet	4	6	6	3	7	3	3	2	✓	1		
Mugiloididae	Weaver	1											
Mullidae	Goatfish	4	1				2	2					
Muraenidae	Moray eel												
Myliobatidae	Eagle and Manta ray										1		
Nemipteridae	Spinecheek												
Ophichthidae	Snake eel			1			1	1					
Opistognathidae	Jawfish												
Orectolobidae	Nurseshark	1	1			1	1						
Ostraconidae	Cowfish												
Paralichthyidae	Flounder	1				2							
Platycephalidae	Flathead	3	2			4							
Plotosidae	Catfish	1		1	2	1							
Poecillidae	Livebearer	3		2	3	4	>2	1					
Polynemidae	Threadfin	3				2							
Pomacentridae	Damselfish	4											
Pomadasyidae		2		2									
Pseudomugilidae	Blue eye			1	2		1						
Rhinobatidae	Guitarfish	1				1							
Scaridae	Parrotfish	1					2						
Scatophagidae	Scat			1		1							
Sciaenidae	Drum	2				4	10	3				9	
Scombridae	Mackerel	2											
Scorpaenidae	Lionfish												
Serranidae	Grouper	6					1						
Siganidae	Rabbitfish	2	1			2	3						
Sillanginidae	Northern whiting	1	2	4		3	1						

Table 11.1 (*cont.*)

	Geographic location	Hong Kong	Solomon Islands	Trinity, Australia	Dampier, Western Australia	Embley estuary, Gulf of Carpentaria, Northern Australia	Alligator Creek, Northern Queensland, Australia	Vellar-Coloroon (Pichavaram), India	Selangor, Malaysia	Florida Bay	South Florida	Indian River Lagoon, Central Florida	Bahia de la Paz, Mexico	Caeté Estuary, Brazil	Gulf of Nicoya, Costa Rica	East Africa
Soleidae	Sole		1	2						1	1					
Sparidae	Porgy			1	1	1	1			3	2	2				
Sphyraenidae	Barracuda		3	1	1	1			2	1	1					
Synanceiidae	Stonefish			1		1				1	1					
Syngnathidae	Seahorse and pipefish		3			1				5	3					
Synodontidae	Lizardfish		1						1		1					
Tachysuridae				2												
Teraponidae	Perch		1	1	2	1		3	1							
Tetraodontidae	Puffer		2	3	3	3	1		3	2					5	
Tetraogidae	Waspfish		1													
Toxotidae	Archerfish		1						1							
Triacanthidae	Triplespine			1												
Trichiuridae	Cutlassfish								1							
Triglidae	Searobin										2					
Zandidae	Moorish idol															

Table 11.2

Fish Families Recorded from Intertidal Mudflats or Estuaries. The Fishes were Investigated when the Habitat was not Exposed to Air (Figures Indicate Number of Species)

	Geographic location	Embley estuary, Northern Australia	Selangor, Malaysia	Ponggol Estuary, Singapore
	Temperature (°C)	25–32	24–28	
	Salinity (‰)	15–35	20–32	
	Depth (m)	-	<5	Intertidal −3 m at low tide
	Reference	Blaber et al., 1989	Chong et al., 1990; Sasekumar et al., 1992	Chua, 1973
Family	**Common name**			
Antennariidae	Frogfish			1
Apogonidae	Cardinalfish		1	3
Ariidae	Sea catfish	1	6	
Atherinidae	Silversides			2
Bagridae	Bagrid catfish		1	
Balistidae	Leatherjack			1
Batrachoididae	Toadfish			1
Belonidae	Needlefish	2	2	1
Carangidae	Jack	1	1	2
Carcharhinidae	Requiem sharks	1		
Centropomidae	Snook	1		2
Chandidae	Milkfish		1	
Chirocentridae	Tang			1
Cichlidae	Tilapia			1
Clupeidae	Sardine	1	4	4
Cynoglossidae	Tonguefish		1	1
Cyprinodontidae	Killifish			1
Dasyatidae	Stingray	6	1	
Dorosomidae				1
Drepaneidae	Sicklefish	1	2	
Echeneidae	Remoras	1		
Engraulidae	Anchovy	1	5	6
Gerreidae	Mojarra	1	1	
Gobiidae	Goby		6	8
Haemulidae	Sweetlip/grunt	2	2	
Hemihamphidae	Halfbeak			2
Lactariidae	False trevallies	1		
Leiognathidae	Ponyfish	2	2	10
Lethrinidae	Emperorfish			1
Lutjanidae	Snapper	2	1	2
Mugilidae	Mullet	4	3	1
Mullidae	Goatfish		1	1
Muraenesocidae	Pike conger		1	

Table 11.2 (*cont.*)

	Geographic location	Embley estuary, Northern Australia	Selangor, Malaysia	Ponggol Estuary, Singapore
Muraenidae	Eel		1	
Myliobatidae	Eagle and manta rays	1		
Ophichthidae	Snake eel		1	
Paralichthyidae	Flounder	1		
Periophthalmidae				2
Platycephalidae	Flathead	2	2	
Plectorhynchidae				1
Plotosidae	Catfish		1	3
Polynemidae	Threadfins	1	2	
Pomadasyidae				1
Psettodidae				1
Rhinobatidae	Guitarfishes	1		
Scaridae	Parrotfish			
Scatophagidae	Scat	1		1
Sciaenidae	Drum	1	5	
Scomberomoridae				1
Scorpaenidae	Lionfish			1
Scyliorhinidae	Cat shark		1	
Serranidae	Grouper			3
Siganidae	Rabbitfish			1
Sillanginidae	Northern whiting		1	2
Soleidae	Sole		1	1
Sparidae	Porgies	1		
Sphyraenidae	Barracuda	1		1
Stromateidae	Butterfish		1	
Syngnathidae	Seahorse and pipefish			2
Synodontidae	Lizardfish			1
Terapontidae	Grunter, tigerperch or thornfish		1	1
Tetraodontidae	Puffer	1	2	
Toxotidae	Archerfish			1
Triacanthidae	Triplespine			1
Trichiuridae	Cutlassfish		1	1
Trygonidae				1

to be expected due to seagrass photosynthesis in daylight and respiration at night. These variables, however, have seldom been recorded in parallel with studies of the resident fish assemblages (Table 11.3).

Fig. 11.2 Seagrass bed in Mullet Bay, Bermuda, being exposed at low tide.

4. TIDEPOOLS

Tidepools are created by either depressions or ridges on bedrock lying between the highest (in very exposed situations even above Extreme High Water Spring Tide) and lowest tides and where wave action and current speed are sufficiently high to prevent sediment build-up (Figures 11.3, 11.4). Physical conditions are different among tidepools even on the same shore. These are determined by their height on the shore with respect to low water, the degree of experienced wave exposure, the internal dimensions and shapes of the pools (Emson, 1985) and degree of exposure to sunlight. Presence of biota, especially macroalgae, in the pool is also an important factor determining pH levels, oxygen and carbon dioxide contents in the water. Macroalgae also create sheltered microhabitats and refugia (Emson, 1985). Again, the physical conditions experienced by tidepools are seldom described in parallel with the resident fish assemblages (Table 11.4).

5. SHALLOW CORAL REEFS

There are three basic types of intertidal or shallow sub-tidal coral reef. First, fringing reefs develop along the inclined shoreline and whose upper levels may be exposed during very low tides. Second, extended shallow reef flats contain a moat surrounded by a reef crest and third, the reef crest itself.

Table 11.3
Fish Families Recorded from Tropical Shallow Seagrass Beds (Figures Indicate Number of Species; ✓ = Presence)

Family	Common name	Philippines	Cairns Harbour, Australia	Embley estuary, Northern Australia	Alligator Creek, Northern Queensland, Australia	Jakarta Bay, Java, Indonesia	Guam	Puerto Rico	Panama	Everglade National Park, South Florida	Indian River Lagoon, Central Florida	Bermuda	St Croix, US Virgin Islands
Geographic location													
Temperature (°C)			22–33	25–30	21–31				27–31		>20		25.3–30.2
Salinity (‰)			20–37	16–34	30–38				21–38	5.5–20	~ <20 to oceanic		2–6
Water depth (m)			<5	<2	0–3	1	3	0.5–1	1–2	0.8–1.2	Next to shoreline	Intertidal	
Reference		Fortes, 1990	Coles et al., 1993	Blaber et al., 1989	Robertson and Duke, 1987	Pollard, 1984	Jones and Chase, 1975	Martin and Cooper, 1981	Weinstein and Heck, 1979	Thayer et al., 1987	Gilmore, 1995	Logan and Cook, 1992	Robblee and Zieman, 1984
Remarks						Dominant fish families	Dominant fish families				(Top 25 species only)		
Acanthuridae	Sailfin tang	1											2
Ambassidae	Glassfish		2				✓	2	2				
Anguillidae	Freshwater eel												
Antennariidae	Frogfish								2				
Apogonidae	Cardinalfish	10	4	3		✓	✓		2				2
Ariidae	Sea catfish		3							1			
Atherinidae	Silverside	1	1	1	1			1	1	1			
Aulostomidae	Trumpetfish								1	1	1	1	
Balistidae	Triggerfish						✓			3			
Batrachoididae	Toadfish								1	1			
Belonidae	Needlefish	3	1			✓			1	1	2		
Blenniidae	Blenny	5				✓	✓						
Bothidae	Lefteye flounder		1					1	2	2			
Callionymidae	Dragonet		2			✓			✓				
Carangidae	Jack	1	3						3	1	1		1

Family	Common name									
Carcharhinidae	Requiem shark	2								
Centriscidae	Shrimpfish	1			1					
Centrogeniidae	False scorpionfish		1							
Centropomidae	Snook	1	2							
Chaetodontidae	Butterfly fish	1	2	2	✓	2	3	1	1	1
Chandidae	Milkfish		3	2			3			
Clinidae	Clinid					4		4		4
Clupeidae	Sardine	1	6	1	3	✓	2	1	3	2	3
Congridae	Garden eel		1				1			
Cynoglossidae	Tonguefish		3					2		
Cyprinodontidae	Killifish						2			
Dactylopteridae	Flying gurnard					1	1	1		
Dactyloscopidae	Sand stargazer					1				
Dasyatidae	Stingray	1	1				1			
Diodontidae	Porcupinefish					1	1		1	1
Eleotridae	Sleeper	3	1				1			
Elopidae	Tarpon	2	2							
Engraulidae	Anchovy	1	8	1	2		2	2	2	1
Ephippidae	Batfish	2	2				1	1		
Exocoetidae	Flyingfish	1					1			
Fistulariidae	Flutemouth	1				1	1		1	1
Gerridae	Mojarra	1	4	4	1	✓	1	5	2	3	1
Ginlymostomatidae	Nurseshark						1			
Gobiidae	Goby	1	8	5	3	✓	4	4	3	2	1
Haemulidae	Sweetlip/grunt	1	5				6	4	2	4
Hemiramphidae	Halfbeak	3	4	1	1				3
Holocentridae	Squirrelfish and soldierfish					2	1		5
Kyphosidae	Ridderfish	1			✓			1
Labridae	Wrasse	7	1		✓		3	3	2	4
Labrisomidae	Labrisomid								1
Lactariidae	False trevally	1					3			
Leiognathidae	Ponyfish	2	8	6	5	✓		3		

Table 11.3 *(cont.)*

Geographic location		Philippines	Cairns Harbour, Australia	Embley estuary, Northern Australia	Alligator Creek, Northern Queensland, Australia	Jakarta Bay, Java, Indonesia	Guam	Puerto Rico	Panama	Everglade National Park, South Florida	Indian River Lagoon, Central Florida	Bermuda	St Croix, US Virgin Islands
Lethrinidae	Emperorfish	7				✓							
Lutjanidae	Snapper	8	4	1		✓		1	7	2		1	2
Monacanthidae	Japanfilefish	4	1	3		✓	✓		3			4	
Monodactylidae	Moonfish	1											
Mugilidae	Mullet	2	3		1						2		
Mugiloididae	Weaver	1											
Mullidae	Goatfish	4	1			✓	✓		1	1		1	
Muraenidae	Eel	1	1					3	1				1
Nemipteridae	Spinecheek	3											
Ophichthidae	Snake eels							2				1	
Opistognathidae	Jawfish	1											1
Ostraciidae	Boxfish, cowfish or trunkfish							1	3			3	3
Ostracionidae	Cowfish	1											
Paralichthyidae	Flounder	1	2	1					2				
Platycephalidae	Flathead	2	6	2		✓							
Plotosidae	Catfish	1											
Polynemidae	Threadfin		5						1				
Pomacentridae	Damselfish	7				✓		1	2			1	
Pomadasyidae								1					
Pseudomugilidae	Blue eye				1								
Scaridae	Parrotfish	1				✓	✓	6	5	1		1	7
Scatophagidae	Scat	1											
Sciaenidae	Drum		2						2	3	2	1	
Scorpaenidae	Lionfish	1	1					1	5			1	
Serranidae	Grouper	1	1	1		✓		1	8				3
Siganidae	Rabbitfish	7	5	4	1	✓	✓		1				1

Family	Common name							
Sillanginidae	Northern whiting	1	1				1	
Soleidae	Sole	1	1		✓		1	2
Sparidae	Porgie	1	1				3	2	2
Sphyraenidae	Barracuda	1	1		1	3	1	2	1
Sphyrnidae	Hammerhead shark	1	1			3	1		2
Sternopygidae	Glass knifefish					1		
Synanceiidae	Stonefish	1	1					
Syngnathidae	Seahorse and pipefish	4	1	2	✓	✓	2	2	6	2	3	3
Synodontidae	Lizardfish	2	1	1			3	1	
Teraponidae	Perch		3	2				
Tetrabranchiidae	Ray-finned fish		1					
Tetraodontidae	Puffer	4	8	3	✓	✓	1	4	1	1	1
Theraponidae	Perch	2		2				
Toxotidae	Archerfish	1	1					
Triacanthidae	Tripodfish		1					
Urolophidae	Round ray					1		
Xenocongridae					✓			
Zanclidae	Moorish idol	1		1				

Fig. 11.3 Tidepool on a wave-exposed shore on Cape d'Aguilar, Hong Kong.

Fig. 11.4 Tidepools can be very shallow, like these on Elbow Beach, Bermuda.

Although the structural complexity of the coral community provides ideal shelter for fish against predators, those that reside in shallow reef flats experience large diurnal fluctuations in physical conditions. Physical properties of these shallow reefs are subjected to (i) semi-diurnal changes in water properties brought about by tidal flows bringing in oceanic water and (ii) diurnal changes in temperature, salinity and circulation (Andrews and Pickard, 1990). The degree of diurnal fluctuation in physical properties varies among reefs, depending on their size, depth, topography, and geographic location (Table 11.5). Diurnal variations in temperature near fringing reefs are high, ranging from 3 °C at the outer edge to 12 °C near the shore. Decreases in salinity, especially in the upper water layer, occurs chiefly during the wet season as a consequence of rainfall and river run-off. Because hermatypic (reef-building) corals have symbiotic zooxanthellae, the reef system is photosynthetic during the day and respires at night. Dissolved oxygen and carbon dioxide levels in the water column, therefore, follow the diurnal photosynthetic and respiration cycle, i.e., dissolved oxygen being maximum ($10.4 \, \mathrm{mg \, l^{-1}}$) when the sun is bright but minimum ($\sim 5 \, \mathrm{mg \, l^{-1}}$) during early evening when there is a heavy demand for oxygen which leveled off at 20:00–22:00 h (Cope, 1986) (Table 11.5).

B. Associated Fish Families

1. MANGROVES

A number of studies provide checklists of tropical mangrove fishes (Table 11.1) but seldom distinguish between true and associated residents. One of the reasons for this is that the migratory patterns of many associated species have yet to be identified. It appears that the majority of fishes spend their juvenile stage in mangroves and are often nursery grounds for pelagic species (Birkeland, 1985; Robertson and Duke, 1987) such as the schoolmaster (*Lutjanus apodus*), gray snapper (*L. griseus*), great barracuda (*Sphyraena barracuda*), and foureye butterfly (*Chaetodon capistratus*) (van der Velde *et al.*, 1992).

Extreme changes in a mangrove's physical parameters limit the number of fish species that can survive such variable conditions. Some taxa are specialized to inhabit mangroves and are true residents, for example, representatives of the Gobiidae and its subfamily Periophthalminae, the Blenidae and the Eleotriidae. The Periophthalminae, i.e., the mudskippers, consists of numerous species of *Periophthalmus* which replace each other in different regions of the same mangrove. Other members of the Gobiidae restricted to mangrove habitats include mangrove gobies, species of *Hemigobius, Gobinellus, Gobius* and *Dormitator* (Por, 1984). Other true residents include

Table 11.4

Fish Families Recorded from Tropical Tidepools (Figures Indicate Number of Species)

Family	Common name	Martins Bay, Barbados (Mahon and Mahon, 1994) — (True resident)	(Partial resident)	Tamarindo, Costa Rica (Weaver, 1970)	Salvador, Brazil (Almeida, 1973)	Hong Kong (Lam, 1986, 1990; Cheung, 1991)	Seychelles Islands (Chotkowski et al., 1999) (Only top 20 families)	Hawaii (Gosline, 1965) (Only top 20 families)
Acanthuridae	Sailfin tang		1					1
Antennariidae	Frogfish		1					
Apogonidae	Cardinalfish	1			1		1	
Atherinidae	Silverside	1						
Blenniidae	Blenny	2		2			5	2
Bothidae	Lefteye flounder				1			
Brotulidae				1	1			
Bythitidae	Viviparous brotula	1		1				
Carangidae	Jack	1						
Chaenopsidae	Blenny	1		1				
Chaetodontidae	Butterfly fish	1		1				
Clinidae	Clinid			6				
Congridae	Garden eel							
Cottidae	Sculpin							
Dactyloscopidae	Sand stargazer				1			
Gobiesocidae	Clingfish	1		3	3			
Gobiidae	Goby	5		3	3	2	5	2
Grammistidae					1			
Haemulidae	Sweetlip/grunt	1						
Holocentridae	Squirrelfish or soldierfish	4			1		1	

Family	Common name						
Kyphosidae	Ridderfish						
Labridae	Wrasse		5				2	1
Labrisomidae	Labrisomid	13	5				2	
Liparidae	Snailfish	1						
Lutjanidae	Snapper	1	1					
Microdesmidae	Wormfish	1	1					
Moringuidae	Worm or spaghetti eel				1			
Mugilidae	Mullet	1	2					
Mullidae	Goatfish	1	1					
Muraenidae	Eel	3	4				3	
Ophichthidae	Snake eels	3						
Ophidiidae	Cusk-eel				1			
Opistognathidae	Jawfish	1	1					
Orectolobidae	Carpet or nurse shark		1					
Plesiopidae	Roundhead	1	1				1	
Pomacanthidae	Angelfish	1	5				2	
Pomacentridae	Damselfish	2	1				2	3
Pomadasyidae			1				1	
Scaridae	Scorpionfish or rockfish	1	1		1		1	
Scorpaenidae	Parrotfish	1	1					
Serranidae	Grouper	2	2	1				
Siganidae	Rabbitfish			1				
Sparidae	Porgie							
Stromateidae	Butterfish		1					
Syngnathidae	Seahorse and pipefish		1	1			1	
Teraponidae	Perch			1				
Tetraodontidae	Puffer	2	2					
Tripterygiidae	Threefin blenny	1	1					
Other families			1					

Table 11.5
Fluctuation in Physical Conditions on Shallow Coral Reefs

Geographical location	Site	Diurnal temperature change (°C)	Diurnal salinity change (‰)	Time	Reference
Pacific atolls					
Johnston Island (mid-Pacific)	Deep reef lagoon and platform reef	Increased by 0.7		Between 0900 h and 1500 h	Wiens, 1962
Arno Atoll, Marshall Islands	Near the beach	Rose from 29 to 34.5		Before sunrise and within a few hours	Wells, 1952
Onotoa Atoll, Kiribati	Near the shore	Increased by 10.5	Increased by 4.7	0500–1200 h (noon)	Cloud, 1952
Murray Islands (Torres Strait, 10°S, 144°E)	Over the reef flat	Increased by 9.8	Increased by 4.0		
Chagos and Peros Banhos, 6°S, Indian Ocean Atolls	Near the shore	Daily mean temperature range up to 12.5 Minimum at 0600 h and maximum at 1800 h, range <0.2		In Sep/Oct	Pugh and Rayner, 1981
Barrier reefs					
Low Isles, GBR Lagoon	A pool on reef flat with depth 0.3–2 m	Range was ~10 from 0200–1430 h	Semi-diurnal change of ~0.35, with maxima at 0100 h and 1430 h, i.e., low tides		Orr, 1933

Location	Notes		Value	Reference
Britomart Reef (GBR)	In lagoon	Diurnal range 0.3		Wolanski and Jones, 1980
Thursday Island and Yule Entrance, northern GBR		Semi-diurnal fluctuation 0.7–0.8		Wolanski and Ruddick, 1981
Heron Island, GBR	Shallow water close to the island	Daily temperature range has monthly mean values between 3–5; maximum is 12 in March; daily maximum range 14		Potts and Swart, 1984
				Endean et al., 1956
Fringing reefs				
Grand Cayman Island, Caribbean	Along the reef flat 3.5–4.5 between 0300–1500 h			Kohn and Helfrich, 1951
Kaneobe Bay, Oahu, Hawaii		0800–1600 h range 0.12 to 0.36		Bathen, 1968
Laurel Reef, 4 km offshore south of Puerto Rico		Early evening and next morning fall by 6–9		Glynn, 1973
Hoi Ha Wan, Hong Kong		~29 in summer ~16 in winter	29–33	Cope, 1986

species of *Scaricchthys* (Blenniidae), *Eleotris* (Eleotridae), *Selenapsis* and *Arius* (Ariidae), *Centropomus* (Centropomidae), *Diapterus* and *Eugerres* (Gerreidae), *Sphaeroides* and *Chelonodon* (Tetraodontidae) and some syngnathids.

2. MUDFLATS

Most studies on fishes have focused on the mangroves at the back of the mudflat or sub-tidal estuaries at the shore's front (Blaber, 2000). Checklists for tropical, shallow, intertidal mudflat fishes are rare. The only fish lists are from the Embley Estuary, Northern Australia and Selangor, Malaysia (Blaber *et al.*, 1989; Chong *et al.*, 1990; Sasekumar *et al.*, 1992). Many other studies have shown the importance of estuaries to fisheries and have sampled sub-tidal areas (Kuo and Shao, 1999). The resident species of intertidal and sub-tidal zones form a small proportion of the total estuarine fish fauna and are usually of small body size and juveniles which include many tropical and subtropical species of Clupeidae, Mugilidae, Sciaenidae, Arridae, Polynemidae, Hemulidae, Gobiidae, Engraulidae, Ambassidae, Antheridae, and Ayngnathidae (Table 11.2). The most common species are *Ambassis productus, A. natalensis, Gilchristella aestuaria, Hyporhamphus knysnaensis* and the gobies *Croilia mossambica, Glossogobius callidua* and the numerous *Periophthalmus* taxa (Blaber, 2000).

3. SEAGRASS BEDS

Shallow and intertidal sparse seagrass areas are dominated by several ubiquitous gobies, e.g., *Drombus triangularis* and *Yongeichthys nebulosus*, and small juveniles of the Gerridae (*Gerres* sp.), Lutjanidae and Sillangini-dae (Blaber, 2000; Table 11.3). The species composition of fish assemblages in seagrass beds changes according to the plant taxa and habitat type (Blaber, 2000).

Seagrass beds serve as an important feeding ground for adult and juvenile coral reef fishes (Robblee and Zieman, 1984). Ogden (1976, 1980) has shown that herbivorous fishes are extremely abundant in the Carribbean, in terms of number of individuals and standing crop, where they actively graze algae and seagrasses during the day. The main group of seagrass grazers in Bermuda are parrot fishes (Scaridae), which move from mangrove to seagrass to reef habitats, probably over a daily cycle. Van der Velde *et al.* (1992) showed that seagrass beds are nursery grounds for reef fishes such as French grunt (*Hemulon flavolineatum*), bluestriped grunt (*H. sciurus*), yellowtail snapper (*Ocyurus chrysurus*), stoplight parrot fish (*Sparisoma viride*), and doctor fish (*Acanthurus chirurgus*). More permanent seagrass residents include eels, wrasses, razor fishes, pipe fishes, and cow fishes. Predators such as barracuda and sennet fish are frequent visitors to seagrass beds and may restrict the size classes of the resident fishes in these seagrass beds to those small enough to find refuge among the blades (Logan and Cook, 1992).

4. TIDEPOOLS

Although there are a number of studies on temperate tidepool fish assemblages (Stepien *et al.*, 1991), tropical systems are seldom investigated. Table 11.4 summarizes tropical and subtropical tidepool fish assemblages. The most important families in tropical tidepools are the Blenniidae, Gobiesocidae, Gobiidae, Labrisomidae, Muraenidae, and Ophichthidae. The list of tidepool fishes in Barbados provided by Mahon and Mahon (1994) has been divided into true and temporary residents (Table 11.4). The major true residents include representatives of Blenniidae, Gobiidae, Labrisomidae, Muraenidae, and Ophichthidae. In Hong Kong, *Bathygobius hongkongensis* is the true resident in tidepools (Lam, 1990; Figure 11.5). There is little information on the ecology of the true residents in tropical tidepools.

Fig. 11.5 A rockpool goby, *Bathygobius hongkongensis* (Gobiidae) in a mid-shore tidepool on Cape d'Aguilar, Hong Kong.

Table 11.6
Fish Families Recorded from Shallow Coral Reefs or Reef Flats (Figures Indicate Number of Species; ✓ = presence)

	Geographic location	South Pacific	Pratas Island, South China Sea	San Blas Archipelago, Caribbean Panama	Sri Lanka	Bahama Island
	Temperature (°C)		22–30	26–32		
	Salinity (‰)		0–5	33–35		
	Depth (m)			Nearshore reef	Shallow reef flat	Patch reef (4–7 m) and crest (1–3 m)
	Reference	Wright, 1993	Chen et al., 1995	Wilson, 2001	Öhman et al., 1997	Alevizon et al., 1985
	Remarks				Major species only	
Family	**Common name**					
Acanthuridae	Sailfin tang	✓	3	3	2	2
Albulidae	Bonefish	✓				
Antennariidae	Frogfish		1			
Aploactinidae	Velvetfish					
Apogonidae	Cardinalfish		7	6		
Aulostomidae	Trumpetfish		1	1		1
Balistidae	Triggerfish	✓	3	2		1
Batrachoidae	Toadfish			1		
Belonidae	Needlefish	✓		1		
Bleniidae	Blenny		6	9		
Bothidae	Lefteye flounder			4		
Bregmacerotidae	Codlet			1		
Caesionidae	Fusilier				2	
Carangidae	Jack	✓		4		
Centropomidae	Snook					
Chaenopsidae	Pike-, tube- and flagblenny			1		
Chaetodontidae	Butterfly fish		12	2	2	4

Family	Common name				
Chandidae	Milkfish				
Congridae	Conger and garden eel	✓	1		1
Coryphaenidae	Dolphinfish	✓			
Cynoglossidae	Tonguefish	✓			
Dactylopteridae	Flying gurnard		1		
Dasyatidae	Stingray	✓	1		
Diodontidae	Porcupinefish		1	2	
Echeneidae	Remora	✓			
Elopidae	Tarpon	✓	1		
Exocoetidae	Flyingfish	✓			
Fistulariidae	Flutemouth	✓			
Gerreidae	Mojarra		1	1	
Gobiesocidae	Clingfish		2	2	
Gobiidae	Goby		11	7	
Grammistidae			1		
Gymnuridae	Ray	✓			
Haemulidae	Grunt	✓	3		3
Hemiramphidae	Halfbeak	✓			
Holocentridae	Squirrelfish or soldierfish	✓	1	3	
Kyphosidae	Ridderfish	✓			
Labridae	Wrasse	✓	17	9	4
Labrisomidae	Blenny			5	
Leiognathidae	Ponyfish	✓			
Lethrinidae	Emperorfish	✓	2		
Lutjanidae	Snapper	✓	2	7	3
Monacanthidae	Japanfilefish			3	
Mugilidae	Mullet	✓		1	
Mullidae	Goatfish	✓	7	1	
Muraenidae	Eel	✓	2	2	2

Table 11.6 (*cont.*)

	Geographic location	South Pacific	Pratas Island, South China Sea	San Blas Archipelago, Caribbean Panama	Sri Lanka	Bahama Island
Myliobatidae	Eagle and manta ray					
Narcinidae	Apron, electric and sleeper ray					
Nemipteridae	Spinecheek	✓	4			
Ostraciidae	Boxfish, cowfish or trunkfish		1			
Percichthyidae	Temperate perch					
Pinguipedidae	Sandperch		1			
Platycephalidae	Flathead	✓				
Plectropomids		✓				
Plotosidae	Catfish					
Polynemidae				1		
Pomacanthidae	Angelfish		1	4	1	
Pomacentridae	Damselfish		19	12	8	
Priacanthidae	Bigeye			1		2
Pseudogrammidae			1			
Sauridae		✓				
Scaridae	Parrotfish	✓	8	3	1	6
Scombridae	Mackerel	✓		3		
Scorpaenidae	Lionfish	✓	3			
Serranidae	Grouper	✓	4	7		
Siganidae	Rabbitfish	✓	2			3
Sillaginidae	Northern whiting					
Soleidae	Sole	✓			1	
Solenostomidae	Ghost pipefish					

Family	Common name			
Sparidae	Porgie			
Sphyraenidae	Barracuda	✓	1	2
Syngnathidae	Seahorse and pipefish		1	2
Synodontidae	Lizardfish		2	1
Teraponidae	Grunter or thornfish			3
Tetraodontidae	Puffer	✓		
Tripterygiidae	Threefin blenny			1
Uranoscopidae	Stargazer			1
Zanclidae	Moorish idol		1	

5. SHALLOW CORAL REEFS

There are 100 fish families known to have coral reef representatives (Leis, 1991). The groups most commonly associated with coral reef environments are (i) three labroid families, i.e., the Labridae, the Scaridae, and Pomacentridae; (ii) three acanthuroid families, i.e., the Acanthuridae, Siganidae, Zanclidae in its single genus *Zanclus*, and (iii) two chaetodontoid families, i.e., the Chaetodontidae and Pomacanthidae. The Gobiidae, which associate with all types of intertidal habitats, are also an important group on shallow reefs (Hixon, 2001). These reef fish assemblages are usually studied in whole coral reef systems (Montgomery, 1990; Bellwood and Wainwright, 2002). Checklists that exclusively include shallow and intertidal reef areas are more limited (Table 11.6).

C. Evolution of Fish in the Intertidal Zone

The discontinuity and variability of the intertidal zone has played an important role in fish evolution. Intertidal fish populations demonstrate a considerable degree of convergent evolution in searches to improve their fitness to the intertidal environment. The intertidal zone is also the place where fish, with considerable phenotypic plasticity, have evolved and spread to more stable environments such as sub-tidal marine and freshwater systems.

It has been shown that tropical intertidal habitats are characterized by fluctuations in physical parameters on diel and monthly scales. These habitats are also subjected to change on larger spatial and temporal scales. For example, the extent of estuarine areas, including mangroves, mudflats, and seagrass beds, are affected by several interacting factors such as sea level rises (and falls), sedimentation by land run-off, and human impacts. It has been shown that mangroves respond to the first two contrasting factors on a geological time scale. A Philippine mangrove developed and then drowned associated with a slow, and then rapid, sea level rise between 6000 and 4500 ^{14}C years BP and until 2500 ^{14}C years BP, respectively (Berdin *et al.*, 2003). Mangroves in Pari State, Amazonia, first occurred at $\sim$7500 ^{14}C years BP followed by a retreat after $\sim$6700 ^{14}C BP when relative sea level was lowered (Behling, 2002). Over a shorter time scale, an estimated loss of approximately one-third of the mangroves due to natural causes and mangrove clearing has occurred over the past 50 years (Alongi, 2002). The dominant factors responsible for this mangrove loss are sea level rise due to global warming, and human disturbance (Field, 1995; Ellison and Farnsworth, 1996; Nicholls *et al.*, 1999; Vilarrubia and Rull, 2002). Changes in the hydrodynamics and channel dimensions of estuarine areas are also caused by sea level rise

(Wolanski and Chappell, 1996). The degree and pattern of patchiness on coral reefs are also the result of sea level changes in the past and their relationship to the continental landmass (Hubbard, 1988).

The intertidal zone is also fragmented on a spatial scale. The formation of each type of intertidal habitat is basically by alternating exposed and sheltered stretches of the coastline due to different degrees of wave action. That is, each habitat type is patchy and non-continuous, for example, rocky shores are commonly separated by stretches of sandy beaches and *vice versa*; mangroves and mudflats occur in sheltered bays only, whereas the occurrence of coral reefs is restricted by latitudinal temperature gradients and within warm oceanic circulation patterns. Those intertidal areas that are isolated to varying degrees, such as rocky shores, are where fish communities undergo adaptation to increase their fitness to the intertidal environment after resulting in convergent evolution (Horn, 1999). This convergent adaptation has been demonstrated for temperate rocky shore fishes with respect to three functional traits, i.e., desiccation tolerance, air breathing and parental care (Horn, 1999).

It is suggested that there is an increase in overall phenotypic plasticity in populations residing in widely fluctuating environments (Soares *et al.*, 1999). Phenotypic plasticity within the genotypes of a population is suggested to be the most important mechanism of adaptation to the environment (Bradshaw, 1965). This has been demonstrated for the intertidal fish *Rivulus marmoratus* which has been shown to demonstrate considerable phenotypic variation in response to biotic (food) and abiotic (temperature and salinity) factors (Lin and Dunson, 1995).

Chiba (1998) provided a model which suggests that characters that promote a high intrinsic growth rate and a low carrying capacity, i.e., small body size, high fecundity and simple body form, tend to evolve in the most unstable environments. During mass extinctions, taxa living in less stable shallow seas have higher extinction rates than those living in the stable deep sea (Jablonski, 1986; Hallam, 1987). The reverse is true, however, during normal times or normal extinctions (Ward and Signor, 1983). Opportunistic life history (*r*-selected) traits are, thus, more favoured in stressful environments such as the intertidal zone.

Paleontological studies imply that the intertidal zone is the place of origin of most fish groups (and other vertebrates), and is the place from where fish conquered the open-marine and freshwater systems, and from where the tetrapods entered the terrestrial realm (Schultze, 1999).

Fossil records show that agnathans appeared in the Ordovician between 510 and 439 MYA in the intertidal zone. These intertidal vertebrates occur more frequently in Lower Palaeozoic (Silurian, 439–409 MYA and

Devonian, 409–363 MYA) rocks. Many early forms of fish, such as placoderms from the Silurian and Early Devonian, mongolepids, which are close relatives of chondrichthyans, and acanthodians from the Early Silurian to the Late Permian, all occur in shallow-marine and intertidal deposits (Goujet, 1984; Karatajute-Talima and Predtechenskyi, 1995; Schultze, 1996a). The first osteichthyans also appeared in the coastal zone of the Late Silurian.

It appears that these early fish moved into freshwaters from the marine environment. Acanthodians known from the early Silurian to the late Permian do not appear in the freshwater environment until the Upper Devonian Escuminac Formation. Xenacanths, a group of chondrichthyans, are common in freshwater and lagoonal marine deposits in the Carboniferous (363–290 MYA) and Permian (290–245 MYA) (Schultze, 1996a). The earliest actinopterygians are represented by scales from Silurian and Lower Devonian marine near-shore deposits (Schultze, 1996a). Actinopterygians had spread to the open marine environment by the Devonian and may have entered freshwater at that time too. During the Carboniferous, most fishes (elasmobranchs, acanthodians, actinopterygians, actinistians, rhipidstians and dipnoans) could be found in both marine and freshwater environments (Schultze, 1999). The Elpistostegalia, the closest relatives of tetrapods, are coastal zone fishes which occur in intertidal sediments of the Upper Devonian (Schultze, 1996b). Elpistostegalia and the early tetrapods show intertidal and supratidal adaptations such as a high orbit position with a median bony support ("eye brow"), comparable to the present day mudskipper *Periophthalmus* (Schultze, 1997).

The oldest known hermatypic Scleractinia occurred in the shallow-water Tethys sea of the Middle Triassic (245–208 MYA) with subsequent reef building beginning in the Late Jurassic (208–146 MYA) and formation of "modern" tropical reefs in the late Cretaceous (Wells, 1956; Rosen, 1988). It is generally agreed that mangrove systems first appeared in the Late Cretaceous (146–65 MYA) and Early Tertiary (65–1.6 MYA) on the shores of the Tethys Sea with subsequent genera diversification in the Late Tertiary (Hutchings and Saenger, 1987; Ellison *et al.*, 1999). The oldest seagrass fossil is also from the Cretaceous, followed by establishment of most modern seagrass genera in the late Eocene (~40 MYA; Hemminga and Duarte, 2000). The early form of reef fishes, Perciformes, stem from the Late Cretaceous (~75 MYA), followed by the earliest record of modern reef-fish families in the Eocene (~50 MYA) (Bellwood, 1996, 1997; Bellwood and Wainwright, 2002). When orders of fish increase noticeably at Eocene, i.e., between ~100 and 50 MYA (Budyko *et al.*, 1985), it was time when global mean air temperature was ~23 °C (Budyko *et al.*, 1985). The global mean air temperature fluctuated around 20 °C at 50 MYA and decreased subsequently to 15 °C at present. The tropical ecosystems, i.e., mangroves, seagrass beds, and coral reefs, should have been more extensive latitudinally at that time. The mangrove and intertidal seagrass bed fishes, similar to coral reef ones, evolved after these extensive tropical ecosystems appeared. Although both intertidal and open-water fishes occurred before the Triassic, the fishes associated with these relatively modern ecosystems are still more likely derived from intertidal and shallow water species rather than open-sea ones, following the rule that decreases in environment stability increase phenotypic plasticity (Soares *et al.,* 1999). One example supporting shallow sea as a place for fish speciation is shown on the geologically young, i.e., ~1–8 MYA, volcanic-produced Azores Archipelago. Among the 462 marine fishes there, including shallow-water, pelagic, and deep-water species, only one shallow-water species is endemic (Morton and Britton, 2000). This endemic fish, therefore, should have been evolved in shallow waters of Azores between 1 and 8 MYA.

D. Potential Impacts of Global Warming on Tropical Shallow Water Environments

Global temperature increases of between 2 and 4 °C are expected over the next half-century due to climate changes (Southward *et al.,* 1995; Somero and Hofmann, 1997). Temperature changes of this magnitude can affect the physiological and ecological states of animals, especially when those changes in temperature occur near their upper thermal limits. Since some aquatic animals live close to their thermal limits, an increase of only a few degrees in water temperature could threaten their survival (Logue *et al.,* 1995).

El Niño events can exacerbate the effects of global warming on aquatic animals, especially during the summer, since it is characterized by warm ocean temperatures in the Equatorial Pacific. For example, the 1997–98 El Niño/Southern Oscillation (ENSO) was unusually large (Izaurralde *et al.,* 1999), resulting in the warmest weather for 150 years (Wilkinson *et al.,* 1999). Temperatures of 3–5 °C above normal were often recorded in the Indian Ocean during the ENSO of 1998 causing bleaching and the death of corals (Wilkinson *et al.,* 1999). Severe coral bleaching was also observed during 1998 around the Andaman Islands, India (Ravindran *et al.,* 1999), and on the Great Barrier Reef, Australia (Berkelmans and Oliver, 1999). There are reports demonstrating that the population dynamics of fish were affected near California during the 1997–98 ENSO (Richards and Engle, 2001; Sánchez-Velasco *et al.,* 2002). In New Zealand, many new fish species were observed when sea surface temperature (SST) was higher than average, such as in the autumn months of 1996, 1998, and 1999 (Francis et al., 1999). Growth reductions in the otoliths coinciding with a strong El Niño event during 1982–83 were also reported in the widow (Sebastes entomelas) and the yellowtail rockfish (S. flavidus) inhabiting the coast of central and northern California (Woodbury, 1999). Forty-one of 44 coral reef fishes (Pomacentridae) in the Galapagos Archipelago had a check in otoliths growth, due to a major change in environmental and/or growth conditions that corresponded to the timing of the 1982–83 El Niño event (Meekan et al., 1999). These reports indicate that warmer ocean temperatures of but a few degrees during the ENSO had major impacts on the survival of corals and population dynamics of fish over a wide sea area.

Reid et al. (1995, 1998) reported that the addition of $2\,°C$ to a low ambient temperature increased growth and protein turnover rates by approximately 16% in juvenile rainbow trout (Oncorhynchus mykiss). However, an increase of $2\,°C$ during the period of peak summer ambient temperature caused a reduction in both growth and protein turnover rates by 20% in the fish. These results demonstrated that ambient water temperatures could significantly affect cellular protein turnover and growth rates in fish and that the effects of additional temperature can be altered depending on the original ambient temperature and, therefore, season (Morgan et al., 2001). There may be a critical threshold temperature below which there are positive effects on growth and protein turnover rates, while negative effects may result above it. If the current environmental temperature is already close to such threshold temperatures, a slight increase in water temperature could have a major impact on fish physiology and metabolism. Indeed, Nakano et al. (2004) found changes in the cellular HSP70 expression patterns and in the growth rate of the Indo-Pacific sergeant (Abudefduf vaigiensis) with an increase in ocean temperature of only $1\,°C$. In a laboratory heat shock experiment, 50% mortality was observed at $32\,°C$, which was only about $2\,°C$ higher than the ocean temperature at the time. These reports propose a possibility that even if the rate of increase in the global yearly average temperature is low, the addition of a few degrees to the maximum daytime temperature during the summer could be fatal for many ectothermic animals, especially in tropical intertidal environments in which fishes are already exposed to considerably high temperatures.

It appears that global warming will cause elimination of temperate species that fail to redistribute themselves, however, the resulting more extensive warmer shallow seas may repeat what had happen in the Eocene. That is, wider latitudinal distribution for tropical ecosystems and fish and increase in fish species diversification through their adaptation to the tropical conditions.

II. RESPIRATORY ADAPTATIONS

As noted in section I.A, tropical intertidal environments are associated with rapid changes in physical parameters such as temperature, pH, salinity, oxygen, and carbon dioxide levels. Increases in temperature during emersion in the daytime greatly affect the metabolic rate of fish. Changes in temperature of the order of only 1 °C can alter the rate of biological reactions by 10% (Hochachka and Somero, 1984; Prosser, 1986; Cossins and Bowler, 1987). The level of ATP demands at the cellular level increases with temperature, requiring higher ATP production in organs such as the liver and muscle. Metabolism is elevated with increases in temperature, such that ATP supply matches demand. Gill ventilation rates and cardiac output increase with temperature to maintain oxygen delivery to the tissues. Oxygen consumption rates respond to changes in temperature in a similar way between intertidal and non-intertidal fishes (Bridges, 1988). In addition, aquatic hypoxia is a common phenomenon in the intertidal environments. The low availability of oxygen aggravates the problem of an increase in energy demand due to increases in temperature. Therefore, fishes inhabiting these environments have to possess adaptations to hypoxia in order to maintain an adequate energy supply to the body.

Many tropical intertidal fishes breathe air, presumably as an adaptation to aquatic hypoxia. However, air-breathing fishes need to deal with the additional problems associated with the water-to-air transition. These problems will be described together with appropriate adaptations below. In this section, respiratory adaptations observed in the gill, buccopharyngeal cavity, skin, and blood will be described. Some speculations about the possible existence of certain intertidal specific adaptations will also be presented (see Chapter 12, this volume).

A. Gill Adaptations

1. MORPHOLOGICAL ADAPTATIONS

During aerial exposure, fish gills face the problem of collapse due to the lack of buoyancy exerted by water, and the problem of secondary lamella coalescence due to water surface tension. Both reduce the effective surface area for gaseous exchange. In addition, a gill faces the danger of desiccation. Certain tropical intertidal fishes possess gills that appear to be well adapted for air breathing. Different fishes occupy different ecological niches. Some explore the high tide area, while others prefer the low tide and hence spend more time submerged. Such species differ in their gill morphologies, which reflect their life styles. Some are well adapted to air breathing, others, less so.

A comparative study of the gills of three mudskippers occupying different ecological niches has demonstrated differences in adaptation (Low *et al.*, 1988). *Periophthalmodon schlosseri* stays in higher zone burrows of the intertidal mudflat than *Boleophthalmus boddaerti*. *Periophthalmus chrysospilos* inhabits the eulittoral zone near the mudflat. In terms of life style, *P. schlosseri* spends the largest amount of time on land, followed by *P. chrysospilos*, and then *B. boddaerti*. It was demonstrated that the secondary lamellar surfaces of the three mudskippers' gills were highly folded, greatly increasing their surface area for gaseous exchange (Low *et al.*, 1988). There were also raised ridges on the gill epithelium that might function as mucus support. The mucus could have two functions: (i) prevent desiccation and (ii) prevent frictional abrasion between the lamellae.

In addition to the above adaptations, *Periophthalmodon schlosseri* has branched gill filaments. This gives rise to a larger surface area and provides more skeletal support from the gill rods as compared with unbranched filaments. Perhaps the most interesting adaptation in the gill of *P. schlosseri* is that it has extensive fusion of the secondary lamellae. Water will be trapped in the spaces between the lamellae formed by this fusion. This could further lower the risk of desiccation during air exposure. Similar secondary lamella fusions have also been found in the bowfin, *Amia calva* (Daxboeck *et al.*, 1981). This fusion, however, renders the gills less suitable for water-breathing by increasing the resistance to flow through them between the secondary lamellae. In comparison to *Boleophthalmus boddaerti*, *P. schlosseri* possesses fewer, shorter filaments and hence a smaller gill area. It was subsequently shown that *P. schlosseri* respires more efficiently on land than in water (Kok *et al.*, 1998; Takeda *et al.*, 1999). The secondary lamellae of *P. chrysospilos*, on the other hand, are not fused, but decreases in size towards the filament tip. This morphology allows a large gill surface area to be in contact with air, even when the gill filaments are stuck together. The gill of *B. boddaerti* is the most well adapted for water breathing. It has the largest surface area and largest number of the longest filaments. Having long filaments also implies that *B. boddaerti* is less well adapted for air breathing. This is because the longer the filament the easier it will collapse and coalesce in air. Kok *et al.* (1998) showed that it respires better in water than in air.

2. BIOCHEMICAL ADAPTATIONS

Fish gills are generally thought of as lactate-consuming organs (Mommsen, 1984). Therefore, the heart-type lactate dehydrogenase (LDH), which oxidizes lactate, is expected to be the dominant isozyme present in the gills. Low *et al.* (1992), however, found that the muscle-type LDH was the dominant isozyme in the gills of the mudskippers, *Periophthalmodon schlosseri* and *Periophthalmus chrysospilos*. This might be because the gill morphology of these two species are not well adapted for water breathing, as described in the previous section (Low et al., 1988; Kok et al., 1998; Takeda et al., 1999). Upon exposure to aquatic hypoxia ($0.8\,\mu lO_2 ml^{-1}$) for 6 hours, there were significant increases in lactate contents in the gills of P. chrysospilos and P. schlosseri, but not in B. boddaerti (Ip and Low, 1990; Ip et al., 1990; Low et al., 1993), which could breathe perfectly well in water. The muscle-type LDH is pyruvate-reducing and facilitates redox balance in tissues when oxygen supply is low. In addition, the activities of many glycolytic enzymes, such as phosphofructokinase-1 (PFK-1) and pyruvate kinase (PK), were found to be higher in the gill of P. schlosseri than B. boddaerti (Low et al., 1993). As a result, despite its lower water-breathing efficiency in comparison with the gill of B. boddaerti, the gill of P. schlosseri could still function during hypoxia by sustaining anaerobic glycolysis. Together with the unique morphology described in the previous session, the high glycolytic capacity of the gill of P. schlosseri could be seen as a package of adaptations for air breathing and hypoxia in tropical intertidal environments.

3. CARBON DIOXIDE EXCRETION IN AIR

Tropical intertidal fishes face the potential problem of carbon dioxide excretion during aerial exposure. Carbon dioxide exists as molecular CO_2 and HCO_3^- in solution, and most of it exists in the latter form over the physiological pH range. Cell membranes are not permeable to the charged HCO_3^-, and thus carbon dioxide has to travel across membranes in the form of molecular CO_2. During aquatic respiration, upon crossing the respiratory epithelial membrane and entering the ambient water, most of the molecular CO_2 becomes HCO_3^- through hydration. This is not the case with aerial respiration, where molecular CO_2 always remains as it is in air. As a result, the body-to-environment PCO_2 gradient is larger in aquatic than in aerial respiratory systems, assuming that other factors such as ventilation are equal. Thus, carbon dioxide excretion might become a challenge for intertidal fishes when they emerge from water. Blood gas data has shown that the total CO_2 content in Periophthalmodon schlosseri increased significantly after aerial exposure for 24 h (Kok et al., 1998).

Although there are data showing elevated blood CO_2 contents, the respiratory exchange ratio (RE) of many intertidal fishes remained the same when breathing air, as when submerged. The REs have been reported to fall within the range of 0.7–1.0 (Bridges, 1988; Martin, 1993, 1995). This implies that these intertidal fishes have the ability to excrete CO_2 even in air. The increase in blood CO_2 content during aerial exposure in mudskippers (Kok et al., 1998) results in elevated PCO_2 levels in the blood and, hence, an increasing body-to-environment PCO_2 gradient maintaining normal carbon dioxide excretion in air. This presumably happens in other intertidal fishes too.

Carbonic anhydrase (CA) associated with the intravascular endothelial membrane has been located recently in the gas exchange organs of amphibians, reptiles, birds, and mammals (Stabenau and Heming, 2003). In teleost fish, on the other hand, only intracellular CA is found in the gills. Although intravascular CA has been identified in the gills of chondrichthyan fish (Wood *et al.*, 1994; Wilson *et al.*, 2000b; Gilmour *et al.*, 2002), it has not been detected in any teleost gill. However, it has been reported that intravascular CA occurs in the air bladder of the bowfin, *Amia calva*, an air-breathing teleost (Gervais and Tufts, 1998). Many tropical intertidal fishes are air-breathers. Some rely heavily on the gill for gaseous exchange even in air, e.g., the mudskipper, *Periophthalmodon schlosseri* (Kok *et al.*, 1998). An interesting question, therefore, is: is there intravascular CA in the gills of these tropical intertidal fishes?

Carbonic anhydrase catalyzes the interconversion of molecular CO_2 and HCO_3^- and plays an important role in carbon dioxide transport and excretion. Carbon dioxide is transported mainly in the form of HCO_3^- in the plasma. At the teleost gill, where there is no intravascular CA, HCO_3^- has to enter the red blood cells (RBC) via a Cl^-/HCO_3^- exchanger, be converted to molecular CO_2 by intracellular CA inside the RBC, and then diffuse across the RBC membrane through the plasma and gill epithelial cells and enter the ambient water (Tufts and Perry, 1998). The rate-limiting step in this carbon dioxide excretion process is thought to be the entry of HCO_3^- into the RBC via the Cl^-/HCO_3^- exchanger (Perry and Gilmour, 1993). The presence of intravascular CA could potentially speed up the above process by bypassing the Cl^-/HCO_3^- exchange step. This has been suggested to be a physiological function of intravascular CA in gas exchange organs (Heming *et al.*, 1993, 1994).

However, as pointed out by Henry and Swenson (2000), the increment in CO_2 excretion caused by the intravascular CA in gas exchange organs cannot be significant. This is because the RBC has more CA activity than the gas exchange organ and more buffering capacity than the plasma. There is likely a limited supply of protons for the HCO_3^- dehydration reaction in the plasma whereas proton supply by hemoglobin inside the RBC is relatively much larger (Heming and Bidani, 1992). Thus, most of the HCO_3^- dehydration reaction would take place inside the RBC despite the availability of CA in the plasma. Intravascular CA activities are more likely to be involved in affecting the pH/CO_2 disequilibrium in the blood leaving the gas exchange organ (Gilmour, 1998).

Regardless of the physiological function of the intravascular CA, the fact that many air-breathing animals possess this enzyme in their gas exchange organ may suggest its presence in the gill, or even skin, of air-breathing intertidal fishes.

B. Buccopharyngeal and Skin Adaptations

In comparison to the gill, the buccopharyngeal cavity and the skin of fish are relatively less studied. The greater number of studies on the gill might be explained by its massive involvement not only in gaseous exchange, but also in nitrogenous waste excretion as well as osmotic and acid-base regulation. Nevertheless, the skin and, in some fish, the buccopharyngeal cavity are heavily involved in respiration (Tamura *et al.*, 1976; Nonnotte and Kirsch, 1978; Steffensen *et al.*, 1981; Meredith *et al.*, 1982; Sacca and Burggren, 1982). Cutaneous gas exchange in vertebrates in general has been extensively reviewed by Feder and Burggren (1985). This session will focus on skin and buccopharyngeal respiration in tropical intertidal fishes.

Respiration through the skin was considered an important adaptation for fishes that frequently leave water (Randall *et al.*, 1981). This is because during emergence the gills are generally thought to be less effective as a respiratory organ due to the collapse of their filaments (this might exclude gills with special adaptations as described in section A.1.). Many tropical intertidal fishes rely heavily on skin respiration when out of water, e.g., *Periophthalmus sobrinus* (Teal and Carey, 1967), *Mnierpes macrocephalus* (Graham, 1973), *Periophthalmus cantonensis* and *Boleophthalmus chinensis* (Tamura *et al.*, 1976), *Blennius pholis* (Pelster *et al.*, 1988), and *Clinocottus analis* (Martin, 1991). It should be noted that the skin itself generally has a high oxygen consumption rate, presumably due to its functions of ionic regulation and mucus production. Nonnotte (1981) found that the oxygen consumption rates of skin from many fish species such as the rainbow trout and crucian carp were either equal to or exceeded the total transfer of oxygen across the skin. On the other hand, the skin oxygen consumption rates of some other fish species such as the shanny and flounder were lower than skin oxygen uptake rates (Graham, 1976; Nonnotte and Kirsch, 1978) and some oxygen must have been supplied from the blood. Therefore, it remains to be determined if cutaneous respiration in tropical intertidal fishes is beneficial to organs other than the skin itself.

It was demonstrated in *Boleophthalmus chinensis* and *Periophthalmus cantonensis* that there was increase in reliance on skin respiration during emergence in comparison to submergence (Tamura *et al.*, 1976). There have been no similar studies of tropical intertidal fishes on the relative importance of cutaneous respiration in air and in water since the work by Tamura *et al.* (1976), although there are studies on freshwater fishes (Sacca and Burggren, 1982). It is not clear if the increase in reliance on cutaneous respiration is an adaptation generally adopted by tropical intertidal air-breathers. It was suggested that recruitment of capillaries in the skin during aerial exposure was the mechanism behind the increase in cutaneous respiration (Sacca and Burggren, 1982).

Studies on the histology of tropical intertidal fish skin suggested that the penetration of capillaries into the epidermis was an adaptation to air breathing. As a result, the air–blood diffusion distance is reduced. In the different species of mudskippers examined, the air–blood diffusion distance in the skin of the head and the dorsal body, which are often exposed in air, falls within the range of 2–6 μm (Al-Kadhomiy and Hughes, 1988; Yokoya and Tamura, 1992; Zhang *et al.*, 2000; Park, 2002). In contrast, the capillaries in the skin of the ventral and caudal area, which are often submerged, are located in the sub-epidermal region (Zhang *et al.*, 2000). The capillaries of *Rivulus marmoratus* in the epidermis of the dorsal head and antero-dorsol body were also located within 1 μm of the epidermis (Grizzle and Thiyagarajah, 1987).

During emergence, there is a potential threat of desiccation in the fish epidermis. The skin surface of amphibious intertidal fishes such as mudskippers and rockskippers are slimy (Whitear and Mittal, 1984; Yokoya and Tamura, 1992). Lillywhite and Maderson (1988) proposed that the layer of mucus secreted onto the epidermis is an adaptation to emergence because it can reduce direct evaporation from the epidermal cells. Goblet cells occur in the skin of *Boleophthalmus pectinirostris* and were thought to be the site of mucus production (Yokoya and Tamura, 1992). In the skin of *Periophthalmus cantonensis* and *Periophthalmus koelreuteri*, however, goblet cells do not occur (Yokoya and Tamura, 1992). Mucus was thought to be produced by superficial epithelial cells in these fish. In addition to the mucous layer, vacuolated cells were demonstrated in the middle layer of the skin of species of *Boleophthalmus* and *Periophthalmus* (Whitear, 1986, 1988; Yokoya and Tamura, 1992). These cells were suggested to contain large amounts of water for further protection from desiccation in addition to the mucus. The buccopharyngeal cavity is also considered to be an important respiratory organ in tropical intertidal fishes because the skin lining is heavily vascularized. Many species possess a large buccopharyngeal volume which allows the fish to hold a large volume of air when in hypoxic water. In addition, it was observed in the mudskipper, *Periophthalmodon schlosseri*, that the opercular bone is large, thin, and can be easily distorted, facilitating increases in buccopharyngeal volume during breath-holding (Aguilar *et al.*, 2000). The capillaries in the buccal cavity of *P. schlosseri* are intraepithelial and diffusion distances between blood and water are <1 μm (Wilson *et al.*, 1999).

In summary, capillaries are found within the epidermis of skin and buccopharyngeal cavity of fish which rely heavily on air breathing. This morphological arrangement decreases the blood–air diffusion distance and, thus increases gaseous exchange efficiency. For skin portions that are frequently exposed to air directly, mucus are generally present to minimize epidermal desiccation.

C. Blood Adaptation

The gaseous function of fish blood is to transport oxygen from the site of uptake to where it is utilized and carbon dioxide in the reverse direction. However, to understand exactly how blood does this and how different parameters affect the processes involved is relatively more difficult. Increases in hemoglobin concentrations and hematocrit values raise the blood oxygen-carrying capacity and augment oxygen entry into and release from the blood. Increases in temperature, pH, organophosphate concentrations and O_2 and CO_2 levels are known to affect the oxygen-affinity of hemoglobin (Weber and Wells, 1989; Nikinmaa, 1990). Regulation of some, or all, of these parameters in order to maintain proper oxygen delivery in a changing external environment is therefore complex.

Although intensive efforts have been spent on investigating fish blood properties and their environmental adaptations, little data on tropical intertidal fishes are available. It is unclear how the blood of tropical intertidal fishes adapts to the ever-changing parameters of their environments. Notwithstanding, since most tropical intertidal fishes are air-breathers, research on the blood of other amphibious fishes may increase our understanding of the blood adaptations in this group. One of the problems faced by all air-breathing fishes is that oxygen acquired from air in the air breathing organ (ABO) would be lost to hypoxic water (which they often encounter) via the gill because the ABO and the gill are in series. Many fishes, however, have shunt systems to allow oxygenated blood to bypass the gills (Ishimatsu and Itazawa, 1983) or the hemoglobin affinity is high reducing oxygen loss from blood to the water across the gills (Daxboeck *et al.*, 1981). It has been observed that the hemoglobin concentrations and hematocrit values are high in air-breathing fishes (Graham, 1997). Hemoglobin concentrations and hematocrit values obtained from the few studies on the blood of the mudskippers *Periophthalmus sobrinus* and *Boleophthalmus boddaerti* were high (blood hemoglobin concentration: 14–17 g%; hematocrit: ~50%) (Pradhan,

1961; Vivekanandan and Pandian, 1979; Manickam and Natarajan, 1985). It has been suggested that blood–oxygen affinity generally decreases with increasing dependence on air breathing (Lenfant and Johansen, 1967; Johansen and Lenfant, 1972). This might be seen as an advantage for oxygen unloading in tissues (oxygen loading is not a problem due to the increase in oxygen supply in air). A comparative study of the oxygen affinity of Hb in water- and air-breathing freshwater fishes in the Amazon supported the theory (Johansen *et al.*, 1978). However, results from another study carried out by Powers *et al.* (1979) appeared to reject the theory. These authors failed to detect any significant difference between the oxygen affinity of Hb in water- and air-breathing fishes. Morris and Bridges (1994) later resolved this apparent contradiction. When the relationship between pH and P_{50} (the PO_2 needed to oxygenate 50% of the Hb), instead of between PCO_2 and P_{50} as in the Powers *et al.* (1979) study, was examined, there existed a clear trend, with an increase in P_{50} values from water- to obligate air-breathing fishes (Morris and Bridges, 1994). However, Graham (1997) pointed out that data from two tropical intertidal fishes, *Gillichthys mirabilis* and *Periophthalmus barbarus*, broke the trend as the former species (a facultative air-breather) showed lower hemoglobin–oxygen affinity than the latter (an amphibious fish). Further comparisons are needed in order to see if there really is a trend.

Perhaps the most interesting, and also the most challenging, topic in dealing with blood adaptation in tropical intertidal fishes is how rapid changes in environmental parameters such as pH, temperature, CO_2, and O_2 levels are tolerated. For example, when challenged with increasing temperatures and decreasing ambient O_2 levels when trapped in a small volume of water, would they decrease their blood concentration of organophosphates (such as ATP and GTP) in order to raise the hemoglobin–oxygen affinity (Wood and Johansen, 1972; Greaney and Powers, 1978; Val *et al.*, 1995; Val, 2000)? Or would they adjust their hemoglobin concentration and hematocrit values to increase their oxygen-carrying capacity? Would they possess hemoglobin that has either small or no Bohr or Root effect so that they can still function properly at high temperatures, hypercapnia and/or low pH? Considering the range of habitats found in the tropical intertidal environment, one might expect diversity in adaptation in the different fishes, each having a set of features suitable to its local habitat. Val (2000) has shown that erythrocytic GTP and ATP concentrations can change in a few minutes in some tropical species and are fast enough to compensate for short-term hypoxia. GTP changes more rapidly than ATP. Whether or not such changes occur in intertidal fishes is unknown.

Another feature that might be of advantage to tropical intertidal fishes is hemoglobin multiplicity. Many fish species possess more than one type of hemoglobin (Riggs, 1970, 1979; Weber, 1990). Anodally resolved hemoglobins (with low isoelectric points, pI) are sensitive to temperature and pH, generally with large Bohr or Root effects. Cathodic hemoglobins (with a high pI) are relatively insensitive to temperature and pH, and may have intrinsically higher oxygen affinities. Brix *et al.* (1998, 1999) have concluded from their work on New Zealand fishes that those fishes in habitats with unstable oxygen values tend to possess cathodic hemoglobins. Since tropical intertidal environments are also characterized by unstable oxygen levels, one might expect fishes there to possess cathodic hemoglobins too. Further work may verify this hypothesis.

In summary, tropical intertidal fishes might have adaptation(s) in their blood in order to maintain a relative constant hemoglobin–oxygen affinity even during rapid environmental changes. It is speculated that the erythrocyte GTP and ATP levels would change rapidly to compensate for the effects exerted by changes in environmental parameters. Hemoglobin multiplicity might also be another adaptation found in this group of fish.

III. PHYSIOLOGICAL AND MOLECULAR ADAPTATIONS TO ELEVATED TEMPERATURES

Generally, fish cannot maintain thermal disequilibrium between the surrounding environment and their body, because of the high specific heat and low oxygen content of water and the associated design of the gills, which facilitates rapid heat exchange across them. Some species of Scombroidei possess vascular heat exchangers which allow them to maintain temperatures in some regions of the body considerably above ambient (Block and Finnerty, 1994) and other fish may show transient increases in muscle temperature associated with high levels of exercise. With these exceptions, the body temperatures of most fishes are kept within ±1 °C of the surrounding water (Hazel and Prosser, 1974). Changes in water temperature can directly affect fish body temperature and, therefore, metabolic processes.

The thermal physiology of fishes is diverse. Fishes inhabiting tidepools experience rapid increases in water temperature when their habitat is isolated from the ocean during low tide periods. Fish can sense changes in ambient temperature through peripheral and central thermoreceptors (Crawshaw, 1977, 1979). When ambient temperature increases, fish can first thermoregulate by selecting cooler waters in the ocean or river (behavioral thermoregulation). However, in shallow waters, such as in a tidepool, the choice a fish can make may be more limited. Often the fish must stay at the high temperature until it decreases naturally, such as at high tide. Such eurythermal species of fish seem to have some physiological features beneficial for survival when subjected to these large thermal fluctuations.

A. Cellular Protein Stability

Enzymes control chemical reactions within living organisms. Enzymes are thermolabile proteins requiring a particular complex 3D-conformation to be active, and are denatured at high temperatures. When temperature increases, the general body metabolic rates generally increase, as mentioned above. However, when the temperature exceeds the threshold temperature of enzyme denaturation, the metabolic rate can decrease as the number of such enzymes increases. The minimum temperature that denatures cellular proteins is well correlated in stenothermal and eurythermal animals (Somero, 1995; Somero *et al.*, 1996). Organisms occupy almost every corner of the globe, the thermal habitat ranging from $-50\,°C$ to $110\,°C$. Organisms, however, have only 20 amino acids from which to compose all their cellular proteins. Fields (2001) reviewed the amazing adaptation in protein stabilities to species-specific thermal habitats. By relatively small modifications to amino acid sequences, organisms maintain the appropriate balance between stability and flexibility of functional proteins within their physiological temperatures. For example, A_4-lactate dehydrogenase (LDH) in warm-adapted species has a lower Michaelis constant (K_m) value for pyruvate than cold-adapted species at one tested temperature, indicating that the A_4-LDH of warm-adapted species has inherently higher stability under higher temperatures. When homologous M_4-LDH from fish species from various thermal habitats are assayed at species normal body temperatures, the K_m values for pyruvate, as well as resting pyruvate concentrations at physiological body temperatures, appeared to be uniform, meaning that the optimal functional temperature for LDH has been evolutionarily adapted to the habitat temperature of the species (Hochachka and Somero, 1984; Moerland, 1995; Fields, 2001). The mechanism underlying cellular protein stability in species of fishes inhabiting a wide range of temperatures has not been fully studied. However, Fields (2001) showed that an extremely eurythermal intertidal goby, *Gillichthys seta*, had a more constant A_4-LDH K_m value for pyruvate at various temperatures, as compared with other temperate and Antarctic fishes, indicating that this species can maintain the enzyme-substrate affinity of A_4-LDH over a wide range of temperatures.

Intracellular pH is one of the most important factors determining the functions of cellular proteins when temperature changes occur (Sartoris *et al.*, 2003), i.e., tissue pH decreases when temperature increases. The temperature-induced adjustment of intracellular pH is generally known to occur by active ion transport in eurythermal species whereas it depends on a passive mechanism in stenothermal species, resulting in slower intracellular pH changes in eurythermal species than in stenothermal ones (Sartoris and

Pörtner, 1997; Pörtner and Sartoris, 1999). Also, the ability for active temperature-dependent pH regulation appears essential for inhabiting thermally unstable environments such as the intertidal zone and estuaries. It has been suggested that a slower pH change is beneficial for species inhabiting a thermally fluctuating environment in order to minimize rapid and frequent pH shifts by fluctuations in water temperatures (Sartoris *et al.*, 2003).

B. Molecular Chaperones

Various organisms, including fish, respond to unfavorable conditions or stress, such as acute increase in temperature, by expressing a small number of specific proteins (heat shock proteins = HSPs or stress proteins = SPs) (Gross, 1998). Heat shock proteins (HSPs) play a critical role in the thermal tolerance of organisms (Lindquist and Craig, 1988; Sanchez and Lindquist, 1990; Prändle *et al.*, 1998; Soto *et al.*, 1999). They function as molecular chaperones in the folding and translocation of newly synthesized proteins, as well as in the repair of damaged proteins denatured by stressors (Hartl, 1996). Correlations between the cellular levels of HSPs and the resistance to stress in animals are now well established by various field and laboratory experiments (Feder and Hofmann, 1999), and it is considered that the cellular level of HSPs is an important factor determining the level of cyto-protection, preventing thermal injury of the cell, and contributes to the whole-body thermal tolerance of the animal. Ulmasov *et al.* (1992) demonstrated a correlation between the range of habitat temperatures and HSP70 levels at the physiological temperature in several species of lizards. Norris *et al.* (1995) also reported that the amount of HSP70 was positively correlated with time of survival at 41 °C in the tropical topminnow, *Poeciliopsis gracilis*. These authors concluded that an increased expression of HSP70 might contribute to an increase in stress response in this species.

Although HSPs have a broad range of functions essential for cellular protein homeostasis, all of them are based on the ability of HSPs to recognize unstructured proteins (Morimoto *et al.*, 1990; Freeman *et al.*, 1999; Santoro, 2000). Since the threshold temperature for denaturation of a cellular protein is different among species (see above), it is possible that the functional importance of HSPs varies depending on the thermal habitat of the species. Indeed, the flexibility of reprogramming in gene expression and in cellular signaling pathway for the expression of HSPs seems to differ among species, depending on its thermal tolerance or ability to acclimate (Horowitz, 2001). The function of HSPs may have been influenced by habitat temperature during evolution, so that it can improve evolutionary fitness by reducing the impact of changes in environmental temperatures on cellular protein stability. Further, synthesizing HSPs as well as maintaining the function of these proteins can represent a major energy demand for the cell (Hawkins, 1985; Houlihan, 1991; Martin *et al.*, 1991). A close relationship between the threshold temperature for increased 70 kDa HSP (HSP70) levels and the thermal preference of fish (*Poeciliopsis*) was demonstrated by Hightower *et al.* (1999).

The functional importance of the cellular heat shock response in intertidal invertebrates has been extensively studied. HSP70 is known to be the most abundant protein induced by heat stress in various organisms and is considered to play an essential role in conferring tolerance and survival under stress (Parsell and Lindquist, 1993). Hofmann and Somero (1996) demonstrated that the northern intertidal mussel, *Mytilus trossulus*, had lower threshold induction temperatures for HSP70 synthesis than the southern species, *M. galloprovincialis*, while the overall intensity of synthesis and induction of HSP70 was greater in the latter than the former species at higher temperatures. Tomanek and Somero (1999) noted that the threshold temperatures for increased and peaked levels of HSP70 in gills were higher in species of marine snails in the higher intertidal zone, compared with those from lower down. Tomanek and Somero (2000) also demonstrated that levels of various HSPs in the mid-intertidal marine snail, *Tegula funebralis*, increased and returned to pre-stress levels faster than the low-intertidal to sub-tidal species, *T. brunnea*. The authors hypothesized that *T. funebralis* may be more capable of repairing protein damage incurred during the mid-day low tide before the occurrence of the next low tide, as compared to *T. brunnea.*

Although the functional importance of HSPs has not been studied in intertidal fishes as much as in invertebrates, Nakano and Iwama (2002) demonstrated that the cellular HSP70 response of two intertidal sculpins correlated with their thermal habitats. The tidepool (*Oligocottus maculosus*) and fluffy sculpins (*O. snyderi*) are common on the intertidal of the West Coast of North America between the Gulf of Alaska and mid-California (Morris, 1960, 1962; Green, 1971) with a similar morphology, ecology, and diet (Nakamura, 1970, 1971). However, the tidepool sculpin occurs in both lower and upper tidepools, while the fluffy sculpin's vertical distribution does not extend past the lower tidepools and the sub-tidal zone during low tides (Green, 1971; Nakamura, 1976a, b). Larger water temperature fluctuations occur in upper than in lower tidepools (Green, 1967; Nakamura, 1976a). Thus, this distinct vertical distribution may reflect a difference in the thermal tolerances of these species. Indeed, the tidepool sculpin has been shown to have a higher resistance to warmer temperatures than the fluffy sculpin (Nakamura, 1970, 1976a). Nakano and Iwama (2002) demonstrated that the tidepool sculpin had higher lethal and induction temperatures for

HSP70; the liver HSP70 level of the tidepool sculpin was less sensitive to changes in water temperatures; and it had higher natural constitutive HSP70 levels than the fluffy sculpin. The authors concluded that the less thermally sensitive tidepool sculpin might enhance its thermal tolerance by having a large pool of cellular HSP70, thus allowing it to inhabit the upper intertidal zone, experiencing relatively large and unpredictable fluctuations in water temperature.

IV. OSMOREGULATION

A. Ion Regulation

Fishes in intertidal habitats face daily changes in salinity, which is usually diel and influenced by the tide. Sudden drops in salinity may also occur because of heavy rainfall and freshwater input from rivers whereas increases in salinity may be caused by excessive evaporation during low tide. Most tropical intertidal fishes, especially estuarine ones, occur in habitats with salinity fluctuations from ~30–33‰ down to 20–25‰ in the rainy season. Salinities of <25‰ may pose osmoregulatory problems for tropical marine fishes. Changes in salinity in intertidal habitats such as estuaries affect the spatial distribution pattern of fish (Martin, 1988; Sheaves, 1998; Prodocimo and Freire, 2001). Most estuarine fishes in the tropics and subtropics are euryhaline and able to cope with salinities down to almost freshwater (<1‰) to at least 35‰ (Blaber, 2000). For example, some tolerant Indo-Pacific euryhaline species are *Elops machnata* (0–155‰), *Pomadasys commersonnii* (0–90‰), *Monodactylus falciformis* (0–90‰), *Mugil cephalus* (0–90‰), *Rhabdosargus sarba* (1–80‰), *Liza macrolepis* (1–75‰), *Siganus vermiculatus* (2–55‰), *Gilchristella aesturia* (0–90‰) and *Ambassis natalensis* (0–52‰) (Whitfield *et al.*, 1981; Gundermann and Popper, 1984; Whitfield, 1996). The latter two have also been recorded from hypersaline conditions whereas the other tropical estuarine species have been recorded from freshwater or 1‰ (Blaber, 2000). Although euryhaline fishes living in intertidal habitats can tolerate a wide salinity range, they do have preferences, for instance, the blenny *Alticus kirki* occurs in habitats with a salinity of between 0 and 67‰ but has a preference for 30–40‰ (Brown *et al.*, 1991). This salinity preference may indicate the minimum energy the fish needs to invest in osmotic regulation. Many studies have reported that ~10% of the total fish energy budget is an osmotic cost and that water salinity influences fish metabolic rate, development, and growth (Plaut, 1999a, b; Bœuf and Payan, 2001).

Osmoregulation in marine fishes takes place in the kidneys, gills, the integument, the gut, urinary bladder, and in the case of elasmobranchs, a specialized salt rectal gland (Rankin *et al.*, 1983; Karnaky, 1998). Hormonal systems are also involved in salt water acclimation (Hanke *et al.*, 1993; Evans, 2002). A review of crossover hormonal systems in controlling the movement of ions across the gill epithelium is provided by Evans (2002). Circulating factors include prolactin, cortisol, growth hormone, and insulin-like growth factor, thyroid hormones, natriuretic peptides, arginine vasotocin, and angiotensin, whereas local effectors include epinephrine/norepinephrine, prostaglandins, nitric oxide, and endothelin (Evans, 2002). Part of the mechanism of ionic regulation in tropical intertidal fishes has been demonstrated for mudskippers (Sakamoto *et al.*, 2000; Sakamoto and Ando, 2002).

Mudskippers can survive 20–100% seawater and tolerate freshwater after acclimation (Clayton, 1993). In *Periophthalmus chrysospilos*, high rates of both Na^+ influx and efflux and an increase in Na^+-K^+ ATPase activity occur at high salinities (30–34‰ *in situ* on the shore in Singapore). Long-term acclimation to high salinity cause an increase in plasma Na^+ and Cl^- concentrations. Regulation of plasma sodium and potassium concentrations was short term but long-term subjection of the fish to high salinities resulted in an increase in plasma Na^+ and K^+ content. Corresponding adjustments were made in intracellular iso-osmotic regulation through an increase in ninhydrin substances (NPS) (Lee *et al.*, 1987). Glutamate dehydrogenase (GDH) levels for amination or deamination also increase in higher salinities in the muscle and liver of *P. chrysospilos* (Chew and Ip, 1990). *Periophthalmus cantonensis* showed a similar increasing response in GDH and free amino acids (FAA) exposed to 80% seawater (Iwata *et al.*, 1981). When *P. chrysospilos* was exposed to a hypotonic environment, nuclear and prolactin cell size increased, indicating prolactin plays an important role in freshwater osmoregulation (Ogasawara *et al.*, 1991).

In brackish waters (8–25‰ *in situ* on the Singapore mudflat), *Boleophthalmus boddaerti* showed no change in NPS or GDH at high salinities (Chew and Ip, 1990). This is because *B. boddaerti* is naturally exposed to the tide and to fluctuations in salinity twice a day, and a constant synthesis and degradation of NPS and FAA for osmoregulation would be inefficient. This mudskipper instead employs behavioral adaptations to avoid low salinities, i.e, they stay in their seawater-filled burrow or move to the mudflat and feed during low tide. In this fish, branchial Na^+ and K^+ activated adenosine triphosphase (Na^+, K^+-ATPase) and HCO_3^- and Cl^- stimulated adenosine triphosphatase (Cl^-, HCO_3^- ATPase) are important for hyperosmotic and hypoosmotic adaptations, respectively (Ip *et al.*, 1991). Hydromineral regulation in this species is also dependent on cortisol secretion (Lee *et al.*,

1991). In *Periophthalmus modestus*, the apical crypts of the chloride cells or mitochondrial-rich cells in the skin under the pectoral fin open in seawater to secrete salt but close in freshwater to tolerate hypotonicity until the tide returns (Sakamoto *et al.*, 2000; Sakamoto and Ando, 2002).

B. Acid-base Regulation

Hypercapnia and hypoxia often occur in tropical intertidal environments during low tide at night due to respiration by living organisms. On the other hand, hypocapnia and hyperoxia occur during low tide in the daytime because of algal photosynthesis. Nakano and Iwama observed the dissolved oxygen level of a tidepool on the west coast of Vancouver Island, Canada, becoming more than 200% DO during daytime, while dropping to less than 10% DO during night (unpublished data). Hypercapnia would cause an increase in PCO_2 and a decrease in pH in the blood of fish because of the direct uptake of CO_2 from the ambient water (Heisler *et al.*, 1976; Toews *et al.*, 1983). Hyperoxia has the same effect on fish blood as environmental hypercapnia (Heisler *et al.*, 1988), but the mechanism by which it acts is different. During hyperoxia, ventilation rate decreases as fish regulate ventilation rate according to O_2 availability. This hyperoxia-induced decrease in ventilation rate causes a decrease in CO_2 excretion and results in an elevation of PCO_2 and a decrease in blood pH. Transition from water- to air-breathing has the same effect as aquatic hypercapnia and hyperoxia (Daxboeck *et al.*, 1981; Heisler, 1982; Ishimatsu and Itazawa, 1983; Pelster *et al.*, 1988). Due to the generally small size of intertidal fishes, blood vessels cannulation is often not possible. Hence, the number of accurate experiments done on acid-base regulation during hypercapnia, hypoxia, and air breathing are few. This is because the acid-base status of fish is sensitive to slight disturbance and cannulation is the only reliable way to obtain fish blood samples with minimal effects. Yet, Pelster *et al.* (1988) showed that after 3 hours of aerial exposure, venous blood pH significantly fell from 7.69 to 7.49 in the rockpool teleost *Blennius pholis*. There was a concomitant increase in PCO_2 and [lactate⁻] in the blood. It is not known if blood pH would later return to normal. It is, however, unlikely that *B. pholis* would stay emerged for so long anyway. The drop in blood pH is tolerated presumably because of the much higher buffering capacity of RBC and other cells, where most enzymatic reactions, which would be affected by a change in pH, occur. The situation in the mudskipper, *P. schlosseri*, is different. When exposed to air for 1 hour, there was a significant increase in blood pH and it remained significantly higher than the submerged value for up to 6 hours in air (Ishimatsu *et al.*, 1999). There was a transient decrease in blood PCO_2 at 1 h aerial exposure, but the value later returned to that of the control. An increase in blood pH might be partially explained by the increase in blood ammonia levels caused by decreased excretion during aerial exposure (Ishimatsu *et al.*, 1999). Again, the increase in blood pH was tolerated in *P. schlosseri*, presumably because of the high intracellular buffering capacity.

Blennius pholis and *P. schlosseri* are two of the few intertidal fish species large enough for cannulation. From the limited data available, it is not yet possible to draw any general conclusions about acid-base regulation in tropical intertidal fish. *Blennius pholis* responded like other facultative air-breathers to aerial exposure whereas *P. schlosseri* responded in an opposite way. It is not known if *P. schlosseri* is unique. Discovery of new, large intertidal fish species would facilitate further research into acid-base regulation.

V. AMMONIA TOLERANCE

Tropical intertidal environments are often associated with high ammonia levels. During low tide, water ammonia can accumulate to high levels due to excretion by living organisms inhabiting isolated water bodies on the shore. Decomposition of organic matter could also contribute to high ammonia levels. For example, the ammonia concentration in water collected from mudflat burrows can be as high as $3 \, \text{mmol} \, \text{l}^{-1}$ (Ip *et al.*, 2004). In addition, during aerial exposure, fish gills are not irrigated. Ammonia excretion is thus impeded because of the absence of ventilatory water which carries away excreted matter. The excreted ammonia accumulates in the boundary water layer on gills and its excretion becomes increasingly difficult. Consequently, ammonia accumulates in the fish body during aerial exposure and animals remaining in these environments have to deal with this toxicant. Ammonia toxicity as well as fish adaptations to the problem have been reviewed (Ip, Chew *et al.*, 2001a; Randall and Tsui, 2002; Chew *et al.*, 2005). However, not all described ammonia-tolerating mechanisms are found in tropical intertidal fishes. This section will highlight those adaptations to ammonia either reported upon or most likely to be observed in this group of fishes.

The mudskipper, *Periophthalmodon schlosseri*, has been reported to actively transport NH_4^+ across its gill against a concentration gradient (Randall *et al.*, 1999). A model explaining the mechanism of active transport has also been proposed (Wilson *et al.*, 2000a). Using immunohistochemical techniques, Na^+/H^+ exchangers (NHEs), a cystic fibrosis transmembrane regulator (CFTR)-like anion channel, a vacuolar-type H^+-ATPase (V-ATPase), and carbonic anhydrase (CA) were located in the apical crypt region of mitochondria-rich (MR) gill cells. The Na^+/K^+-ATPase and a $Na^+/K^+/2Cl^-$ cotransporter were found to be associated with the basolateral membrane of MR cells (Wilson *et al.*, 2000a). It was suggested that NH_3 diffuses into the MR cell, where it combines with H^+ to form NH_4^+. The H^+ is produced by hydration of CO_2, which also results in the formation of HCO_3^-. This HCO_3^- is thought to leave the cell via the apical CFTR-like anion channel. NH_4^+ formed could then be eliminated via the NHEs on the apical membrane. The Na^+/K^+-ATPase on the basolateral membrane is thought to participate in ammonia excretion only against a concentration gradient. The roles of V-ATPase and $Na^+/K^+/2Cl^-$ cotransporter in this process still remain unclear.

Alteration of the membrane permeability to ammonia has been proposed as an adaptation to high environmental ammonia (Ip *et al.*, 2004). Ammonia exists in two forms when in solution: NH_3 and NH_4^+. Biological membranes are generally permeable to NH_3, but not NH_4^+. Therefore, NH_3 can enter the fish body and exerts its effects on the animal. The permeability of a membrane to NH_3 is related to its fluidity (Lande *et al.*, 1995). The fluidity of a membrane is, in turn, influenced by lipid composition (Zeidel, 1996). *Periophthalmodon schlosseri* has high cholesterol contents in its skin $(4.5\,\mu\mathrm{mol\,g^{-1}})$ and is likely to have low fluidity and low permeability to NH_3 (Ip *et al.*, 2004). In addition, upon exposure to 30 $\mathrm{mmol^{-1}}$ NH_4Cl pH 7 for 6 days, the skin cholesterol level increased significantly. Sphingomyelin content, a high level of which causes low fluidity, also increased significantly (Ip *et al.*, 2004). This clearly is an adaptation to the problem of high environmental ammonia.

Tropical intertidal fishes that frequently depart water have to deal with the problem of body ammonia accumulation due to a reduction in its excretion. When subjected to aerial exposure in total darkness, *P. schlosseri* and *B. boddaerti* showed a reduction in proteolysis and amino acid catabolism (Lim *et al.*, 2001). There were also decreases in ammonia excretion and steady state free amino acids in various tissues (Lim *et al.*, 2001). These mudskippers became less active when exposed to air in total darkness. This decrease in activity could contribute to the observed reduction in proteolysis and amino acid catabolism, resulting in a decrease in ammonia production thereby alleviating the ammonia accumulation problem.

Proteins and amino acids are the main energy source for fish (Moon and Johnston, 1981). One of the end products of complete amino acid catabolism is ammonia. As mentioned above, during aerial exposure, ammonia excretion is impeded and it is accumulated in the body. Activities during emergence require energy and thus would potentially aggravate the ammonia problem in air-exposed tropical intertidal fishes. The mudskipper, *P. schlosseri*, deals with this problem by carrying out partial amino acid catabolism when it is active on land (Ip, Lim *et al.*, 2001). The amino groups of amino acids are transferred to pyruvate, catalyzed by alanine aminotransferase, leading to the formation of alanine. The α-ketoglutarate thus formed is fed into the Krebs cycle and undergoes partial catabolism. The overall result is the derivation of energy without the production of ammonia from amino acids. Alanine accumulation was observed in *P. schlosseri* after exposure to air under a 12h:12h dark–light regime (Ip, Lim *et al.*, 2001).

Ammonia is known to be detoxified to glutamine in the fish brain (Arillo *et al.*, 1981; Iwata, 1988; Mommsen and Walsh, 1992; Peng *et al.*, 1998). The same phenomenon was observed in brains of tropical intertidal fishes. For example, when exposed to sub-lethal ammonia concentrations, glutamine levels in the brains of *P. schlosseri* and *B. boddaerti* increased to 28 and $15 \mu mol\, g^{-1}$, respectively (Peng *et al.*, 1998). Glutamine is formed from glutamate and ammonia, catalyzed by glutamine synthetase (GS). Glutamate, in turn, is formed from α-ketoglutarate and ammonia, catalyzed by glutamate dehydrogenase (GDH). This strategy has an advantage over detoxification to urea because it is less energy demanding for each mole of ammonia detoxified (Ip, Chew *et al.*, 2001b). The two mudskippers, however, do not accumulate glutamine in other tissues such as the liver and muscle during ammonia loading conditions (Peng *et al.*, 1998).

In addition to detoxifying ammonia to glutamine, urea formation has also been observed in certain intertidal fishes (Mommsen and Walsh, 1992; Rozemeijer and Plaut, 1993). The Gulf toadfish, *Opsanus beta*, turned from ammonotelic to ureotelic when subjected to confinement and crowding (Walsh *et al.*, 1994). The increase in urea production was suggested to be a result of ammonia detoxification to glutamine, which serves as a substrate for carbamoyl phosphate synthetase III (CPS III), a urea cycle enzyme (Wood *et al.*, 1995). There was also an increase in GS mRNA as well as protein levels in the liver during such conditions (Kong *et al.*, 2000). In the leaping blenny, *Alticus kirki*, more than 50% of its total nitrogenous waste excretion was in the form of urea during aerial exposure (Rozemeijer and Plaut, 1993). The mechanism behind this high rate of urea formation has, however, not been investigated.

VI. SULFIDE TOLERANCE

Sulfide, including undissolved H_2S, the bisulfide ion HS^-, and the sulfide anion S^{2-}, is an important environmental factor for a variety of organisms in shallow marine habitats. It is extremely toxic for most fish species even in the low $\mu mol\, l^{-1}$ range (Smith *et al.*, 1977). In the marine environment, sulfide is produced by the anaerobic decomposition of organic matter and mostly from bacterial sulfate reduction in sediments (Jørgensen and Fenchel, 1974). Salt marshes, enclosed bays and estuaries are characterized by micromolar to millimolar sulfide concentrations in sediments (Bagarinao, 1992). Species inhabiting intertidal environments are, therefore, exposed frequently to considerable concentrations of sulfide (Grieshaber and Völkel, 1998). Anthropogenic inputs of organic carbon in the form of pollution from land also lead to increases in hydrogen sulfide in streams and coastal habitats (Bagarinao, 1992). Bagarinao (1992: Table 1) gives some sulfide values recorded from shallow marine habitats. Most of the values are from pore water. For tropical examples, sulfide concentration of pore water under mangrove roots and unvegetated area in Exuma, Bahamas was between 1100 and 4125 μM (Nickerson and Thibodeau, 1985); that of brine seep in East Flower Garden, Gulf of Mexico was up to 2200 μM (Brooks et al., 1979; Powell et al., 1983) and that of intertidal Spartina seagrass bed sediment from Dauphin Island, Alabama, USA was between $\sim$1 mmol l^{-1}, up to 8 mmol l^{-1} (Lee et al., 1996). Some portion of the sediment sulfide diffuses into the overlying water column and thus affects fish (Fenchel, 1969). Sulfide concentrations in salt marshes show diel fluctuations (Ingvorsen and Jorgensen, 1979), and vary with season, being typically low in winter and high in summer and early fall, in consonance with seasonal fluctuations in rates of organic matter production and sulfate reduction (Bagarinao, 1992).

Acute and lethal effects of sulfide are mainly due to inhibition of cytochrome c oxidase by reversibly binding at the heme site to cytochrome aa_3 (Nicholls, 1975). Apart from being a respiratory toxicant, sulfide uptake through respiratory epithelia can severely affect red blood cell (RBC) functions (see Chapter 7, this volume). Within RBCs, sulfide inhibits several enzymes such as glutathione peroxidase, super-oxide dismutase and catalase (Khan et al., 1987). Oxygen transport can also be impaired by the formation of sulfhemoglobin (SHb), in which FeII SHb binds O_2 with a much lower affinity than Hb (Carrico et al., 1978). Sulfide also affects ion transport across fish RBC membranes by inhibiting Na$^+$, K$^+$-ATPase (Völkel et al., 2001).

Tolerance levels to sulfide concentrations are better documented for freshwater fish species, e.g., the Northern pike Esox lucius (Adelman and Smith, 1970; Oseid and Smith, 1972), the Channel catfish Ictalurus punctatus (Torrans and Clemens, 1982), Amazon fishes (Affonso and Waichman, 1996), the River catfish Mystus nemurus (Hoque et al., 1998) and temperate marine species (Bagarinao and Vetter, 1989). The only tropical intertidal species studied is Boleophthalmus boddaerti, which has 96 h LC$_{50}$ values of 0.567 mM sulfide (Kuah et al., in press a). In general, intertidal species exhibit higher sulfide tolerance than pelagic ones, i.e., fish living in sulfide-rich habitats are more tolerant to hydrogen sulfide (Bagarinao and Vetter, 1989, 1992; Brauner et al., 1995). In temperate ecosystems, total sulfide levels at 50% mortality of intertidal marsh inhabitants, e.g., the California killifish,

Fundulus parvipinnis, the long-jawed mudsucker, *Gillichthys mirabilis* and *Mugil cephalus*, were >1200 μM whereas those of outer bay and open coast species were between 30 and 950 μM. *F. parvipinnis* and *G. mirabilis* also exhibit high sulfide tolerances, with 96 h LC_{50} values of 700 and 525 μmol l^{-1}, respectively (Bagarinao and Vetter, 1989).

Although tropical intertidal marine habitats are environments where high sulfide concentrations may occur, studies on sulfide toxicity and the tolerances of inhabiting fishes have been demonstrated only for the Tarpon, *Megalops atlanticus*, from South Florida estuaries, and *Boleophthalmus boddaerti* from Malaysia. Metamorphic larvae of *Megalops atlanticus* recruit to estuarine areas after 2–3 months and juveniles occur in shallow pools in which H_2S concentrations can be 250 ppb (Abel *et al.*, 1987). In laboratory conditions, they survived stepwise increase in total sulfide up to 232 μM. Tarpon decrease their gill ventilation rates in response to increased sulfide (up to 150.9 μM) and acidic pH but increase air-breathing frequency (Geiger *et al.*, 2000). There are other indications that elevated sulfide concentration may induce air-breathing activity in some fishes. *Gillichthys mirabilis* turns to air breathing at sulfide concentrations >700 μM (Bagarinao and Vetter, 1989). A mangrove forest fish, *Rivulus marmoratus*, will leap from the water and begin to respire cutaneously when H_2S levels reach 123 μM (Abel *et al.*, 1987). Armored catfish *Hoplosternum littorale* apply air breathing as an adaptation to H_2S (Brauner *et al.*, 1995). In a study of estuarine fishes by Bagarinao and Vetter (1989), only 2 out of 11 fish species could survive exposure to 1.5 mM sulfide; one of these is *G. mirabilis*, a species capable of air breathing (Todd and Ebeling, 1966).

There are several other possible physiological adaptations in fish to high sulfide levels. For example, a shift to anaerobic glycolysis, immobilization of sulfide-binding proteins, and enzymatic oxidation by mitochondria. It has been demonstrated that the Hb of three sulfide-tolerant fish species, *Fundulus parvipinnis, Gillichthys mirabilis* and the common carp *Cyprinus carpio*, exhibit a low susceptibility to SHb formation (Bagarinao and Vetter, 1992; Völkel and Berenbrink, 2000). Bagarinao and Vetter (1989, 1990) investigated sulfide adaptation in sulfide-tolerant *F. parvipinnis,* including the role of anaerobic metabolism, and hemoglobin and mitochondria characteristics. *Fundulus parvipinnis* blood neither catalyzes sulfide oxidation significantly nor binds sulfide at low environmental concentrations. Exposure to 200 μM and 700 μM sulfide over several days causes significant increases in lactate concentrations, indicating a shift to anaerobic glycolysis. It is, therefore, suggested that the high sulfide tolerance of the Killifish can be explained by mitochondrial sulfide oxidation (Bagarinao and Vetter, 1993). Bagarinao and Vetter (1990) also showed that liver mitochondria of *F. parvipinnis* oxidized sulfide to thiosulfate and produce ATP in the process whereas mitochondria of the pelagic, sulfide-sensitive Sanddab oxidized sulfide at a much lower rate. It is concluded that mitochondria with a sufficiently high affinity for sulfide and high capacity for sulfide oxidation are the chief detoxification mechanism through which fish erythrocytes prevent intracellular sulfide from reaching extracellular concentrations. Most sulfide-tolerant invertebrates studied so far also exhibit sulfide oxidation as one major detoxification mechanism with thiosulfate being the main oxidation product (Grieshaber and Völkel, 1998).

The sulfide tolerance of *Boleophthalmus boddaerti*, however, is not due to the presence of a sulfide-insensitive cytochrome oxidase, nor is it effected by a shift to anaerobic fermentation (Kuah *et al.*, in press b). Sulfide detoxification occurs mainly in the liver, in which different mechanisms involving cysteine aminotransferase (CAT), 3-mercaptopyruvate sulfurtransferase (MPST), and rhodanese, were used to detoxify sulfide in normoxia or hypoxia, producing mainly sulfate and sulfane sulfur, respectively (Kuah *et al.*, in press a, b).

The ability of predatory juvenile fishes, e.g., the tarpon, to withstand a combination of unfavourable intertidal conditions, including high sulfide levels, probably allows them to take advantage of the mangrove environment during their critical juvenile periods. These habitats contain abundant prey such as poeciliids, cyprinodontids, and juvenile mullet (Lewis *et al.*, 1983). They, therefore, are provided with a rich supply of prey that allows them to grow rapidly and obtain size refuge from predators (Geiger *et al.* 2000). It appears probable that a similar ecological role of sulfide adaptation also applies in tropical intertidal systems, in which there are sulfide-rich mangroves or mudflats, plenty of invertebrate prey, and small resident fishes.

VII. REPRODUCTION

A. Sexual Status

Tropical intertidal and shallow water fishes exhibit a considerable array of reproductive strategies to cope with habitats experiencing changing physical conditions. It appears there is a general pattern of sexual strategy within those teleost groups, which occupy a certain habitat type. For example, most tropical estuarine fishes are gonochoristic, that is, separate sexes, and each individual remains the same sex throughout its life, with but a few exceptions of successive hermaphroditism, i.e., sex change (Blaber,

2000). Protandrous sex reversal is demonstrated by some economically important species of Centropomidae, e.g., *Lates calcarifer* and *Centropomus undecimalis*, Clupeidae, e.g., *Tenualosa toil* and *T. macrura*, Platycephalidae, and Sparidae, e.g., *Acanthopagrus australis*, whereas protogyny occurs in most *Epinephelus* species (Serranidae) and *Rivulus marmoratus* (Rivulidae) (Blaber, 2000).

Sex change is usually under social control and is more common on coral reefs. Protogyny has been demonstrated for a number of reef fish such as the Angelfish *Centropyge potteri*. Some species of parrotfish (Scaridae) and wrasses (Labridae) are diandric. Populations of these species may consist of juveniles (immatures), gonochoristic females (females incapable of changing sex), hermaphroditic females (females that will change sex later), primary males (gonochoristic males incapable of changing sex), and secondary males (derived from sex-reversed females). Gonochoristic species occur in the Chaetodontidae. These butterfly fishes mate and have a long-term pair bonding with a single partner (Jobling, 1995).

Sexual plasticity in hermaphroditic reef fishes has been shown to be controlled by morphological changes (dimorphism) in the number and size of gonadotropin releasing hormone (GnRH) and arginine vasotocin (AVT) neurons present in the forebrain's preoptic area (POA) (Foran and Bass, 1999; Bass and Grober, 2001). The number and size of these neurons is altered either when these fishes change sex or when they change from non-territorial/non-courting male to territorial/courting male. Unidirectional sex transformation and POA neuron change is seen in tropical reef fishes like the Bluehead wrasse *Thalassoma bifasciatum*, the Anemonefish *Amphiprion melanopus*, and the Saddle wrasse *Thalassoma duperrey*. Reversible sex change in the marine goby *Trimma okinawae* was associated with reversible changes in AVT cell size (Grober and Sunobe, 1996).

According to the size-advantage hypothesis (Ghiselin, 1969; Warner, 1975), there is a greater advantage to being large in one sex than in the other and that selection would favor sex change provided that the cost of the change is not high. This hypothesis thus predicts that protogyny allows large males to achieve many matings and protandry promotes population fecundity by larger females (Turner, 1993). The killifish, *Rivulus marmoratus*, which inhabits small pools in mangroves where it is subject to wide fluctuations in physical factors, e.g., salinity and temperature extremes and high sulfide levels (section VI) and where these variations are different from locality to locality, is a diandric protogynous hermaphrodite. This unusual reproductive strategy may have evolved in response to the problem of finding mates and preserving the genotypes best suited to cope with the particular local environment (Lin and Dunson, 1999; Blaber, 2000).

B. Reproductive Behavior

Resident intertidal rocky shore fishes show territorial reproductive behaviors, i.e., the male selects a spawning site to which he attracts females, fertilizes the eggs and cares for them in the nest until they hatch (Gibson, 1982). Most observations on this have been made in temperate to subtropical waters (Gibson, 1982, 1993; Coleman, 1999; DeMartini, 1999), with few tropical studies.

Time available for mating is limited to the low tide period for those individuals breeding out of water and to daylight hours, as this behavior relies on visual communication. The attraction phase of the courtship behavior is usually in the form of body movement, such as vertical looping, as in reef fishes *Tripterygion* spp. (Tripterygiidae) and *Istiblennius zebra* (Blenniidae), and leaping into the air, as in mudskippers *Periophthalmus chrysospilos* and *P. kalolo* (MacNae, 1968; Magnus, 1972; Phillips, 1977; Wirtz, 1978). Blennies, *Alticus saliens* and *Entomacrodus vermicularis*, display courtship behavior to attract females to the nest (Abel, 1973). Hole-dwelling blennies also display sharp color patterns on their head to attract a mate (Gibson, 1993). It has also been demonstrated that subtropical mature male blennies *Blennius pavo* secrete pheromones from a special gland on the anal fin rays to attract females (Zander, 1975). Losey (1969) showed that other mature males of *Hypsobelnnius* are attracted by pheromones produced by courting males, whereas egg-guarding males and females do not respond. This reaction was considered to promote the social facilitation of courtship by attracting males which had not yet mated and possibly enhancing their sexual receptivity. Such a sexual facilitating pheromone is thus of selective advantage to the species in that it increases the number of heterosexual contacts which might lead to pair formation and promotes spawning synchrony (Losey, 1969).

Nest-holding and sneaking tactics occur in tropical Blenniidae and Gobiidae males. In *Trypterygion* spp. (Blenniidae), small males without territories gather around a spawning pair. It was suggested that these "satellite males" may attempt to fertilize the eggs of the spawning female thereby producing offspring without having to guard them (Wirtz, 1978). In the rocky or cobble shore gobiid fish *Bathygobius fuscus*, nest holders are always >55 mm whereas smaller individuals play both roles of nest holder and sneaker. Smaller males employ nest holding tactics with the removal of the large nest holders from the population. It is suggested that the sneaking males switch their tactics according to social status (Taru et al., 2002). Sneaking male tactics are also observed in the reef-associated gobiid *Eviota prasina* and the temperate *Pomatoschistus microps* (Sunobe and Nakazono, 1999; Taru et al., 2002).

Table 11.7
Tropical Fishes Exhibiting Lunar Periodicity in Reproduction

Taxon	Cycle	Spawning time	Habitat	Egg incubation site	Reference
Cyprinodontidae					
Fundulus grandis	Semilunar	Spawn on high spring tide	Estuary	Plants or sand	Greeley and MacGregor, 1983
F. similes	Semilunar	Spawn on high spring tide	Estuary	Sand	Greeley, 1982
Lutjanidae					
Lutjanus fulvus	Lunar		Coral reef	Scattered	Randall and Brock, 1960
Pomacentridae					
Centropyge potteri	Lunar	Spawn in the evenings of the week before full moon	Coral reef	Scattered	Lobel, 1978
Pomacentridae					
Microspathodon chrysurus	Lunar	Guard eggs in nests from several days before full moon until several days after new moon, i.e., spawn between just before full moon and just before new moon	Coral reef	Nest	Pressley, 1980

Pomacentrus nagasakiensis	Lunar	Spawn between just before full moon and just before new moon	Rock, et cetera	Nest	Moyer, 1975
Labridae					
Thalassoma duperrey	Semilunar (?)	Spawn on spring tides	Coral reef	Scattered	Ross, 1983
T. lucasanum	Semilunar (?)	Spawn on spring tides	Coral reef	Scattered	Warner, 1982
Polynemidae					
Polydactylus sexfilis	Lunar	Spawning peaks in evening for several days before or after the last quarter moon, i.e., spawn on edding tides	Shore zone	Scattered (pelagic)	May et al., 1979
Blenniidae					
Istiblennius enosimae	Semilumnar	Spawn in neap tides	Rocky tidepool		Sunobe et al., 1995
Sciaenidae					
Sciaenops ocellatus	Semilunar	Peak spawning on nights of the full or new moon	Estuary		Peters and McMichael, 1987
Siganidae					
Siganus guttatus	Lunar	Spawn between the new and full moon for the whole year in the Philippines and between April and September in Okinawa	Mangrove		Hara et al., 1986; Rahman et al., 2000

Parental care, especially egg guarding, occurs in many fishes breeding in rocky intertidal habitats (Almada and Serrao Santos, 1995; Coleman, 1999). Such parental care, usually fraternal, is associated frequently with external fertilization and male nest attachment. It is suggested that such behavior ensures the male has a high probability of genetic relatedness to the offspring he is guarding and site attachment enables the male to attract a succession of females and fertilize their eggs while guarding those from a previous spawning (Gibson, 1982). The function of egg care is to protect them against predators, keep them clean and provide an adequate supply of oxygen for development. The female, relieved of parental care, is allowed to feed, grow, and increase fecundity. This behavioral pattern has been demonstrated for species of Gobiidae and Blenniidae such as *Boleophthalmus dussumierei, Periophthalmus kalolo, Bathygobius fuscus*, and *Eviota abax* (Magnus, 1972; Gibson, 1982; Taru and Sunobe, 2000; Taru *et al.*, 2002). *Periophthalmus schlosseri* larvae develop in a protected moist air space in the spawning chamber of a water-filled J-shaped burrow. The air is carried through the burrow water in the buccopharyngeal cavity of the adult fish. This behavioral accumulation of air in the burrows provides a significant oxygen reservoir for burrow-dwelling fish and for developing embryos (Ishimatsu *et al.*, 1998). Air has been detected in the spawning chambers of mudskippers, including *Boleophthalmus dussumieri, Oxuderces dentatus, Periophthalmus chrysospilos*, and *Scartelaos histophorus* (Ishimatsu *et al.*, 1998). Burrow water also has high ammonia levels that may deter other animals from entering the burrow (Ip *et al.*, 2004). In *Periophthalmus cantonensis*, however, neither parents guard the eggs (Kobayaski *et al.*, 1971). Male parental care studies for subtropical and temperate rocky shore species are more extensive and documented for representatives of the Blennidae, Batrachoididae, Cottidae, Gobiidae, Gobiesocidae, Hexagrammidae, Pholidae, and Strichaeidae (Gibson, 1982; Coleman, 1999; Swenson, 1999).

Some intertidal fishes also show various levels of maternal care. For example, several species of Clinidae from South Africa are viviparous (Prochazka, 1994) and females of *Leuresthes tennis* (Atherinidae) from California bury their eggs in sand (Clark, 1938). It appears that tropical intertidal fishes seldom provide care past the point of hatching with the following exceptions. Mouthbrooding is demonstrated by cichlids (Cichlidae). External egg carrying and brood-pouch egg carrying is found in representative of the Syngnathidae, the pipefishes and seahorses (Vincent *et al.*, 1995). Post-hatching care is demonstrated by the Plainfin midshipman (*Porichthys notatus*) and possibly its relatives on temperate intertidal shores.

There is a close relationship between reproductive hormonal levels and behavior (Liley and Stacey, 1983). Testicular androgen and gonadotropin stimulate male sexual behavior in a variety of teleost species

(Stacey, 1993). The reproductive endocrine status is mediated through the action of GnRH. In nesting male demoiselles *Chromis dispilus*, elevated gonadotropin-II (dGtH-II) levels are detected during periods of reproductive display but are low during the brooding stage. Plasma testosterone levels also showed a similar pattern whereas plasma 11-ketotesterone and 17,20β-dihydroxy-4-pregnen-3-one levels tended to remain high during early phases of egg brooding (Pankhurst and Peter, 2002).

In the tropics, hatching eggs also show adaptation to an intertidal life. In *Periophthalmus sobrinus* the eggs require a moist atmosphere but not immersion to achieve complete development. The larvae hatch rapidly when submerged by the rising tide (Brillet, 1975). They are positively phototaxic and this helps them find their way out of the burrows in which the eggs are laid.

C. Lunar Synchronization

Lunar-synchronized spawning has been well documented for temperate intertidal spawners and coral reef fishes (Taylor, 1984, 1990; DeMartini, 1999). Tropical intertidal and reef fish examples are identified in Table 11.7. It appears that the spawning cycle exhibited by each species has its own adaptive significance. For example, the primary reproductive adaptation in atheriniforms, e.g., *Fundulus*, is oviposition in the high intertidal zone. Lunar-synchronized spawning is, therefore, a secondary feature of this spawning mechanism. Pressley (1980) observed that the spawning activity of damselfishes, *Microspathodon chrysurus* and *Pomacentrus nagasakiensis*, does not coincide with a particular phase of the spring tide cycle, but corresponds to the time of greatest moonlight. It is suggested that moonlight might improve conditions for nocturnal egg care and make it easier for adults or larvae to escape from predators (Taylor, 1984). *Thalassoma duperrey, T. lucasanum*, and *P. sexfilus* (Table 11.7) spawn on either high or ebbing tides and produce pelagic eggs. This results in egg dispersal off-shore on ebbing tides which is a potential effective antipredator adaptation (Taylor, 1984). Male *Istiblennius enosimae* guard the eggs until the embryos hatch at following spring tides. It was suggested that this semilunar spawning cycle guarantees the maximum dispersal of newly hatched embryos away from natal tidepools (Sunobe *et al.*, 1995).

VIII. FUTURE STUDIES

Although the tropical intertidal environment is one of the harshest habitats to live in, knowledge about the physiological adaptation of such organisms to it is far from sufficient. Most studies on fish physiology have been conducted on temperate species. For this reason we have relied on observations on temperate fishes to draw conclusions about the probable responses of tropical intertidal fishes to temperature increases. Nonetheless, there have been some studies on tropical intertidal fishes such as the mudskippers, rockskippers, and blennies. In fact, studies on the nitrogen metabolism of mudskippers have provided new insights into our understanding of fish (Ip, Chew *et al.*, 2001a, b; Ip, Lim *et al.*, 2001; Chew *et al.*, 2005).

Knowing the massive mortality of some tropical fishes that could result from a mere 2 °C elevation in temperature (Nakano *et al.*, 2004), we suggest that tropical intertidal fishes could be influenced most by global warming. Tropical intertidal fish species can be, therefore, good animal models for further understanding of physiological and molecular adaptations to an extreme thermal environment, and estimation of the realistic impact of elevated temperatures, caused by natural factors as well as human activities.

IX. CONCLUSIONS

Life for fish in the intertidal zone is harsh, since the tidal cycle, in combination with changes in weather conditions, can bring rapid and high fluctuation in the physical and chemical conditions of the water in isolated tidepools and shallow seas. In the tropics, environmental stress on the intertidal shore is more severe especially during low tide. Fishes evolved in this variable environment demonstrate physiological and biochemical adaptation to fluctuations in oxygen, water temperature, salinity, and high levels of ammonia and sulfide. They also have respiratory and reproductive adaptations to cope with the intertidal environment.

ACKNOWLEDGMENTS

We thank Professor Brian Morton, Dr Michael Martin, and anonymous reviewers for their comments on the drafts of the manuscript. This chapter was supported by a grant from City University of Hong Kong, Project 9660001-740.

REFERENCES

Abel, D. C., Koenig, C. C., and Davis, W. P. (1987). Emersion in the mangrove forest fish *Rivulus marmoratus*: A unique response to hydrogen sulphide. *Env. Biol. Fish.* **18**, 67–72.
Abel, E. F. (1973). Zur Oko-Ethologie des amphibisch lebenden Fisches *Alticus saliens* (Foster) und von *Entomacrodus vermiculatus* (Val.) (Blennioidea, Salariidae) unter besondere Berucksichtigung des Fortpflanzungsverhaltens. *Sber. Öst. Akad. Wiss. Abteil I* **181**, 137–153.

Achituv, Y., and Dubinsky, Z. (1990). Evolution and Zoogeography of coral reefs. *In* "Ecosystems of the World. 25: Coral Reefs" (Dubinsky, Z., Ed.), pp. 1–9. Elsevier Science, Amsterdam.

Adelman, I. R., and Smith, L. L. (1970). Effect of hydrogen sulphide on Northern pike eggs ands sac fry. *Trans. Am. Fish. Soc.* **3**, 501–509.

Affonso, E. G., and Waichman, A. V. (1996). Hydrogen sulfide tolerance in Amazon fish. *In* "The Physiology of Tropical Fish Symposium Proceedings, International Congress on the Biology of Fishes, San Francisco State University, July 14–18, 1996" (Val, A., Randall, D., and MacKinlay, D., Eds.), pp. 75–79. Physiology Section, American Fisheries Society, United States.

Aguilar, N. M., Ishimatsu, A., Ogawa, K., and Khoo, K. H. (2000). Aerial ventilatory responses of the mudskipper, *Periophthalmodon schlosseri*, to altered aerial and aquatic respiratory gas concentrations. *Comp. Biochem. Physiol.* **127A**, 285–292.

Al-Kadhomiy, N. K., and Hughes, G. M. (1988). Histological study of different regions of the skin and gills in the mudskipper, *Boleophthalmus boddaerti* with respect to their respiratory function. *J. Mar. Biol. Assn UK* **68**, 413–422.

Alevizon, W., Richardson, R., Pitts, P., and Servis, G. (1985). Coral zonation and patterns of community structure in Bahamian reef fishes. *Bull. Mar. Sci.* **36**, 304–318.

Almada, V. C., and Serrao Santos, R. (1995). Parental care in the rocky intertidal: A case study of adaptation and exaptation in Mediterranean and Atlantic blennies. *Rev. Fish. Biol. Fish.* **5**, 23–37.

Almeida, V. G. (1973). New records of tidepool fishes from Brazil. *Papéis Avulsos Zool. S. Paulo* **26**, 187–191.

Alongi, D. M. (2002). Present status and future of the world's mangrove forest. *Environ. Conserv.* **29**, 331–349.

Andrews, J. C., and Pickard, G. L. (1990). The physical oceanography of coral reef systems. *In* "Coral Reefs" (Dubinsky, Z., Ed.), pp. 11–48. Elsevier Science, Amsterdam.

Ansell, A. D., and Gibson, R. N. (1990). Patterns of feeding and movement of juvenile flatfishes on an open sandy beach. *In* "Trophic Relationships in the Marine Environment" (Barnes, M. and Gibson, R. N., Eds.), pp. 191–207. Aberdeen University Press, Aberdeen.

Arillo, A., Margiocco, C., Medlodia, F., Mensi, P., and Schenone, G. (1981). Ammonia toxicity mechanism in fish: Studies on rainbow trout (*Salmo gairdneri* Rich.). *Ecotoxicol. Environ. Saf.* **5**, 316–325.

Austin, H. M. (1971). A survey of the ichthyofauna of the mangroves of western Puerto Rico during December, 1967 – August, 1968. *Carib. J. Sci.* **11**, 27–39.

Bagarinao, T. (1992). Sulfide as an environmental factor and toxicant: Tolerance and adaptations of aquatic organisms. *Aquatic Toxicol.* **24**, 21–62.

Bagarinao, T., and Vetter, R. D. (1989). Sulfide tolerance and detoxification in shallow-water marine fishes. *Mar. Biol.* **103**, 251–262.

Bagarinao, T., and Vetter, R. D. (1990). Oxidative detoxification of sulphide by mitochondria of the California killifish *Fundulus parvipinnis* and the speckled sanddab Citharichthys stigmaeus. *J. Comp. Physiol. B* **160**, 519–527.

Bagarinao, T., and Vetter, R. D. (1992). Sulfide-haemoglobin interactions in the sulphide-tolerant salt marsh resident, the California killifish Fundulus parvipinnis. *J. Comp. Physiol. B* **162**, 614–624.

Bagarinao, T., and Vetter, R. D. (1993). Sulphide tolerance and adaptation in the California killifish, *Fundulus parvipinnis*, a salt marsh resident. *J. Fish Biol.* **42**, 729–748.

Barletta, M., Saint-Paul, U., Barletta-Bergan, A., Ekau, W., and Schories, D. (2000). Spatial and temporal distribution of *Myrophis punctatus* (Ophichthidae) and associated fish fauna in a northern Brazilian intertidal mangrove forest. *Hydrobiologia* **426**, 65–74.

Bass, A. H., and Grober, M. S. (2001). Social and neural modulation of sexual plasticity in toleost fish. *Brain Behav. Evol.* **57**, 293–300.

Bathen, K. H. (1968). A descriptive study of the physical oceanography of Kaneohe Bay, Oahu, Hawaii. *Hawaii Inst. Mar. Biol. Tech. Rep.* **14**, 1–353.

Behling, H. (2002). Impact of the Holocene sea-level changes in coastal, eastern and Central Amazonia. *Amazoniana* **17**, 41–52.

Bellwood, D. R. (1996). The Eocene fishes of Monte Bolca: The earliest coral reef fish assemblage. *Coral Reefs* **15**, 11–19.

Bellwood, D. R. (1997). Reef fish biogeography: Habitat association, fossils and phylogenies. *Proc. 8th Int. Coral Reef Symp.* **1**, 379–384.

Bellwood, D. R., and Wainwright, P. C. (2002). The history and biogeography of fishes on coral reefs. *In* "Coral Reef Fishes" (Sale, P. F., Ed.), pp. 5–32. Elsevier Science, California, USA.

Berdin, R. D., Siringan, F. P., and Maeda, Y. (2003). Holocene relative sea-level changes and mangrove response in southwest Bohol, Philippines. *J. Coast. Res.* **19**, 304–313.

Berkelmans, R., and Oliver, J. K. (1999). Large-scale bleaching of corals on the Great Barrier Reef. *Coral Reefs* **18**, 55–60.

Birkeland, C. (1985). Ecological interactions between mangroves, seagrass beds and coral reefs. UNEP Regional Seas Reports and Studies No. 73. UNEP, Nairobi.

Blaber, S. J. M. (1980). Fishes of the Trinity Inlet system of North Queensland with notes on the ecology of fish faunas of tropical Indo-Pacific estuaries. *Aust. J. Mar. Freshwat. Res.* **31**, 137–146.

Blaber, S. J. M. (2000). *Tropical Estuarine Fishes. Ecology, Exploitation and Conservation.* Blackwell Science, London.

Blaber, S. J. M., and Milton, D. A. (1990). Species composition, community structure and zoogeography of fishes of mangrove estuaries in the Solomon Islands. *Mar. Biol.* **105**, 259–267.

Blaber, S. J. M., Brewer, D. T., and Salini, J. P. (1989). Species composition and biomasses of fishes in different habitats of a tropical Northern Australian estuary: Their occurrence in the adjoining sea and estuarine dependence. *Estuar. Coast. Shelf Sci.* **29**, 509–531.

Blaber, S. J. M., Young, J. W., and Dunning, M. C. (1985). Community structure and zoogeographic affinities of the coastal fishes of the Dampier region of north-west Australia. *Aust. J. Mar. Freshwat. Res.* **36**, 247–266.

Block, B. A., and Finnerty, J. R. (1994). Endothermy in fishes: A phylogenetic analysis of constraints, predispositions, and selection pressures. *Environ. Biol. Fish.* **40**, 283–302.

Bœuf, G., and Payan, P. (2001). How should salinity influence fish growth? *Comp. Biochem. Physiol. C* **130**, 411–423.

Bradshaw, A. D. (1965). Evolutionary significance of phenotypic plasticity in plants. *Adv. Genetics* **13**, 115–155.

Brauner, C. J., Ballantyne, C. L., Randall, D. J., and Val, A. L. (1995). Air breathing in the armoured catfish (Hoplosternum littorale) as an adaptation to hypoxic, acidic and hydrogen sulphide rich waters. *Can. J. Zool.* **73**, 739–744.

Bridges, C. R. (1988). Respiratory adaptations in intertidal fish. *Am. Zool.* **28**, 79–96.

Bridges, C. R. (1993). Ecophysiology of intertidal fish. *In* "Fish Ecophysiology" (Rankin, J. C., and Jensen, F. B., Eds.), pp. 375–400. Chapman & Hall, London.

Brillet, C. (1975). Relations entre territoire et comportement aggressif chez *Periophthalmus sobrinus* Eggert (Pisces, Periophthalmidae) au laboratoire et en milieu natural. *Z. Tierpsychol.* **39**, 283–331.

Brix, O., Clements, K. D., and Wells, R. M. G. (1998). An ecophysiological interpretation of hemoglobin multiplicity in three herbivorous marine teleosts species from New Zealand. *Comp. Biochem. Physiol. A* **121,** 189–195.

Brix, O., Clements, K. D., and Wells, R. M. G. (1999). Haemoglobin components and oxygen transport in relation to habitat distribution in triplefin fishes (Tripterygiidae). *J. Comp. Physiol. B* **169,** 329–334.

Brooks, J. M., Bright, T. J., Bernard, B. B., and Schwab, C. R. (1979). Chemical aspects of a brine pool at the East Flower Garden Bank, northwest Gulf of Mexico. *Limnol. Oceanogr.* **24,** 735–745.

Brown, C. R., Gordon, M. S., and Chin, H. G. (1991). Field and laboratory observations on microhabitat selection in the amphibious Red Sea rockskipper fish *Alticua kirki* (Family Blenniidae). *Mar. Behav. Physiol.* **19,** 1–13.

Budyko, M. I., Ronov, A. B., and Yanshin, A. L. (1985). "History of the Earth's Atmosphere." Springer-Verlag, Berlin.

Carrico, R. J., Blumberg, W. E., and Peisach, J. (1978). The reversible binding of oxygen to sulfhaemoglobin. *J. Biol. Chem.* **253,** 7212–7215.

Charpy-Roubaud, C., and Sournia, A. (1990). The comparative estimation of phytoplanktonic and microphytobenthic production in the oceans. *Mar. Microb. Food Webs* **4,** 31–57.

Chen, J., Shao, K., and Lin, C. (1995). A checklist of reef fishes from Tingsha Tao (Pratas Island), South China Sea. *Acta Zool. Taiwan* **6,** 13–40.

Cheung, P. S. (1991). An intertidal survey of Cape d'Aguilar, Hong Kong with special reference to the ecology of high-zoned rock pools. MPhil thesis, University of Hong Kong, Hong Kong.

Chew, S. F., and Ip, Y. K. (1990). Differences in the responses of two mudskippers, *Boleophthalmus boddaerti* and *Periophthalmus chrysospilos* to changes in salinity. *J. Exp. Zool.* **256,** 227–231.

Chew, S. F., Wilson, J. M., Ip, A. Y. K., and Randall, D. J. (2005). Nitrogen excretion and defense against ammonia toxicity. *In* "Fish Physiology" (Val, A. L., Almeida-Val, V. M. F., and Randall, D. J., Eds.), Vol. 22, pp. 307–395. Academic Press, New York.

Chiba, S. (1998). A mathematical model for long-term patterns of evolution: Effects of environmental stability and instability on macroevolutionary patterns and mass extinctions. *Paleobiology* **24,** 336–348.

Chong, V. C., Sasekumar, A., Leh, M. U., and D'Cruz, R. (1990). The fish and prawn communities of a Malaysian coastal mangrove system, with comparisons to adjacent mudflat and inshore waters. *Estuar. Coast. Shelf Sci.* **31,** 703–722.

Chotkowski, M. A., Buth, D. G., and Prochazka, K. (1999). Systematics of intertidal fishes. *In* "Intertidal Fishes: Life in Two Worlds" (Horn, M. H., Martin, K. L. M., and Chotkowski, M. A., Eds.), pp. 297–355. Academic Press, San Diego, CA.

Chua, T. E. (1973). An ecological study of the Ponggol Estuary in Singapore. *Hydrobiologia* **43,** 505–533.

Clark, F. N. (1938). Grunion in southern California. *Calif. Fish Game* **24,** 49–54.

Clayton, D. A. (1993). Mudskippers. *Ocean. Mar. Biol. Ann. Rev.* **31,** 507–577.

Cloud, P. E. (1952). Preliminary report on the geology and marine environment of Onotoa Atoll, Gilbert Islands. *Atoll Res. Bull.* **12,** 1–73.

Coleman, R. M. (1999). Parental care in intertidal fishes. *In* "Intertidal Fishes: Life in Two Worlds" (Horn, M. H., Martin, K. L. M., and Chotkowski, M. A., Eds.), pp. 165–180. Academic Press, San Diego, CA.

Coles, R. G., Lee-Long, W. J., Watson, R. A., and Derbyshire, K. J. (1993). Distribution of seagrasses, and their fish and penaeid prawn communities, in Cairns harbour, a tropical estuary, Northern Queensland, Australia. *Aust. J. Mar. Freshwat. Res.* **44,** 193–210.

Cope, M. (1986). Seasonal, diel and tidal hydrographic patterns, with particular reference to dissolved oxygen, above a coral community at Hol Ha Wan, Hong Kong. *Asian Mar. Biol.* **3**, 59–74.

Cossins, A. R., and Bowler, K. (1987). "Temperature Biology of Animals." Chapman & Hall, London.

Crawshaw, L. I. (1977). Physiological and behavioral reactions of fishes to temperature change. *J. Fish. Res. Board Can.* **34**, 730–734.

Crawshaw, L. I. (1979). Responses to rapid temperature change in lower vertebrate ectotherms. *Am. Zool.* **19**, 225–237.

Daxboeck, C., Barnard, D. K., and Randall, D. J. (1981). Functional morphology of the gills of the bowfin *Amia calva*, with special reference to their significance during air exposure. *Respir. Physiol.* **43**, 349–364.

DeMartini, E. E. (1999). Intertidal spawning. *In* "Intertidal Fishes: Life in Two Worlds" (Horn, M. H., Martin, K. L. M., and Chotkowski, M. A., Eds.), pp. 143–164. Academic Press, San Diego, CA.

Duarte, C. M. (2000). Benthic ecosystems: Seagrasses. *In* "Encyclopedia of Biodiversity" (Levin, S. L., Ed.), Vol. 5, pp. 255–268. Academic Press, San Diego, CA.

Ellison, A. M., and Farnsworth, E. J. (1996). Anthropogenic disturbance of Caribbean mangrove ecosystems: Past impacts, present trends, and future predictions. *Biotropica* **28**, 549–565.

Ellison, A. M., Farnsworth, E. J., and Merkt, R. E. (1999). Origins of mangrove ecosystems and the mangrove biodiversity anomaly. *Global Ecol. Biogeogr.* **8**, 95–115.

Emson, R. H. (1985). Life history patterns in rock pool animals. *In* "The Ecology of Rocky Coasts" (Moore, P. G., and Seed, R., Eds.), pp. 220–222. Hodder and Stoughton, London.

Endean, R., Stephenson, W., and Kenny, R. (1956). The ecology and distribution of intertidal organisms on certain islands of the Queensland coast. *Aust. J. Mar. Freshwat. Res.* **7**, 317–342.

Evans, D. H. (2002). Cell signalling and ion transport across the fish gill epithelium. *J. Exp. Zool.* **293**, 336–347.

Feder, M. E., and Burggren, W. W. (1985). Cutaneous gas exchange in vertebrates: Design, patterns, control and implications. *Biol. Rev.* **60**, 1–45.

Feder, M. E., and Hofmann, G. E. (1999). Heat-shock proteins, molecular chaperones, and the stress response: Evolutionary and ecological physiology. *Annu. Rev. Physiol.* **61**, 243–282.

Fenchel, T. (1969). The ecology of marine microbenthos IV. Structure and function of the benthic ecosystem, its chemical and physical factors and the microfauna communities with special reference to the ciliated protozoa. *Ophelia* **6**, 1–182.

Field, C. D. (1995). Impact of expected climate-change on mangroves. *Hydrobiologia* **295**, 75–81.

Fields, P. A. (2001). Review: Protein function at thermal extremes: Balancing stability and flexibility. *Comp. Biochem. Physiol. A* **129**, 417–431.

Foran, C. M., and Bass, A. H. (1999). Preoptic GnRH and AVT: Axes for sexual plasticity in toleost fish. *Gen. Comp. Endocrin.* **116**, 141–152.

Fortes, M. D. (1990). "Seagrasses: A Resource Unknown in the ASEAN Region." International Center for Living Aquatic Resources Management, Manila, Philippines.

Francis, M. P., Worthington, C. J., Saul, P., and Clements, K. D. (1999). New and rare tropical and subtropical fishes from northern New Zealand. *N. Z. J. Mar. Freshwat. Res.* **33**, 571–586.

Freeman, M. L., Borrelli, M. J., Meredith, M. J., and Lepock, J. R. (1999). On the path to the heat shock response: Destabilization and formation of partially folded protein intermediates, a consequence of protein thiol modification. *Free Radic. Biol. Med.* **26**, 737–745.

Geiger, S. P., Torres, J. J., and Crabtree, R. E. (2000). Air breathing and gill ventilation frequencies in juvenile tarpon, *Megalops atlanticus*: Responses to changes in dissolved oxygen, temperature, hydrogen sulphide, and pH. *Environ. Biol. Fish.* **59,** 181–190.

Gervais, M. R., and Tufts, B. L. (1998). Evidence for membrane-bound carbonic anhydrase in the air bladder of bowfin (*Amia calva*), a primitive air-breathing fish. *J. Exp. Biol.* **201,** 2205–2212.

Ghiselin, M. T. (1969). The evolution of hermaphrodism among animals. *Q. Rev. Biol.* **44,** 180–208.

Gibson, R. N. (1982). Recent studies on the biology of intertidal fishes. *Oceanogr. Mar. Biol. Ann. Rev.* **20,** 363–414.

Gibson, R. N. (1993). Intertidal teleosts: Life in a fluctuating environment. *In* "Behaviour of Teleost Fishes" (Pitcher, T. J., Ed.), 2nd edn., pp. 513–536. Chapman & Hall, London.

Gilmore, R. G. (1995). Environmental and biogeographic factors influencing ichthyofaunal diversity: Indian River Lagoon. *Bull. Mar. Sci.* **57,** 153–170.

Gilmour, K. M. (1998). Causes and consequences of acid-base disequilibria. *In* "Fish Physiology" (Perry, S. F., and Tufts, B., Eds.), Vol. 17, pp. 321–348. Academic Press, New York.

Gilmour, K. M., Shah, B., and Szebedinszky, C. (2002). An investigation of carbonic anhydrase activity in the gills and blood plasma of brown bullhead (*Ameiurus nebulosus*), longnose skate (*Raja rhina*), and spotted ratfish (*Hydrolagus colliei*). *J. Comp. Physiol. B* **172,** 77–86.

Glynn, P. W. (1973). Ecology of a Caribbean coral reef. The *Porites* reef flat biotope: Part I: Meteorology and hydrography. *Mar. Biol.* **20,** 297–318.

Gosline, W. A. (1965). Vertical zonation of inshore fishes in the upper water layers of the Hawaiian Islands. *Ecology* **46,** 823–831.

Goujet, D. (1984). "Les Poissons Placodermes du Spitsberg: Arthrodires Dolichothoraci de la Formation de Wood Bay (Dévonien Inferieur)." Cahiers de Paléontologie, Centre National de la Recherche Scientifique, Paris.

Graham, J. B. (1973). Terrestrial life of the amphibious fish *Mnierpes macrocephalus*. *Mar. Biol.* **23,** 83–91.

Graham, J. B. (1976). Respiratory adaptations of marine air-breathing fishes. *In* "Respiration of Amphibious Vertebrates" (Hughes, G. M., Ed.), pp. 165–187. Academic Press, New York.

Graham, J. B. (1997). "Air-breathing Fishes: Evolution, Diversity, and Adaptation." Academic Press, San Diego, CA.

Greaney, G. S., and Powers, D. A. (1978). Allosteric modifiers of fish hemoglobins: *In vitro* and *in vivo* studies of the effect of ambient oxygen and pH on erythrocyte ATP concentrations. *J. Exp. Zool.* **203,** 339–350.

Greeley, M. S., Jr. (1982). Tide-controlled reproduction in the longnose killifish, Fundulus similes. *Am. Zool.* **22,** 870.

Greeley, M. S., Jr., and MacGregor, R. M., III. (1983). Annual and semilunar reproductive ecycles of the gulf killifish, *Fundulus grandis*, on the Alabama gulf coast. *Copeia* **1983,** 711–718.

Green, J. M. (1967). A field study of the distribution and behavior of *Oligocottus maculosus* Girard, a tidepool cottid of the Northeast Pacific Ocean. PhD thesis, University of British Columbia, Vancouver.

Green, J. M. (1971). Local distribution of *Oligocottus maculosus* Girard and other tidepool cottids of the west coast of Vancouver Island, British Columbia. *Can. J. Zool.* **49,** 1111–1128.

Grieshaber, M. K., and Völkel, S. (1998). Animal adaptations for tolerance and exploitation of poisonous sulphide. *Annu. Rev. Physiol.* **60,** 33–53.

Grizzle, J. M., and Thiyagarajah, A. (1987). Skin histology of *Rivulus ocellatus marmoratus*: Apparent adaptation for aerial respiration. *Copeia* **1987**, 237–240.

Grober, M. S., and Sunobe, T. (1996). Serial adult sex change involves rapid and reversible changes in forebrain neurochemistry. *Neuroreport* **7**, 2945–2949.

Gross, M. (1998). "Life on the Edge: Amazing Creatures Thriving in Extreme Environments." Plenum Press, New York.

Gundermann, N., and Popper, D. M. (1984). Notes on the Indo-Pacific mangal fishes and on mangrove related fisheries. *In* "Hydrobiology of the Mangal" (Por, F. D. and Dor, I., Eds.), pp. 201–206. W. Junk, The Hague.

Hallam, A. (1987). End-Cretaceous mass extinction event: Argument for terrestrial causation. *Science* **238**, 1237–1242.

Hanke, W., Hegab, S. A., Assem, H., Berkowsky, B., Gerhard, A., Gupta, O., and Reiter, S. (1993). Mechanisms of hormonal action on osmotic adaptation in teleost fish. *In* "Fish Ecotoxicology and Ecophysiology" (Braunbeck, T., Hanke, W., and Segner, H., Eds.), Proceedings of an International Symposium, Heidelberg, September 1991, pp. 315–325. VCH Verlagsgesellschaft mbH, Weinheim, Germany.

Hara, S., Duray, M. N., Parazo, M., and Taki, Y. (1986). Year-round spawning and seed production of the rabbitfish, *Siganus guttatus*. *Aquaculture* **59**, 259–272.

Hartl, F. U. (1996). Molecular chaperones in cellular protein folding. *Nature* **381**, 571–580.

Hawkins, A. J. S. (1985). Relationships between the synthesis and breakdown of protein, dietary absorption and turnovers of nitrogen and carbon in the blue mussel, *Mytilus edulis* L. *Oecologia* **66**, 42–49.

Hazel, J. R., and Prosser, C. L. (1974). Molecular mechanisms of temperature compensation in poikilotherms. *Physiol. Rev.* **54**, 620–677.

Heald, E. J., and Odum, W. E. (1970). The contribution of the mangrove swamps to Florida fisheries. *Proc. Gulf Carib. Fish. Inst.* **22**, 130–135.

Heisler, N. (1982). Intracellular and extracellular acid-base regulation in the tropical freshwater teleost fish *Synbranchus marmoratus* in response to the transition from water breathing to air breathing. *J. Exp. Biol.* **99**, 9–28.

Heisler, N., Toews, D. P., and Holeton, G. F. (1988). Regulation of ventilation and acid-base status in the elasmobranch *Scyliorhinus stellaris* during hyperoxia-induced hypercapnia. *Respir. Physiol.* **71**, 227–246.

Heisler, N., Weitz, H., and Weitz, A. M. (1976). Hypercapnia and resultant bicarbonate transfer processes in an elasmobranch fish. *B. Eur. Physiopath. Res.* **12**, 77–85.

Heming, T. A., and Bidani, A. (1992). Influence of proton availability on intracapillary CO_2-HCO_3^--H^+ reactions in isolated rat lungs. *J. Appl. Physiol.* **72**, 2140–2148.

Heming, T., Stabenau, E., Vanoye, C., Magahadasi, H., and Bidani, A. (1994). Roles of intra- and extracellular carbonic anhydrase in alveolar-capillary CO_2 equilibration. *J. Appl. Physiol.* **77**, 697–705.

Heming, T., Vanoye, C., Stabenau, E., Roush, E., Fierke, C., and Bidani, A. (1993). Inhibitor sensitivity of pulmonary vascular carbonic anhydrase. *J. Appl. Physiol.* **75**, 1642–1649.

Hemminga, M. A., and Duarte, C. M. (2000). Seagrass Ecology. Cambridge University Press, Cambridge.

Henry, R. P., and Swenson, E. R. (2000). The distribution and physiological significance of carbonic anhydrase in vertebrate gas exchange organs. *Respir. Physiol.* **121**, 1–12.

Hightower, L. E., Norris, C. E., diIorio, P. J., and Fielding, E. (1999). Heat shock responses of closely related species of tropical and desert fish. *Am. Zool.* **39**, 877–888.

Hixon, M. A. (2001). Coral reef fishes. *In* "Encyclopedia of Ocean Sciences" (Steele, A. C., Turekian, J. H., and Thorpe, S. A., Eds.), Vol. 1, pp. 538–542. Academic Press, San Diego, CA.

Hochachka, P. W., and Somero, G. N. (1984). "Biochemical Adaptation." Princeton University Press, Princeton, NJ.

Hofmann, G. E., and Somero, G. N. (1996). Interspecific variation in thermal denaturation of proteins in the congeneric mussels *Mytilus trossulus* and *M. galloprovincialis*: Evidence from the heat shock response and protein ubiquitination. *Mar. Biol.* **126**, 65–75.

Hoque, M. T., Yusoff, F. M., Law, A. T., and Syed, M. A. (1998). Effect of hydrogen sulphide on liver somatic index and Fulton's condition factor in *Mystus nemurus*. *J. Fish Biol.* **52**, 23–50.

Horn, M. H. (1999). Convergent evolution and community convergence: Research potential using intertidal fishes. *In* "Intertidal Fishes: Life in Two Worlds" (Horn, M. H., Martin, K. L. M., and Chotkowski, M. A., Eds.), pp. 356–372. Academic Press, San Diego, CA.

Horowitz, M. (2001). Heat acclimation: Phenotypic plasticity and cues to the underlying molecular mechanisms. *J. Therm. Biol.* **26**, 357–363.

Houlihan, D. F. (1991). Protein turnover in ectotherms and its relationships to energetics. *Adv. Comp. Env. Physiol.* **7**, 1–43.

Hubbard, D. K. (1988). Controls of modern and fossil reef development. Common ground for biological and geological research. *Proc. Int. Coral Reef Symp. 6th* **1**, 243–252.

Hutchings, P., and Saenger, P. (1987). *Ecology of Mangroves*. University of Queensland Press, St. Lucia, Australia.

Ingvorsen, K., and Jorgensen, B. B. (1979). Combined measurements of oxygen and sulfide in water samples. *Limnol. Oceanogr.* **24**, 390–393.

Ip, Y. K., and Low, W. P. (1990). Lactate production in the gills of the mudskipper *Periophthalmodon schlosseri* exposed to hypoxia. *J. Exp. Zool.* **253**, 99–101.

Ip, Y. K., Lee, C. G. L., Low, W. P., and Lam, T. J. (1991). Osmoregulation in the mudskipper, *Boleophthalmus boddaerti* I. Responses of branchial cation activated and anion stimulated adenosine triphosphatases to changes in salinity. *Fish Physiol. Biochem.* **9**, 63–68.

Ip, Y. K., Low, W. P., Lim, A. L. L., and Chew, S. F. (1990). Changes in lactate content in the gills of the mudskippers *Periophthalmus chrysospilos* and *Boleophthalmus boddaerti* in response to environmental hypoxia. *J. Fish Biol.* **36**, 481–487.

Ip, Y. K., Chew, S. F., and Randall, D. J. (2001a). Ammonia toxicity, tolerance, and excretion. *In* "Fish Physiology. Vol. 20: Nitrogen Excretion" (Wright, P. A., and Anderson, P. M., Eds.), pp. 109–148. Academic Press, San Diego, CA.

Ip, Y. K., Chew, S. F., Leong, I. W. A., Jin, Y., and Wu, R. S. S. (2001b). The sleeper *Bostrichthys sinensis* (Teleost) stores glutamine and reduces ammonia production during aerial exposure. *J. Comp. Physiol. B* **171**, 357–367.

Ip, Y. K., Lim, C. B., Chew, S. F., Wilson, J. M., and Randall, D. J. (2001). Partial amino acid catabolism leading to the formation of alanine in *Periophthalmodon schlosseri* (mudskipper): A strategy that facilitates the use of amino acids as an energy source during locomotory activity on land. *J. Exp. Biol.* **204**, 1615–1624.

Ip, Y. K., Randall, D., Kok, T. K., Bazaghi, C., Wright, P. A., Ballantyne, J. S., Wilson, J. M., and Chew, S. F. (2004). The giant mudskipper *Periophthalmodon scholsseri* facilitates active NH_4^+ excretion by increasing acid excretion and having a low NH_3 permeability in the skin. *J. Exp. Biol.* **207**, 787–801.

Ishimatsu, A., and Itazawa, Y. (1983). Blood oxygen levels and acid-base status following air exposure in an air-breathing fish, *Channa argus*: The role of air ventilation. *Comp. Biochem. Physiol. A* **74**, 787–793.

Ishimatsu, A., Aguilar, N. M., Ogawa, K., Hishida, Y., Takeda, T., Oikawa, S., Kanda, T., and Khoo, K. H. (1999). Arterial blood gas levels and cardiovascular function during varying environmental conditions in a mudskipper, *Periophthalmodon schlosseri*. *J. Exp. Biol.* **202**, 1753–1762.

Ishimatsu, A., Hishida, Y., Takita, T., Kanda, T., Oikawa, S., Takeda, T., and Khoo, K. H. (1998). Mudskippers store air in their burrows. *Nature* **391**, 237–238.

Iwata, K. (1988). Nitrogen metabolism in the mudskipper, *Periophthalmus cantonensis*: Changes in free amino acids and related compounds in various tissues under conditions of ammonia loading, with special reference to its high ammonia tolerance. *Comp. Biochem. Physiol. A* **91**, 499–508.

Iwata, K., Kakuta, I., Ikeda, M., Kimoto, S., and Nada, N. (1981). Nitrogen metabolism in the mudskipper, *Periophthalmus cantonensis*: A role of free amino acids in detoxification of ammonia produced during its terrestrial life. *Comp. Biochem. Physiol. A* **86**, 589–596.

Izaurralde, R. C., Rosenberg, N. J., Brown, R. A., Legler, D. M., Tiscareño, M., and Srinivasan, R. (1999). Modeled effects of moderate and strong 'Los Niños' on crop productivity in North America. *Agric. Forest Meteorol.* **94**, 259–268.

Jablonski, D. (1986). Background and mass extinctions: The alternation of macroevolutionary regimes. *Science* **231**, 129–133.

Jobling, M. (1995). "Environmental Biology of Fishes." Chapman & Hall, London.

Johansen, K., and Lenfant, C. (1972). A comparative approach to adaptability of O_2.Hb affinity. *In* "Oxygen Affinity of Hemoglobins and Red Cell Acid Base Status" (Astrup, P. and Rorth, M., Eds.), pp. 750–780. Academic Press/Munksgaard, Copenhagen.

Johansen, K., Mangum, C. P., and Lykkeboe, G. (1978). Respiratory properties of the blood of Amazon fishes. *Can. J. Zool.* **56**, 898–906.

Jones, R. S., and Chase, J. A. (1975). Community structure and distribution of fishes in an enclosed high island lagoon in Guam. *Micronesica* **11**, 127–148.

Jørgensen, B. B., and Fenchel, T. (1974). The sulphur cycle of a marine sediment model system. *Mar. Biol.* **24**, 189–201.

Karatajute-Talima, V., and Predtechenskyi, N. (1995). The distribution of the vertebrates in the Late Ordovician and Early Silurian palaeobasins of the Siberian Platform. *Bull. Mus. Natl Hist. Nat. 4 Sér., Sect. C: Sci. Terre* **27**, 39–55.

Karnaky, K. J., Jr. (1998). Osmotic and ionic regulation. *In* "The Physiology of Fish" (Evans, D. H., Ed.), 2nd edn., pp. 157–176. CRC Press, Boca Raton, FL.

Khan, A. A., Schuler, M. M., and Coppock, R. W. (1987). Inhibitory effects of various sulphur compounds on the activity of bovine erythrocyte enzymes. *J. Toxicol. Environ. Health* **22**, 481–490.

Kneib, R. T. (1987). Predation risk and the use of intertidal habitats by young fishes and shrimp. *Ecology* **68**, 379–386.

Kobayaski, T., Dotsu, Y., and Takita, T. (1971). Nest and nesting behavior of the mud skipper, *Periophthalmus cantonensis* in Ariake Sound. *Bull. Fac. Fish. Nagasaki Univ.* **32**, 27–40.

Kohn, A., and Helfrich, P. (1951). Primary productivity of a Hawaiian coral reef. *Limnol. Oceanogr.* **2**, 241–251.

Kok, W. K., Lim, C. B., Lam, T. J., and Ip, Y. K. (1998). The mudskipper *Periophthalmodon schlosseri* respires more efficiently on land than in water and vice versa for *Boleophthalmus boddaerti*. *J. Exp. Zool.* **280**, 86–90.

Kong, H. Y., Kahatapitiya, N., Kingsley, K., Salo, W. L., Anderson, P. M., Wang, Y. X., and Walsh, P. J. (2000). Induction of carbamoyl phosphate synthetase III and glutamine synthetase mRNA during confinement stress in gulf toadfish (*Opsanus beta*). *J. Exp. Biol.* **203**, 311–320.

Kuah, S. S. L., Chew, S. F. and Ip, Y. K. (In press a). The mudskipper *Boleophthalmus boddaerti* does not undergo anaerobic energy metabolism when exposed to 0.4 mM sulphide in normoxia or 0.2 mM sulphide in hypoxia. *J. Exp. Biol.*

Kuah, S. S. L., Chew, S. F. and Ip. Y. K. (In press b). The mudskipper *Boleophthalmus boddaerti* activates different mechanisms to detoxify sulphide in normoxia or hypoxia, producing mainly sulphate and sulfane sulphur, respectively. *J. Exp. Biol.*

Kuo, S., and Shao, K. (1999). Species composition of fish in the coastal zones of the Tsengwen Estuary, with descriptions of five new records from Taiwan. *Zoological Studies* **38**, 391–404.

Lam, C. (1986). A new species of *Bathygobius* (Pisces: Gobiidae) from Hong Kong. *Asian Mar. Biol.* **3**, 75–87.

Lam, C. (1990). Intertidal gobies (Pisces: Gobiidae) from Hong Kong. *In* "Proceedings of the Second International Marine Biological Workshop: The Marine Flora and Fauna of Hong Kong and Southern China, Hong Kong, 1986" (Morton, B., Ed.), pp. 673–690. Hong Kong University Press, Hong Kong.

Lande, M. B., Donovan, J. M., and Zeidel, M. L. (1995). The relationship between membrane fluidity and permeabilities to water, solute, ammonia, and protons. *J. Gen. Physiol.* **106**, 67–84.

Larkum, A. W. D., and den Hartog, C. (1989). Evolution and biogeography of seagrasses. *In* "Biology of Seagrasses" (Larkum, A. W. D., McComb, A. J., and Shepherd, S. A., Eds.), pp. 112–156. Elsevier, New York.

Lee, C. G. L., Low, W. P., and Ip, Y. K. (1987). Na^+, K^+ and volume regulation in the mudskipper, *Periophthalmus chrysospilos*. *Comp. Biochem. Physiol. A* **87**, 439–448.

Lee, C. G. L., Low, W. P., Lam., T. J., Munro, A. D., and Ip, Y. K. (1991). Osmoregulation in the mudskipper, *Boleophthalmus boddaerti* II. Transepithelial potential and hormonal control. *Fish Physiol. Biochem.* **9**, 69–75.

Lee, R. W., Kraus, D. W., and Doeller, J. E. (1996). Oxidation of sulphide by *Spartina alterniflora* roots. *Limnol. Oceanogr.* **44**, 1155–1159.

Leis, J. M. (1991). The pelagic stage of reef fishes: The larval biology of coral reef fishes. *In* "The Ecology of Fishes on Coral Reefs" (Sale, P. F., Ed.), pp. 183–230. Academic Press, San Diego.

Lenfant, C., and Johansen, K. (1967). Respiratory adaptations in selected amphibians. *Resp. Physiol.* **2**, 247–260.

Lewis, R. R., III, Gilmore, R. G., Jr., Crewz, D. W., and Odum, W. E. (1983). Mangrove habitat and fishery resources in Florida. *In* "Florida Aquatic Habitat and Fishery Resources" (Seaman, W., Jr., Ed.), pp. 281–336. Florida Chapter of the American Fisheries Society, Eustis.

Ley, J. A., McIvor, C. C., and Montague, C. L. (1999). Fishes in mangrove prop-root habitats of northeastern Florida Bay: Distinct assemblages across an estuarine gradient. *Estuar. Coast Shelf Sci.* **48**, 701–723.

Liley, N. R., and Stacey, N. E. (1983). Hormones, pheromones, and reproductive behaviour in fish. *In* "Fish Physiology. Volume 9: Reproduction, Part B, Behavior and Fertility Control" (Hoar, W. S., Randall, D. J., and Donaldson, E. M., Eds.), pp. 1–63. Academic Press, London.

Lillywhite, H. B., and Maderson, P. F. A. (1988). The structure and permeability of integument. *Am. Zool.* **28**, 945–962.

Lim, C. B., Chew, S. F., Anderson, P.M, and Ip, Y. K. (2001). Mudskippers reduce the rate of protein and amino acid catabolism in response to terrestrial exposure. *J. Exp. Biol.* **204**, 1605–1614.

Lin, H. J., and Dunson, W. A. (1999). Phenotypic plasticity in the growth of the self-fertilizing hermaphroditic fish *Rivulus marmoratus*. *J. Fish Biol.* **54**, 250–266.

Lin, H. C., and Dunson, W. A. (1995). An explanation of the high strain diversity of a self-fertilizing hermaphroditic fish. *Ecology* **76**, 593–605.

Lindquist, S., and Craig, E. A. (1988). The heat shock response. *Ann. Rev. Genet.* **22**, 631–677.

Little, C., Reay, P. J., and Grove, S. J. (1988). The fish community of an East African mangrove creek. *J. Fish. Biol.* **32**, 729–747.

Lobel, P. S. (1978). Diel, lunar and seasonal periodicity in the reproductive behaviour of the pomacanthid fish, *Centropyge potteri*, and some other reef fishes in Hawaii. *Pacif. Sci.* **32**, 193–207.

Logan, A., and Cook, C. B. (1992). Seagrass beds. *In* "A Guide to the Ecology of Shoreline and Shallow-Eater Marine Communities of Bermuda" (Thomas, M. L. H., and Logan, A., Eds.), pp. 69–92. Wm. C. Brown, Iowa.

Logue, J., Tiku, P., and Cossins, A. R. (1995). Heat injury and resistance adaptation in fish. *J. Therm. Biol.* **20**, 191–197.

Losey, G. S., Jr. (1969). Sexual pheromone in some fishes of the genus Hypsoblennius Gill. *Science N. Y.* **163**, 181–183.

Low, W. P., Chew, S. F., and Ip, Y. K. (1992). Differences in electrophoretic patterns of lactate dehydrogenases from the gills, hearts and muscles of three mudskippers. *J. Fish Biol.* **40**, 975–977.

Low, W. P., Lane, D. J. W., and Ip, Y. K. (1988). A comparative study of terrestrial adaptations of the gills in three mudskippers – *Periophthalmus chrysopilos, Boleophthalmus boddaerti*, and *Periophthalmodon schlosseri*. *Biol. Bull.* **175**, 434–438.

Low, W. P., Peng, K. W., Phuan, S. K., Lee, C. Y., and Ip, Y. K. (1993). A comparative study on the responses of the gills of two mudskippers to hypoxia and anoxia. *J. Comp. Physiol. B* **163**, 487–494.

MacNae, W. (1968). A general account of the fauna and flora of mangrove swamps and forests in the Indo-West-Pacific region. *Adv. Mar. Biol.* **6**, 73–270.

MacNae, W., and Kalk, M. (1962). The ecology of the mangrove swamps of Inhaca Island, Mozambique. *J. Ecol.* **50**, 19–34.

Maeda-Martinez, A, Contreras, S., and Maravilla, O. (1982). Fish diversity and abundance in three mangrove areas of the Bahia de Paz, Mexico. *Trans. Cibcasio.* **6**, 138–151.

Magnus, D. B. E. (1972). On the ecology of the reproduction behaviour of *Periopthalmus kalolo* Lesson at the east African coast. *Proc. Int. Congr. Zool.* **17**, Theme 6, 1–5.

Mahon, R., and Mahon, S. D. (1994). Structure and resilience of tidepool fish assemblage at Barbados. *Environ. Biol. Fish.* **41**, 171–190.

Manickam, P., and Natarajan, G. M. (1985). Observations on the blood parameters of two air breathing mudskippers of South India. *Ind. J. Curr. Biosci.* **2**, 19–22.

Martin, F. D., and Cooper, M. (1981). A comparison of fish faunas found in pure stands of two tropical Atlantic seagrasses, *Thalassia festidinum* and *Syringodium filiforme*. *Northeast Gulf Sci.* **5**, 31–37.

Martin, J., Langer, T., Boteva, R., Schramel, A., Horwich, A. L., and Hartl, F. U. (1991). Chaperonin-mediated protein folding at the surface of groEL through a "molten globule" like intermediate. *Nature* **352**, 36–42.

Martin, K. L. M. (1991). Facultative aerial respiration in an intertidal sculpin, *Clinocottus analis* (Scorpaeniformes: Cottidae). *Physiol. Zool.* **64**, 1342–1355.

Martin, K. L. M. (1993). Aerial release of CO_2 and respiratory exchange ratio in intertidal fishes out of water. *Environ. Biol. Fishes* **37**, 189–196.

Martin, K. L. M. (1995). Time and tide wait for no fish: Intertidal fishes out of water. *Environ. Biol. Fishes* **44**, 165–181.

Martin, T. J. (1988). Interaction of salinity and temperature as a mechanism for spatial separation of three co-existing species of Ambassidae (Cuvier) (Teleostei) in estuaries on the south east coast of Africa. *J. Fish Biol.* **33**(Suppl. A), 9–15.

May, R. C., Akiyama, G. S., and Santerre, M. T. (1979). Lunar spawning of the threadfin, *Polydactylus sexfilis*, in Hawaii. *US Natl. Mar. Service Fish. Bull.* **76**, 900–904.

Meekan, M. G., Wellington, G. M., and Axe, L. (1999). El Niño-Southern Oscillation events produce checks in the otoliths of coral reef fishes in the Galápagos Archipelago. *Bull. Mar. Sci.* **64**, 383–390.

Meredith, A. S., Davies, P. S., and Forster, K. E. (1982). Oxygen uptake by the skin of the Canterbury mudfish, *Neochanna burrowsius*. *N. Z. J. Zool.* **9**, 387–390.

Moerland, T. S. (1995). Temperature: Enzyme and organelle. *In* "Biochemistry and Molecular Biology of Fishes" (Hochachka, P. W., and Mommsen, T. P., Eds.), pp. 57–71. Elsevier Science, Amsterdam.

Mommsen, T. P. (1984). Metabolism of the fish gill. *In* "Fish Physiology. Volume 10: Gills" (Hoar, W. S., and Randall, D. J., Eds.), pp. 203–238. Academic Press, New York.

Mommsen, T. P., and Walsh, P. J. (1992). Biochemical and environmental perspectives on nitrogen metabolism in fishes. *Experentia* **48**, 583–593.

Montgomery, W. L. (1990). Zoogeography, behaviour and ecology of coral-reef fishes. *In* "Ecosystem of the World, Volume 25, Coral Reefs" (Dubinsky, Z., Ed.), pp. 329–364. Elsevier Science Publishers Amsterdam.

Moon, T. W., and Johnston, I. A. (1981). Amino acid transport and interconversions in tissues of freshly caught and food-deprived plaice, *Pleuronectes platessa* L. *J. Exp. Biol.* **19**, 653–663.

Morgan, I. J., McDonald, D. G., and Wood, C. M. (2001). The cost of living for freshwater fish in a warmer, more polluted world. *Global Change Biol.* **7**, 345–355.

Morimoto, R. I., Tissières, A., and Georgopoulos, C. (1990). The stress response, function of the proteins, and perspective. *In* "Stress Proteins in Biology and Medicine" (Morimoto, R. I., Tissières, A., and Georgopoulos, C., Eds.), pp. 1–36. Cold Spring Harbor Laboratory Press, New York.

Moring, J. R. (1990). Seasonal absence of fishes in tidepools of a boreal environment (Maine, USA). *Hydrobiologia* **194**, 163–168.

Morris, R. W. (1960). Temperature, salinity, and southern limits of three species of Pacific cottid fishes. *Limnol. Oceanogr.* **5**, 175–179.

Morris, R. W. (1962). Distribution and temperature sensitivity of some eastern Pacific cottid fishes. *Physiol. Zool.* **34**, 217–227.

Morris, S., and Bridges, C. R. (1994). Properties of respiratory pigments in bimodal breathing animals: Air and water breathing by fish and crustaceans. *Am. Zool.* **34**, 216–228.

Morris, S., and Taylor, A. C. (1983). Diurnal and seasonal variation in physico-chemical conditions with intertidal rock pools. *Estuar. Coast. Shelf Sci.* **17**, 339–355.

Morton, B., and Britton, J. C. (2000). The origins of the coastal and marine flora and fauna of the Azores. *Oceanogr. Mar. Biol. Ann. Rev.* **38**, 13–84.

Morton, B., and Lee, C. N. W. (2003). The biology and ecology of juvenile horseshoe crabs along the northwestern coastline of the New Territories, Hong Kong: Prospects and recommendations for conservation. Final Report. The Swire Institute of Marine Science, The University of Hong Kong, Hong Kong.

Moyer, J. T. (1975). Reproductive behaviour of the damselfish *Pomacentrus nagasakiensis* at Miyakejima, Japan. *Japan J. Ichthy.* **22**, 151–163.

Nakamura, R. (1970). The Comparative Ecology of Two Sympatric Tidepool Fishes, *Oligocottus maculosus* (Girard) and *Oligocottus snyderi* (Greeley). PhD thesis. University of British Columbia, Vancouver.

Nakamura, R. (1971). Food of two cohabiting tide-pool cottidae. *J. Fish. Res. Bd. Canada* **28**, 928–932.

Nakamura, R. (1976a). Temperature and the vertical distribution of two tidepool fishes (*Oligocottus maculosus, O. snyderi*). *Copeia* **1976**, 143–152.

Nakamura, R. (1976b). Experimental assessment of factors influencing microhabitat selection by the two tidepool fishes *Oligocottus maculosus* and *O. snyderi. Mar. Biol.* **37**, 97–104.

Nakano, K., and Iwama, G. K. (2002). The 70-kDa heat shock protein response in two intertidal sculpins, *Oligocottus maculosus* and *O. snyderi*: Relationship of hsp70 and thermal tolerance. *Comp. Biochem. Physiol. A* **133**, 79–94.

Nakano, K., Takemura, A., Nakamura, S., Nakano, Y., and Iwama, G. K. (2004). Changes in the cellular and organismal stress responses of the subtropical fish, the Indo-Pacific sergeant (*Abudefduf vaigiensis*), due to the 1997–98 El Niño/Southern Oscillation. *Environ. Biol. Fish.* **70**(4), 321–329.

Nicholls, P. (1975). The effect of sulphide on cytochrome aa3. Isosteric and allosteric shrifts of the reduced α-peak. *Biochem. Biophys. Acta* **396**, 24–35.

Nicholls, R. J., Hoozemans, F. M. J., and Marchand, M. (1999). Increasing flood risk and wetland losses due to global sea-level rise: Regional and global analyses. *Global Environ. Change* **9**, S69–S87.

Nickerson, N. H., and Thibodeau, F. R. (1985). Association between pore water sulfide concentrations and the distribution of mangroves. *Biogeochem.* **1**, 183–192.

Nikinmaa, M. (1990). "Vertebrate Red Blood Cells: Adaptations of Function to Respiratory Requirements." Springer-Verlag, Berlin.

Nonnotte, G. (1981). Cutaneous respiration in 6 fresh-water teleosts. *Comp. Biochem. Physiol.* **70A**, 541–543.

Nonnotte, G., and Kirsch, R. (1978). Cutaneous respiration in seven sea-water teleosts. *Respir. Physiol.* **35**, 111–118.

Norris, C. E., diIorio, P. J., Schultz, R. J., and Hightower, L. E. (1995). Variation in heat shock proteins within tropical and desert species of Poeciliid fishes. *Mol. Biol. Evol.* **12**, 1048–1062.

Odum, W. E., and Heald, E. J. (1972). The detritus food web of an estuarine mangrove community. *In* "Estuarine Research" (Cronin, L. E., Ed.), pp. 265–286. Academic Press, New York.

Ogasawara, T., Ip, Y. K., Hasegawa, S., Hagiwra, Y., and Hirano, T. (1991). Changes in prolactin cell activity in the mudskipper, *Periophthalmus chrysospilos*, in response to hypotonic environment. *Zool. Sci.* **8**, 89–95.

Ogden, J. C. (1976). Some aspects of plant-herbivore relationships on Caribbean reefs and seagrass beds. *Aquat. Bot.* **2**, 103–116.

Ogden, J. C. (1980). Faunal relationships in Caribbean seagrassbeds. *In* "Handbook of Seagrass Biology: An Ecosystem Perspective" (Philips, R. C., and McRoy, C. P., Eds.), pp. 173–198. Garland STPM Press, New York and London.

Öhman, M. C., Rajasuriya, A., and Ólafsson, E. (1997). Reef fish assemblages in north-western Sri Lanka: Distribution patterns and influences of fishing practices. *Environ. Biol. Fish.* **49**, 45–61.

Orr, A. P. (1933). Variations in some physical and chemical conditions on and near Low Isles Reef. *Sci. Rep. Great Barrier Reef Expedition, 1928–29, Br. Mus. (Natl Hist.)* **2**, 87–98.

Oseid, D. M., and Smith, L. L. (1972). Swimming endurance and resistance to copper and malathion of bluegills treated by long-term exposure to sublethal levels of hydrogen sulphide. *Trans. Am. Fish. Soc.* **4**, 620–625.

Pankhurst, N. W., and Peter, R. E. (2002). Changes in plasma levels of gonadal steroids and putative gonadotropin in association with spawning and brooding behaviour of male demoiselles. *J. Fish Biol.* **61**, 394–404.

Park, J. Y. (2002). Structure of the skin of an air-breathing mudskipper, *Periophthalmus magnuspinnatus*. *J. Fish Biol.* **60,** 1543–1550.

Parsell, D. A., and Lindquist, S. (1993). The function of heat-shock proteins in stress tolerance: Degradation and reactivation of damaged proteins. *Ann. Rev. Genet.* **27,** 437–496.

Pelster, B., Bridges, C. R., and Grieshaber, M. K. (1988). Physiological adaptations of the intertidal rockpool teleost *Blennius pholis* L., to aerial exposure. *Respir. Physiol.* **71,** 355–374.

Peng, K. W., Chew, S. F., Lim, C. B., Kuah, S. S. L., Kok, W. K., and Ip, Y. K. (1998). The mudskippers *Periophthalmodon schlosseri* and *Boleophthalmus boddaerti* can tolerate environmental NH_3 concentrations of 446 and 36 μM, respectively. *Fish Physiol. Biochem.* **19,** 59–69.

Perry, S. F., and Gilmour, K. (1993). An evaluation of factors limiting carbon dioxide excretion by trout red blood cells *in vitro*. *J. Exp. Biol.* **180,** 39–54.

Peters, K. M., and McMichael, R. H. (1987). Early life history of the Red Drum *Sciaenops ocellatus* (Pisces Sciaenidae), in Tampa Bay, Florida. *Estuaries* **10,** 92–107.

Phillips, R. R. (1977). Behavioural field study of the Hawaiian rockskipper *Istiblennius zebra* (Teleostei, Blenniidae) I. Ethogram. *Z. Tierpsychol.* **32,** 1–22.

Plaut, I. (1999a). Effects of salinity acclimation on oxygen consumption in the freshwater blenny, *Salaria fluviatilis*, and the marine peacock blenny, *S. pavo*. *Mar. Freshwat. Res.* **50,** 655–659.

Plaut, I. (1999b). Effects of salinity on survival, osmoregulation and oxygen consumption in the intertidal blenny, *Parablennius sanguinolentus*. *Copeia* **3,** 775–779.

Pollard, D. A. (1984). A review of ecological studies on seagrass fish communities, with particular reference to recent studies in Australia. *Aquat. Bot.* **18,** 3–42.

Por, F. D. (1984). Editor's note on mangal fishes of the world. *In* "Hydrobiology of the Mangal" (Por, F. D., and Dor, I., Eds.), pp. 207–209. W. Junk, The Hague.

Pörtner, H. O., and Sartoris, F. J. (1999). Invasive studies of intracellular acid-base parameters: Quantitative analyses during environmental and functional stress. *In* "Regulation of Tissue pH in Plants and Animals" (Taylor, E. W., Egington, S., and Raven, J. A., Eds.), pp. 68–98. Cambridge University Press, Cambridge.

Potts, D. C., and Swart, P. K. (1984). Water temperature as an indicator of environmental variability on a coral reef. *Limnol. Oceanogr.* **29,** 504–513.

Powell, E. N., Bright, T. J., Woods, A., and Gittings, S. (1983). Meiofauna and the thiobios in the East Flower Garden brine seep. *Mar. Biol.* **73,** 269–283.

Powers, D. A., Fyhn, H. J., Fyhn, U. E. H., Martin, J. P., Garlick, R. L., and Wood, S. C. (1979). A comparative study of the oxygen equilibria of blood from 40 genera of Amazonian fishes. *Comp. Biochem. Physiol. A* **62,** 67–85.

Pradhan, V. (1961). A study of blood of a few Indian fishes. *Proc. Indian Acad. Sci.* **54,** 251–256.

Prändle, R., Hinderhofer, K., Eggers-Schumacher, G., and Schöffl, F. (1998). HSF3, a new heat shock factor from *Arabidopsis thaliana*, derepresses the heat shock response and confers thermotolerance when overexpressed in transgenic plants. *Mol. Genet. Genom.* **258,** 269–278.

Pressley, P. H. (1980). Lunar periodicity in the spawning of yellowtail damselfish, *Microspathodon chrysurus*. *Environ. Biol. Fish.* **5,** 153–159.

Prochazka, K. (1994). The reproductive biology of intertidal klipfish (Perciformes: Clinidae) in South Africa. *S. Afr. J. Zool.* **29,** 244–251.

Prodocimo, V., and Freire, C. A. (2001). Ionic regulation in aglomerular tropical estuarine pufferfishes submitted to sea water dilution. *J. Expt. Mar. Biol. Ecol.* **262,** 243–253.

Prosser, C. L. (1986). "Adaptational Biology: Molecules to Organisms." John Wiley, New York.

Pugh, D. T., and Rayner, R. F. (1981). The tidal regimes of three Indian Ocean atolls and some ecological implications. *Estuar. Coast. Mar. Sci.* **13**, 289–407.

Rahman, M. S., Takemura, A., and Takano, K. (2000). Lunar synchronization of testicular development and plasma steroid hormone profiles in the golden rabbitfish. *J. Fish Biol.* **57**, 1065–1074.

Randall, D. J., and Tsui, T. K. N. (2002). Ammonia toxicity in fish. *Mar. Pollut. Bull.* **45**, 17–23.

Randall, D. J., Burggren, W. W., Farrell, A. P., and Haswell, M. S. (1981). "The Evolution of Air Breathing in Vertebrates." Cambridge University Press, Cambridge.

Randall, D. J., Wilson, J. M., Peng, K. W., Kok, T. W. K., Kuah, S. S. L., Chew, S. F., Lam, T. J., and Ip., Y. K. (1999). The mudskipper, *Periophthalmodon schlosseri*, actively transports NH_4^+ against a concentration gradient. *Am. J. Physiol.* **277**, R1562–R1567.

Randall, J. E., and Brock, V. E. (1960). Observations on the ecology of the epinepheline and lutjanid fishes of the Society Islands, with emphasis on food habitats. *Trans. Am. Fish. Soc.* **89**, 9–16.

Rankin, J. C., Henderson, I. W., and Brown, J. A. (1983). Osmoregulation and the control of kidney function. *In* "Control Processes in Fish Physiology" (Rankin, J. C., Pitcher, T. J., and Duggan, R. T., Eds.), pp. 66–88. Croom Helm, Manuka, Australia.

Ravindran, J., Raghukumar, C., and Raghukumar, S. (1999). Disease and stress-induced mortality of corals in Indian reefs and observations on bleaching of corals in the Andamans. *Curr. Sci.* **76**, 233–237.

Reid, S. D., Dockray, J. J., Linton, T. K., McDonald, D. G., and Wood, C. M. (1995). Effects of a summer temperature regime representative of a global warming scenario on growth and protein synthesis in hardwater- and softwater-acclimated juvenile rainbow trout (*Oncorhynchus mykiss*). *J. Therm. Biol.* **20**, 231–244.

Reid, S. D., Linton, T. K., Dockray, J. J., McDonald, D. G., and Wood, C. M. (1998). Effects of chronic sublethal ammonia and a simulated summer global warming scenario: protein synthesis in juvenile rainbow trout (*Oncorhynchus mykiss*). *Can. J. Fish. Aquat. Sci.* **55**, 1534–1544.

Richards, D. V., and Engle, J. M. (2001). New and unusual reef fish discovered at the California Channel Island during the 1997–1998 El Niño. *Bull. South. Calif. Acad. Sci.* **100**, 175–185.

Riggs, A. (1970). Properties of fish hemoglobins. *In* "Fish Physiology. Vol. 4: The Nervous System, Circulation, and Respiration" (Hoar, W. S., and Randall, D. J., Eds.), pp. 209–252. Academic Press, New York.

Riggs, A. (1979). Studies of the hemoglobins of Amazonian fishes: An overview. *Comp. Biochem. Physiol.* **62A**, 257–272.

Robblee, M. B., and Zieman, J. C. (1984). Diel variation in the fish fauna of a tropical seagrass feeding ground. *Bull. Mar. Sci.* **34**, 335–345.

Robertson, A. I. and Alongi, D. M. (Eds.) (1992). *In* "Tropical Mangrove Ecosystems." American Geophysical Union, Washington, DC.

Robertson, A. I., and Duke, N. C. (1987). Mangroves as nursery sites: Comparisons of the abundance and species composition of fish and crustaceans in mangroves and other nearshore habitats in tropical Australia. *Mar. Biol.* **96**, 193–205.

Robertson, A. I., and Duke, N. C. (1990). Recruitment, growth and residence time of fishes in a tropical Australian mangrove system. *Estuar. Coast. Shelf Sci.* **31**, 723–743.

Rojas, M. J. R., Pizarro, J. F., and Castro, V. (1994). Diversidad abundancia íctica en areas de manglar en el Golfo de Nicoya, Costa Rica. *Rev. Biol. Trop.* **42**, 663–672.

Rosen, B. R. (1988). Progress, problems and patterns in the biogeography of reef corals and other tropical marine organisms. *Helgol. Wiss. Meeresunters* **42**, 269–301.

Ross, R. M. (1983). Annual, semilunar and diel reproductive rhythms in the Hawaiian labrid *Thalassoma duperry.* *Mar. Biol.* **72**, 311–318.

Rozemeijer, M. J. C., and Plaut, I. (1993). Regulation of nitrogen excretion of the amphibious blenniidae *Alticus kirki* (Guenther, 1868) during emersion and immersion. *Comp. Biochem. Physiol. A* **104,** 57–62.

Sacca, R., and Burggren, W. W. (1982). Oxygen uptake in air and water in the air-breathing redfish *Calamoichthys calabaricus*: Role of skin, gills and lungs. *J. Exp. Biol.* **97,** 179–186.

Sakamoto, T., and Ando, M. (2002). Calcium ion triggers rapid morphological oscillation of chloride cells in the mudskipper, *Periophthalmus modestus*. *J. Comp. Physiol. B* **172,** 435–439.

Sakamoto, T., Yokota, S., and Ando, M. (2000). Rapid morphological oscillation of mitochondrion-rich cell in estuarine mudskipper following salinity changes. *J. Exp. Zool.* **286,** 666–669.

Sanchez, Y., and Lindquist, S. L. (1990). HSP104 required for induced thermotolerance. *Science* **248,** 1112–1115.

Sánchez-Velasco, L., Valdez-Holguín, J. E., Shirasago, B., Cisneros-Mata, M. A., and Zarata, A. (2002). Changes in the spawning environment of *Sardinops caeruleus* in the Gulf of California during El Niño 1997–1998. *Estuar. Coast. Shelf Sci.* **54,** 207–217.

Santoro, M. G. (2000). Heat shock factors and the control of the stress response. *Biochem. Pharmacol.* **59,** 55–63.

Sartoris, F. J., and Pörtner, H. O. (1997). Temperature dependence of ionic and acid-base regulation in boreal and arctic *Crangon crangon* and *Pandalus borealis*. *J. Exp. Mar. Biol. Ecol.* **211,** 69–83.

Sartoris, F. J., Bock, C., and Pörtner, H. O. (2003). Temperature-dependent pH regulation in eurythermal and stenothermal marine fish: An interspecies comparison using ^{31}P-NMR. *J. Thermal. Biol.* **28,** 363–371.

Sasekumar, A., Chong, V. C., Leh, M. U., and D'Cruz, R. (1992). Mangroves as a habitat for fish and prawns. *Hydrobiologia* **247,** 195–207.

Schultze, H. P. (1996a). Terrestrial biota in coastal marine deposits: Fossil-Lagerstätten in the Pennsylvanian of Kansas, USA. *Palaeogeogr. Palaeoclimatol. Palaeoecol.* **119,** 255–273.

Schultze, H. P. (1996b). The elpistostegid fish *Elpistostege*, the closest the Miguasha fauna comes to a tetrapod. In "Devonian Fishes and Plants of Miguasha, Quebec, Canada" (Schultze, H. P., and Cloutier, R., Eds.), pp. 316–327. Verlag Dr. Friedrich Pfiel, Munich.

Schultze, H. P. (1997). Umweltbedingungen beim Übergang von Fisch zu Tetrapode. *Sitzungsber. Gesell. Naturforsch. Freunde Berlin* **36,** 59–77.

Schultze, H. P. (1999). The fossil record of the intertidal zone. In "Intertidal Fishes: Life in Two Worlds" (Horn, M. H., Martin, K. L. M., and Chotkowski, M. A., Eds.), pp. 373–392. Academic Press, San Diego, CA.

Sheaves, M. J. (1998). Spatial patterns in estuarine fish faunas in tropical Queensland: A reflection of interaction between long-term physical and biological processes. *Mar. Freshwat. Res.* **47,** 827–830.

Sheppard, C., Price, A., and Roberts, C. (1992). "Marine Ecology of the Arabian Region" Academic Press, London.

Smith, L., Kruszyna, H., and Smith, R. P. (1977). The effect of methemoglobin on the inhibition of cytochrome c oxidase by cynide, sulphide or azide. *Biochem. Pharmacol.* **26,** 2247–2250.

Soares, A. G., Scapini, F., Brown, A. C., and McLachlan, A. (1999). Phenotypic plasticity, genetic similarity and evolutionary inertia in changing environments. *J. Moll. Stud.* **65,** 136–139.

Somero, G. N. (1995). Proteins and temperature. *Annu. Rev. Physiol.* **57,** 43–68.

Somero, G. N., and Hofmann, G. E. (1997). Temperature thresholds for protein adaptation: When does temperature start to 'hurt'? In "Global Warming: Implications for Freshwater and Marine Fish" (Wood, C. M., and McDonald, D. G., Eds.), SEB Seminar Series 61, pp. 1–24. Cambridge University Press, Cambridge.

Somero, G. N., Dahlhoff, E., and Lin, J. J. (1996). Stenotherms and eurytherms: Mechanisms establishing thermal optima and tolerance ranges. In "Animals and Temperature" (Johnston, I. A., and Bennett, A. F., Eds.), pp. 53–78. Cambridge University Press, Cambridge.

Soto, A., Allona, I., Collada, C., Guevara, M., Casado, R., Rodriguez-Cerezo, E., Aragoncillo, C., and Gomez, L. (1999). Heterologous expression of a plant small heat-shock protein enhances Escherichia coli viability under heat and cold stress. Plant Physiol. 120, 521–528.

Southward, A. J., Hawkins, S. J., and Burrows, M. T. (1995). Seventy years' observations of changes in distribution and abundance of zooplankton and intertidal organisms in the western English Channel in relation to rising sea temperature. J. Ther. Biol. 20, 127–155.

Stabenau, E. K., and Heming, T. (2003). Pulmonary carbonic anhydrase in vertebrate gas exchange organs. Comp. Biochem. Physiol. A 136, 271–279.

Stacey, N. E. (1993). Hormones and reproductive behaviour in teleosts. In "Control Processes in Fish Physiology" (Rankin, J. C., Pitcher, T. J., and Dugan, R. T., Eds.), pp. 117–129. Croom Helm, Manuka, Australia.

Steffensen, J. F., Lomholt, J. P., and Johansen, K. (1981). The relative importance of skin oxygen uptake in the naturally buried plaice, Pleuronectes platessa, exposed to graded hypoxia. Respir. Physiol. 44, 269–272.

Stepien, C. A., Phillips, H., Adler, J. A., and Mangold, P. J. (1991). Biogeographic relationships of a rocky intertidal fish assemblage in an area of Cold Water Upwelling off Baja California, Mexico. Pac. Sci. 45, 63–71.

Sunobe, T., and Nakazono, A. (1999). Alternative mating tactics in the gobiid fish Eviota prasina. Ichthyol. Res. 46, 212–215.

Sunobe, T., Ohta, T, and Nakazono, A. (1995). Mating system and spawning cycle in the blenny, Istiblennius enosimae, at Kagoshima, Japan. Environ. Biol. Fish. 43, 195–199.

Swenson, R. O. (1999). The ecology, behaviour, and conservation of the tidewater goby, Eucyclogobius newberryi. Environ. Biol. Fish. 55, 99–114.

Takeda, T., Ishimatsu, A., Oikawa, S., Kanda, T., Hishida, Y., and Khoo, K. H. (1999). Mudskipper Periophthalmodon schlosseri can repay oxygen debts in air but not in water. J. Exp. Zool. 284, 265–270.

Tamura, S. O., Morii, H., and Yuzuriha, M. (1976). Respiration of the amphibious fishes Periophthalmus cantonensis and Boleophthalmus chinensis in water and on land. J. Exp. Biol. 65, 97–107.

Taru, M., and Sunobe, T. (2000). Notes on reproductive ecology of the gobiid fish Eviota abax at Kominato, Japan. Bull. Mar. Sci. 66, 507–512.

Taru, M., Kanda, T., and Sunobe, T. (2002). Alternative mating tactics of the gobiid fish Bathygobius fuscus. J. Ethology 20, 9–12.

Taylor, M. H. (1984). Lunar synchronization of fish reproduction. Trans. Am. Fish. Soc. 113, 484–493.

Taylor, M. H. (1990). Estuarine and intertidal teleosts. In "Reproductive seasonality in Teleosts: Environmental Influences" (Munro, A. D., Scott, A. P., and Lam, T. J., Eds.), pp. 109–124. CRC Press, Boca Raton, FL.

Teal, J. M., and Carey, F. G. (1967). Skin respiration and oxygen debt in the mudskipper Periophthalmus sobrinus. Copeia 1967, 677–679.

Thayer, G. W., Colby, D. R., and Hettler, W. F., Jr. (1987). Utilization of red mangrove prop habitat by fishes in South Florida. Mar. Ecol. Progr. Ser. 35, 25–38.

Todd, S. E., and Ebeling, A. W. (1966). Aerial respiration in the longjaw mudsucker Gillichthys mirabilis (Teleostei: Gobiidae). Biol. Bull. 130, 265–288.

Toews, D. P., Holeton, G. F., and Heisler, N. (1983). Regulation of the acid-base status during environmental hypercapnia in the marine teleost fish *Conger conger*. *J. Exp. Biol.* **107**, 9–20.

Tomanek, L., and Somero, G. N. (1999). Evolutionary and acclimation-induced variation in the heat-shock responses of congeneric marine snails (genus *Tegula*) from different thermal habitats: Implications for limits of thermotolerance and biogeography. *J. Exp. Biol.* **202**, 2925–2936.

Tomanek, L., and Somero, G. N. (2000). Time course and magnitude of synthesis of heat-shock proteins in congeneric marine snails (genus *Tegula*) from different tidal heights. *Physiol. Biochem. Zool.* **73**, 249–256.

Torrans, E. L., and Clemens, H. P. (1982). Physiological and biochemical effects of acute exposure of fish to hydrogen sulfide. *Comp. Biochem. Physiol. C* **71**, 183–190.

Truchot, J. P., and Duhamel-Jouve, A. (1980). Oxygen and carbon dioxide in the marine intertidal environment: Diurnal and tidal changes in rockpools. *Resp. Physiol.* **39**, 241–254.

Tufts, B., and Perry, S. F. (1998). Carbon dioxide transport and excretion. *In* "Fish Physiology. Vol. 17: Fish Respiration" (Perry, S. F., and Tufts, B., Eds.), pp. 229–281. Academic Press, New York.

Turner, G. F. (1993). Teleost mating behaviour. *In* "Behaviour of Teleost Fishes" (Picher, T. J., Ed.), 2nd edn., pp. 307–331. Chapman & Hall, London.

Ulmasov, K. A., Shammakov, S., Karaev, K., and Evgen'ev, M. B. (1992). Heat shock proteins and thermoresistance in lizards. *Proc. Natl Acad. Sci. USA* **89**, 1666–1670.

Val, A. L. (2000). Organic phosphates in the red blood cells of fish. *Comp. Biochem. Physiol.* **125A**, 417–435.

Val, A. L., Lessard, J., and Randall, D. (1995). Effects of hypoxia on rainbow trout (*Oncorhynchus mykiss*): Intraerythrocytic phosphates. *J. Exp. Biol.* **198**, 305–310.

Van der Velde, G., Gorissen, M. W., Den Hartog, C., van 't Hoff, T, and Meijer, G. J. (1992). Importance of the Lac-lagoon (Bonaire, Netherlands Antilles) for a selected number of reef fish species. *Hydrobiologia* **247**, 139–140.

Vance, D. J. (1999). Distribution of shrimp and fish associated with the mangrove forest of Mai Po Marshes Nature Reserve, Hong Kong. *In* "The Mangrove Ecosystem of Deep Bay and the Mai Po Marshes, Hong Kong, Proceedings of the International Workshop on the Mangrove Ecosystem of Deep Bay and the Mai Po Marshes, Hong Kong, 3–20 September 1993" (Lee, S. Y., Ed.), pp. 23–32. Hong Kong University Press, Hong Kong.

Vilarrubia, T. V., and Rull, V. (2002). Natural and human disturbance history of the Playa Medina mangrove community (Eastern Venezuela). *Caribb. J. Sci.* **38**, 66–76.

Vincent, A. C. J., Berglund, A., and Ahnesjo, I. (1995). Reproductive ecology of five pipefish species in one seagrass meadow. *Environ. Biol. Fish.* **44**, 347–361.

Vivekanandan, E., and Pandian, T. J. (1979). Erythrocyte count and haemoglobin concentration of some tropical fishes. *J. Madurai Kamaraj Univ. (Sci.)* **8**, 71–75.

Völkel, S., and Berenbrink, M. (2000). Sulphaemoglobin formation in fish: A comparison between the haemoglobin of the sulphide-sensitive rainbow trout (*Oncorhynchus mykiss*) and of the sulphide-tolerant common carp (*Cyprinus carpio*). *J. Exp. Biol.* **203**, 1047–1058.

Völkel, S., Berenbrink, M., Heisler, N., and Nikinmaa, M. (2001). Effects of sulfide on K^+ pathways in red blood cells of crucian carp and rainbow trout. *Fish Physiol. Biochem.* **24**, 213–223.

Walsh, P. J., Tucker, B. C., and Hopkins, T. E. (1994). Effects of confinement/crowding on ureogenesis in the Gulf toadfish *Opsanus beta*. *J. Exp. Biol.* **191**, 195–206.

Ward, P. D., and Signor, P. W. (1983). Evolutionary tempo in Jurassic and Cretaceous ammonites. *Paleobiology* **9**, 183–198.

Warner, R. R. (1975). The adaptive significance of sequential hermaphrodism in animals. *Am. Nat.* **109**, 61–82.

Warner, R. R. (1982). Mating systems, sex change and sexual demography in the rainbow wrasse, *Thalassoma lucasanum*. *Copeia* **1982**, 653–661.

Weaver, P. L. (1970). Species diversity and ecology of tidepool fishes in three Pacific coastal areas of Costa Rica. *Rev. Biol. Trop.* **17**, 165–185.

Weber, R. E. (1990). Functional significance and structural basis of multiple hemoglobins with special reference to ectothermic vertebrates. *In* "Animal Nutrition and Transport Processes. Volume 6: Transport, Respiration and Excretion: Comparative and Environmental Aspects" (Truchot, J. P., and Lahlou, B., Eds.), pp. 58–75. Karger, Basle.

Weber, R. E., and Wells, R. M. G. (1989). Hemoglobin structure and function. *In* "Comparative Pulmonary Physiology" (Wood, S. C., Ed.), pp. 279–310. Marcel Dekker, New York.

Weinstein, M. P., and Heck, K. L., Jr. (1979). Ichthyofauna of seagrass meadows along the Caribbean coast of Panama and the Gulf of Mexico: Composition, structure and community ecology. *Mar. Biol.* **50**, 97–107.

Wells, J. W. (1952). The coral reefs of Arno Atoll, Marshall Islands. *Atoll Res. Bull.* **9**, 1–14.

Wells, J. W. (1956). Scleractinia. *In* "Treatise on Invertebrate Paleontology, Part F, Coelenterata" (Moore, R. C., Ed.), pp. F328–F444. Geological Society of America and University of Kansas Press, Lawrence, Kansas.

Whitear, M. (1986). The skin of fishes including cyclostomes. *In* "Biology of the Integument" (Bereiter-Hahn, J., Matoltsy, A. G., and Richards, K. S., Eds.), pp. 8–38. Springer-Verlag, Berlin.

Whitear, M. (1988). Variation in the arrangement of tonofilaments in the epidermis of teleost fish. *Biol. Cell* **64**, 85–92.

Whitear, M., and Mittal, A. K. (1984). Surface secretions of the skin of *Blennius (Lipophrys) pholis*. *J. Fish Biol.* **25**, 317–331.

Whitfield, A. K. (1996). A review of estuarine ichthyology in South Africa over the past 50 years. *Trans. R. Soc. S. Afr.* **51**, 79–89.

Whitfield, A. K., Blaber, S. J. M., and Cyrus, D. P. (1981). Salinity ranges of some southern African fish species occurring in estuaries. *S. Afr. J. Zool.* **16**, 151–155.

Wiens, H. J. (1962). "Atoll Environment and Ecology." Yale University Press, New Haven, CT.

Wilkinson, C., Lindén, O., Cesar, H., Hodgson, G., Rubens, J., and Strong, A. E. (1999). Ecological and socioeconomic impacts of 1998 coral mortality in the Indian Ocean: An ENSO impact and a warning of future change? *Ambio* **28**, 188–196.

Wilson, D. T. (2001). Patterns of replenishment of coral-reef fishes in the nearshore waters of the San Blas Archipelago, Caribbean Panama. *Mar. Biol.* **139**, 735–753.

Wilson, J. M., Kok, T. W. K., Randall, D. J., Vogl, W., and Ip, Y. K. (1999). Fine structure of the gill epithelium of the terrestrial mudskipper, *Periophthalmodon schlosseri*. *Cell Tissue Res.* **298**, 345–356.

Wilson, J. M., Randall, D. J., Donowitz, M., Vogl, A. W., and Ip, Y. K. (2000a). Immunolocalization of ion-transport proteins to branchial epithelium mitochondria-rich cells in the mudskipper (*Periophthalmodon schlosseri*). *J. Exp. Biol.* **203**, 2297–2310.

Wilson, J. M., Randall, D. J., Vogl, A. W., Harris, J., Sly, W. S., and Iwama, G. K. (2000b). Branchial carbonic anhydrase is present in the dogfish, *Squalus acanthias*. *Fish Physiol. Biochem.* **22**, 329–336.

Wirtz, P. (1978). The behaviour of the Mediterranean Tripterygion species (Pisces, Blennioidei). *Z. Tierpsychol.* **48**, 142–174.

Wolanski, E., and Chappell, J. (1996). The response of tropical Australian estuaries to a sea level rise. *J. Mar. Syst.* **7**, 267–279.

Wolanski, E., and Jones, M. (1980). Water circulation around Britomart Reef, Great Barrier Reef, during July 1979. *J. Mar. Freshwat. Res.* **31**, 415–430.

Wolanski, E., and Ruddick, B. (1981). Water circulation and shelf waves in the northern Great Barrier Lagoon. *Aust. J. Mar. Freshwat. Res.* **32,** 721–740.

Wood, C. M., Hopkins, T. E., Hogstrand, C., and Walsh, P. J. (1995). Pulsatile urea excretion in the ureagenic toadfish *Opsanus beta*: An analysis of rates and routes. *J. Exp. Biol.* **198,** 1729–1741.

Wood, C. M., Perry, S. F., Walsh, P. J., and Thomas, S. (1994). HCO_3. dehydration by the blood of an elasmobranch in the absence of a Haldane effect. *Respir. Physiol.* **98,** 319–337.

Wood, S. C., and Johansen, K. (1972). Adaptation to hypoxia by increased HbO_2 affinity and decreased red cell ATP concentration. *Nature* **237,** 278–279.

Woodbury, D. (1999). Reduction of growth in otoliths of widow and yellowtail rockfish (*Sebastes entomelas* and *S. flavidus*) during the 1983 El Niño. *Fish. Bull.* **97,** 680–689.

Wright, A. (1993). Shallow water reef-associated finfish. *In* "Nearshore Marine Resources of the South Pacific" (Wright, A., and Hill, L., Eds.), pp. 203–284. Institute of Pacific Studies, Suva, Fiji.

Yokoya, S., and Tamura, O. S. (1992). Fine structure of the skin of the amphibious fishes, *Boleophthalmus pectinirostris* and *Periophthalmus cantonensis*, with special reference to the location of blood vessels. *J. Morphol.* **214,** 287–297.

Zander, C. D. (1975). Secondary sex characters of Belnnioid fishes (Perciformes). *Pubbl. Staz. Zool. Napoli* **39**(Suppl.), 717–727.

Zeidel, M. L. (1996). Low permeabilities of apical membranes of barrier epithelia: What makes watertight membranes watertight? *Am. J. Physiol.* **271,** F243–F245.

Zhang, J., Taniguchi, T., Takita, T., and Ali, A. B. (2000). On the epidermal structure of *Boleophthalmus* and *Scartelaos* mudskippers with reference to their adaptation to terrestrial life. *Ichthyol. Res.* **47,** 359–366.

Zhou, H. (2001). Meiofaunal community structure and dynamics in a Hong Kong mangrove. PhD thesis, University of Hong Kong, Hong Kong.

12

HYPOXIA TOLERANCE IN CORAL REEF FISHES

GÖRAN E. NILSSON
SARA ÖSTLUND-NILSSON

I. INTRODUCTION

Coral reef fishes are not known for their hypoxia tolerance, and coral reefs are not generally thought of as hypoxic habitats. However, recent investigations on Australia's Great Barrier Reef paint a different picture. We will here discuss some examples of hypoxic coral reef habitats and hypoxia-tolerant coral reef inhabitants, and we will even suggest that hypoxia tolerance is a widespread phenomenon among coral reef fishes.

Because coral reefs have the highest biodiversity of any marine habitat, finding hypoxia tolerance there could mean that there is an exceptional wealth of hypoxia adaptations waiting to be explored in this ecosystem. The best studied hypoxia tolerant vertebrates, which include North American freshwater turtles, carps, and goldfish (Lutz *et al.*, 2003), have all evolved their hypoxia tolerance to allow overwintering in hypoxic habitats at temperatures close to 0 °C. By contrast, coral reef fishes live at a water temperature near 30 °C. This is not far from the body temperature of homeothermic vertebrates like mammals. Thus, to disclose the mechanisms that coral reef fishes have evolved to allow hypoxic survival could be of particular relevance for biomedical hypoxia and ischemia research, besides deepening our understanding of the coral reef ecosystem.

The Physiology of Tropical Fishes: Volume 21
FISH PHYSIOLOGY

II. THE EPAULETTE SHARK: A HYPOXIA-TOLERANT TROPICAL ELASMOBRANCH

Our first, and best known, example of hypoxia tolerance and hypoxia on a coral reef is that of the epaulette shark (*Hemiscyllium ocellatum*) on Heron Island – a small and low coral cay situated close to the southern end of the Great Barrier Reef. In 1966, Kinsey and Kinsey reported measurements of water oxygen levels on the reef platform surrounding Heron Island. At nocturnal low tides, they found that the water [O_2] could fall below 30% of air saturation (about 2.1 mg $O_2 l^{-1}$). A prerequisite for this hypoxia is that the huge (*ca.* 3 × 10 km) reef platform becomes cut off from the surrounding ocean at low tides, and essentially becomes a very large tide-pool. When this happens on calm nights, with little water movements, the respiration of the coral, and all associated organisms, leads to a drastic fall in oxygen levels. More recent measurements show that water [O_2] levels can fall to 18% of air saturation on the Heron Island reef platform (Routley *et al.*, 2002).

During these hypoxic episodes, the epaulette shark stays on the reef platform. Wise *et al.* (1998) were the first to show that this shark was hypoxia tolerant, surviving in water with an [O_2] of 5% of air saturation for at least 3.5 hours without any impairment of neural functions like righting reflex, ventilation, and rhythmic swimming. Subsequent studies have revealed that hypoxia is tolerated without any delayed phase of neuronal death (Renshaw and Dyson, 1999), a common event in mammals subjected to hypoxia. Moreover, the epaulette shark also survives complete anoxia for about an hour at temperatures close to 30 °C, without any considerable drop in brain [ATP] (Renshaw *et al.*, 2002).

A. Natural Hypoxic Preconditioning on the Reef

When the epaulette shark encounters hypoxia on the reef, this will not happen abruptly. Initially during a period of spring tides, when the tides become lower and lower on subsequent nights, the epaulette shark should experience longer and longer periods of hypoxia (Figure 12.1). This should allow the shark to gradually acclimate to longer and longer hypoxic episodes over a period of a few days. Interestingly, this situation becomes a natural parallel to a hypoxic treatment scheme that has been termed hypoxic preconditioning in biomedical science. In mammals, exposures to relatively mild hypoxia has been found to reduce the detrimental effects of more severe subsequent hypoxic or ischemic insults in both the brain and heart (Dirnagl *et al.*, 2003).

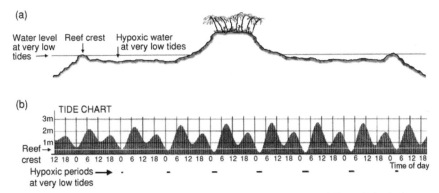

Fig. 12.1 Hypoxic preconditioning on a coral reef platform like that of Heron Island. At very low tides, the water on reef platform gets cut off from the surrounding ocean (a). If this happens at night, the respiration of the reef organisms will make the water hypoxic. The tide chart (b) shows that the hypoxic episodes will increase in length during subsequent nights in periods of low tide, providing the inhabitant of the reef platform with a natural scheme of hypoxic preconditioning.

Experiments on the epaulette shark show that repeated exposure to hypoxia in the laboratory has a profound effect on the rate of oxygen consumption (VO_2) and the critical oxygen concentration ($[O_2]_{crit}$) of this shark (Figure 12.2a–d). The $[O_2]_{crit}$ is the concentration below which the fish is unable to maintain a resting VO_2 that is independent of the ambient $[O_2]$ (Beamish, 1964). The sharks acclimated to hypoxia this way had a mean VO_2 of $59.5 \, mg \, O_2 \, kg^{-1} \, h^{-1}$ while that of non-acclimated sharks was $83.4 \, mg$ $O_2 \, kg^{-1} \, h^{-1}$. Probably as a consequence of this, the mean $[O_2]_{crit}$ of the acclimated sharks was 25% of air saturation while that of non-acclimated sharks was 32% of air saturation (Routley *et al.*, 2002).

B. Mechanisms of Hypoxia Tolerance in the Epaulette Shark

So far, experiments on the epaulette shark have yielded both interesting and surprising results. The epaulette shark shows no change in the cerebral blood flow when exposed to severe hypoxia (5% of air saturation) for 2 hours (Söderström, Renshaw *et al.*, 1999). This was an unexpected finding in light of the fact that hypoxia has a highly stimulatory effect on brain blood flow in all other vertebrates examined, including mammals, crocodiles, turtles, and teleost fish (Söderström, Renshaw *et al.*, 1999; Söderström, Nilsson *et al.*, 1999). However, the brain blood flow in the epaulette shark was at least maintained during hypoxia, despite a fall blood pressure, suggesting that hypoxia induces cerebral vasodilation (Söderström, Renshaw *et al.*, 1999).

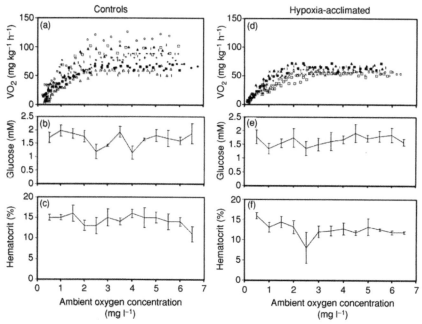

Fig. 12.2 Effect of falling ambient [O_2] on the rate of O_2 consumption (VO_2; a, d), blood glucose (b, e), and hematocrit (c, f) in control epaulette sharks (a–c) and hypoxia-acclimated epaulette sharks (d–f). The different symbols in (a) and (d) represent individual sharks while other values are means ± SEM from 10 (control) or 7 (acclimated) animals. Acclimated animals had been exposed to hypoxia (5% air saturation) for 2 hours twice-daily on four subsequent days. (Data from Routley *et al.*, 2002.)

In contrast to the hyperglycemic response of many vertebrates, including teleost fish, exposed to hypoxia, the blood glucose level remains constant during hypoxia in the epaulette shark (Figure 12.2b–f) (Routley *et al.*, 2002). This is also the case in another shark, the dogfish, *Scyliorhinus canicula* (Butler *et al.*, 1979), and may reflect an elasmobranch trait. Moreover, the erythrocyte content of the epaulette shark blood was found to be relatively low (a hematocrit of about 15%), and unlike many other vertebrates, hematocrit did not increase in response to hypoxia (Figure 12.2c), not even after repeated hypoxic exposures (Figure 12.2f) (Routley *et al.*, 2002).

There are also similarities. As in other vertebrates, adenosine appears to play an important role in the hypoxia tolerance of the epaulette shark. Adenosine is an inhibitory neuromodulator in vertebrates that, for example, functions to reduce ATP consumption in brain during hypoxia or anoxia (Lutz *et al.*, 2003, for review). When a cell becomes energy deficient, and is unable to match ATP use with ATP production, adenosine is formed from dephosphorylation of the high-energy phosphorylated adenosines:

$$ATP \Rightarrow ADP \Rightarrow AMP \Rightarrow Adenosine$$

Renshaw *et al.* (2002) found that epaulette sharks in anoxia lose their righting reflex after about 36 minutes (but they immediately recover in normoxia). Twenty-four hours later, the same sharks were injected with either saline or the adenosine receptor blocker aminophylline and again exposed to anoxia. The saline controls now showed a 56% decrease in the time taken to lose the righting reflex (possibly a preconditioning effect), while conversely, the aminophylline-treated sharks displayed a 46% increase in this variable. Moreover, while control sharks maintained their brain ATP level during anoxia, ATP fell significantly in brains of aminophylline-treated animals. Since also a 3.5-fold increase in brain adenosine levels was found after anoxia, these results suggest that during anoxia, adenosine acts as a retaliatory signal to initiate neural metabolic depression, manifested as a loss of righting reflex. This should allow the epaulette shark brain to reduce ATP consumption and maintain cellular ATP levels. Thus, while a loss of the righting reflex in anoxia-exposed fish is generally thought to be a sign of neural energy deficiency and an immediate threat to survival, this study indicates that in the epaulette shark, the loss of the righting reflex is actually a survival strategy to save energy, probably mediated by elevated adenosine levels and speeded up by hypoxic preconditioning.

Elasmobranchs may be especially fortunate when it comes to surviving hypoxia. A recent study has suggested that elasmobranchs (including the epaulette shark) have a relatively low rate of cerebral ATP consumption compared to other vertebrates, as indicated by a very low activity of Na^+/K^+ ATPase – the main ATP consumer in brain (Nilsson *et al.*, 2000). The mass specific brain Na^+/K^+ ATPase activity of sharks and rays was found to be only one-third of that of teleosts.

Besides explaining why elasmobranchs can afford brains that are generally much bigger than those of teleosts (Nilsson *et al.*, 2000), the low brain Na^+/K^+ ATPase activity of elasmobranchs also indicates that it should be less of a feat for an elasmobranch to be hypoxia-tolerant. The major threat of hypoxia lies in the difficulty of keeping the ATP production high enough to match ATP consumption. The reason is simple: anaerobic ATP production is much less efficient than aerobic ATP production. The higher the rate of ATP use by an organ, the faster that organ will run into energy deficiency. When the ATP level starts to fall, the cell will become unable to maintain ion balance and a catastrophic cascade of degenerative processes will start. Because of its high specific rate of energy consumption, the brain is generally the most hypoxia-sensitive organ (Lutz *et al.*, 2003, for review).

The same logic applies to temperature: ATP consumption increases exponentially with temperature, so surviving hypoxia at the high temperature of a tropical coral reef is much more of a challenge than surviving hypoxia in cold temperate waters. Thus, possessing a brain with a relatively low rate of energy consumption, like sharks do, should extend the time the animal can endure a period of hypoxia and a reduced ability to produce ATP. This raises the question: are there teleost fishes that can tolerate hypoxia living in the warm waters of coral reefs?

III. WIDESPREAD HYPOXIA TOLERANCE IN CORAL REEF FISHES

While studying the respiratory consequences of mouthbrooding in two species of cardinalfish (*Apogon leptacanthus* and *A. fragilis*) on Lizard Island, on the Northern portion of the Great Barrier Reef, we found that these fishes had an $[O_2]_{crit}$ just below 20% of air saturation (Östlund-Nilsson and Nilsson, 2004). To us, this was an unexpectedly low $[O_2]_{crit}$ for fishes living in a tropical coral reef habitat. First, to our knowledge, severe hypoxia had never been reported in this habitat. Secondly, the ability to maintain O_2 uptake in hypoxia is not uncomplicated in seawater at such a high temperature (30 °C), due to the combined effects of a low solubility of O_2 in warm seawater, and a high rate of oxygen consumption of a small fish at such a high temperature. In all animals, VO_2 increases with body temperature and decreases with body mass, which in the case of the cardinal fishes was 1–2 g.

A comparison can be made with tropical freshwater fishes for which there are data available. Some African cichlid species, including tilapia, *Oreochromis niloticus*, have $[O_2]_{crit}$ of about 20% of air saturation at 25 °C. However, these species are relatively large fishes adapted to tropical freshwater habitats which regularly become severely hypoxic (Verheyen *et al.*, 1994; Chapman *et al.*, 1995). Indeed, these cichlids are renowned for their hypoxia tolerance.

Our first hypothesis was that the low $[O_2]_{crit}$ measured in the cardinalfishes could reflect an adaptation for mouthbrooding, allowing the fish to strip more oxygen from the water when faced with a reduced ability to ventilate the gills. In cardinal fishes, the males carry the fertilized eggs in their mouth for 1–2 weeks (Thresher, 1984; Okuda, 1999). In the species we studied, the egg clutch could make up 25% of the fish weight. Clearly such a mouth full of eggs could represent a significant ventilatory problem. We decided to test this hypothesis by comparing $[O_2]_{crit}$ values of several species of cardinalfish (which are all mouthbrooders) with other fish species

Fig. 12.3 Branching coral at the reef near Lizard Island Research Station – a habitat where all fish examined display a considerable hypoxia tolerance. Depth 3 m. (Photo G. E. Nilsson.)

(non-mouthbrooders) in the same habitat. The habitat was 2–5 m deep reefs with a high abundance of branching coral (Figure 12.3), situated in the lagoon outside the Lizard Island Research Station.

The $[O_2]_{crit}$ of the nine cardinalfish species studied varied between 17% and 34% of air saturation (Table 12.1), but surprisingly, all of the other fishes sampled from the same habitat, representing 31 species from seven families, showed strikingly low $[O_2]_{crit}$ values, varying from 13% to 32% (Nilsson and Östlund-Nilsson, 2004). Low $[O_2]_{crit}$ were, for example, displayed by several species of damselfishes (Pomacentridae) (Table 12.1), a large and well-known family of often colorful fishes, and one of the dominating fish groups on coral reefs around the world. Also the gobiids and blennids examined displayed low $[O_2]_{crit}$ values. Several of the fishes did not show any signs of distress or loss of coordination until the O_2 level fell below 5–10% of air saturation, indicating high anaerobic capacities. This hypoxia tolerance was combined with high metabolic rates. Most of the fishes studied weighed less than 10 g and had a resting VO_2 of 200–700 mg $O_2\,kg^{-1}\,h^{-1}$, which is several times higher than that of fishes in cold temperate water.

Obviously, we could dismiss our initial hypothesis. Cardinalfishes in general do not display a $[O_2]_{crit}$ that is lower than other species in the same habitat, indicating that their low $[O_2]_{crit}$ is not an adaptation to mouthbrooding.

Table 12.1

Hypoxia Tolerance at 30 °C of Fishes in the Lagoon at Lizard Island Research Station, Great Barrier Reef

Family/Species	N	Weight (g)	Normoxic VO_2 (mg kg^{-1}h^{-1})	$[O_2]_{crit}$ (% of air saturation)	$[O_2]_{out}$ (% of air saturation)
Cardinalfishes (Apogonidae)					
Apogon compressus	4	7.0 ± 1.2	179 ± 67	19 ± 5	6.7 ± 1.9
Apogon cyanosoma	1	2.2	259	30	
Apogon doederleini	1	4.4	288	31	
Apogon exostigma	1	3.7	218	26	11.4
Apogon fragilis	14	1.9 ± 0.1	255 ± 17	17 ± 1	7.2 ± 1.0
Apogon leptacanthus	14	1.5 ± 0.1	239 ± 19	19 + 1	7.0 + 1.2
Archamia fucata	1	5.8	225	34	
Cheilodipterus quinquelineatus	2	1.8–7.4	244–263	23–31	7.2–11.1
Sphaeramia nematoptera	1	7.3	131	17	10.0
Damselfishes (Pomacentridae)					
Acanthochromis polyacanthus	1	15.4	197	26	6.5
Chromis atripectoralis	5	8.4 ± 2.5	358 ± 84	22 ± 2	8.8 ± 0.8
Chromis viridis	6	2.5 ± 1.1	555 ± 108	23 ± 1.2	7.4 ± 0.9
Chrysiptera flavipinnis	1	2.4	384	30	12.0
Dascyllus aruanus	3	4.1 ± 1.3	306 ± 37	19 ± 0	5.9 ± 0.6
Neoglyphidodon melas	6	32.1 ± 8.8	216 ± 32	25 ± 2	5.6 ± 0.7
Neoglyphidodon nigroris	6	14.9 ± 2.4	162 ± 21	22 ± 3	8.9 ± 1.5
Neopomacentrus azysron	1	3.2	493	32	
Pomacentrus ambionensis	4	12.6 ± 1.7	201 ± 11	22 ± 4	7.1 ± 2.0
Pomacentrus bankanensis	1	7.8	237	19	
Pomacentrus coelestis	6	7.8 ± 3.0	387 ± 85	22 ± 4	9.3 ± 1.0
Pomacentrus lepidogenys	5	3.1 ± 0.6	516 ± 73	31 ± 2	12.5 ± 1.1
Pomacentrus moluccensis	4	5.2 ± 4.0	397 ± 85	25 ± 3	10.4 ± 1.9
Pomacentrus philippinus	2	2.2–6.9	320–348	26–33	9.3–10.5
Gobies (Gobiidae)					
Amblygobius phalaena	1	2.4	333	21	2.8
Asteropteryx semipunctatus	1	1.4	403	26	1.4
Gobiodon histrio	10	1.2 ± 0.2	248 ± 31	18 ± 1	2.8 ± 0.5
Blennies (Blenniidae)					
Atrosalarias fuscus	3	7.3 ± 1.9	208 ± 34	18 ± 2	1.6 ± 0.7
Atrosalarias fuscus juvinile	1	0.29	552	13	1.5
Filefishes (Monacanthidae)					
Paramonacanthus japonicus	1	1.7	486	23	9.5
Breams (Nemipteridae)					
Scolopsis bilineata juvenile	1	1.9	375	28	12.8
Wrasses (Labridae)					
Halichoeres melanurus	1	1.8	394	25	6.8
Labroides dimidiatus juvenile	3	0.56 ± 0.09	736 ± 35	24 ± 5	7.8 ± 0.9

Normoxic VO_2 = rate of O_2 consumption at a water $[O_2]$ >70% air saturation; $[O_2]_{crit}$ = critical $[O_2]$, below this level VO_2 starts falling and is no longer independent of ambient $[O_2]$; $[O_2]_{out}$ = $[O_2]$ at which the fish showed signs of agitation or balance problems. Values for 3 or more fish are means ± SD. Taxonomy follows Randall *et al.* (1997). From Nilsson and Östlund-Nilsson (2004).

Fig. 12.4 Extreme low tide at a shallow reef near Lizard Island Research Station. At night, this could be a hypoxic environment. (Photo Lizard Island Research Station.)

Why do virtually all fishes in this habitat show a $[O_2]_{crit}$ that is much lower than the O_2 levels that they, at a first glance, can be expected to encounter? Possibly this is related to the fact that the same species may occur also on more shallow reefs which, like the reef platform around Heron Island, becomes cut off from the surrounding ocean during low tides.

Indeed, shallow parts of the reef around Lizard Island can get partially air exposed, with the resultant formation of tidepools, during exceptionally low tides (Figure 12.4). So far, however, we have not been able to explore the reef at such low tides. It would be particularly interesting to measure oxygen levels in the tidepools on calm nights, and examine the composition of the fish fauna that may remain there.

Another possible explanation for the hypoxia tolerance displayed by at least some of the species examined at Lizard Island is that they move deep into the branching coral to feed or hide from predators. If they do this at night, as many night divers reportedly have seen, they may enter a micro-habitat which becomes hypoxic due to coral respiration. To examine this hypothesis further, we decided to conduct a case study by taking a closer look at the respiratory characteristics of a true coral-dweller.

A. The Most Cowardly Fish on the Reef: The Broad-Barred Goby

The broad-barred goby, *Gobiodon histrio* (Figure 12.5), is arguably an exceptionally cowardly fish. This species secretes a poisonous mucus, and is therefore probably inedible to most predators. Its bright green color with red markings could possibly serve as a warning to predators, and fish fed with

Fig. 12.5 The broad-barred goby, *Gobiodon histrio*. (Photo G. E. Nilsson.)

pieces of *Gobiodon* have been found to die within a few minutes (Schubert *et al.*, 2003). Nevertheless, *G. histrio* spends virtually its whole adult life in 5–10 mm wide spaces formed between the branches of *Acropora* corals (preferentially *A. nasuta*), a shelter that should be inaccessible to most predators. Moreover, the need to leave the coral to find a mating partner is minimized by its ability to change sex. Thus, if two individuals of the same sex end up in the same coral, one of them will change its sex unless it can readily move to another suitable coral very near by (Munday *et al.*, 1998).

This extreme habitat fidelity makes the broad-barred goby ideal for studying whether a coral habitat demands hypoxia tolerance. In particular, we expected that the water between branches of coral becomes hypoxic during calm nights, due to the combined effects of the nocturnal stop in photosynthesis, continued respiration of the coral (and associated organisms), and lack of water movements. Low nocturnal oxygen levels are indicated by physiological studies revealing night-time hypoxia in coral tissue (Jones and Hoegh-Guldberg, 2001).

Another interesting problem for this goby is that the whole coral colony may be out of the water at times. At Lizard Island on the Great Barrier Reef, it had been observed that *A. nasuta* corals may be out of the water for 1–4 hours during the exceptionally low tides that occur about 30 times every year. Therefore, we set out to measure the water oxygen level in *A. nasuta* corals at night in calm water (simulated in a large outdoor tank at Lizard Island Research Station), and to use closed respirometry to examine the ability of *G. histrio* to tolerate hypoxia and air exposure (Nilsson *et al.*, 2004).

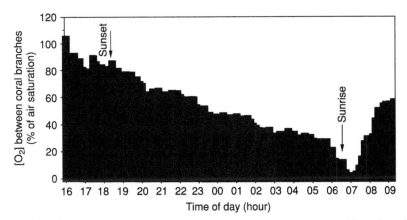

Fig. 12.6 Tracing of the oxygen level inside the coral *Acropora nasuta*, the habitat where the broad-barred goby spends its whole adult life.

The results indicated that the coral home of *G. histrio* can become severely hypoxic under calm conditions at night. The average $[O_2]$ minimum reached between branches in the coral (Figure 12.6) could be as low as 3% for a short period of time in the early morning.

The respirometric measurements showed that *G. histrio* has a $[O_2]_{crit}$ of 18% of air saturation, and that it can tolerate at least 2 hours at even lower O_2 levels, not losing its equilibrium until water $[O_2]$ falls below 3%. Furthermore, during air exposure, which was tolerated for up to 4.5 hours, *G. histrio* upheld a rate of O_2 consumption that was 60% of that in water. Possible mechanisms utilized for oxygen uptake in air by *G. histrio* include the circulation of a small droplet of water over the gills and respiration through the skin. There are other examples of gobies, particularly estuarine gobies, that show air-breathing abilities (Graham, 1976; Gee and Gee, 1995; Martin, 1995). The best known gobiid air-breathers are probably mudskippers, some of which show specialized respiratory epithelia in the buccal cavity (Al-Kadhomiy and Hughes, 1988).

Clearly, the ability of this goby to tolerate such severe hypoxia is likely to be a prerequisite for it to be able to remain in its coral shelter during hypoxic episodes, which are most likely to occur during calm nights. The air-breathing abilities of *G. histrio* should serve the same function during the extremely low tides which render its coral home completely air-exposed.

This is to our knowledge the first documented case of predator avoidance through hypoxia tolerance in a coral-dwelling fish. However, another goby, *Valenciennea longipinnis*, that lives in burrows in sandy areas near coral reefs, appears to be quite tolerant to hypoxia (Takegaki and Nakazono, 1999), and this may be a prerequisite for it to stay in its burrow, which at times can become hypoxic.

IV. CONCLUSIONS

Hypoxia in coral reef habitats, and hypoxia tolerance among coral reef fishes, are probably much more common phenomena than generally thought. Few scientists have, however, set out to study hypoxia in the paradisiacal setting of a coral reef. The only relatively well studied example of a hypoxia-tolerant reef fish is the epaulette shark on Heron Island, which becomes preconditioned to hypoxia during subsequent nights of nocturnal low tides.

Recent studies indicate that hypoxia tolerance is widespread among coral reef teleosts, including damselfishes, cardinalfishes, gobiids, and blennids. At daytime, the damselfishes and cardinalfishes are seen hovering in the well oxygenated water above the coral, but at night, they may encounter hypoxia as they go down further into the coral, probably to avoid nocturnal predators. Hypoxia may also be encountered on the reef by fishes that remain in temporary pools at nocturnal low tides. In both cases, the combination of a stop in photosynthesis and lack of water movements can make these habitats severely hypoxic.

A case study on a coral-dwelling goby indicates that hypoxia tolerance, as well as air breathing, are prerequisites for this species to stay in the shelter of their coral homes indefinitely. However, our knowledge of these phenomena are still very incomplete, and we have at present a very inadequate understanding of the role of hypoxia in coral reef habitats and of hypoxia adaptations in coral reef inhabitants.

REFERENCES

Al-Kadhomiy, N. K., and Hughes, G. M. (1988). Histological study of different regions of the skin and gills in the mudskipper, *Boleophthalmus boddarti* with respect to their respiratory function. *J. Mar. Biol. Ass. U.K.* **68,** 413–422.

Beamish, F. W. H. (1964). Seasonal temperature changes in the rate of oxygen consumption of fishes. *Can. J. Zool.* **42,** 189–194.

Butler, P. J., Taylor, E. W., and Davison, W. (1979). The effect of long term, moderate hypoxia on acid base balance, plasma catecholamines and possible anaerobic end products in the unrestrained dogfish. *Scyliorhinus canicula. J. Comp. Physiol. B.* **132,** 297–303.

Chapman, L. J., Kaufman, L. S., Chapman, C. A., and McKenzie, F. E. (1995). Hypoxia tolerance in twelve species of East African cichlids: Potential for low oxygen refugia in Lake Victoria. *Conservation Biol.* **9**, 1274–1288.

Dirnagl, U., Simon, R. P., and Hallenbeck, J. M. (2003). Ischemic tolerance and endogenous neuroprotection. *Trends Neurosci.* **26**, 248–254.

Gee, J. H., and Gee, P. A. (1995). Aquatic surface respiration, buoyancy control and the evolution of air-breathing. *J. Exp. Biol.* **198**, 79–89.

Graham, J. B. (1976). Respiratory adaptations of marine air-breathing fishes. *In* "Respiration of Amphibious Vertebrates" (Hughes, G. M., Ed.), pp. 165–187. Academic Press, New York.

Jones, R. J., and Hoegh-Guldberg, O. (2001). Diurnal changes in the photochemical efficiency of the symbiotic dinoflagellates (Dinophyceae) of corals: Photoreception, photoinactivation and the relationship to coral bleaching. *Plant Cell Environ.* **24**, 89–99.

Kinsey, D. W., and Kinsey, B. E. (1966). Diurnal changes in oxygen content of the water over the coral reef platform at Heron Island. *Aust. J. Mar. Freshw. Res.* **1**, 23–24.

Lutz, P. L., Nilsson, G. E., and Prentice, H. M. (2003)."The Brain without Oxygen," 3rd edn. Kluwer, Dordrecht.

Martin, K. L. M. (1995). Time and tide wait for no fish: Intertidal fishes out of water. *Environ. Biol. Fish.* **44**, 165–181.

Munday, P. L., Caley, M. J., and Jones, G. P. (1998). Bi-directional sex change in a coral-dwelling goby. *Behav. Ecol. Sociobiol.* **43**, 371–377.

Nilsson, G. E., and Östlund-Nilsson, S. (2004). Hypoxia in paradise. Widespread hypoxia tolerance in coral reef fishes. *Proc. R. Soc. B. (Biol. Lett. Suppl.)* **271**, S30–S33.

Nilsson, G. E., Hobbs, J.-P., Munday, P. L., and Östlund-Nilsson, S. (2004). Coward or braveheart: Extreme habitat fidelity through hypoxia tolerance in a coral-dwelling goby. *J. Exp. Biol.* **207**, 33–39.

Nilsson, G. E., Routley, M. H., and Renshaw, G. M. C. (2000). Low mass specific brain Na^+K^+ ATPase activity in elasmobranch and teleost fishes: Its implications for the large brain size of elasmobranchs. *Proc. R. Soc. Ser. B* **267**, 1335–1339.

Okuda, N. (1999). Sex roles are not always reversed when the potential reproductive rate is higher in females. *Am. Nat.* **153**, 540–548.

Östlund-Nilsson, S., and Nilsson, G. E. (2004). A mouth full of eggs: Respiratory consequences of mouthbrooding in cardinalfishes. *Proc. R. Soc. Ser. B* **271**, 1015–1022.

Randall, J. E., Allen, G. R., and Steene, R. C. (1997). "Fishes of the Great Barrier Reef and Coral Sea," 2nd edn. Crawford House Press, Bathurst.

Renshaw, G. M. C., and Dyson, S. E. (1999). Increased nitric oxide synthase in the vasculature of the epaulette shark brain following hypoxia. *Neuroreport* **10**, 1707–1712.

Renshaw, G. M. C., Kerrisk, C. B., and Nilsson, G. E. (2002). The role of adenosine in the anoxic survival of the epaulette shark. *Hemiscyllium ocellatum. Comp. Biochem. Physiol.* **131B**, 133–141.

Routley, M. H., Nilsson, G. E., and Renshaw, G. M. C. (2002). Exposure to hypoxia primes the respiratory and metabolic responses of the epaulette shark to progressive hypoxia. *Comp. Biochem. Physiol.* **131A**, 313–321.

Schubert, M., Munday, P. L., Caley, M. J., Jones, G. P., and Llewellyn, L. E. (2003). The toxicity of skin secretions from coral-dwelling gobies and their potential role as a predator deterrent. *Environ. Biol. Fish.* **67**, 359–367.

Söderström, V., Nilsson, G. E., Renshaw, G. M. C., and Franklin, C. E. (1999). Hypoxia stimulates cerebral blood flow in the estuarine crocodile (*Crocodylus porosus*). *Neurosci. Lett.* **267**, 1–4.

Söderström, V., Renshaw, G. M. C., and Nilsson, G. E. (1999). Brain blood flow and blood pressure during hypoxia in the epaulette shark (*Hemiscyllium ocellatum*), a hypoxia tolerant elasmobranch. *J. Exp. Biol.* **202**, 829–835.

Takegaki, T., and Nakazono, A. (1999). Responses of the egg-tending gobiid fish *Valenciennea longipinnis* to the fluctuation of dissolved oxygen in the burrow. *Bull. Mar. Sci.* **65**, 815–823.

Thresher, R. E. (1984). "Reproduction in Reef Fishes." T. F. H. Publications, Neptune City, NJ.

Verheyen, R., Blust, R., and Decleir, W. (1994). Metabolic rate, hypoxia tolerance and aquatic surface respiration of some lacustrine and riverine African cichlid fishes. *Comp. Biochem. Physiol.* **107A**, 403–411.

Wise, G., Mulvey, J. M., and Renshaw, G. M. C. (1998). Hypoxia tolerance in the epaulette shark (*Hemiscyllium ocellatum*). *J. Exp. Zool.* **281**, 1–5.

INDEX

A

OTHER VOLUMES IN THE FISH PHYSIOLOGY SERIES

Printed and bound by CPI Group (UK) Ltd, Croydon, CR0 4YY

08/05/2025

01864951-0005